MW00358798

# *MECHATRONICS*

**SABRI CETINKUNT**
*University of Illinois at Chicago*

**JOHN WILEY & SONS, INC.**

Associate Publisher *Daniel Sayre*
Acquisition Editor *Joseph Hayton*
Senior Production Editor *Sujin Hong*
Marketing Manager *Frank Lyman*
Design Director *Harry Nolan*
Senior Designer *Kevin Murphy*
Senior Illustration Editor *Sigmund Malinowski*
Editorial Assistant *Mary Moran-McGee*

This book was set in *Times Roman* by *TechBooks* and printed and bound by *Hamilton Printing.* The cover was printed by *Phoenix Color.*

The book is printed on acid-free paper. ∞



To order books or for customer service, please call 1-800-CALL WILEY (225-5945).

***Library of Congress Cataloging in Publication Data*:**
Cetinkunt, Sabri.
Mechatronics/Sabri Cetinkunt.
p. cm.
Includes bibliographical references and index.
ISBN-13 978-0-471-47987-1 (cloth)
1. Mechatronics. I. Title.
TJ163.12.C43 2006
621—dc22 2005031908

Printed in the United States of America

10 9 8 7 6 5

# CONTENTS

*PREFACE* **vii**

**CHAPTER 1** *INTRODUCTION TO MECHATRONICS* **1**

1.1 Introduction **1**
1.2 Case Study: Modeling and Control of Combustion Engines **13**
1.2.1 Diesel Engine Components **14**
1.2.2 Engine Control System Components **20**
1.2.3 Engine Modeling with Lug Curve **22**
1.2.4 Engine Control Algorithms: Engine Speed Regulation Using Fuel Map and a Proportional Control Algorithm **26**
1.3 Problems **26**

**CHAPTER 2** *CLOSED-LOOP CONTROL* **29**

2.1 Components of a Digital Control System **30**
2.2 The Sampling Operation and Signal Reconstruction **32**
2.2.1 Sampling: A/D Operation **32**
2.2.2 Sampling Circuit **32**
2.2.3 Mathematical Idealization of the Sampling Circuit **34**
2.2.4 Signal Reconstruction: D/A Operation **39**
2.2.5 Real-Time Control Update Methods and Time-Delay **42**
2.2.6 Filtering and Bandwidth Issues **44**
2.3 Open-Loop Control versus Closed-Loop Control **46**
2.4 Performance Specifications for Control Systems **49**
2.5 Time Domain and s-Domain Correlation of Signals **51**
2.5.1 Selection of Pole Locations **52**
2.5.2 Step Response of a Second-Order System **52**
2.5.3 Standard Filters **56**
2.5.4 Steady-State Response **56**
2.6 Stability of Dynamic Systems **58**
2.6.1 Bounded Input–Bounded Output Stability **59**
2.7 The Root Locus Method **60**
2.8 Basic Feedback Control Types **64**
2.8.1 Proportional Control **67**
2.8.2 Derivative Control **68**
2.8.3 Integral Control **69**
2.8.4 PI Control **70**
2.8.5 PD Control **72**
2.8.6 PID Control **73**
2.9 Translation of Analog Control to Digital Control **74**
2.9.1 Finite Difference Approximations **76**
2.10 Problems **78**

**CHAPTER 3** *MECHANISMS FOR MOTION TRANSMISSION* **81**

3.1 Introduction **81**
3.2 Rotary-to-Rotary Motion Transmission Mechanisms **84**
3.2.1 Gears **84**
3.2.2 Belt and Pulley **85**
3.3 Rotary-to-Translational Motion Transmission Mechanisms **87**
3.3.1 Lead Screw and Ball Screw Mechanisms **87**
3.3.2 Rack-and-Pinion Mechanism **89**
3.3.3 Belt and Pulley **90**
3.4 Cyclic Motion Transmission Mechanisms **91**
3.4.1 Linkages **91**
3.4.2 Cams **92**
3.5 Shaft Misalignments and Flexible Couplings **101**
3.6 Actuator Sizing **102**
3.6.1 Inertia Match Between Motor and Load **108**
3.7 Homogeneous Transformation Matrices **110**
3.8 Problems **119**

**CHAPTER 4** *MICROCONTROLLERS* **123**

4.1 Embedded Computers versus Nonembedded Computers **123**
4.1.1 Design Steps of an Embedded Microcontroller-Based Mechatronic System **125**

4.1.2 Microcontroller Development Tools **125**
4.1.3 Microcontroller Development Tools for PIC 18F452 **127**
4.2 Basic Computer Model **129**
4.3 Microcontroller Hardware and Software: PIC 18F452 **133**
4.3.1 Microcontroller Hardware **133**
4.3.2 Microprocessor Software **137**
4.3.3 I/O Peripherals of PIC 18F452 **139**
4.4 Interrupts **145**
4.4.1 General Features of Interrupts **145**
4.4.2 Interrupts on PIC 18F452 **147**
4.5 Problems **152**

**CHAPTER 5** *ELECTRONIC COMPONENTS FOR MECHATRONIC SYSTEMS* **153**

5.1 Introduction **153**
5.2 Basics of Linear Circuits **153**
5.3 Equivalent Electrical Circuit Methods **156**
5.3.1 Thevenin's Equivalent Circuit **157**
5.3.2 Norton's Equivalent Circuit **157**
5.4 Impedance **160**
5.4.1 Concept of Impedance **160**
5.4.2 Amplifier: Gain, Input Impedance, and Output Impedance **163**
5.4.3 Input and Output Loading Errors **164**
5.5 Semiconductor Electronic Devices **166**
5.5.1 Semiconductor Materials **166**
5.5.2 Diodes **168**
5.5.3 Transistors **172**
5.6 Operational Amplifiers **183**
5.6.1 Basic Op-Amp **184**
5.6.2 Common Op-Amp Circuits **188**
5.7 Digital Electronic Devices **201**
5.7.1 Logic Devices **201**
5.7.2 Decoders **202**
5.7.3 Multiplexer **202**
5.7.4 Flip-Flops **204**
5.8 Digital and Analog I/O and Their Computer Interface **206**
5.9 D/A and A/D Converters and Their Computer Interface **208**
5.10 Problems **214**

**CHAPTER 6** *SENSORS* **217**

6.1 Introduction to Measurement Devices **217**
6.2 Measurement Device Loading Errors **220**
6.3 Wheatstone Bridge Circuit **222**
6.3.1 Null Method **223**
6.3.2 Deflection Method **223**
6.4 Position Sensors **225**
6.4.1 Potentiometer **225**
6.4.2 LVDT, Resolver, and Syncro **227**
6.4.3 Encoders **232**
6.4.4 Hall Effect Sensors **237**
6.4.5 Capacitive Gap Sensors **238**
6.4.6 Magnetostriction Position Sensors **239**
6.4.7 Sonic Distance Sensors **240**
6.4.8 Photoelectric Distance and Presence Sensors **241**
6.4.9 Presence Sensors: ON/OFF Sensors **243**
6.5 Velocity Sensors **245**
6.5.1 Tachometers **245**
6.5.2 Digital Derivation of Velocity from Position Signal **247**
6.6 Acceleration Sensors **248**
6.6.1 Inertial Accelerometers **249**
6.6.2 Piezoelectric Accelerometers **252**
6.6.3 Strain-Gauge-Based Accelerometers **254**
6.7 Strain, Force, and Torque Sensors **254**
6.7.1 Strain Gauges **254**
6.7.2 Force and Torque Sensors **256**
6.8 Pressure Sensors **259**
6.8.1 Displacement-Based Pressure Sensors **260**
6.8.2 Strain-Gauge-Based Pressure Sensor **261**
6.8.3 Piezoelectric-Based Pressure Sensor **262**
6.8.4 Capacitance-Based Pressure Sensor **262**
6.9 Temperature Sensors **263**
6.9.1 Temperature Sensors Based on Dimensional Change **264**
6.9.2 Temperature Sensors Based on Resistance **264**
6.9.3 Thermocouples **265**
6.10 Flow Rate Sensors **267**
6.10.1 Mechanical Flow Rate Sensors **267**
6.10.2 Differential Pressure Flow Rate Sensors **269**
6.10.3 Thermal Flow Rate Sensors: Hot Wire Anemometer **271**
6.10.4 Mass Flow Rate Sensors: Coriolis Flow Meters **272**
6.11 Humidity Sensors **272**
6.12 Vision Systems **273**
6.13 Problems **277**

**CHAPTER 7** *ELECTROHYDRAULIC MOTION CONTROL SYSTEMS* **281**

7.1 Introduction **281**
7.1.1 Fundamental Physical Principles **294**

7.1.2 Analogy Between Hydraulic and Electrical Components **296**
7.1.3 Energy Loss and Pressure Drop in Hydraulic Circuits **299**
7.2 Hydraulic Pumps **301**
7.2.1 Types of Positive Displacement Pumps **303**
7.2.2 Pump Performance **307**
7.2.3 Pump Control **313**
7.3 Hydraulic Actuators: Hydraulic Cylinder and Rotary Motor **320**
7.4 Hydraulic Valves **324**
7.4.1 Pressure Control Valves **326**
7.4.2 Example: Multifunction Hydraulic Circuit with Poppet Valves **330**
7.4.3 Flow Control Valves **332**
7.4.4 Example: A Multifunction Hydraulic Circuit Using Post-Pressure Compensated Proportional Valves **337**
7.4.5 Directional Flow Control Valves: Proportional and Servo Valves **339**
7.4.6 Mounting of Valves in a Hydraulic Circuit **351**
7.4.7 Performance Characteristics of Proportional and Servo Valves **352**
7.5 Sizing of Hydraulic Motion System Components **359**
7.6 EH Motion Axis Natural Frequency and Bandwidth Limit **371**
7.7 Linear Dynamic Model of a One-Axis Hydraulic Motion System **373**
7.7.1 Position Controlled Electrohydraulic Motion Axes **375**
7.7.2 Load Pressure Controlled Electrohydraulic Motion Axes **378**
7.8 Nonlinear Dynamic Model of a Hydraulic Motion System **379**
7.9 Current Trends in Electrohydraulics **381**
7.10 Case Studies **384**
7.10.1 Case Study: Multifunction Hydraulic Circuit of a Caterpillar Wheel Loader **384**
7.11 Problems **388**

**CHAPTER 8** *ELECTRIC ACTUATORS: MOTOR AND DRIVE TECHNOLOGY* **393**

8.1 Introduction **393**
8.1.1 Steady-State Torque-Speed Range, Regeneration, and Power Dumping **395**
8.1.2 Electric Fields and Magnetic Fields **399**
8.1.3 Permanent Magnetic Materials **412**
8.2 Solenoids **423**
8.2.1 Operating Principles of Solenoids **423**
8.2.2 DC Solenoid: Electromechanical Dynamic Model **426**
8.3 DC Servo Motors and Drives **430**
8.3.1 Operating Principles of DC Motors **431**
8.3.2 Drives for DC Brush-Type and Brushless Motors **438**
8.4 AC Induction Motors and Drives **447**
8.4.1 AC Induction Motor Operating Principles **448**
8.4.2 Drives for AC Induction Motors **454**
8.5 Step Motors **461**
8.5.1 Basic Stepper Motor Operating Principles **463**
8.5.2 Step Motor Drives **468**
8.6 Switched Reluctance Motors and Drives **471**
8.6.1 Switched Reluctance Motors **471**
8.6.2 SR Motor Control System Components: Drive **475**
8.7 Linear Motors **478**
8.8 DC Motor: Electromechanical Dynamic Model **481**
8.8.1 Voltage Amplifier Driven DC Motor **484**
8.8.2 Current Amplifier Driven DC Motor **485**
8.8.3 Steady-State Torque-Speed Characteristics of a DC Motor under Constant Terminal Voltage **486**
8.8.4 Steady-State Torque-Speed Characteristics of a DC Motor and Current Amplifier **486**
8.9 Energy Losses in Electric Motors **488**
8.9.1 Resistance Losses **489**
8.9.2 Core Losses **490**
8.9.3 Friction and Windage Losses **491**
8.10 Problems **491**

**CHAPTER 9** *PROGRAMMABLE LOGIC CONTROLLERS* **495**

9.1 Introduction **495**
9.2 Hardware Components of PLCs **498**
9.2.1 PLC, CPU, and I/O Capabilities **498**
9.2.2 Opto-Isolated Discrete Input and Output Modules **502**
9.2.3 Relays, Contactors, Starters **503**
9.2.4 Counters and Timers **505**
9.3 Programming of PLCs **505**
9.3.1 Hardwired Seal-In Circuit **509**
9.4 PLC Control System Applications **510**

9.5 PLC Application Example: Conveyor and Furnace Control 511
9.6 Problems 514

**CHAPTER 10** *PROGRAMMABLE MOTION CONTROL SYSTEMS* 515

10.1 Introduction 515
10.2 Design Methodology for PMC Systems 520
10.3 Motion Controller Hardware and Software 521
10.4 Basic Single-Axis Motions 522
10.5 Coordinated Motion Control Methods 526
10.5.1 Point-to-Point Synchronized Motion 527
10.5.2 Electronic Gearing Coordinated Motion 528
10.5.3 CAM Profile and Contouring Coordinated Motion 531
10.5.4 Sensor-Based Real-Time Coordinated Motion 532
10.6 Coordinated Motion Applications 532
10.6.1 Web Handling with Registration Mark 532
10.6.2 Web Tension Control Using Electronic Gearing 535
10.6.3 Smart Conveyors 539
10.7 Problems 544

**APPENDIX A** *TABLES* 547

**APPENDIX B** *MODELING AND SIMULATION OF DYNAMIC SYSTEMS* 549

B.1 Modeling of Dynamic Systems 549
B.2 Complex Variables 550
B.3 Laplace Transforms 552
B.3.1 Definition of Laplace Transform 552
B.3.2 Properties of the Laplace Transform 554
B.3.3 Laplace Transforms of Some Common Functions 558
B.3.4 Inverse Laplace Transform: Using Partial Fraction Expansions 562
B.4 Fourier Series, Fourier Transforms, and Frequency Response 566
B.4.1 Basics of Frequency Response: Meaning of Frequency Response 571
B.4.2 Relationship Between the Frequency Response and Transfer Function 572
B.4.3 s-Domain Interpretation of Frequency Response 573
B.4.4 Experimental Determination of Frequency Response 574
B.4.5 Graphical Representation of Frequency Response 574
B.5 Transfer Function and Impulse Response Relation 574
B.6 Convolution 579
B.7 Review of Differential Equations 581
B.7.1 Definitions 581
B.7.2 System of First-Order O.D.E.s 581
B.7.3 Existence and Uniqueness of the Solution of O.D.E.s 582
B.8 Linearization 583
B.8.1 Linearization of Nonlinear Functions 583
B.8.2 Linearization of Nonlinear First-Order Differential Equations 585
B.8.3 Linearization of Multidimensional Nonlinear Differential Equations 586
B.9 Numerical Solution of O.D.E.s and Simulation of Dynamic Systems 588
B.9.1 Numerical Methods for Solving O.D.E.s 589
B.9.2 Numerical Solution of O.D.E.s 589
B.9.3 Time Domain Simulation of Dynamic Systems 591
B.10 Details of the Solution for Example on Page 162: RL and RC Circuits 600
B.11 Problems 604

*BIBLIOGRAPHY* 607

*INDEX* 611

# PREFACE

This book covers the fundamental scientific principles and technologies that are used in the design of modern computer-controlled machines and processes. Today, the technical background necessary for an engineer to design an automated machine, component, or process is very different from that of 30 years ago. The underlying difference is the availability of embedded computers used to control such machines. An automated machine designed 30 years ago would have complicated linkages and cams to define the coordinated motion relationship between different stations. Today, such relationships are defined in computer control software. A computer controlled electromechanical system designer not only needs to know proper mechanical design principles, but also needs to know embedded computer control hardware and software, sensors in order to measure variables of interest, and actuation technologies.

Many computer-aided design tools in all of these areas (i.e., mechanical design, embedded controller) make it possible for a designer to be knowledgeable in all of these areas to the extent that he or she can use them effectively in the design. This book should be useful to senior undergraduate or first-year graduate-level students as well as practicing engineers. Its purpose is to present all the technical background needed in designing an automated machine or process. These technical areas cover traditionally different engineering disciplines, namely mechanical, aerospace, chemical, electrical, and computer engineering. The book has enough material for two semester courses. If it is used for one semester course only, it is advised that Chapters 1 through 6 be covered first, then some of the selected chapters can be covered. Chapters 10 and 11 may be assigned as a self-study or left as a reference for students. If time permits, these chapters may be used as a basis for comprehensive lab projects where all aspects of the mechatronics field are brought together in modern design projects. The reader should be prepared to refer to other good reference books for more details in each topic covered. Because a large number of topics are covered under the topic of mechatronics, the depth of coverage had to be limited in one book.

The emphasis is the view of a design engineer: What does one need to know about a component or subsystem in order to effectively use it in a design? While covering the fundamental physical principles in each area, we skip historic perspectives and long reviews, and go straight into the discussion of relevant technology in its current state-of-art form. We avoid long derivations or proofs. However, proper references are provided where the details of the derivations and proofs can be found. In this book, we do not try to find all answers to the questions with equations and numbers. Quite often, we rely on "rule of thumb" design guides and justify their validity with reasonable physics-based discussions. Good design requires good understanding of the fundamental principles and good judgment. Examples throughout the text and the problem assignments at the end of each chapter are intended to make the student think of the design issues as opposed to requiring the student to make some numerical calculations. Therefore, the reader should be prepared to consult other reference books and especially supplier web pages to find a good solution (among multiple possible solutions) to a problem.

At the references section, we also provide information on the major suppliers of different products. A modern mechatronics engineer is a systems integration designer. It is rarely the case that all of the system components are designed from scratch for a design project. Quite often, the designer selects components and subsystems, and then properly designs their custom hardware and software integration. Publisher web site at www.wiley.com/college/cetinkunt provides various lab experiments involving microcontroller-based electromechanical design experiments, and some brief review material on Matlab/Simulink, C/C++ programming language.

The material in this book is a result of the courses I have taught at the University of Illinois at Chicago over the past five years as well as the experience I gained in working with various companies over the years in many research and development projects. I am indebted to many people with whom I have worked and who taught me most of the material covered here. I have had the good fortune of having worked with many outstanding, bright, talented young students: U. Pinsopon, A. Egelja, M. Cobo, C. Chen, S. Haggag, G. Larsen, S. Ku, T. Hwang, F. Riordan, D. Norlen, D. Alstrom, J. Woloszko, M. Nakamura, S. Velamakanni, D. Vecchiato, and M. Bhanabhagvanwala. I also would like to acknowledge the following colleagues who over the years shared their expertise and educated me in many aspects of the field: R. Ingram, M. Hopkins, J. Aardema, J. Krone, J. Schimpf, J. Mount, M. Sorokine, M. Vanderham, S. Kherat, S. Anwar, M. Guven of Caterpillar Inc., D. Wohlsdorf of Sauer-Danfoss, L. Schrader of Parker, H. Yamamoto of Neomax, D. Hirschberger of Moog Gmbh, G. Al-ahmad of Hydraforce, W. Fisher of OilGear, M. Brown, P. Eck, T. Klikuszowian of Abbott Labs, and J. Gamble of Magnet-Schultz, C. Wilson of Delta Tau, C. Johnson, A. Donmez of National Institute of Standards and Technology, and R. Cesur of Servo Tech. I would like to thank my editor Joseph Hayton, editorial assistant Mary Moran, and senior production editor Sujin Hong at John Wiley & Sons for their patience and kind guidance throughout the process of writing this book.

The following faculty has reviewed this edition in various stages: Hon Zhang–Rowan University, Michael Goldfarb–Vanderbuilt University, George Chiu–Purdue University, Sandford Meek–University of Utah, Ji Wang–San Jose State University, Kazuo Yamazaki–University of California at Davis, and Mark Nagurka–Marquette University.

*Sabri Cetinkunt*
Chicago, Illinois
November 2005

CHAPTER 1

# INTRODUCTION TO MECHATRONICS

## 1.1 INTRODUCTION

The mechatronics field consists of the synergistic integration of three distinct traditional engineering fields for the system level design process. These three fields are:

1. Mechanical engineering, where the word "mecha" is taken from
2. Electrical or electronics engineering, where the part of the word "tronics" is taken from
3. Computer science

The mechatronics field is not simply the sum of these three major areas, but rather the field defined as the intersection of these areas when taken in the context of systems design (Fig. 1.1). It is the current state of evolutionary change of the engineering fields that deals with the design of controlled electromechanical systems. The word mechatronics was first coined by engineers at Yaskawa Electric Company [1, 2]. Virtually every modern electromechanical system has an embedded computer controller. Therefore, computer hardware and software issues (in terms of their application to the control of electromechanical systems) are part of the field of mechatronics. Had it not been the widespread availability of the low-cost microcontrollers for the mass market, the field of mechatronics as we know it today would not have existed [2a]. The availability of embedded microprocessors for the mass market at an ever-reducing cost and increasing performance makes possible the use of computer control in thousands of consumer products.

The old model for an electromechanical product design team includes:

1. Engineer(s) who designs the mechanical components of a product
2. Engineer(s) who designs the electrical components such as actuators, sensors, and amplifiers, as well as design the control logic and algorithms
3. Engineer(s) who designs the computer hardware and software implementation to control the product in real time

A mechatronics engineer is trained to do all of these three functions. In addition, the design process is not sequential from mechanical design, followed by electrical and computer control system designs, but rather all aspects (mechanical, electrical, and computer control) of design are done simultanenously for optimal product design. Clearly, mechatronics is not a new engineering discipline, but is rather the current state of the evolutionary process of engineering disciplines needed in design of electromechanical systems. The end product of a mechatronics engineer's work is a working prototype of an embedded computer-controlled electromechanical device or system. This book covers the fundamental technical topics needed to enable an engineer to accomplish such designs. We define the word *device* as a stand-alone product that serves a function such as a microwave oven,

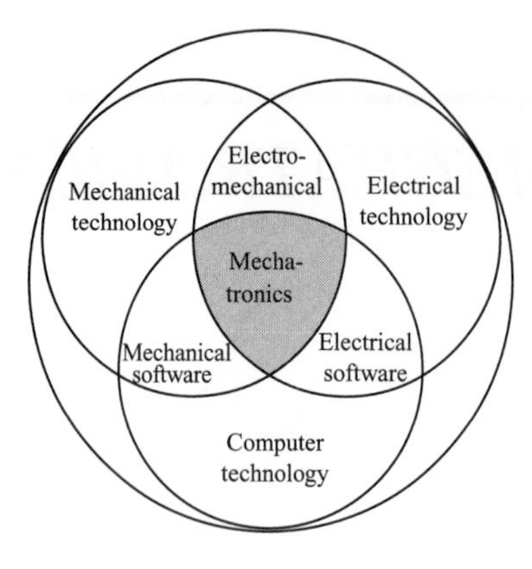


**FIGURE 1.1:** The field of mechatronics: intersection of mechanical, electrical, and computer science.

whereas a *system* may be a collection of multiple devices such as an automated robotic assembly line.

As a result, this book has sections in mechanical design of various mechanisms used in automated machines and robotic applications. Such mechanisms are over a century old designs and the basic designs are still used in modern applications. Mechanical design forms the "skeleton" of the electromechanical product, upon which the rest of the functionalities are built (such as "eyes," "muscles," "brains"). These mechanisms are discussed in terms of their functionality and common design parameters. Detailed stress or force analysis of these mechanisms is omitted, as they are covered in traditional stress analysis and machine design courses.

The analogy between a human-controlled system and a computer-controlled system is shown in Fig. 1.2. If a process is controlled and powered by a human operator, the operator observes the behavior of the system (i.e., using visual observation), makes a decision regarding what action to take, and then, using his muscular power, a particular control action is taken. One could view the outcome of a decision-making process as a low-power control or decision signal, and the action of the muscles as the actuator signal which is the amplified version of the control (or decision) signal. The same functionalities of a system can be automated by use of a digital computer as shown in the same figure.

The sensors replace the eyes, actuators replace the muscles, and the computer replaces the human brain. Every computer-controlled system has these four basic functional blocks:

1. Process to be controlled
2. Actuators
3. Sensors
4. Controller (i.e., digital computer)

The microprocessor and digital signal processing ($\mu$P/DSP) technology had two types of impact in the control world:

1. Replaced the *existing* analog controllers
2. Prompted *new* products and designs such as fuel-injection systems, active suspension, home temperature control, microwave ovens, and auto-focus cameras

Every mechatronic system has some sensors to measure the status of the process variables. The sensors are the "eyes" of a computer-controlled system. We study most common types of sensors used in electromechanical systems for the measurement of temperature,

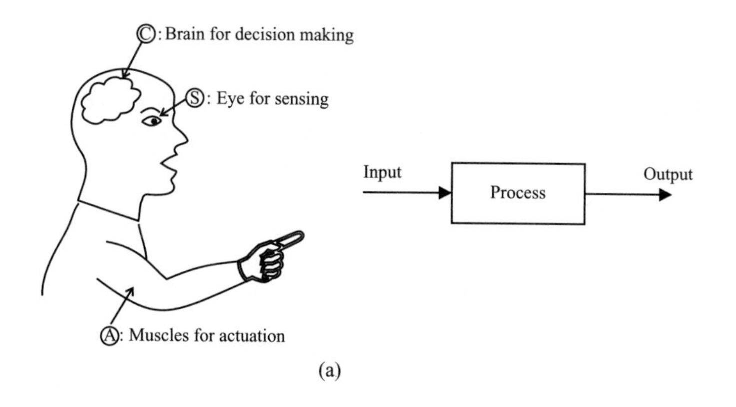


(a)

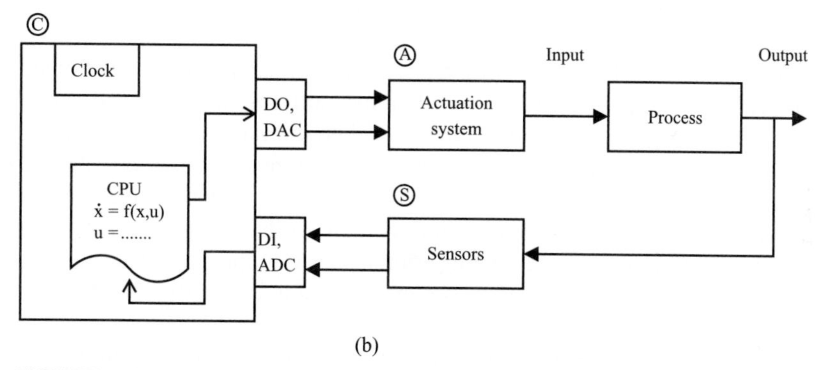


(b)

**FIGURE 1.2:** Manual and automatic control system analogy: (a) human controlled and (b) computer controlled.

pressure, force, stress, position, speed, acceleration, flow, etc. (Fig. 1.3). This list does not attempt to cover every conceivable sensor available in the current state of art, but rather makes an attempt to cover all major sensor categories, their working principles, and typical applications in design.

Actuators are the "muscles" of a computer-controlled system. We focus in depth on the actuation devices that provide high-performance control as opposed to simple ON/OFF actuation devices. In particular, we discuss the hydraulic and electric power actuators in

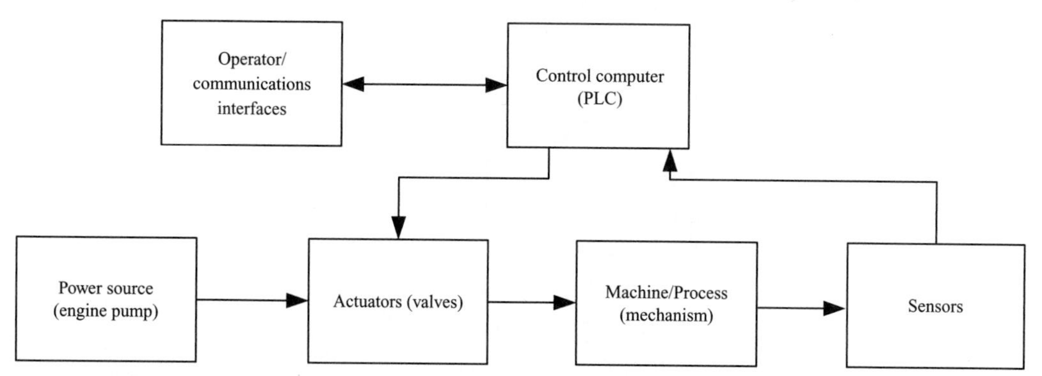


**FIGURE 1.3:** Main components of any mechatronic system: mechanical structure, sensors, actuators, decision-making component (microcontroller), power source, and human/supervisory interfaces.

detail. Pneumatic power (compressed air power) actuation systems are not discussed. They are typically used in low-performance, ON/OFF type control applications (although, with advanced computer control algorithms, even they are starting to be used in high-performance systems). The component functionalities of pneumatic systems are similar to those of hydraulic systems. However, the construction detail of each is quite different. For instance, both hydraulic and pneumatic systems need a component to pressurize the fluid (pump or compressor), a valve to control the direction, amount, and pressure of the fluid flow in the pipes, and translation cylinders to convert the pressurized fluid flow to motion. The pumps, valves, and cylinders used in hydraulic systems are quite different than those used in pneumatic systems.

Hardware and software fundamentals for embedded computers, microprocessors, and digital signal processors (DSP) are covered with applications to the control of electromechanical devices in mind. Hardware I/O interface, microprocessor hardware architecture, and software concepts are discussed. The basic electronics circuit components are discussed because they form the foundation of interface between the digital world of computers and analog real world. It is important to note that the hardware interfaces and embedded controller hardware aspects are largely standard and do not vary greatly from one application to another. On the other hand, the software aspects of mechatronics designs are different for every product. The development tools used may be same, but the final software created for the product (also called the application software) is different for each product. It is not uncommon that over 80% of engineering effort in the development of a mechatronic product is spent on the software aspects only. Therefore, the importantance of software, especially as it applies to embedded systems, cannot be overemphasized.

Mechatronic devices and systems are products of the natural evolution of automated systems. We can view this evloution as having three major phases:

1. Completely mechanical automatic systems (before and early 1900)
2. Automatic devices with electronic components such as using relays, transistors, op-amps (early 1900–1970)
3. Computer-controlled automatic systems (1970s–present)

Early automatic control systems performed an automated function completely with mechanical means. For instance, a water level regulator for a water tank uses a float connected to a valve via a linkage (Fig. 1.4). The desired water level in the tank is set by the adjustment of the float height or the linkage arm length connecting it to the valve. The float opens and closes the valve in order to maintain the desired water level. Another

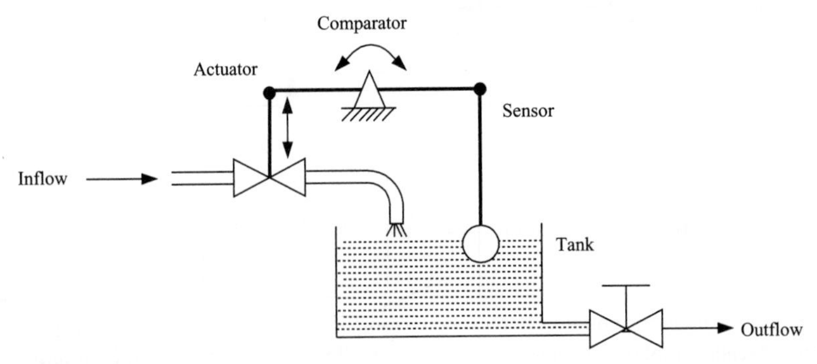


**FIGURE 1.4:** A completely mechanical closed-loop control system for liquid level regulation.

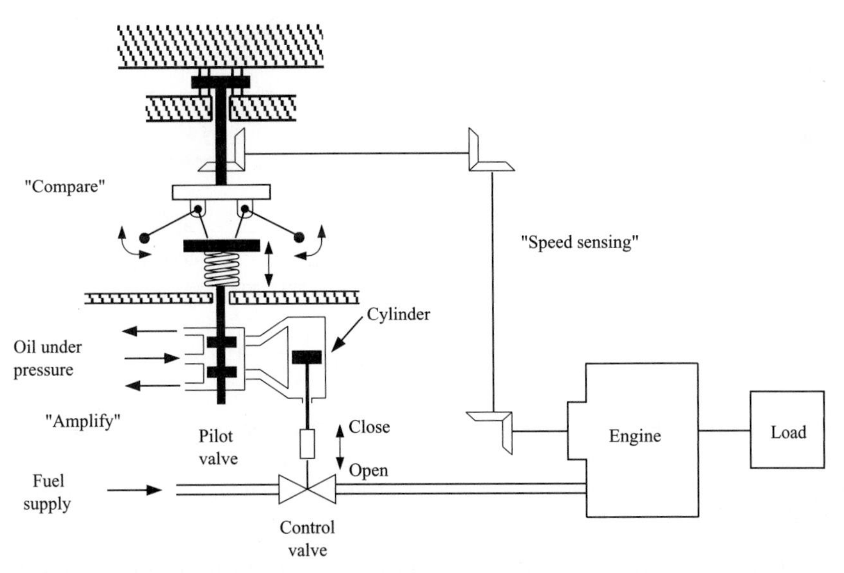


**FIGURE 1.5:** Mechanical "governor" concept for automatic engine speed control using all mechanical components.

classic automatic control system that is made of completely mechanical components (no electronics) is the Watt's flyball governor, which is used to regulate the speed of an engine (Fig. 1.5). The same concept is still used in some engines today. The engine speed is regulated by controlling the fuel control valve on the fuel supply line. The valve is controlled by a mechanism that has a desired speed setting using the bias in the spring in the flywheel mechanism. The actual speed is measured by the flyball mechanism. The higher the speed of the engine, the more the flyballs move out due to centrifugal force. The difference between the desired speed and actual speed is turned into control action by the movement of the valve which controls a small cylinder which is then used to control the fuel control valve. In today's engines, the fuel rate is controlled directly by an electrically actuated injector. The actual speed of the engine is sensed by an electrical sensor (i.e., tachometer, pulse counter, encoder) and an embedded computer controller decides how much fuel to inject based on the difference between the desired and actual engine speed (Fig. 1.8).

Analog servo controller using operational amplifiers made the second major change in mechatronic systems. Now the automated systems no longer had to be all mechanical. An operational amplifier is used to compare a desired response (presented as an analog voltage) and a measured response by an electrical sensor (also presented as a voltage) and actuates an electrical device (solenoid or electric motor) based on the difference. This brought about many electromechanical servo control systems (Figs. 1.6 and 1.7). Figure 1.6 shows a web handling machine with tension control. The wind-off roll runs at a speed that may vary. The wind-up roll is to run such that no matter what the speed of the web motion, a certain tension is to be maintained on the web. Therefore, a displacement sensor on the web is used to indirectly sense the web tension because the sensor measures the displacement of a spring. The measured tension is then compared to the desired tension (command signal in the figure) by an operational amplifier. The operational amplifier sends a speed or current command to the amplifier of the motor based on the tension error. Modern tension control systems use a digital computer controller in place of the analog operation amplifier controller. In addition, the digital controller may use a speed sensor from the wind-off roll or from the web on the incoming side in order to react to tension changes faster and improve the dynamic performance of the system.

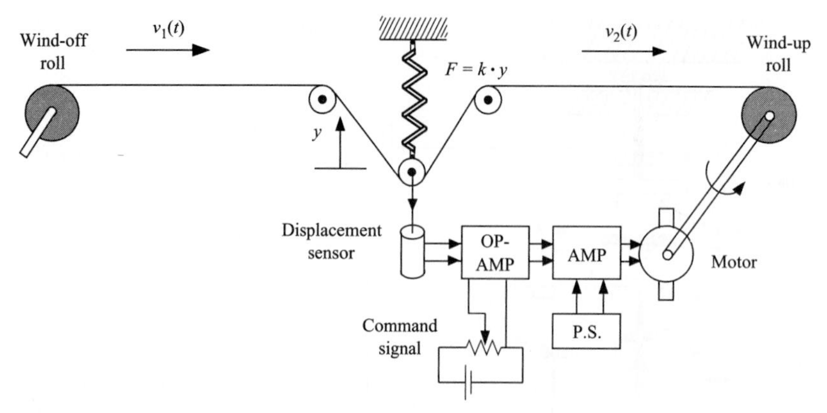


**FIGURE 1.6:** A web handling motion control system. The web is moved at high speed while maintaining a desired tension. The tension control system can be considered a mechatronic system where the control decision is made by an analog op-amp, which can be replaced by a digital computer.

Figure 1.7 shows a temperature control system that can be used to heat a room or oven. The heat is generated by the electric heater. Heat is lost to outside through the walls. A thermometer is used to measure the temperature. An analog controller has the desired temperature setting. Based on the difference between the set and measured temperature, the op-amp turns ON or OFF the relay which turns the heater ON/OFF. In order to make sure the relay does not turn ON and OFF due to small variations about the set temperature, the op-amp would normally have a hysteresis functionality implemented on its circuit. More details on the relay control with hysteresis will be discussed in later chapters.

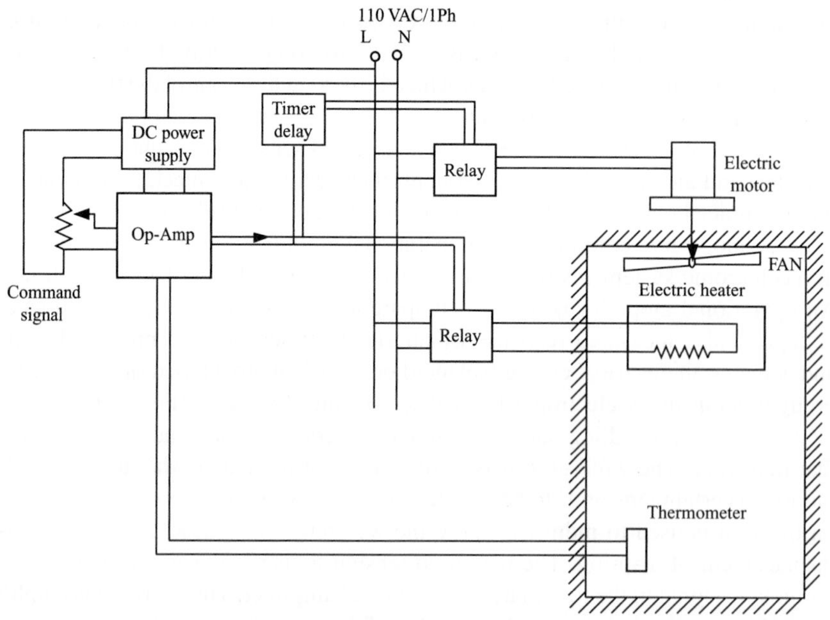


**FIGURE 1.7:** A furnace or room temperature control system and its components using analog op-amp as controller. Notice that a fan driven by an electric motor is used to force the air circulation from the heater to the room. A timer is used to delay the turn ON and turn OFF time of the fan motor by a specified amount of time after the heater is turned ON or OFF. A microcontroller-based digital controller can replace the op-amp and timer components.

Finally, with the introduction of microprocessors into the control world in late 1970s, the programmable control and intelligent decision making were introduced to the automatic devices and systems. Digital computers not only duplicated the automatic control functionality of previous mechanical and electromechanical devices, but also brought about new possibilities of device designs that were once not possible. The control functions incorporated into the designs included not only the servo control capabilities but also many operational logic, fault diagnostics, component health monitoring, network communication, and nonlinear, optimal, and adaptive control strategies (Fig. 1.3). Many such functions were practically impossible to implement using analog op-amp circuits. With digital controllers, such functions are rather easy to implement. It is only a matter of coding these functionalities in software. The difficulty is in knowing what to code that works.

The automotive industry (largest industry in the world) has transformed itself both in terms of its products (the content of the cars) and the production methods of its products since the introduction of microprocessors. Use of microprocessor-based embedded controllers significantly increased the robotics-based programmable manufacturing processes, such as assembly lines, CNC machine tools, and material handling. This changed the way the cars are made, reducing the needed labor and increasing the productivity. The product itself, cars, have also changed significantly. Before the widespread introduction of 8-bit and 16-bit microcontrollers into the embedded control mass warket, the only electrical components in a car were radio, starter, alternator, and battery charging system. Engine, transmission, and brake subsystems were all controlled by mechanical or hydromechanical means. Today the engine on a modern car has a dedicated embedded microcontroller that controls the timing and amount of fuel injection in an optimized manner based on the load, speed, temperature, and pressure sensors in real time. Thus, it improves the fuel efficiency, reduces emissions, and increases performance (Fig. 1.8). Similarly, automatic transmission is controlled by an embedded controller. Braking systems include ABS (anti-lock braking systems), TCS (traction-control systems), and DVSC (dynamic vehicle stability control) systems which use dedicated microcontrollers to modulate the control of brake and engine

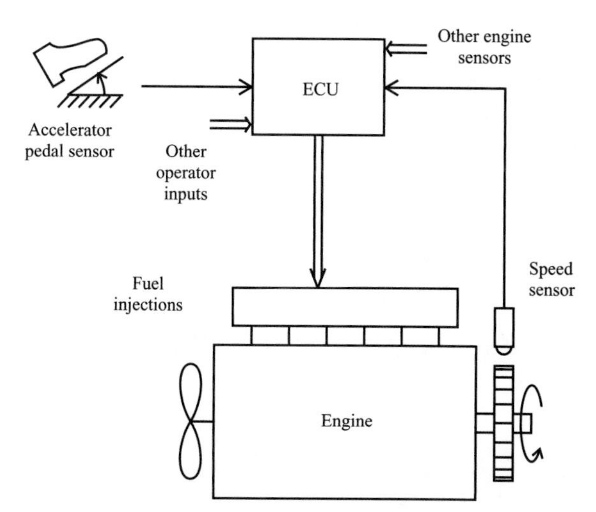


**FIGURE 1.8:** Electronic "governor" concept for engine control using embedded microcontrollers. Electronic control unit (ECU) decides on fuel injection timing and amount in real time based on sensor information.

in order to maintain better control of the vehicle. It is estimated that an average car today has over 30 embedded microprocessor-based controllers on board. This number continues to increase as more intelligent functions are added to cars. It is clear that traditionally all mechanical devices in cars have now become computer-controlled electromechanical devices which we call mechatronic devices. Therefore, the new generation of engineers must be well versed in the technologies that are needed in the design of modern electromechanical devices and systems. The field of mechatronics is defined as the integration of these areas to serve this type of modern design process.

Robotic manipulator is a good example of a mechatronic system. Low-cost, high-computational power and wide availability of digital signal processors (DSP) and microprocessors energized the robotics industry in late 1970s and early 1980s. The robotic manipulators, the reconfigurable, programmable, multi-degrees of freedom motion mechanism have been applied in many manufacturing processes, and many more applications are being developed, including robotic assisted surgery. The main subsystems of a robotic manipulator serves as a good example of mechatronic system. A robotic manipulator has four major subsystems (Fig. 1.3), and every modern mechatronic system has the same subsystem functionalities:

1. A mechanism to transmit motion from actuator to tool
2. Actuator (i.e., a motor and power amplifer, a hydraulic cylinder and valve) and power source (i.e., DC power supply, internal combustion engine and pump)
3. Sensors to measure the motion variables
4. Controller (DSP or microprocessor) along with operator user interface devices and communication capabilities to other intelligent devices

Let us consider an electric servo motor-driven robotic manipulator with three axes. The robot would have a predefined mechanical structure, i.e., cartesian, cylindrical, spherical, SCARA type robot (Figs. 1.9, 1.10, 1.11). Each of the three electric servo motors (i.e., brush type DC motor with integrally mounted position sensor such as encoder or stepper motor with separately mounted position sensor) drives one of the axes. There is a separate power amplifier for each motor which controls the current (hence torque) of the motor. A DC power supply provides a DC bus at a constant voltage and derives it from a standard AC line. The DC power supply is sized to support all three motor amplifiers.

The power supply, amplifer, and motor combination forms the actuator subsystem of a motion system. The sensors in this case are used to measure the position and velocity of each motor so that this information is used by the axis controller to control the motor through

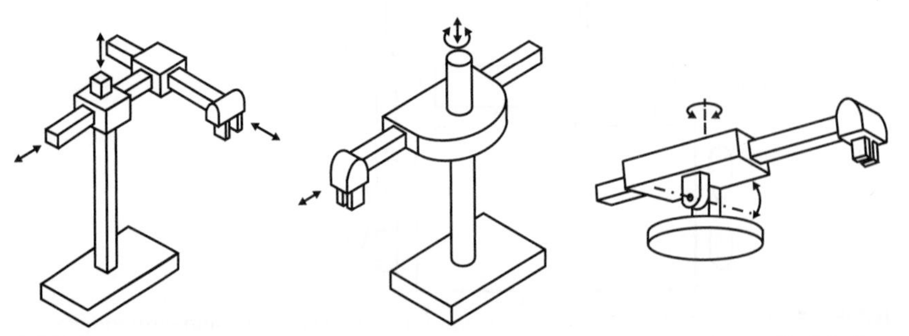

**FIGURE 1.9:** Three major robotic manipulator mechanisms: cartesian, cylindrical, and spherical coordinate axes.

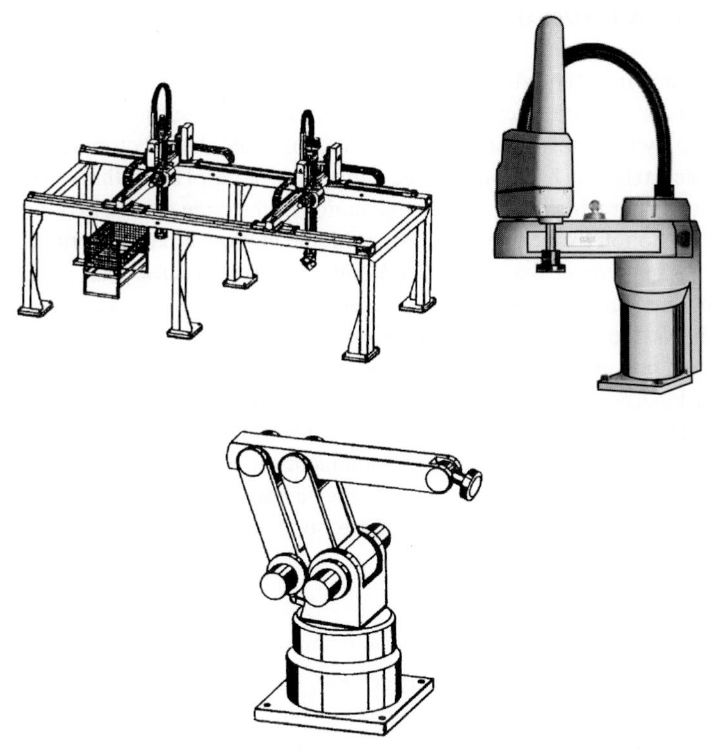

**FIGURE 1.10:** Gantry, SCARA, and parallel linkage drive robotic manipulators.

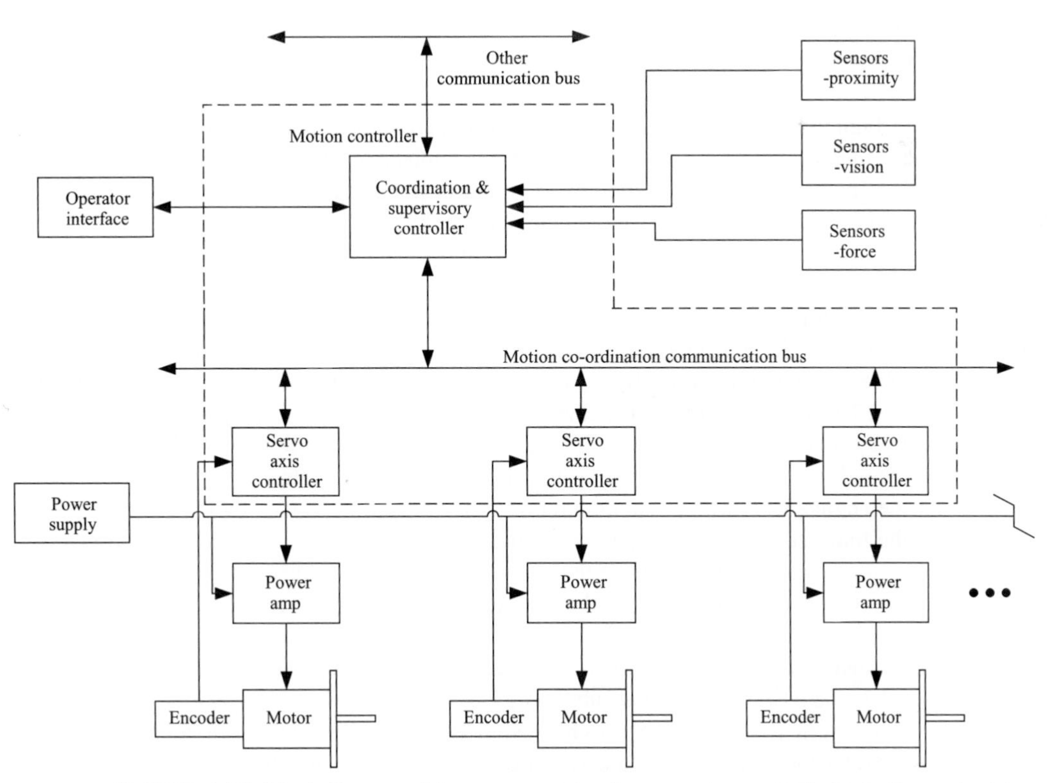


**FIGURE 1.11:** Block diagram of the components of a computer-controlled robotic manipulator.

the power amplifier in a closed-loop configuration. Other external sensors not directly linked to the actuator motions, such as a vision sensor or a force sensor or various proximity sensors, are used by the supervisory controller to coordinate the robot motion with other events. While each axis may have a dedicated closed-loop control algorithm, there has to be a supervisory controller which coordinates the motion of the three motors in order to generate a coordinated motion by the robot, i.e., straight line motion, circular motion. The hardware platform to implement the coordinated and axis level controls can be based on a single DSP/microprocessor or it may be distributed over multiple processors as shown. Figure 1.11 shows the components of a robotic manipulator in block diagram form. The control functions can be implemented on a single DSP hardware or a distributed DSP hardware. Finally, just as no man is an island, no robotic manipulator is an island. A robotic manipulator must communicate with a user and other intelligent devices to coordinate its motion with the rest of the manufacturing cell. Therefore, it has one or more other communication interfaces, typically over a common fieldbus (i.e., DeviceNET, CAN, ProfiBus, Ethernet). The capabilities of a robotic manipulator are quantified by the following:

1. Workspace: volume and envelope that the manipulator end effector can reach
2. Number of degrees of freedom that determines the positioning and orientation capabilities of the manipulator
3. Maximum load capacity, determined by the actuator, transmission components, and structural component sizing
4. Maximum speed (top speed) and small motion bandwidth
5. Repeatability and accuracy of end effector positioning
6. Manipulator's physical size (weight and volume it takes)

Figure 1.12 shows the power flow in modern construction equipment. The power source in most mobile equipment is an internal combustion engine, which is a diesel engine in large power applications. The power is hydromechanically transmitted from engine to transmission, brake, steering, implement, and cooling fan. All subsystems get their power in hydraulic power form from a group of pumps mechanically connected to the engine. These pumps convert mechanical power to hydraulic power. In automotive type designs, the power from engine to transmission gear mechanism is linked via a torque converter. In other designs, the transmission may be a hydrostatic design where the mechanical power is converted hydraulic power by a pump and then back to mechanical power by hydraulic motors. This is the case in most excavator designs. Notice that each major subsystem has its own electronic control module (ECM). Each ECM deals with the control of the subsystem and possibly communicates with a machine level master controller. For instance, ECM for the engine deals with maintaining an engine speed commanded by the operator pedal. As the load increases and engine needs more power, the ECM automatically commands more fuel to the engine to regulate the desired speed. The transmission ECM deals with the control of a set of solenoid actuated pressure valves in order to select the desired gear ratio. Steering ECM controls a valve which controls the flow rate to a steering cylinder. Similarly, other subsystem ECMs control electrically controlled valves and other actuation devices to modulate the power used in that subsystem.

The agricultural industry uses harvesting equipment where the equipment technology has the same basic components used in automotive industry. Therefore, automotive technology feeds and benefits the agricultural technology. Using global positioning systems (GPS)

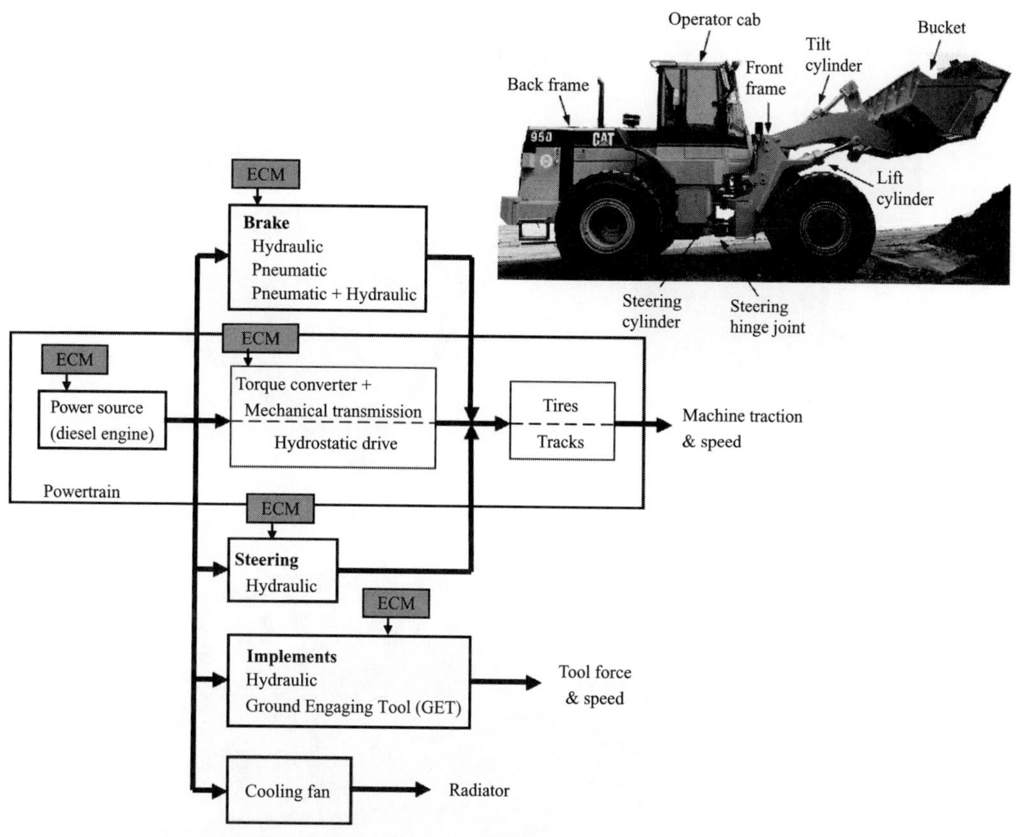


**FIGURE 1.12:** Block diagram of controlled power flow in a construction equipment. Power flow in automotive applications is similar. Notice that modern construction equipment has electronic control modules (ECMs) for most major subsystems such as engine, transmission, brake, steering, and implement subsystems.

and land mapping for optimal utilization, large-scale farming has started to be done by autonomous harvesters where the machine is automatically guided and steered by GPS systems. Farmlands are fertilized in an optimal manner based on previously collected satellite maps. For instance, the planning and execution of an earth moving job, such as road building or a construction site preparation or farming, can be done completely under the control of GPSs and autonomously driven machines without any human operators on the machine. However, safety concerns have so far delayed the introduction of such autonomous machine operations. The underlying technologies are relatively mature for autonomous construction equipment and farm equipment operation (Fig. 1.13).

The chemical process industry involves many large-scale computer-controlled plants. The early application of computer-controlled plants were based on a large central computer controlling most of the activities. This is called the *centralized control* model. In recent years, as the microcontrollers became more powerful and low cost, the control systems for large plants are designed using many layers of hierarchy of controllers. In other words, the control logic is distributed physically to many microcomputers. Each microcomputer is physically closer to the sensors and actuators for which it is responsible. Distributed controllers communicate with each other and higher level controllers over a standard communication network. There may be a separate communication network at each layer of the hierarchical control

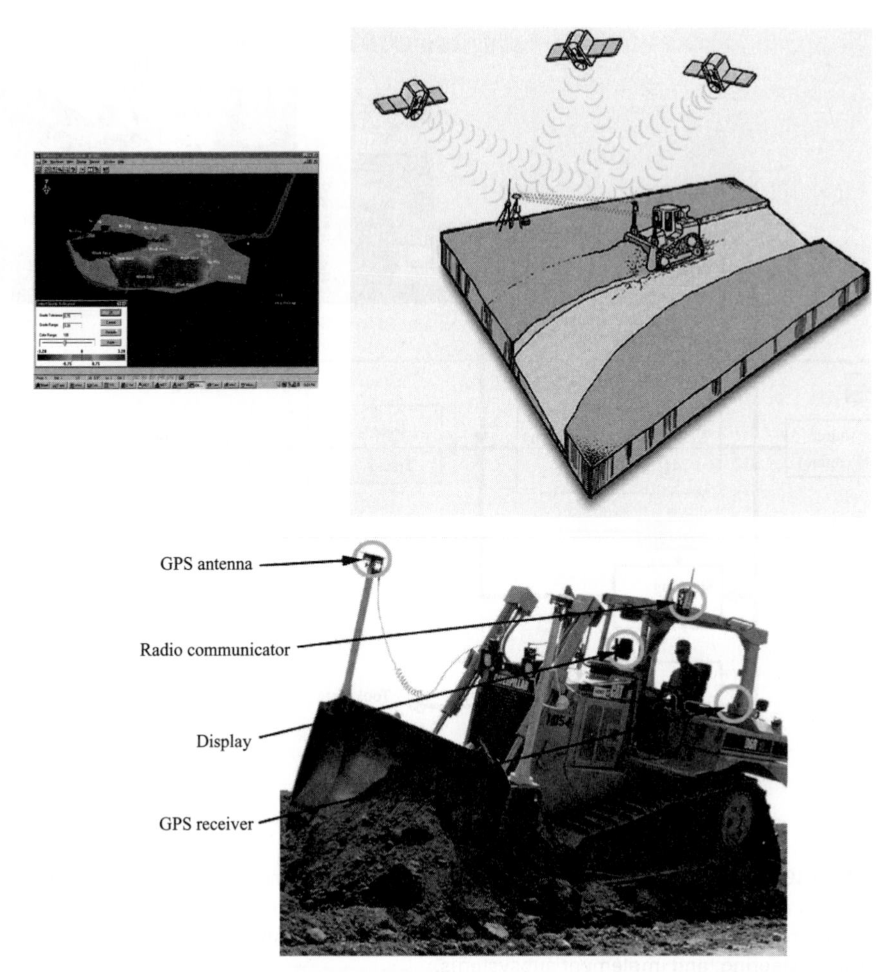


**FIGURE 1.13:** Semiautonomous construction equipment operation using global positioning system (GPS), local sensors, and on-vehicle sensors for closed-loop subsystem control.

system. The typical variables of control in process industry are fluid flow rate, temperature, pressure, mixture ratio, fluid level in tank, and humidity.

Energy management and control of large buildings is a growing field of applications of optimized computer control. Home appliances are more and more microprocessor controlled, instead of being just electromechanical appliances. For instance, old ovens used relays and analog temperature controllers to control the electric heater in the oven. The new ovens use a microcontroller to control the temperature and timing of the oven operation. Similar changes occur in many other appliances used in homes, such as washers and dryers.

Microelectromechanical systems (MEMS) and MEMS devices incorporate all of the computer-control, electrical, and mechanical aspects of the design directly on the silicon substrate in such a way that it is impossible to discretely identify each functional component. Finally, the application of mechatronic design in medical devices, such as surgery assistive devices, robotic surgery, intelligent drills, is perhaps one of the most promising fields in this century.

## 1.2 CASE STUDY: MODELING AND CONTROL OF COMBUSTION ENGINES

The internal combustion engine is the power source for most of the mobile equipment applications including automotive, construction, and agricultural machinery. As a result, it is an essential component in most mobile equipment applications. Here we discuss the modeling and basic control concepts of internal combustion engines from a mechatronics engineering point of view. This case study may serve as an example of how a dynamic model and a control system should be developed for a computer-controlled electromechanical system. Basic modeling and control of any dynamic system invariably involves use of Laplace transforms, which is covered in the next chapter. As a result, detailed analysis using Laplace transforms is minimized here.

We will discuss the basic characteristics of a diesel engine from a mechatronics engineer's point of view. Any modeling and control study should start with a good physical understanding of how a system works. We identify the main components and subsystems. Then each component is considered in terms of its input and output relationship in modeling. For control system design purposes, we identify the necessary sensors and controlled actuators. With this guidance, we study:

1. Engine components—basic mechanical components of the engine
2. Operating principles and performance—how energy is produced (converted from chemical energy to mechanical energy) through the combustion process
3. Electronic control system components—actuators, sensors, and electronic control module (ECM)
4. Dynamic models of engine from a mechatronics engineer's point of view
5. Control algorithms—basic control algorithms and various extensions in order to meet fuel efficiency and emission requirements

The engine converts chemical energy of fuel to mechanical energy through the combustion process. In mobile equipment, subsystems derive their power from the engine. There are two major categories of internal combustion engines: (1) Clerk (two-stroke) cycle engine and (2) Otto (four-stroke) cycle engine. In a two-stroke cycle engine, there is a combustion in each cylinder once per revolution of the crankshaft. In a four-stroke cycle engine, there is a combustion in each cylinder once every two revolution of the crankshaft. Only four-stroke cycle engines are discussed here.

Four-stroke cycle internal combustion engines are also divided into two major categories: (1) gasoline engines and (2) diesel engines. The fundamental difference between them is in the way the combustion is ignited every cycle in each cylinder. Gasoline engines use a spark plug to start the combustion, whereas the combustion is self-ignited in diesel engines as a result of the high temperature rise (typical temperature levels in the cylinder toward the end of the compression cylce are around the 700°C–900°C range) due to the large compression ratio. If the ambient air temperature is very low (i.e., extremely cold conditions), the temperature rise in the cylinder of a diesel engine due to the compression of air–fuel mixture may not be high enough for self-ignition. Therefore, diesel engines have electric heaters to pre-heat the engine block before starting the engine in a very cold environment.

The basic mechanical design and size of the engine define an envelope of maximum performance (speed, torque, power, and fuel consumption). The specific performance of an engine within the envelope of maximum performance is customized by the engine controller. The decision block between the sensory data and fuel injection defines a particular performance within the bounds defined by the mechanical size of the engine. This decision block includes considerations of speed regulation, fuel efficiency, and emission control.

### 1.2.1 Diesel Engine Components

The main mechanical components of a diesel engine are located on the engine block (Fig. 1.14). The engine block provides the frame for the combustion chambers where each combustion chamber is made of a cylinder, a piston, one or more intake valves and exhaust valves, and a fuel injector. The power obtained from the combustion process is converted to the reciprocating linear motion of the piston. The linear motion of the piston is converted to a unidirectional, continuous rotation of the crankshaft through the connecting rod. In the case of a spark-ignited engine (gasoline engine), there would also be a spark plug to generate ignition. The main difference between gasoline engine and diesel engine is in the way the combustion process is started. In gasoline engines, the combustion in the compressed air–fuel mixture is started by the ignition generated by a spark plug in each cycle. In diesel engines, the combustion is self-ignited by the temperature rise as a result of the high compression ratio. The compression ratio of diesel engines is in the 1:14–1:24 range, while the gasoline engine compression ratio range is about half that.

Normally, there are multiple cylinders (i.e., 4, 6, 8, 12) where each cylinder operates with a different crankshaft phase angle from each other in order to provide nonpulsating power. An engine power capacity is determined primarily by the number of cylinders, volume of each cylinder (piston diameter and stroke length), and compression ratio. Figure 1.15 shows the engine block and its surrounding subsystems: throttle, intake manifold, exhaust manifold, turbo charger, and charge cooler. In most diesel engines, there is no physical throttle valve. A typical diesel engine does not control the inlet air; it takes the available air and controls the injected fuel rate, while some diesel engines control both the inlet air (via the throttle valve) and the injected fuel rate.

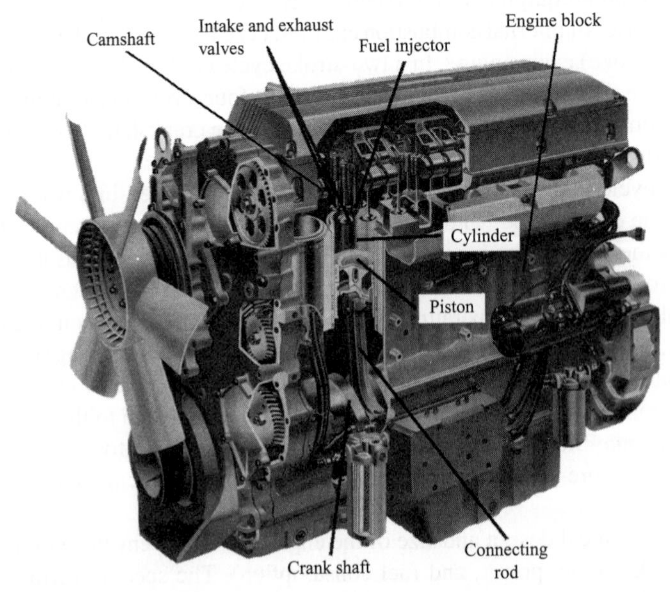


**FIGURE 1.14:** Mechanical components of an engine: engine block, cylinders, pistons, connecting rod, crankshaft, camshaft, intake valves, exhaust valves, and fuel injectors.

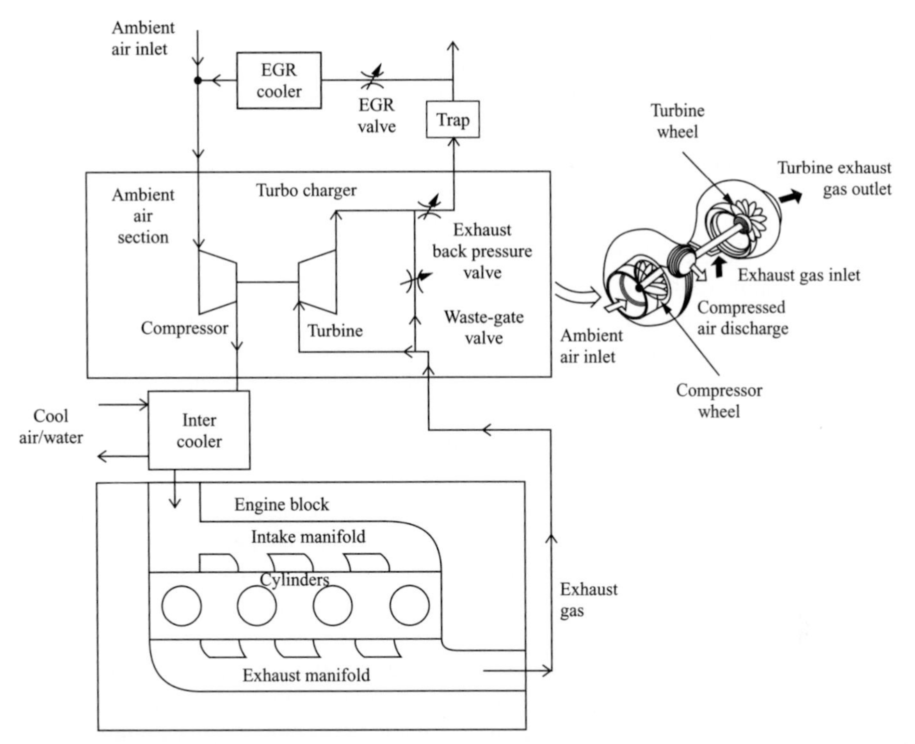


**FIGURE 1.15:** The engine and its surrounding subsystems: intake manifold, exhaust manifold, turbo charger with waste-gate valve, charge inter cooler, exhaust gas recirculation (EGR), and trap or catalytic converter.

The surrounding subsystems support the preparation of air and fuel mixture before the combustion and exhausting it. Timing of the intake valve, exhaust valve, and injector is controlled either by mechanical means or by electrical means. In completely mechanically controlled engines, a mechanical *camshaft* coupled to the *crankshaft* by a timing belt with a 2:1 gear ratio is used to control the timing of these components, which is periodic with two revolutions of the crankshaft. *Variable valve control* systems incorporate a mechnically controlled lever which adjusts the phase of the camshaft sections in order to vary the timing of the valves. Similar phase adjustment mechanisms are designed into individual fuel injectors as well [2b, 2c]. In electronically controlled engines some or all of these components (fuel injectors, intake and exhaust valves) are each controlled by electrical actuators (i.e., solenoid actuated valves). In today's diesel engines, the injectors are electronically controlled, while the intake valves and exhaust valves are controlled by the mechanical camshaft. In the fully electronically controlled engine, also called *camless engine*, the intake and exhaust valves are also electronically controlled.

*Turbo charger* (also called *super charger*) and *charge cooler* (also called *inter cooler*, or *after cooler*) are passive mechanical devices that assist in the efficiency and maximum power output of the engine. The turbo charger increases the amount of air pumped ("charged") into the cylinders. It gets the necessary energy to perform the pumping function from the exhaust gas. The turbo charger has two main components: (1) the turbine and (2) the compressor, which are connected to the same shaft. Exhaust gas rotates the turbine, and it in turns rotates the compressor which performs the pumping action. By making partial use of the otherwise wasted energy in the exhaust gas, the turbo charger pumps more air, which in turn means more fuel can be injected for a given cylinder size. Therefore, an engine can generate more power from a given cylinder size using a turbo charger. An engine without a turbo charger is called the *naturally aspirated engine*. The turbo charger gain is a function of the turbine speed, which is related to the engine speed. Therefore, some turbo chargers have variable blade orientation or moving nozzle [called *variable geometry turbo chargers (VGT)*] to increase the turbine gain at low speed and reduce it at high speed (Fig. 1.15).

While the main purpose of the turbo charger is to increase the amount of inlet air pumped into the cylinders, it is not desirable to increase the inlet boost pressure beyond a maximum value. Some turbo charger designs incorporate a *waste-gate* valve for that purpose. When the boost pressure sensor indicates that the pressure is above a certain level, the electronic control unit opens a solenoid actuated butterfly type valve at the waste-gate. This routes the exhaust gas to bypass the turbine to the exhaust line. Hence the name "waste-gate" because it wastes the exhaust gas energy. This reduces the speed of the turbine and the compressor. When the boost pressure drops below a value, the waste-gate valve is closed again and the turbo charger operates in its normal mode.

Another feature of some turbo chargers is the *exhaust back pressure device*. Using a butterfly type valve, the exhaust gas flow is restricted and hence the exhaust back pressure is increased. As a result, the engine experiences larger exhaust pressure resistance. This leads to faster heating of the engine block. This is used for rapid warming of the engine under cold starting conditions.

In some turbo charger designs, in order to reduce the cylinder temperature, *charge cooler* (also called *inter cooler*) is used between the turbo charger's compressor output and the intake manifold. The turbo charger's compressor outputs air with temperature as high as 150°C. The ideal temperature for inlet air for a diesel engine is around 35°C to 40°C. The charge cooler performs the cooling function of the intake air so that air density can be increased. High air temperature reduces the density of the air (hence, the air–fuel ratio) as well as increases the wear in the combustion chamber components.

The exhaust gas recirculation (EGR) mixes the intake air with a controlled amount of exhaust gas for combustion. The main advantage of the EGR is the reduction of NOx content in the emission. However, EGR results in more engine wear and increases smoke and particulate content in the emission.

Fuel is injected to the cylinder by cam-actuated (mechanically controlled) or solenoid-actuated (electrically controlled) injectors. The solenoid actuation force is amplified by hydraulic means in order to provide the necessary force for the injectors. Figure 1.16 shows an electrically controlled fuel injector system where a hydraulic oil pressure line is used as the amplifier stage between the solenoid signal and injection force. Notice that the main components of the fuel delivery system are the:

1. Fuel tank
2. Fuel filter
3. Fuel pump
4. Pressure regulator valve
5. High-pressure oil pump
6. Fuel injectors

The fuel pump maintains a constant fuel pressure line for the injectors. In common-rail (CR) and electronic unit injector (EUI) type fuel injection systems (such as the hydraulic electronic unit injector (HEUI) by Caterpillar and Navistar Inc.), a high-pressure oil pump in conjunction with a pressure-regulating valve is used to provide the

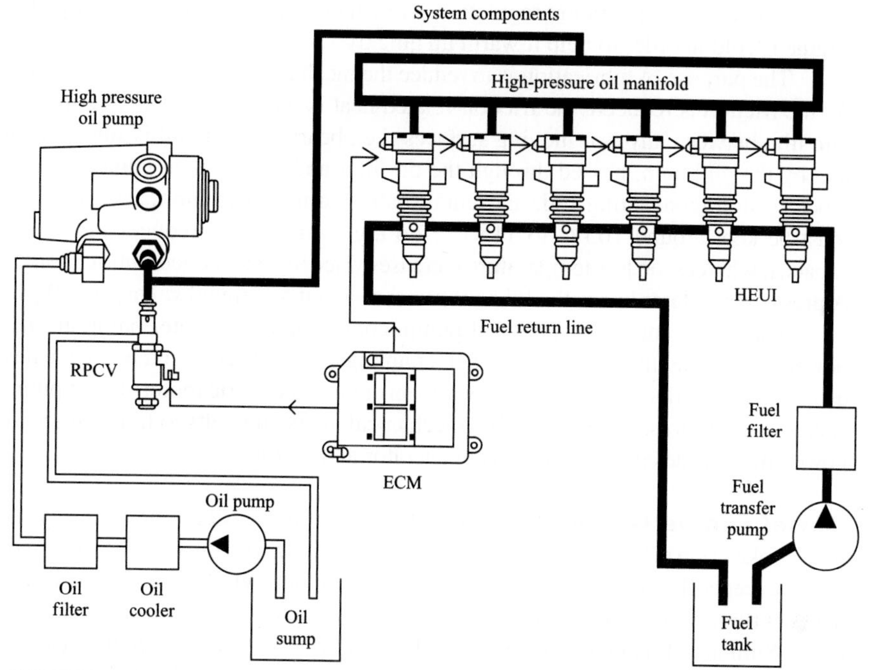


**FIGURE 1.16:** An example of an electronically controlled fuel injection system used in diesel engines: HEUI fuel injection system by Caterpillar Inc. and Navistar Inc.

high-pressure oil line to act as an amplifier line for the injectors. Hydraulic oil is the same oil used for engine lubrication. The injectors are controlled by the low-power solenoid signals coming from the electronic control module (ECM). The motion of the solenoid plunger is amplified by the high-pressure oil line to provide the higher power levels needed for fuel injector. The typical actuation timing accuracy of an HEUI injector is around 5 $\mu$sec.

There are four different fluids involved in any internal combustion engine:

1. Fuel for combustion
2. Air for combustion and cooling
3. Oil for lubrication
4. Water-coolant mixture for cooling

Each fluid (liquid or gas) circuit has a component to store, condition (filter, heat, or cool), move (pump), and direct (valve) it within the engine.

The cooling and lubrication systems are closed-circuit systems which derive their power from the crankshaft via a gearing arrangement to the coolant pump, and the oil pump. Reservoir, filter, pump, valve, and circulation lines are very similar to other fluid circuits. The main component of the cooling system is the radiator. It is a heat exchanger where the heat from the coolant is removed to the air through a series of convective tubes or cores. The coolant is used not only to remove heat from the engine block, but also to remove heat from the intake air at after cooler (inter cooler) as well as to remove heat from the lubrication oil. Finally the heat is dissipated out to the environment at the radiator. The radiator fan provides forced air for higher heat exhange capacity. Typically, the cooling system includes a temperature regulator valve which directs the coolant flow path when the engine is cold in order to help it warm up quickly.

The purpose of lubrication is to reduce the mechanical friction between two surfaces. As the friction is reduced, the friction-related heat is reduced. The lubrication oil forms a thin film between any two moving surfaces (i.e., bearings). The oil is sucked from the oil pan by the oil pump, passed through the oil filter and cooler, then guided to the cylinder block, piston, connecting rod, and crankshaft bearings. The lubrication oil temperature must be kept around 105°C to 115°C. Too high temperature reduces the load handling capacity, whereas too low temperature increases viscosity and reduces lubrication capability. A pressure regulator keeps the lube oil pressure about a nominal value (40–50 psi range).

The fuel pump, lubrication oil pump, cooling fan, and water pump all derive their power from the crank with gear and belt couplings. The current trend in engine design is to use electric generators to transfer power from engine to electric motor-driven pumps for the subsystems. That is, instead of using mechanical gears and belts to transmit and distribute power, the new designs use electrical generator and motors.

***Diesel Engine Operating Principles*** Let us consider one of the cylinders in a four-stroke cycle diesel engine (Fig. 1.17). Other cylinders go through the same sequence of cycles except offset by a crankshaft phase angle. In a four-cylinder diesel engine, each cylinder goes through the same sequence of four-stroke cycles offset by 180 degrees of crankshaft angle. Similarly, this phase angle is 120 degrees for a six-cylinder engine, and 90 degrees for an eight-cylinder engine. The phase angle between cylinders is (720 degrees)/(number of cylinders). During the intake stroke, the intake valve opens and the exhaust valve closes. As the piston moves down, the air is sucked into the cylinder until the piston reaches the bottom dead center (BDC). The next stroke is the compression stroke during which the intake valve closes and as the piston moves up, the air is compressed. The fuel injection (and spark ignition in the SI engine) is started at some position before the piston reaches the top dead center (TDC).

The combustion, and the resulting energy conversion to the mechanical energy, are accomplished during the expansion stroke. During that stroke, the intake valve and exhaust valve are closed. Finally, when the piston reaches the BDC position and starts to move up, the exhaust valve opens to evacuate the burned gas. This is called the exhaust stroke. The cycle ends when the piston reaches the TDC position.

This four-stroke cycle repeats for each cylinder. Note that each cylinder is in one of these strokes at any given time. For the purpose of illustrating the basic operating principle, we stated above that the intake and exhaust valves open and close at the end or beginning of each cycle. In an actual engine, the exact opening and closing position of these valves, as well as the fuel injection timing and duration, are a little different than the BDC or TDC positions of the piston.

It is indeed these intake and exhaust valve timings as well as the fuel injection timing (start time, duration, and injection pulse shape) decisions that are made by the electronic control module (ECM) in real time. The control decisions are made relative to the crankshaft angular position based on a number of sensory data. The timing relative to the crankshaft position may be varied as a function of engine speed in order to optimize the engine performance. The delay from the time current pulse sent to the injector and the time that combustion is fully developed is in the order of 15 degrees of crankshaft angle. The typical shape of the pressure in the cylinder during a four-stroke cycle is also shown in the Fig. 1.17. The maximum combustion pressure is in the range of 30 bar (3 Mpa) to 160 bar (16 Mpa). Notice that even though the pressure in the cylinder is positive, the torque contribution of each cylinder as a result of this pressure is positive during the cycle when

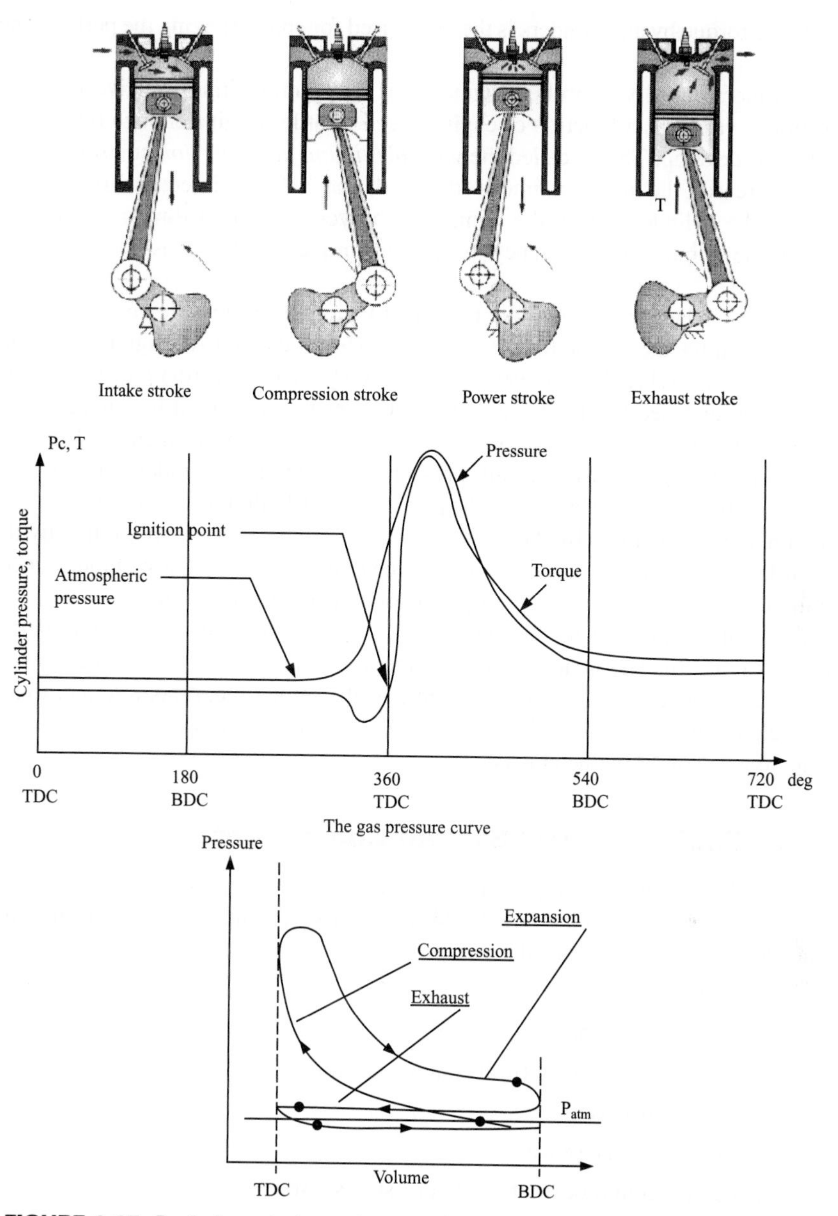


**FIGURE 1.17:** Basic four-stroke cycle operation of a diesel engine: intake, compression, expansion, exhaust stroke. The pressure in a cylinder is a function of crankshaft position. Other cylinders have almost identical pressures as a function of crankshaft angle, except with a phase angle. Compression ratio is the ratio of the cylinder volume at BDC ($V_{BDC}$) to the cylinder volume at TDC ($V_{TDC}$). Notice that during compression stroke, the cylinder pressure opposes the crank motion; hence, the effective torque is negative. During expansion stroke, the cylinder pressure supports the crank motion; hence, the effective torque is positive.

the piston is moving down (the pressure is helping the motion, and the net torque contribution to the crankshaft is positive) and negative during the cycle when the piston is moving up (the pressure is opposing the motion, and the net torque contribution to the crankshaft is negative). As result, the net torque generated by each cylinder oscillates as a function of crankshaft angle with a period of two revolutions. The mean value of that generated torque by all cylinders is the value used for characterizing the performance of the engine.

In electronically controlled engines, the fuel injection timing relative to the crankshaft position is varied as a function of engine speed in order to give enough time for the combustion to develop. This is called the *variable timing fuel injection* control. As the engine speed increases, the injection time is advanced; that is, fuel is injected earlier relative to the TDC of the cylinder during the compression cycle. Injection timing has significant effect on the combustion efficiency, hence the torque produced, as well as the emission content.

It is standard in the literature to look at the cylinder pressure versus the combustion chamber volume during the four-stroke cycle. The so-called p-v diagram shape in general is shown in Fig. 1.17. The net energy developed by the combustion process is proportional to the area enclosed by the p-v diagram. In order to understand the shape of the torque generated by the engine, let us look at the pressure curve as a function of the crankshaft (Fig. 1.17) and superimpose the same pressure curve for other cylinders with the appropriate crankshaft phase angle. The sum of the pressure contribution from each cylinder is the total pressure curve generated by the engine (Fig. 1.17). The pressure multiplied by the piston top surface area is the net force generated. The effective moment arm of the connecting rod multiplied by the force gives the torque generated at the crankshaft.

The net change in the acceleration is the net torque divided by the inertia. If the inertia is large, the transient variations in the net torque will result in smaller acceleration changes, and hence smaller speed changes. At the same time, it takes longer to accelerate or decelerate the engine to a different speed. These are the advantages and disadvantages of the flywheel used on the crankshaft.

### 1.2.2 Engine Control System Components

There are three groups of components of an engine control system: (1) sensors, (2) actuators, and (3) electronic control module (ECM) (Fig. 1.18). The number and type of sensors used in an electronic engine controller vary from manufacturer to manufacturer. The following is a typical list of sensors used:

1. Accelerator pedal position sensor
2. Throttle position sensor (if the engine has throttle)
3. Engine speed sensor
4. Air mass flow rate sensor
5. Intake manifold (boost) absolute pressure sensor
6. Atmospheric pressure sensor
7. Manifold temperature sensor
8. Ambient air temperature sensor
9. Exhaust gas oxygen (EGO) sensor
10. Knock detector sensor (piezoaccelerometer sensor)

The controlled actuators in an engine (outputs) are:

1. Fuel injector actuation: injection timing, duration (injected fuel amount, also called fuel rate), and pulse shape control
2. Ignition sparks: timing (only in spark-ignited gasoline engines, not used in diesel engines)

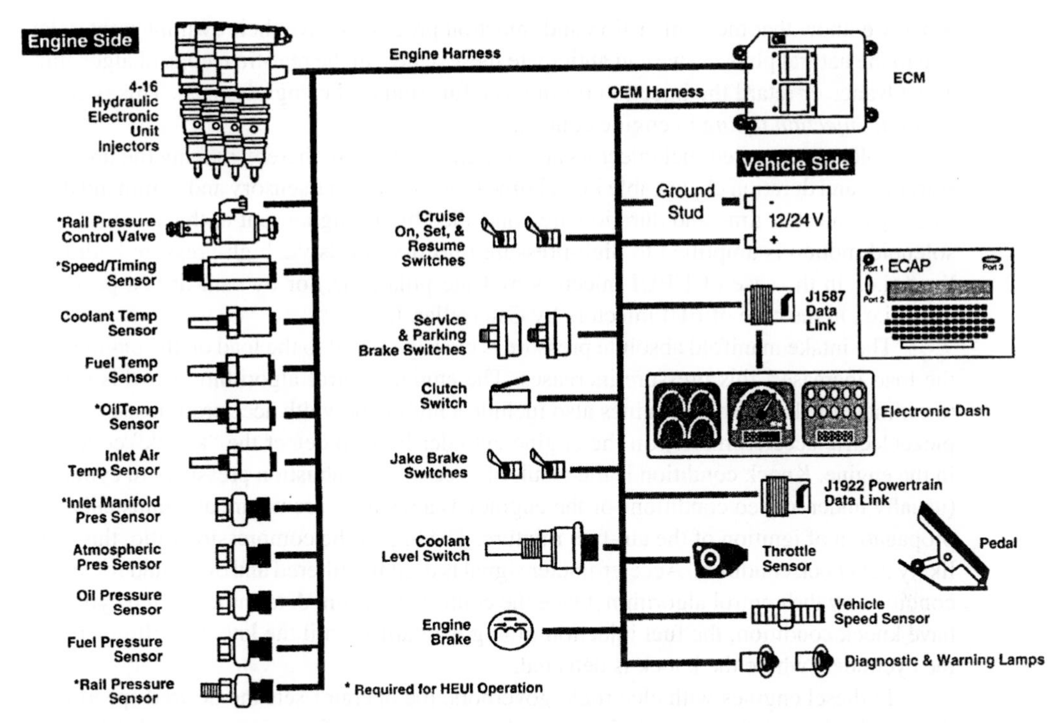


**FIGURE 1.18:** Control system components for a modern engine controller: sensor inputs, ECM (electronic control module), and outputs to actuators.

3. Exhaust gas recirculation (EGR) valve
4. Idle air control (IAC) valve, which may not be present in all engine designs

ECM is the digital computer hardware which has the interface circuitry for the sensors and actuators, and runs the engine control algorithm (engine control software) in real time. The engine control algorithm implements the control logic that defines the relationship between the sensor signals and actuator control signals. The main objectives of engine control are:

1. Engine speed control
2. Fuel efficiency
3. Emission concerns

In its simplest form, the engine control algorithm controls the fuel injector in order to maintain a desired speed set by the accelerator pedal position sensor.

The actively controlled variables by ECM are the injectors (when and how much fuel to inject is indicated by an analog signal per injector) and the RPCV valve which is used to regulate the pressure of the amplification oil line. As a result, for a six-cylinder, four-stroke cycle diesel engine, the engine controller has seven control outputs: six outputs (one for each injector solenoid) and one output for the RPCV valve (Fig. 1.16). Notice that at 3000-rpm engine speed, 36 degrees of crankshaft rotation takes only about 2.0 msec, which is about the window of opportunity to complete the fuel injection. Controlling the injection start time with an accuracy of 1 degree of crankshaft position requires 55.5-microsecond repeatability in the fuel injection control system timing. Therefore, accuracy of controlling the injection start time and duration at different engine speeds is clearly very important.

Since we know that the combustion and injection processes have their own inherent delay due to the natural physics, we can anticipate these delays in the real-time control algorithm, and advance or retard the injection timing as a function of the engine speed. This is called *variable injection timing* in engine control.

Solenoid-actuated fuel injectors are digitally controlled, thereby making the injection start time and duration changeable in real time based on various sensory and command data. The injection start time and duration are controlled by the signal sent to the solenoid. The solenoid motion is amplified to high-pressure injection levels via high-pressure hydraulic lines (i.e., in the case of HEUI injectors by Caterpillar, Inc.) or by cam-driven push rod arms (i.e., in the case of EUI injectors by Caterpillar, Inc.).

The intake manifold absolute pressure is closely related to the load on the engine—as the load increases, this pressure increases. The engine control algorithm uses this sensor to estimate the load. Some engines also include a high-bandwidth acceleration sensor (i.e., piezoelectric accelerometer) on the engine cylinder head to detect the "knock" condition in the engine. Knock condition is the result of excessive combustion pressures in cylinders (usually under loaded conditions of the engine) as a result of premature and unusually fast propagation of ignition of the air–fuel mixture. The higher the compression ratio, the more likely the knock condition. Accelerometer signal is digitally filtered and evaluated for knock condition by the control algorithm. Once the control algorithm determines which cylinders have knock condition, the fuel injection timing is retarded until the knock is eliminated in the cyclinders where the knock is detected.

In diesel engines with electronic governors, the operator sets the desired speed with the pedal which defines the desired speed as percentage of maximum speed. Then the electronic controller modulates the fuel rate up to the maximum rate in order to maintain that speed. The engine operates along the vertical line between the desired speed and the lug curve (Fig. 1.19). If the load at that speed happens to be larger than the maximum torque the engine can provide at that speed, the engine speed drops and torque increases until the balance between load torque and engine torque is achieved. In most gasoline engines, the operator pedal command is a desired engine torque. The driver closes the loop on the engine speed by observing and reacting to the vehicle speed. When "cruise control" is activated, then the electronic controller regulates the engine fuel rate in order to maintain the desired vehicle speed.

### 1.2.3 Engine Modeling with Lug Curve

If we neglect the transient response delays in the engine performance and the oscillations of engine torque within one cycle (two revolution of crank angle), the steady-state performance of an engine can be described in terms of its mean (average) torque over each cycle, power, and fuel efficiency as a function of engine speed (Fig. 1.19). The most important of these three curves that defines the capabilities of an engine is the torque-speed curve. This curve is also called the "lug curve" due to its shape. As the speed is reduced down from the rated speed, the mean torque generated by the engine increases under constant fuel rate conditions. Hence, if load increases to slow down the engine, the engine inherently increases torque to overcome the load. In order to define the lug curve model for an engine, we need a table of torque versus the engine speed for maximum fuel rate. A linear interpolation between intermediate points is satisfactory for initial analysis. The table should have, in minimum, the low idle, high idle, peak torque, and rated speed points (four data points). The points under that curve are achieved by lower fuel rates. As the fuel injection rate decreases from its maximum value, the lug curve does not necessarily scale linearly with it. At very low fuel rates, the combustion process works in such a way that the shape of the lug curve

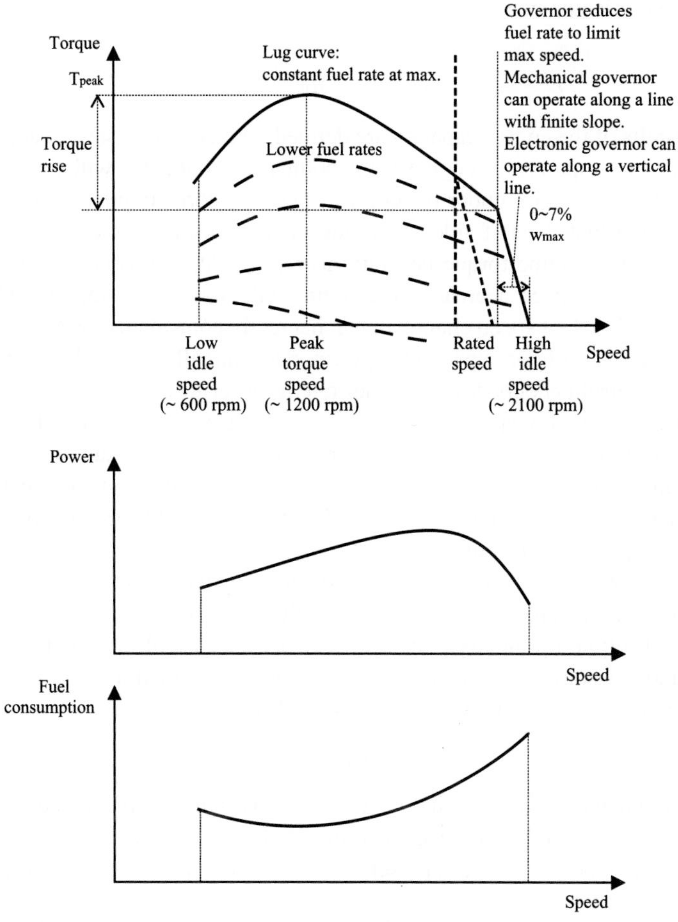


**FIGURE 1.19:** Steady-state engine performance: torque (lug), power, and fuel efficiency as function of engine speed.

becomes quite different (Fig. 1.19). The best fuel rate is accomplished when the engine speed is around the middle of low and high idle speeds.

The torque defined by the lug curve is the *mean effective* steady-state torque capacity of the engine as a function of speed at maximum fuel rate. This curve does not consider the cyclic oscillations of net torque as a result of the combustion process. The actual torque output of the engine is oscillatory as a function of the crankshaft angle within each four-stroke (intake-compression-combustion-exhaust) per two-revolution cycle. The role of the flywheel is to reduce the oscillations of engine speed. Each cylinder has one cycle of net torque contribution (which has both positive and negative portions) per two revolutions, and the torque functions of each cyclinder are phase shifted from each other.

A given mechanical engine size defines the boundaries of this curve:

1. Maximum engine speed (high idle speed) determined by the friction in the bearings and combustion time needed
2. Lug portion of the curve corresponding to torque-speed relation when the maximum fuel rate is injected to the engine. This is determined by the heat capacity of the engine

as well as the injector size.

$$T_{eng}^{lug} = f_o^{lug}(w_{eng}) \qquad \text{for} \qquad u_{fuel} = u_{fuel}^{max} \tag{1.1}$$

Once the limits of the engine capacity are defined by mechanical design and sizing of the components, we can operate the engine at any point under that curve with the governor—the engine contoller. Any point under the engine lug curve corresponds to a fuel rate and engine speed. Notice that the left side of the lug curve (speeds below the peak torque) is unstable because the engine torque capacity decreases. At speeds below this point, if the load increases as the speed decreases, the engine will stall. The slope of the high idle speed setting can be 3–7% of speed change in mechanically controlled engines, or it can be made perpendicular (almost zero speed change) in electronically controlled engines. The governor (engine controller) cuts down the fuel rate in order to limit the maximum speed of the engine. The sharp, almost vertical line on the lug curve (steady-state torque-speed curve) is the result of the governor action to limit the engine speed by changing the fuel rate from its maximum value at the rated speed down to smaller values. Hence, electronically controlled engines can maintain the same speed as the load conditions vary from zero to maximum torque capacity of the engine.

The parameterized version of the lug curve (torque versus engine speed), where there is a curve defined for each value of fuel rate, is called the *torque map*. The engine is modeled as a torque source (output: torque) as a function of two input variables: fuel rate and engine speed. Such a model assumes that the necessary air flow is provided in order to satisfy the fuel rate to torque transfer function (Figs. 1.19 and 1.20).

$$T_{eng} = f_0(u_{fuel}, w_{eng}) \tag{1.2}$$

Notice that, when $u_{fuel} = u_{fuel}^{max}$, the function $f_0(\cdot, \cdot)$ represents the lug curve. In order to define the points below the lug curve, data points need to be specified for different values of fuel rate (Fig. 1.19). An analytical representation of such a model can be expressed as

$$T_{eng} = \left( \frac{u_{fuel}}{u_{fuel}^{max}} \right) \cdot f_0 \left( u_{fuel}^{max}, w_{eng} \right) \tag{1.3}$$

Clearly, if the engine capacity limits are to be imposed on the model, the lug curve limits can be added as a nonlinear block in this model. The stardard lug curve defines the steady-state torque-speed relationship for an engine when the maximum fuel rate is injected. This is also called the *rack stop* limit of the torque-speed curve. Under the governor control (mechanical or electronic engine controller), any point (speed, torque) under that curve can be accomplished by a fuel rate that is less than the maximum fuel rate. In order to impose the engine lug curve limits, the following must be defined:

1. Torque-speed (the lug curve) for maximum fuel rate ($T_{eng}(w_{eng})$ when $u_{fuel} = u_{fuel}^{max}$)
2. The parameterized torque-speed curves for different fuel rates less than the maximum fuel rate

This model and the lug curve model do not include the transient behavior due to the combustion process. The simplest transient (dynamic) model to account for the delay between the fuel rate and generated torque can be included in the form of a first-order filter. In other words, there is a filtering type delay between the torque obtained from the lug curve and the actual torque developed by the engine for a given fuel rate and engine speed,

$$\tau_e \frac{dT_{eng}(t)}{dt} + T_{eng}(t) = T_{eng}^{lug}(t) \tag{1.4}$$

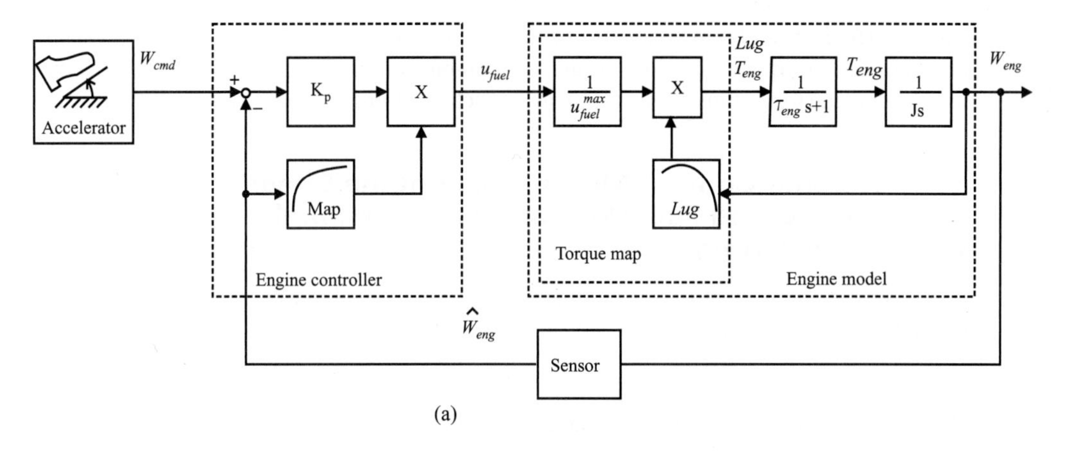


(a)

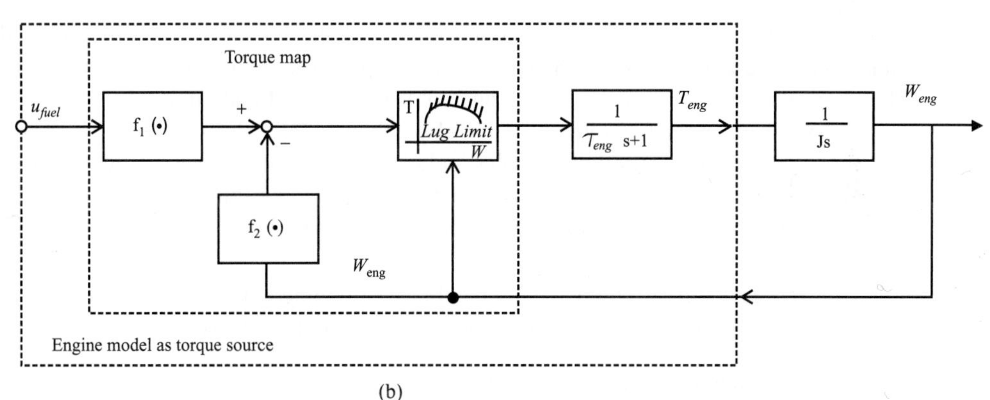


(b)

**FIGURE 1.20:** Engine model and its closed-loop control. (a) Engine is modeled as a relationship between the fuel rate and generated torque. The time delay and filtering effects due to combustion and injection are taken into account with first-order filter dynamics with a time constant of $\tau_{eng}$. Torque pulsation due to individual cylinder combustion is also neglected; however, it can be included by using such a model for individual cylinders. This model assumes that the lug curve scales linearly with the fuel injection rate. (b) This model assumes that the torque can be expressed as the difference of two functions where one function output depends on engine speed and the other on the fuel rate.

and its Laplace transform (readers who are not familiar with Laplace transforms can skip related material in the rest of this section without loss of continuity),

$$T_{eng}(s) = \frac{1}{(\tau_{eng} \cdot s + 1)} \cdot T_{eng}^{lug}(s) \tag{1.5}$$

where $\tau_{eng}$ represents the time constant of the combution to torque generation process, $T_{eng}^{lug}$ is the torque prediction based on lug curve, and $T_{eng}$ is the torque produced including the filtering delay.

***Special Case: A Simple Engine Model*** Simpler versions of this model can be used to represent engine steady state dynamics as follows:

$$T_{eng} = f_1(u_{fuel}) - f_2(w_{eng}) \tag{1.6}$$

where $T_{eng}$ is the torque generated by the engine, $u_{fuel}$ is the injected fuel rate, $w_{eng}$ is engine speed, $f_0(\cdot, \cdot)$ and represents the nonlinear mapping function between the two independent variables (fuel rate and engine speed) and the generated torque. The function $f_1(\cdot)$ represents the fuel rate to torque generation through the combustion process, and $f_2(\cdot)$ represents the load torque due to friction in the engine as a function of engine speed (Fig. 1.20).

### 1.2.4 Engine Control Algorithms: Engine Speed Regulation Using Fuel Map and a Proportional Control Algorithm

A very simple engine control algorithm may decide on the fuel rate based on the accelerator pedal position and engine speed sensors as follows (Fig. 1.20):

$$u_{fuel} = g_1(w_{eng}) \cdot K_p \cdot (w_{cmd} - w_{eng}) \tag{1.7}$$

where $g_1(\cdot)$ is the fuel rate look-up table as a function of engine speed and $K_p$ is a gain multiplying the speed error.

An actual engine control algorithm is more complicated and uses more sensory data and embedded engine data in the form of look-up tables, estimators, and various logic functions such as cruise control mode and cold start mode. In addition, control algorithm decides not only on the fuel rate ($u_{fuel}$), but also on the injection timing relative to the crank shaft position. Other controlled variables include the exhaust gas recirculation (EGR) and idle air control valves. However, simple engine control algorithms like this are useful in various stages of control system development in vehicle applications.

***Emission Issues*** A fundamental challenge in engine control comes from emission requirements. There are five major emission concerns in the exhaust smoke:

1. $CO_2$ content due to global warming effects
2. CO—health concerns (colorless, odorless, tasteless, but when inhaled in 0.3% of air can result in death within 30 minutes)
3. $NO_x$ content (NO and $NO_2$)
4. HC when combined with $NO_2$ becomes harmful to health
5. PM (particulate matter) solids and liquids present in the exhaust gas

The fuel injection timing has a major influence on the quality of combustion and hence the exhaust gas composition. The challenge is to design engines and controllers that will create efficient combustion while reducing all of the undesirable emission components in the exhaust gas.

## 1.3 PROBLEMS

1. Consider the mechanical closed-loop control system for the liquid level shown in Fig. 1.4.

**(a)** Draw the block diagram of the whole system showing how it works to maintain a desired liquid level.
**(b)** Modify this system with an electromechanical control system involving a digital controller. Show the components in the modified system and explain how they would function.
**(c)** Draw the block diagram of the digital control system version.

2. Consider the mechanical governor used in regulating the engine speed in Fig. 1.5.

**(a)** Draw the block diagram of the system and explain how it works.

**(b)** Assume you have a speed sensor on the crack shaft, an electrically actuated valve in place of the valve which is actuated by the fly-ball mechanism, and a microcontroller. Modify the system components for digital control version and draw the new block diagram.

**3.** Consider the web tension control system shown in Fig. 1.6.

**(a)** Draw the block diagram of the system and explain how it works.
**(b)** Assume analog electronic circuit of op-amp and command signal source is replaced by a microcontroller. Draw the new components and block diagram of the system. Explain how the new digital control system would work.
**(c)** Discuss what would be different in the real-time control algorithm if the microcontroller were to control the speed of the wind-off roll's motor instead of the wind-up roll's motor.

**4.** Consider the room temperature control system shown in Figure 1.7. The electric heater and fan motor is turned ON or OFF depending on the actual room temperature and desired room temperature.

**(a)** Draw the block diagram of the control system and explain how it works.
**(b)** Replace the op-amp, command signal source, and timer components with a microcontroller. Explain using pseudocode the main logic of the real-time software that must run on the microcontroller.

CHAPTER 2

# CLOSED-LOOP CONTROL

**THIS CHAPTER** contains the fundamental material on closed-loop control systems. Before one uses feedback, that is, closed-loop control, one should explore the option of open-loop control. We will address the following questions:

- What are the advantages and disadvantages of closed-loop control versus open-loop control?
- Why should we use feedback control instead of open-loop control?
- In what cases open-loop control may be better than closed-loop control?

A control system is designed to *make a system do what we want it to do*. Therefore, a control system designer needs to know the desired behavior or performance expected from the system. The performance specifications of a control system must cover certain fundamental characteristics such as stability, quality of response, and robustness. Despite the great variety and richness of the control theory, more than 90% of the feedback controller in practice are the proportional-integral-derivative (PID) type. Due to its wide usage in practice, PID control is considered a fundamental controller type. The PID control is discussed in the last section of this chapter.

The control decisions can be made either by an analog control circuit, in which case the controller is called an *analog controller*, or by a digital computer in which case the controller is called a *digital controller*. In analog control, the control decision rules are designed into the analog circuit hardware. In digital control, the control decision rules are coded in software. This software code implementing the control decisions is called the *digital control algorithm*.

The main advantages of digital control over analog control are as follows:

1. *Increased flexibility:* changing the control algorithm is a matter of changing the software. Making software changes in digital control is much easier than changing analog circuit design in analog control.
2. *Increased level of decision-making capability:* implementing nonlinear control functions, logical decision functions, conditional actions to be taken, learning from experience can all be programmed in software. Building analog controllers with these capabilities would be a prohibitive task, if not impossible.

It is important to identify the place of the control of dynamic systems in the big picture of control systems. Real-world control systems involve many discrete event controls involving sequencing and logic decisions. Discrete event control refers to the control logic based on sensors which provide only ON/OFF signals (i.e., limit switches, proximity sensors), and use actuators which have only a two-level state, ON/OFF (i.e., pneumatic cylinders controlled by an ON/OFF solenoid, relays). The sequence controllers use sensors and actuators which have only an ON/OFF state, and the control algorithm is a logic between the ON/OFF sensors and ON/OFF actuators. Such controls are generally implemented using programmable logic controllers (PLC) in the automation industry. The servo control loops may be part of such a control system. Closed-loop servo control is often a subsystem of the logic control systems where servo and logic control are hierarchically organized.

A control system is called *closed loop* if the control decisions are made based on some sensor signals. If the control decisions do not take any sensor signal of the controlled variables into account and decisions are made based on some predefined sequence or operator commands, such a control system is called *open loop*. It has been long recognized that using feed back information (sensor signals) about the controlled variable in determining the control action provides robustness against the changing conditions and disturbances.

## 2.1 COMPONENTS OF A DIGITAL CONTROL SYSTEM

Let us consider the control of a process using (1) analog control (Fig. 2.1) and (2) digital control (Fig. 2.2). The only difference is the controller box. In analog control, all of the signals are continuous, whereas in digital control the sensor signals must be converted to digital form, and the digital control decisions must be converted to analog signals to send to the actuation system.

The basic components of a digital controller are shown in Fig. 2.2:

1. Central processing unit (CPU) for implementing the logic and mathematical control algorithms (decision-making process)
2. Discrete state input and output devices (i.e., switches and lamps)
3. Analog to digital converter (A/D or PWM input) to convert the sensor signals from analog to digital signals
4. Digital to analog converter (D/A or PWM output) to convert the control decisions made by the control algorithm in the central processing unit (CPU) to the analog signal form so that it can be commanded to actuation system for amplification

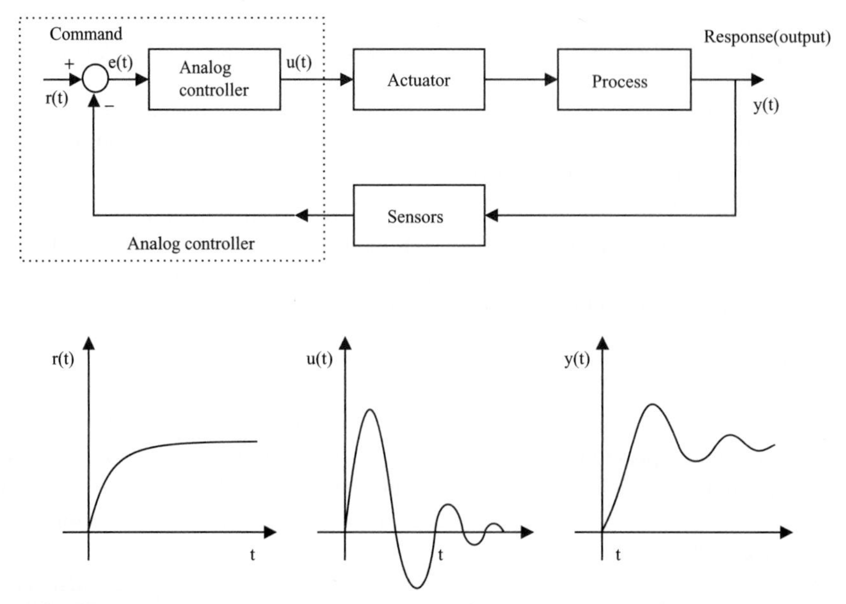


**FIGURE 2.1:** Analog closed-loop control system and the nature of the signals involved.

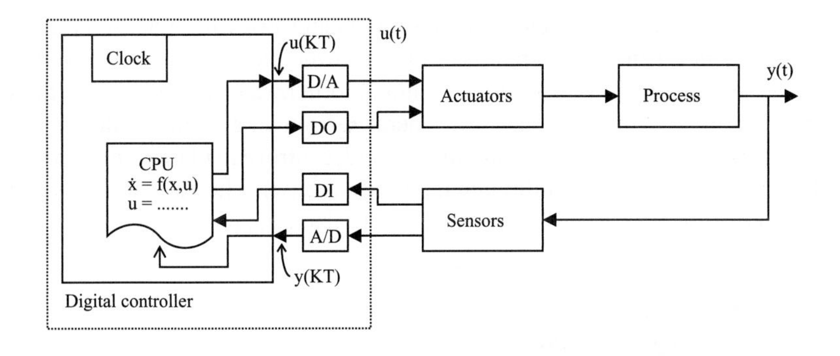


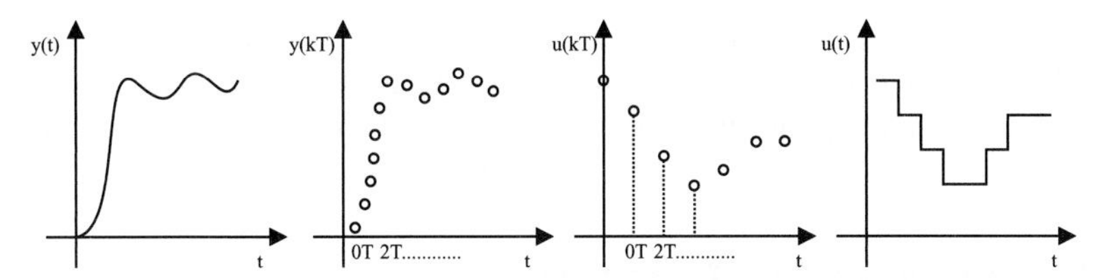


**FIGURE 2.2:** Digital closed-loop control system and the nature of the signals involved.

5. Clock for controlling the operation of the digital computer. The digital computer is a discrete device, and its operations are controlled by the clock cycle. The clock is for the computer what the heart is for a body.

In Figure 2.2, it is shown that the signals travel from sensors to the control computer in analog form. Similiarly, the control signals from the controller to the amplifier/actuator travel in analog form. The conversion of signal from analog to digital form (A/D converter) occurs at the control computer end. Similarly, the conversion of digital signal to analog signal occurs at the control computer end (D/A converter) and travels to the actuators in analog form. Recent trends in the computer-controlled systems are such that the analog to digital, and digital to analog conversion occurs at the sensor and actuator point. Such sensors and actuators are marketed as "smart sensors" and "smart actuators." In this approach, the signal travels from sensor point to control computer, and from control computer to the actuator point in digital form. Especially, the use of fiber-optic transmission medium provides very high signal transmission speed with high noise immunity. It also simplifies the interface problems between the computers, sensors, and amplifiers. In either case, digital input and output (DI/DO), A/D, and D/A operations are needed in a computer control system as interface between the digital world of computers and the analog world of real systems. The exact location of the digital and analog interface functions can vary from application to application.

Let us consider the operations performed by the components of a digital control computer and their implications compared to analog control:

1. Time delay associated with signal conversion (at A/D and D/A) and processing (at CPU)
2. Sampling
3. Quantization
4. Reconstruction

The digital computer is a discrete-event device. It can work with finite samples of signals. The sampling rate can be programmed based on the clock frequency. Every sampling period, the sensor signals are converted to digital form by the A/D converter (the sampling operation). If the command signals are generated from an external analog device, it also must be sampled. During the same sampling period, control calculations must be performed, and the result must be sent out through the D/A converter. The A/D and D/A conversions are finite precision operations. Therefore, there is always a quantization error.

## 2.2 THE SAMPLING OPERATION AND SIGNAL RECONSTRUCTION

Due to the fact that the controller is a digital computer, the following additional problems are introduced in a closed-loop control system: time delay associated with signal conversion and processing, sampling, quantization error due to finite precision, and reconstruction of signals.

### 2.2.1 Sampling: A/D Operation

In this section, we will focus on the sampling only and its implications. We will consider the sampling operation in the following order:

1. physical circuit of the sampler
2. mathematical model of sampling
3. implications of sampling

### 2.2.2 Sampling Circuit

Consider the sample and hold circuit shown in Fig. 2.3. When the switch is turned ON, the output will track the input signal. This is the sampling operation. When the switch is turned OFF, output will stay constant at the last value. This is the hold operation.

While the switch is ON, the output voltage is

$$\overline{y}(t) = \frac{1}{C}\int_0^t i(\tau)\,d\tau \tag{2.1}$$

where

$$i(t) = \frac{y(t) - \overline{y}(t)}{R} \tag{2.2}$$

Taking the Laplace transforms of the differential equations and substituting the value of $i$ from the second equation gives the input–output transfer function of the sample and hold circuit.

$$\overline{y}(s) = \frac{1}{(RCs + 1)} y(s) \tag{2.3}$$

While the switch is OFF, $i(t) = 0$; $\overline{y}$ remains constant (hold operation).

Let $T$ be sampling period, $T_0$ is the portion of $T$ for which the switch stays ON, and $T_1$ is the remaining portion of the sampling period during which the switch stays OFF (Fig. 2.4). If the input signal $y(t)$ changes as a step function, the output signal will track it according to the solution of the transfer function in response to the step input. Figure 2.4 shows the typical response of a realistic sample and hold circuit of an A/D converter.

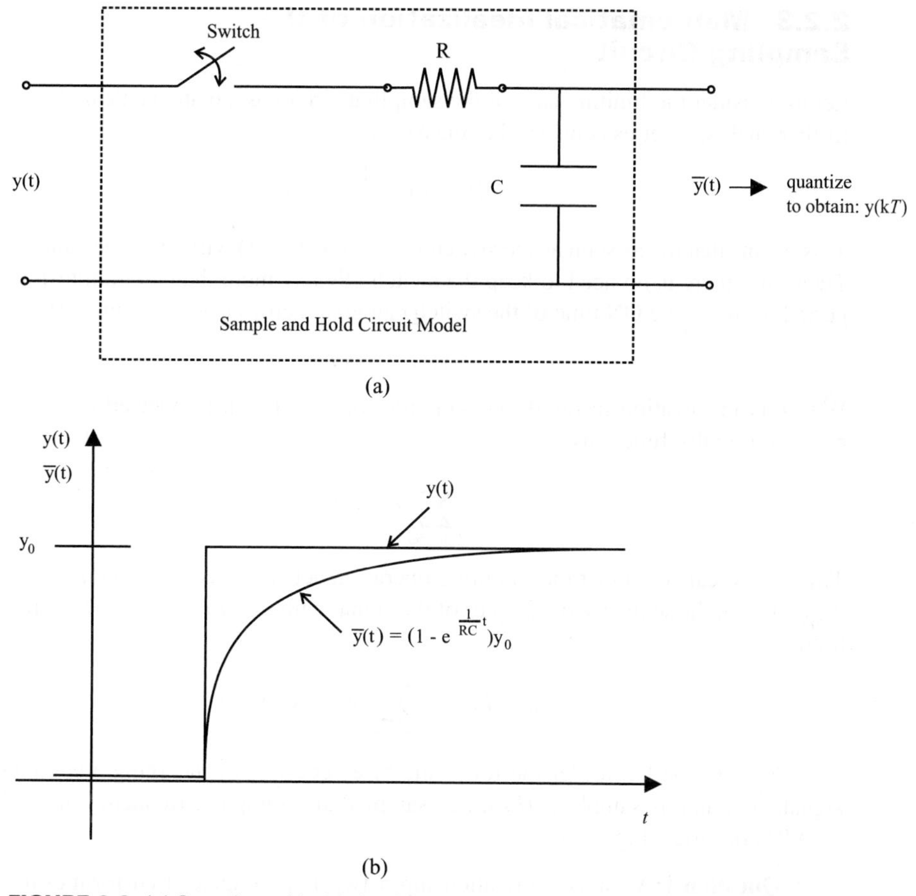


**FIGURE 2.3:** (a) Sample and hold circuit model and (b) response of the sampled voltage output.

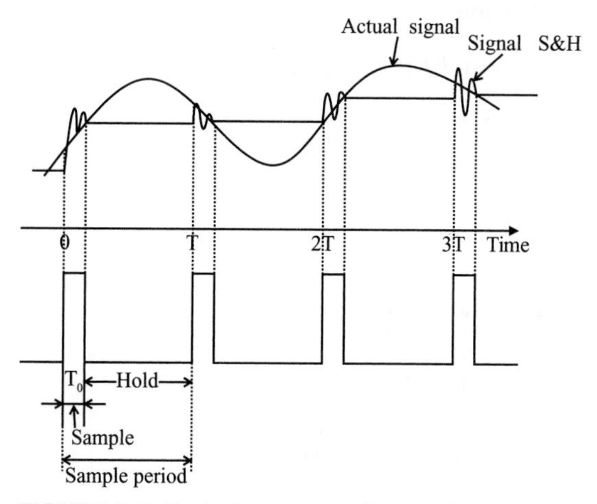


**FIGURE 2.4:** Typical response of a sample and hold circuit.

### 2.2.3 Mathematical Idealization of the Sampling Circuit

Let us consider the limiting case of the sampling circuit as a mathematical idealization for further analysis. Let us consider that the $RC$ value goes to zero.

$$RC \rightarrow 0; \frac{1}{RC} \rightarrow \infty$$

This means that the as soon as the switch is turned ON, $\overline{y}(t)$ will reach the value of the $y(t)$. Therefore, there is no need to keep the switch ON any more than an infinitesimally small period of time. The ON time of the switch can go to zero, $T_0 \rightarrow 0^+$ (Fig. 2.4).

$$y(t) \simeq y(kT) \tag{2.4}$$

With this idealization in mind, the sampling operation can be viewed as a sequence of periodic impulse functions.

$$\sum_{k=-\infty}^{+\infty} \delta(t - kT)$$

This is also called so that the sampling operation acts as a "comb" function (Fig. 2.5). If we represent the sequence of samples of the signal with $\{y(kT)\}$, the following relationship holds:

$$\{y(kT)\} = \sum_{k=-\infty}^{\infty} \delta(t - kT)y(t) \tag{2.5}$$

Now we will consider the following three questions concerning a continuous time signal, $y(t)$, and its samples, $y(kT)$; i.e., sampled at a sampling frequency, $w_s = 2\pi/T$ by an A/D converter (Fig. 2.6).

- **Question 1:** What is the relationship between the Laplace transform of the samples and the Laplace transform of the original continuous signal?

$$L\{y(kT)\} \ ? \ L\{y(t)\}$$

- **Question 2:** What is the relationship between Fourier transform of the samples and the Fourier transform of the original continuous signal? Shannon's sampling theorem provides the answer to this question:

$$F\{y(kT)\} \ ? \ F\{y(t)\}$$

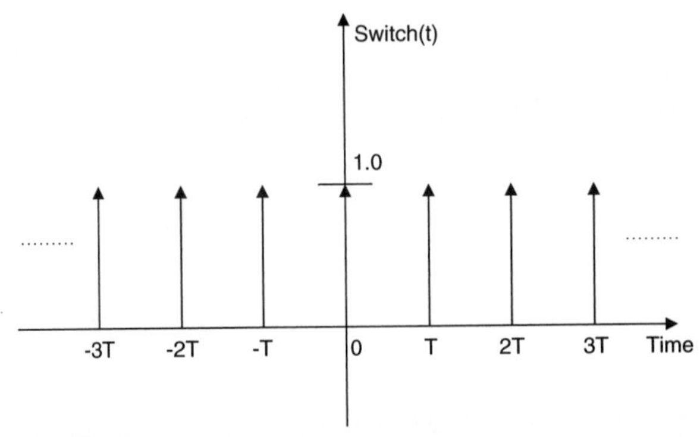


**FIGURE 2.5:** Idealized mathematical model of the sampling operation via a "comb" function.

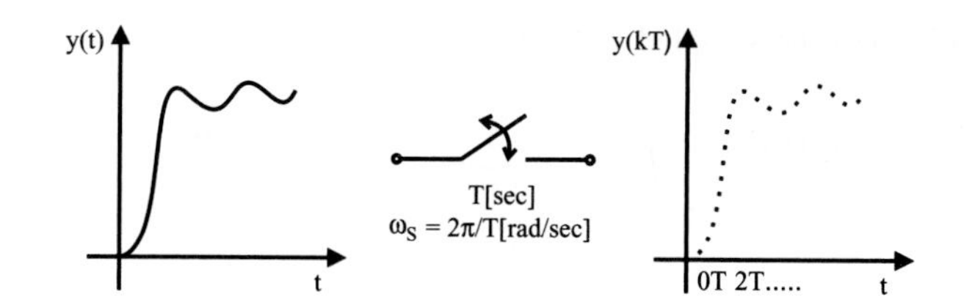


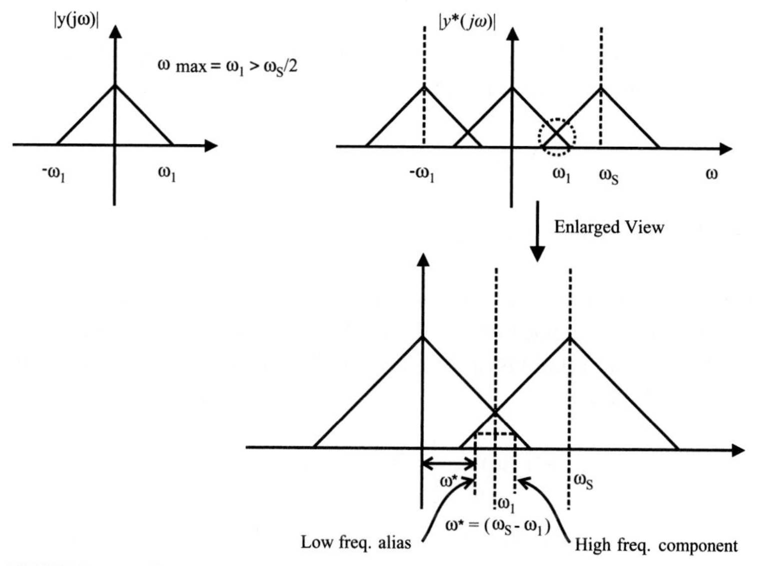


**FIGURE 2.6:** Sampling of a continuous signal, and the frequency domain relationship between the original signal and the sampled signal.

- **Question 3:** Point out a least three implications of the sampling theorem that comes out of the relationship derived in question 3.

Let us address each of these questions in order.

***Question 1*** Notice that because the sampling "comb" function is periodic, it can be expressed as a sum of Fourier series,

$$\sum_{k=-\infty}^{\infty} \delta(t-kT) = \sum_{n=-\infty}^{\infty} C_n e^{j\frac{2\pi}{T}nt} \tag{2.6}$$

where

$$C_n = \frac{1}{T}\int_{-\frac{T}{2}}^{\frac{T}{2}} \sum \delta(t-kT) e^{-j\frac{2\pi}{T}\cdot n\cdot t} dt = \frac{1}{T} \tag{2.7}$$

Therefore,

$$\sum_{k=-\infty}^{\infty} \delta(t-kT) = \frac{1}{T}\sum_{n=-\infty}^{\infty} e^{j(\frac{2\pi}{T})n\cdot t} \tag{2.8}$$

The Laplace transform of the sampled signal is (two-sided Laplace transform, see [16])

$$L\{y(kT)\} = \int_{-\infty}^{\infty} y(t) \sum_{n=-\infty}^{\infty} \delta(t - kT)e^{-st}\,dt$$

$$= \int_{-\infty}^{\infty} y(t)\frac{1}{T} \sum_{n=-\infty}^{\infty} e^{j\frac{2\pi}{T}nt} e^{-st}\,dt$$

$$= \frac{1}{T} \sum_{n=-\infty}^{\infty} \int_{-\infty}^{\infty} y(t)e^{-(s-j\frac{2\pi}{T}n)t}\,dt \tag{2.9}$$

The relationship between the Laplace transform of the sampled signal and the Laplace transform of the original continuous signal is

$$Y^*(s) = \frac{1}{T} \sum_{n=-\infty}^{\infty} Y(s - jw_s n) \tag{2.10}$$

where $w_s = \frac{2\pi}{T}$ is the sampling frequency and $T$ is the sampling period.

***Question 2*** The Fourier transform of a signal can be obtained from the Laplace transform by substituting $jw$ in place of $s$ in the Laplace transform of the function. Therefore, we obtain the following relationship between the Fourier transforms of the sampled and continuous signal (Fig. 2.6),

$$Y^*(jw) = \frac{1}{T} \sum_{n=-\infty}^{\infty} Y(jw - jw_s n) \tag{2.11}$$

where $Y^*(jw)$ – Fourier transform of sampled signal, and $Y(jw - jw_s n)$ – Fourier transform of the original signal.

The frequency content of the sampled signal is the frequency content of the original signal plus the same content shifted in the frequency axis by integer multiples of the sampling frequency. In addition, the magnitude of the frequency content is scaled by the sampling period. The physical interpretation of the above relation is the famous sampling theorem, also called the Shannon's sampling theorem.

***Sampling Theorem*** In order to recover the original signal from its samples, the sampling frequency, $w_s$, must be at least two times the highest frequency content, $w_{max}$, of the signal,

$$w_s \geq 2 \cdot w_{max} \tag{2.12}$$

***Question 3*** We now consider various implications of the sampling operation.

**(i) Aliasing** Aliasing is the result of violating the sampling theorem, that is,

$$w_s < 2 \cdot w_{max} \tag{2.13}$$

The high-frequency components of the original signal show up in the sampled signal as if they are low-frequency components (Fig. 2.7). This is called the *aliasing*. The aliasing frequency that shows up on the samples of the signal is

$$w^* = \left| \left(w_1 + \frac{w_s}{2}\right) mod(w_s) - \left(\frac{w_s}{2}\right) \right| \tag{2.14}$$

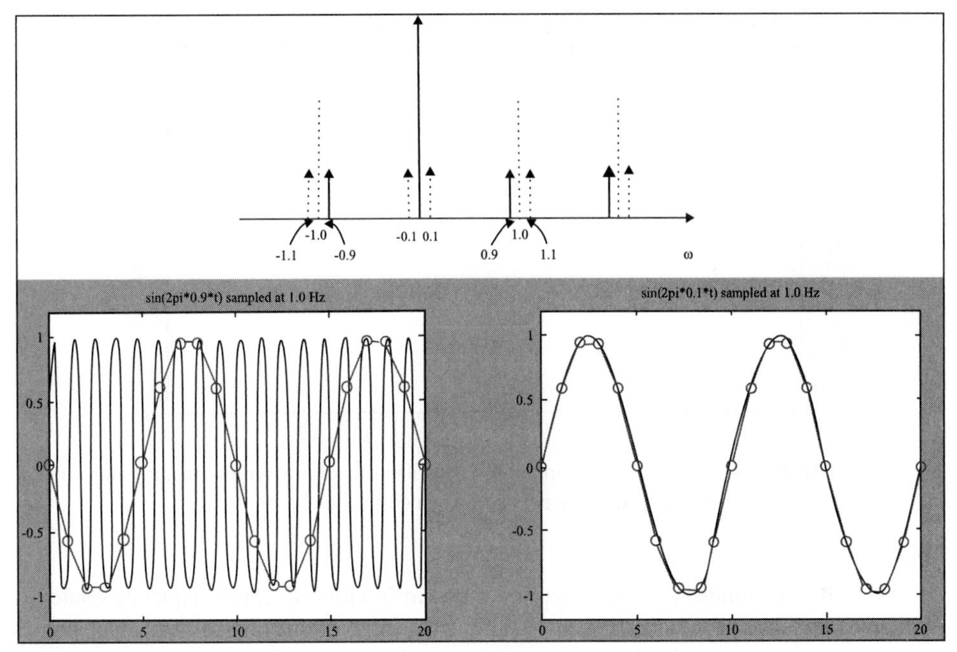


**FIGURE 2.7:** Example of aliasing as a result of sampling operation and misleading picture of the sampled signal.

where $w_1$ is the frequency content of the original signal. For simplicity, we consider a specific frequency content for the original signal. If the sampling theorem is violated (the sampling frequency is less than twice the highest frequency content of the original signal),

1. Reconstruction of original signal from its samples is impossible.
2. High-frequency components look like low-frequency components.

Let us consider two sinusoidal signals with frequencies 0.1 Hz and 0.9 Hz. If we sample both of the signals at $w_s = 1$ Hz, the sampling theorem is not violated in sampling the first signal, but it is violated in sampling the second signal. Due to the aliasing, the samples of the 0.9 Hz sinusoidal signal will look like the samples of the 0.1 Hz signal (Fig. 2.6). Figure 2.9 shows the two cases,

1. Continuous signal $sin(2\pi(0.9)t)$ and sampling frequency $w_s = 1.0$ Hz.
2. Continuous signal $sin(2\pi(0.1)t)$ and sampling frequency $w_s = 1.0$ Hz.

High-frequency signal 0.9 Hz looks like a 0.1 Hz signal when sampled at a 1.0 Hz rate as a result of the aliasing,

$$
\begin{aligned}
w^* &= \left| \left( w_1 + \frac{w_s}{2} \right) mod(w_s) - \left( \frac{w_s}{2} \right) \right| \\
w^* &= |(0.9 + 0.5)\, mod(1.) - 0.5| = |0.4 - 0.5| \\
w^* &= 0.1 \text{ Hz}
\end{aligned}
\tag{2.15}
$$

Another example is a sinusoidal signal with $w_1 = 3$ Hz, and sampled values of it at $w_s =$ 4 Hz. The sampling theorem is violated. The samples will show a 1 Hz oscillation which does not exist in the original signal.

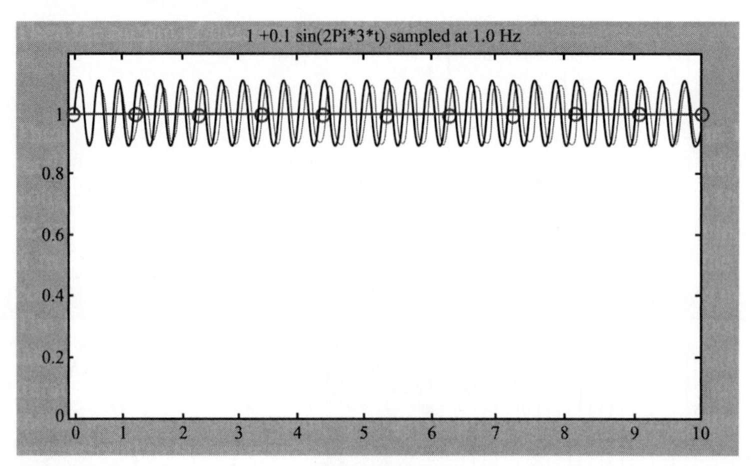


**FIGURE 2.8:** Hidden oscillations in the samples of a signal when the sampling rate is an integer multiple of a frequency content in the continuous signal.

In summary, if the sampling theorem is violated, high-frequency content of signal shows up as a low-frequency (aliasing frequency) content in the sampled signal as a result of aliasing. The aliasing frequency is given by equation 2.14.

**(ii) Hidden oscillations:** If the original signal has frequency content which is an exact integer multiple of the sampling frequency (sampling theorem is violated), then there could be hidden oscilations. In other words, the original signal would have high-frequency oscillations, whereas the sampled signal would not show them at all (Fig. 2.8). If $w_{signal} = n \cdot w_s, n = 1, 2, \ldots.$. When $n = \frac{1}{2}$, with correct phase of the sampling time to the oscillation frequency, hidden oscillations are also possible.

**(iii) Beat phenomenon:** Beat phenomenon is observed when two signals with very close frequency content with very close magnitudes are added. The result looks like two signals (one slowly varying, the other quickly varying) are multiplied.

This phenomenon occurs as a result of sampling operation when the sampling frequency is just a little larger than the highest frequency content of the signal (Fig. 2.9). Notice that the sampling theorem is not violated. Consider the following signal,

$$u(t) = A\ cos(w_1 t) + B\ cos(w_2 t) \tag{2.16}$$

$$= A\ cos((w_2 - w_1)/2)t) \cdot cos((w_2 + w_1)/2)t) \tag{2.17}$$

If $w_1$ is very close to $w_2$, i.e., $w_1 = 5$, $w_2 = 5.2$, then the so-called beat phenomenon occurs:

$$u(t) = A\ cos(w_{beat}t) \cdot cos(w_{ave}t) \tag{2.18}$$

where $w_{beat} = (w_2 - w_1)/2$, $w_{ave} = (w_2 + w_1)/2$. The same effect occurs when we sample a signal with a frequency which is very closed to the minimum sampling frequency required, yet sampling theorem is not violated. Let us consider a sinusoidal signal,

$$y(t) = sin(2\pi 0.9t) \tag{2.19}$$

and sample it with sampling frequency of $w_s = 1.9$ Hz, which satifies the sampling theorem requirements. However, since sampling a signal effectively shifts the original signal frequency content in frequency domain in integer multiples of sampling frequency, it results

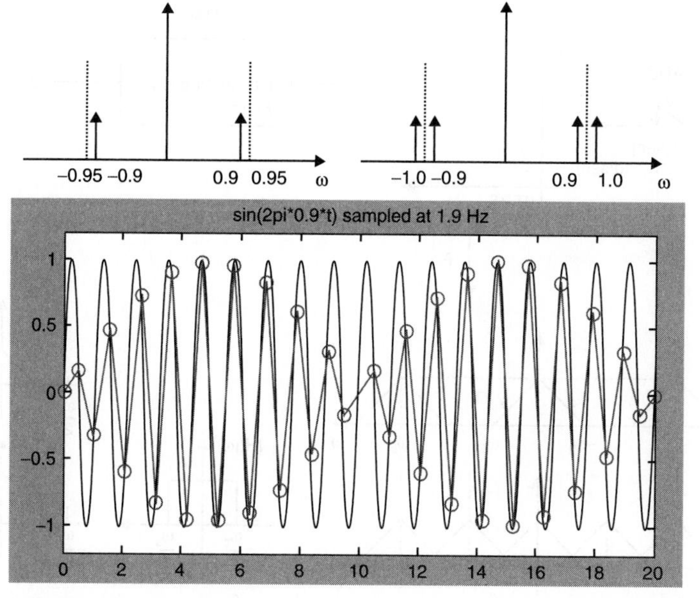


**FIGURE 2.9:** Beat phenomenon and its explanation in frequency and time domain. Notice that the sampling theorem is not violated. The sampling frequency is very close to the minimum requirement.

in adding two very close frequency content. The result is the sampled signal shows a beat phenomemon (Fig. 2.9).

## 2.2.4 Signal Reconstruction: D/A Operation

A D/A converter is used to convert a digital number to an analog voltage signal. It is also referred to as the signal reconstruction device. Let us consider that we sample a continuous signal through an A/D converter, then send that signal out without any modification through a D/A converter. The difference between the original analog signal (input to the A/D converter) and the analog signal output from the D/A converter is an undesired distortion due to sampling, quantization, time-delay, and reconstruction errors (Fig. 2.1). For instance, in communication systems, the analog voice signal is sampled, transmitted digitally over the phone lines, and converted back to the analog voice signal at the other end of the phone line. The goal there is to be able to reconstruct the original signal as accurately as possible.

We know that the sampled signal frequency content is the original frequency content plus the same content shifted in frequency axis by integer multiples of the sampling frequency. In order to recover the frequency content of the original signal, we need an ideal filter which has a square gain, and zero phase transfer function (Fig. 2.10). Clearly, if there is aliasing, it is impossible to recover the original signal even with an ideal reconstruction filter. Here we will consider the following two rescontruction D/A converters:

1. Ideal reconstruction filter (Shannon's Reconstruction),
2. ZOH (Zero-Order-Hold).

In order to focus on D/A functionality and its ability to reconstruct original signal from its samples, let us assume that the time-delay and quantization errors are negligable and that the sampled signal is sent out to D/A converter without any modification.

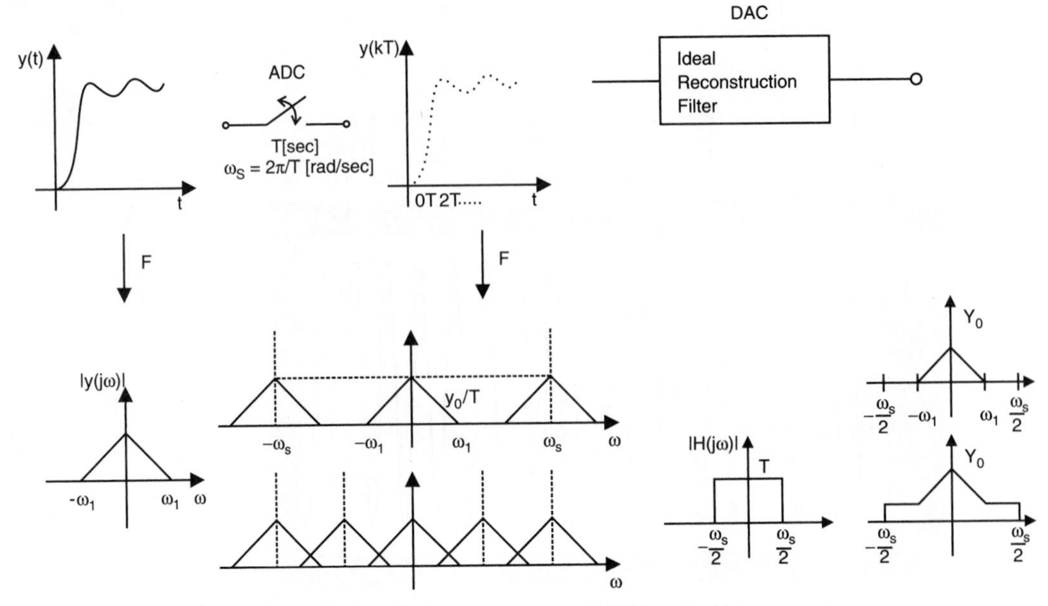


**FIGURE 2.10:** Signal reconstruction with an ideal D/A converter.

**(i) Ideal Reconstruction Filter D/A Converter** In order to recover the original signal from the sampled signal, we need to recover the original frequency content of the signal from the frequency content of the sampled signal (Fig. 2.10). Therefore, we need an ideal filter, an *ideal reconstruction filter*, which has the following frequency response,

$$\begin{cases} |H(j\omega)| = \begin{cases} T; & w \in \left[-\frac{\omega_s}{2}, \frac{\omega_s}{2}\right] \\ 0; & otherwise \end{cases} \\ \angle H(j\omega) = 0; \ \forall\omega \end{cases} \tag{2.20}$$

Let us take the inverse Fourier transform of this filter transfer function to determine what kind of impulse response such a filter would have:

$$\begin{aligned} F^{-1}(H(jw)) &= \frac{1}{2\pi}\int_{-\frac{\pi}{T}}^{\frac{\pi}{T}} T \cdot e^{jwt} \cdot dw \\ &= \frac{T}{2\pi}\int_{-\frac{\pi}{T}}^{\frac{\pi}{T}} e^{j\omega t} \cdot dw \end{aligned} \tag{2.21}$$

$$h(t) = \frac{2}{w_s t} \cdot sin\left(\frac{w_s t}{2}\right) \tag{2.22}$$

This is the impulse response of an ideal reconstruction filter. Notice that the impulse response of the ideal reconstruction filter is noncausal (Fig. 2.11).

In order to practically implement it, one must introduce time-delay into the system large enough compared to sampling period. It cannot be implemented in closed-loop control systems due to stability problems the time-delay would cause. The original signal could, in theory if not in practical applications, be reconstructed from its samples as follows:

$$y(t) = \sum_{k=-\infty}^{\infty} y(kt) \cdot sinc\frac{\pi(t-kT)}{T} \tag{2.23}$$

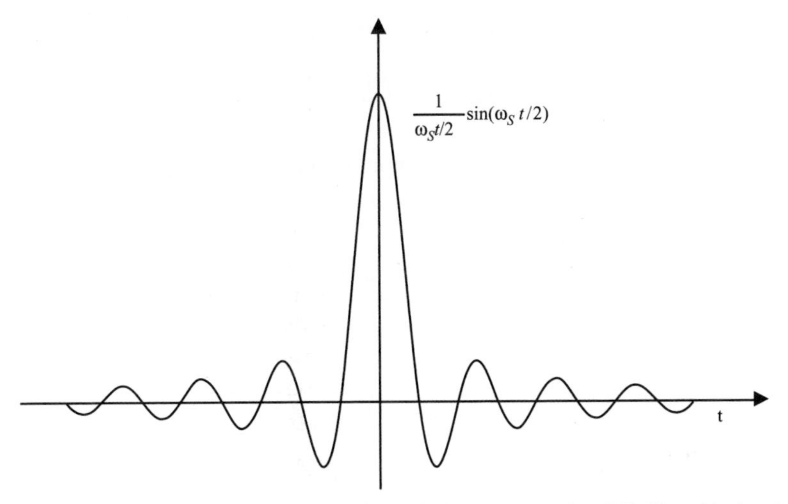


**FIGURE 2.11:** Impulse response of an ideal reconstruction D/A filter. Notice that it is a noncausal filter.

**(ii) D/A: Zero-Order-Hold (ZOH)** Great majority of D/A converters operate as zero-order-hold functions. The signal is kept at the last value until a new value comes. The change between the two value is a step change. Let us try to obtain a transfer function for a zero-order-hold (ZOH) D/A converter. To this end, consider that a single unit pulse is sent to the ZOH D/A (Fig. 2.12). The output of the ZOH D/A would be a single pulse with unit magnitude and duration of a single sampling period. The transfer function of it is an

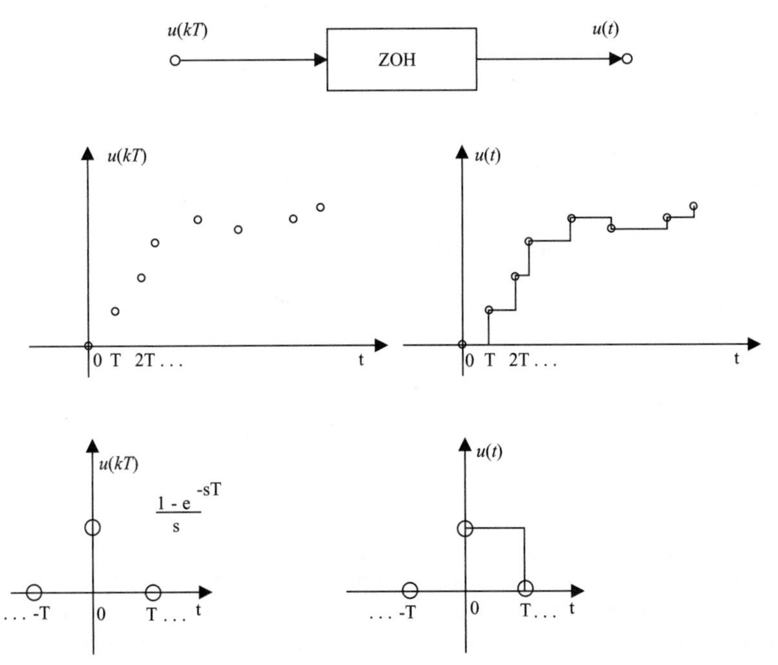


**FIGURE 2.12:** Zero-order-hold (ZOH) type practical D/A converter, its operation, and transfer function.

integrator response to an impulse minus the same response delayed by a sampling period,

$$\mathrm{ZOH}(s) = \left(\frac{1 - e^{-sT}}{s}\right) \tag{2.24}$$

The frequency response (filter transfer function) of the ZOH D/A can be obtained from the above equation by substituting $jw$ for $s$ variable, and after some algebraic manipulations it can be shown that the frequency response of a ZOH D/A is

$$\begin{aligned}
\mathrm{ZOH}(j\omega) &= \left(\frac{1 - e^{-j\omega T}}{j\omega}\right) \\
&= e^{-\frac{j\omega T}{2}} \left(\frac{e^{\frac{j\omega T}{2}} - e^{-\frac{j\omega T}{2}}}{jw}\right) \\
&= T \cdot e^{-\frac{j\omega T}{2}} \frac{\sin\frac{\omega T}{2}}{\frac{\omega T}{2}} \\
&= e^{-\frac{j\omega T}{2}} \cdot T \cdot \left(sinc\frac{\omega T}{2}\right)
\end{aligned} \tag{2.25}$$

Clearly, compared to equation 2.20, which represents the ideal reconstruction filter transfer function, the transfer function of the ZOH type D/A converter is different than the ideal case, but it is a practical one.

### 2.2.5 Real-Time Control Update Methods and Time-Delay

Time-delay is an important issue in control systems. Time-delay in the feedback loop can cause instability because it introduces phase lag. There is inherent time-delay in digital control. The A/D and D/A conversion takes a finite amount of time to complete. The execution of the control calculations takes a finite amount of time. The sampling period is determined by the sum of the time periods that these operations take. The sampling period is a good indication of the time-delay introduced into the loop due to the digital implementation. If the sampling frequency is much higher than the bandwidth of the closed-loop system (i.e., 50 times faster), the influence of time-delay due to digital sampling period will not be significant. As the sampling frequency gets closer to the control system bandwidth (i.e., 2 times), the time-delay associated with sampling rate can create very serious stability and performance problems. Figure 2.13 shows two different implementations of a closed-loop system in terms of sampling and control update timing.

A control system will have periodic sampling intervals. The sampling period can be programmed using a clock. After every sampling period is passed, an interrupt is generated. The real-time control software can be divided into two groups:

1. Foreground program
2. Background program

Normally, CPU runs the background program handling operator input–output operations, checks error and alarm conditions, and checks other process inputs and outputs (I/O) not used in closing the control loop, but used for other logic and sequencing functions. The foreground program is the one that is executed every time sampling clock generates an interrupt. When a new interrupt is generated every sampling period, the CPU saves the

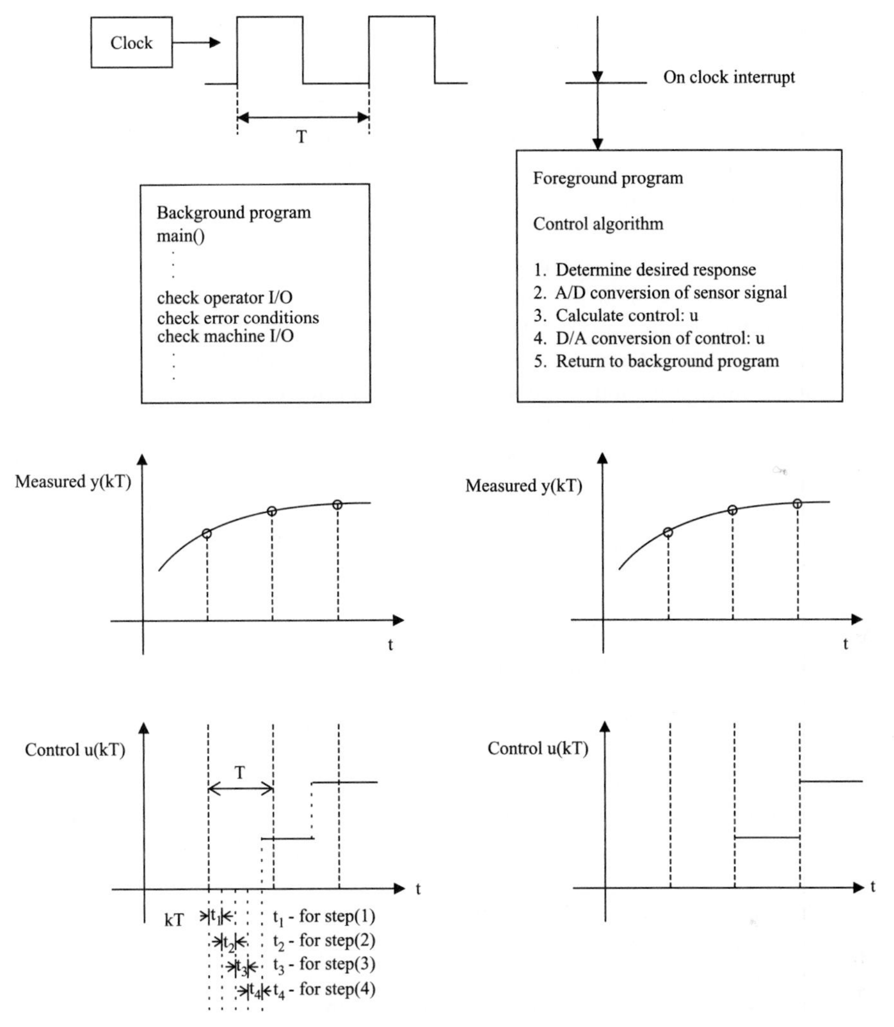


**FIGURE 2.13:** Effective time-delay due to signal conversion, processing, and update methods in a digital control system.

status of what it is doing in the background program, and jumps to the foreground program as soon as possible. In the foreground program, it handles the closed-loop control update for the current sampling period. This involves the following tasks:

1. Determine the desired response (sample from A/D if necessary)
2. Sample the A/D convertor for the sensor signal
3. Calculate the control action
4. Send the control action to D/A
5. Return to the background program

This implementation, and timing of the sequence of operations are shown in the Fig. 2.13 on the left-hand side. Note that A/D conversion is an iterative process, and the conversion times may vary from one cycle to next. As a result, the effective update period for control signals can be a little different from one cycle to the next due to variations in the A/D conversion. The period for which control signal is held constant is

$$T_u = T - T_{c_k} + T_{c_{k+1}} \tag{2.26}$$

where $T_{c_k} = (t_1 + t_2 + t_3 + t_4)_k$ and $T_{c_{k+1}} = (t_1 + t_2 + t_3 + t_4)_{k+1}$ are the total time spent in the foreground program execution during sampling interval $k$, and $k+1$, respectively. Due to the variations in the A/D conversion time, $t_2$, the effective update period $T_u$ for control action may vary from one sample to next. Hence, control system may not have a truly constant sampling period. This is a potential problem which may or may not be significant depending on the application. However, this implementation minimizes the time-delay between measured sensor signal and the corresponding control action since the control action is updated as soon as it is available. If minimizing the time-delay associated with digital controller is more important than maintaining a truly constant sampling frequency, this implementation is appropriate to use. Another implementation is shown on the right side of Fig. 2.13. The sequence of operations in the foreground program is a little different.

1. Send the control action calculated from the previous period to D/A
2. Determine the desired response (sample from A/D, if necessary)
3. Sample the A/D convertor for the sensor signal
4. Calculate the control action and keep it for the next sampling period
5. Return to the background program

The difference is that sending out a control signal to the D/A converter is the first thing done every sampling period. The control signal sent out is the signal calculated during the previous sampling period. The signal calculated during this period is sent to a D/A converter at the beginning of the next sampling period. The advantage of this implementation is that the control signal is updated in truly fixed sampling period. The sampling period must be long enough to complete the foreground program every sampling period. The disadvantage is that the effective time-delay associated with digital processing of the control signal is longer than the previous implementation. The time-delay is at most one sampling period long. If the sampling frequency is much larger than the bandwidth of the closed-loop system, that should not be a serious problem. As the sampling frequency gets closer to the bandwidth of the closed-loop system, the larger time delay in the second implementation compared to the first implementation may have serious performance degrading consequences.

### 2.2.6 Filtering and Bandwidth Issues

The sampling theorem requires that the sampling frequency be at least twice the highest frequency content of the signal being sampled. However, real-world signals always have some level of noise in them. The frequency of the noise component of the signal is generally very high. Therefore, it would not be practical to use very fast sampling rates in order to handle the noise and avoid an aliasing problem. Furthermore, the noise content of the signal is not something we would like to capture; it is an unwanted component. Therefore, signals are generally passed through anti-aliasing filters before sampling at the A/D circuit. This is called *pre-filtering*. The purpose of anti-aliasing filters is to attenuate the high-frequency noise components and pass the low-frequency components of the signal (Fig. 2.14). The bandwidth of the anti-aliasing filter (also called the noise filter) should be such that it should not pass much of the signal beyond 1/2 of the sampling frequency. An ideal filter would have frequency response characteristics similar to the ideal reconstruction filter. It would pass identically all the frequency components up to 1/2 of the sampling frequency,

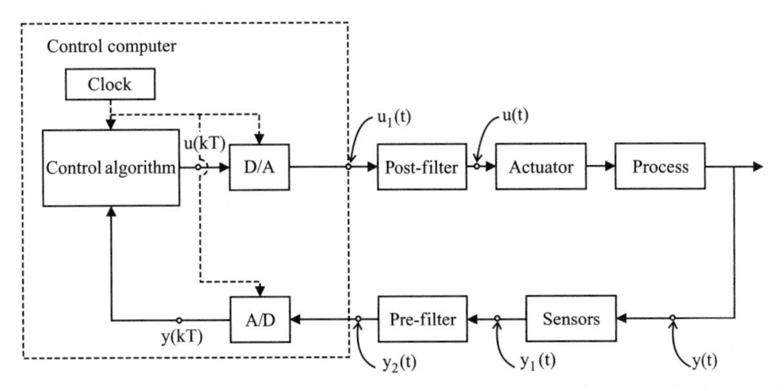


**FIGURE 2.14:** Filtering and bandwidth issues in a digital closed-loop control system.

and cancel the rest. However, it is not practical to use such filters in control and data-acqusition systems.

Generally, a second-order or higher order filter is used as a noise filter. Typically filter transfer function for a second-order filter is

$$G_F(s) = \frac{w_n^2}{s^2 + 2\xi_n w_n s + w_n^2}$$

If a higher order filter is desired, multiple second-order filters can be cascaded is series. The exact parameters of the noise filter ($\xi_n$, $w_n$) are selected based on the class of the filter. The Butterworth, ITAE, and Bessel filter are popular filter parameters used for that purpose [12].

Similarly, since the output of the D/A converter is a sequence of step changes, it may be useful to smooth the control signal before it is applied to the amplification stage. The same type of noise filters can be used as the *post-filtering* device to reduce the high-frequency content of the control signal.

The pre-filters and post-filters add time-delay into the closed-loop system due to their finite bandwidth. In order to use the pre- and post-filters for noise cancellation and smoothing purposes without significantly affecting the closed-loop system bandwidth, the following general guidelines should be followed. We have four frequencies of interest:

1. Closed-loop system bandwidth, $w_{cls}$
2. Sampling frequency, $w_s$
3. Pre- and post-filter bandwidth, $w_{filter}$
4. The maximum frequency content of the signal presented to the sampling and hold circuit of A/D converter, $w_{signal}$

In order to ensure, pre- and post-filtering does not affect the closed-loop system bandwidth, the filters must have about 10 times or more higher bandwidth than the closed-loop system bandwidth. The filter bandwidth is also a good estimate of the highest frequency content allowed to enter the sampling circuit.

$$w_{filter} \approx w_{signal} \approx 5 \text{ to } 10 * w_{cls}$$

From the sampling theorem,

$$w_s \geq 2 * w_{signal}$$

In practice, the sampling frequency should be much larger than the minimum requirement imposed by the sampling theorem, i.e.,

$$w_s \approx 5 \text{ to } 20 * w_{signal}$$

Therefore, the magnitude relation between the four frequencies of interest is

$$w_s \approx 5 \text{ to } 20 * w_{signal} \approx 5 \text{ to } 20 * w_{filter} \approx 25 \text{ to } 200 * w_{cls} \tag{2.27}$$

## 2.3 OPEN-LOOP CONTROL VERSUS CLOSED-LOOP CONTROL

Open-loop control means control decisions are made without making use of any measurement of the actual response of the system. Open-loop control decisions do not need a sensor. Closed-loop control means control decisions are made based on the measurements of the actual system response. The actual response is fed back to the controller, and the control decision is made based on this feed-back signal and the desired response. In the comparision, we will take the real-world issues into consideration, namely, the *disturbances, changes in process dynamics, and sensor noise*. These are common real-world problems faced in different degrees of significance in every control system. Consider a general dynamic process nominally modeled by, *G(s)*, controlled by on open-loop and a closed-loop controller (Fig. 2.15).

The response for the open-loop control case is

$$y(s) = D(s)G(s)r(s) + G(s)w(s)$$

and for the closed-loop control case is

$$y(s) = \frac{D(s)G(s)}{1 + D(s)G(s)}r(s) + \frac{G(s)}{1 + D(s)G(s)}w(s) - \frac{D(s)G(s)}{1 + D(s)G(s)}v(s)$$

In real-world systems, the following issues exist and must be addressed:

1. **Disturbances (*w*(*s*)):** There are always disturbances which are not under our control. They exist and cause error in the system response. For instance, the wind acts as a disturbance on an airplane changing its flight direction. Low outside temperature and the heat loss due to it from the walls of a heated house act as a disturbance on the control system since the outside temperature is not under our control, yet it affects the temperature of the house.

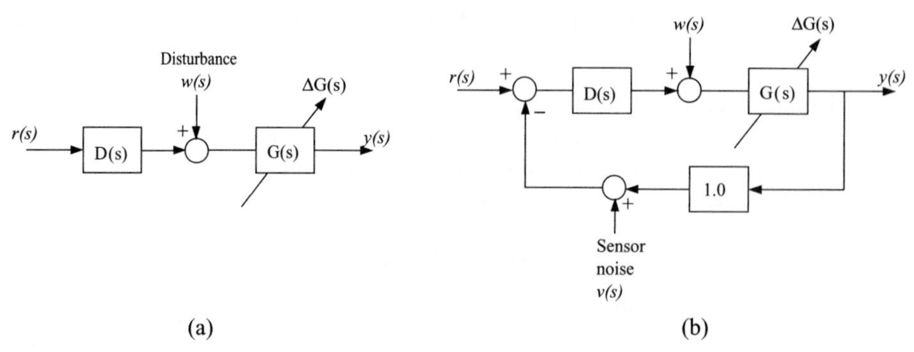


**FIGURE 2.15:** Open-loop versus closed-loop control system comparison.

2. **Variations in process dynamics ($\Delta G(s)$):** The dynamics of the process may change structurally or parametrically. Structural changes in the dynamics imply drastic significant changes, such as the change in the dynamics of an aircraft due to loss of an engine or a wing. However, the parametric changes imply less-significant, more smooth, nondrastic changes, such as the change in the weight of an aircraft as the fuel is being consumed, or due to opening of the wing control surfaces.
3. **Sensor noise ($v(s)$):** Closed-loop control requires the measurement of the actual response (the controlled variable). The sensor signals always have some noise in the measurement. The noise is included in the control decisions and hence affects the overall performance of the system.

Let us consider the effect of these three groups of real-world problems in system performance under open-loop and closed-loop control [Fig. 2.15(a)]. In open-loop control, the effect of disturbance is

$$y_w(s) = G(s)w(s)$$

the effect of process dynamic variations is

$$\begin{aligned} y(s) &= (G_0(s) + \Delta G(s))r(s) \\ &= G_0(s)r(s) + \Delta G(s)r(s) \end{aligned}$$

Since no feedback sensors are needed in open-loop control, there is no sensor noise problem.

The response due to disturbance $G(s)\cdot w(s)$ and due to process dynamic variations $\Delta G(s)\cdot r(s)$ are not wanted and is considered error. Open-loop control has no mechanism to correct for these errors. The error is proportional to the disturbance magnitude and the changes in the process dynamics. On one hand, if the process dynamics is well known and no variations occur, and there is no disturbance or the nature of disturbance is well known, the open-loop control can provide excellent performance. On the other hand, if process dynamics vary or there are disturbances whose nature is not known or not repeatable, open-loop control has no mechanism to reduce the effect of them.

In closed-loop control, there is an added component, the sensor, to provide feedback measurement about the actual output of the process. By definition, feedback control action is generated based on the error between desired and actual output. Hence, error is an inherent part of such a design. Let us consider the output of the system shown in Fig. 2.15(b).

$$y(s) = \frac{D(s)G(s)}{1 + D(s)G(s)}r(s) + \frac{G(s)}{1 + D(s)G(s)}w(s) - \frac{D(s)G(s)}{1 + D(s)G(s)}v(s)$$

Ideally we would like $y(s) = r(s)$ and no output result because of $w(s)$, variations in $G(s)$, and $v(s)$. Let us consider the effects of disturbance, dynamic process variations, and sensor noise.

1. **Disturbance effect:** $w(s)$

$$y_w(s) = \frac{G(s)}{1 + D(s)G(s)}w(s)$$

The response due to the disturbance, $y_w(s)$, is desired to be small or zero if possible. If we can make $D(s)G(s) \gg G(s)$ and $D(s)G(s) \gg 1$ for the frequency range where $w(s)$ is significant, then the response due to disturbances in that frequency range would be small. If the response due to disturbances is small or zero, the control system is said to *have good disturbance rejection* or *be insensitive to the disturbances*. Another way of stating this result is that it is desirable to have a controller with large gain.

**2. Variation of process dynamics:** $G(s) = G_0(s) + \Delta G(s)$. Consider only the command signal and process dynamic variations, and let us analyze the effect of the variations in the process dynamics on the response.

$$y_r(s) = \frac{D(s)(G_0(s) + \Delta G(s))}{1 + D(s)(G_0(s) + \Delta G(s))} r(s)$$
$$= \frac{D(s)G_0(s)}{1 + D(s)(G_0(s) + \Delta G(s))} r(s) + \frac{D(s)\Delta G(s)}{1 + D(s)(G_0(s) + \Delta G(s))} r(s)$$

The second term is the main contribution of process dynamic variations to the output. In order to make the effect of this small on the system response, the following condition must hold:

$$D(s)G(s) \gg D(s) \quad \text{and} \quad D(s)G(s) \gg 1$$

If the response due to changes in the process dynamics is small, the control system is called *insensitive to the variations in process dynamics*, which is a desired property.

So far, disturbance rejection capability requires

$$D(s)G(s) \gg G(s) \quad \text{and} \quad D(s)G(s) \gg 1$$

and insensitivity to process dynamic variations requires

$$D(s)G(s) \gg D(s) \quad \text{and} \quad D(s)G(s) \gg 1$$

Therefore, the conditions $\{D(s)G(s) \gg G(s),\ D(s)G(s) \gg D(s),\ \text{and}\ D(s)G(s) \gg 1\}$ mean that the loop gain must be well balanced between the controller and the process in order to have good disturbance rejection and be insensitive to process dynamics variations.

**3. Sensor noise effect:** Sensor noise is a problem for closed-loop control systems only. The open-loop control does not need feedback sensors, and therefore it does not have sensor noise problem. Let us consider the response, $y_v(s)$, of a closed-loop system due to sensor noise, $v(s)$.

$$y_v(s) = -\frac{D(s)G(s)}{1 + D(s)G(s)} v(s)$$

In order to make $y_v(s)$ small, $D(s)G(s) \ll 1$ must be, which is contradictory to the loop gain requirements for previous two properties: (1) disturbance rejection and (2) insensitivity to variations in process dynamics. *This is the fundamental design conflict of feedback control systems.* Robustness against disturbances and variations in process dynamics require large loop gain balanced between controller and process, $D(s)G(s) \gg 1$, while robustness against sensor noise requires $D(s)G(s) \ll 1$. These are conflicting requirements, and both cannot be satisfied at the same time for all frequencies.

In practice, the control engineering problems are generally such that disturbances and variations in process dynamics are slowly varying and have low-frequency content, whereas sensor noise has high-frequency content. If a given control problem has this frequency separation property between various uncertainties, then a controller can be designed such that $D(s)G(s) \gg 1, D(s), G(s)$ for low-frequency range so that the system has good robustness against disturbances and variations in process dynamics, and $D(s)G(s) \ll 1$ for high-frequency range so that sensor noise is also rejected. *This is the basic feedback control system design compromise.* If there is no such frequency separation between disturbance, variations in process dynamics, and sensor noise, no feedback controller can be designed to provide robustress against all of these real-world problems.

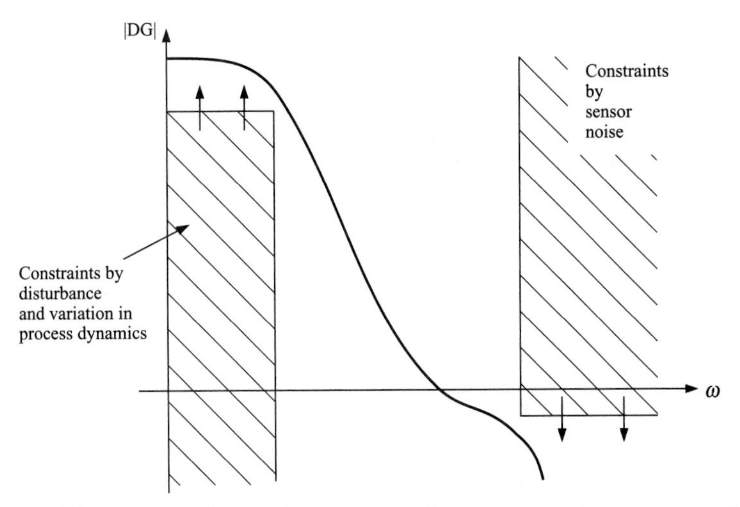


**FIGURE 2.16:** Desired performance specification for control systems in frequency domain.

In summary, the loop transfer function of a typical well-designed control system has the following desired shape as a function of frequency: It should be as large as possible at low frequencies to provide robustness against disturbance and variations in process dynamics, and it should be as small as possible at high frequencies to reject sensor noise (Fig. 2.16). Furthermore, it should cross the 0-db magnitude by about −20 db/*decade* slope in $20\ log_{10}|D(jw)G(jw)|$ versus $log_{10}w$ plot in order to have a good stability margin.

So far, we compared the advantages and disadvantages of closed-loop control versus open-loop control. The main advantage of feedback control over open-loop control is that it increases the robustness of the system against the disturbances and variations in the process dynamics. The general characteristics of control systems are discussed in terms of the shape of the loop transfer function in order to provide good robustness against these undesirable real-world problems of control systems. However, sensor noise or sensor failure can make a closed-loop system unstable. If the process dynamics do not vary much and the disturbances are well known, open-loop control may be a better choice than closed-loop control. Open-loop control does not suffer from potential stability problems associated with sensor failures.

## 2.4 PERFORMANCE SPECIFICATIONS FOR CONTROL SYSTEMS

The performance desired from a control system can be described under three groups:

1. Response quality
   **(a)** transient response
   **(b)** steady-state response
2. Stability
3. Robustness of the system stability and response quality against various uncertainties such as disturbances, process dynamic variations, and sensor noise

The main advantage of feedback control over open-loop control is its ability to reduce the effect of disturbances and process dynamic variations on the quality of system response. In other words, the main advantage of feedback control is the robustness it provides against various uncertainties.

A basic feedback control system and typical uncertainties associated with it are shown in Fig. 2.15(b). As we discussed in the previous section, the total response of the system due to command, disturbance, and sensor noise (for $H = 1$ case) is

$$y = \frac{DG}{1+DG}r + \frac{G}{1+DG}w - \frac{DG}{1+DG}v$$

The goal of the control is to make $y(t)$ equal to $r(t)$. Therefore $DG \gg 1$ should be in general. If $DG \gg G$, the effect of disturbance, $w$, is reduced. However, the sensor noise directly contributes to the output, $y$. In order to track $r$ and reject disturbance, $w$, we want $DG \gg 1$ (large), but in order to reject sensor noise, we want $DG \ll 1$ (small). This is the basic dilemma of feedback control design. A compromise is reached by the following engineering judgment: disturbance, $w(t)$, is generally of low-frequency content; whereas sensor noise $v(t)$ is high-frequency content. Therefore if we design a controller such that $DG \gg 1$ around low-frequency regions to reject disturbances, and $DG \ll 1$ around high-frequency regions to reject sensor noise, the closed-loop system has good robustness against the uncertainties.

Robustness of the closed-loop system (CLS) is closely related to the gain of loop transfer function as a function of frequency (Fig. 2.16). Therefore the robustness properties are best conveyed in frequency domain. In general, a loop transfer function should have a large gain at low frequency in order to reject low-frequency disturbances and slow variations in process dynamics, and low loop gain at high frequency in order to reject sensor noise. The $s$-plane pole-zero representation of a transfer function does not convey gain information. Hence, robustness properties are not well conveyed by $s$-plane pole-zero structure of transfer function.

Stability requirements are equally well described in $s$-plane as well as frequency domain. In $s$-plane, all the CLS poles must be on the left-hand plane. In frequency domain, the gain margin and phase margin must be large enough to provide sufficient stability margin. Desired relative stability margin from a CLS can be expressed either in terms of gain and phase margin in frequency domain or in terms of the distance of CLS poles from the imaginary axis in the $s$-plane.

Finally, the response quality must be specified. The response of a dynamic system can be divided into two parts: (1) transient response part and (2) steady-state response part.

Transient response is the immediate response of the system when it is commanded new desired output. The steady-state response is the response of the system after sufficient amount of time has passed. The steady-state response quality is quantified by the error between the desired and actual output after long enough time have passed for the system to respond. Clearly, the response of a dynamic system depends on its input. The standard input signal used in defining the transient response characteristics of a dynamic system is the step input. By using a standard test signal, various competing controller and process designs can be compared in terms of their performance (Fig. 2.17).

In general a CLS step response looks like the response shown in (Fig. 2.17). The transient response to step command can be characterized by a few quantitative measures of the response, such as the maximum percent overshoot, the time it takes for the response to settle down within certain percentage of the final value.

The transient response to a step input is typically described by maximun percent overshoot, $P.O.\%$, and settling time, $t_s$, the time it takes for the output to settle down to within $\pm 2\%$ or $\pm 1\%$ of the desired output.

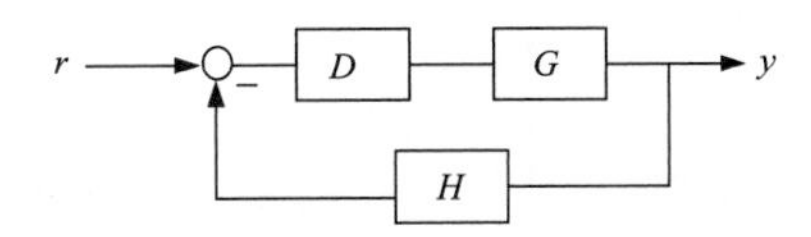


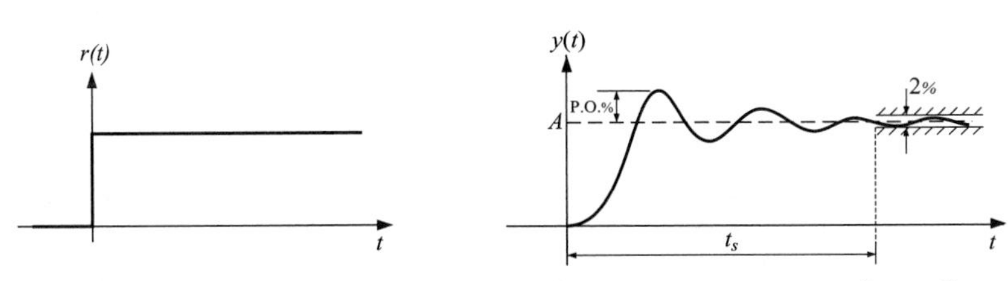


**FIGURE 2.17:** A typical closed-loop control system step response shape. Quantification of transient response characteristics to step input can done using percent overshoot and settling time.

For a second-order system, there is a one-to-one relationship between ($P.O.\%$, $t_s$) and the damping ratio and natural frequency ($\xi$, $\omega_n$) of the second-order system (Figs. 2.17 and 2.18). It can be shown that

$$P.O.\% = e^{\frac{-\pi\xi}{\sqrt{1-\xi^2}}}$$

$$t_s = \begin{cases} \frac{4}{\xi\omega_n}, & \pm 2\% \\ \frac{4.6}{\xi\omega_n}, & \pm 1\% \end{cases}$$

Therefore, if it is desired that the CLS step response should not have more than a certain amount of $P.O.\%$, and should settle down within $\pm 2\%$ of final value within $t_s$ (sec), the designer must seek a controller which will make the closed-loop system behave like (or similiar to) a second-order system whose two poles are given by the above relationships.

## 2.5 TIME DOMAIN AND s-DOMAIN CORRELATION OF SIGNALS

The response of a linear time invariant dynamic system, $y(t)$, as a result of an input signal, $r(t)$, can be calculated using the Laplace transforms

$$y(s) = G(s)r(s)$$

The $y(s)$ can be expanded to its partial fraction expansion (P.F.E.), which has the general form as

$$y(s) = \frac{1}{s} + \sum_{i=1}^{m} \frac{A_i}{s+\sigma_i} + \sum_{j=1}^{m} \frac{B_j}{s^2 + 2\alpha_j s + \left(\alpha_j^2 + \omega_i^2\right)}$$

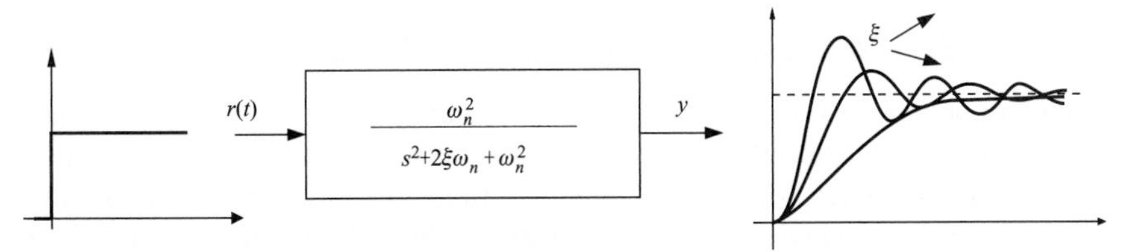


**FIGURE 2.18:** A standard second-order system model and its step response. The step response is determined by the damping ratio ($\xi$) and natual frequency ($\omega_n$).

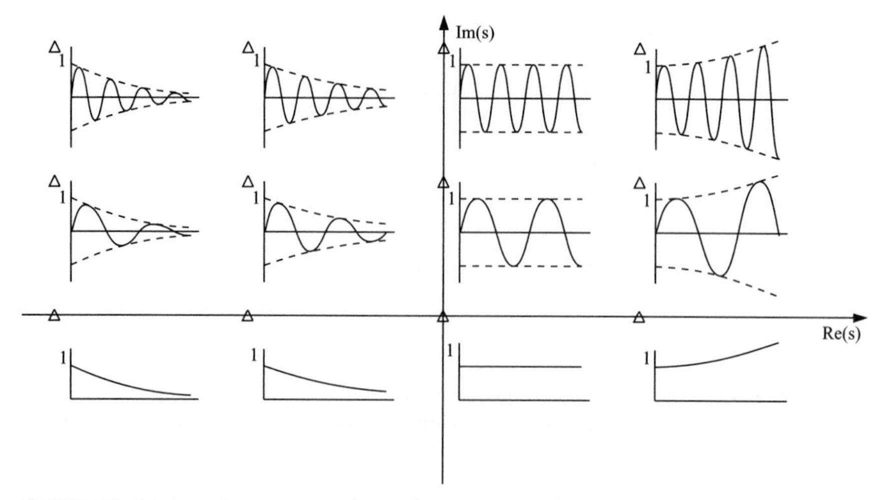


**FIGURE 2.19:** Impulse response for various root locations in the $s$-plane.

$$A_i, B_j \;-\; \text{residue of P.F.E. of } G(s)r(s)$$

The time domain response, $y(t)$, can be obtained by taking the inverse Laplace transform of each term of the P.F.E.,

$$y(t) = 1(t) + \sum_{i=1}^{m} A_i e^{-\sigma_i t} + \sum_{j=1}^{m} B_j e^{-\alpha_j t}\left(\frac{1}{\omega_j}\sin\omega_j t\right)$$

The correlation between the time domain response (impulse response) and the poles of the transfer function is shown in Fig. 2.19.

## 2.5.1 Selection of Pole Locations

The feedback control changes the dynamics of the open-loop system to the desired form by closed-loop control action. For linear systems, closed-loop control changes the locations of the poles. The control effort required is proportional to the amount of movement of pole locations (the difference between the open-loop and closed-loop pole locations). Large pole movements will require unnecessarily large actuators. The desired pole locations can be selected to approximate the step response behavior of a dominant second-order model or the pole locations of some standard filters such as Bessel and Butterworth filters. The second-order system parameters ($\omega_n$, $\xi$) can be selected fairly accurately to satisfy $t_{settling}$ and $P.O.\%$ specifications by designing the CLS such that it has dominant second-order system poles and the rest of poles are further to the left in $s$-plane (Fig. 2.20).

## 2.5.2 Step Response of a Second-Order System

Step response is the standard signal used in evaluating the transient response characteristics of a control system. Specifically, the step response behavior is summarized by the maximum percent overshoot ($P.O.\%$) and the amount of time it takes for the output to settle to within 1 or 2% of the commanded step output (settling time, $t_s$), the time it takes for the output to reach 90% of command rise time $t_r$, and to reach maximum value (peak time, $t_p$). The ($P.O.\%$, $t_s$, $t_r$, $t_p$) all are related to the pole locations of a second-order system. We will

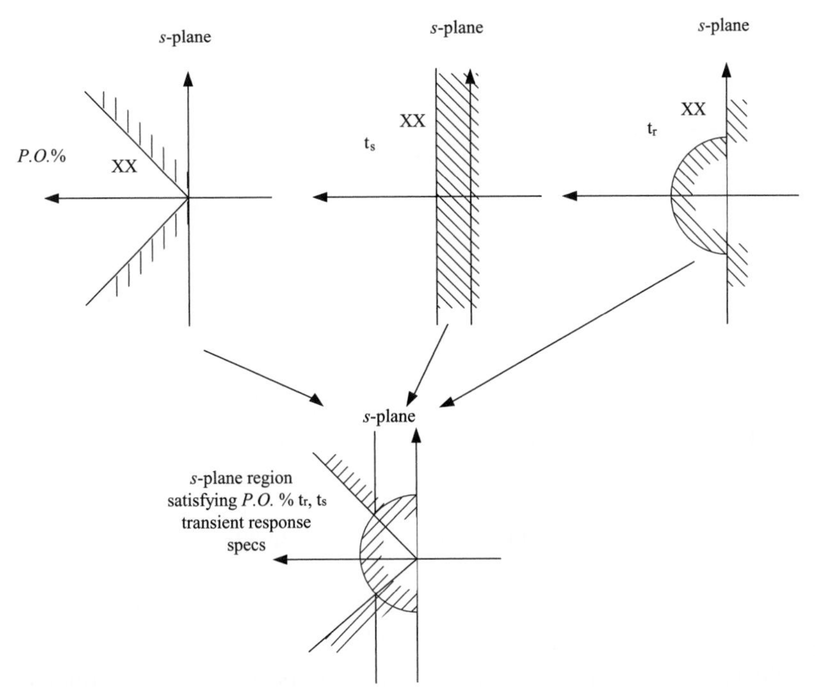


**FIGURE 2.20:** Desired response performance specifications using s-domain pole locations.

consider the step response of a second-order system of the form

$$\frac{\omega_n^2}{s^2 + 2\xi\omega_n s + \omega_n^2}$$

Further, we will consider the step response when there is an additional pole and zero:

$$\frac{\left(\frac{s}{a} + 1\right)}{\left(\frac{s}{b} + 1\right)} \frac{\omega_n^2}{s^2 + \xi\omega_n s + \omega_n^2}$$

The utility of this is that given a transient response specifications for a control system (*P.O.%*, $t_s$) we can determine where the dominant poles should be in order to meet the specifications. Even though our system may not be second-order, second-order system pole-zero locations can provide a good starting point in design, especially if the higher order system can be made to have dominant second-order dynamics.

Consider the step response of a second-order system (Fig. 2.21):

$$m\ddot{x} + c\dot{x} + kx = f$$

let $f(t) = kr(t)$

$$\ddot{x} + \frac{c}{m}\dot{x} + \frac{k}{m}x = \frac{k}{m}r$$

Let $\frac{c}{m} = 2\xi\omega_n$, $\frac{k}{m} = \omega_n^2$ and take the Laplace transform of the differential equation with zero initial conditions,

$$\frac{x(s)}{r(s)} = \frac{\omega_n^2}{s^2 + 2\xi\omega_n s + \omega_n^2}$$

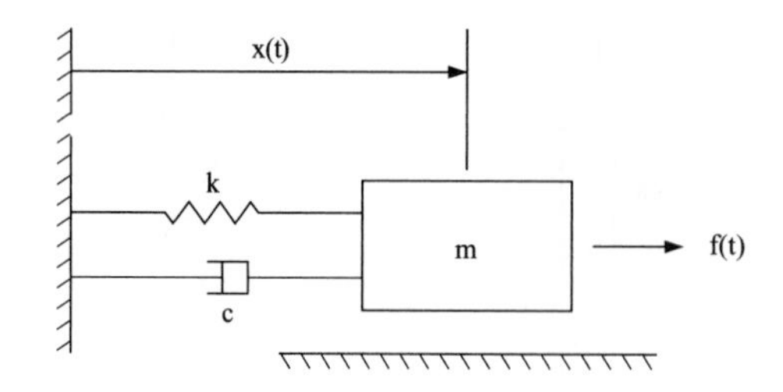


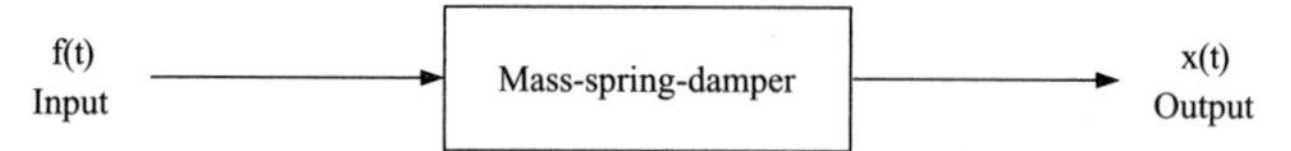


**FIGURE 2.21:** Mass-spring system dynamics and its position control.

If $r(t)$ is a step input, $r(s) = \frac{1}{s}$, the response $x(s)$ is given by

$$x(s) = \frac{\omega_n^2}{s^2 + 2\xi\omega_n s + \omega_n^2} \cdot \frac{1}{s}$$

Using the partial fraction expansions and taking inverse Laplace transform, the response can be found as

$$x(t) = 1 - e^{-\xi\omega_n t}\left(cos\sqrt{1-\xi^2}\omega_n t + \frac{\xi}{\sqrt{1-\xi^2}} sin\sqrt{1-\xi^2}\omega_n t\right) \quad \text{for } 0 \le \xi < 1 \text{ range}$$

The maximum overshoot occurs at the first time instant, $t_p$, where the derivative of $x$ is zero. Once $t_p$ is found, then the maximum value of response at that time can be evaluated and percent overshoot can be determined.

$$\frac{dx(t)}{dt} = 0 \Rightarrow \text{ find } t_p : t_p = \frac{2.5}{\omega_n}$$

Then

$$P.O.\% = \frac{x(t_p) - 1}{1} \times 100 = e^{\frac{-\pi\xi}{\sqrt{1-\xi^2}}}$$

The settling time is the time it takes for response to settle within $\pm 1$ or $\pm 2\%$ of final value, and it can be shown that

$$t_s = \begin{cases} \frac{4.6}{\xi\omega_n}; & \pm 1\% \\ \frac{4.0}{\xi\omega_n}; & \pm 2\% \end{cases}$$

Therefore, given a $(P.O.\%, t_s)$ specification, the corresponding second-order system pole locations $(-\xi w_n \pm \sqrt{(1-\xi^2)}w_n$ can be directly obtained.

***Effect of an Additional Zero*** Let us consider a second-order system with a real zero (Fig. 2.22). The system is the same as a standard second-order system with two complex conjugate poles and d.c. gain of 1, with an additional zero on the real axis.

$$G(s) = \left(\frac{s}{a} + 1\right) \frac{\omega_n^2}{s^2 + 2\xi\omega_n s + \omega_n^2}$$

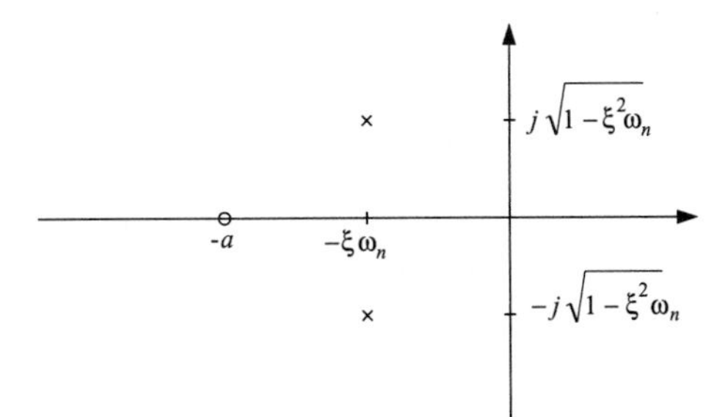


**FIGURE 2.22:** Second-order system with two complex conjugate poles and a real zero.

Let $\omega_n = 1$; $a = \alpha\,\xi\omega_n$, the transfer function can be expressed as

$$G(s) = \frac{\left(\frac{s}{\alpha\xi} + 1\right)}{s^2 + 2\xi s + 1} = \frac{1}{s^2 + 2\xi s + 1} + \frac{1}{\alpha\xi}\frac{s}{s^2 + 2\xi s + 1}$$

Notice that the effect of zero is to add the derivative of the step response to the second-order system response by an amount proportional to $\frac{1}{\alpha\xi}$. Clearly if $\alpha$ is large, $a$ is far to the left of $\xi\omega_n$, and the influence of addition zero is not much. As $\alpha$ gets smaller, $a$ gets closer to the $\xi\omega_n$ area, and $(1/\alpha\xi)$ grows. Hence, the influence of zero on the response increases. The main effect of zero as it gets close to $\xi\omega_n$ value is to increase the percent overshoot. If the zero is on the right half $s$-plane (nonmininum phase transfer function) initial value of step response goes in the opposite direction. This is illustrated in the figures below (Figs. 2.23 and 2.24).

Step response of a second-order system with a zero on left half plane has the form

$$x_{step}(t) + \frac{1}{\alpha\xi}\dot{x}_{step}(t)$$

and response with a zero on the right half $s$-plane [make the $(s/a + 1)$ term $(s/a - 1)$ in the above equation and the result follows (Fig. 2.24)] has the form

$$x_{step}(t) - \frac{1}{\alpha\xi}\dot{x}_{step}(t)$$

***Effect of an Additional Pole*** The system becomes third order when an additional pole exists in addition to the two complex conjugate poles.

$$\left(\frac{1}{\frac{s}{b} + 1}\right)\left(\frac{1}{\frac{s^2}{\omega_n^2} + \frac{2\xi}{\omega_n}s + 1}\right)$$

If the $b = \alpha\xi\omega_n$ is 3 to 5 times to the left of $\xi\omega_n$, the effect of an additional pole on the step response is negligible. As $b$ gets close to $\xi\omega_n$, the effect is to increase the rise time, hence slow the response of the system (Fig. 2.24).

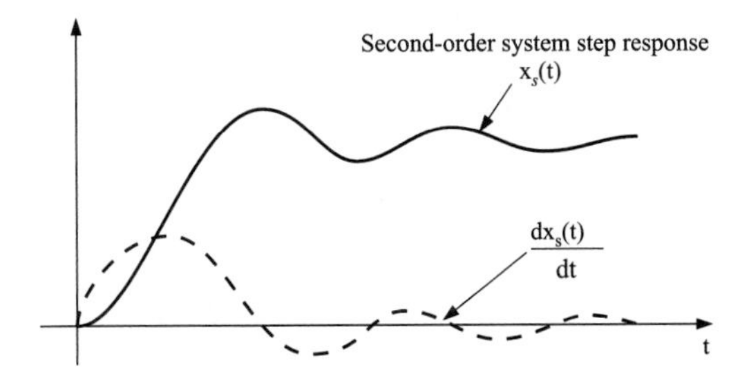


**FIGURE 2.23:** Step response of a second-order system and the derivative of its response.

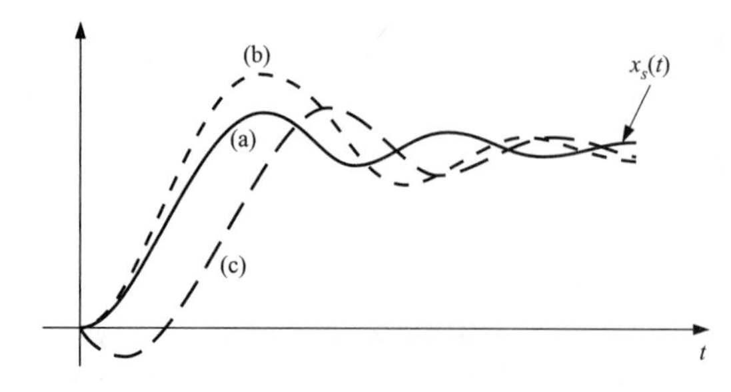


**FIGURE 2.24:** Step response of a second-order system and the effect of a zero: (a) without zero, (b) zero on the left-hand plane, and (c) zero on the right-hand plane.

### 2.5.3 Standard Filters

Most dynamic systems are higher order than second order. However, a higher order system behaves very similiar to a standard second-order system if the poles are distributed on the $s$-plane such that there are two dominant poles and the rest of the poles and zeros are far (3–5 times) to the left of these dominant poles. Instead of choosing a dominant second-order system model as a design goal, we can choose other higher order system models (Fig. 2.25). Among the popular standard filters, which can be used as the goal for a control system performance, are Bessel, Butterworth, and ITAE filters [13].

The step response, and frequency response characteristics of these filters are well tabulated in standard textbooks in control systems and digital signal processing fields. They may be used as a reference to describe the desired performance (hence, desired pole-zero locations) for a closed-loop control design problem. Note that the more the open-loop poles must be moved to desired closed-loop pole locations, the larger the control action requirements. This may quickly saturate existing actuators and result in poor transient performance or require unnecessarily large actuators on the system. The dominant closed-loop system poles (also called the bandwidth) should be as large as possible with sufficient damping on the $s$-plane, but it should be a balanced choice between desired speed of response and required actuator size and control effort.

### 2.5.4 Steady-State Response

Steady-state response is usually characterized by the steady-state error between a desired output and the actual output.

$$y(\cdot) = \frac{D(\cdot)G(\cdot)}{1 + D(\cdot)G(\cdot)} r(\cdot)$$

The error between the desired and actual response is

$$\begin{aligned} e(\cdot) &= r(\cdot) - y(\cdot) \\ &= \left(1 - \frac{D(\cdot)G(\cdot)}{1 + D(\cdot)G(\cdot)}\right) r(\cdot) \\ e(\cdot) &= \frac{1}{1 + D(\cdot)G(\cdot)} r(\cdot) \end{aligned}$$

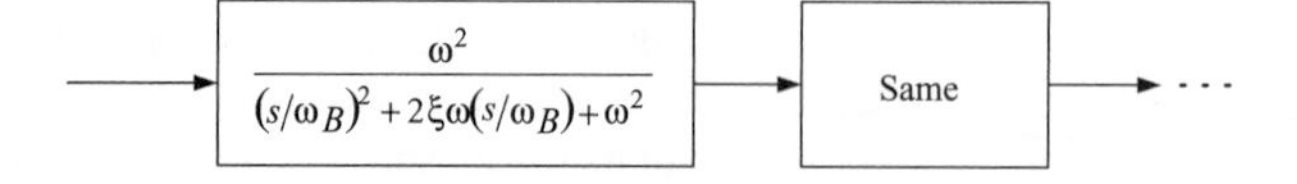


**FIGURE 2.25:** Second-order filters.

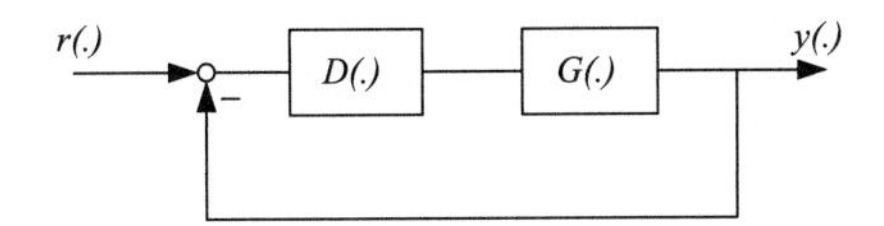


**FIGURE 2.26:** Block diagram of a standard feedback control system.

In $s$-domain the steady-state error (for continuous time systems)

$$e(s) = \frac{1}{1 + D(s)G(s)} r(s)$$

Steady-state value of the error as time goes to infinity can be determined using the final value theorem of Laplace transforms, provided that the $e(s)$ has stable poles and at most has one ploe at the origin, $s = 0$,

$$\lim_{t \to \infty} e(t) = \lim_{s \to 0} s\, e(s) = \lim_{s \to 0} s \frac{1}{1 + D(s)G(s)} r(s)$$

Let us consider the following cases:

1. The loop transfer function, $D(s)G(s)$, has $N$ number of poles at the origin $s = 0$,

$$D(s)G(s) = \frac{\prod_{i=1}^{m}(s + z_i)}{s^N \prod_{i=1}^{n}(s + p_i)}; \quad N = 0, 1, 2, \ldots.$$

2. The commanded signal is step, ramp, or parabolic signals (Fig. 2.26),

$$r(s) = \frac{1}{s}, \frac{A}{s^2}, \frac{B}{2s^3}$$

Now we will consider the steady-state error of a closed-loop system in response to a step, ramp, and parabolic command signals where the loop transfer function $D(s)G(s)$ has $N(N = 0, 1, 2)$ poles at the origin (Fig. 2.27).

1. $N = 0$;

   **(a)** $\lim_{t \to \infty} e_{step}(t) = \lim_{s \to 0} s \dfrac{1}{1 + \dfrac{\prod(s + z_i)}{\prod(s + p_i)}} \dfrac{1}{s} = \dfrac{1}{1 + D(0)G(0)} = \dfrac{1}{1 + K_p}$

| r(t) \ N | 0 | 1 | 2 |
|---|---|---|---|
| 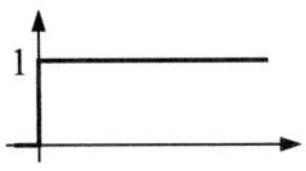 step (1) | $\frac{1}{1+K_p}$ | 0 | 0 |
| 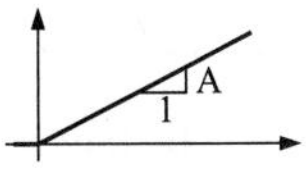 ramp (slope A/1) | $\infty$ | $\frac{A}{K_v}$ | 0 |
| 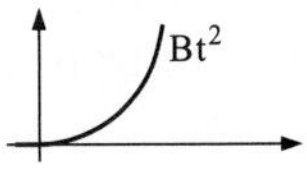 parabola ($Bt^2$) | $\infty$ | $\infty$ | $\frac{B}{2K_a}$ |

**FIGURE 2.27:** Steady-state error of a feedback control system in response to various command signals depends on the number of poles at the origin of the loop transfer function.

(b) $\lim_{t\to\infty} e_{ramp}(t) = \lim_{s\to 0} s \dfrac{1}{1+\dfrac{\prod(s+z_i)}{\prod(s+p_i)}} \dfrac{A}{s^2} = \dfrac{A}{0} \Rightarrow \infty$

(c) $\lim_{t\to\infty} e_{parab}(t) = \lim_{s\to 0} s \dfrac{1}{1+\dfrac{\prod(s+z_i)}{\prod(s+p_i)}} \dfrac{B}{2s^3} = \dfrac{1}{0} \Rightarrow \infty$

2. $N = 1$

(a) $\lim_{t\to\infty} e_{step}(t) = \lim_{s\to 0} s \dfrac{1}{1+\dfrac{1}{s}\dfrac{\prod(s+z_i)}{\prod(s+p_i)}} \dfrac{1}{s} = 0$

(b) $\lim_{t\to\infty} e_{ramp}(t) = \lim_{s\to 0} s \dfrac{1}{1+\dfrac{1}{s}\dfrac{\prod(s+z_i)}{\prod(s+p_i)}} \dfrac{A}{s^2} = \lim_{s\to 0} \dfrac{A}{sD(s)G(s)} = \dfrac{A}{K_v}$

(c) $\lim_{t\to\infty} e_{parab}(t) = \lim_{s\to 0} s \dfrac{1}{1+\dfrac{1}{s}\dfrac{\prod(s+z_i)}{\prod(s+p_i)}} \dfrac{B}{2s^3} = \dfrac{1}{0} \Rightarrow \infty$

3. $N = 2$

(a) $\lim_{t\to\infty} e_{step}(t) = \lim_{s\to 0} s \dfrac{1}{1+\dfrac{1}{s^2}\dfrac{\prod(s+z_i)}{\prod(s+p_i)}} \dfrac{1}{s} = 0$

(b) $\lim_{t\to\infty} e_{ramp}(t) = \lim_{s\to 0} s \dfrac{1}{1+\dfrac{1}{s^2}\dfrac{\prod(s+z_i)}{\prod(s+p_i)}} \dfrac{A}{s^2} = 0$

(c) $\lim_{t\to\infty} e_{parab}(t) = \lim_{s\to 0} s \dfrac{1}{1+\dfrac{1}{s^2}\dfrac{\prod(s+z_i)}{\prod(s+p_i)}} \dfrac{B}{2s^3} \lim_{s\to 0} \dfrac{B/2}{s^2 D(s)G(s)} = \dfrac{B}{2K_a}$

Notice that the d.c gain ($D(0)G(0)$) and the number of poles that the loop transfer function has at the origin are important factors in determining the steady-state error. It is convenient to define three constants to describe the steady-state error behavior of a closed-loop system: $K_p$ the position error constant, $K_v$ velocity error constant, and $K_a$ acceleration error constant.

$$
\begin{aligned}
K_p &= \lim_{s\to 0} D(s)G(s) \\
K_v &= \lim_{s\to 0} sD(s)G(s) \\
K_a &= \lim_{s\to 0} s^2 D(s)G(s)
\end{aligned}
$$

## 2.6 STABILITY OF DYNAMIC SYSTEMS

Stability of a control system is always a fundamental requirement. In fact, not only is it required that the system be stable, but it must be stable against uncertainties and reasonable variations in the system dynamics. In other words, it must have a good stability robustness. The stability of a dynamic system can be defined in two general terms (Fig. 2.28):

1. In terms of input–output magnitudes
2. In terms of the stability around an equilibrium point

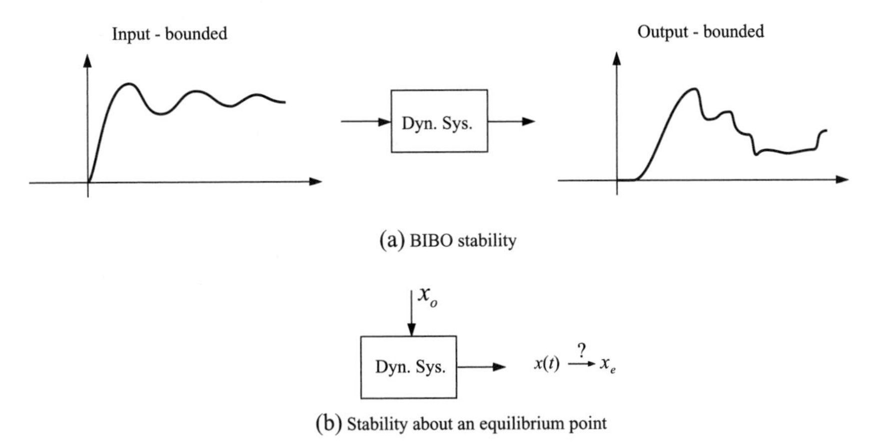


**FIGURE 2.28:** Definition of two different stability notions.

A dynamic system is said to be bounded input–bounded output (BIBO) stable if the response of the system stays bounded for every bounded input. This definition is referred to as *input–output stability* or *BIBO stability*. The stability of a dynamic system can also be defined only in terms of its equilibrium points and initial conditions without any reference to input. This definition is called the *stability in the sense of Lyapunov* or *Lyapunov stability*.

## 2.6.1 Bounded Input–Bounded Output Stability

***Definition*** A dynamic system (linear or nonlinear) is said to be bounded input–bounded output (BIBO) stable if for every bounded input, the output is bounded.

For a linear time invariant (LTI) system, the output to any input can be calculated as

$$y(t) = \int_{-\infty}^{t} h(\tau)u(t-\tau)\,d\tau$$

where $h(t)$ is the impulse response of the LTI system. If the input is bounded, there must exist a constant $M$ such that

$$|u(t)| \leq M < \infty$$

Hence,

$$\begin{aligned}|y(t)| &= |\int_{-\infty}^{t} h(\tau)u(t-\tau)\,d\tau| \\ &\leq \int_{-\infty}^{t} |h(\tau)||u(t-\tau)|d\tau \\ &\leq M \int_{-\infty}^{t} |h(\tau)|d\tau\end{aligned}$$

For the LTI system to be BIBO stable, $y(t)$ must be bounded for all $t$ as $t \to \infty$. Therefore, for $y(t)$

$$\lim_{t\to\infty} |y(t)| \leq M \int_{-\infty}^{\infty} |h(z)|dz$$

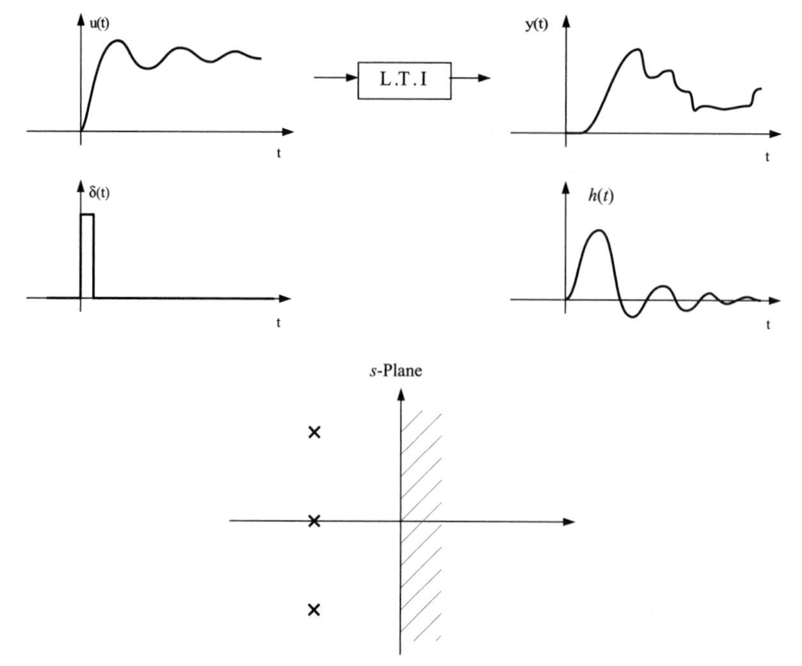


**FIGURE 2.29:** Bounded input–bounded output stability of linear time invariant systems.

to be bounded, the following expression must be bounded,

$$\int_{-\infty}^{\infty} |h(z)|dz$$

This requires that

$$h(t) \to 0 \text{ as } t \to \infty$$

In conclusion, if a LTI system is BIBO stable, this means that its impulse response goes to zero as time goes to infinity. The opposite is also true. If the impulse response of a LTI system goes to zero as time goes to infinity, the LTI system is BIBO stable (Fig. 2.29).

***Remark*** Impulse response going to zero as time goes to infinity means that all the poles of the dynamic system are on the left half $s$-plane. Therefore, for LTI systems, BIBO stability means that all the poles have negative real part on $s$-plane. The following expression summarizes the BIBO stability for LTI systems.

$$\{BIBO\ stable\} \Leftrightarrow \{h(t) \to 0 \text{ as } t \to \infty\} \Leftrightarrow \{\forall R_e(p_i) < 0\}$$

## 2.7 THE ROOT LOCUS METHOD

The root locus method is a graphical method for plotting the roots of an algebraic equation as one or more parameters vary. It is used primarily in studying the effect of variations in one parameter of a control system on the locations of closed-loop system poles. In the next section we will use the root locus method in order to understand the characteristics of PID type closed-loop controllers. The basic mathematical functionality is to *find* and *plot* the roots of an algebraic equation for various values of a parameter. The solution of algebraic

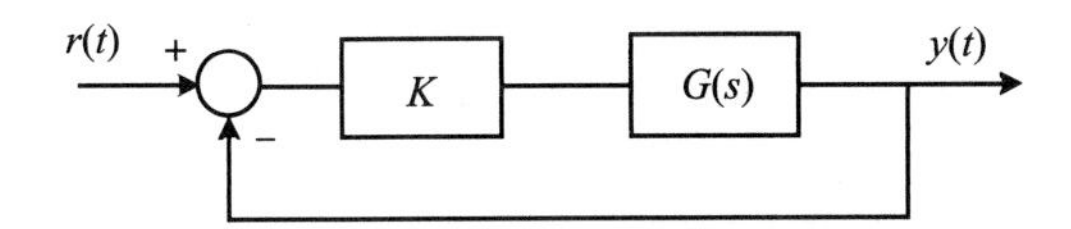


**FIGURE 2.30:** Basic root locus problem.

equations can be done easily by numerical means using a digital computer. Solving it for various values of one or more parameters is nothing more than implementing the numerical procedure in an iteration loop, i.e., FOR or DO loop in a high-level programming language. This is certianly a tool which became more and more effective with the availablity of CAD tools for control system design. A control engineer must, however, always keep in mind that the basic principle about computers is *garbage in–garbage out*. Therefore, it is very important that a designer be able to quickly verify in general terms a computer calculation with hand calculations or analysis. The graphical hand sketching rules of the root locus method provides such a tool. Understanding graphical root locus method not only provides a way of quickly checking the results of a computer calculations, but also develops very valuable insights to the control system design.

Let us consider an algebraic equation, i.e., a polynomial of degree $n$,

$$a_0 s^n + a_1 s^{n-1} + \cdots + a_{n-1} s + a_n = 0 \tag{2.28}$$

which has $n$ roots $(s_1, s_2, \ldots., s_n)$. If the value of any of $a_i$ changes, there will be a new set of $n$ roots $(s_1, s_2, \ldots., s_n)$. If a particular parameter $a_i$ varies from one minimum value to another value, an $n$ set of roots can be solved for every value of the parameter $a_i$. If we plot the roots on the complex $s$-plane, we end up with the graphical representation of the *locus of the roots* of the algebraic equation as one of the parameters in the equation varies. This is the basic functionality of the root locus method. Clearly, this can be done numerically by a computer algorithm.

Let us consider the feedback control system shown in Fig. 2.30. The closed-loop system transfer function is

$$\frac{y(s)}{r(s)} = \frac{KG(s)}{1 + KG(s)} \tag{2.29}$$

where the poles of the closed-loop system are given by the roots of the denominator,

$$\Delta_{cls}(s) = 1 + KG(s) \tag{2.30}$$

The standard root locus analysis problem involves the sketch of the locus of the roots of this equation as $K$ varies from zero to infinity. From the above general discussion, it is clear that the root locus method is not limited to that type of problem. It can address any problem which involves finding roots of an algebraic equation as any one or more parameters vary. Consider another example as shown in Fig. 2.31, where $a$ is a parameter. We would like

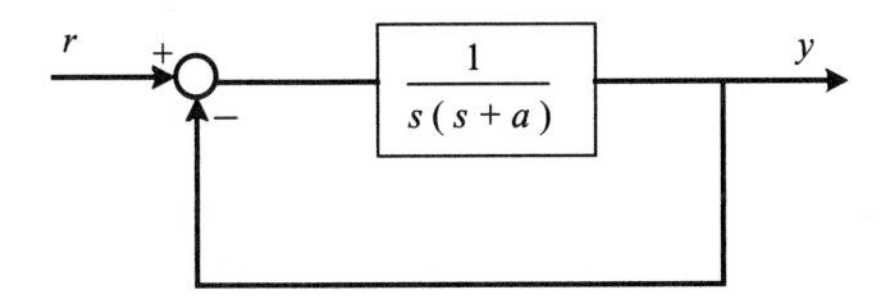


**FIGURE 2.31:** An example: closed-loop transfer function poles as parameter *a* varies.

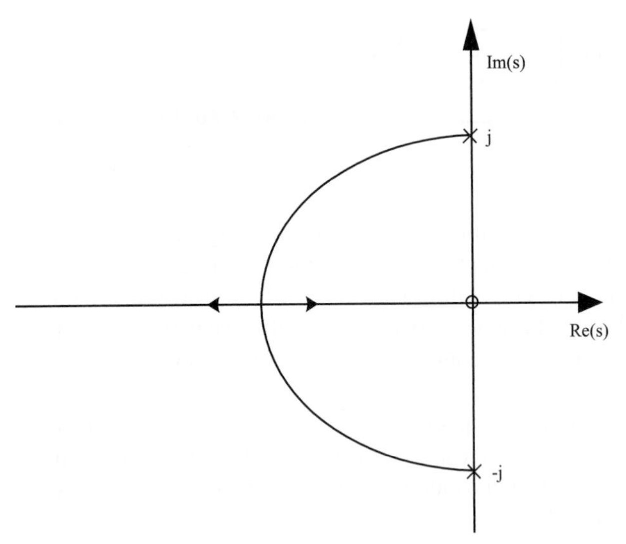


**FIGURE 2.32:** Value of gain is parameterized along the root locus curves. The particular value of gain can be calculated in order to be at a selected point on the root locus.

to study the locations of closed-loop system poles as the parameter $a$ varies from zero to infinity.

$$\frac{y(s)}{r(s)} = \frac{1}{1 + s(s + a)} = \frac{1}{s^2 + as + 1} \tag{2.31}$$

The characteristic equation is

$$\Delta_{cls}(s) = s^2 + as + 1 = 0 \tag{2.32}$$

which can be expressed in standard root locus formulation form suitable for graphical sketching as

$$1 + a\frac{s}{s^2 + 1} = 0 \tag{2.33}$$

The graphical root locus method rules are developed for sketching the roots of a polynominal equation as one parameter varies (Fig. 2.32). The polynomial equation is always expressed in the form of

$$1 + (parameter) \cdot \frac{(numerator)}{(denominator)} \tag{2.34}$$

Therefore, the locus of roots can be studied as a function of any parameter in the closed-loop system, not just the gain of the loop transfer function. Let us assume that we are interested in studying the locus of the roots of the following equation as parameter $b$ varies from zero to infinity,

$$s^3 + 6s^2 + bs + 8 = 0 \tag{2.35}$$

This problem can be expressed in a form suitable for the application of the root locus sketching rules as follows:

$$1 + b\frac{s}{s^3 + 6s^2 + 8} = 0 \tag{2.36}$$

Matlab provides the *rlocus(...)* function for root locus. The *rlocus()* function is overloaded and can accept different parameters. In principle, it takes the loop transfer function and assumes that a parameter $K$ is in the feedback loop. Therefore, in order to use the *rlocus()* function to analyze the roots of a transfer function or albegraic equation as one parameter varies, the equivalent root locus problem must be formed before calling the *rlocus()* function. Below are some samples of calls to the *rlocus()* function.

```
rlocus(sys) ; /* given LTI system, plot closed loop poles
                 as K varies from 0 to infinity
rlocus(sys, K) ; /*.............for the values of the
                    parameter given in the vector K*/
[R]=rlocus(sys,K) ; /* Stores the closed loop roots in the
                       R for numerical reference. */

sys = tf(num,den) ; /* sys can be formed by tf, ss, zpk
                       function calls */
sys = zpk(z,p,k) ; /* sys= ( K (s-z1)(s-z2).../)(s-p1).
                      (s-p2)....) */
sys = ss(A,B,C,D) ; /* G(s) = C (sI-A)^-1 B + D */
```

***Root Locus Sketching Rules*** In this section we list the rules used in approximate hand sketching of root locus. The derivation of the rules are given in many classic textbooks in control systems.

1. Put the problem in the $1 + KG(s) = 0$ form, where $K$ is the varying parameter. Mark the poles of $G(s) = n(s)/d(s)$ by $x$ and the zeros by $o$ on the $s$-plane. Notice that root locus begins at $x$'s for $K = 0$ and ends at $o$'s as $K \longrightarrow \infty$. The $x$'s and $o$'s represent the asymptotic location of closed-loop pole locations. If $K$ is indeed the loop transfer function gain, the o's are also the zeros of the open- and closed-loop systems. Otherwise, they only represent the asymptotic location of closed-loop system poles.
2. Mark the part of the real axis to the left of odd number of poles and zeros as part of the root locus. All the points on the real axis to the left of odd number of poles and zeros satisfy the angle criteria, hence are part of the root locus. Notice that the root locus is the collection of points on the $s$-plane that makes the phase angle of the transfer function equal to 180°, since

$$K = -\frac{1}{G(s)} ; \tag{2.37}$$

Since $K$ is a positive real number, $G(s)$ complex function will have negative real values at the collection of $s$-points that makes up the root locus. In other words, the phase angle of $G(s)$ at all points that are part of the root locus is 180°.

$$G(s) = \frac{1}{K} = |\frac{1}{K}| \cdot e^{j180^\circ} \tag{2.38}$$

3. If there are $n$ poles, and $m$ zeros, $m$ of the poles ends up at the $m$ zero locations as parameter $K$ goes to infinity. The remaining $n - m$ poles go to infinity along asymptotes. If $n - m = 1$, then the one extra pole goes to infinity along the negative real axis. If $n - m \geq 2$, then they go to infinity along the asymptotes defined by the asymptote center and angles as follows:

$$\sigma = \frac{\sum p_i - \sum z_i}{n - m} \tag{2.39}$$

$$\psi_l = \frac{180 + l360}{n - m} ; \; l = 0, 1, 2, \ldots, n - m - 1. \tag{2.40}$$

The quick hand sketches of root locus allow the designer to quickly check the computer analysis results for correctness and provides valuable insight in controller design.

## 2.8 BASIC FEEDBACK CONTROL TYPES

Figure 2.33 shows the three basic feedback control actions: (1) proportional, (2) integral, and (3) derivative control actions. Figure 2.34 shows the input–output behavior of these control types. In practical terms, proportional control action is generated based on the current error, the integral control action is generated based on the past error, and the derivate control action is generated based on the anticipated future error. Integral of the error can be interpreted as the past information about it. Derivative of the error can be interpreted as a measure of future error. Assume that the error signal entering the control blocks has a trapeziodal form. The control actions generated by the proportional, integral, and derivative actions are shown in Fig. 2.34. Proportional-integral-derivative (PID) control has control decision blocks which take into account the past, current, and future error. In a way, it covers all the history of error. Therefore, most practical controllers are either a form of the PID controller or have the properties of a PID controller.

The block diagram of a textbook standard PID controller is shown in Fig. 2.35. The control algorithm can be expressed in both continuous (analog) time domain (which can be implemented using op-amps) and in discrete (digital) time domain (which can be implemented usign a digital computer in software). At any given time, $t$, the control signal, $u(t)$, is determined as a function,

$$u(t) = K_p e(t) + K_I \int_0^t e(\tau)d\tau + K_D \dot{e}(t) \tag{2.41}$$

Proportional control (P)

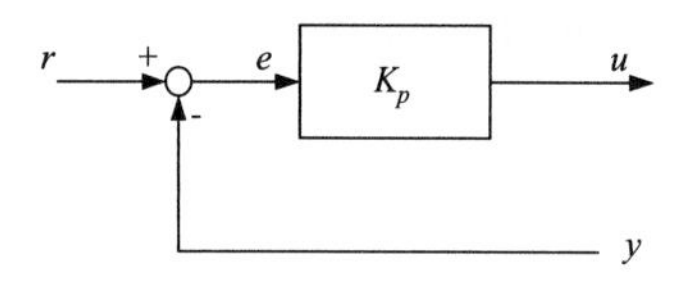


$$u(t) = K_p e(t)$$

$$D(s) = \frac{u(s)}{e(s)} = K_p$$

Integral control (I)

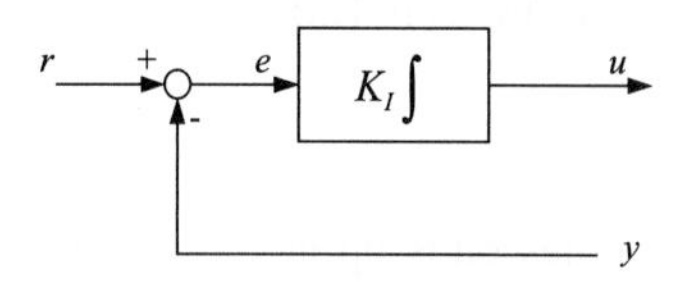


$$u(t) = K_I \int_0^t e(\tau)d\tau$$

$$D(s) = \frac{u(s)}{e(s)} = \frac{K_I}{s}$$

Derivative control (D)

r + e $K_D \frac{d(\ )}{dt}$ u - y

$$u(t) = K_D \frac{d}{dt}(e(t))$$

$$D(s) = \frac{u(s)}{e(s)} = K_D s$$

**FIGURE 2.33:** Basic feedback control actions: proportional control, integral control, and derivative control.

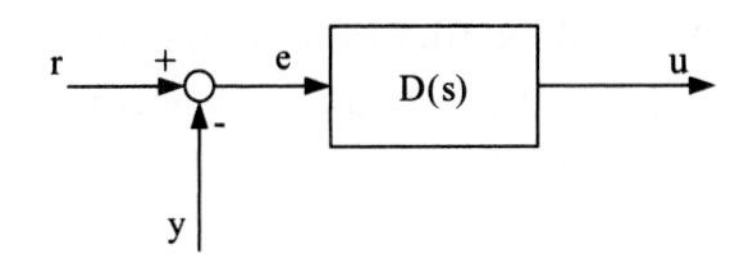


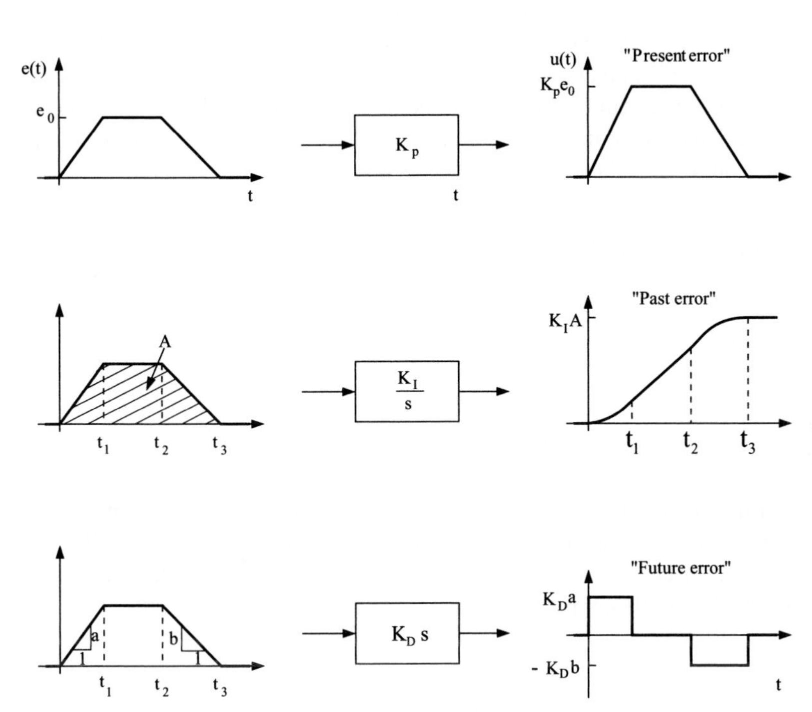


**FIGURE 2.34:** Illustration of the input–output behavior of basic feedback control actions: proportional, integral, and derivative control.

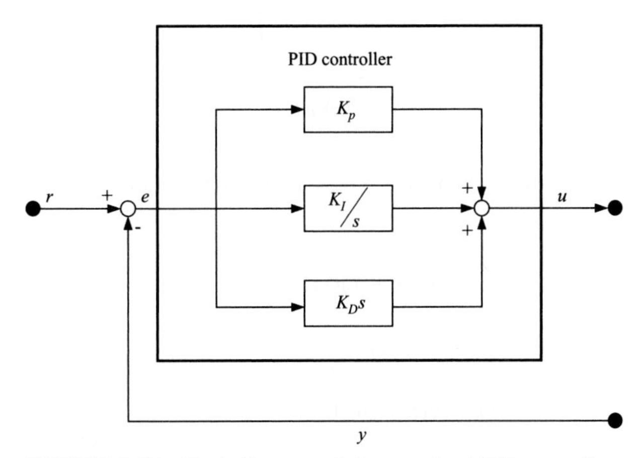


**FIGURE 2.35:** Block diagram of the standard PID controller.

which shows that the control signal is function of the error between the commanded and measured output signals, $e(t)$ at time $t$, as well as the derivative of the error signal, $\dot{e}(t)$, and the integral of the error signal since the control loop is enabled ($t = 0$), $\int_0^t e(\tau)d\tau$.

The discrete time approximation of the PID control algorithm can be implemented by finite difference approximation to the derivative and integral functions. In digital implementation, the control signal can be updated in periodic intervals, $T$, also called the sampling period. The control is updated at integer multiples of the sampling period. The value of the signal is kept constant in between each update period. At any update instant $k$, time is $t = kT$, and previous update instant is $t - T = kT - T$, next update instant is $t = kT + T$, and so on, the control signal can be expressed as $u(t) = u(kT)$,

$$u(kT) = K_p \cdot e(kT) + K_I \cdot u_I(kT) + K_D \left( \frac{(e(kT) - e(kT - T))}{T} \right) \tag{2.42}$$

where

$$u_I(kT) = u_I(kT - T) + e(kT) \cdot T \tag{2.43}$$

$$u_I(0) = 0.0; \text{ at the initialization} \tag{2.44}$$

Let us take the Laplace transform of the continuous time domain (analog) version of the PID control and analyze its effect on a controlled system. Basically, the same results apply for discrete time version (digital implementation) provided the sampling period is short enough (high sampling frequency) relative to the bandwidth of the closed-loop system.

$$u(s) = \left( k_p + K_I \frac{1}{s} + K_D s \right) e(s) \tag{2.45}$$

$$D(s) = K_p + K_I \frac{1}{s} + K_D s \tag{2.46}$$

$$= K_p \left( 1 + \frac{1}{T_I s} + T_D s \right) \tag{2.47}$$

$$K_I = \frac{K_p}{T_I}, \quad K_D = K_p T_D$$

Consider a second-order mass-force system to study its behavior under various forms of PID control (Fig. 2.36).

$$m\ddot{x}(t) = f(t) - f_d(t)$$

$$\ddot{x}(t) = \frac{1}{m} f(t) - \frac{1}{m} f_d(t)$$

$$\ddot{x}(t) = u(t) - w_d(t)$$

$$s^2 x(s) = u(s) - w_d(s)$$

$$x(s) = \frac{1}{s^2} u(s) - \frac{1}{s^2} w_d(s)$$

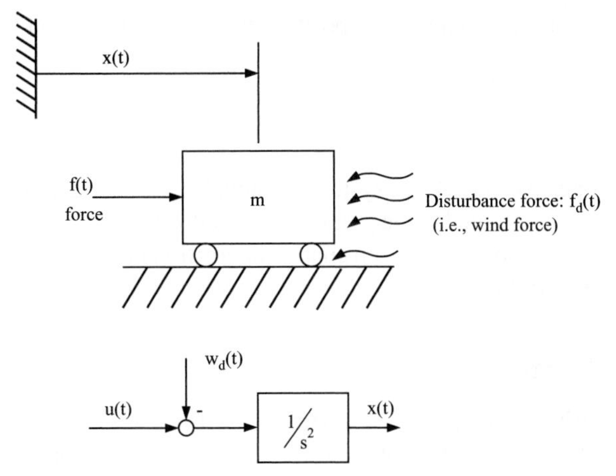


**FIGURE 2.36:** Mass-force system model: (a) model components, and (b) block diagram.

## 2.8.1 Proportional Control

Let us consider input and output relationship, and do not consider disturbance for the purpose of studying proportional control properties only (Fig. 2.37),

$$\frac{x(s)}{u(s)} = \frac{1}{s^2} \tag{2.48}$$

If the control action $u(t)$ is decided on by a proportional control based on the error between the desired position, $x_d(t)$ and the actual measured position, $x(t)$,

$$u(t) = K_p(x_d(t) - x(t)) \tag{2.49}$$

$$u(s) = K_p(x_d(s) - x(s)) \tag{2.50}$$

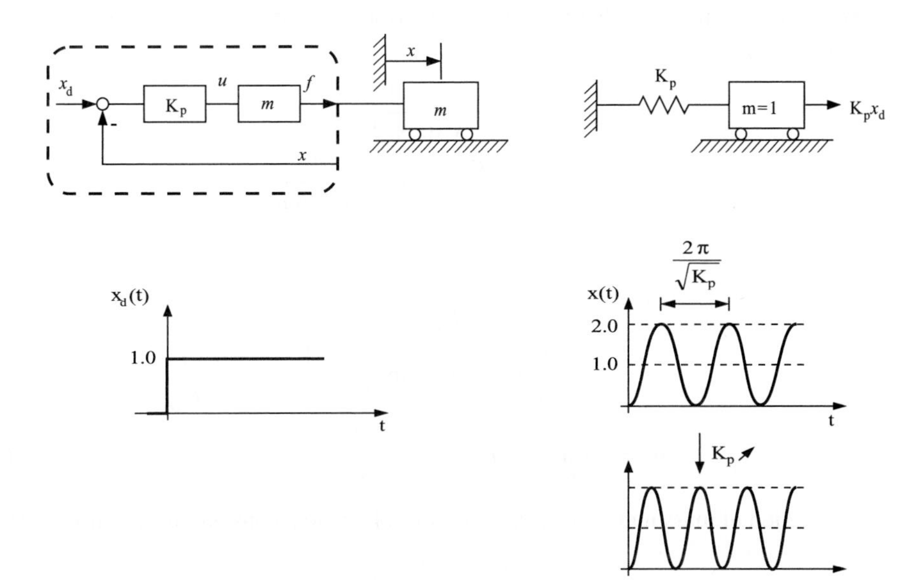


**FIGURE 2.37:** Mass-force system with position feedback control and its step response.

The CLS transfer function from the commanded position to the actual position under the proportional control is

$$x(s) = \frac{K_p}{s^2 + K_p} x_d(s) \tag{2.51}$$

The proportional control alone on a mass-force system is equivalent to adding a spring to the system where the spring constant is equal to the proportional feedback gain, $K_p$, (Fig. 2.37). The response of this system to a commanded step change in position is shown in Fig. 2.37. Figure 2.39(a) shows the CLS root locus as $K_p$ varies from zero to infinity.

### 2.8.2 Derivative Control

Let us consider only the derivative control on the same mass-force system. Assume that the control is proportional to the derivative of position which means proportional to the velocity,

$$u(t) = -K_D \dot{x} \tag{2.52}$$

$$u(s) = -K_D s x(s) \tag{2.53}$$

$$s^2 x(s) = -K_D s x(s) \tag{2.54}$$

$$s(s + K_D) x(s) = 0 \tag{2.55}$$

If we consider disturbance in the model, the transfer function from the disturbance (i.e., wind force) to the position of the mass can be determined as

$$m\ddot{x} = f(t) - f_d(t) \tag{2.56}$$

$$\ddot{x} = u(t) - w_d(t) \tag{2.57}$$

$$s(s + K_D) x(s) = -w_d(s) \tag{2.58}$$

$$x(s) = \frac{1}{s(s + K_D)} (-w_d(s)) \tag{2.59}$$

Let us consider the case that the disturbance is a constant step function, $w_d = \frac{1}{s}$ and the resultant response is

$$x(s) = -\frac{1}{s} \frac{1}{s(s + K_D)} \tag{2.60}$$

$$= \frac{a_1}{s^2} + \frac{a_2}{s} + \frac{a_3}{(s + K_D)} \tag{2.61}$$

where

$$a_1 = \lim_{s \to 0} s^2 x(s) = \frac{1}{K_D} \tag{2.62}$$

$$a_2 = \lim_{s \to 0} \frac{d}{ds} [s^2 x(s)] = \lim_{s \to 0} \frac{-1}{(s + K_D)^2} = -\frac{1}{K_D^2} \tag{2.63}$$

$$a_3 = \lim_{s \to K_D} [s + K_D] x(s) = \frac{1}{K_D^2} \tag{2.64}$$

The position of mass under the derivative control due to a constant disturbance force is

$$x(t) = \frac{1}{K_D} t - \frac{1}{K_D^2} 1(t) + \frac{1}{K_D^2} e^{-K_D t} \tag{2.65}$$

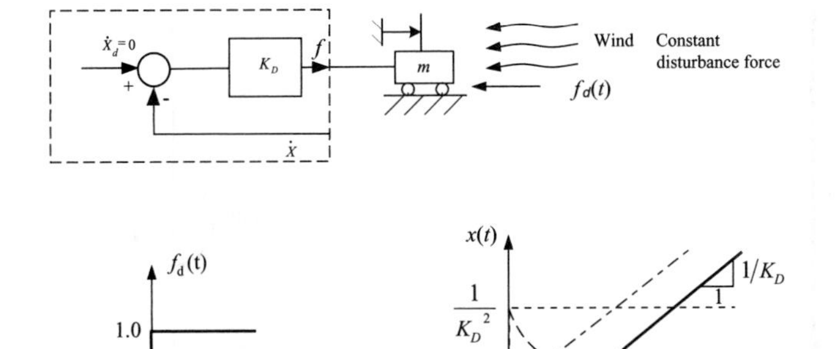


**FIGURE 2.38:** Mass-force system under velocity feedback control, and response to a constant disturbance.

This example shows that the derivative feedback control alone would not be able to reject a constant disturbance acting on a second-order mass-force system. Derivative feedback introduces damping into the closed-loop system poles (Fig. 2.38) and increases the stability margin.

### 2.8.3 Integral Control

Now let's consider the case where the control action is based on the integral of position error,

$$u(t) = K_I \int_0^t [x_d(\tau) - x(\tau)]d\tau \tag{2.66}$$

$$u(s) = \frac{K_I}{s}[x_d(s) - x(s)] \tag{2.67}$$

Substituting this into mass-force model (equation 2.48),

$$s^2 x(s) = u(s) = \frac{K_I}{s}[x_d(s) - x(s)] \tag{2.68}$$

$$s^3 x(s) + K_I x(s) = K_I x_d(s) \tag{2.69}$$

$$(s^3 + K_I)x(s) = K_I x_d(s) \tag{2.70}$$

The closed-loop system poles are given by (Fig. 2.39):

$$\Delta_{cl}(s) = 1 + K_I \frac{1}{s^3} \tag{2.71}$$

Figure 2.39(b) shows the locus of CLS poles for various values of $K_I$ as it takes on values from 0 to $\infty$. The integral control alone would result in an unstable mass-force system. It tends to destabilize the system. However, the main purpose of integral control is to reject the disturbances and reduce the steady-state error, as will be shown in the next section.

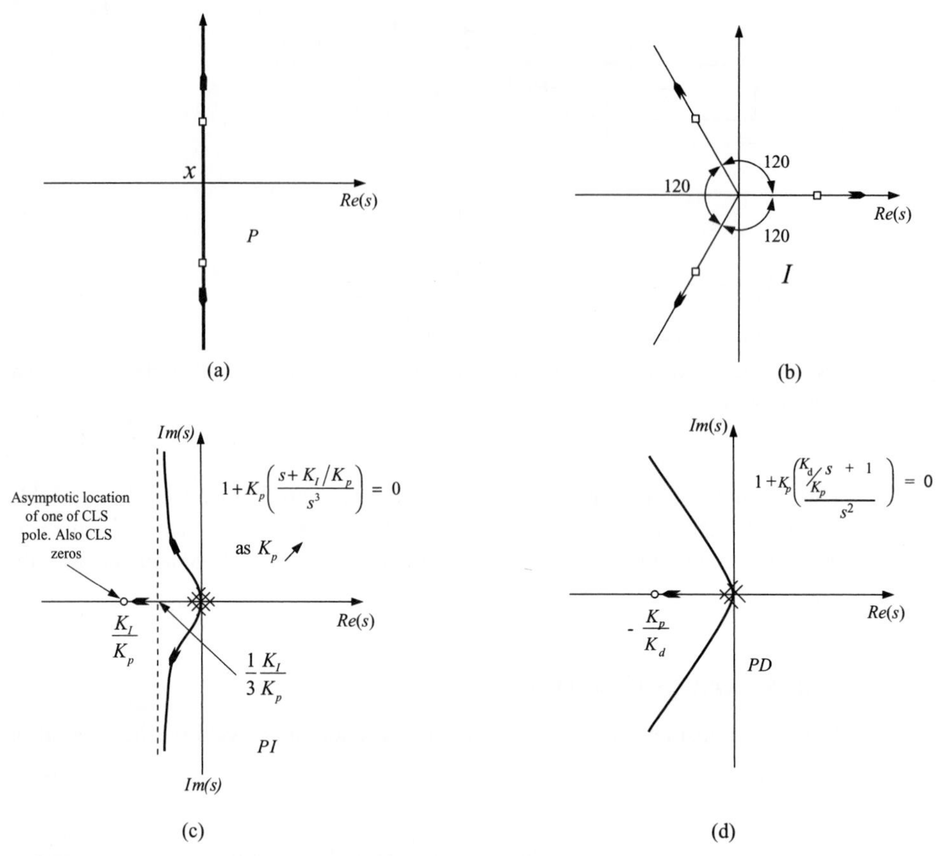


**FIGURE 2.39:** Locus of the poles of the system under P, I, PI, PD control (closed-loop system poles) as the gain increases from zero to infinity.

## 2.8.4 PI Control

Let us consider the mass-force system under a proportional plus integral (PI) control. The control algorithm is

$$u(t) = K_p(x_d(t) - x(t)) + K_I \int_0^t (x_d(\tau) - x(\tau))\, d\tau \tag{2.72}$$

and its Laplace transform is

$$u(s) = K_p(x_d(s) - x(s)) + K_I \frac{1}{s}(x_d(s) - x(s)) \tag{2.73}$$

$$= \left(K_p + \frac{K_I}{s}\right)(x_d(s) - x(s)) \tag{2.74}$$

Let's take the Laplace transform of the mass-force system and substitute the PI controller for $u(s)$

$$\ddot{x} = u(t) - w_d(t)$$

$$s^2 x(s) = u(s) - w_d(s) = \left(K_p + \frac{K_I}{s}\right)(x_d(s) - x(s)) - w_d(s) \tag{2.75}$$

After a few algebraic manipulations, the transfer function between position, desired position, and disturbance force is found as

$$x(s) = \frac{K_p(s) + K_I}{s^3 + K_p s + K_I} x_d(s) - \frac{s}{s^3 + K_p s + K_I} w_d(s) \tag{2.76}$$

Consider the case in which the commanded position is zero, $x_d(t) = 0$, and there is a constant step disturbance, $w_d(s) = \frac{1}{s}$. Any nonzero response due to the disturbance would be an error.

$$x(s) = -\frac{s}{s^3 + K_p s + K_I} \frac{1}{s}$$

$$x(s) = -\frac{1}{s^3 + K_p s + K_I} \tag{2.77}$$

If $\Delta_{cls}(s) = s^3 + K_p s + K_I$ has all stable roots, the response of the system will be zero despite a constant disturbance.

Using the final value theorem,

$$\begin{aligned} \lim_{t \to \infty} e(t) = e_{ss}(\infty) &= \lim_{s \to 0} s e(s) \\ &= \lim_{s \to 0} s \frac{1}{s^3 + K_p s + K_I} \qquad (2.78) \\ e_{ss}(\infty) &= 0 \end{aligned}$$

The steady-state error due to a constant disturbance is zero under the PI type control. If there is no integral control action, $K_I = 0$, the steady-state error would have been

$$e(s) = \frac{s}{s(s^2 + K_p)} \frac{1}{s}$$

$$\lim_{s \to 0} s e(s) = \frac{1}{s^2 + K_p} = \frac{1}{K_p} \to \neq 0$$

Therefore it is clear that it is the integral of position error used in feedback control which enables the control system to reject the constant disturbance and keep $x(t) = x_d(t)$ in steady state. Transient response to a step command change in desired position under no disturbance condition is given by

$$w_d(t) = 0$$

$$x_d(t) = 1(t)$$

$$\begin{aligned} x(s) &= \frac{K_p s + K_I}{s^3 + K_p s + K_I} x_d(s) \\ &= \frac{\frac{1}{s^2}\left(K_p + \frac{K_I}{s}\right)}{1 + \frac{1}{s^2}\left(K_p + \frac{K_I}{s}\right)} \end{aligned}$$

The closed-loop system has a zero at

$$-\frac{K_I}{K_p}$$

and three poles at the roots of the following equation:

$$s^3 + K_p s + K_I = 0$$

$$1 + \frac{1}{s^2}\left(K_p + \frac{K_I}{s}\right) = 0$$

Let us study the locus of the roots of this equation for various values of $K_P, K_I$ [Fig. 2.39(c)]. CLS dominant pole damping is limited by $\left(\frac{1}{3}\frac{K_I}{K_p}\right)$. If we want to add more damping, we must use other control actions, such as derivative (D) control, which is considered next.

### 2.8.5 PD Control

Now we will consider the characteristics of mass-force of system under proportional plus derivative (PD) control. The PD control algorithm is given by

$$u(t) = K_p(x_d(t) - x(t)) - K_D\dot{x}(t) \tag{2.79}$$

$$u(s) = K_p(x_d(s) - x(s)) - K_D\dot{x}(s) \tag{2.80}$$

Substituting this into the mass-force model,

$$s^2x(s) = u(s) - w_d(s) \tag{2.81}$$

$$(s^2 + K_Ds + K_p)x(s) = K_px_d(s) - w_d(s) \tag{2.82}$$

We will consider the dominant response to a step command in desired position, and steady-state response to a constant disturbance.

**(i)** Transient response in $s$-domain

$$x(s) = \frac{K_p}{s^2 + K_Ds + K_p}(1/s) \tag{2.83}$$

and the time domain response is found by taking the inverse Laplace transform,

$$x(t) = 1 - e^{-\xi\omega_n t}\frac{1}{\sqrt{1-\xi^2}}\,sin\left(\sqrt{1-\xi^2}\,\omega_n t + \phi\right) \tag{2.84}$$

where

$$K_p = \omega_n^2 \tag{2.85}$$

$$K_D = 2\xi\omega_n \tag{2.86}$$

$$\phi = \tan^{-1}\left(\frac{\sqrt{1-\xi^2}}{\xi}\right) \tag{2.87}$$

Since PD control gains determine the natural frequency and the damping ratio of the closed-loop system, PD control can be efficient in shaping transient response.

**(ii)** Zero command, constant step disturbance case - $x_d = 0;\;\; w_d \neq 0;\;\; w_d(s) = \frac{1}{s}$. The response due to step disturbance is given by

$$x(s) = -\frac{1}{s^2 + K_Ds + K_p}\frac{1}{s} \tag{2.88}$$

$$x(s) = e(s) = -\frac{1}{s^2 + K_Ds + K_p}\frac{1}{s} \tag{2.89}$$

Any nonzero response due to the disturbance is indeed an error. The magnitude of the error under PD control is

$$\lim_{t\to\infty} e(t) = \lim_{s\to 0} s(s) \tag{2.90}$$

$$= \lim_{s\to 0}(-s)\frac{1}{s^2 + K_Ds + K_p}\frac{1}{s} \tag{2.91}$$

$$= \frac{1}{K_p} \tag{2.92}$$

Therefore, the PD control alone cannot provide zero steady-state error in the presence of constant disturbance.

### 2.8.6 PID Control

PID control is basically a PD control plus PI control. It combines the capabilities of PD and PI control. PD control is primarily used to shape transient response and stabilize the system. The D (derivative) action introduces damping into the closed-loop system. If the steady-state error is constant, hence its derivative is zero, the derivate action has no influence in the steady-state response. PI control is used to reduce the steady-state error and improve disturbance rejection capability. Almost all practical controllers exhibit the features of PID control. They have control action components which deal with the present error (proportional-P control), past error using the integral of error (integral-I control), and the future error using the anticipatory nature of derivative (D-control). There are many different implementations of PID control. One possible implementation of PID control is shown below. In this implementation, the derivative action is only applied to the feedback signal, not on the error. Sometimes this may be preferable if the command signal has jump discontinuities such as step changes.

$$u(t) = K_p(x_d(t) - x(t)) + K_I\left(\int_0^t (x_d(\tau) - x(\tau))\right)d\tau - K_D\dot{x}(t) \tag{2.93}$$

$$u(s) = K_p(x_d(s) - x(s)) + \frac{K_I}{s}(x_d(s) - x(s)) - K_D s x(s) \tag{2.94}$$

$$= -\left(K_p + \frac{K_I}{s} + K_D s\right)x(s) + \left(K_p + \frac{K_I}{s}\right)x_d(s) \tag{2.95}$$

If we use the PID controller for position control of the mass-force system,

$$s^2 x(s) = u(s) - w_d(s) \tag{2.96}$$

$$\left\{s^2 + \left(K_p + \frac{K_I}{s} + K_D s\right)\right\}x(s) = \left(K_p + \frac{K_I}{s}\right)x_d(s) - w_d(s) \tag{2.97}$$

$$(s^3 + K_D s^2 + K_p s + K_I)x(s) = (K_p s + K_I)x_d(s) - s w_d(s) \tag{2.98}$$

The closed-loop system transfer function for the mass-force system under the PID control is

$$x(s) = \frac{(K_p s + K_I)}{(s^3 + K_D s^2 + K_p s + K_I)}x_d(s) - \frac{s}{(s^3 + K_D s^2 + K_p s + K_I)}w_d(s) \tag{2.99}$$

Let us consider the behavior of this system for two different conditions: (i) commanded input is zero, but there is a constant unit magnitude disturbance, $x_d(t) = 0$; $w_d(t) = 1(t)$; and (ii) there is a unit magnitude step command, but no disturbance, $x_d(t) = 1(t)$; $w_d(t) = 0$.

**(i)** $x_d(t) = 0.0$; $w_d(t) = 1(t)$: The nonzero response due to disturbance is an unwanted response and can be considered as error,

$$x(t) = e(t) \tag{2.100}$$

$$x(s) = e(s) \tag{2.101}$$

$$= -\frac{s}{(s^3 + K_D s^2 + K_p s + K_I)}\frac{1}{s} \tag{2.102}$$

In steady state, the position of the mass-force system under the PID control

$$\lim_{t\to\infty} x(t) = \lim_{s\to 0} x(s) = -\frac{s}{(s^3 + K_D s^2 + K_p s + K_I)} = 0 \tag{2.103}$$

is zero as commanded despite the constant magnitude disturbance force. If there is no integral action, $k_I = 0$, the response is

$$e(\infty) = \frac{1}{K_p}$$

finite and inversely proportional to the proportional feedback gain. Notice the importance of the integral (I) action in rejecting the disturbance. The integral action makes the system type I with respect to the disturbace entering the system after the controller block.

**(ii)** $x_d(t) = 1(t)$; $w_d(t) = 0$: Let us study the step response of the system when there is no disturbance,

$$x_d(t) = 1(t); \quad \textit{step function} \tag{2.104}$$

$$x(s) = \frac{K_p s + K_I}{(s^3 + K_D s^2 + K_p s + K_I)} \frac{1}{s} \tag{2.105}$$

The PID controller can be designed as a cascade of PD and PI controllers. Design PD control first to set shape the transient response, then design the PI control to shape the steady-state response. The PD control introduce a zero to the open- and closed-loop transfer functions. Therefore, it has a tendency to pull the root locus to the left side of $s$-plane, hence has a stabilizing effect on the closed-loop system. PI control introduces a zero close to origin and a pole at the origin. Generally, the PI controller zero is placed closer to the origin relative to the other poles and zeros of the system. The result of placing the zero of PI control closer to the pole at the origin is that it will not influence the transient response much which was shaped by the PD control, but will still increase the *type* of the loop transfer function by one. Therefore, PI controller primarily influences the steady-state error. The resultant PID control parameters as function of PI and PD controller parameters can be found as follows:

$$D(s) = K_P \left(1 + \frac{1}{T_I s} + T_D s\right) \tag{2.106}$$

$$= K_P^* \left(1 + \frac{1}{T_I^* s}\right)(1 + T_D^* s) \tag{2.107}$$

where

$$K_P = K_P^*(1 + T_D^*/T_I^*) \tag{2.108}$$

$$T_I = T_I^* + T_D^* \tag{2.109}$$

$$T_D = (T_I^* T_D^*)/(T_I^* + T_D^*) \tag{2.110}$$

## 2.9 TRANSLATION OF ANALOG CONTROL TO DIGITAL CONTROL

A controller can be completely analyzed and designed using continuous time methods. The resultant controller is an analog controller which can also be implemented in hardware using op-amps. The controller can be approximated with a digital controller which would be implementated using a digital computer. The fundamental tool is the approximation of differentiation by finite differences.

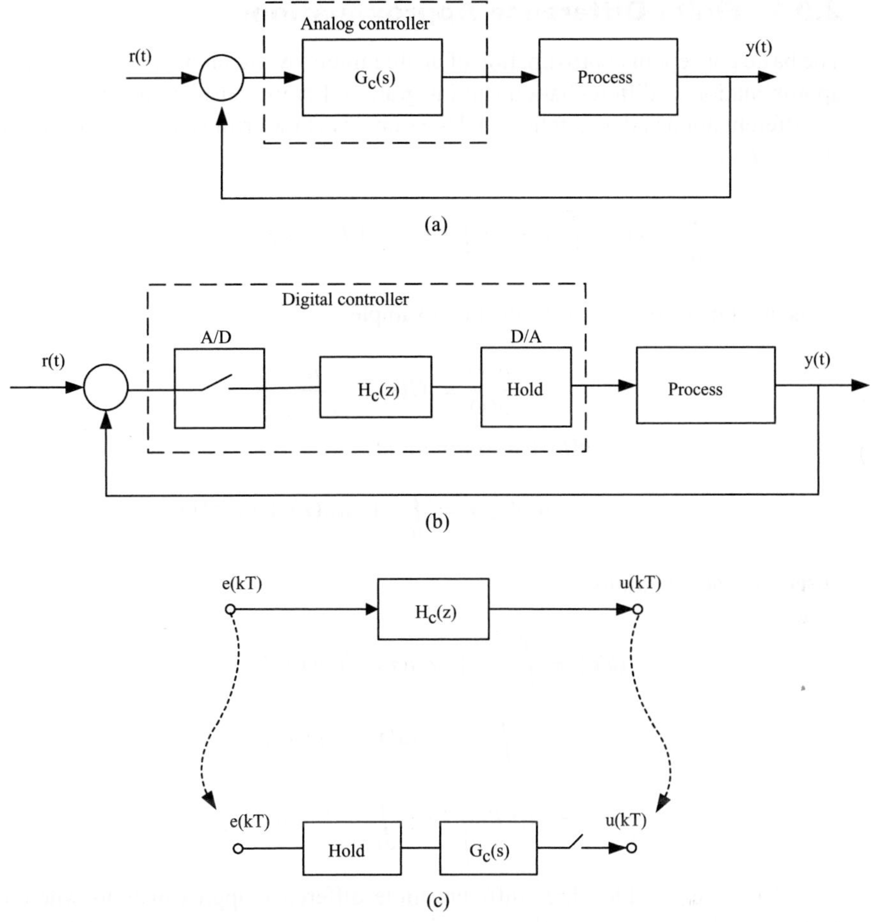


**FIGURE 2.40:** Digital approximation to analog controllers.

The basic problem is the following: given $G_c(s)$ (analog controller transfer function), find $H_c(z)$ (digital controller transfer function) such that closed-loop system (CLS) under digital controller performs as close as possible to the CLS under analog controller performance (Fig. 2.40).
The finite-difference approximations considered are:

- Forward difference approximation
- Backward difference approximation
- Trapezoidal approximation

There are other methods such as *trapezoidal approximation with frequency prewarping*, *zero-order hold (ZOH) equivalent approximation*, *pole-zero mapping*, *first-order equivalence*. These are not discussed here.

It should be quickly noted that as the sampling rate gets very large relative to the bandwidth of the controller (i.e., 20–50 times larger), the differences between different approximation methods gets insignificant. Likewise, if the sampling frequency is not very large relative to the bandwidth of the controller (i.e., 2–4 times larger than the controller bandwidth), the differences between various approximations get significant.

### 2.9.1 Finite Difference Approximations

The basic concept in approximation of analog filters by digital filters is the finite difference approximation of differentiation and integration. Let us consider an error signal, $e(t)$, and its differentiation and integration, and the samples of the error signal $\{\ldots, e(kT-T), e(kT), e(kT+T), \ldots\}$,

$$\frac{d}{dt}[e(t)], \left(\int_{t_o}^{t} e(\tau)d\tau\right) \Longleftrightarrow (e(kT), e(kT-T), \ldots) \tag{2.111}$$

Consider a first-order transfer function example

$$\frac{u(s)}{e(s)} = G(s) = \frac{a}{s+a} \tag{2.112}$$

$$\dot{u}(t) + au(t) = ae(t) \tag{2.113}$$

$$u(t)|_{t=kT} = \int_0^{kT} [-au(\tau) + ae(\tau)]d\tau \tag{2.114}$$

Discretize the integration

$$u(kT) = \int_0^{kT-T} [-au(\tau) + ae(\tau)]d\tau + \int_{kT-T}^{kT} [-au(\tau) + ae(\tau)]d\tau \tag{2.115}$$

$$u(kT) = u(kT-T) + \int_{kT-T}^{kT} [-au(\tau) + ae(\tau)]d\tau \tag{2.116}$$

Now we consider three different finite difference approximations where each one makes a different approximation to the integration term in the above equation.

**(i)** Forward difference approximation:

$$u(kT) = u(kT-T) + T[-au(kT-T) + ae(kT-T)] \tag{2.117}$$

$$u(kT) = (1-aT)\,u(kT-T) + a\,Te(kT-T) \tag{2.118}$$

Notice that this equation can be easily implemented in software on a digital computer. At every control sampling period, all is needed the value of the output at the previous cycle and the error. The algorithm involves two multiplication and one addition operations.

In order to develop a more generic relationship between the analog and digital controller approximate conversion, we take the Z-transform of the above difference equation,

$$[1-(1-aT)z^{-1}]u(z) = aTz^{-1}e(z) \tag{2.119}$$

Notice that a single sampling period of delay in a signal adds a $z^{-1}$ to the transform of the signal. Likewise, an advance of a single sampling period in a signal adds a $z$ to the transform of a signal. Using that principle, it is easy to take the Z-transform of difference equations and inverse Z-transform of z-domain transfer functions to obtain difference equations.

For real-time algorithmic implementation, we need the difference equation form of the controller.

$$\frac{u(z)}{e(z)} = \frac{aTz^{-1}}{1-(1-aT)z^{-1}}$$
$$= \frac{a.T}{z-(1-aT)}$$
$$= \frac{a}{((z-1)/T)+a} \tag{2.120}$$

Notice the subsititution relationship between $s$ and $z$ using this approximation:

$$\frac{a}{s+a} \longrightarrow \frac{a}{\frac{z-1}{T}+a} \tag{2.121}$$

$$s \longrightarrow \frac{z-1}{T}; \quad z = sT+1 \tag{2.122}$$

**(ii)** Backward difference approximation: Another possible approximation is to use the backward difference rule

$$u(kT) = u(kT-T) + T[-au(kT) + ae(kT)] \tag{2.123}$$

Again, the above equation is in a form suitable for real-time implementation in software. In order to obtain a more generic relationship for this type of approximation, let us take the Z-transform of the above equation,

$$u(z) = z^{-1}u(z) - Tau(z) + Tae(z) \tag{2.124}$$

$$(1 + Ta - z^{-1})u(z) = Tae(z) \tag{2.125}$$

$$\frac{u(z)}{e(z)} = \frac{Ta}{1+Ta-z^{-1}} = \frac{zTa}{z-1+Taz}$$
$$= \frac{a}{\frac{z-1}{zT}+a} \tag{2.126}$$

The backward approximation is equivalent to the following substitution between $s$ and $z$:

$$\frac{a}{s+a} \longrightarrow \frac{a}{\frac{z-1}{Tz}+a} \tag{2.127}$$

$$s \longrightarrow \frac{z-1}{Tz} \tag{2.128}$$

**(iii)** Trapezoidal approximation (Tustin's method, bilinear transformation): Finally, we will consider the trapezoidal rule approximation among the finite difference approximations to the integration,

$$u(kT) = u(kT-T) + \frac{T}{2}[-a[u(kT-T) + u(kT)]$$
$$+ [a(e(kT-T) + e(kT)]] \tag{2.129}$$

Similarly, we take the Z-transform of the above equation,

$$zu(z) = u(z) + \frac{T}{2}[-a(1+z)u(z) + a(1+z)e(z)] \tag{2.130}$$

$$[z + \frac{Ta}{2}(1+z) - 1]u(z) = \frac{T \cdot a}{2}(1+z)e(z) \tag{2.131}$$

$$\begin{aligned}\frac{u(z)}{e(z)} &= \frac{\frac{T}{2} \cdot a(1+z)}{(z-1) + \frac{T \cdot a}{2}(1+z)} \\ &= \frac{a}{\frac{2}{T}\frac{z-1}{z+1} + a}\end{aligned} \tag{2.132}$$

The equivalent substitution relationship between $s$ and $z$ is

$$\frac{a}{s+a} \longleftrightarrow \frac{a}{\frac{2}{T}\frac{z-1}{z+1} + a} \tag{2.133}$$

$$s \longleftrightarrow \frac{2}{T}\frac{z-1}{z+1} \tag{2.134}$$

Summary of finite difference based digital approximation of analog filters is given next.

| Method | Approximation |
|---|---|
| *FWD–rule* | $s \longleftarrow \frac{z-1}{T}$ |
| *BWD–rule* | $s \longleftarrow \frac{z-1}{Tz}$ |
| *Trapezoidal–rule* | $s \longleftarrow \frac{2}{T}\frac{z-1}{z+1}$ |

(2.135)

## 2.10 PROBLEMS

**1.** Consider a time domain signal $y(t) = 1.0 \cdot sin(2\pi \cdot 10t)$ which is periodic. Select proper sampling frequencies to illustrate the following sampling effects, and plot the sampled results in time domain as well as the Fourier series plot the results in Bode plot form of the original signal and the Fourier transform of the sampled signal.

**(a)** A sampling period that is fast enough for accurate sampling and does not violate the sampling theorem.
**(b)** A sampling period that illustrates an aliasing problem as a result of sampling.
**(c)** A sampling period that illustrates the beat phenomenon as a result of sampling.
**(d)** A sampling period that illustrates the hidden oscillations problem as a result of sampling.

Explain your time domain results with the frequency content of the sampled signals.

**2.** Consider a mass-force system. Let $m = 5$ kg. A controller decides on the force using a PD controller on position error. Select the PD controller gains such that the step response of the closed-loop system has no more than 5% overshoot, and the settling time is less than 2.0 sec. Confirm your results with a Simulink or Matlab simulation.

**3.** What is the effect of switching the sign on the controller, that is, $u(s) = (K_p + K_d s)(x(s) - x_{cmd}(s))$? This can easily happen in practice by swapping the signal input lines to the analog PD controller between the command signal and the sensor signal or by switching the polarity of the controller output (sign change once in the controller). What happens if we switch the polarity of both the command signal and sensor signal as well as the output polarity of the controller (sign change twice in the controller)? Use a simple mass-force system model under a PD control algorithm to simulate your analysis and present results.

**4.** Consider a mass-force system. Let $m = 5$ kg. A controller decides on the force using a PID (P or PI) controller on *velocity* error. How can we guarantee finite steady state following error in response to: (a) a ramp velocity command signal, and (b) a parabolic veloctiy command signal? Confirm your results with a Simulink or Matlab simulation. Let velocity commands be $r(t) = 10 \cdot t$ and $r(t) = 10 \cdot t^2$. It is assumed that the feedback sensor provides the velocity measurement of the mass. The controller acts on the velocity error.

**5.** Consider the dynamics of a DC motor speed control system and its current mode amplifier. Consider that the motor torque-speed transfer function is a first-order filter and the current mode amplifier input–output dynamics is also a first-order filter.

$$\frac{w(s)}{T(s)} = \frac{100}{0.02s + 1} \tag{2.136}$$

$$T(s) = 10 \cdot i(s) \tag{2.137}$$

$$\frac{i(s)}{i_{cmd}(s)} = \frac{10}{0.005s + 1} \tag{2.138}$$

Consider that the current command is generated by an analog controller (PID type) using motor speed as feedback signal and commanded velocity signals.

**(a)** Ignore the filtering effect of the current amplifier, and determine the locus of closed-loop poles (root locus) of the closed-loop control system under three different controllers: P, PD, PI. In each case vary proportional gain as the root locus parameter and select different values of derivative and integral gains.

$$i_{cmd}(s) = K_p \cdot (w_{cmd}(s) - w(s)); \; P\ control \tag{2.139}$$

$$= (K_p + K_d s) \cdot (w_{cmd}(s) - w(s)); \; PD\ control \tag{2.140}$$

$$= \left(K_p + \frac{K_i}{s}\right) \cdot (w_{cmd}(s) - w(s)); \; PI\ control \tag{2.141}$$

**(b)** Repeat the same root locus analysis by including the amplifier dynamics in the analysis. Discuss your results in terms of the effect of selecting different controller types (P, PD, PI), their gains, and the effect of additional filter type dynamics in the control loop.

**6.** Consider the dynamics of a DC motor speed control system and its current mode amplifier. Consider that the motor torque-speed transfer function is a first-order filter and the current mode amplifier input–output dynamic is also a first-order filter.

$$\frac{w(s)}{T(s)} = \frac{K_m}{\tau_m s + 1} \tag{2.142}$$

$$T(s) = K_T \cdot i(s) \tag{2.143}$$

$$\frac{i(s)}{i_{cmd}(s)} = \frac{K_i}{\tau_a s + 1} \tag{2.144}$$

Assume that there is a load torque acting as a disturbance on the motor and that it is in the form of (i) step function and (ii) a ramp function. If we want to make sure that the steady-state speed error due to the load torque is zero, what kind of controller is required in order to deal with each type of disturbance (load) torque? Why can't a PD type controller do the job?

**7.** Given a analog PD controller,

$$G_c(s) = K_p + K_d \cdot s \tag{2.145}$$

**(a)** Obtain its digital PD controller equivalent for sampling period $T$ using all of the digital approximation methods discussed above (eqn. 2.135).

**(b)** Let $K_p = 1.0$, $K_d = 0.7$. Plot the Bode diagrams of the analog and the digital approximations for three different values of $T = 1/5$ sec., 1/50 sec, 1/500 sec.

**8.** Given a analog PI controller,

$$G_c(s) = K_p + K_i \cdot (1/s) \tag{2.146}$$

**(a)** Obtain its digital PI controller equivalent for sampling period $T$ using all of the digital approximation methods discussed above (eqn. 2.135).

**(b)** Let $K_p = 1.0$, $K_i = 0.1$. Plot the Bode diagrams of the analog and the digital approximations for three different values of $T$ = 1/5 sec., 1/50 sec, 1/500 sec.

**9.** Given a analog PID controller,

$$G_c(s) = K_p + K_i \cdot (1/s) + K_d \cdot s \tag{2.147}$$

**(a)** obtain its digital PID controller equivalent for sampling period $T$ using all of the digital approximation methods discussed above (eqn. 2.135).

**(b)** Let $K_p = 1.0$, $K_i = 0.1$, $K_d = 0.7$. Plot the Bode diagrams of the analog and the digital approximations for three different values of $T$ = 1/5 sec., 1/50 sec, 1/500 sec.

CHAPTER 3

# MECHANISMS FOR MOTION TRANSMISSION

## 3.1 INTRODUCTION

Every computer-controlled mechanical system that involves motion is built around a basic frame of a mechanism which is used to transmit the motion generated by the actuators to the desired tool. The actuators provide purely rotary or purely linear motion. For instance, a rotary electric motor can be viewed as a rotary motion source while a hydraulic or pneumatic cylinder can be viewed as a linear motion source. In general, it is not practical to place the actuator exactly at the location where the motion of a tool is needed. Therefore, a motion transmission mechanism is needed between the actuator and the tool. Motion transmission mechanisms perform two different roles:

1. Transmit motion from actuator to tool when the actuator cannot be designed into the same location as the tool with the desired motion type,
2. Increase or reduce torque and speed between input and output shafts while maintaining the power conservation between input and output (output power is input power minus the power losses).

The most common motion transmission mechanisms fit into one of three major categories:

1. Rotary-to-rotary motion transmission mechanisms (gears, belts and pulleys),
2. Rotary-to-translational motion transmission mechanisms (lead-screw, rack-pinion, belt-pulley), and
3. Cyclic motion transmission mechanisms (linkages and cams).

Common to all of these mechanisms is that an input shaft displacement is related to the output shaft displacement with a fixed mechanical relationship. During the conversion, there is inevitable loss of power due to friction. However, for analysis purposes here, we will assume ideal motion transmission mechanisms with 100% efficiency.

The efficiency of a motion transmission mechanism is defined as the ratio between the output power and input power,

$$\eta = \frac{P_{out}}{P_{in}} \tag{3.1}$$

The efficiency can vary from 75 to 95% range for different types of motion transmission mechanisms. If we assume perfect efficiency, then

$$P_{in} = P_{out} \tag{3.2}$$

which is a convenient relationship in determining the input–output relationships.

The mechanical construction of the mechanism determines the ratio of input displacement to output displacement, which is called the *effective gear ratio*. The effective gear ratio is not influenced by efficiency. If a mechanism is not 100% efficient, the loss is a percentage

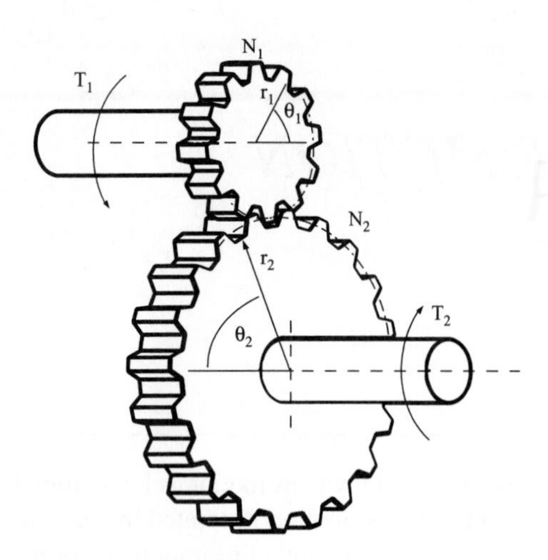


**FIGURE 3.1:** Rotary-to-rotary motion conversion mechanism: gear mechanism.

of the torque or force transmitted. In other words, let us consider a simple gear arrangement (Fig. 3.1) with a gear ratio of $N = \Delta\theta_{in}/\Delta\theta_{out}$,

$$P_{out} = \eta \cdot P_{in} \tag{3.3}$$

$$T_{out} \cdot \dot{\theta}_{out} = \eta \cdot T_{in} \cdot \dot{\theta}_{in} \tag{3.4}$$

and regardless of the efficiency,

$$N = \frac{\dot{\theta}_{in}}{\dot{\theta}_{out}} \tag{3.5}$$

Hence,

$$T_{out} = \eta \cdot N \cdot T_{in} \tag{3.6}$$

where the torque output is reduced by the efficiency of the mechanism.

The effective gear ratio of a mechanism is determined using the energy equations. Kinetic energy of the tool on the output is expressed in terms of output speed. Then, the output speed is expressed as a function of the input speed. Since both expressions represent the same kinetic energy, the effective gear ratio is obtained. For the following discussion, we refer to the load inertia ($J_l$) and load torque ($T_l$) as being the applied on the output shaft. In other words, $J_l = J_{out}$, $T_l = T_{out}$. The reflected values of these on the input shaft side are referred to as $J_{in,eff} = J_{in}$, $T_{in,eff} = T_{in}$. For instance, for a rotary gear reducer, let $KE_l$ be the kinetic energy of the load with inertia $J_l$ and output speed $\dot{\theta}_{out}$. In order to provide such an energy, the actuator must provide a kinetic energy plus the losses. Hence,

$$KE_l = \eta \cdot KE_{in} \tag{3.7}$$

$$KE_l = \frac{1}{2} \cdot J_l \cdot \dot{\theta}_{out}^2 \tag{3.8}$$

$$= \eta \cdot \frac{1}{2} \cdot J_{in,eff} \cdot \dot{\theta}_{in}^2 \tag{3.9}$$

$$= \frac{1}{2} \cdot J_l \cdot (\dot{\theta}_{in}/N)^2 \tag{3.10}$$

Hence, the effective reflected inertia is

$$J_{in,eff} = \frac{1}{\eta \cdot N^2} \cdot J_l \tag{3.11}$$

In summary, an ideal motion transmission mechanism (efficiency is 100%, $\eta = 1.0$) has the following reflection properties between its input and output shafts:

$$\dot{\theta}_{in} = N \cdot \dot{\theta}_{out} \tag{3.12}$$

$$J_{in,eff} = \frac{1}{N^2} \cdot J_l \tag{3.13}$$

$$T_{in,eff} = \frac{1}{N} \cdot T_l \tag{3.14}$$

where $N$ is the effective gear ratio. The efficiency factor of the motion transmission mechanism is often taken into account by a relatively large safety factor. If the efficiency factor is to be explicitly included in the actuator sizing calculations, then the following relations hold:

$$\dot{\theta}_{in} = N \cdot \dot{\theta}_{out} \tag{3.15}$$

$$J_{in,eff} = \frac{1}{N^2 \cdot \eta} \cdot J_l \tag{3.16}$$

$$T_{in,eff} = \frac{1}{N \cdot \eta} \cdot T_l \tag{3.17}$$

It is also of interest to determine the reflection (transmitted) motion, inertia, and torque from input shaft to output shaft direction. The relationships are then,

$$\theta_{out} = \frac{1}{N} \cdot \theta_{in} \tag{3.18}$$

$$J_{out,eff} = \frac{N^2}{\eta} \cdot J_{in} \tag{3.19}$$

$$T_{out,eff} = \frac{N}{\eta} \cdot T_{in} \tag{3.20}$$

It is important to note that in either direction of power transmission, the efficiency factor is in the denominator of the equations which indicates loss of power due to transmission efficiency in either direction.

A motion transmission mechanism is characterized by the following parameters:

1. *Effective gear ratio:* The main characteristic of a motion transmission mechanism is its gear ratio. This is sometimes called the *effective gear ratio* because the motion conversion may not necessarily be performed by gears.
2. *Efficiency:* Efficiency of a real transmission mechanism is always less than 100%. For most gear mechanisms, forward and back drive efficiencies are the same, except the lead-screw and ball-screw type mechanisms. In such mechanisms, it is appropriate to talk about rotary-to-linear motion conversion efficiency (forward efficiency) $\eta_f$, and linear-to-rotary motion conversion efficiency (backward efficiency) $\eta_b$. For ball-screw, the typical values of the efficiency coefficients are $\eta_f = 0.9$, $\eta_b = 0.8$, and for lead-screws they vary as function of the lead angle. The *lead angle* is defined as the angle the lead helix makes with a line perpendicular to the axis of rotation. As the lead (linear distance traveled per rotation) increases, so does the lead angle, hence the efficiency.
3. *Backlash:* There is always an effective *backlash* in motion transmission mechanisms. The backlash is given in units of *arc minutes* = 1/60 *degrees* in rotary mechanisms, and linear distance in translational mechanisms. Notice that backlash directly affects the positioning accuracy. If the position sensor is connected to the motor, not to the load, it will not be able to measure the positioning error due to backlash. Therefore, if backlash is large enough to be a concern for positioning accuracy, there has to be a position sensor connected to the load in order to measure the true position, including the effect of backlash. In such systems, it is generally necessary to use two position sensors (dual

sensor feedback or dual loop control): one position sensor connected to the motor and the other to the load. The motor-connected sensor is primarily used in velocity control loop to maintain closed loop stability, whereas the load-connected sensor is primarily used for accurate position sensing and control. Without the motor-connnected sensor, the closed loop system may be unstable. Without the load-connected sensor, desired positioning accuracy cannot be achieved. In systems where the backlash is much smaller than the positioning accuracy required, the backlash can be ignored.

4. *Stiffness:* The transmission components are not perfectly rigid. They have finite stiffness. The stiffness of the transmission box between input and output shaft is rated with a *torsional or translational stiffness* parameter.
5. *Break-away friction:* This friction torque (or force) is an estimated value and highly function of the lubrication condition of the moving components. This is the minimum torque or force needed at the input shaft to move the mechanism.
6. *Back driveability:* Motion conversion mechanisms involve two shafts, input shaft and output shaft. In normal mode of operation, the motion and torque in the input shaft are transmitted to the output shaft with a finite efficiency. Back driveability refers to the transmission of power in the opposite direction; that is, the motion source is provided at the output shaft and transmitted to the input shaft. Most spur gear, belt, and pulley type mechanisms are back driveable with the same efficiency in both directions. Rotary-to-linear motion conversion mechanisms such as lead-screw and ball-screws have different efficiencies and are not necessarily back driveable. Ball screws are considered back driveable for all cases. The back driveabilitiy of lead screw depends on the lead angle. If the lead angle is below a certain value (i.e., 30°), the back driveability may be in question. Furthermore, it also highly depends on the lubrication condition of the mechanism because it affects the friction force that must be overcome in order to back drive the mechanism. Worm-gear mechanisms are not back driveable. There are applications which benefit from that, i.e., while raising a very heavy load, if there is power failure in the input shaft of the motor, the mechanism is not suppose to back drive under gravitational force. It is required to hold the load in position. In short, when back driveability is required, two variables are of interest: (1) $\eta_b$ efficiency in the back drive direction, and (2) $F_{fric}$ friction force to overcome in order to initiate motion, which is highly dependent on the lubrication conditions.

## 3.2 ROTARY-TO-ROTARY MOTION TRANSMISSION MECHANISMS

### 3.2.1 Gears

Gears are used to increase or decrease the speed ratio between the input and output shaft. The effective gear ratio is obvious (Fig. 3.1). Assuming that the gears do not slip, the linear distance traveled by each gear at the contact point is the same:

$$s_1 = s_2 \tag{3.21}$$

$$\Delta\theta_1 \cdot r_1 = \Delta\theta_2 \cdot r_2 \tag{3.22}$$

$$N = \frac{\Delta\theta_1}{\Delta\theta_2} \tag{3.23}$$

$$N = \frac{r_2}{r_1} \tag{3.24}$$

Since the pitch of each gear must be the same, the number of teeth on each gear is proportional to their radius,

$$N = \frac{\dot{\theta}_1}{\dot{\theta}_2} = \frac{N_2}{N_1} = \frac{r_2}{r_1} \tag{3.25}$$

where $N_1$ and $N_2$ represent the number of gear teeth on each gear. It can be shown that for an ideal gear box (100% power transmission efficiency),

$$P_{out} = P_{in} \tag{3.26}$$

$$T_{out} \cdot \dot{\theta}_{out} = T_{in} \cdot \dot{\theta}_{in} \tag{3.27}$$

Hence,

$$N = \frac{N_2}{N_1} = \frac{\dot{\theta}_{in}}{\dot{\theta}_{out}} = \frac{T_{out}}{T_{in}} \tag{3.28}$$

The reflection of inertia and torque from output shaft to the input shaft can be determined by using the energy and work relationships. Let the rotary inertia of the load on the output shaft be $J_l$ and the load torque be $T_l$. Let us express the kinetic energy of the load

$$KE = \frac{1}{2} \cdot J_l \cdot \dot{\theta}_{out}^2 \tag{3.29}$$

$$= \frac{1}{2} \cdot J_l \cdot (\dot{\theta}_{in}/N)^2 \tag{3.30}$$

$$= \frac{1}{2} \cdot J_l \cdot \frac{1}{N^2} (\dot{\theta}_{in})^2 \tag{3.31}$$

$$= \frac{1}{2} \cdot J_{in,eff} (\dot{\theta}_{in})^2 \tag{3.32}$$

where the reflected inertia (inertia of the load seen by the input shaft) is

$$J_{in,eff} = \frac{J_l}{N^2} \tag{3.33}$$

Similarly, let us determine the effective load torque seen by the input shaft. The work done by a load torque $T_l$ over an output shaft displacement $\Delta\theta_{out}$ is

$$W = T_l \cdot \Delta\theta_{out} \tag{3.34}$$

$$= T_l \cdot \frac{\Delta\theta_{in}}{N} \tag{3.35}$$

$$= T_{in,eff} \cdot \Delta\theta_{in} \tag{3.36}$$

The effective reflective torque on the input shaft as a result of the load torque on the output shaft is

$$T_{in,eff} = \frac{T_l}{N} \tag{3.37}$$

The same concept of kinetic energy and work of the tool is used in all of the other mechanisms to determine the reflected inertia (or mass) and torque (or force) between output and input shafts.

### 3.2.2 Belt and Pulley

The gear ratio of a belt-pulley mechanism is the ratio between the input and output diameters. Assuming no slip between the belt and pulleys on both shafts, the linear displacement along

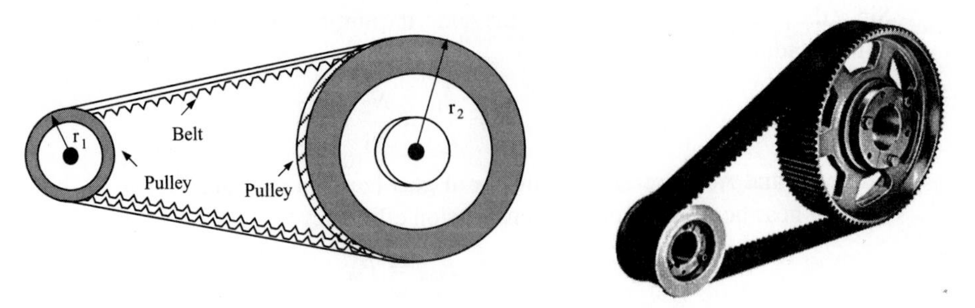


**FIGURE 3.2:** Rotary-to-rotary motion conversion mechanism: timing belt and toothed pulley.

the belt and both pulleys should be equal (Fig. 3.2),

$$x = \Delta\theta_1 \cdot r_1 = \Delta\theta_2 \cdot r_2 \tag{3.38}$$

The effective gear ratio is by definition

$$N = \frac{\Delta\theta_1}{\Delta\theta_2} \tag{3.39}$$

$$= \frac{r_2}{r_1} \tag{3.40}$$

$$= \frac{d_2}{d_1} \tag{3.41}$$

The inertia and torque reflection between input and output shaft has the same relationship as the gear mechanisms.

***Example*** Consider a spur gear mechanism with a gear ratio of $N = 10$. Assume that the load inertia connected to the output shaft is solid steel with diameter $d = 3.0$ in, length $l = 2.0$ in. The friction-related torque at the load is $T_l = 200$ lb.in. The desired speed of the load is 300 rev/min. Determine the necessary speed at the input shaft as well as reflected inertia and torque.

The necessary speed at the input shaft is related to the output shaft speed by a kinematic relationship defined by the gear ratio,

$$\dot{\theta}_{in} = N \cdot \dot{\theta}_{out} = 10 \cdot 300\,[\text{rpm}] = 3000\,[\text{rpm}] \tag{3.42}$$

The inertia and torque experienced at the input shaft due to the load alone (which we call the reflected inertia and the reflected torque) are

$$J_{in,eff} = \frac{1}{N^2} \cdot J_l \tag{3.43}$$

$$T_{in,eff} = \frac{1}{N} \cdot T_l \tag{3.44}$$

The mass moment of inertia of the cylindrical load,

$$J_l = \frac{1}{2} \cdot m \cdot (d/2)^2 \tag{3.45}$$

$$= \frac{1}{2} \cdot \rho \cdot \pi \cdot (d/2)^2 \cdot l \cdot (d/2)^2 \tag{3.46}$$

$$= \frac{1}{2} \cdot \rho \cdot \pi \cdot l \cdot (d/2)^4 \tag{3.47}$$

$$= \frac{1}{2} \cdot (0.286/386) \cdot \pi \cdot 2.0 \cdot (3/2)^4 \text{ [lb in sec}^2] \tag{3.48}$$

$$= 0.0118 \text{ [lb in sec}^2] \tag{3.49}$$

where $g = 386$ in./sec$^2$, the gravitational acceleration, is used to convert the weight density to mass density. Hence, the reflected inertia and torque are

$$J_{in,eff} = \frac{1}{10^2} \cdot 0.0118 = 0.118 \times 10^{-3} \text{ [lb in sec}^2] \tag{3.50}$$

$$T_{in,eff} = \frac{1}{10} \cdot 200 = 20.0 \text{ [lb in]} \tag{3.51}$$

## 3.3 ROTARY-TO-TRANSLATIONAL MOTION TRANSMISSION MECHANISMS

The rotary-to-translational motion transmission mechanisms convert rotary motion to linear translational motion. Translational motion is also referred to as linear motion. Both terms will be used interchangeably in the following discussions. In addition, torque input is converted to force at the output. It should be noted that all of the rotary-to-translational motion transmission mechanisms discussed here are back driveable, meaning that they also make the conversion in the reverse direction.

### 3.3.1 Lead Screw and Ball Screw Mechanisms

Lead-screw and ball-screw mechanisms are the most widely used precision motion conversion mechanisms which transfer rotary motion to linear motion (Fig. 3.3). The lead screw is basically a precision threaded screw and nut. In a screw and nut pair, typically we turn the nut and it advances linearly on the stationary screw. In the lead screw, used as a motion conversion mechanism, the nut is not allowed to make rotation around the screw, but supported by linear (translational) bearings to move. The screw is rotated (i.e., by an electric motor). Since the screw is not allowed to travel, and can only rotate, the nut moves translationally along the screw. The tool is connected to the nut. Hence, the rotary motion of the motor connected to the screw is converted to the translational motion of the nut and the tool.

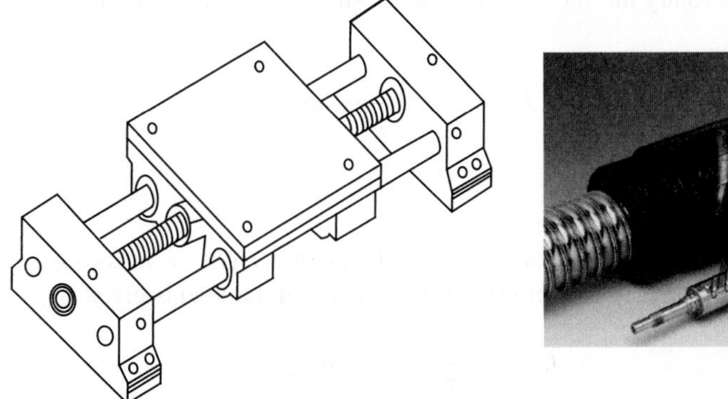

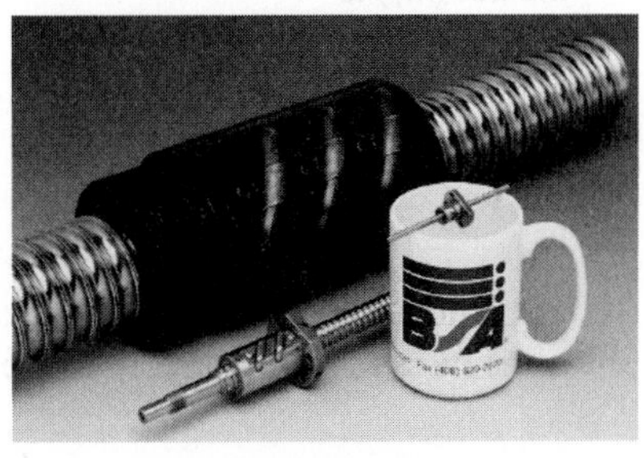

**FIGURE 3.3:** Rotary-to-translational motion conversion mechanism: lead-screw or ball-screw with linear guide bearings.

Ball screw design uses precision ground spherical balls in the grove between the screw and nut threads to reduce backlash and friction in the motion transmission mechanism. Any lead screw has a finite backlash typically in the order of micrometer range. Preloaded springs on a set of spherical bearing type balls are used to reduce the backlash. Hence, they are called ball screws. Most XYZ table type positioning devices use ball screws instead of lead screws. However, the load carrying capacity of ball screws are less than that of the lead screws since the contact between the moving parts (lead and nut) is provided by point contacts of balls. On the other hand, a ball screw has less friction than a lead screw.

The kinematic motion conversion factor, or effective gear ratio, of a lead screw is characterized by its pitch, $p$ [rev/in] or [rev/mm]. Therefore, for a lead screw having a *pitch* of $p$, which is the inverse of the distance traveled for one turn of the thread called *lead*, $l = 1/p$ [in/rev] or [rev/mm], the rotational displacement ($\Delta\theta$ in units of [rad]) at the lead shaft and the translational displacement ($\Delta x$) of the nut is related by

$$\Delta\theta = 2\pi \cdot p \cdot \Delta x \tag{3.52}$$

$$\Delta x = \frac{1}{2\pi \cdot p} \cdot \Delta\theta \tag{3.53}$$

The effective gear ratio may be stated as

$$N_{ls} = 2\pi \cdot p \tag{3.54}$$

Let us determine the inertia and torque seen by the input end of the lead screw due to a load mass and load force on the nut. We follow the same method as before and use energy–work relations. The kinetic energy of the mass $m_l$ at a certain speed $\dot{x}$ is

$$KE = \frac{1}{2} m_l \cdot \dot{x}^2 \tag{3.55}$$

Noting the above motion conversion relationship,

$$\dot{x} = \frac{1}{2\pi \cdot p} \cdot \dot{\theta} \tag{3.56}$$

Then,

$$KE = \frac{1}{2} m_l \cdot \frac{1}{(2\pi \cdot p)^2} \cdot \dot{\theta}^2 \tag{3.57}$$

$$= \frac{1}{2} J_{eff} \cdot \dot{\theta}^2 \tag{3.58}$$

Then, the effective rotary inertia seen at the input shaft ($J_{eff}$) due to a translational mass on the nut ($m_l$) is

$$J_{eff} = \frac{1}{(2\pi \cdot p)^2} \cdot m_l \tag{3.59}$$

$$= \frac{1}{N_{ls}^2} \cdot m_l \tag{3.60}$$

It should be noted that $m_l$ is in units of mass (not weight, $weight = mass \cdot g$), and the $J_{eff}$ is the mass moment of inertia. Therefore, if the weight of the load is given, $W_l$,

$$J_{eff} = \frac{1}{(2\pi \cdot p)^2} \cdot (W_l/g) \tag{3.61}$$

$$= \frac{1}{N_{ls}^2} \cdot (W_l/g) \tag{3.62}$$

The lead screw has very large effective gear ratio effect. Notice that for $p = 2, 5, 10$, the net gear ratio is $N = 4\pi, 10\pi, 20\pi$, respectively, and the inertial reflection of a translational mass is a factor determined by the square of the effective gear ratio.

Let us determine the reflected torque at the input shaft due to a load force, $F_l$. The work done by a load force during a incremental displacement is

$$Work = F_l \cdot \Delta x \tag{3.63}$$

The corresponding rotational displacement is

$$\Delta x = \frac{1}{2\pi \cdot p} \cdot \Delta\theta \tag{3.64}$$

Hence,

$$Work = F_l \cdot \Delta x \tag{3.65}$$

$$= F_l \cdot \frac{1}{2\pi \cdot p} \cdot \Delta\theta \tag{3.66}$$

$$= T_{eff} \cdot \Delta\theta \tag{3.67}$$

The equivalent torque seen at the input shaft ($T_{eff}$) of the lead screw due to the load force $F_l$ at the nut is

$$T_{eff} = \frac{1}{2\pi \cdot p} \cdot F_l \tag{3.68}$$

$$= \frac{1}{N} \cdot F_l \tag{3.69}$$

***Example*** Consider a ball screw motion conversion mechanism with a pitch of $p = 10$ rev/in. The mass of the table and workpiece is $W_l = 1000$ lb, and the resistance force of the load is $F_l = 1000$ lb. Determine the reflected rotary inertia and torque seen by a motor at the input shaft of the ball screw.

$$J_{eff} = \frac{1}{(2\pi \cdot 10)^2 [\text{rad/in}]^2} \cdot (1000/386)\ [\text{lb/(in/sec}^2)] \tag{3.70}$$

$$= 6.56 \times 10^{-3}\ [\text{lb.in.}\ \text{sec}^2] \tag{3.71}$$

The torque that is reflected on the input shaft due to the load force $F_l$ is

$$T_{eff} = \frac{1}{2\pi \cdot p} \cdot F_l \tag{3.72}$$

$$= \frac{1}{2\pi\,[\text{rad/rev}] \cdot 10\,[\text{rev/in}]} \cdot 1000\ [\text{lb}] \tag{3.73}$$

$$= \frac{100}{2\pi}\ [\text{lb.in}] \tag{3.74}$$

$$= 15.91\ [\text{lb.in}] \tag{3.75}$$

### 3.3.2 Rack-and-Pinion Mechanism

A rack-and-pinion mechanism is an alternative rotary-to-linear motion conversion mechanism (Fig. 3.4). The pinion is the small gear. Rack is the translational (linear) component. It is similar to a gear mechanism where one of the gear is a linear gear. The effective gear

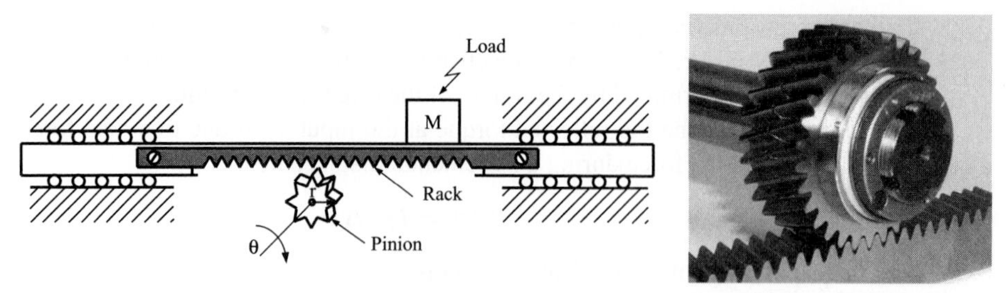


**FIGURE 3.4:** Rotary-to-translational motion conversion mechanism: rack-and-pinion mechanism. The advantange of the rack-and-pinion mechanism over the lead-screw mechanism is that the translational motion range can be very long. The lead-screw length is limited by the torsional stiffness. In a rack-and-pinion mechanism since the translational part does not rotate, it does not have the reduced torsional stiffness problem due to the long length.

ratio is calculated or measured from the assumption that there is no slip between the gears,

$$\Delta x = r \cdot \Delta\theta \tag{3.76}$$

$$\Delta\dot{x} = r \cdot \Delta\dot{\theta} \tag{3.77}$$

$$V = r \cdot w \tag{3.78}$$

where $\Delta\theta$ is in radian units. Hence, the effective gear ratio is

$$N = \frac{1}{r}; \quad \textit{if} \quad \Delta\theta \textit{ is in } [rad] \textit{ units} \quad \textit{or} \quad w \textit{ is in } [rad/sec] \textit{ units} \tag{3.79}$$

$$= \frac{1}{2\pi \cdot r}; \quad \textit{if} \quad \Delta\theta \textit{ is in } [rev] \textit{ units} \quad \textit{or} \quad w \textit{ is in } [rev/sec] \textit{ units} \tag{3.80}$$

The same mass and force reflection relations we developed for lead screws apply for the rack-pinion mechanisms. The only difference is the effective gear ratio.

### 3.3.3 Belt and Pulley

Belt and pulley mechanisms are used both as rotary-to-rotary and rotary-to-linear motion conversion mechanisms depending on the output point of interest. If the load (tool) is connected to the belt and used to obtain linear motion, then it acts as a rotary-to-linear motion conversion device (Fig. 3.5). The relations we developed for a rack-and-pinion mechanism identically apply here. This mechanism is widely used in low inertia, low load force, and high bandwidth applications such as coil winding machines.

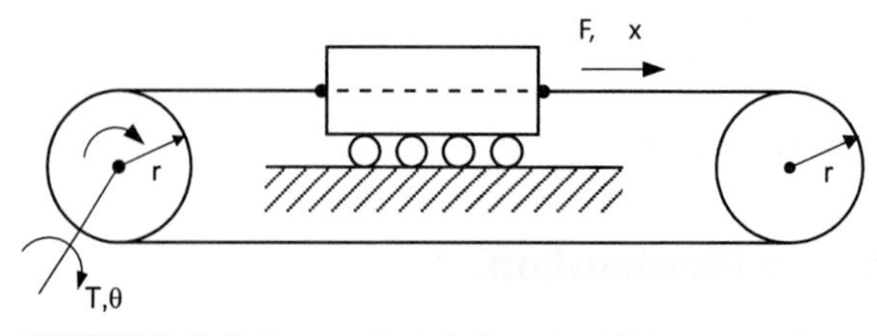


**FIGURE 3.5:** Rotary-to-translational motion conversion mechanism: belt-pulley mechanism where both pulleys have the same diameter. The output motion is taken from the belt as the translational motion.

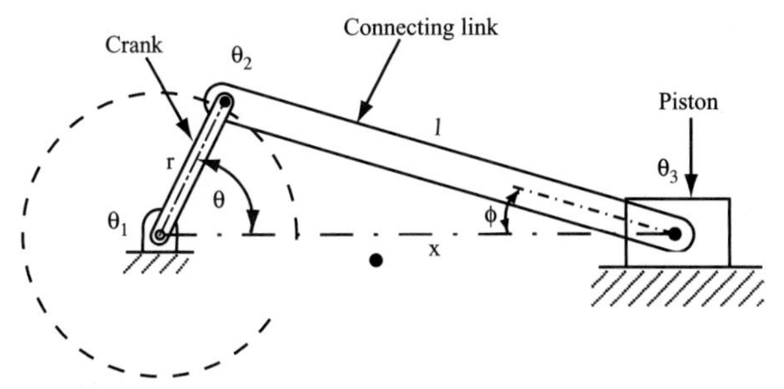


**FIGURE 3.6:** Rotary-to-translational motion conversion mechanism: slider-crank mechanism. In the internal combustion engine, the mechanism is used as translational- (piston motion is input)-to-rotary motion (crank shaft rotation is output) conversion mechanism.

## 3.4 CYCLIC MOTION TRANSMISSION MECHANISMS

### 3.4.1 Linkages

Linkages are generally one degree of freedom, kinematically closed chain, robotic manipulators. The motion of one member (output link) of the linkage is a periodic function of the motion of another linkage member (input link). The most common linkages include:

1. Slider-crank mechanism (i.e., used in internal combustion engines, Fig. 3.6)
2. Four-bar mechanism (Fig. 3.7)

The slider-crank mechanism is most widely recognized because it is used in every internal combustion engine. It can be used to convert the translational displacement of slider to the rotary motion of the crank or vice versa. The length of the crank arm and the length of the connecting link determines the geometric relationship between the slider translational displacement and crank angular rotation.

Four-bar linkage can convert the cyclic motion of the input arm (i.e., full rotation of the input arm shaft) to the limited range cyclic rotation of the output shaft. By selectively choosing the input and output shaft, as well as the lengths of the linkages, different motion conversion functions are obtained. The motion control of such linkages is rather simple,

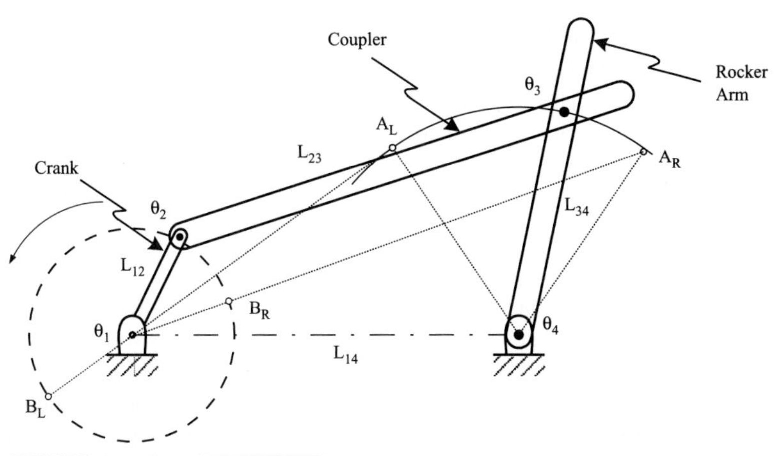


**FIGURE 3.7:** Four-bar mechanism.

because they are mostly used with a constant speed input shaft motion, and the resultant motion is obtained at the output shaft.

From the kinematics of the slider-crank mechanism (Fig. 3.6), the following relations can be derived:

$$x = r\ cos\theta + l\ cos\phi \tag{3.81}$$

$$l\ sin\phi = r\ sin\theta \tag{3.82}$$

$$cos\phi = [1 - sin^2\phi]^{1/2} \tag{3.83}$$

$$= \left[1 - \left(\frac{r}{l}\ sin\theta\right)^2\right]^{1/2} \tag{3.84}$$

$$x = r\ cos\theta\ +\ l\ \left[1 - \left(\frac{r}{l}\ sin\theta\right)^2\right]^{1/2} \tag{3.85}$$

$$\dot{x} = -\,r\ \dot{\theta}\left[sin\theta\ +\ \frac{r}{2l}\frac{sin(2\theta)}{[1-(\frac{r}{l}sin\theta)^2]^{1/2}}\right] \tag{3.86}$$

where $r$ is the radius of the crank (length of the crank link), $l$ is the length of the connecting arm, $x$ is the displacement of the slider, $\theta$ is the angular displacement of the crank, and $\phi$ is the angle between the connecting arm and displacement axis. The position and speed of piston motion and crank motion are related by the above geometric relations. Acceleration relation can be obtained by taking the time derivative of the speed relation [7].

***Example*** Consider a crank-slider mechanism with the following geometric parameters: $r = 0.30$ m, $l = 1.0$ m. Consider the simulation of a condition that the crank shaft rotates at a constant speed $\dot{\theta}(t) = 1200$ rpm. Plot the displacement of the slider as function of crank shaft angle from 0 to 360 degrees of rotation (for one revolution) and the linear speed of the slider.

Since we have the geometric relationship between the crank shaft angle and speed versus the slider position and speed (equations 3.85 and 3.86), we simply substitute for $r$ and $l$ the above values, and calculate the $x$ and $\dot{x}$ for $\theta = 0\ to\ 2\pi\,[rad\,]$ and $\dot{\theta} = 1200 \cdot (1/60) \cdot 2\pi$ rad/sec. A Matlab or Simulink environment can be conveniently used to generate the results (Fig. 3.8).

### 3.4.2 Cams

Cams convert the rotary motion of a shaft into translational motion of a follower (Fig. 3.9). The relationship between the translational motion and rotary motion is not a fixed gear ratio, but a nonlinear function. The nonlinear cam function is determined by the machined cam profile. A cam mechanism has three major components:

1. Input shaft
2. The cam
3. The follower

If the input shaft axis of the cam is parallel to the follower axis of motion, such cams are called *axial* cams. In this case, cam function is machined into the cylindrical surface along the axis of input shaft rotation. If they are perpendicular to each other, such cams are called *radial* cams and the cam function may be machined either along the outside diameter or face diameter. All cam profiles can be divided into periods called *rise, dwell,* and *fall* in various combinations. For instance, a cam profile can be designed such that for one revolution of the input shaft, the follower makes a cyclic motion that is various combinations of *rise, fall,* and *dwell*, i.e., *rise, dwell, fall, dwell*, or *rise, fall, dwell*. In addition, a cam function may be designed such that the follower makes single or multiple cycles per revolution of the input shaft, i.e., one, two, three, or four follower cycles per input shaft revolution. It is common to design rise and fall periods of the cam as symmetric. During the dwell period, the follower is stationary. Therefore, during the dwell period the follower position is constant, speed and acceleration are zero. If symmetric cam functions are used for rise and fall periods, then we are only concerned with the cam function design for the rise period.

A cam profile is defined by the following relationship:

$$x = f(\theta); \quad 0 < \theta < 2\pi \tag{3.87}$$

where $x$ is the displacement of the follower, and $\theta$ is the rotation of the input shaft. The cam function is periodic. The cam follower motion repeats for every revolution of the input shaft.

In addition to selecting an appropriate cam function, there are three other important parameters to consider in cam design (Fig. 3.9):

1. ***Pressure angle:*** Measured as the angle between the follower motion axis and an axis perpendicular to the common tangent line at the contact point between cam and follower (Fig. 3.9). A cam should be machined such that the pressure angle stays less than about 30 degrees in order to make sure the side loading force on the follower is not too high.

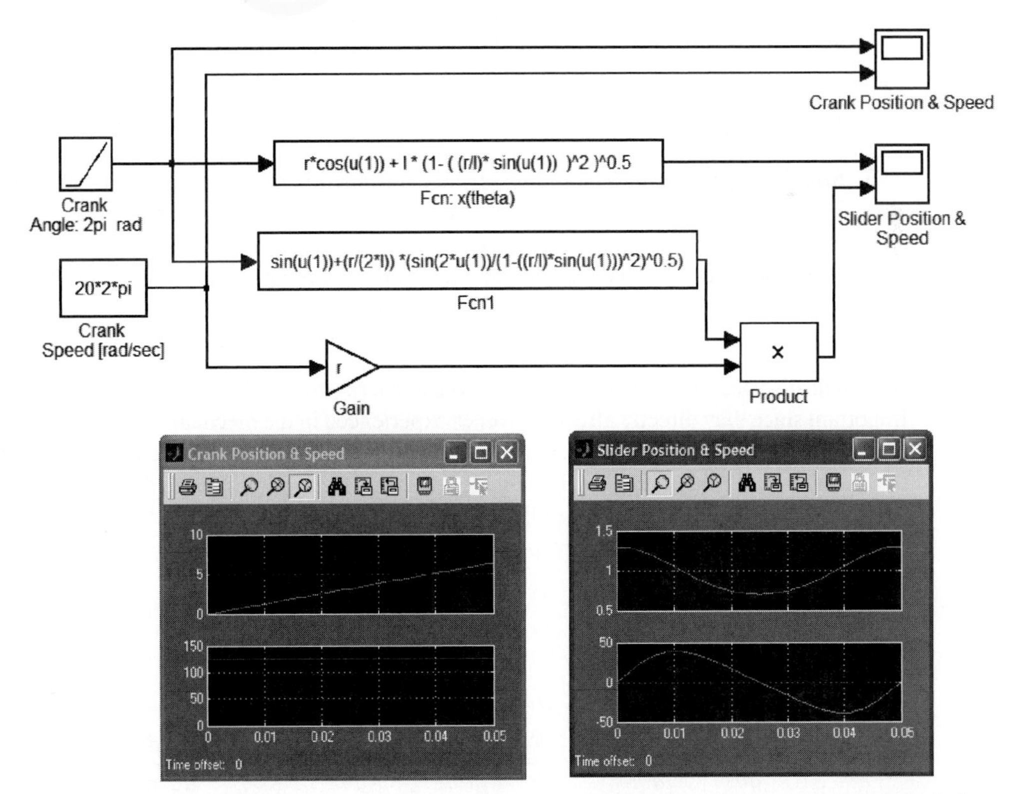


**FIGURE 3.8:** Simulation result of a slider crank mechanisms: $r = 0.3$ m, $l = 1.0$ m, speed of crank shaft is constant at $\dot{\theta} = 1200$ rpm. The resulting slider position and speed functions are shown in the figure.

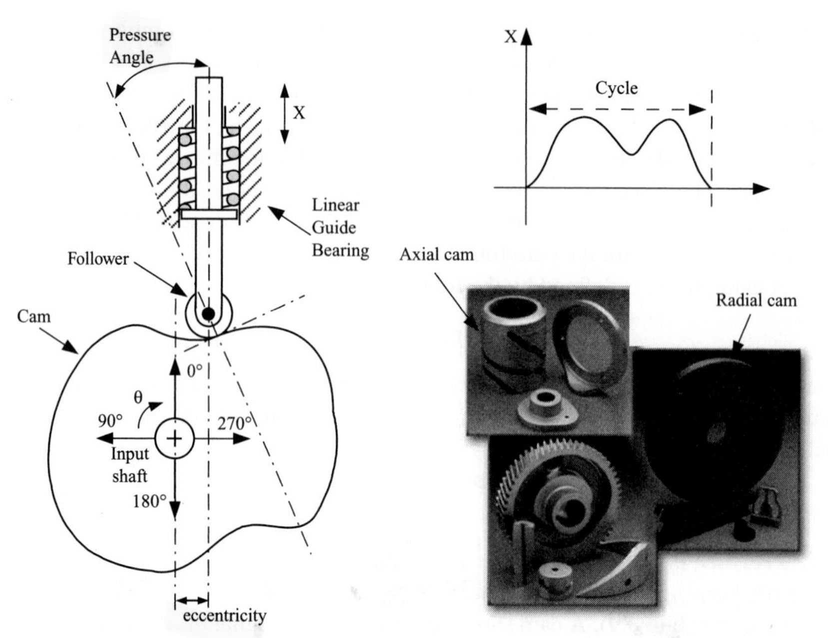


**FIGURE 3.9:** Rotary-to-translational motion conversion mechanism: cam mechanism.

2. ***Eccentricity:*** The offset distance between the follower axis and cam rotation axis in the direction perpendicular to the cam motion. By increasing eccentricity, we can reduce the effective pressure angle, hence the side loading forces on the follower. However, as the eccentricity increases, the cam gets larger and less compact [7].
3. ***Radius of curvature:*** The radius of the cam function curvature along its periphery. The radius of curvature should be continuous function of angular position of cam input shaft. Any discontinuity in the radius of curvature is essentially reflected as a nonsmooth cam surface. In general, the radius of curvature should be at least 2 to 3 times larger than the radius of the follower. The main considerations are the continuity and ability of the follower to maintain contact on the cam at all times.

The displacement, velocity, acceleration, and jerk (time derivative of acceleration) are important since they directly affect the forces experienced in the mechanism. In cam design, quite often we focus on the first, second, and third derivatives of the cam function with respect to $\theta$ instead of $t$, time. In the final analysis, we are interested in the time derivative values since they determine the actual speed, acceleration, and jerk. The relationships between the derivatives of cam function with respect to $\theta$ and $t$ are as follows:

$$\frac{dx}{d\theta} = \frac{df(\theta)}{d\theta}; \quad \frac{dx}{dt} = \frac{dx}{d\theta}\dot{\theta} \tag{3.88}$$

$$\frac{d^2x}{d\theta^2} = \frac{d^2f(\theta)}{d\theta^2}; \quad \frac{d^2x}{dt^2} = \frac{d^2x}{d\theta^2}\dot{\theta}^2 + \frac{dx}{d\theta}\ddot{\theta} \tag{3.89}$$

$$\frac{d^3x}{d\theta^3} = \frac{d^3f(\theta)}{d\theta^3}; \quad \frac{d^3x}{dt^3} = \frac{d^3x}{d\theta^3}\dot{\theta}^3 + \frac{d^2x}{d\theta^2}(3 \cdot \dot{\theta} \cdot \ddot{\theta}) + \frac{dx}{d\theta}\frac{d\ddot{\theta}}{dt} \tag{3.90}$$

Notice that when $\dot{\theta} = constant$, then $\ddot{\theta} = 0$, and $d\ddot{\theta}/dt = 0$. When the input shaft of the cam speed is constant, which is the case in most applications, then the relationship be-

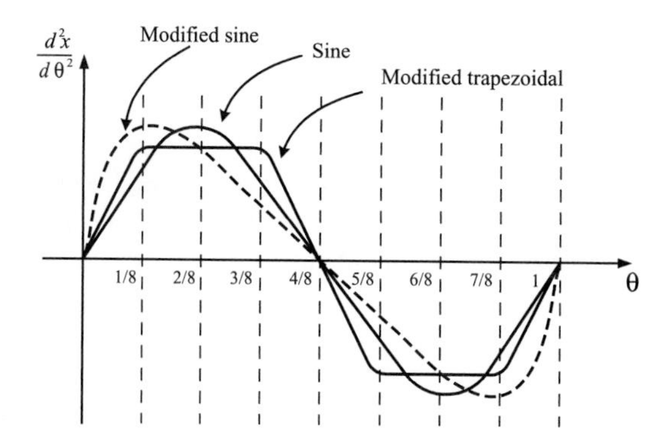


**FIGURE 3.10:** Commonly used cam profiles. The acceleration function is shown as function of the driving axis. Sinusoidal, modified sin, and modified trapezoidal functions are common cam profiles.

tween the time derivatives and derivatives with respect to $\theta$ of the cam function are as follows:

$$\frac{dx}{d\theta} = \frac{df(\theta)}{d\theta}; \quad \frac{dx}{dt} = \frac{dx}{d\theta}\dot{\theta} \tag{3.91}$$

$$\frac{d^2x}{d\theta^2} = \frac{d^2 f(\theta)}{d\theta^2}; \quad \frac{d^2x}{dt^2} = \frac{d^2x}{d\theta^2}\dot{\theta}^2 \tag{3.92}$$

$$\frac{d^3x}{d\theta^3} = \frac{d^3 f(\theta)}{d\theta^3}; \quad \frac{d^3x}{dt^3} = \frac{d^3x}{d\theta^3}\dot{\theta}^3 \tag{3.93}$$

Significant effort is made in selecting the $f(\theta)$ cam function in order to shape the first, second, and even the third derivative of it so that desired results (i.e., minimize vibrations) are obtained from the cam motion. *As a general rule in cam design, the cam function should be chosen so that the cam function, first and second derivatives (displacement, speed, and acceleration functions) are continuous and the third derivative (jerk function) discontinuities (if any) are finite.* In general these continuity conditions are applied to cam function derivatives with respect to $\theta$. If the input cam speed is constant, the same continuity conditions are then satified by the time derivatives as well. For instance, a cam function with trapezoidal shape results in discontinuous velocity and infinite accelerations. Therefore, pure trapezoidal cam profiles are almost never used. The alternatives for the selection of functions that meet the continuity requirements in the cam function are many. Some functions are defined in analytical form (parts of the cam function are portions of the sinusoidal functions joined to form a continuous function), and some are custom developed by experimentation and stored in numerical form. The most common cam functions are as follows (Fig. 3.10).

1. *Cycloidal displacement* cam function has acceleration curve for the rise portion that is sinusoidal with single frequency,

$$\frac{d^2x(\theta)}{d\theta^2} = C_1 \cdot sin(f_1\theta) \tag{3.94}$$

where $C_1$ is constant related to the displacement range of $x$ and $f_1$ is determined by the portion of input shaft rotation to complete the *rise portion of cam motion*. For instance, if rise motion is completed in $\theta_{rise} = 1/2$ rev $= \pi$ rad,

$$f_1\theta_{rise} = 2\pi \tag{3.95}$$

$$f_1 = \frac{2\pi}{\theta_{rise}} \tag{3.96}$$

$$= \frac{2\pi}{\pi} \tag{3.97}$$

$$= 2 \tag{3.98}$$

Similarly if rise motion is completed in 1/3 rev $= 2\pi/3$ rad, then $f_1 = 3$. Both displacement and velocity functions are also sinusoidal functions. The speed and displacement functions can be readily obtained by integrating the acceleration curve,

$$\frac{dx(\theta)}{d\theta} = -C_1 \frac{1}{f_1} \cdot cos(f_1 \theta)|_0^\theta \tag{3.99}$$

$$= \frac{C_1}{f_1} - C_1 \frac{1}{f_1} \cdot cos(f_1 \theta) \tag{3.100}$$

$$x(\theta) = \left[ \frac{C_1}{f_1}\theta - C_1 \frac{1}{f_1^2} \cdot sin(f_1 \theta) \right] \Bigg|_0^\theta \tag{3.101}$$

$$= \frac{C_1}{f_1}\theta - C_1 \frac{1}{f_1^2} \cdot sin(f_1 \theta); \quad for\, 0 \leq \theta \leq \theta_{rise} \tag{3.102}$$

Let us assume that the rise portion is to occur in 1/2 revolution of the cam input shaft. Then, $f_1 = 2$. Let the rise distance be $x_{rise} = 0.2$ m. Then the constant $C_1$ can be determined by evaluating the displacement cam function at the end of the rise cycle for $\theta = \pi$,

$$0.2 = \frac{C_1}{2}\pi \tag{3.103}$$

$$C_1 = \frac{2 \cdot 0.2}{\pi} \tag{3.104}$$

Assuming that cam has symmetric rise and down portions without any dwell portion, the complete period of motion for the cam is $\theta = 0$ to $2 \times \theta_{rise}$. Acceleration, speed, and displacement curves for the *down portion* (also called the fall portion) of the cam are all mirror images of the rise portion. Mirror image and original function relationship are as follows. The original function for rise motion is $x_{rise}$,

$$x_{rise}(\theta) = f(\theta); \quad 0 \leq \theta \leq \theta_{rise} \tag{3.105}$$

then, the mirror image (let us call that $x_{down}$)

$$x_{down}(\theta) = f(2\theta_{rise} - \theta)\,; \quad \theta_{rise} \leq \theta \leq 2\theta_{rise} \tag{3.106}$$

Hence, the mirror image of the displacement cam function $x(\theta)$ is $x(2\theta_{rise} - \theta)$,

$$x(\theta) = \frac{C_1}{f_1}(2\theta_{rise} - \theta) - C_1 \frac{1}{f_1^2} \cdot sin(f_1(2\theta_{rise} - \theta)); \quad \theta_{rise} \leq \theta \leq 2\theta_{rise} \tag{3.107}$$

Figure 3.11 shows the acceleration, speed, and displacement curves for one cycle of the cam motion. If there were a dwell portion for the input shaft rotation range of $[\theta_{rise}, \theta_{rise} + \theta_{dwell}]$, then the dwell portion of the cam function following the rise portion would be as follows:

$$x(\theta) = x_{rise}(\theta_{rise}); \quad \theta_{rise} \leq \theta \leq \theta_{rise} + \theta_{dwell} \tag{3.108}$$

$$\frac{dx(\theta)}{d\theta} = 0 \tag{3.109}$$

$$\frac{d^2x(\theta)}{d\theta^2} = 0 \tag{3.110}$$

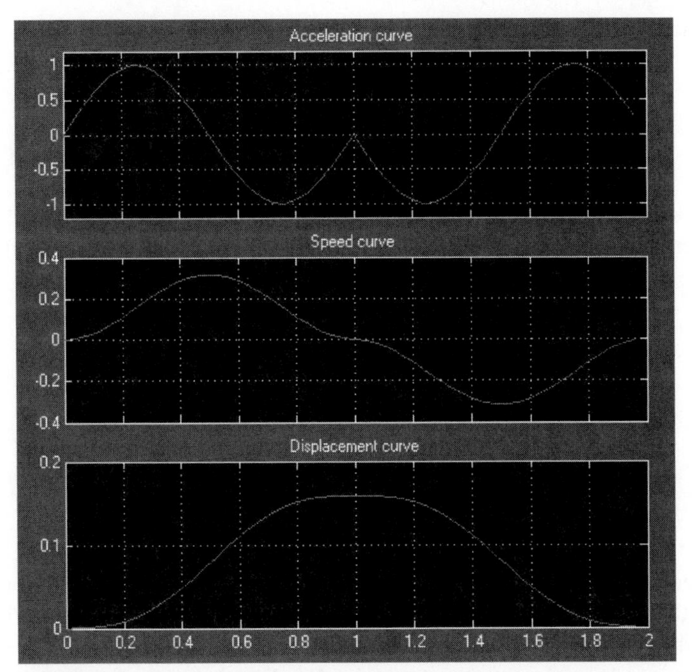


**FIGURE 3.11:** Sinusoidal acceleration cam profile: acceleration, speed, and displacement function for a cam design where half of the shaft revolution is used for rise motion and the other half is used for down motion symmetrically.

and for the down portion of the cam,

$$x(\theta) = \frac{C_1}{f_1}(2\theta_{rise} + \theta_{dwell} - \theta) - C_1\frac{1}{f_1^2} \cdot sin(f_1(2\theta_{rise} + \theta_{dwell} - \theta)); \quad (3.111)$$

$$\theta_{rise} + \theta_{dwell} \leq \theta \leq 2\theta_{rise} + \theta_{dwell} \quad (3.112)$$

$$\frac{dx(\theta)}{d\theta} = -\frac{C_1}{f_1} + C_1\frac{1}{f_1} \cdot cos(f_1(2\theta_{rise} + \theta_{dwell} - \theta)) \quad (3.113)$$

$$\frac{d^2x(\theta)}{d\theta^2} = C_1 \cdot sin(f_1(2\theta_{rise} + \theta_{dwell} - \theta)) \quad (3.114)$$

2. *Modified sine* function is a modified version of the first function, cycloidal displacement function. Compared to the original version, this version results in lower peak acceleration and speed values while maintaining similar cam displacement function. This cam function uses at least two frequencies of the sinusoidal profile. Pieces of the two sine functions are combined to form a smooth cam function.

$$\frac{d^2x(\theta)}{d\theta^2} = C_1 \cdot sin(f_1\theta) + C_2 \cdot sin(f_2\theta) \quad (3.115)$$

where $C_1$, $C_2$ constants would have nonzero values in some segments and zero values in other segments of the cam. In other words, in some segments of the cam function one of the section of the sinusoidal function is used, in other segments the other sinudoidal function is used with its appropriate segment connecting to the preceeding and the following curve. Let $\theta_{rise}$ be the period of input shaft rotation for the rise period. The $f_1$, $f_2$ are two constant frequencies (Fig. 3.10).

$$f_1 = \frac{2\pi}{\theta_{rise}/2} = \frac{4\pi}{\theta_{rise}} \tag{3.116}$$

$$f_2 = \frac{2\pi}{3\theta_{rise}/2} = \frac{4\pi}{3\theta_{rise}} \tag{3.117}$$

The acceleration function is constructed from the segments of two sinusoidal functions as follows:

$$\frac{d^2x(\theta)}{d\theta^2} = A_o \cdot sin\left(\frac{2\pi}{(\theta_{rise}/2)}\theta\right) ; \quad for \quad 0 \le \theta \le \frac{1}{8}\theta_{rise} \tag{3.118}$$

$$\frac{d^2x(\theta)}{d\theta^2} = A_o \cdot sin\left(\frac{2\pi}{(3\theta_{rise}/2)}\theta + (\pi/3)\right) ; \quad for \quad \frac{1}{8}\theta_{rise} \le \theta \le \frac{7}{8}\theta_{rise} \tag{3.119}$$

$$\frac{d^2x(\theta)}{d\theta^2} = A_o \cdot sin\left(\frac{2\pi}{(\theta_{rise}/2)}\theta - 2\pi\right) ; \quad for \quad \frac{7}{8}\theta_{rise} \le \theta \le \theta_{rise} \tag{3.120}$$

The acceleration function like this describes the movement of the follower for the rise period. If the follower cycle is made of *rise and fall* without any dwell periods, the complete displacement cycle is performed over an input shaft rotation of $\theta = 2\theta_{rise}$. Displacement and speed curves are determined by directly integrating the acceleration curve with zero initial condition for speed and zero initial condition for displacement at the beginning of integration.

The fall motion curves are simply obtained using a mirror image of the rise motion curves. During dwell portion (if used in the cam design), the cam follower displacement stays constant, speed and acceleration are zero.

**3.** *Modified trapezoidal* acceleration function modifies the standard trapezoidal acceleration function around the points where the acceleration changes slope. The acceleration function is smoothed out with a smooth function, such as a sinusoidal function, hence eliminating the jerk discontinuity. If purely trapezoidal acceleration profiles are used, there would be discontinuity in the jerk function when the acceleration function slope changes from finite value to zero value (constant acceleration). In order to reduce the effect of this in terms of vibrations in the mechanism, the acceleration profile function includes sinusoidal functions at the corners of the curve and constant acceleration functions in other segments of the cam. Let $\theta_{rise}$ be the period of input shaft rotation for the rise period. The modified trapezoildal acceleration profile is formed by constructing a function where the pieces of it are obtained from a sinusoidal function with peroid $\theta_{rise}/2$ and constant acceleration function.

Let us consider a modified trapezoidal acceleration function. The acceleration function has positive and negative (symmetric to the positive) portions. The variable portions of the acceleration function are implemented with portions of the sinusoidal, and constant acceleration portions are implemented with a constant value (Fig. 3.10),

$$\frac{d^2x}{d\theta^2} = A_o \cdot sin(2\pi\theta/(\theta_{rise}/2)); \quad for \quad 0 \le \theta \le \theta_{rise}/8 \tag{3.121}$$

$$\frac{d^2x}{d\theta^2} = A_o; \quad for \quad \frac{1}{8}\theta_{rise} \le \theta \le \frac{3}{8}\theta_{rise} \tag{3.122}$$

$$\frac{d^2x}{d\theta^2} = A_o \cdot sin(2\pi\theta/(\theta_{rise}/2) - \pi); \quad for \quad \frac{3}{8}\theta_{rise} \le \theta \le \frac{5}{8}\theta_{rise} \tag{3.123}$$

$$\frac{d^2x}{d\theta^2} = -A_o; \quad for \quad \frac{5}{8}\theta_{rise} \leq \theta \leq \frac{7}{8}\theta_{rise} \tag{3.124}$$

$$\frac{d^2x}{d\theta^2} = A_o \cdot sin(2\pi\theta/(\theta_{rise}/2) - 2\pi); \quad for \quad \frac{7}{8}\theta_{rise} \leq \theta \leq \theta_{rise} \tag{3.125}$$

Again, the acceleration functions describe half of the displacement cycle of the follower. For a symmetric return function, a mirror image of the same acceleration function is implemented in the cam profile.

**4.** *Polynominal functions* of many different types, i.e., including third, fourth, fifth, sixth, or seventh order polynomials are used for cam functions.

$$x(\theta) = C_0 + C_1\theta + C_2\theta^2 + C_3\theta^3 + \cdots + C_n\theta^n \tag{3.126}$$

Typically, the input shaft is driven at a constant speed, and output follower motion is obtained as a cyclic function of input shaft rotation. Before the development of computer-controlled programmable machines, almost all of the automated machines had one large power source driving a shaft. Many different cams generated desired motion profiles at individual stations from the same shaft. Hence, all the stations are synchronized to a master shaft. The synchronization is fixed as a result of the machined profiles of cams. If a different automated control is needed, cam profiles have to be physically changed. As programmable computer-controlled machines replace the fixed automation machines, the use of mechanical cams has been decreasing (Fig. 10.2 in Chap. 10). Instead, cam functions are implemented in software to synchronize independently controlled motion axes. If a different type of synchronization is required, all we need is to change the cam software. In mechanical cam synchronized systems, we would have to physically change the cam.

***Example*** Consider a cam with its follower as shown in Fig. 3.9. Let us consider the displacement, speed, acceleration, and jerk profiles for modified sine cam functions. Assume that the input shaft of the cam runs at a constant speed; hence $\ddot{\theta} = 0$, $\frac{d}{dt}\ddot{\theta} = 0.0$. Assume that nominal operating speed of input shaft of the cam is $\dot{\theta} = 2\pi \cdot 10$ [rad/sec] ($\dot{\theta} = 10$ [rev/sec] $= 600$ [rev/min]), the total displacement of the cam follower is 2.0 in, and period of cam motion for a complete up-down cycle is once per revolution of the input shaft. Since the input shaft speed is constant, the time derivatives and derivatives with respect to input shaft angle are related to each other as

$$\dot{x} = \frac{dx}{d\theta}\dot{\theta}; \quad \ddot{x} = \frac{d^2x}{d\theta^2}(\dot{\theta})^2; \quad \frac{d}{dt}\ddot{x} = \frac{d^3x}{d\theta^3}(\dot{\theta})^3 \tag{3.127}$$

We will assume that the cam has rise and fall periods without any dwell period in between them, hence $\theta_{rise} = \theta_{down} = \pi$. For one half cycle (rise period), the cam profile is defined in three sections as a function of input shaft displacement which are regions $0 \leq \theta \leq \frac{1}{8}\theta_{rise}$, $\frac{1}{8}\theta_{rise} \leq \theta \leq \frac{7}{8}\theta_{rise}$, $\frac{7}{8}\theta_{rise} \leq \theta \leq \theta_{rise}$ (equation 3.118–3.120).

It can be shown that the *modified sine cam function* in the rise region is as follows [$cam_1(\theta)$, equation (3.118)]:

$$\frac{d^2x(\theta)}{d\theta^2} = A_o \cdot sin\left(\frac{4\pi}{\theta_{rise}}\theta\right); \quad 0 \leq \theta \leq \frac{1}{8}\theta_{rise} \tag{3.128}$$

$$\frac{dx(\theta)}{d\theta} = \left(\frac{A_o\theta_{rise}}{4\pi}\right) \cdot \left(1 - cos\left(\frac{4\pi}{\theta_{rise}}\theta\right)\right) \tag{3.129}$$

$$x(\theta) = \frac{A_o\theta_{rise}}{4\pi} \cdot \left(\theta - \frac{\theta_{rise}}{4\pi}sin\left(\frac{4\pi}{\theta_{rise}}\theta\right)\right) \tag{3.130}$$

Likewise, the cam functions for the other period of motion are defined as follows [$cam_2(\theta)$, equation (3.119)]:

$$\frac{d^2x(\theta)}{d\theta^2} = A_o \cdot sin\left(\frac{4\pi}{3\theta_{rise}}\theta + \frac{\pi}{3}\right); \quad \frac{1}{8}\theta_{rise} \leq \theta \leq \frac{7}{8}\theta_{rise} \tag{3.131}$$

$$\frac{dx(\theta)}{d\theta} = V_{s1} - A_o \cdot \left(\frac{3\theta_{rise}}{4\pi}\right)\left(cos\left(\frac{4\pi}{3\theta_{rise}}\theta + \frac{\pi}{3}\right)\right) \tag{3.132}$$

$$x(\theta) = X_{s1} + V_{s1} \cdot \theta - A_o \cdot \left(\frac{3\theta_{rise}}{4\pi}\right)^2 \cdot \left(sin\left(\frac{4\pi}{3\theta_{rise}}\theta + \frac{\pi}{3}\right)\right) \tag{3.133}$$

and similarly, for the period of $\frac{7}{8}\theta_{rise} \leq \theta \leq \theta_{rise}$, [$cam_3(\theta)$]

$$\frac{d^2x(\theta)}{d\theta^2} = A_o \cdot sin(4\pi\theta/\theta_{rise} - 2\pi); \quad \frac{7}{8}\theta_{rise} \leq \theta < \theta_{rise} \tag{3.134}$$

$$\frac{dx(\theta)}{d\theta} = V_{s2} - A_o \cdot \left(\frac{\theta_{rise}}{4\pi}\right)\left(cos\left(\frac{4\pi}{\theta_{rise}}\theta - 2\pi\right)\right) \tag{3.135}$$

$$x(\theta) = X_{s2} + V_{s2} \cdot \theta - A_o \cdot \left(\frac{\theta_{rise}}{4\pi}\right)^2 \cdot \left(sin\left(\frac{4\pi}{\theta_{rise}}\theta - 2\pi\right)\right) \tag{3.136}$$

where the constants $X_{s1}, X_{s2}, V_{s1}, V_{s2}$ are determined to meet the continuity requirements at the boundaries of the cam function sections. Using the total travel range of follower, we can determine the constant $A_o$ as function of the total travel range specified. Below are the five equations from which the above five constants can be calculated.

$$cam_1\left(\frac{1}{8}\theta_{rise}\right) = cam_2\left(\frac{1}{8}\theta_{rise}\right) \tag{3.137}$$

$$cam_2\left(\frac{7}{8}\theta_{rise}\right) = cam_3\left(\frac{7}{8}\theta_{rise}\right) \tag{3.138}$$

$$\frac{d}{d\theta}\left(cam_1\left(\frac{1}{8}\theta_{rise}\right)\right) = \frac{d}{d\theta}\left(cam_2\left(\frac{1}{8}\theta_{rise}\right)\right) \tag{3.139}$$

$$\frac{d}{d\theta}\left(cam_2\left(\frac{7}{8}\theta_{rise}\right)\right) = \frac{d}{d\theta}\left(cam_3\left(\frac{7}{8}\theta_{rise}\right)\right) \tag{3.140}$$

$$cam_3\left(\theta_{rise}\right) = x_{rise} \tag{3.141}$$

The result of the solution of the above equations for five unknowns can be shown to be

$$[A_o, X_{s1}, V_{s1}, X_{s2}, V_{s2}] = [1.273, 0.443, 0.318, 1.000, 0.318]$$

The cam function is the combination of equations 3.130, 3.133, and 3.136 and their mirror images, ordered in reverse, and $\theta$ substituted by $2\theta_{rise} - \theta$ as independent variable in the cam functions. The cam function is then

$$cam_1(\theta) = \frac{A_o\theta_{rise}}{4\pi} \cdot \left(\theta - \frac{\theta_{rise}}{4\pi}sin\left(\frac{4\pi}{\theta_{rise}}\theta\right)\right); \quad 0 \leq \theta \leq \frac{1}{8}\theta_{rise} \tag{3.142}$$

$$cam_2(\theta) = X_{s1} + V_{s1} \cdot \theta - A_o\left(\frac{3\theta_{rise}}{4\pi}\right)^2 \cdot \left(sin\left(\frac{4\pi}{3\theta_{rise}}\theta + \frac{\pi}{3}\right)\right);$$

$$\frac{1}{8}\theta_{rise} \leq \theta \leq \frac{7}{8}\theta_{rise} \tag{3.143}$$

$$cam_3(\theta) = X_{s2} + V_{s2} \cdot \theta - A_o \left(\frac{\theta_{rise}}{4\pi}\right)^2 \cdot \left(sin\left(\frac{4\pi}{\theta_{rise}}\theta - 2\pi\right)\right);$$

$$\frac{7}{8}\theta_{rise} \leq \theta \leq \theta_{rise} \tag{3.144}$$

$$cam_4(\theta) = cam_3(2\theta_{rise} - \theta); \quad \theta_{rise} \leq \theta \leq \frac{9}{8}\theta_{rise} \tag{3.145}$$

$$cam_5(\theta) = cam_2(2\theta_{rise} - \theta); \quad \frac{9}{8}\theta_{rise} \leq \theta \leq \frac{15}{8}\theta_{rise} \tag{3.146}$$

$$cam_6(\theta) = cam_1(2\theta_{rise} - \theta); \quad \frac{15}{8}\theta_{rise} \leq \theta \leq 2\theta_{rise} \tag{3.147}$$

For the purpose of manufacturing a cam, we only need the displacement functions for a complete revolution of the input shaft.

## 3.5 SHAFT MISALIGNMENTS AND FLEXIBLE COUPLINGS

Mechanical systems always involve two or more shafts to transfer motion. There is always a finite accuracy with which the two shafts can be aligned in the axial direction. Any shaft misalignments will result in loads on the bearings and cause vibration, hence reducing the life of the machinery. In order to reduce the vibration and life-reducing effects of shaft misalignment, flexible couplings are used between shafts (Fig. 3.12).

There are two main categories of flexible shaft couplings:

1. Couplings for large power transfer between shafts and motors
2. Couplings for precision motion transfer at low powers between shafts and motors

High-precision motion systems include motors with very low friction and yet very delicate bearings. Such motors are very sensitive to shaft misalignments. The bearing failure of the motor as a result of excessive shaft misalignment is a very common reliability problem. Therefore, in most high-performance servo motor applications, the motor shaft is coupled to the load via a flexible coupling. The flexible couplings provide the ability to make the system more tolerant to the shaft misalignments. However, it comes at the cost of reduced stiffness of the mechanical system. Therefore, designers must make sure that the stiffness of the coupling does not interfere with the desired motion bandwidth (especially in variable speed and cyclic positioning applications).

Couplings are rated by the following parameters:

1. Maximum and rated torque capacity
2. Torsional stiffness
3. Maximum allowed axial misallignment
4. Rotary inertia and mass of the coupling
5. Input and output shaft diameters
6. Input/output shaft connection method (set screw, clamped with keyway)
7. Design type (bellow or helical coupling)

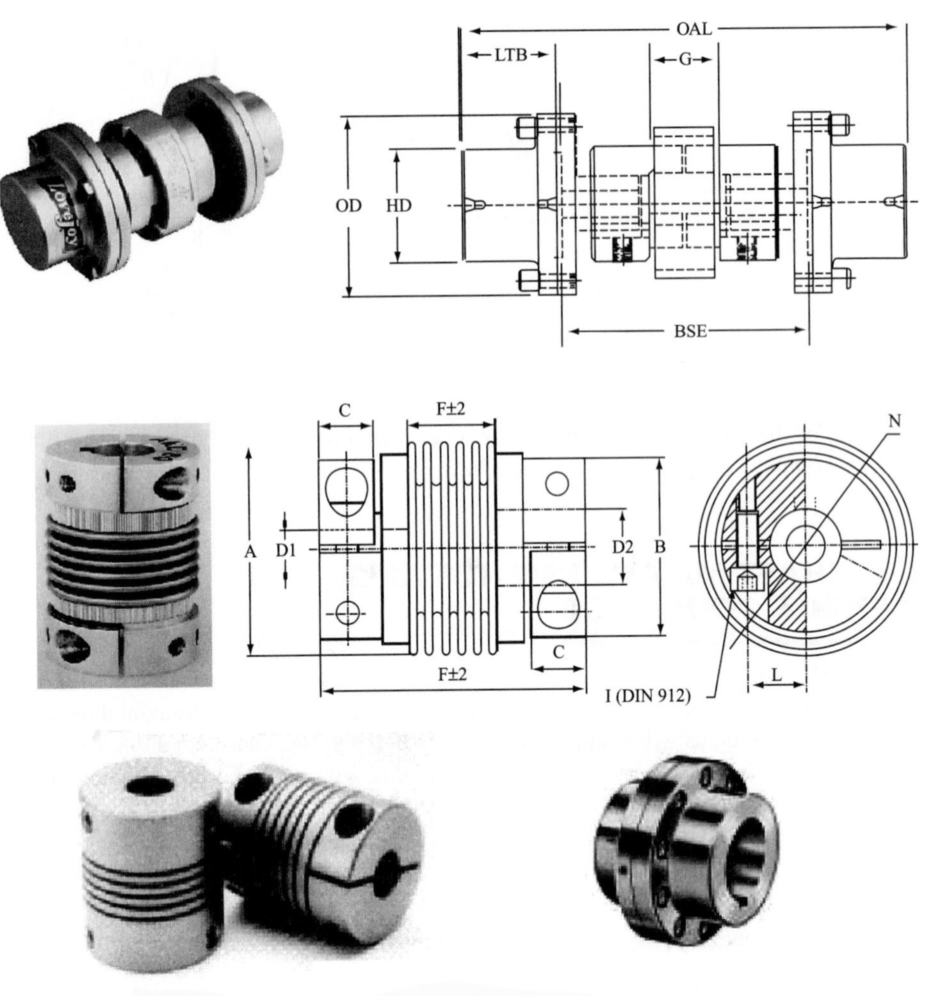


**FIGURE 3.12:** Flexible couplings are used between connecting shafts in motion transmission mechanisms to compensate for the shaft misalignments and protect bearings.

## 3.6 ACTUATOR SIZING

Every motion axis is powered by an actuator. The actuator may be an electric, hydraulic, or pneumatic power based. The size of the actuator refers to its power capacity and must be large enough to be able to move the axis under the given inertial and load force/torque conditions. If the actuator is undersized, the axis will not be able to deliver the desired motion, i.e., cannot deliver desired acceleration or speed levels. If the actuator is oversized, it will cost more and the motion axis will have slower bandwidth since as the actuator size gets larger (larger power levels), the bandwidth gets slower as a general rule. Therefore, it is important to properly size the actuators for a motion axis with a reasonable margin of safety. Along with determining the proper actuator size for an application, the size of a gear mechanism needs to be determined unless the actuator is directly coupled to the load (Fig. 3.13). The focus of this section is on sizing a rotary electric motor type actuator. The same concepts can be used for other types of actuators.

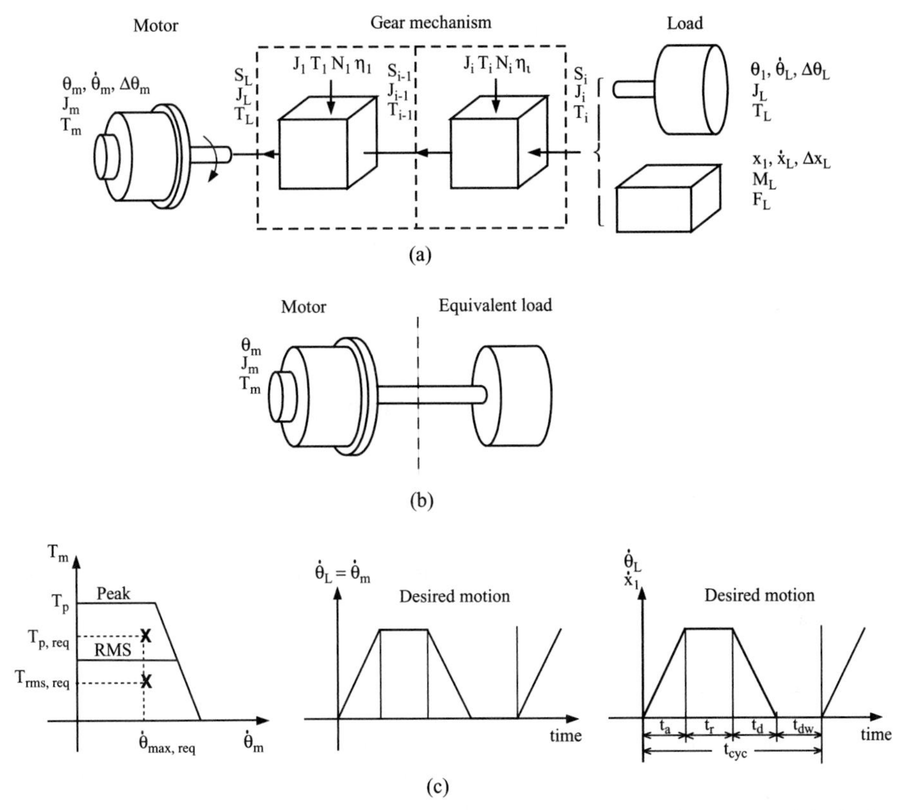


**FIGURE 3.13:** Actuator sizing: load (inertia, torque/force), gear mechanism, and desired motion must be specified.

The question of actuator sizing is the question of determining the following requirements for an axis under worst operating conditions (i.e., largest expected inertia and resistive load):

1. Maximum torque (also called peak torque) required, $T_{max}$
2. Rated (continuous or root mean squared, RMS) torque required, $T_r$
3. Maximum speed required, $\dot{\theta}_{max}$
4. Positioning accuracy required, $\Delta\theta$
5. Gear mechanism parameters: gear ratio, its inertial and resistive load (force/torque), stiffness, backlash characteristics

Once the torque requirements are determined, the amplifier current and power supply requirements are directly determined from them.

In general, accuracy and maximum speed requirements of the load dictate the gear ratio. Below, we will assume that a gear mechanism with an appropriate gear ratio is selected and focus on determining the actuator size. For a given application, the load motion requirements specify the desired positioning accuracy and maximum speed. Let us call that $\Delta x$ and $\dot{x}_{max}$. The desired positioning accuracy and maximum speed at the actuator (i.e., rotary motor) is determined by

$$\Delta\theta = N \cdot \Delta x \tag{3.148}$$

$$\dot{\theta} = N \cdot \dot{x}_{max} \tag{3.149}$$

A gear ratio range is defined by the minimum accuracy and the maximum speed requirement; i.e., in order to provide the desired accuracy ($\Delta x$) for a motor with a given actuator positioning accuracy ($\Delta\theta$), the gear ratio must be

$$N \geq \frac{\Delta\theta}{\Delta x} \tag{3.150}$$

In order to provide the desired maximum speed for a given maximum speed capacity of the motor (not to exceed the maximum speed capability of the motor), the gear ratio must be smaller than or equal to

$$N \leq \frac{\dot{\theta}_{max}}{\dot{x}_{max}} \tag{3.151}$$

Hence, the acceptable gear ratio range is defined by the accuracy and maximum speed requirement,

$$\frac{\Delta\theta}{\Delta x} \leq N \leq \frac{\dot{\theta}_{max}}{\dot{x}_{max}} \tag{3.152}$$

Some of the most commonly used motion conversion mechanisms (also called the gear mechanism) are shown in Fig. 3.14. Notice that the gear mechanism adds inertia and possible load torque to the motion axis in addition to performing the gear reduction role and coupling the actuator to the load. In precision positioning applications, the first requirement that must be satisfied is the accuracy.

The actuator needs to generate torque/force in order to move two different categories of inertia and load (Fig. 3.13):

1. Load inertia and force/torque (including the gear mechanism).
2. Inertia (and any resistive force) of the actuator itself. For instance, an electric motor has a rotor with finite inertia and that inertia is important to how fast the motor can accelerate and decelerate in high cycle rate automated machine applications. Similarly, a hydraulic cylinder has a piston and large rod which has mass.

The torque/force and motion relationship for each axis is determined by Newton's second law. Let us consider it for a rotary actuator. The same relationships follow for translational actuators by replacing the rotary inertia with mass, torque with force, and angular acceleration with translational acceleration [$\{J_T, \ddot{\theta}, T_T\}$ replace with $\{m_T, \ddot{x}, F_T\}$],

$$J_T \cdot \ddot{\theta} = T_T \tag{3.153}$$

where $J_T$ is the total inertia reflected on the motor axis, $T_T$ is the total net torque acting on the motor axis, and $\ddot{\theta}$ is angular acceleration. The *reflected* inertia or torque means the equivalent inertia or torque seen at the motor shaft after the gear reduction affect is taken into account.

There are three issues to determine for the actuator sizing (Fig. 3.13),

1. Determine the net inertia, Fig. 3.14 (it may be a function of the position of the motion conversion mechanism, Fig. 3.9)

| | Mechanism Characteristics | | | | Input Characteristics | | | Output Characteristics | | |
|---|---|---|---|---|---|---|---|---|---|---|
| Mechanism Type | $n_1$ | $\eta_1$ | $J_1$ | $T_1$ | $S_2$ | $J_2$ | $T_2$ | $S_1$ | $J_1$ | $T_1$ |
| Gear 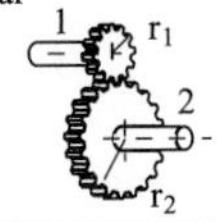 | $\frac{r_2}{r_1}$ | $\leq 1.0$ | $J_{r1}$<br>$J_{r2}$ | $Tc.sgn(\dot{\theta}_1)$<br>$+c\,\dot{\theta}_1$ | $\theta_2$ | $J_2$ | $T_2$ | $\theta_1 = n\,\theta_2$<br>$= \left(\frac{r_2}{r_1}\right)\theta_2$ | $J_{r1} + \left(\frac{J_{r2}}{n^2\eta}\right)$ | $T_i + \frac{1}{n\eta}T_2$ |
| Belt-Pulley 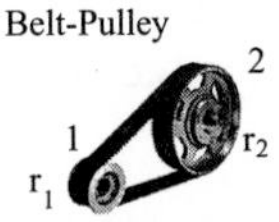 | $\frac{r_2}{r_1}$ | $\leq 1.0$ | $J_{r1}$<br>$J_{r2}$<br>$W_{belt}$ | $Tc.sgn(\dot{\theta}_1)$<br>$+c\,\dot{\theta}_1$ | $\theta_2$ | $J_2$ | $T_2$ | $\theta_1 = n\,\theta_2$<br>$= \left(\frac{r_2}{r_1}\right)\theta_2$ | $J_{r1} + \left(\frac{J_{r2}+J^2}{n^2\eta}\right)$<br>$+ \frac{1}{2}\left(\frac{W_{belt}}{g}\right)(r_1^2 + r_2^2)$ | $T_i + \frac{1}{n\eta}T_2$ |
| Ball Screw or Cam 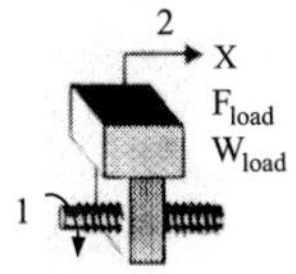<br>$X = \theta/2\pi p$ | $2\pi p$ | $\leq 1.0$ | $J_{lead}$<br>$W_{load}$ | $Tc.sgn(\dot{\theta}_1)$<br>$+c\,\dot{\theta}_1$ | X | $W_{load}$ | $F_{load}$ | $\theta_1 = n\,X$<br>$= (2\pi p)\,X$ | $J_{load} + \left(\frac{W_{load}}{g}\right)\left(\frac{1}{n^2\eta}\right)$ | $T_i + \frac{1}{n\eta}F_{load}$ |
| Rack Pinion  | $\frac{1}{r_p}$ | $\leq 1.0$ | $J_{pinion}$<br>$W_{rack}$ | $Tc.sgn(\dot{\theta}_1)$<br>$+c\,\dot{\theta}_1$ | X | $W_{load}$ | $F_{load}$ | $\theta_1 = n\,X$<br>$= \left(\frac{1}{r_p}\right)X$ | $J_{pinion}$<br>$+ \left(\frac{W_{rack}}{g}\right)\left(\frac{1}{n^2\eta}\right)$<br>$+ \left(\frac{W_{load}}{g}\right)\left(\frac{1}{n^2\eta}\right)$ | $T_i + \frac{1}{n^2\eta}F_{load}$ |
| Conveyor Belt 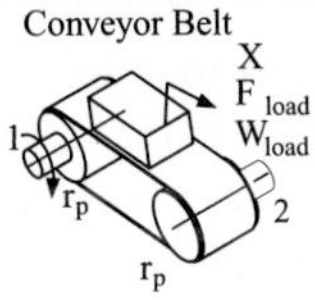 | $\frac{1}{r_p}$ | $\leq 1.0$ | $J_{p1}$<br>$J_{p2}$<br>$W_{belt}$ | $Tc.sgn(\dot{\theta}_1)$<br>$+c\,\dot{\theta}_1$ | X | $W_{load}$ | $F_{load}$ | $\theta_1 = n\,X$<br>$= \left(\frac{1}{r_p}\right)X$ | $J_{p1} + J_{p2}$<br>$+ \left(\frac{W_{belt} + W_{load}}{g}\right)\left(\frac{1}{n^2\eta}\right)$ | $T_i + \frac{1}{n^2\eta}F_{load}$ |

**FIGURE 3.14:** Commonly used motion conversion (gear) mechanisms and their input–output relationships.

2. Determine the net load torques (it may be a function of the position of the motion conversion mechanism and speed)
3. Specify the desired motion profile (Figs. 3.13, 3.15)

Let us discuss the first item, detemination of inertia. Total inertia is the inertia of the rotary actuator and the reflected inertia,

$$J_T = J_m + J_{l,eff} \tag{3.154}$$

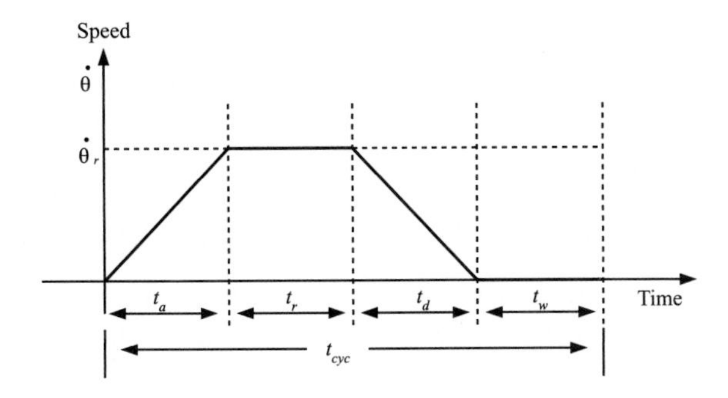


**FIGURE 3.15:** A typical desired velocity profile of a motion axis in programmable motion control applications such as automated assembly machines and robotic manipulators.

where $J_{l,eff}$ includes all the load inertias reflected on the motor shaft. For instance, in the case of a ball-screw mechanism, this includes the inertia of the flexible coupling ($J_c$) between the motor shaft and ball-screw, ball-screw inertia ($J_{bs}$), and load mass inertia ( due to $W_l$),

$$J_{l,eff} = J_c + J_{bs} + \frac{1}{(2\pi p)^2}(W_l/g) \tag{3.155}$$

Notice that the total inertia that the actuator has to move is sum of the load (including motion transmission mechanism) and the inertia of the moving part of the actuator itself.

The total torque is the difference between the torque generated by the motor ($T_m$) minus the resistive load torques on the axis ($T_l$),

$$T_T = T_m - T_{l,eff} \tag{3.156}$$

where $T_l$ represents the sum of all external torques. If the load torque is in the direction of assisting the motion, it will be negative, and net result will be the addition of two torques. The $T_l$ may include friction ($T_f$), gravity ($T_g$), process-related torque and forces (i.e., an assembly application may require the mechanism to provide a desired force pressure, $T_a$), and nonlinear motion related forces/torques, if any (i.e., Coriolis forces and torques, $T_{nl}$),

$$T_l = T_f + T_g + T_a + T_{nl} \tag{3.157}$$

$$T_{l,eff} = \frac{1}{N} \cdot T_l \tag{3.158}$$

Notice that the friction torque may have a constant and speed dependent component to represent the Coulomb and viscous friction, $T_f(\dot{\theta})$.

For actuator sizing purposes, these torques should be considered for the worst possible case. However, care should be exercised that too much safety margin in the worst-case assumptions can lead to unnecessarily large actuator sizing. Once the friction, gravitational loading, task-related forces, and other nonlinear force coupling effects in articulated mechanisms are estimated, the mechanism kinematics is used to determine the reflected forces on the actuator axis. This reflection is a constant ratio for simple motion conversion mechanisms such as gear reducers, belt-pulley, lead screw. For more complicated mechanisms such as linkages, cams, and multidegrees of freedom mechanisms, the kinematic reflection relations for the inertias and forces are not constant. Again, these relationships can be handled using worst-case assumptions in simpler forms, or using more detailed nonlinear kinematic model of the mechanism (see Sect. 3.7).

Finally, we need to know the desired motion profile of the axis as a function of time. Generally, we assume a worst-case cyclic motion. The most common motion profile used is a trapezoidal velocity profile as a function of time (Fig. 3.15). The typical motion includes a constant acceleration period, then a constant speed period, then a constant decelleration period, and a dwell (zero-speed) period.

$$\dot{\theta} = \dot{\theta}(t); \quad 0 \leq t \leq t_{cyc} \tag{3.159}$$

Once the inertias, load torques, and desired motion profile are known, the required torque as a function of time during a cycle of the motion can be determined from

$$T_m(t) = J_T \cdot \ddot{\theta}(t) + T_{l,eff} \tag{3.160}$$

$$= J_T(\theta) \cdot \ddot{\theta}(t) + T_{f,eff}(\dot{\theta}) + T_{g,eff}(\theta) + T_{a,eff}(t) + T_{nl,eff}(\theta, \dot{\theta}) \tag{3.161}$$

Notice that before we calculate the torque requirements, we need to guess the inertia of the actuator itself which is not known yet. Therefore, this calculation may need to be iterated few times.

Once the required torque profile is known as a function of time, two sizing values are determined from it: the maximum and root-mean square (RMS) value of the torque,

$$T_{max} = max(T_m(t)); \tag{3.162}$$

$$T_r = T_{rms} = \left(\frac{1}{t_{cyc}} \int_0^{t_{cyc}} T_m(t)^2 \, dt\right)^{1/2} \tag{3.163}$$

From the desired motion profile specification, we determine the maximum speed the actuator must deliver using the kinematic relations. In order to design an optimal motion control axis, the actuator sizing and the motion conversion mechanism (effective gear ratio) should be considered together. It may be possible that a very small gear ratio may require a motor with very large torque requirement, and yet run at very low speeds, hence make use of a small part of the power capacity of the motor. An increased gear ratio would then require a smaller torque motor and that motor would operate at higher speeds on the average, hence making greater use of the available power of the motor.

Once the torque requirements are known, the drive current and power supply voltage requirements can be directly determined for a given electric motor-driven system. Similarly, for hydraulic actuators, once the force requirements are determined, we would pick the supply pressure and determine the diameter of the linear cylinder. The speed requirement would determine the flow rate. Once these are known, the size of the valve and pump can be determined.

### Actuator Sizing Algorithm (Figs. 3.13 and 3.14)

1. Define the geometric relationship between the actuator and load. In other words, select the type of motion transmission mechanism between the motor and the load ($N$).
2. Define the inertia and torque/force characteristics of the load and the transmission mechanism, i.e., define the inertia of the tool as well as the inertia of the gear reducer mechanisms ($J_l$, $T_l$).
3. Define the desired cyclic motion profile in the form of load speed versus time ($\dot{\theta}_l(t)$).
4. Using the reflection equations developed above, calculate the reflected load inertia and torque/forces ($J_{l,eff}$, $T_{l,eff}$) that will effectively act on the actuator shaft as well as the desired motion at the actuator shaft ($\dot{\theta}_m(t) = \dot{\theta}_{in}(t)$).
5. Guess an actuator/motor inertia from an available list (or make the first calculation with zero motor inertia assumption) and calculate the torque history, $T_m(t)$, for the desired motion cycle. Then calculate the peak torque and RMS torque from $T_m(t)$.
6. Check if the actuator meets the required performance in terms of peak and RMS torque, and maximum speed capacity ($T_p$, $T_{rms}$, $\dot{\theta}_{max}$). If the above selected actuator/motor from the available list does not meet the requirements (i.e., too small or too large), repeat the previous step by selecting a different motor.
7. Continuous torque rating of most electric servo motors is given for 25°C ambient temperature and an aluminum face mount for heat dissipation considerations. If the nominal ambient temperature is different than 25°C, the continuous (RMS) torque capacity of the electric motor should be derated using the following equation for a temperature:

$$T_{rms} = T_{rms}(25°C)\sqrt{(155 - Temp°C)/130} \tag{3.164}$$

If the $T_{rms}$ rating is exceeded, the temperature of the motor winding will increase proportionally. If the temperature rise is above the rated temperature for the winding insulation of the motor, the motor will be damaged permanently.

### 3.6.1 Inertia Match Between Motor and Load

The ratio of the motor's rotor inertia and the reflected load inertia is always a concern in high-performance motion control applications. It is a rule of thumb that the ratio of motor inertia to load inertia should be between one-to-one and up to one-to-ten:

$$\frac{J_m}{J_l/N^2} = \frac{1}{1} \sim \frac{1}{10} \tag{3.165}$$

The one-to-one match is considered the optimal match. Below, we show that one-to-one inertial match is *optimal* only in ideal cases in which the motor drives a purely inertial load and that this inertia ratio results in *minimum heating of the motor*.

Let us consider the case that the motor is coupled to a purely inertial load through an effective gear ratio. The torque and motion relationship is

$$T_m(t) = (J_m + \frac{1}{N^2} J_l) \cdot \ddot{\theta}_m \tag{3.166}$$

$$= (J_m + \frac{1}{N^2} J_l) \cdot N \cdot \ddot{\theta}_l \tag{3.167}$$

The minimal heating occurs when the required torque is minimized, because torque is proportional to current and heat generation is related to current ($P_{elec} = R \cdot i^2$, where $P_{elec}$ is the electric power dissipated at the motor windings due to its electrical resistance $R$ and current $i$). The minimum torque occurs at the gear ratio where the derivative of $T_m$ with respect to $N$ is equal to zero,

$$\frac{d}{dN}(T_m) = \left(J_m + \frac{1}{N^2} J_l\right) \ddot{\theta}_l + \left(\frac{-2N}{N^4} J_l\right) \cdot N \ddot{\theta}_l \tag{3.168}$$

$$= \ddot{\theta}_l \cdot \left(J_m - \frac{1}{N^2} J_l\right) \tag{3.169}$$

$$= 0 \tag{3.170}$$

Therefore, the optimal gear ratio between the motor and a purely inertial load which minimizes the torque requirements (hence, the heating), is

$$J_m = \frac{1}{N^2} J_l \tag{3.171}$$

$$= J_{l,reflected} \tag{3.172}$$

It is important to note that this ideal inertia match (1:1) between motor's rotor inertia and reflected load inertia is optimal only for purely inertial loads. In applications where the load may be dominated by friction or other application related load torque or forces, the ideal inertia match may not be a good design.

***Example*** Consider a rotary motion axis driven by an electric servo motor. The rotary load is directly connected to the motor shaft without any gear reducer. The rotary load is a solid cylindrical shape made of steel material, $d = 3.0$ in, $l = 2.0$ in, $\rho = 0.286$ lb/in$^3$. The desired motion of the load is a periodic motion (Fig. 3.13). The total distance to be traveled is 1/4 of a revolution. The period of motion is $t_{cyc} = 250$ msec, and the dwell portion of it is $t_{dw} = 100$ msec., and the remaining part of the cycle time is equally divided between acceleration, constant speed, and deceleration periods, $t_a = t_r = t_d = 50$ msec. Determine the required motor size for this application.

This example matches the ideal model shown in Fig. 3.14. The load inertia is

$$J_l = \frac{1}{2} \cdot m \cdot (d/2)^2 \tag{3.173}$$

$$= \frac{1}{2} \cdot \rho \cdot \pi \cdot (d/2)^2 \cdot l \cdot (d/2)^2 \tag{3.174}$$

$$= \frac{1}{2} \cdot \rho \cdot \pi \cdot l \cdot (d/2)^4 \tag{3.175}$$

$$= \frac{1}{2} \cdot (0.286/386) \cdot \pi \cdot 2.0 \cdot (3/2)^4 \text{ [lb in sec}^2\text{]} \tag{3.176}$$

$$= 0.0118 \text{ [lb in sec}^2\text{]} \tag{3.177}$$

Let us assume that we will pick a motor which has an rotor inertia same as the load so that there is an ideal load and motor inertia match, $J_m = 0.0118$ [lb $\cdot$ in $\cdot$ sec$^2$]. The acceleration, top speed, and deceleration rates are calculated from the kinematic relationships,

$$\theta_a = \frac{1}{2}\dot{\theta} \cdot t_a = \frac{1}{4} \cdot \frac{\pi}{2} \tag{3.178}$$

$$\dot{\theta} = 2 \cdot \theta_a / t_a = 2 \cdot (1/4) \cdot (\pi/2) \text{ [rad]}/(0.05 \text{ [sec]}) \tag{3.179}$$

$$= 80\pi/16 \text{ [rad/sec]} = 40/16 \text{ [rev/sec]} = 2400/16 \text{ [rev/min]} = 150 \text{ [rev/min]} \tag{3.180}$$

$$\ddot{\theta}_a = \dot{\theta}_a / t_a = (80\pi/16)(1/0.05) = 1600\pi/16 \text{ [rad/sec}^2\text{]} = 100\pi \text{ [rad/sec}^2\text{]} \tag{3.181}$$

$$\ddot{\theta}_r = 0.0 \tag{3.182}$$

$$\ddot{\theta}_d = -100\,\pi \text{ [rad/sec}^2\text{]} \tag{3.183}$$

$$\ddot{\theta}_{dw} = 0.0 \tag{3.184}$$

The required torque to move the load through the desired cyclic motion can be calculated as follows:

$$T_a = (J_m + J_l) \cdot \ddot{\theta} = (0.0118 + 0.0118) \cdot (100\pi) = 7.414 \text{ [lb in]} \tag{3.185}$$

$$T_r = 0.0 \tag{3.186}$$

$$T_d = (J_m + J_l) \cdot \ddot{\theta} = (0.0118 + 0.0118) \cdot (-100\pi) = -7.414 \text{ [lb in]} \tag{3.187}$$

$$T_{dw} = 0.0 \tag{3.188}$$

Hence, the peak torque requirement is

$$T_{max} = 7.414 \text{ [lb in]} \tag{3.189}$$

and the RMS torque requirement is

$$T_{rms} = \left( \frac{1}{0.250} \left( T_a^2 \cdot t_a + T_r^2 \cdot t_r + T_d^2 \cdot t_d + T_{dw}^2 \cdot t_{dw} \right) \right)^{1/2} \tag{3.190}$$

$$= \left( \frac{1}{0.250} (7.414^2 \cdot 0.05 + 0.0 \cdot 0.05 + (-7.414)^2 \cdot 0.05 + 0.0 \cdot 0.1) \right)^{1/2} \tag{3.191}$$

$$= 4.689 \text{ [lb in]} \tag{3.192}$$

Therefore, a motor which has rotor inertia of about 0.0118 [lb in $sec^2$], maximum speed capabaility of 150 [rev/min] or better, and peak and RMS torque rating in the range of 8.0 [lb in] and 5.0 [lb in] range would be sufficient for the task.

This design may be improved further by the following consideration. The top speed of the motor is only 150 rpm. Most electric servo motors runs in the 1500 rpm to 5000 rpm range where they deliver most of their power capacity, while maintaining a constant torque capacity up to these speeds. As a result, it is reasonable to consider a gear reducer between the motor shaft and the load in the range of 10:1 to 20:1 and repeat the sizing calculations. This will result in a motor that will run at a higher speed and will have lower torque requirements.

## 3.7 HOMOGENEOUS TRANSFORMATION MATRICES

The geometric relationships in simple one degree of freedom mechanisms can be derived using basic vector algebra. The derivation of geometric relations for multi-degrees of freedom mechanisms, such as the robotic mechanisms, is rather difficult using three-dimensional vector algebra. The so-called $(4 \times 4)$ homogeneous transformation matrices are very powerful matrix methods to describe the geometric relations [9–13]. They are used to describe the geometric relations of a mechanism between the absolute values of:

1. Displacement variables
2. The relations between the incremental changes in displacements
3. Force and torque transmission through the mechanism

*The position and orientation of a three-dimensional object with respect to a reference frame can be uniquely described by the position coordinates of a point on it (three components of position information in three-dimensional space) and the orientation of a coordinate frame fixed to the object (described by three angles).* The position coordinates are associated with a point and are unique for a given point with respect to a reference coordinate frame. Orientation is associated with an object, not a point. The best way to describe the position and orientation of an object is to attach a coordinate frame to the object, and describe the orientation and the origin coordinates of the attached coordinate frame with respect to a reference frame. For instance, the position and orientation of a tool held by the gripper of a robotic manipulator can be described by a coordinate frame attached to the tool (Fig. 3.16). The position coordinates of the origin of the attached coordinate frame and its orientation with respect to another reference frame also describes the position and orientation of the tool.

The transformation of an object between any two different orientations can be accomplished by a sequence of three independent rotations. However, the number and sequence of rotation angles to go from one orientation to another are not unique. There are many possible rotation combinations to make a desired orientation change. For instance, an orientation change between any two different orientation of two coordinate frames can be accomplished by a sequence of three angles such as

1. roll, pitch, and yaw angles
2. Z, Y, X Euler angles
3. X, Y, Z Euler angles

There are 24 different possible combinations of a seqeunce of three angles to go from one orientation to another [11]. *Finite rotations are not commutative. Infinitesimal rotations are*

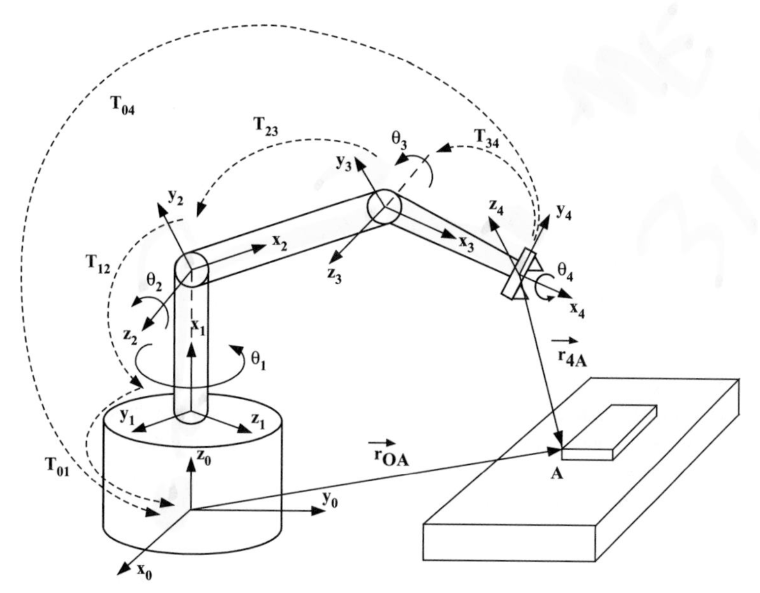


**FIGURE 3.16:** Multi-degree of freedom mechanisms: a robotic manipulator with four joints. We use this example to illustrate how to describe the position and orientation of one coordinate frame with respect to another. If we attach a coordinate frame to a workpiece, we can describe the position and orientation of it with respect to other coordinate frames through the coordinate frame attached to it.

*commutative.* That means the order of a sequence of finite rotations makes a difference in the final orientation. For instance, making a 90 degree rotation about $X$ axis followed by another 90 degree rotation about $Y$ axis results in a different orientation than that of making a 90 degree rotation about $Y$ axis followed by another 90 degree rotation about $X$ axis. However, if the rotations are infinitesimal, the order does not matter.

The $4 \times 4$ homogeneous transformation matrices describe the position of a point on an object and the orientation of the object in three-dimensional space using a $4 \times 4$ matrix. The first $3 \times 3$ portion of the matrix is used to define the orientation of a coordinate frame fixed to the object with respect to another reference coordinate frame. The last column of the matrix is used to describe the position of the origin of the coordinate frame fixed to the object with respect to the origin of the reference coordinate frame. The last row of the matrix is [0001]. A general $4 \times 4$ homogeneous transformation matrix $T$ has the following form (Fig. 3.17):

$$T = \begin{bmatrix} e_{11} & e_{12} & e_{13} & x_A \\ e_{21} & e_{22} & e_{23} & y_A \\ e_{31} & e_{32} & e_{33} & z_A \\ 0 & 0 & 0 & 1 \end{bmatrix} \tag{3.193}$$

It describes the position and orientation of the coordinate frame A with respect to the coordinate frame 0. The columns of the $(3 \times 3)$ portion of the matrix which contain the orientation information are the cosine angles between the unit vectors of the coordinate frames. Notice that, even though we know that the orientation of one coordinate frame with respect to another can be described by three angles, the general form of the rotation portion of the $(4 \times 4)$ transformation matrix (the $(3 \times 3)$ portion) requires nine parameters. However, they are not all independent. There are six constraints between them, leaving three independent

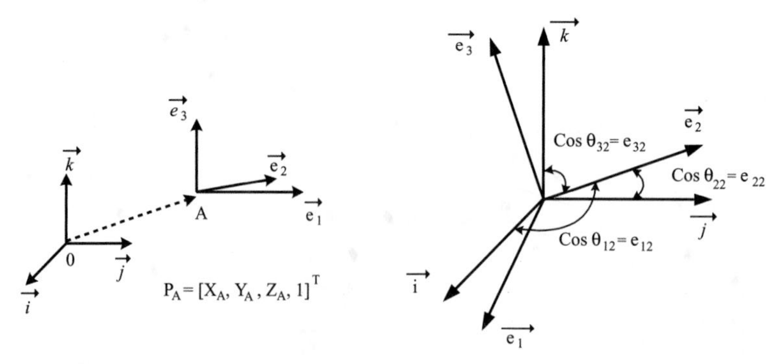


**FIGURE 3.17:** Use of (4 × 4) coordinate transformation matrices to describe the kinematic (geometric) relationships between different objects in three-dimensional space. A point is described with respect to a reference coordinate frame by its three position coordinates. An object is described by a coordinate frame fixed to it: position of its origin and its orientation with respect to the reference coordinate frame.

variables (Fig. 3.17). The six constraints are

$$\vec{e}_1 \cdot \vec{e}_1 = 1.0 \tag{3.194}$$

$$\vec{e}_2 \cdot \vec{e}_2 = 1.0 \tag{3.195}$$

$$\vec{e}_3 \cdot \vec{e}_3 = 1.0 \tag{3.196}$$

$$\vec{e}_1 \cdot \vec{e}_2 = 0 \tag{3.197}$$

$$\vec{e}_1 \cdot \vec{e}_3 = 0 \tag{3.198}$$

$$\vec{e}_2 \cdot \vec{e}_3 = 0 \tag{3.199}$$

where

$$\vec{e}_i = e_{1i}\vec{i} + e_{2i}\vec{j} + e_{3i}\vec{k}; \quad i = 1, 2, 3 \tag{3.200}$$

are the unit vectors along each of the axes of the attached coordinate frame expressed in terms of its components in the unit vectors of the other coordinate frame. The use of cosine angles in describing the orientation of one coordinate frame with respect to another one is a very convenient way to determine the elements of the matrix.

The 4 × 4 homogeneous transformation matrices are the most widely accepted and powerful (if not computationally most efficient) method to describe kinematic relations. The algebra of transformation matrices follows the basic matrix algebra. Let us consider the coordinate frames numbered 1, 2, 3, 4 and a point $A$ on the object where the position coordinates of the point with respect to the third coordinate frame is described by $r_{4A}$ (Fig. 3.16). The description of the coordinate frame 4 (position of its origin and orientation) can be expressed as

$$T_{04} = T_{01} \cdot T_{12} \cdot T_{23} \cdot T_{34} \tag{3.201}$$

where $T_{02}$, $T_{12}$, $T_{23}$, and $T_{34}$ are the description of the origin position coordinates and orientations of the axes of coordinate frames 1 with respect to 0, 2 with respect to 1 and that of 3 with respect to 2. The position coordinate vector of point $A$ can be expressed with respect to coordinates 2 and 3 as follows (Fig. 3.16):

$$r_{3A} = T_{34} \cdot r_{4A} \tag{3.202}$$

$$r_{2A} = T_{23} \cdot r_{3A} = T_{23} \cdot T_{34} \cdot r_{4A} \tag{3.203}$$

$$r_{1A} = T_{12} \cdot r_{2A} = T_{12} \cdot T_{23} \cdot T_{34} \cdot r_{4A} \tag{3.204}$$

$$r_{0A} = T_{01} \cdot r_{1A} = T_{01} \cdot T_{12} \cdot T_{23} \cdot T_{34} \cdot r_{4A} \tag{3.205}$$

where, $r_{4A} = [x_{4A} \quad y_{4A} \quad z_{4A} \quad 1]^T$. The $r_{3A}, r_{2A}, r_{1A}, r_{0A}$ are similarly defined. Notice that $T_{12}$ is the description (position coordinates of the origin and the orientation of the axes of coordinate system 2 with respect to coordinate system 1) of coordinate system 2 with respect to coordinate system 1. Then the reverse description, that is, the description of coordinate system 1 with respect to coordinate system 2 is the inverse of the previous transformation matrix,

$$T_{21} = T_{12}^{-1} \tag{3.206}$$

The (4 × 4) transformation matrix has a special form, and the inversion of the matrix also has a special result. Let

$$T_{12} = \begin{bmatrix} R_{12} & p_A \\ ------ & \\ 0 \;\; 0 \;\; 0 & 1 \end{bmatrix} \tag{3.207}$$

Then the inverse of this matrix can be shown as

$$T_{12}^{-1} = \begin{bmatrix} R_{12}^T & -R_{12}^T \cdot p_A \\ ------ & \\ 0 \;\; 0 \;\; 0 & 1 \end{bmatrix} \tag{3.208}$$

Also notice that the order of transformations is important (multiplication of matrices are dependent on the order):

$$T_{12} \cdot T_{23} \neq T_{23} \cdot T_{12} \tag{3.209}$$

For a general purpose multi-degrees of freedom mechanism, such as a robotic manipulator, the relationship between a coordinate frame attached to the tool (the coordinates of its origin and orientation) with respect to a fixed reference frame at the base can be expressed as a seqeunce of transformation matrices where each transformation matrix is a function of one of the axis position variables. For instance, for a four degrees of freedom robotic manipulator, the coordinate system at the wrist joint can be described with respect to base as (Fig. 3.16)

$$T_{04} = T_{01}(\theta_1) \cdot T_{12}(\theta_2) \cdot T_{23}(\theta_3) \cdot T_{34}(\theta_4) \tag{3.210}$$

where $\theta_1, \theta_2, \theta_3, \theta_4$ are the position variables of axes driven by motors. Any given position vector relative to the forth coordinate frame ($r_{4A}$) can be expressed with respect to base as follows:

$$r_{0A} = T_{04} \cdot r_{4A} \tag{3.211}$$

$$= T_{01}(\theta_1) \cdot T_{12}(\theta_2) \cdot T_{23}(\theta_3) \cdot T_{34}(\theta_4) \cdot r_{4A} \tag{3.212}$$

where $r_{4A} = [x_{4A}, y_{4A}, z_{4A}, 1]^T$, the three coordinates of the point $A$ with respect to the coordinate frame 4.

The Denavit–Hartenberg method [12,13] defines a standard way of attaching coordinate frames to a robotic manipulator such that only four numbers (one variable, three constant parameters) are needed per one-degree of freedom joint to represent the kinematic relationships.

In generic terms, the relationship between the coordinates of the tool and the joint displacement variables can be expressed as

$$\underline{x} = \underline{f}(\underline{\theta}) \tag{3.213}$$

The vector variable $\underline{x}$ represents the cartesian coordinates of the tool (i.e., position coordinates $x_P, y_P, z_P$ in a given coordinate frame and oriention angles where three angles can be used to describe the orientation). The description of the position coordinates of a point with respect to a given reference frame is unique. However, the orientation of an object with respect to a reference coordinate frame can be described by many different possible combinations of angles. Hence, the orientation description is not unique. The vector variable $\underline{\theta}$ represents the joint variables of the robotic manipulator, i.e., for a six joint robot $\underline{\theta} = [\theta_1, \theta_2, \theta_3, \theta_4, \theta_5, \theta_6]^T$.

The $\underline{f}(\underline{\theta})$ is called the *forward kinematics* of the mechanism, which is a vector nonlinear function of joint variables,

$$\underline{f}(\underline{\theta}) = [f_1(\underline{\theta}), f_2(\underline{\theta}), f_3(\underline{\theta}), f_4(\underline{\theta}), f_5(\underline{\theta}), f_6(\underline{\theta})]^T \tag{3.214}$$

The inverse relationship, that is the geometric function which defines the axis positions as a function of tool position and orientation, is called the *inverse kinematics* of the mechanism,

$$\underline{\theta} = \underline{f}^{-1}(\underline{x}) \tag{3.215}$$

The inverse kinematics function may not be possible to express in one analytical closed form for every mechanism. It must be determined for each special mechanism on a case-by-case basis. For a six revolute joint manipulator, a sufficient condition for the existence of the inverse kinematic solution in analytical form is that three consecutive joint axes must intersect at a point. Forward and inverse kinematic functions of a mechanism relate the joint positions to the tool positions.

The *differential* relationships between joint axis variables and tool variables are obtained by taking the differential of the forward kinematic function. The resultant matrix that relates the differential values of joint and tool position variables (in other words, it relates the velocities of joint axes and tool velocity) is called the *jacobian matrix* of the mechanism (Fig. 3.16).

$$\dot{\underline{x}} = \frac{d\underline{f}(\underline{\theta})}{d\underline{\theta}} \cdot \dot{\underline{\theta}} \tag{3.216}$$

$$= J\dot{\underline{\theta}} \tag{3.217}$$

where the $J$ matrix is called the jacobian of the mechanism. Each element of the Jacobian matrix is defined as

$$J_{ij} = \frac{\partial f_i(\theta_1, \theta_2, \theta_3, \theta_4, \theta_5, \theta_6)}{\partial \theta_j} \tag{3.218}$$

where $i = 1, 2, \ldots, m$ and $j = 1, 2, \ldots, n$, where $n$ is the number of joint variables. For a six degrees of freedom mechanism, $n = m = 6$. If the mechanism has less than six degrees of freedom, the jacobian matrix is not a square matrix. Likewise, the inverse of the jacobian, $J^{-1}$ relates the changes in the tool position to the changes in the axis displacements,

$$\dot{\underline{\theta}} = J^{-1} \cdot \dot{\underline{x}} \tag{3.219}$$

If the jacobian is not invertible in certain positions, these positions are called the *geometric singularities* of the mechanism. It means that at these locations, there are some directions that the tool cannot move no matter what the change is in the joint variables. In other words, no joint axis variable combination can generate a motion in certain directions at a singularity point. A robotic manipulator may have many singularity points in its workspace. The geometric singularity is directly function of the mechanical configuration of the manipulator. There are two groups of singularities:

1. **Workspace boundary:** A given manipulator has a finite span in three-dimensional space. The locations that the manipulator can reach is called the *workspace*. At the boundary of workspace, the manipulator tip cannot move out because it reached its limits of reach. Hence, all points in workspace boundary are singularity points since at these points there are directions along which manipulator tip cannot move.
2. **Workspace interior points:** These singularity points are inside the workspace of the manipulator. Such singularity points depend on the manipulator geometry and generally occur when two or more joints line up.

The same jacobian matrix also describes the relationship between the torques/forces at the controlled axes and the force/torque exprienced at the tool. Let the tool force be $\underline{Force}$ and the corresponding tool position differential displacement be $\delta \underline{x}$. The differential work done is

$$Work = \underline{\delta x}^T \cdot \underline{Force} \tag{3.220}$$

Note that the jacobian relationship from eqn. (3.217)

$$\underline{\delta x} = J \cdot \underline{\delta \theta} \tag{3.221}$$

and the equivalent work done by the corresponding torques at the controlled axes can be expressed as

$$Work = \underline{\delta \theta}^T \cdot \underline{Torque} \tag{3.222}$$

$$= \underline{\delta x}^T \cdot \underline{Force} \tag{3.223}$$

$$= (J \cdot \delta \theta)^T \cdot \underline{Force} \tag{3.224}$$

Hence, the force-torque relationship between the tool and joint variables is

$$\underline{Torque} = J^T \cdot \underline{Force} \tag{3.225}$$

and the inverse relationship is

$$\underline{Force} = (J^T)^{-1} \cdot \underline{Torque} \tag{3.226}$$

Notice that, at the *singular configurations* (geometric singularities) of the mechanism, those configurations of the mechanism at which the inverse of the jacobian matrix does not exist, there are some force directions at the tool which does not result in any change in the axes torques. They only result in reaction forces in the linkage structure, but not in the actuation axes. Another interpretation of this result is that there are some directions of tool motion that we cannot generate force no matter what combination of torques are applied at the joints.

There are different methods for the calculation of the jacobian matrix for a mechanism [8–14]. The inverse jacobian matrix can be either obtained analytically in symbolic form or calculated numerically off-line or on-line (in real time). However, real-time numerical inverse calculations present a problem both in terms of the computational load and the possible numerical stability problems around the singularities of the mechanism. The decision regarding the jacobian matrix and its inverse computations in real time should be made on a mechanism by mechanism basis.

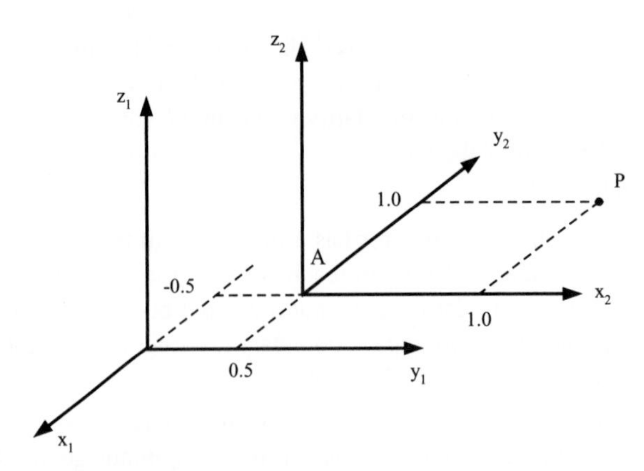


**FIGURE 3.18:** Describing the position and orientation of one coordinate frame with respect to the other.

***Example*** Consider two coordinate frames numbered 1 and 2 as shown in Fig. 3.18. Let the origin coordinates of second coordinate frame have the following coordinates, $r_{1A} = [x_{1A} \quad y_{1A} \quad z_{1A} \quad 1]^T = [-0.5 \quad 0.5 \quad 0.0 \quad 1]^T$. Orientations of axes are such that $X_2$ is parallel to $Y_1$, $Y_2$ is parallel but in opposite direction to $X_1$, and $Z_2$ has the same direction as $Z_1$. Determine the vector description of point $P$ with respect to coordinate frame 1 whose coordinates are given in second frame as $r_{2P} = [x_{2P} \quad y_{2P} \quad z_{2P} \quad 1]^T = [1.0 \quad 1.0 \quad 0.0 \quad 1]^T$.

Using the homogeneous transformation matrix relationship between coordinate frames 1 and 2,

$$r_{1P} = T_{12} \cdot r_{2P} \tag{3.227}$$

where the $T_{12}$ is described by the orientation of the second coordinate frame and position coordinates of its origin with respect to first coordinate frame.

$$T_{12} = \begin{bmatrix} e_{11} & e_{12} & e_{13} & x_{1A} \\ e_{21} & e_{22} & e_{23} & y_{1A} \\ e_{31} & e_{32} & e_{33} & z_{1A} \\ 0 & 0 & 0 & 1 \end{bmatrix} \tag{3.228}$$

$$= \begin{bmatrix} 0.0 & -1.0 & 0.0 & -0.5 \\ 1.0 & 0.0 & 0.0 & 0.5 \\ 0.0 & 0.0 & 1.0 & 0.0 \\ 0 & 0 & 0 & 1 \end{bmatrix} \tag{3.229}$$

Notice that the orientation portion of the matrix is the coefficients of the relationships between the unit vectors,

$$\vec{i} = cos\theta_{11} \cdot \vec{e}_1 + cos\theta_{12} \cdot \vec{e}_2 + cos\theta_{13} \cdot \vec{e}_3 \tag{3.230}$$

$$= e_{11} \cdot \vec{e}_1 + e_{12} \cdot \vec{e}_2 + e_{13} \cdot \vec{e}_3 \tag{3.231}$$

$$= 0.0 \cdot \vec{e}_1 + (-1.0) \cdot \vec{e}_2 + 0.0 \cdot \vec{e}_3 \tag{3.232}$$

$$\vec{j} = cos\theta_{21} \cdot \vec{e}_1 + cos\theta_{22} \cdot \vec{e}_2 + cos\theta_{23} \cdot \vec{e}_3 \tag{3.233}$$

$$= e_{21} \cdot \vec{e}_1 + e_{22} \cdot \vec{e}_2 + e_{23} \cdot \vec{e}_3 \tag{3.234}$$

$$= 1.0 \cdot \vec{e}_1 + (0.0) \cdot \vec{e}_2 + (0.0) \cdot \vec{e}_3 \tag{3.235}$$

$$\vec{k} = cos\theta_{31} \cdot \vec{e}_1 + cos\theta_{32} \cdot \vec{e}_2 + cos\theta_{33} \cdot \vec{e}_3 \tag{3.236}$$

$$= e_{31} \cdot \vec{e}_1 + e_{32} \cdot \vec{e}_2 + e_{33} \cdot \vec{e}_3 \tag{3.237}$$

$$= (0.0) \cdot \vec{e}_1 + (0.0) \cdot \vec{e}_2 + (1.0) \cdot \vec{e}_3 \tag{3.238}$$

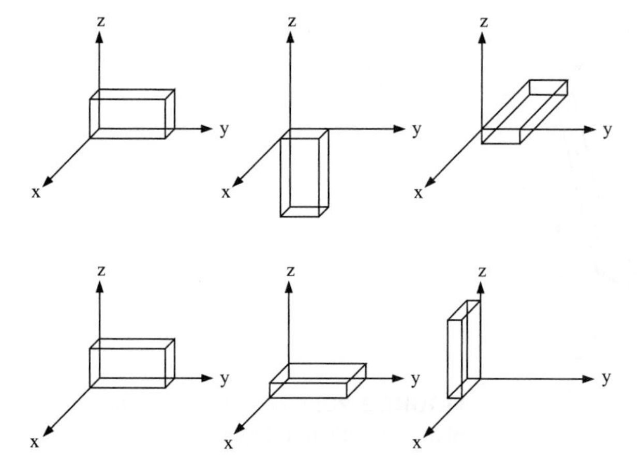


**FIGURE 3.19:** Two finite rotation sequences performed in reverse order. The final orientations are different. Finite rotations are not commutative.

Hence,

$$r_{1P} = T_{12} \cdot r_{2P} \tag{3.239}$$

$$= [-1.5 \quad 1.5 \quad 0.0 \quad 1]^T \tag{3.240}$$

***Example*** The purpose of this example is to illustrate graphically that the order of a sequence of finite rotations is important. If we change the order of rotations, the final orientation is different (Fig. 3.19). In other words, $T_1 \cdot T_2 \neq T_2 \cdot T_1$. Let $T_1$ represent a rotation about the $X$ axis by 90°, and $T_2$ represent a rotation about the $Y$ axis by 90°. Figure 3.19 shows the sequence of both $T_1$ followed by $T_2$ and $T_2$ followed by $T_1$. The resulting final orientations are different because the order of rotations are different. Finite rotations are not commutative. We can show this algebraically for this case.

$$T_1 = \begin{bmatrix} 1.0 & 0.0 & 0.0 & 0.0 \\ 0.0 & 0.0 & 1.0 & 0.0 \\ 0.0 & -1.0 & 0.0 & 0.0 \\ 0 & 0 & 0 & 1 \end{bmatrix} \tag{3.241}$$

$$T_2 = \begin{bmatrix} 0.0 & 0.0 & -1.0 & 0.0 \\ 0.0 & 1.0 & 0.0 & 0.0 \\ 1.0 & 0.0 & 0.0 & 0.0 \\ 0 & 0 & 0 & 1 \end{bmatrix} \tag{3.242}$$

Clearly, $T_1 \cdot T_2 \neq T_2 \cdot T_1$.

***Example*** Consider the geometry of a two-link robotic manipulator (Fig. 3.20). The geometric parameters (link lengths $l_1$ and $l_2$) and joint variables are shown in the figure.

1. Derive the tip position $P$ coordinates as function of joint variables (forward kinematic relations).
2. Derive the jacobian matrix that relates the joint velocities to tip position coordinate velocities.
3. If there is a load with weight $W$ at the tip, determine the necessary torques at joints 1 and 2 in order to balance the load.

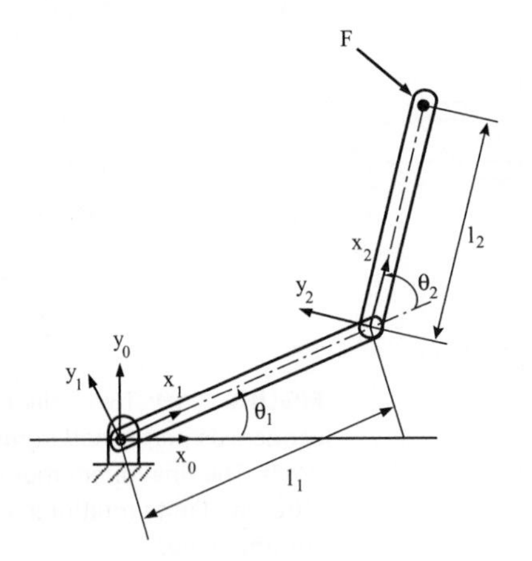


**FIGURE 3.20:** Kinematic description of a two-joint planar robotic manipulator.

In this example, we are asked to determine the following relations:

$$\underline{x} = \underline{f}(\underline{\theta}) \tag{3.243}$$

$$\underline{\dot{x}} = J(\theta)(\underline{\dot{\theta}}) \tag{3.244}$$

$$\underline{Torque} = J^T(\theta)(\underline{Force}) \tag{3.245}$$

Let us attach three coordinate frames to the manipulator. Coordinate frame 0 is attached to the base and fixed, coordinate frame 1 is attached at joint 1 to link 1 (moves with link 1), coordinate frame 2 is attached at joint 2 to link 2 (moves with link 2). The position vector of the tip with respect to coordinate frame 2 is simple, $r_{2P} = [l_2 \quad 0 \quad 0 \quad 1]^T$. The transformation matrices between the three coordinate frames are functions of $\theta_1$ and $\theta_2$ as follows:

$$T_{01} = \begin{bmatrix} cos(\theta_1) & -sin(\theta_1) & 0.0 & 0.0 \\ sin(\theta_1) & cos(\theta_1) & 0.0 & 0.0 \\ 0.0 & 0.0 & 1.0 & 0.0 \\ 0 & 0 & 0 & 1 \end{bmatrix} \tag{3.246}$$

$$T_{12} = \begin{bmatrix} cos(\theta_2) & -sin(\theta_2) & 0.0 & l_1 \\ sin(\theta_2) & cos(\theta_2) & 0.0 & 0.0 \\ 0.0 & 0.0 & 1.0 & 0.0 \\ 0 & 0 & 0 & 1 \end{bmatrix} \tag{3.247}$$

Hence, the tip vector description in base coordinates,

$$r_{0P} = [x_{0P} \quad y_{0P} \quad 0 \quad 1]^T = T_{01} \cdot T_{12} \cdot r_{2P} \tag{3.248}$$

The vector components of $r_{0P}$ can be expressed as individual functions for more clarity,

$$x_{0P} = x_{0P}(\theta_1, \theta_2; l_1, l_2) = l_1 \cdot cos(\theta_1) + l_2 \cdot cos(\theta_1 + \theta_2) \tag{3.249}$$

$$y_{0P} = y_{0P}(\theta_1, \theta_2; l_1, l_2) = l_1 \cdot sin(\theta_1) + l_2 \cdot sin(\theta_1 + \theta_2) \tag{3.250}$$

$$z_{0P} = 0.0 \tag{3.251}$$

The jacobian matrix for this case can simply be determined by taking the derivative of the forward kinematic relations with respect to time and express the equation in the matrix form to find the jacobian matrix.

$$\dot{x}_{OP} = \frac{d}{dt}(x_{OP}(\theta_1, \theta_2; l_1, l_2)) = J_{11}\dot{\theta}_1 + J_{12}\dot{\theta}_2 \tag{3.252}$$

$$\dot{y}_{OP} = \frac{d}{dt}(y_{OP}(\theta_1, \theta_2; l_1, l_2)) = J_{21}\dot{\theta}_1 + J_{22}\dot{\theta}_2 \tag{3.253}$$

where it is easy to show that the elements of the jacobian matrix $J$

$$J = \begin{bmatrix} J_{11} & J_{12} \\ J_{21} & J_{22} \end{bmatrix} \tag{3.254}$$

$$= \begin{bmatrix} -l_1 \cdot sin(\theta_1) - l_2 \cdot sin(\theta_1 + \theta_2) & -l_2 \cdot sin(\theta_1 + \theta_2) \\ l_1 \cdot cos(\theta_1) + l_2 \cdot cos(\theta_1 + \theta_2) & l_2 \cdot cos(\theta_1 + \theta_2) \end{bmatrix} \tag{3.255}$$

The torque needed to balance a weight load, $F = [F_x \ F_y]^T = [0 \ -W]^T$, at the tip is determined by, which is in opposite direction to eqn. (3.245)

$$\begin{bmatrix} Torque_1 \\ Torque_2 \end{bmatrix} = -\begin{bmatrix} J_{11} & J_{21} \\ J_{12} & J_{22} \end{bmatrix} \begin{bmatrix} 0.0 \\ -W \end{bmatrix} \tag{3.256}$$

which shows the necessary static torque at each joint to balance a weight at the tip for different positions of the manipulator. Notice that since the jacobian matrix is $2 \times 2$, it is relatively simple to obtain the inverse jacobian in analytical form.

$$J^{-1} = \frac{1}{l_1 \cdot l_2 \cdot sin(\theta_2)} \begin{bmatrix} l_2 cos(\theta_1 + \theta_2) & l_2 \cdot sin(\theta_1 + \theta_2) \\ -l_1 \ cos(\theta_1) - l_2 \ cos(\theta_1 + \theta_2) & l_1 \ sin(\theta_1) - l_2 \cdot sin(\theta_1 + \theta_2) \end{bmatrix} \tag{3.257}$$

Notice that when $\theta_2 = 0.0$, the mechanism is at a singular point, which is indicated by the $sin(\theta_2)$ term in the denominator of the inverse jacobian equation above.

## 3.8 PROBLEMS

**1.** Consider the gear reducer shown in Fig. 3.1. Let the diameter of the gear on input shaft be equal to $d_1 = 2.0$ in and witdh $w_1 = 0.5$ in. Assume that the gear is made of material steel and that it is a solid frame without any holes. The ouput gear has the same width and material, and the gear reduction from input to output is $N = 5$, ($d_2 = 10.0$ in). The length and diameters of the shafts that extend to the sides of the gears are $d_{s1} = 1.0$ in, $d_{s2} = 1.0$ in. and $l_{s1} = 1.0$ in., $l_{s2} = 1.0$ in. Let us consider that there is a net load torque of $T_{L2} = 50$ lb/in at the output shaft.

**(a)** Determine the net rotary inertia reflected on the input shaft due to two gears and two shafts.
**(b)** Determine the necessary torque at the input shaft to balance the load torque.
**(c)** If the input shaft is actuated by a motor that is controlled with 1/10 degree accuracy, what is the angular positioning accuracy that can be provided at the output shaft?

**2.** Repeat the same analysis and calculations for a belt and toothed pulley mechanism. The gear ratios and the shaft sizes are the same. The load torque in the output shaft is the same. Neglect the inertia of the belt. Comment on the functional similarities. Also discuss practical differences between the two mechanisms.

**3.** Consider a linear positioning system using a ball-screw mechanism. The ball screw is driven by an electric servo motor. The ball screw is made of steel, has length of $l_{ls} = 40$ in, and diameter of $d_{ls} = 2.5$ in. The pitch of the lead is $p = 4$ rev/in (or the lead is 0.25 in/rev). Assume that the lead-screw mechanism is in a vertical direction and moving a load of 100 lbs against the gravity up and down in z-direction.

**(a)** Determine the net rotary inertia reflected on the input shaft of the motor.
**(b)** Determine the necessary torque at the input shaft to balance the weight of the load due to gravity.
**(c)** If the input shaft is actuated by a motor that is controlled with 1/10° accuracy, what is the translational positioning accuracy that can be provided at the output shaft?

**4.** For the problem 3, consider that the typical cyclic motion that the workpiece is to make is defined by a trapezoidal velocity profile. The load is to be moved a distance of 1.0 in. in 300 msec, wait there for 200 msec, and then move in the reverse direction. This motion is repeated continuously. Assume that the 300-msec motion time is equally divided between acceleration, run, and deceleration times, $t_a = t_r = t_d = 100$ msec (Fig. 3.15).

**(a)** Calculate the necessary torque (maximum and continuous rated) and maximum speed required at the motor shaft. Select an appropriate servo motor for this application.
**(b)** If an incremental position encoder is used on the motor shaft for control purposes, what is the required minimum resolution in order to provide a tool positioning accuracy of 40 $\mu$ in?
**(c)** Select a proper flexible coupling for this application to be used between the motor shaft and the lead screw. Include the inertia of the flexible coupling in the inertia calculations and motor sizing calculations. Repeat step 1.

**5.** Repeat the same analysis and calculations of Problems 3 and 4, if a rack-pinion mechanism were used to convert the rotary motion of motor to a translational motion of the tool.

**(a)** First determine the rack and pinion gear that gives the same gear ratio (from rotary motion to translational motion) as the ball-screw mechanism.
**(b)** Discuss the differences between the ball-screw, rack-pinion, and belt-pulley (translational version) mechanisms.

**6.** Given a four-bar linkage (Fig. 3.7), derive the geometric relationship between the following motion variables of the mechanism. Let the link lengths be $l_1, l_2, l_3, l_4$.

**(a)** Input is the angular position ($\theta_1(t)$) of link 1, output is the angular position of link 3, ($\theta_3(t)$). Find $\theta_3 = f(\theta_1)$. Plot $\theta_3$ for one revolution of link 1 as function of $\theta_1$.
**(b)** Determine the $x$ and $y$ coordinates of the tip of the link 3 during the same motion cycle. Plot the results on the x–y plane (path of the tip of link 3 during one revolution of link 1).

**7.** Consider the cam and follower mechanism shown in Fig. 3.9. The follower arm is connected to a spring. The follower is to make an up-down motion once per revolution of the cam. The travel range of the follower is to be 2.0-in. total.

**(a)** Select a modified trapezoidal cam profile for this task.
**(b)** Assume the input shaft to the cam is driven at 1200 rpm constant speed. Calculate the maximum linear speed and linear accelerations experienced at the tool tip.
**(c)** Let the stiffness of the spring be $k = 100$ lb/in and the mass of the follower and the tool it is connected to $m_f = 10$ lb. Assume the input shaft motion is not affected by the dynamics of the follower and tool. The input shaft rotates at constant speed at 1200 (rev/min). Determine the net force function at the follower and tool assembly during one cycle of the motion and plot the result. Notice that the

$$F(t) = m_f \ddot{x}(t) + k \cdot x(t) \tag{3.258}$$

and $x(t)$, $\ddot{x}(t)$ are determined by the input shaft motion and cam function. What happens if the net force $F(t)$ becomes negative? One way to assume that $F(t)$ does not become negative is to use a preloaded spring. What is the preloading requirement to ensure $F(t)$ is always positive during the planned motion cycle? Preload spring force can be taken into account in the above equation as follows:

$$F(t) = m_f \ddot{x}(t) + k \cdot x(t) + F_{pre} \tag{3.259}$$

where $F_{pre} = k \cdot x_o$ is a constant force due to the preloading of the spring. This force can be set to a constant value by selection of spring constant and initial compression.

**8.** Consider an electric servo motor and a load it drives through a gear reducer (Fig. 3.13).

**(a)** What is the generally recommended relationship between the motor inertia and reflected load inertia?
**(b)** What is the optimal relationship in terms of minimizing the heating of the motor?
**(c)** Derive the relationship for the optimal relationship.

**9.** Consider the coordinate frames $x_2y_2z_2$ attached to link 2 and $x_3y_3z_3$ attached to link 3 in Fig. 3.16. Let the length of link 2 be $l_2$.

**(a)** Given that joint 2 and joint 3 axes are parallel to each other ($z_2$ is parallel to $z_3$), write the $4 \times 4$ transformation matrix that describes the coordinate frame 3 with respect to coordinate frame 2; that is, $T_{23} =?$
**(b)** Determine $T_{32}$, the description of coordinate frame 2 with respect to coordinate frame 3 (Hint: $T_{32} = T_{23}{}^{-1}$).

**10.** Consider the problem in Fig. 3.20. Let $l_1 = 0.5$ m, $l_2 = 0.25$ m, and $F$ be the weight of a payload mass of m = 100 kg, then $F = 9.81 \times 100$ N in the vertical negative direction of $Y_0$. Determine the joint torques necessary to hold the load in position.

**(a)** $\theta_1 = 0°, \theta_2 = 0°$
**(b)** $\theta_1 = 30°, \theta_2 = 30°$
**(c)** $\theta_1 = 90°, \theta_2 = 0°$

**11.** The objective of this problem is to illustrate the problem of *backlash* in motion control systems and how to deal with it. Consider the closed loop position control system shown in Fig. 10.4. Assume that the gear motion transmission mechanism is a lead-screw type (Fig. 3.3). Further, let us model components as follows:

- motor dynamics as inertia only ($J_m$) with no damping, and current to torque gain is $K_T$,
- position sensor connected to the motor that gives $N_{s1}$ [count/rev] number of counts per revolution,
- amplifier as a voltage to current gain ($K_a$) with all filtering effects neglected,
- lead-screw and the load it carries is modeled with its effective gear ratio ($N = 1/(2\pi p)$) where $p$ [rev/mm] is the pitch and mass $m$ (rotary inertia of the lead screw is neglected). Assume there is load force acting on the inertia ($F_l$). In addition, consider that the lead screw has a backlash of $x_b$, which we will assume is a constant value.
- control algorithm is implemented with an analog op-amp as a form of PID controller.

Assume the following numerical values for the system components: $J_m = 10^{-5} kg \cdot m^2$, $K_T = 2.0\, Nm/A$, $N_{s1} =$ 2000 count/rev, $K_a = 2A/V$, $p =$ 0.5 rev/mm, $m =$ 100 kg, $F_l = 0.0N$, $x_b =$ 0.1 mm. For simplicity, use the following relationship for the total inertia acting on the motor (although during the period of motion when backlash is in effect and the lead-screw is not moving the nut, the load inertia is not coupled to the motor, but we will neglect that) as well as motor torque and transmitted force to the moving mass,

$$J_t = J_m + \frac{1}{(2\pi p)^2} \cdot m$$

$$T_l = \frac{1}{2\pi p} \cdot F_l$$

**(a)** If the desired positioning accuracy of the load is 0.001 mm, draw a control system block diagram and sensors to achieve that. (Hint: Due to backlash, we must have a load-coupled position sensor. Let the resolution of that sensor be called $N_{S2}$ [counts/mm]. Further, we should have measurement accuracy that is 2 to 5 times better than desired positioning accuracy.)

**(b)** Develop a dynamic model of the closed loop control system (i.e., using Simulink). Simulate the motion in response to a rectangular pulse, i.e., initial position and commanded position are at zero until $t = 1.0$ sec, then a step position command of 1.0 mm, back to zero position command at $t = 3.0$ sec, and continue simulations until $t = 5.0$ sec. Use the motor-coupled position sensor for velocity loop with a P-only gain, and load-coupled position sensor in the position loop with a PD-type control. Adjust the gains in order to achieve good response.

**(c)** What happens if you use only the motor-coupled position sensor, not the load-coupled position sensor? Show your claim with simulation results. Modify component parameters if necessary to illustrate your point.

**(d)** What happens if you use only the load-coupled position sensor, not the motor-coupled position sensor? Show your claim with simulation results. Modify component parameters if necessary to illustrate your point.

CHAPTER 4

# MICROCONTROLLERS

**MOST** of the discussions in this chapter are based on PIC 18F452 microcontroller. The following manuals can be downloaded from http://www.microchip.com and should be used as a reference as part of this chapter:

1. PIC 18FXX2 Data Sheet (Users' Manual)
2. MPLAB IDE V6.xx Quick Start Guide
3. MPLAB C18 C Compiler Getting Started
4. MPLAB C18 C Compiler Users' Guide
5. MPLAB C18 C Compiler Libraries

## 4.1 EMBEDDED COMPUTERS VERSUS NONEMBEDDED COMPUTERS

The digital computer is the brain of a mechatronic system. As such, it is called the *controller* when used for the control function of an electromechanical system. Any computer with proper I/O interface devices (digital and analog I/O) and software tools can be used as a controller. For instance, a desktop PC can be used as a process controller by adding an I/O expansion board and control software to it. Clearly, there are many hardware components on a desktop PC (a nonembedded computer) that are not needed for process control functions. An embedded computer uses only the needed hardware and software components and is much smaller than a nonembedded computer, such as a desktop PC. An embedded computer used as the controller of a mechatronic system is referred to as the *embedded controller*. A microcontroller is the main building block of an embedded computer. In year 2000, the world market for 8-bit microcontroller was about 4 billion units. For 16-bit and 32-bit microcontrollers combined, it was about 1 billion units.

The main differences between *embedded* and *nonembedded* (i.e., desktop) computer-control systems are as follows (Fig. 4.1):

1. Embedded computers are generally used in real-time applications. Therefore, they have hard real-time requirements. *Hard* real-time requirement means that certain tasks must be completed within a certain amount of time, or the computer must react to an external event within a certain time. Otherwise, the consequences may be very serious. The consequences of not meeting the real-time response requirements in a desktop application are not as serious.
2. Embedded computers are not general-purpose computing machines, but have more specialized architectures and resources. For instance, a desktop computer would have hard disk drive, floppy disk drive, CD/DVD drive, tape drive for permanent data

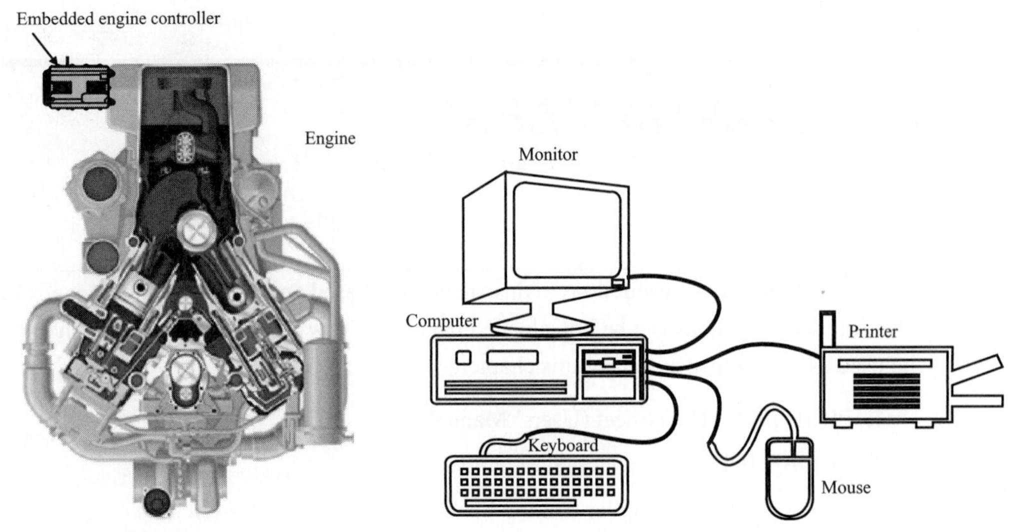


**FIGURE 4.1:** Comparison of an embedded computer and a nonembedded computer. The embedded computer has just enough resources for the application, must operate in harsh environments, has smaller physical size, and has hard real-time requirements.

storage, whereas an embedded computer may have battery backed RAM or flash or ROM memory to store just the application software. Embedded microcontrollers have limited resources in terms of power (i.e., it may be powered by a battery), memory and CPU speed. Embedded computers are dedicated to specific tasks. They do not store general purpose programs such as word processors, graphics programs, etc.

3. Embedded controllers and I/O interfaces are more integrated in the chip design level than general-purpose computers; i.e., I/O interface channels such as analog to digital (ADC) and digital-to-analog (DAC) converters are integrated into the microchip hardware.
4. The physical size of the embedded computer is typically required to be very small, which may be dictated by the application.
5. Embedded controllers operate in extreme environmental conditions (i.e., an embedded controller for a diesel engine must operate under conditions of large temperature and vibration variations).
6. Embedded computers invariably incorporate a *watchdog timer* circuit to reset the system in case of a failure.
7. Embedded microcontrollers may have dedicated debugging circuit on the chip so that the timing of all the I/O signals can be checked and application program debugged on the target hardware. Debugging tools (hardware and software) are a very important part of an embedded system development suite. Unlike desktop applications, the real-time embedded applications must be debugged and I/O signals must be checked for worst-case conditions.
8. The developer must know the details of the embedded system hardware (bus architecture, registers, memory map, interrupt system) because the application software development is influenced by the hardware resources of the embedded computer.
9. Interrupts play a very important role in almost all embedded controller applications. It is through the interrupts that the embedded controller interacts with the controlled process and reacts to events in real time.

10. As the complexity increases, embedded systems require a real-time operating system (RTOS). RTOS provides already tested I/O and resource management software tools. For instance, Ethernet communication may be provided by RTOS functions instead of writing it by ourselves. Furthermore, RTOS provides task scheduling, guaranteed interrupt latency, and resource availability.

Perhaps the most significant factors that differentiate embedded computers from desktop computers are the *real-time requirement*, *limited resources*, and *smaller physical size*. The real-time performance of an embedded system is defined by how fast it responds to interrupts and how fast it can switch tasks.

### 4.1.1 Design Steps of an Embedded Microcontroller-Based Mechatronic System

Design of a microcontroller-based mechatronic system includes the following steps. These steps include the microcontroller and its interface to the electromechanical system. We assume the electromechanical system is already designed.

*Step 1.* ***Specifications***—define the purpose and function of the device. What are the required inputs (sensors and communication signals) and required outputs (actuators and communication signals)? This is the hardware requirement of microcontroller. What are the logical functions required? This is the software requirement of a microcontroller.

*Step 2.* ***Selection***—select a proper microcontroller that has the necessary I/O capabilities to meet the hardware requirements as well as CPU and memory capabilities to meet the software requirements.

*Step 3.* ***Selection***—select a proper development tool set for the microcontroller, i.e., PC-based development software tools and hardware tools for debugging.

*Step 4.* ***Selection***—identify electronic components and subsystems necessary to interface the microcontroller to the mechatronic system.

*Step 5.* ***Design***—complete schematics for the hardware interface circuit, including each component identification and its connections in the interface circuit.

*Step 6.* ***Design***—write the pseudocode of the application software structure, its modules and flow chart.

*Step 7.* ***Implement***—build and test the hardware.

*Step 8.* ***Implement***—write the software.

*Step 9.* ***Implement***—test and debug the hardware and the software.

### 4.1.2 Microcontroller Development Tools

Typical development tools used in an embedded controller development environment are grouped into two categories: (1) hardware tools and (2) software tools (Fig. 4.2). The hardware tools include:

1. A desktop PC to host most of the development software tools
2. The target processor (i.e., evaluation board or the final target microcontroller hardware)
3. Debugging tools such as ROM emulator, logic analyzer
4. EEPROM/EPROM/Flash writer tools

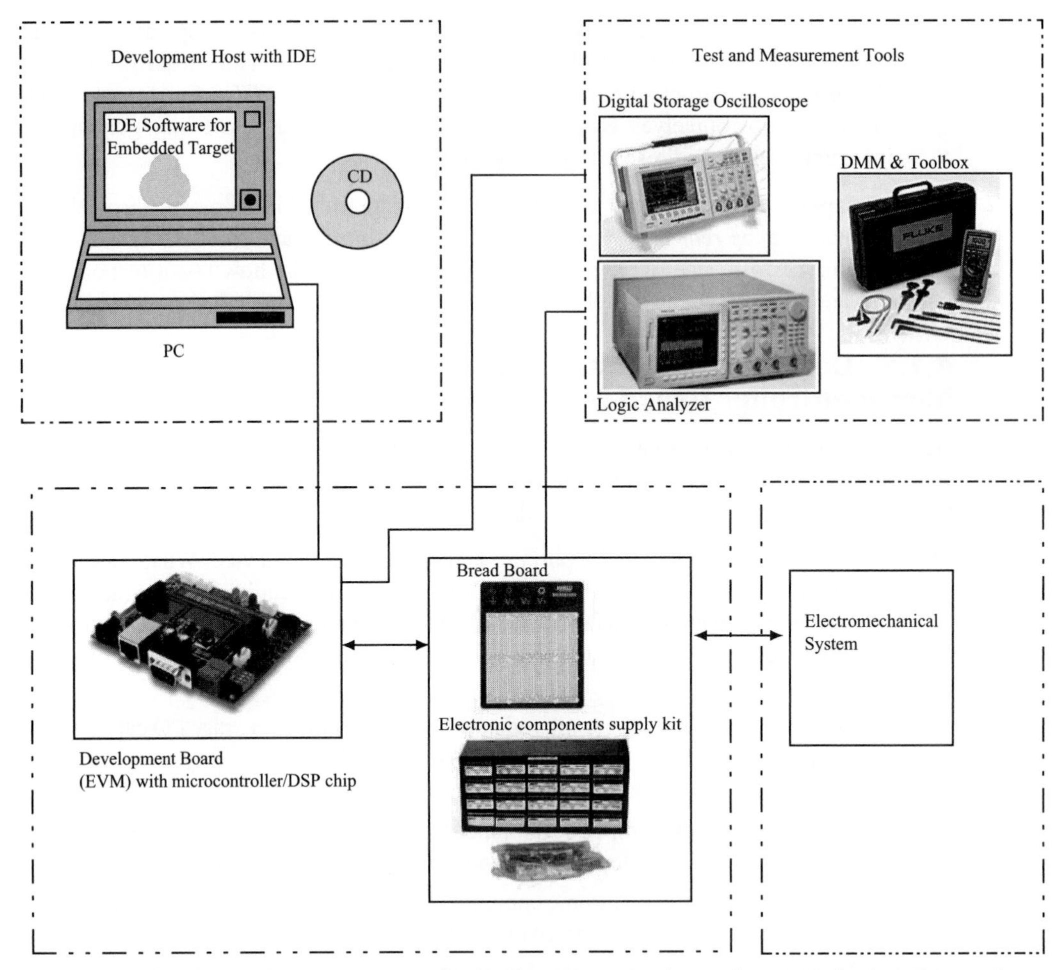


**FIGURE 4.2:** The components of a development setup for a microcontroller-based control system: PC as host development environment including the development software tools for the microcontroller, communication cable, microcontroller board, breadboard, test and measurement tools, and electronic components supply kit.

The software tools include the target processor specific

1. compiler, linker, and debugger
2. real-time operating system (not required)

The compiler includes a *start-up code* for the embedded system to boot-up the system on RESET, check resource integrity, and load the application program from a known location and start its execution. Furthermore, compiler generates *relocatable code*, which is then used by the "linker" to specifically locate it in physical memory based on the selections made in the application file for the linker. This file is prepared by the developer to properly use the target system memory resources. The linker decides where in memory to place the program code and the data.

In particular, the debugging tools deserve special attention. The debugging requirements for embedded applications are significantly more stringent that those of desktop applications. The reason is that the consequences of a failure are much more serious in real-time systems.

The simplest form of debugging tool includes a debug kernel on the target system which communicates with a more comprehensive debugging software on the host computer. More recent embedded controllers include on-chip debugging circuitry which improves the debugging capability. An ROM emulator allows us to use RAM in place of ROM during the development phase so that the host can change the target code and write to the target hardware quickly.

### 4.1.3 Microcontroller Development Tools for PIC 18F452

The development board for PIC 18F452 is the PICDEM 2 Plus Demo board plus various peripheral hardware devices that allow the PIC 18F452 microcontroller to effectively interface with electromechanical devices. These include proper power supply through a 9-V AC/DC adapter and regulator, a 4-MHZ clock, RS-232 interface, LCD, four LEDs, and prototyping area.

The MPLAB ICD 2 module is a low-cost debug and development tool that connects between the PC and the PICDEM 2 Plus Demo board (or the designers target board) allowing direct in-circuit debugging of PIC 18F452 microcontroller (Fig. 4.2). Programs can be executed in real time or single-step mode, watch variables established, break points set, memory read/writes accomplished, and more. It is also used as a development programmer for the microcontrollers to download and save the code in the microcontroller memory.

Hardware development tools are:

1. PC as the host for development tools
2. Development board (PICDEM 2 Plus Demo board), which has the microcontroller (PIC 18F452), its power supply, support circuits, and space for breadboard for custom application specific hardware interface development
3. Communication cable and in-circuit debugger hardware (MPLAB ICD 2)

Software development tools that run on a PC and interfaced to the board via the communication cable are:

1. MPLAB IDE V.6xx integrated development environment (IDE), which includes editor, assembler (MPASM), linker (MPLINK), debugger, and software simulator (SIM) for the PIC chip.
2. MPLAB C18 C-complier (works under MPLAB IDE V.6xx).

The PIC microcontroller can be programmed in a number of ways, using both C-language and assembly language. Our lab experiments require the use of the MPLAB IDE together with the MPLAB C18 ANSI-compliant C compiler installed on a PC. The language tool suite also consists of the MPASM assembly language interpreter, the MPLINK object linker and debugger. The C18 complier supports mixing assembly language and C language instructions in the same file. A block of assembly code included in a C source file must be labeled by:

```
_asm

  ....
_endasm
```

The basic sequences for programming the PIC, using a PC as the development tool, are as follows:

1. On the PC, create a new project in the MPLAB IDE environment. Set up the project environment by selecting the target PIC microcontroller, specify directories to access libraries, add source codes to project, and specify project name. For each project, the following configurations must be made in the MPLAB IDE environment:
   - **(a)** Select the target PIC microcontroller: `MPLAB IDE>Configure>Select Device: PIC18F452`
   - **(b)** Configure project options: `MPLAB IDE>Project>Select Language Toolsuite: Microchip C18 Toolsuite`
   - **(c)** Configure project options: `MPLAB IDE>Project>Set Language Tool Locations: MPLAB C18 C Compiler and .... >Set Language Tool Locations: MPLINK Object Linker`
   - **(d)** Configure project build options: `MPLAB IDE >Project>Build Options..>Project`, then set up the options in the window under different tabbed pages (options: General, Assembler, Compiler, Linker options)
   - **(e)** Add source files to the project: on project window, right click on "Source Files" and select "Add Files"
   - **(f)** Select a script file for the linker: on project window, right click on "Linker Scripts," and select "Add Files," and then select file "18f452.lkr" from the "lkr" directory.
2. Write the program source code in C language, using the built-in editor or any ASCII text editor, and save it as *filename.c*. Add other relevant files to your project.
3. Build (compile and link) the project in the MPLAB IDE environment. This converts the high-level C code to the corresponding hex files which contain the binary coded machine instructions: `MPLAB IDE>Project>Build All`.
4. First debug the program using the software simulator (SIM, provided as part of MPLAB IDE) for the PIC chip on the PC. This is a nonreal time simulation of the PIC chip. Once the program is debugged using the nonreal time simulator (MPLAB SIM), it can be transferred to the PIC microcontroller. If programming in C language, the program must be re-built using a different linker script file (*18f452i.lkr* file for building a program to run on the actual PIC 18F452 chip, instead of *18f452.lkr*, which is used for debugging the program on the PC). Then transfer the program from the PC to the PIC board through the MPLAB IDE environment and communication cable (MPLAB ICD 2):

   ```
   MPLAB IDE>Programmer>Select programmer and
   ...>Settings and ...>Connect.
   ```

   Debugging commands for nonreal time software debugger (MPLAB SIM) and hardware debugger MPLAB ICD2 are largely the same. Typical debugging commands to find errors in the program are similar to debugging commands for other high-level programming languages. Typically we must be able to:
   - **(a)** Run, Halt, Continue the program execution.
   - **(b)** Single step, Step into the function, Step over functions.
   - **(c)** Set up break points at various lines in the program. When the program reaches that line, the program execution will halt.
   - **(d)** Set up "watch window" to view the values of selected variables; i.e., when the program stops at a break point, values of various variables can be examined to check for errors.

**(e)** Set up the ability to trace (also called code trace) all operations of the program for a finite number of clock cycles or until a break point is reached. This is a very thorough view of what happens in the program.

**5.** Give the run command to the PIC chip from the PC. Debug the code on the PIC chip with software and hardware debugging tools (i.e., MPLAB ICD2). When debugging a program, disable the watchdog timer (WDT). Watchdog timer can be enabled/disabled from the IDE menus. Otherwise, while the program is paused for debugging, the WDT will reset the processor.

## 4.2 BASIC COMPUTER MODEL

Let us consider the operation of a basic computer using a human analogy (Fig. 4.3). As shown in the Fig. 4.3, the human has the brain to process information, eyes to read, hands to reach various components, and fingers to write. There is also a clock. On the desk, there is an deck of cards which has the instructions to follow, a chalk, an eraser, a blackboard, input–output trays, and two pockets with one card each for quick access to read/write things on.

The analogy between this human model and a computer is as follows:

```
brain                        --- CPU,
wall clock                   --- clock,
deck of instruction cards    --- read only memory (ROM)
chalk-eraser-black-board     --- random access memory (RAM)
pocket cards                 --- accumulators (also called
                                 registers)
input-output tray            --- I/O devices
eyes, hands and arms         --- bus to access resources
                                 (read/write)
```

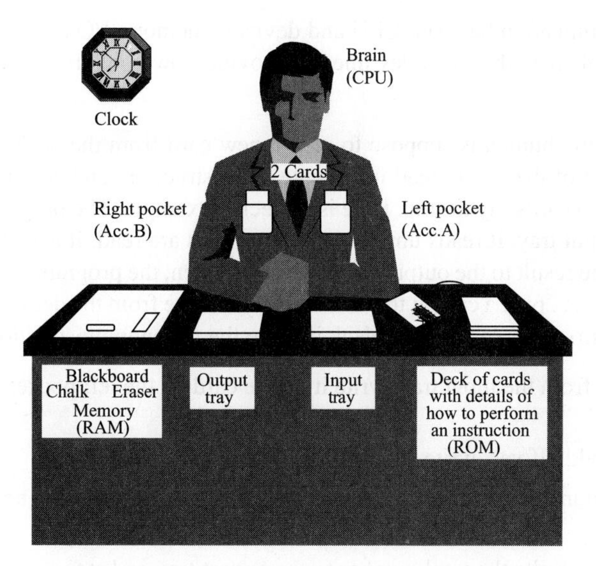


**FIGURE 4.3:** Basic computer and human analogy.

There are seven basic components of a computer:

1. CPU, which is the brain of the computer. The CPU is made of a collection of arithmetic/logic units and registers. For instance, every CPU has:
   (a) A program counter (PC) register, which holds the address of the next instruction to be fetched from the memory.
   (b) An instruction decoder register, which interprets the fetched instruction and passes on the appropriate data to other registers.
   (c) An arithmetic logic unit (ALU), which is the "brain inside the brain," executes the mathematical and logical operations.
2. Clock—a computer executes even the simplest instruction at the tick of a new clock cycle. The clock is like the heart of a human body. Nothing happens in the computer without the clock.
3. ROM—read-only memory, which contains information on how to execute the basic instruction set. It can only be read from and cannot be written to. It maintains data when power is lost. EPROM is an erasable programmable ROM. An EPROM chip has a window where the memory can be erased by exposing it to ultraviolet light and reprogrammed. EEPROM is electrically erasable programmable ROM where the memory can be rewritten by electrical signals in the communication interface without any ultraviolet light.
4. RAM memory—random access memory serves as an erasable blackboard where the information can be read from and written to. Data are lost if power is lost. *Static* RAM and *dynamic* RAM are two common RAM type memory, Static RAM stores data in flip-flop circuits and does not require a refresh-write cycle to hold the data as long as power is not lost. Dynamic RAM requires periodic refresh-write cycle to hold the data even when power is maintained.
5. Registers are a few specific memory locations which can be accessed faster than the other RAM memory locations.
6. I/O devices—every computer must interact with a user and external devices to perform a useful function. It must be able to read in information from the outside world, process it, then output information to the outside world. An I/O device interface chip is also called *peripheral interface adaptor* (PIA).
7. Bus to allow communication between CPU and devices (memory, I/O devices). The bus includes a set of lines that includes lines to provide power, address, data, and control signals.

Let us assume that the human is suppose to pick a new card from the deck (ROM) at the tick of every minute of the clock, read it, execute the instruction, and continue this process until the instruction cards say to stop. Here is a specific example. The program is to read numbers from the input tray. It reads until five odd numbers are read. It adds the five odd numbers and writes the result to the output tray at the end. Then, the program execution stops. Minute: 1, 2, 3, 4, . . . . . pick a card at the tick of every minute from the deck of cards (ROM). Assume that the first four cards in the deck include the following instructions:

***Card 1:*** Read a number from the input tray, write it on the card on the left pocket (Accumulator A).

***Card 2:*** Is the number odd? If yes, write it to the black board.

***Card 3:*** Are there five numbers on the black board? If yes - go to Card 4, if no- go to Card 1.

***Card 4:*** Add all numbers, write the total number to the output tray and stop.

Notice the following characteristics in a computer program:

- Normally the program instructions are executed sequentially.
- Order of execution can be changed using the conditional statements.
- CPU, clock, ROM, RAM and accumulators, I/O are the key components of a basic computer operation.

The digital computer is a collection of many ON/OFF switches. The transistor switches are so small that 1,000 $\times$ 1,000 array of them (1,000,000 transistor switches) can be built on a single chip. A combination of transistor switches can be used to realize various logic (i.e., AND, OR, XOR) functions as well as mathematical operations ($+, -, *, /$).

Every CPU has an instruction set that it can understand. This is called the *basic instruction set* or the *machine instructions*. Each instruction has a unique binary code that tells the CPU what operation to perform and programmable operands for the source of data. Some microprocessors are designed to have a smaller instruction set. They have fewer instructions but execute faster than general-purpose microprocessors. Such microprocessors are called *reduced instruction set computers* (RISC).

*Assembly language* is a set of mnemonic commands corresponding to the basic instruction set. The mnemonic commands make it easier for programmers to remember the instructions instead of trying to remember their binary code. A program written in assembly language must be converted to machine instructions before it can be run on a computer. This is done by the *assembler*.

All of the instructions in a program written in a high-level language must first be reduced to the combinations of the basic instruction set the CPU understands. This is done by the compiler and the linker which translate the high-level instructions to low-level machine instructions. Basic instruction set is microprocessor specific. The build process (compile and link) of C18 compiler generates various files including the executable file with extension "*.hex." In addition, the filename extensions "*.map" contains the list of variable names and the allocated memory address for them. The filename extension "*.lst" contains the machine code (disassembled) generated for each line of the assembly and C-code in a program. These files can be useful in the debugging process.

High-level programming languages try to hide the processor specific details from the programmer. This allows the programmer to program different microprocessors with a single high-level language. In microcontroller applications, it is generally necessary to understand the hardware and assembly language capabilities of a particular microcontroller in order to fully utilize its capabilities. A computer program minimally needs the following machine instructions:

- Access memory (and I/O devices), and perform read/write operations between the CPU registers and the memory (and I/O devices). These operations are typically called

  ```
  LOAD address
  STORE address
  ```

- Mathematical (add, subtract, multiply) and logical operations, such as

  `ADD address` (implied to be added to the content of one the accumulators),
  `SUB address` (implied to be subtracted from the content of one the accumulators),
  `AND, OR, NOT, JUMP, CALL, RETURN`

In high-level languages, data structures (variables, structures, classes, etc.) are used to manage the information. Assembly language does not provide data structures to manage

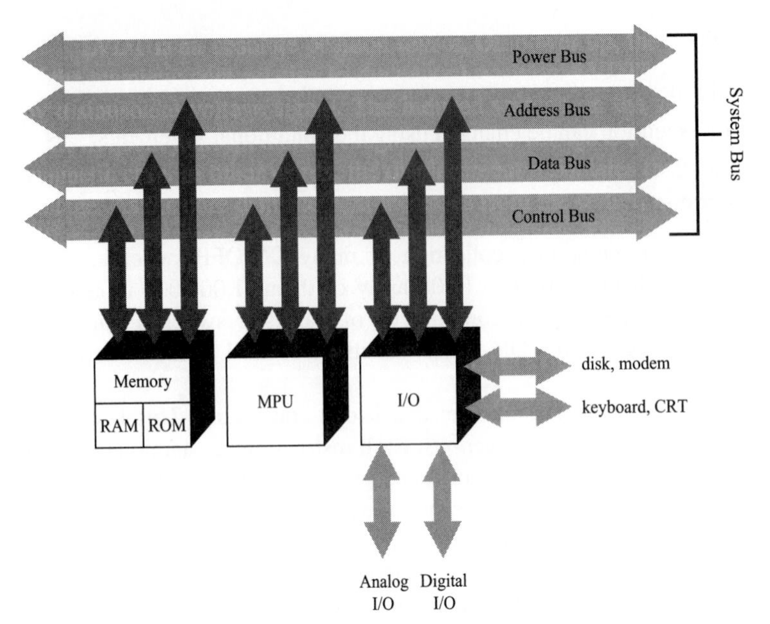


**FIGURE 4.4:** General functional grouping of lines in a computer bus: power bus, address bus, data bus, and control bus.

data in a program. It is up to the programmer to define variables and allocate proper memory address space for them. Assembly language provides access instructions to the memory and I/O devices as a source and destination for data. In addition it provides operators (mathametical, logical, etc.) as well as various decision-making instructions such as GOTO, JUMP, CALL, RETURN. Using these instructions, we operate on the data in memory and I/O devices.

The computer CPU accesses the resources (ROM, RAM, I/O devices) through a bus. The hardware connection between the components of a computer is the *bus*. The bus has four basic groups of lines (Fig. 4.4).

1. power bus
2. control bus
3. address bus
4. data bus

The power bus provides the power for the components to operate. Each line carries a TTL level signal: 0 VDC or 5 VDC (OFF or ON). The control bus indicates whether the CPU wants to read or write. The address bus selects a particular device or memory location by specifying its address. The data bus carries the data between the devices. Notice that if address bus is 16 lines (16-bit), there can be $2^{16} = 65{,}536$ distinct addresses the CPU can access.

An I/O operation between the CPU and the memory (or I/O devices) involves the following steps at the bus interface level:

- CPU places the address of memory or I/O device on the address bus
- CPU sets the appropriate control lines on the control bus to indicate whether it is input or output operation

- CPU reads the data bus if it is an input operation, or write the data to the data bus if it is a write operation

In short, performing I/O between the CPU and the other computer devices involves controlling all three components of the bus: (1) address bus, (2) control bus, and (3) data bus. In a high-level language such as C, we have a line of code as follows:

```
x = 5.0 ;
```

The compiler generates the necessary code to assign a memory address for x, place the address of it in the address bus, turn on the write control lines on the control bus, and write the data "5.0" on the data bus.

## 4.3 MICROCONTROLLER HARDWARE AND SOFTWARE: PIC 18F452

The term *microprocessor* refers to the central processing unit (CPU) and its on-chip memory. The term *microcontroller* ($\mu$C) refers to a chip that integrates the microprocessor and many I/O device interfaces such as ADC, DAC, PWM, digital I/O, and the communication bus (Fig. 4.5). A microcontroller is an integrated microprocessor chip with many I/O interfaces. It results in small foot print (physical size) and low cost. In microcontroller applications, the amount of memory typically needed is in the order of tens of kilobytes, as opposed to hundreds of megabytes memory commonly available in desktop applications. As a result, the cost of microcontrollers is less compared to general-purpose computers, which makes them good candidates for embedded controller applications.

### 4.3.1 Microcontroller Hardware

As users of microcontrollers in mechatronic systems, we need to understand the hardware of microcontrollers. We will study that from the inside out. In the discussion, we are interested in the functionality of the microcontroller components as opposed to how they are designed or manufactured.

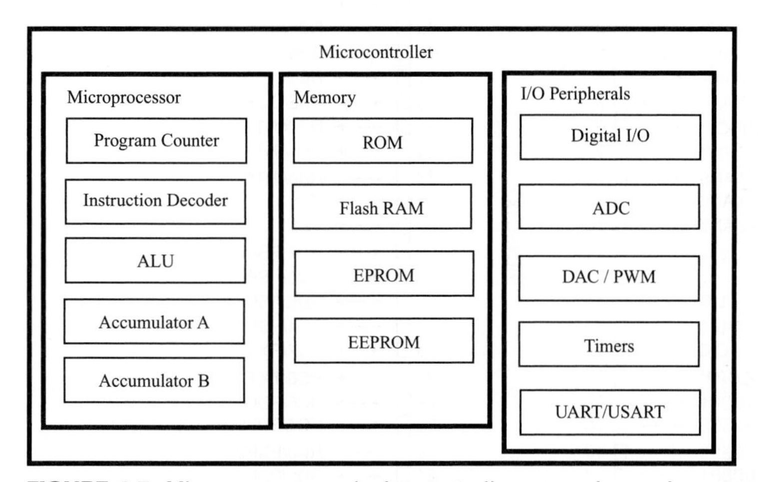


**FIGURE 4.5:** Microprocessor and microcontroller comparison: microcontroller includes a microprocessor and I/O peripherals on the same chip.

The main hardware features that one needs to understand for any microcontroller or digital signal processor (DSP) are as follows:

1. Pin-out on the chip that identifies the role of each pin
2. Registers in the CPU that define the "brain" structure and internal workings of the microprocessor
3. Bus structure which defines how the CPU communicates with the rest of the memory and I/O resources
4. Support chips such as real-time clock, watch-dog timer, interrupt controller, programmable timers/counters, analog to digital converter (ADC), and pulse width modulation (PWM) module

A microcontroller is a single-chip integrated circuit (IC) computer. PIC stands for peripheral interface controller. PIC is a trade name given to a series of microcontrollers manufactured by Microchip Technology, Inc. We will be using the *PIC 18F452* chip for the laboratory experiments. The PIC 18F452 has five bidirectional input–output ports, named Ports A to E. Port A is 7-bit-wide port while Port E is 3 bits wide (Fig. 4.6). The other ports, B, C, D, are all 8-bit ports. Port pins are labeled RA0 through RA6, RB0 through RB7, RC0 through RC7, RD0 through RD7, and RE0 through RE2. Hence total number of pins at ports A through E is $7 + 8 + 8 + 8 + 3 = 34$ out of 40 pin DIP package of PIC 18F452 chip (Fig. 4.6). The remaining pins are used for $V_{DD}$ (two pins), $V_{SS}$ (two pins), *OSC1/CLK1, MCLR/V*$_{PP}$. Most of the pins are software configurable for one of multiple functions between general-purpose I/O and peripheral I/O. Using registers in the chip, one function is selected for each pin under software control.

The PIC 18F452 is an 8-bit microcontroller based on the Harvard architecture. It is available in 40-pin DIP, 44-pin PLCC, and 44-pin TQFP packages. In the 44-pin packages, four of the pins are not used and are labeled as NC (Fig. 4.6). The Harvard architecture separates the program memory and the data memory. During a single instruction cycle, both program instructions and data can be accessed simultaneously. The PIC 18F452 contains a RISC processor. This particular RISC processor has a 75-word basic instruction set.

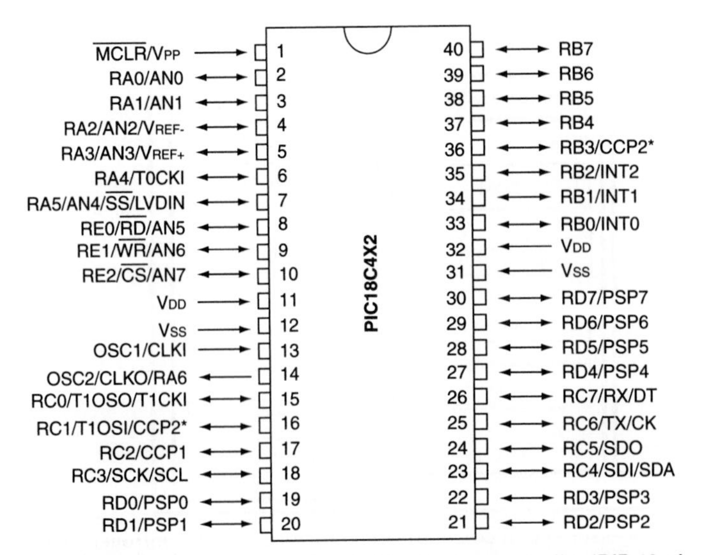


**FIGURE 4.6:** Pin diagrams of PIC 18F452 microcontroller (DIP 40-pin model shown).

The PIC 18F452 chip is compatible with clock speeds between 4 MHz and 40 MHz. The clock speed is determined by two factors:

1. ***Hardware:*** the external crystal oscillator or ceramic resonators along with a few capacitors and resistors,
2. ***Software:*** configuration register settings to select the operating mode of the processor.

PIC 18F452 supports up to 31 levels of stack, which means that up to 31 subroutine calls and interrupts can be nested. This limitation is removed by software handling of interrupts by the C18 compiler. Stack space in memory holds the return address from the function calls and from interrupt service routines. Stack space is neither part of the program memory nor data memory.

The program counter (PC) register is 21 bits long (Fig. 4.7). Hence, the address space for program memory can be up to 2 MB. It is flash memory based and has 32 KB of FLASH program memory (addresses 0000h through 7FFFh) which is 16K 2-byte (single-word) instruction space (Fig. 4.7). Data memory is implemented in RAM and EEPROM. Data RAM has 4096 (4-KB) bytes. EEPROM memory space is 256 bytes. In C-language using C18 compiler, using "rom" directive, data variables can be explicitly directed to be allocated in program memory space if data memory space is not sufficient for a given application. Similarly, "ram" directive in data declarations allocates memory in data memory.

```
rom char c ;
rom int n ;

ram float x ;
```

Some important locations in the program memory map are as follows:

1. RESET vector is at address 0x0000h.
2. High-priority interrupt vector is at address 0x0008h.
3. Low-priority interrupt vector is at address 0x0018h.

RESET condition is the start-up condition of the processor. When the processor is RESET, program counter content is set to 0x0000h; hence, the program branches to this address to get the address of the instruction to execute. C-compiler places the beginning address of **main()** function at this location. Hence, on RESET, the **main()** function of a C-program is where the program execution starts. The source of RESET can be:

1. ***Power-on-reset (POR)*:** When on-chip $V_{DD}$ rise is detected, POR pulse is generated. On power-up, internal power-up timer (PWRT) provides a fixed time-out delay to keep the processor in RESET state so that $V_{DD}$ rises to an acceptable and stable level. After the PWRT time-out, oscillator start-up timer (OST) provides another time-delay of 1024 cycles of oscillator period.
2. ***MCLR reset during normal operation or during SLEEP*:** MCLR input pin can be used to RESET the processor on demand.
3. ***Brown-out reset (BOR)*:** When the $V_{DD}$ voltage goes below a specified voltage level for more than a certain amount of time (both parameters are programmable), the BOR reset is activated automatically.
4. ***Watchdog timer (WDT) reset*:** WDT can be used to reset the processor if it times-out during normal operation in order to clear the state of the processor or it can also be used to wake-up the processor from SLEEP mode. In WDT reset during normal operation,

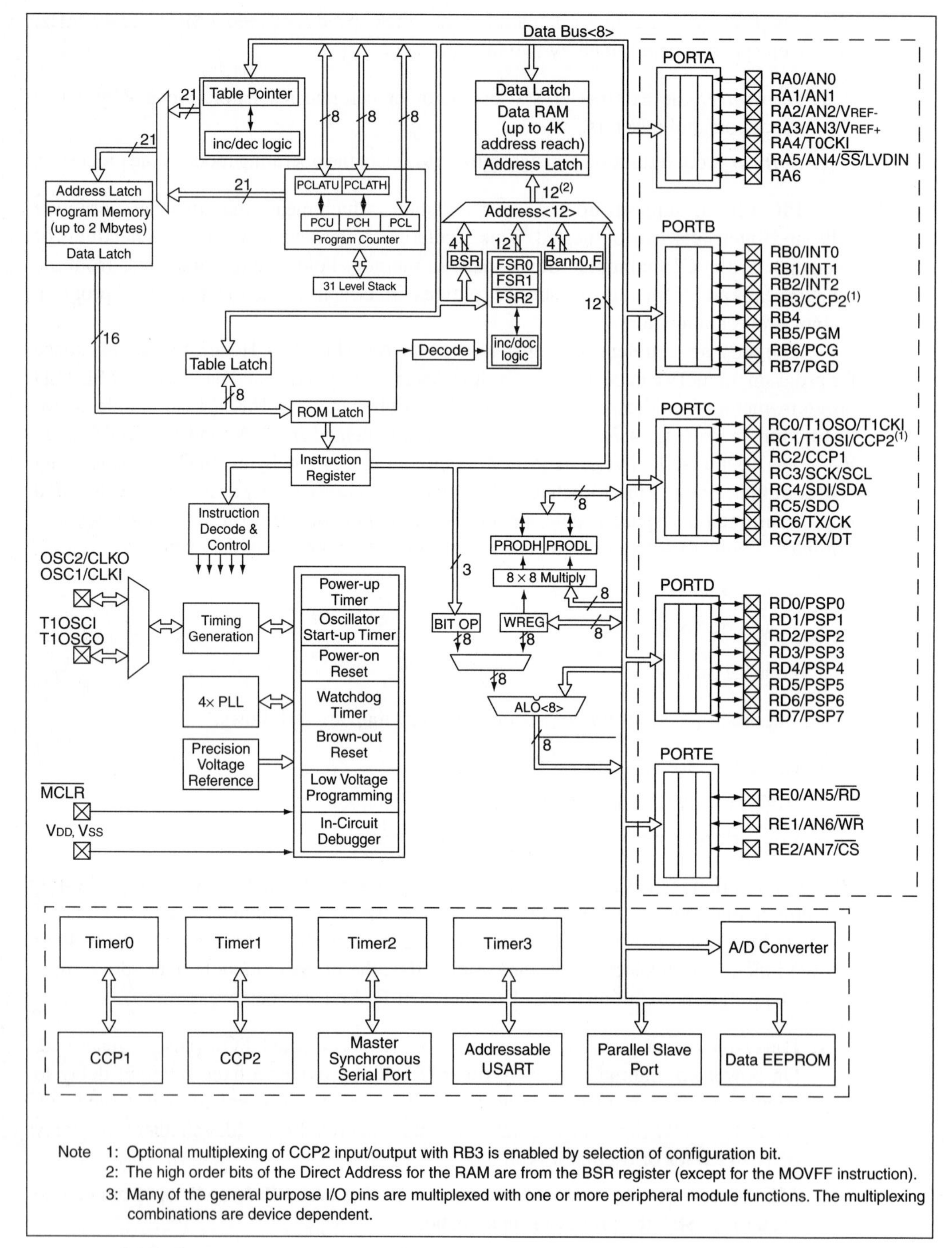


Note 1: Optional multiplexing of CCP2 input/output with RB3 is enabled by selection of configuration bit.
2: The high order bits of the Direct Address for the RAM are from the BSR register (except for the MOVFF instruction).
3: Many of the general purpose I/O pins are multiplexed with one or more peripheral module functions. The multiplexing combinations are device dependent.

**FIGURE 4.7:** Block diagram of PIC18F452: registers, bus, peripheral devices, and ports (Fig. 1-2 from PIC 18Fxx2 Data Sheet).

program counter is initialized to 0x0000h, whereas in WDT wake-up mode, program counter is incremented by 2 to continue to instructions where it was left.

5. RESET instruction, stack full and stack underflow conditions result in software initiated reset conditions. However, the RESET condition due to stack overflow or underflow condition can be disabled in software.

When PIC 18F452 executes SLEEP instruction, the processor is held at the beginning of an instruction cycle. The processor can be woken-up from SLEEP mode by either external RESET, Watchdog timer reset, or an external interrupt. RCON register, which is part of the special function registers (SFRs), holds the information that allows the application software to determine the source of RESET.

## 4.3.2 Microprocessor Software

***Addressing Modes*** Data memory space in PIC 18F452 is accessed by a 12-bits-long address space, which means that it is 4096 bytes long. This space is organized as 16 banks of 256 bytes of memory (Fig. 4.8). Banks 0–14 are used as General Purpose Registers (GPRs). GPRs can be used for general data storage. The upper half of bank 15 (F80h-FFFh) is used as Special Function Registers (SFR). Lists of all registers of PIC 18F452 are provided in the PIC 18F452 Users' Manual. SFRs are used for configuration, control, and status information of the microcontroller. GPRs are used for application program data storage. Most features of the microcontroller are configured by properly setting the SFRs. EEPROM data memory space is 256 bytes and is accessed indirectly through SFRs. SFRs are in two groups: one group is associated with core functions of CPU, the other group is associated with peripheral I/O functions. There are four SFRs involved in accessing EEPROM: EECON1, EECON2,

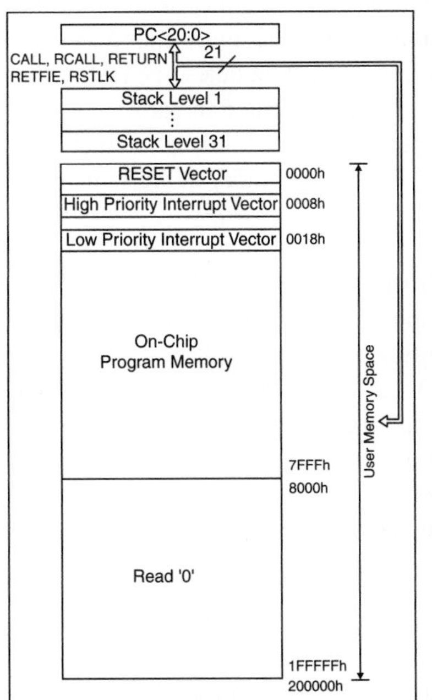


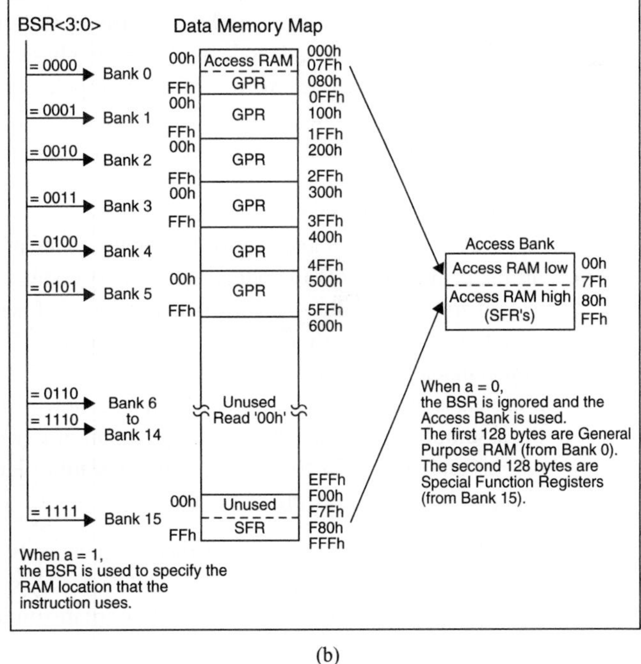


(a) (b)

**FIGURE 4.8:** Memory map of PIC 18F452: (a) program memory map (Fig. 4-2 from PIC 18Fxx2 Data Sheet), (b) data memory map (Fig. 4-7 from PIC 18Fxx2 Data Sheet).

EEDATA, and EEADR. EEDATA and EEADR registers are used to hold the data and address for EEPROM. EECON1 and EECON2 registers control the access to EEPROM.

There are two addressing modes: (1) *direct addressing* and (2) *indirect addressing*. In *direct addressing mode*, Bank Select Register (BSR which is at FE0h address in SFR section of Bank 15 in data RAM) bits 3:0 (4 bits) are used to select one of 16 banks, and the bits 11:4 (8 bits) from the opcode are used to select one of 256 memory locations in the selected bank.

In *indirect addressing mode*, the address of the data memory is not fixed in the instruction. Rather, the address is obtained from the content of a file select register (FSR) as a pointer to a RAM data memory location. The instruction simply contains a pointer to the address location where the address is to be obtained. Since the address is obtained through a pointer indirectly, it is called the indirect addressing mode. Notice that the content of the memory location which holds the address can be modified by the program; hence, different memory locations can be accessed based on program logic. Indirect addressing uses File Select Register (FSR), which is a 12-bit register and Indirect File Operand Register (INDFn). Each FSRx register is a pointer. There are three FSR registers FSR0, FSR1, FSR2 and associated INDF0, INDF1, INDF2 registers.

***Instruction Set*** PIC 18Fxxx chip instruction set has 75 *instructions* (see PIC 18F452 Users' Manual for the list and description of the basic instruction set). All, except three instructions, are 16-bit, single-word instructions. Three instructions are double word (32-bit). Each instruction has a unique binary code. When the microcontroller decodes the instruction code, the dedicated hardware circuit performs a function, i.e., increment the value in accumulator, add two numbers, compare two numbers. An instruction cycle is equal to four oscillator cycles. These four oscillator cycles per instruction cycle are called Q1, Q2, Q3, and Q4 cycles. Therefore, if a 4-MHz oscillator is used, an instruction cycle is 1 $\mu$sec.

Obtaining an instruction from memory is called *fetch*. Decoding it to determine what to do is called *decode*. And performing the operation of the instruction is called *execute*. Each one of these fetch, decode, and execute phase of the instruction takes one instruction cycle. The fetch-decode-execute cyles are pipelined, meaning executed in parallel, so that each instruction executes in one instruction cycle. In other words, while one instruction is decoding and executing, another instruction is fetched from the memory at the same time. Single-word instructions typically take one instruction cycle, and two-word instructions take two instruction cycles to execute.

The CPU understands only the binary codes of its instruction set. All programs must ultimately be reduced to a collection of binary codes supported by its instruction set, specific to that microcontroller. However, it is clear that writing programs with binary codes is very tedious. *Assembly language* is a collection of three to five characters that are used as *mnemonics* corresponding to each instruction. Using assembly language instructions is much easier than using binary code for instructions. The *assembler* is a program that converts the assembly language code to the equivalent binary code (machine code) instructions that can be understood by the CPU. At the assembly language level, the instructions can be grouped into the following categories:

1. Data access and movement instructions used to read, write, and copy data between locations in the memory space. The memory space locations can be registers, RAM, peripheral device registers. Examples of instructions in this category are: MOV, PUSH, POP.
2. Mathematical operations: add, subtract, multiply, increment, decrement, shift left, shift right. Example instructions include ADDWF, SUBWF, MULWF, INCF, DECF, RLCF, RRCF.

3. Logic operations such as AND, OR, inclusive OR. Example PIC 18F452 instructions include ANDWF, IORWF, XORWF.
4. Comparison operations: Example instructions include CPFSEQ (=), CPFSGT (>), CPFSLF (<).
5. Program flow control instructions to change the order of program execution based on some logical conditions: GOTO, CALL, JUMP, RETURN.

In order to effectively use assembly language, one must very closely understand the hardware and software architecture (pinout, bus, registers, addressing modes) of a particular microprocessor. High-level languages, such as C, attempt to relieve the programmer from the details of tedious programming in assembly language. For instance, in order to perform a long mathematical expression in assembly language, we would have to write a sequence of add, subtract, and multiply operations, whereas in C we can directly type the expression in a single statement. Although the basic mathematical and logical operations are expressed the same way for different processors in C, specific hardware capabilities of a processor cannot be made totally generic. For instance, one must understand the interrupt structure in a particular microcontroller in order to be able to make use of it even with C programming language. Quite often, I/O-related operations require access to certain registers that are specific to each microcontroller design. Therefore, a good programmer needs to understand the architecture of a particular microcontroller even if a high-level language such as C is used as the programming language.

### 4.3.3 I/O Peripherals of PIC 18F452

The input/output (I/O) hardware connection to the outside world is provided by the pins of the microcontroller. Figure 4.6 shows the pins of PIC 18F452 and their functions.

***I/O Ports*** There are five ports on PIC 18F452: PORTA (7-pin), PORTB (8-pin), PORTC (8-pin), PORTD (8-pin), and PORTE (3-pin). Functionality of each pin is software selectable among two or three functions (i.e., select between digital I/O or external interrupt 0 for pin RBO or select one of three functions, among digital I/O, analog input 2, AID reference input for pin RA2). Each port has three registers for its operation. They are:

- TRISx register, where 'x' is the port name A, B, C, D, or E. This is the *data direction* or *setup register*. The value of the TRISx register for any port determines whether that port acts as an input (i.e., reads the value present on the port pins to the data register of the port) or an output (writes contents of the port data register to the port pins). Setting a TRISx register bit (= 1) makes the corresponding PORTx pin an input. Clearing a TRISx register bit (= 0) makes the corresponding PORTx pin an output.
- PORTx register. This is the data register for the port. When we want to read the status of the input pins, this register is read. When we write to this register, its content is written to the output latch register.
- LATx register (output latch). The content of the latch register is written to the port. The data latch (LAT register) is useful for read-modify-write operations on the value that the I/O pins are driving.

As an example, the code below sets all pins of PORT C as output, and all pins of PORT B as input.

```
TRISC = 0; /* Binary 00000000 */
PORTC = value ; /* write value variable to port C
```

```
TRISB = 255; /* Binary 11111111 */
value = PORTB; /* read the data in port B to variable
                  value.
```

MPLAB C18 compiler provided C-library functions to set up and use I/O ports. Port B interrupt-on-change and pull-up resistor functions are enabled/disabled using these functions.

```
#include <portb.b>

OpenPORTB (PORTB_CHANGE_INT_ON & PORTB_PULLUPS_ON) ;
                      /* Configure interrupts and internal
                         pull-up resistors on PORTB */
ClosePORTB ();        /* Disable ...........................
                         ...................... */

OpenRBxINT (PORTB_CHANGE_INT_ON & RISING_EDGE_INT &
  PORTB_PULLUPS_ON);
                      /* Enable interrupts for PORTB pin x */
CloseRBxINT () ;      /* Disable ........................ */

EnablePullups() ;     /* Enable the internal pull-up
                         resistors on PORTB */
DisablePullups();     /* Disable ...........................
                         .......... */
```

***Capture/Compare/Pulse Width Modulation (PWM)*** The capture/compare/PWM (CCP) module is used to perform one of three functions: (1) capture, (2) compare, or (3) generate PWM signal. PIC 18F452 has two CCP modules: (1) CCP1 and (2) CCP2. The operation of both these modules is identical, with the exception of special event trigger. The operation of CCP involves the following:

1. Input/output pin: RC2/CCP1 and RC1/CCP2 pins for CCP1 and CCP2, respectively. For capture function, this pin must be configured as input; for compare and PWM functions, it must be configured as output.
2. Timer source: In capture and compare mode, TIMER1 or TIMER3, and in PWM mode TIMER2 can be selected as the timer source.
3. Registers: Used to configure and operate the CCP1/CCP2 modules.
   **(a)** CCP1CON (or CCP2CON) register: 8-bit register used to configure the CCP1 (CCP2) module and select which operation is desired (capture, compare or PWM).
   **(b)** CCPR1 (CCPR2) register: 16-bit register used as the data register. In capture mode, it holds the captured value of TIMER1 or TIMER3 when the defined event occurs in RC2/CCP1 pin. In compare mode, it holds the data value being constantly compared to the TIMER1 and TIMER3 value, and when they match, output pin status is changed or interrupt is generated (the action taken is programmable via the CCP1CON or CCP2CON register settings). In PWM mode, CCP1 (CCP2) pin provides the 10-bit PWM output signal. PWM signal period information is encoded in the PR2, and duty cycle in CCP1CON bits 5:4 (2 bits) and CCPR1 (8 bits) registers which makes up the 10 bit PWM duty cycle resolution. TIMER2 and register settings are used to generate the desired PWM signal.
   **(c)** PR2 register: Used to set the PWM period in microseconds.
   **(d)** T2CON register: Used to select the prescale value of TIMER2 for the PWM function.

In *capture mode*, the CCPR1 (CCPR2) register captures the 16-bit value of TIMER1 or TIMER3 (depending on which one of them has been selected in the setup) when an external event occurs on pin RC2/CCP1 (RC1/CCP2). These pins must be configured as input in capture mode. CCP1CON (CCP2CON) register bits are set to select the capture mode to be one of the following: every falling edge, every rising edge, every fourth rising edge, every 16th rising edge. When a capture occurs, the interrupt request flag bit CCP1IF (PIR register, bit 2) is set and must be cleared in software. If another capture interrupt occurs before the CCPR1 register value is read, it will be overwritten by the new captured value. C18 compiler library provides the following functions to implement capture.

```
#include <capture.h>
#include <timers.h>

OpenCapture1(C1_EVERY_4_RISE_EDGE & CAPTURE_INT_OFF) ;
  /* Configure capture 1 module: capture at every 4th
     rising edge of capture 1 pin signal, no interrupt on
     capture. */

OpenTimer3(TIMER_INT_OFF & T3_SOURCE_INT) ;
  /* Timer3 is the source clock to capture on trigger. */

while (!PIR1bits.CCP1IF) ; /* Wait until Capture module 1
                              has captured */
result = ReadCapture1() ;  /* Read the captured value */

....                       /* Process captured data */

if(!CapStatus.Cap1OVF)     /* Check (if needed) if there
                              was any overflow condition*/
{
  ...                      /* Further processing of
                              captured data if needed */
}
```

In *compare mode*, the 16-bit CCPR1 (CCPR2) register value is continuously compared to the TIMER1 or TIMER3 register value. When they match, RC2/CCP1 pin (RC1/CCP2 pin) is controlled to either high, low, toggle or unchanged state. In addition, interrupt flag bit CCP1IF (or CCP2IF) is set. The choice of action on the compare match event is made by proper settings of CCP1CON (CCP2CON) registers. In compare mode, RC2/CCP1 (RC1/CCP2) pin must be configured for output. It is possible to select to generate interrupt on the "compare match" event. In this case, output pin status is not changed. That action is left to the interrupt service routine (ISR).

In *PWM mode*, the CCP1 pin produces PWM output with 10-bit resolution. The CCP1 pin is multiplexed with the PORTC data latch, and the output is obtained from the PORTC-2 (RC2) pin. The TRISC-2 bit must be cleared to make the CCP1 pin an output. PWM output has two parameters: (1) *period* which defines the frequency of PWM signal and (2) *duty cycle* which defines the percentage of ON-time of the signal within each period. The PWM period obtained is a function of the frequency of the oscillator used to drive the PIC 18F452.

Example register values are given in the code below. The code configures the CCP2 port for PWM of 0.256 ms period (frequency 3.9 KHz) and a duty cycle of 25%.

```
PR2 = 255;      /* Sets the period for the PWM1 output
                   channel in microseconds*/
CCP1CON = 12; /* Activates the PWM mode in the CCP
                   register*/
```

```
CCP1RL = 63;   /* Sets the duty cycle for the PWM output */
TRISC = 0;     /* Configures PortC for output */
T2CON = 62;    /* Configures Timer 2 for prescale value of
                  1:1 and postscale 1:8*/
```

The PR2 register sets the PWM period, while the duty cycle is set by CCP1RL and CCP1CON < 5 : 4 > bits. PWM period is given by

$$PWM_{period} = (PR2 + 1) \times 4 \times (Clock\ Period) \times (Timer\ Prescale\ Value) \quad (4.1)$$

$$= 256 \times 4 \times \frac{1}{4\ \text{MHz}} \times 1 = 0.256\ \text{ms} \quad (4.2)$$

$$PWM_{dutycycle} = CCPR1L : CCP1CON < 5:4 > \times \quad (4.3)$$

$$(ClockPeriod) \times (Timer\ Prescale\ Value) \quad (4.4)$$

$$= 252 \times \frac{1}{4\ \text{MHz}} \times 1 = 0.063\ \text{ms} \quad (4.5)$$

C library functions used to set up and operate the PWM outputs (PWM1 or PWM2) are shown below:

```
#include <pwm.h>
#include <timers.h>

OpenTimer2(TIMER_INT_OFF & T2_PS_1_4 & P2_POST_1_8) ;
          /* Setup TIMER2: disable interrupt, set pre and
             post scalers to 4 and 8 */

OpenPWM1(char period) ;
          /* Enable and setup PWM1 module output signal
             "period" (8-bit value). */
...       /* PWM period=(period+1)*4*Tosc*(TIMER2
             Prescaler)*/

SetDCPWM1(unsigned int duty_cycle) ;
          /* Set "duty cycle" of PWM output signal
             (10-bit value) */
          /* High Time of PWM = (duty_cycle *Tosc) */
...
ClosePWM1() ; /* Disable PWM1 output module */
```

***Analog-to-Digital Converter (ADC)*** The analog-to-digital converter (A/D or ADC) allows conversion of an analog input signal to a digital number. The ADC on the PIC 18F452 has a 10-bit range and eight multiplexed input channels. Unused analog input channel pins can be configured as digital I/O pins. Analog voltage reference is software selected to be either the chip positive and negative supply ($V_{DD}$ and $V_{SS}$) or the voltage levels at RA3/$V_{REF+}$ and RA2/$V_{REF-}$ pins.

There are two main registers to control the operation of ADC:

1. ADCON0 and ADCON1 registers: to configure the ADC and control its operation, i.e., select which channel to sample, set the conversion rate, format of the conversion result, enable ADC and start conversion.
2. ADRESH and ADRESL registers: to hold the ADC converted data in 10-bit format in this two 8-bit registers.

A 10-bit ADC conversion takes $12 \times T_{AD}$ time period. The $T_{AD}$ time can be software selected to be one of seven possible values: $2 \times Tosc$, $4 \times Tosc$, $8 \times Tosc$, $16 \times Tosc$, $32 \times Tosc$, $64 \times Tosc$, or internal A/D module RC oscillator. The selection should be made so that there is a minimum of $T_{AD} = 1.6$ $\mu$sec conversion time available.

ADC must be set up with proper register value settings. Then when ADCON0 bit:2 is set, the ADC conversion process starts. At that point, sampling of the signal is stopped, the charge capacitor holds the sampled voltage, and ADC conversion process converts the analog signal to a digital number using successive approximation method. The conversion process takes about $12 \times T_{AD}$ time periods. When an ADC conversion is complete:

1. The result is stored in ADRESH and ADRESL registers.
2. ADCON0 bit:2 is cleared, indicating ADC conversion is done.
3. ADC interrupt flag bit is set.

The program can determine whether ADC conversion is complete or not by either checking the ADCON bit:2 (cleared) or waiting for ADC conversion complete interrupt if it is enabled. The sampling circuit is automatically connected to the input signal again, and the charge capacitor tracks the analog input signal. Next, conversion is started again by setting the ADCON0 bit:2.

The ADC conversion process can be started by the special event trigger of the CCP2 module. In order to use this, CCP2CON bits 3:0 must be set to 1011 and ADCON0 bit 0 must be set. TIMER1 (or TIMER3) period controls the ADC conversion frequency.

The ADC can be configured for use either by directly writing to the registers or by using C library functions present in the C-18 compiler. The C functions OpenADC(..), ConvertADC (), BusyADC(), and ReadADC() are used to configure the ADC module, begin the conversion process, and read the result of conversion. Examples of their use are given below:

```
#include <p18f452.h>
#include <adc.h>
#include <stdlib.h>

OpenADC(ADC_FOSC_RC & ADC_RIGHT_JUST & ADC_1ANA_0REF,
ADC_CH0 & ADC_INT_OFF);
  /* Enable ADC in specified configuration. */
  /* Select: clock source, format, ref. voltage source,
     channel, enable/disable interrupt on ADC-conversion
     completion*/

ConvertADC();       /* Start the ADC conversion process*/
while ( BusyADC() ) ; /* Wait until conversion is
                         complete: return 1 if busy, 0
                         if not.*/
result = ReadADC();    /* Read converted voltage; store in
                         variable 'result'. 10-bit
                         conversion result will be stored
                         in the least or most significant
                         10-bit portion of result
                         depending on how ADC was
                         configured in OpenADC(...) call.
                         */
closeADC();           /* Disable ADC converter*/
```

Here, the arguments in the `OpenADC(..)` function are used to configure the ADC module to use the specified clock sources, reference voltages, and select the ADC channel. Detailed description of arguments is provided in the C-18 users guide.

***Timers and Counters*** In real-time applications, there are many cases in which different tasks need to be performed at different periodic intervals. For instance, in an industrial control application, the status of doors in a building may need to be checked every minute; the parking lot gate status may need to be checked every hour. The best way to generate such periodic interrupts with different frequencies is to use a programmable timer/counter chip. A timer/counter chip operates as a timer or as a counter. In timer mode of operation, it counts the number of instruction cycles; hence, it can be used as a time measurement peripheral. The clock source can be an internal clock or external clock. In counter mode, it counts the number of signal state transitions at a defined pin, i.e., at rising edge or falling edge.

PIC 18F452 chip supports four timers/counters: TIMER0, TIMER1, TIMER2, and TIMER3. TIMER0 is an 8-bit or 16-bit software selectable, TIMER1 is a 16-bit, TIMER2 is an 8-bit timer and an 8-bit period measurement, and TIMER3 is a 16-bit timer/counter. Timer/counter operation is controlled by setting up appropriate register bits, reading from and writing to certain registers. Timer operation setup requires enable/disable, source of signal, type of operation (timer or counter), etc. When a timer overflows, interrupt is generated. The timer operation is controlled by a set of registers:

1. INTCON register is used to configure interrupts in the microcontroller.
2. TxCON register is used to configure TIMERx, where 'x' is 0, 1, 2, or 3 for the corresponding timer. Configuration of a timer involves the selection of the timer source signal (internal or external), pre- or post-scaling of the source signal, source signal edge selection, and enable/disable timer.
3. TRISA register port may be used to select an external source for the TIMER at port A pin 4 (RA4).
4. TMRxL and TMRxH are 8-bit register pairs that make up the 16-bit timer/counter data register. By using pre-scaler, the maximum range and resolution of time periods that can be measured by timers can be software controlled.

TIMER2 sets the interrupt status flag when its data register is equal to the PR2 register, whereas TIMER0, TIMER1, and TIMER3 set the interrupt status flag when there is overflow from maximum count to zero (from FFh to 00h in 8-bit mode, and from FFFFh to 0000h in 16-bit mode). The registers used to indicate the interrupt status are INTCON, PIR1, and PIR2. The interrupt can be masked by clearing the appropriate timer interrupt mask bit of the timer interrupt control registers.

Counter mode is selected by setting the T0CON register bit 5 to 1. In counter mode, TIMER0 increments on every state transition (on either falling or rising edge) at pin RA4/T0CKI.

C library functions used to set up and operate timers are (the following functions are available for all timers on the PIC microcontroller, TIMER0, TIMER1, TIMER2, and TIMER3),

```
#include <timers.h>

OpenTimer0(char config) ; /* Open and configure timer 0:
                             enable interrupt, select
                             8/16-bit mode, clock source,
                             prescale value */

...
result = ReadTimer0() ;    /* read the timer 0*/
...
```

```
WriteTimer0(data) ;          /* write to timer register 0 */
...
CloseTimer0() ;              /* close (disable) timer 0 and
                                its interrupts */
```

Often, a certain amount of time-delay is needed in the program logic. C-library provide programmable time delay functions, where the minimum delay unit is one instruction cycle. Therefore, the actual real-time delay accomplished by these functions depends on the processor operating speed.

```
#include <delays.h>
....
Delay1TCY(); /* Delay 1 instruction cycle */
Delay10TCYx(unsigned char unit);
             /* unit=[1,255], Delay period = 10*unit
                instruction cycle */
Delay100TCYx(unsigned char unit);
             /* unit=[1,255], Delay period = 100*unit
                instruction cycle */
Delay1KTCY(unsigned char unit);
             /* unit=[1,255], Delay period = 1000*unit
                instruction cycle */
Delay10KTCY(unsigned char unit);
             /* unit=[1,255], Delay period = 10000*unit
                instruction cycle */
....
```

***Watchdog Timers*** Watchdog timer (WDT) is a hardware timer, which is used to reboot or take a predefined action if it expires. The watchdog timer keeps a "watch eye" on the system performance. If something gets stuck, watchdog timer can be used to reset everything. Watchdog timer is essential in embedded controllers. The WDT counts down from a programmable preset value. If it reaches zero before the software resets the counter to preset value, it is assumed that something is stuck. Then the processor's reset line is asserted.

The watchdog timer (WDT) on the PIC chip is an on-chip RC oscillator. It works even if the main clock of the CPU is not working. WDT can be enabled/disabled, and the time-out period can be changed under software control. When the WDT times-out, it generates a RESET signal which can be used to restart the CPU or wake up the CPU from SLEEP mode. There are three registers involved in configuration and use of WDT on PIC 18F452: CONFIG2H, RCON, and WDTCON. If CONFIG2H register bit 0 (WDTEN bit) is 1, WDT cannot be disabled by other software. If this bit is cleared, WDT can be enabled/disabled by WDTCON register, bit 0 (1 for enable, 0 for disable). When WDT times-out, RCON register bit 3 (TO bit) is cleared. The time-out period is determined in hardware, and extended by post-scaler in software (CONFIG2H register, bits 3:1).

## 4.4 INTERRUPTS

### 4.4.1 General Features of Interrupts

An interrupt is an event which stops the current task the microprocessor is executing, and directs it to do something else. And when that task is done, the microprocessor resumes the original task. An interrupt can be generated by two different sources: (1) hardware interrupts (external), and (2) software (internal) generated interrupt with an instruction in assembly language, i.e., **INT n**. When an interrupt occurs, the CPU does the following:

1. Finishes currently executing statement.
2. Saves the status, flags, and registers in the stack so that it can resume its current task later.
3. Checks the interrupt code, looks at the interrupt service table (vector) to determine the location of the the *interrupt service routine* (ISR), a function executed when interrupt occurs.
4. Branches to the ISR and executes it.
5. When ISR is done, restores the original task from the stack and continues.

When an interrupt is generated, the main task stops after some housekeeping tasks, and the address location of the ISR for this interrupt needs to be determined. This information is stored in the interrupt service vector. The table has default ISR addresses. If you like to assign a different ISR for an interrupt number, the old address should be saved, then a new ISR address should be written. Later, when the application terminates, the old ISR address should be restored.

In a given computer control system, there can be more than one interrupt source and they may happen at the same time. Therefore, different interrupts need to be assigned different priority levels to determine which one is more important than the other. Higher priority interrupts can stop lower priority interrupts. Therefore, nested interrupts can be generated (Fig. 4.9).

The time between the generation of interrupt to the time the program branches to the ISR for the interrupt (after saving current state of microprocessor and housekeeping operations, taking care of other higher priority interrupts) is called the *interrupt latency time*. Smaller interrupt latency time is generally desired. Due to multiple interrupt sources and priorities, it may be impossible to calculate the worst possible interrupt latency time since it may depend on the number of higher priority interrupts generated at the same time.

In critical applications, if interrupting the current process is not tolerable, interrupts can be disabled temporarily with appropriate assembly instructions or C functions.

Let us assume that there are two interrupt sources: interrupt 1 and interrupt 2 (Fig. 4.9). Let us assume that interrupt 2 has higher priority than interrupt 1. While the main program

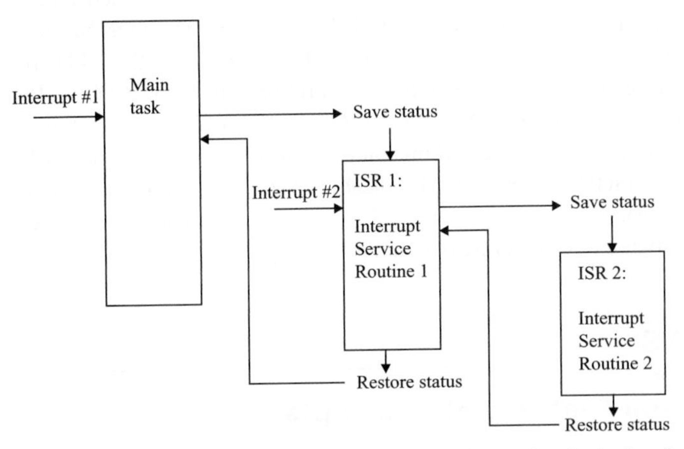


**FIGURE 4.9:** Nesting of interrupts is possible through priority level management.

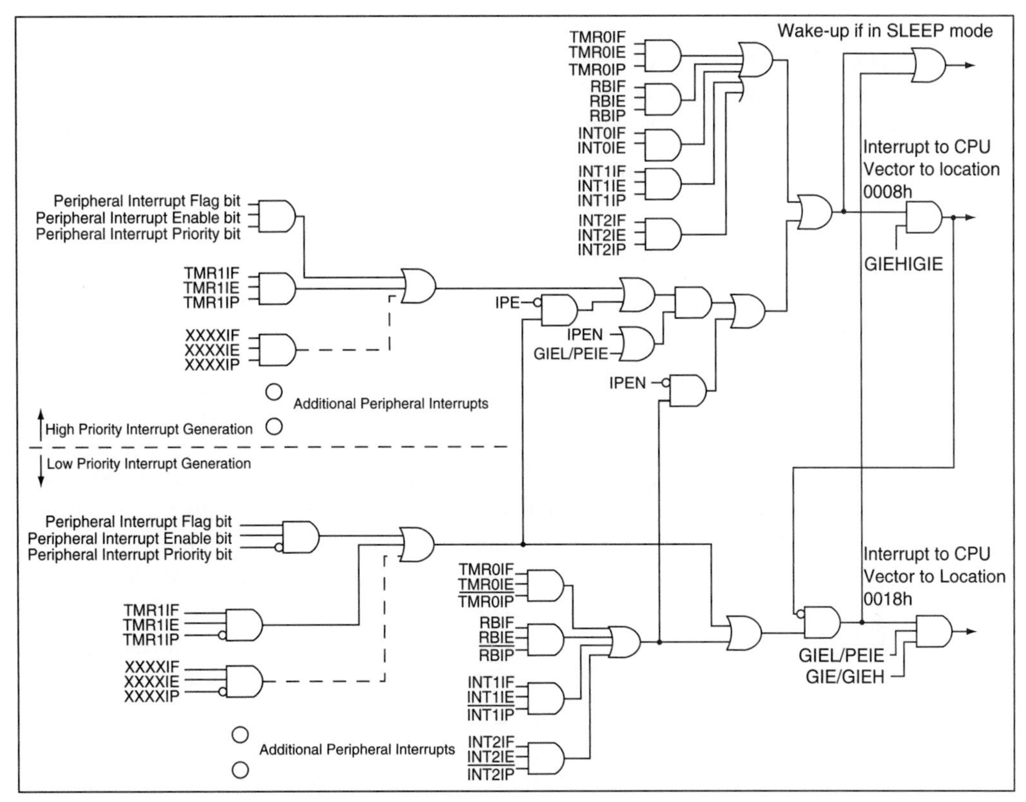


**FIGURE 4.10:** Interrupts and its logic in PIC 18F452 (Fig. 8-1 from PIC 18Fxx2 Data Sheet).

is executing, let us assume that interrupt 1 occurred. The processor will save the status of microprocessor, determine the interrupt number, look in the *interrupt service vector table* for the address of the *interrupt service routine (ISR)* for this particular interrupt, and jump to that ISR-1. Let us further assume that while it is executing ISR-1, interrupt 2 occurs. The processor will save the status, and jump to ISR-2. When ISR-2 is done, processor returns to ISR-1, restores the previous state, and continues the ISR-1 from where it was interrupted. When ISR-1 is finished, it will restore the state of the main task before the interrupt 1 occurs, and continue the main task.

## 4.4.2 Interrupts on PIC 18F452

PIC 18F452 supports external and internal interrupts (Fig. 4.10). There are 18 interrupt sources. In addition, external interrupts can be defined to be active in rising or falling edge of input signal.

It supports two levels of interrupt priority: high priority and low priority. Each interrupt source is assigned one of the two priority levels by setting appropriate register bits. The high-priority interrupt vector is at 000008h, and the low-priority interrupt vector is at 000018h. The address of ISR routine for each interrupt category must be stored at these addresses. High-priority interrupts override any low-priority interrupt in progress. Only one high-priority interrupt can be active at a given time, and cannot be nested. Low-priority interrupts can be nested by setting the INTCON register bit 6 (GIEL bit = 1) at the beginning of the ISR.

There are ten registers used to control the interrupts in PIC 18F452: RCON, INTCON, INTCON2, INTCON3, PIR1, PIR2, PIE1, PIE2, IPR1, and IPR2. Each interrupt source, except INT0 which is always a high-priority interrupt, has three bits for enable/disable and priority setup, and to determine interrupt status (also called *flag* or *request*) register space:

1. One bit to enable/disable the interrupt (enable bit)
2. One bit to set the priority level (priority bit)
3. One bit to indicate the current status of interrupt (flag bit)

These bits are distributed over the ten 8-bit registers referenced above.

Interrupt sources include:

1. Three external interrupt pins: RB0/INT0, RB1/INT1, RB2/INT2 can be programmed to be edge-triggered interrupts on rising or falling edge by setting appropriate bits in the above 10 registers. Both global interrupt enable and individual interrupt enable for a specific interrupt source must be enabled for the interrupt to be enabled. When interrupt occurs, the corresponding interrupt status bit (also called interrupt flag) is set (INTxF), which can be used to determine the source of the interrupt in ISR. INT0 is always high priority, whereas priority of INT1 and INT2 can be programmed as high or low. These interrupts can wake up the processor from SLEEP state.
2. PORTB pins 4,5,6,7 can be programmed to generate interrupt on change-of-state. PORTB pins 4–7 change-of-state interrupts can be enabled or disabled as well as priority is set as a group under program control.
3. ADC conversion done is another interrupt which is generated when an ADC conversion is completed. When ADC conversion complete interrupt is generated (assuming it is enabled), PIR1-bit 6 is set. The logic to trigger the start of the ADC conversion must be handled separately (i.e., after reading the current ADC conversion at the ISR routine, or a periodically performed ADC conversion based on a TIMER generated signal).
4. TIMER interrupts (interrupt-on-overflow) are generated by TIMER0, TIMER1, TIMER3, and when TIMER2 data register = PR2 register. When timer interrupts are enabled and timer overflow occurs, resulting in interrupt generation, bits in corresponding registers are set that can be used in the ISR routine to determine the source of the interrupt (INTCON bit 1 for TIMER0, PIR1-bit 0 for TIMER1, PIR1-bit 1 for TIMER2, and PIR2-bit 1 for TIMER3).
5. CCP1, CCP2 interrupts (once enabled and priorities are selected to be either as high or low) are generated as follows: In capture mode, a TIMER1 register capture occurred, and in compare mode a TIMER1 register match occurred.
6. Communication peripheral interrupts (master synchronous serial port, addressable USART, parallel slave port, and EEPROM write interrupt).

For external interrupts at INT0-INT2 (RB0-RB2) pins or PORTB pins 4-7 (RB3-RB7) input change interrupt, expected interrupt latency time is three to four instruction cycles. Notice that PIC 18F452 groups all interrupts into two categories and provides two ISR addresses: low-priority group with ISR address stored at 000018h, and high-priority group with ISR address stored at 000008h. Within each group, application software must be implemented by testing interrupt flag bits (bits contained in the 10 registered listed above) to determine which one of the interrupt sources in a high- or low-priority group has triggered the interrupt.

The RESET signal can be considered as a nonmaskable interrupt. The address for the program code to execute on the RESET signal is stored at the vector address 0000h.

The following example C-code shows the use of an external interrupt and its software handling in PIC 18F452 microcontroller. First, interrupt service vector locations for high priority and low priority must be set up to point (or branch) to the corresponding ISR routine. Then each ISR function should be defined. The ISR routine should be proceeded by a `#pragma interrupt ISR_name` to indicate that the function is an ISR, not just a regular C-function. Furthermore, the ISR cannot have any argument or return type.

```
#pragma code HIGH_INTERRUPT_VECTOR = 0x8
    /* Place the following code starting at address 0x8,
       which is the location for high priority interrupt
       vector */
  ...
  ...
#pragma code
    /* Restore the default program memory allocation. */
#pragma code LOW_INTERRUPT_VECTOR = 0x18
    /* Place the following code starting at address 0x18,
       which is the location for low priority interrupt
       vector */
  ...
  ...
#pragma code
    /* Restore the default program memory allocation. */

#pragma interrupt High_ISR_name
void High_ISR_name(void)
{
  ....
}

#pragma interruptlow Log_ISR_name
void Low_ISR_name(void)
{
  ....
}
```

***Example: Interrupt Handling with PIC 18F452 on PICDEM 2 Plus Demo Board*** This example requires:

1. ***Hardware:*** The PICDEM 2 Plus Demo board with PIC 18F452 microcontroller chip and MPLAB ICD 2 (in-circuit debugger) hardware.
2. ***Software:*** MPLAB IDE v6.xx and MPLAB C18 development tools.

The INT0 pin is connected to a switch which is an external interrupt. It is a high-priority interrupt. Therefore, when this interrupt occurs, the program will branch to the interrupt vector table location 0x0008h and get the address of the interrupt service routine to branch. By default, this address contains the address of interrupt service routine called `high_ISR`. Hence, when INT0 occurs, the program will branch to the `high_ISR` function, after saving the contents of certain registers and return address in the stack. We can type in the code for the interrupt service routine in `high_ISR` or call another function from there and just have the "call" or "goto" statement in the `high_ISR` function. In addition, we need to set up certain registers for the interrupt to function properly, i.e., enable interrupt, define the active state of interrupt. Interrupt service routine (ISR) must have no input arguments and no return data.

```
#include <p18f452.h> /* Header file for PIC 18F452
                        register declarations */
#include <portb.h>   /* For RB0/INT0 interrupt */

/* Interrupt Service Routine (ISR) logic */

void toggle_buzzer(void) ;

#pragma code HIGH_INTERRUPT_VECTOR = 0x8
                     /* Specify where the program address
                        to be stored */
void high_ISR(void)
{
  _asm
    goto toggle_buzzer
  _endasm
}
#pragma code         /* Restore the default program
                        memory allocation. */

#pragma interrupt toggle_buzzer
void toggle_buzzer(void)
{
  CCP1CON = ~CCP1CON & 0x0F
                     /* Toggle state of buzzer: OFF if
                        ON, ON if OFF */
  INTCONbits.INT0IF = 0
                     /* Clear flag to avoid another
                        interrupt due to same event*/
}

/* Setup of the ISR */

void EnableInterrupt(void)
{
RCONbits.IPEN = 1 ; /* Enable interrupt priority levels*/
INTCONbits.GIEH = 1 ; /* Enable high priority interrupts*/
}

void InitializeBuzzer(void)
{
  T2CON = 0x05 ;           /* postscale 1:1, Timer2 ON,
                              prescaler 4 */
  TRISCbits.TRISC2 = 0 ; /* configure CCP1 module for
                              buzzer operation */
  PR2 = 0x80 ;             /* initialize PWM period */
  CPPR1l = 0x80 ;          /* ......... PWM duty cycle */
}
```

```
void main (void)
{

EnableInterrupts() ;
InitializeBuzzer() ;

OpenRB0INT(PORTB_CHANGE_INT_ON & PORTB_PULLUPS_ON &
  FALLING_EDGE_INT) ;
  /* Enable RB0/INT0 interrupt, configure RB0 pin for
     interrupt, trigger interrupt on falling edge */

CCP1CON = 0x0F ; /* Turn ON buzzer */

while (1) ; /* Wait indefinitely. */
            /* when the interrupt occurs, the
               corresponding ISR will be executed */
}
```

***Example: Timer0 Interrupt*** Below is a sample code which sets up Timer0 to generate interrupt. Timer0 is defined as a low-priority interrupt. The interrupt vector table location for low-priority interrupts is 0 x 18. As in the previous example, a shell form of low-priority ISR for low-priority interrupt address is placed at the 0 x 18 vector address, and the "goto" statement is used to direct the program execution to the Timer0 ISR.

```
#include <p18f452.h> /* Header file for PIC 18F452
                        register declarations */
#include <timer.h>   /* For timer interrupt */

/* Interrupt Service Routine (ISR) logic */

void Timer0_ISR(void) ;

#pragma code LOW_INTERRUPT_VECTOR = 0x18
                     /* Specify where the program address
                        to be stored */

void Low_ISR(void)
{
  _asm
    goto Timer0_ISR
  _endasm
}

#pragma code         /* Restore the default program
                        memory allocation. */

#pragma interruptlow Timer0_ISR
void Timer0_ISR(void)
{
  static unsigned char led_display = 0 ;

  INTCONbits.TMR0IF = 0 ;
  led_display = ~led_display & 0x0F/* toggle LED
  display */
  PORTB = led_display ;
```

```
}

/* Setup of the ISR */

void EnableInterruptTimer0(void)
{
  TRISB = 0 ;
  PORTB = 0 ;
  OpenTimer0 (TIMER_INT_ON & T0_SOURCE_INT & T0_16BIT) ;
  INTCONbits.GIE = 1 ;
}

void main (void)
{

  EnableInterruptTimer0() ;

  while (1) ; /* Wait indefinitely. */
              /* when the interrupt occurs, the
                 corresponding ISR will be executed */
}
```

## 4.5 PROBLEMS

**1.** In a PIC 18F452 microcontroller, how is RESET condition generated and what sequence of events happens under that condition?

**2.** What is the role of "Watchdog timer"? How does it work in PIC 18F452?

**3.** How many hardware interrupt priorities are supported in PIC 18F452? Describe how these interrupts are enabled/disabled, how an interrupt is handled, and the sequence of events that occur in the microcontroller when an interrupt event occurs.

**4.** List the input/output (I/O) ports available on a PIC 18F452 microcontroller. Give examples of C-code to show how they are set up and used in real-time application software.

**5.** What is the role of capture/compare/PWM port on a PIC 18F452 microcontroller? Describe how this port is set up and used in different applications.

**6.** Discuss the differences between "polling-driven" versus "interrupt-driven" programming. Give examples for polling-driven and interrupt-driven applications. What are the internal and external interrupt sources to the microcontroller?

**7.** Discuss analog to digital converter (ADC) inputs available on PIC 18F452 chip (how many channels, resolution of each channel) and how it is controlled in software (how it is configured and used for analog signal to digital signal conversion).

CHAPTER 5

# ELECTRONIC COMPONENTS FOR MECHATRONIC SYSTEMS

## 5.1 INTRODUCTION

Analog and digital electronic components are an integral part of mechatronic devices. Most commonly used analog and digital electronic devices are discussed in this chapter. We start with linear circuits, semiconductors, and discuss electronic switching devices (diodes and transistors). This is followed by the discussion of analog operational amplifiers (op-amps). Examples of proportional, derivative, integral, low-pass and high-pass filters using op-amps are discussed. Digital electronic circuit components are discussed last. The focus of the discussions is on the input–output functionality of the devices, not the detailed modeling of the input–output dynamics nor the details of its construction.

## 5.2 BASICS OF LINEAR CIRCUITS

Basic passive components of electrical circuits are resistor, capacitor, and inductor (Fig. 5.1). By passive components, we mean that the component can be connected to a circuit with two terminal points without requiring its own power supply. Basic variables in an electrical circuit are voltage (potential difference analogous to pressure in hydraulic systems) and current (flow of electrons, analogous to fluid flow). Current is the rate of charge flow through a conductor in a circuit,

$$i(t) = \frac{dQ(t)}{dt} \tag{5.1}$$

where $Q(t)$ is the amount of electrical charge that passes through the conductor. Smallest unit of electrical charge is the electrical charge of an electron or of a proton. The amount of electrical charge of an electron ($e^-$) and that of proton ($p^+$) is the same, except that they are opposite sign,

$$|e^-| = |p^+| = 1.60219 \times 10^{-19}[\text{C}] \tag{5.2}$$

The passive components define relationships between current and voltage potential difference between their two nodes. An ideal *resistor* has a potential difference between its two ends proportional to the current passing through it. The proportionality constant is called the resistance, $R$:

$$V_{12}(t) = R \cdot i(t) \tag{5.3}$$

This is called the *Ohm's Law.* The resistance of a component is a function of the material property, that is, the *resistivity* ($\rho$) and its geometry. As the cross section of the conductor

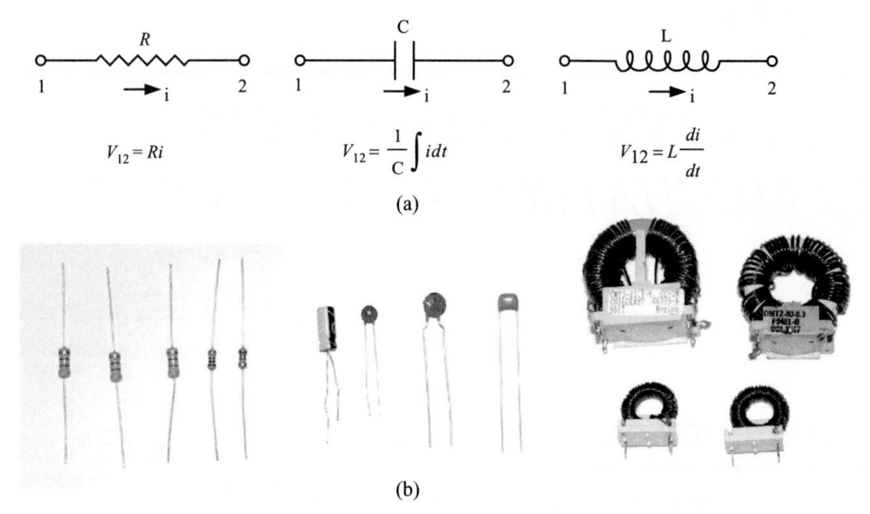


**FIGURE 5.1:** Basic passive components of electrical circuits: (a) symbols of resistance (R), capacitance (C), inductance (L); (b) pictures of resistors, capacitors and inductors.

increases, resistance to the flow of charges reduces, as they have more room to move through. As the length of the conductor increases, resistance increases. Hence, the resistance of a conductor is

$$R = \rho \frac{l}{A} \tag{5.4}$$

where $\rho$ is the resistivity of the material ($\rho = 1.7 \times 10^{-8}\,\Omega \cdot m$ for copper, $\rho = 2.82 \times 10^{-8}\,\Omega \cdot m$ for aluminum, $\rho = 0.46\,\Omega \cdot m$ for germanium, $\rho = 640\,\Omega \cdot m$ for silicon, $\rho = 10^{13}\,\Omega \cdot m$ for hard rubber, $\rho = 10^{10}$ to $10^{14}\,\Omega \cdot m$ for glass), $l$ is the length, and $A$ is the cross-sectional area of the conductor. It is also important to point out that the resistivity of materials, hence the resistance, is a function of temperature. The variation of resistance as function of temperature varies from material to material. In general, many materials, but not all, have the following resistance and temperature relationship,

$$R(T) = R_0[1 + \alpha(T - T_0)] \tag{5.5}$$

where $R(T)$ is the resistance at temperature $T$, $R_0$ is the resistance at temperature $T_0$, and $\alpha$ is the rate of change in resistance as function of temperature. *Superconductors* are materials that have almost zero resistance below a certain critical temperature, $T_c$. For each superconductor, there is a critical temperature $T_c$ below which the above relationship does not hold. The resistance suddenly drops to almost zero at $T \leq T_c$. This critical temperature for mercury is about $T_c = 4.2°K$.

There are two major types of resistors: wire wound and carbon types. Wire wound resistors are used for high current applications such as regenerative power dumping as heat in motor control. Carbon type resistors are used in low current signal processing applications where the resistance value is color coded on the componenet using four color code. The maximum power rating of a resistor defines the maximum power it can dissipate without damage,

$$P = V_{12} \cdot i = V_{12} \cdot (V_{12}/R) = R \cdot i^2 = \frac{V_{12}^2}{R} < P_{max} \tag{5.6}$$

which defines the maximum voltage drop or maximum current that can be present across the resistor.

A *capacitor* stores electric charges, hence creates an electric field (electric voltage) across its terminals. It is analogous to a water storage tank in that it stores a certain amount of water. Water pressure at tank outlet is function of water height in the tank. It is made of two conducting materials separated by an insulating material such as air, vacuum, glass, rubber, paper. The two conductor surfaces store equal but opposite charges. As a result, there is a voltage potential across the capacitor. The voltage potential is proportional to the amount of charge stored and the characteristics of the capacitor. There are four major capacitor types: mica, ceramic, paper/plastic film, and electrolytic. Electrolytic-type capacitors are polarized. Therefore, the positive terminal must be connected to positive side of the circuit.

An ideal capacitor generates a voltage potential difference between its two nodes proportional to the stored electrical charge (integral of current conducted)

$$V_{12}(t) = \frac{Q}{C} = \frac{1}{C}\int_0^t i(\tau) \cdot d\tau \tag{5.7}$$

where $C$ is called the capacitance. An ideal capacitor induces a 90° phase angle between its voltage and current across its terminals due to the integral function. When a capacitor is fully charged, it stores a finite amount of charge,

$$Q = \int_0^t i(\tau) \cdot d\tau = C \cdot V_{12} \tag{5.8}$$

The size of a capacitor is indicated by the amount of charge it can store (capacitance, $C$). A given capacitor can hold voltage up to a certain amount, $V_{max}$ (called the rated voltage). Capacitance ($C$) is a measure of how much charge ($Q$) the capacitor can hold for a given voltage potential, ($V$). A capacitor cannot hold a voltage potential above a maximum value, $V_{max}$. If the rated voltage is exceeded, the capacitor breaks down and cannot hold the charges. Construction principle of a capacitor includes two conductors separated by insulating material. The capacitance is proportional to the surface area of the capacitor (conductors inside the capacitor) which holds the charge and is inversely proportional to the distance between them. The proportionality constant is the permittivity of the medium that separates the conductors, $\epsilon$,

$$C = \frac{\epsilon \cdot A}{l} = \frac{\kappa\epsilon_0 \cdot A}{l} \tag{5.9}$$

where $\kappa$ is the dielectric constant and $\epsilon_0$ is the permittivity of free space. The dielectric constant for some materials is as follows: $\kappa = 1.00$ for vacuum, $\kappa = 3.4$ for nylon, $\kappa = 5.6$ for glass, $\kappa = 3.7$ for paper. Clearly, as the dielectric constant of the insulating material increases, the capacitance of the capacitor increases. The capacitance is proportional to the surface area of the capacitor plates and inversely proportional to the distance between them. The maximum voltage that the capacitor can handle before breakdown, called *breakdown voltage* or *rated voltage*, increases with the increase in dielectric constant of the insulator. The breakdown voltage of the capacitor also increases with the distance parameter $l$.

Energy stored in a capacitor can be calculated as follows:

$$dW = V \cdot dQ = \frac{Q}{C}dQ \tag{5.10}$$

$$W = \frac{1}{2}\frac{Q^2}{C} \tag{5.11}$$

and the $W \leq W_{max}$, where $W_{max}$ is the maximum energy storage capacity of the capacitor,

$$W_{max} = \frac{1}{2}C \cdot V_{max}^2 \tag{5.12}$$

Notice that capacitors block the DC voltage and pass the AC voltage. In other words, the DC voltage will build a potential difference in the capacitor until they are equal, provided

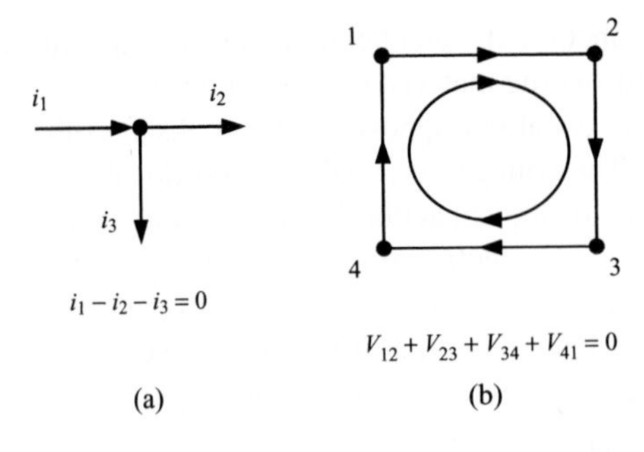


**FIGURE 5.2:** Kirchhoff's electric circuit laws: (a) *current law:* Algebraic sum of currents at any node is zero. (b) *Voltage law:* Algebraic sum of voltages in a loop is zero.

that the DC source voltage is below the break down voltage of the capacitor. The AC voltage simply alternates the charge and discharge of the capacitor and is passed, not blocked, in a circuit.

An ideal *inductor* generates a potential difference proportional to the rate of change of current passing through it

$$V_{12} = L \cdot \frac{di(t)}{dt} \tag{5.13}$$

where $L$ is called the inductance. Notice that, regardless of the direction of current ($i > 0$ or $i < 0$), the voltage across the inductor is proportional to the rate of change of the current. Let $V_{10}$ and $V_{20}$ be the voltages at point 1 and 2 across the inductor with reference to ground. Then,

$$V_{10} > V_{20} \quad \text{when} \quad \frac{di}{dt} > 0$$

$$V_{10} < V_{20} \quad \text{when} \quad \frac{di}{dt} < 0$$

The inductor is made by a coil of a conductor around a core, like a solenoid. The core can be a magnetic or an insulating material. The inductance value, $L$, is a function of permeability of the core material, number of turns in the coil, cross sectional area and length. *Permeability* of a material is a measure of its ability to conduct electromagnetic fields. It is analogous to the electrical conductivity. If the material composition of the space around an inductor changes, i.e., due to motion of device components in a solenoid, the inductance changes.

In deriving the equation for electric circuits, two main relationships are used (Kirchhoff's laws): *1. current law, 2. voltage law* (Fig. 5.2). Current law states that the algebraic sum of currents at a node is equal to zero. Voltage law states that the sum of voltages in a loop is equal to zero. For instance, in Fig. 5.2 current law states that

$$i_1 - i_2 - i_3 = 0 \tag{5.14}$$

and voltage law states that

$$V_{12} + V_{23} + V_{34} + V_{41} = 0 \tag{5.15}$$

## 5.3 EQUIVALENT ELECTRICAL CIRCUIT METHODS

Quite often, we need to reduce a two-terminal circuit with multiple components into an equivalent simpler circuit with a voltage source and an impedance or a current source and an impedance. For now, assume that *impedance* is a generalized form of *resistance*. Two of the most well known equivalent circuit analysis methods are discussed below. These methods are useful in determining input and output loading errors in coupled electrical circuits, i.e., measurement errors introduced due to the effect of a measuring device in an electrical circuit.

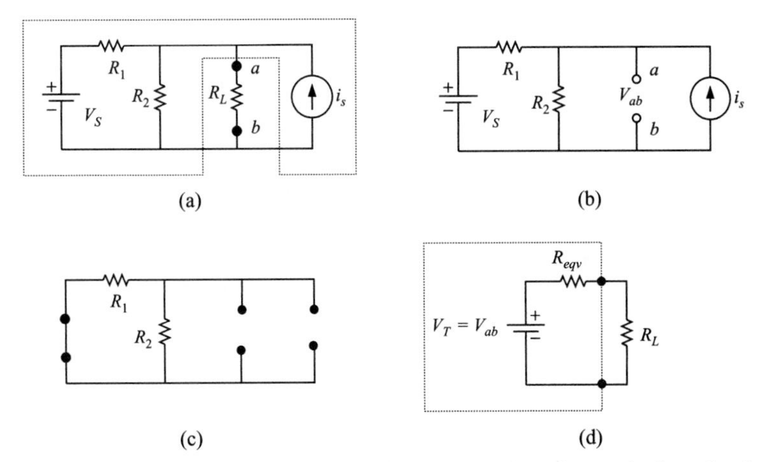


**FIGURE 5.3:** Thevenin's equivalent circuit procedure. An equivalent circuit consists of a voltage source and an equivalent resistor in series.

## 5.3.1 Thevenin's Equivalent Circuit

Thevenin's equivalent circuit consists of a voltage source in series with an equivalent resistor. Any section of a linear circuit with multiple resistors (plus voltage and current sources) components can be replaced with Thevenin's equivalent circuit. Consider the circuit shown in Fig. 5.3. Our objective is to determine the equivalent voltage source and series resistance value for the circuit shown in the dotted line. Hence, we can examine the interaction between the load resistance $R_L$ and the rest of the circuit. Thevenin's equivalent circuit analysis requires the calculation of two parameters: $V_T$ and $R_{eqv}$. For more general circuits involving AC voltage sources, capacitors, and inductors in addition to resistors, the equivalent circuit parameters are $V_T(jw)$ and $Z_{eqv}(jw)$, which are a complex generalization of voltage source and resistor. The $Z_{eqv}(jw)$ is called the impedance, which is discussed in the next section.

The following are the standard steps:

1. Identify the subcircuit whose Thevenin's equivalent circuit is to be determined. Identify the two point terminals ($a$ and $b$) out of the subcircuit. Remove the load resistor between the two points [Fig. 5.3 (a)].
2. Calculate the open-circuit voltage between $a$ and $b$, $V_{ab}$, [Fig. 5.3 (b)].
3. Turn OFF all other voltage sources (short-circuit voltage sources, i.e., zero source) and disconnect all current sources (open-circuit the current sources), and calculate the equivalent series resistor in the circuit (excluding the $R_L$), $R_{eqv}$ [Fig. 5.3 (c)]. The equivalent resistance of the circuit inside the dotted box (as viewed from $V_{ab}$) is represented by $R_{eqv}$ and also called the *output resistance* of the subcircuit.
4. The circuit inside the dotted box can be viewed as a voltage source ($V_T = V_{ab}$) plus a series resistance $R_{eqv}$ the source has in addition to the load [Fig. 5.3 (d)].

As far as the two points $a$ and $b$ are concerned, the circuit inside the dotted block is equivalent to a voltage source $V_T = V_{ab}$ and series resistor $R_{eqv}$.

## 5.3.2 Norton's Equivalent Circuit

Norton's equivalent circuit consists of a current source in parallel with an equivalent resistor (Fig. 5.4). The equivalent resistor is the same as the resistor calculated in Thevenin's equivalent circuit. The value of the current source ($i_N$) is the same as the value of the current that would flow in a short circuit of the output terminals, that is, to replace the output resistance

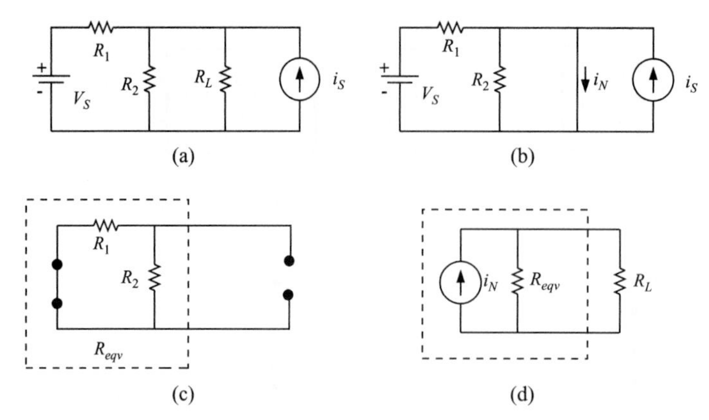


**FIGURE 5.4:** Norton's equivalent circuit procedure. Equivalent circuit consists of a current-source and an equivalent resistor in parallel.

between points $a$ and $b$ with short circuit connection. The procedure for obtaining Norton's equivalent circuit is essentially the same as the procedure for Thevenin's equivalent circuit. The conversions can be made between equivalent Thevenin's circuit and Norton's circuit; i.e., between two points in a circuit any voltage source with a series resistance can be replaced by an equivalent current source and a parallel resistor, and vice versa. It can be shown that

$$V_T = i_N \cdot R_{eqv} \tag{5.16}$$

The steps to follow to find the Norton's equivalent circuit are as follows (Fig. 5.4):

1. Find the Norton source current ($i_N$) by removing the load resistor from the original circuit and calculate the current through the short wire jumping across the open connection points where the load resistor used to be [Fig. 5.4 (b)].
2. Find the Norton resistance by removing all power sources in the original circuit (voltage sources shorted and current sources open) and calculate the total resistance between the open connection points (excluding the load resistance, [Fig. 5.4 (c)]).
3. Draw the Norton equivalent circuit with the Norton equivalent current source in parallel with the Norton equivalent resistance [Fig. 5.4 (d)]. The load resistor re-attaches between the two open points of the equivalent circuit.

***Example*** In the example shown in [Fig. 5.3 (d)], let the following values for the parameters be

$$V_s = 10\ \text{V}, \quad R_1 = 7\ \Omega, \quad R_2 = 3\ \Omega, \quad i_s = 5\ \text{A} \tag{5.17}$$

Then, it is easy to show that [Fig. 5.3 (b)]

$$V_s = R_1 \cdot i_1 + R_2 \cdot (i_1 + i_s) \tag{5.18}$$

$$10\ \text{V} = 7 \cdot i_1 + 3 \cdot (i_1 + 5) \tag{5.19}$$

$$= 10 \cdot i_1 + 15 \tag{5.20}$$

$$i_1 = -0.5\ \text{A} \tag{5.21}$$

The voltage potential $V_{ab}$ is

$$V_{ab} = 3 \cdot (-0.5 + 5.0) = 13.5\ \text{V} \tag{5.22}$$

The equivalent resistance of the circuit after the voltage sources are short-circuited, and current sources are open-circuited [Fig. 5.3 (c)],

$$\frac{1}{R_{eqv}} = \frac{1}{7} + \frac{1}{3} \tag{5.23}$$

$$R_{eqv} = \frac{7 \cdot 3}{7+3} \tag{5.24}$$

$$= 2.1\ \Omega \tag{5.25}$$

Thevenin's equivalent circuit of this example is represented by a voltage source $V_T = 13.5$ V and equivalent resistor $R_{eqv} = 2.1\ \Omega$ [Fig. 5.3 (d)].

For instance, the current on the load resistor $R_L$ can be calculated as function of the load resistance,

$$i_R = \frac{V_T}{R_L + R_{eqv}} \tag{5.26}$$

$$= \frac{13.5}{R_L + 2.1} \tag{5.27}$$

$$i_R = i_R(R_L\ ;\ V_T, R_{eqv}) \tag{5.28}$$

It is instructive to obtain the Norton equivalent of the same circuit and confirm that $V_T = R_{eqv} \cdot i_N$. Referring to Fig. 5.4 (b), the current $i_N$ is sum of the current due to current source and voltage source,

$$i_N = i_s + i_1 \tag{5.29}$$

Notice that no current will pass through $R_2$ since there is zero resistance path parallel to it; therefore,

$$i_1 = \frac{V_s}{R_1} \tag{5.30}$$

Hence,

$$i_N = 5\ \text{A} + \frac{10}{7} A = \frac{45}{7}\ \text{A} \tag{5.31}$$

Referring to Fig. 5.4 (c), the equivalent resistance in the dotted block as seen by the load is again

$$\frac{1}{R_{eqv}} = \frac{1}{R_1} + \frac{1}{R_2} \tag{5.32}$$

$$= \frac{1}{7} + \frac{1}{3} \tag{5.33}$$

$$R_{eqv} = \frac{7 \cdot 3}{7+3} = 2.1\ \Omega \tag{5.34}$$

and it is confirmed that the relationship between Thevenin's and Norton's equivalent circuits is

$$V_T = R_{eqv} \cdot i_N \tag{5.35}$$

$$= 2.1 \cdot \frac{45}{7} = 13.5\ \text{V} \tag{5.36}$$

## 5.4 IMPEDANCE

### 5.4.1 Concept of Impedance

The term *impedance* is a generalization of the resistance. If a circuit consists of a potential source and a resistor, the impedance and the resistance are the same. Impedance is a generalized version of resistance when a circuit contains other components such as capacitive and inductive components. The input–ouput relationship between voltage and current in a resistor circuit is

$$\frac{V(t)}{i(t)} = R \tag{5.37}$$

Let us take the Fourier transform of this,

$$\frac{V(jw)}{i(jw)} = R \tag{5.38}$$

where in frequency domain, the input–output ratio is a constant, $R$, not a function of frequency, $w$, $j = \sqrt{-1}$. The *impedance* of a resistor is a real constant.

Let us examine an RL circuit [Fig. 5.5 (a)], and consider the voltage–current relationship in frequency domain,

$$V(t) = R \cdot i(t) + L\frac{di(t)}{dt} \tag{5.39}$$

$$\frac{V(jw)}{i(jw)} = (R + jwL) \tag{5.40}$$

where $Z(jw)$ is defined as the *impedance* as follows:

$$Z(jw) = \frac{V(jw)}{i(jw)} = (R + jwL) \tag{5.41}$$

which can be viewed as generalized resistance that is a complex function of frequency. The complex function of frequency has a magnitude and phase for each frequency.

*Admittance*, $Y(jw)$, is defined as the inverse of impedance,

$$Y(jw) = \frac{1}{Z(jw)} = Re\,(Y(jw)) + jIm\,(Y(jw)) \tag{5.42}$$

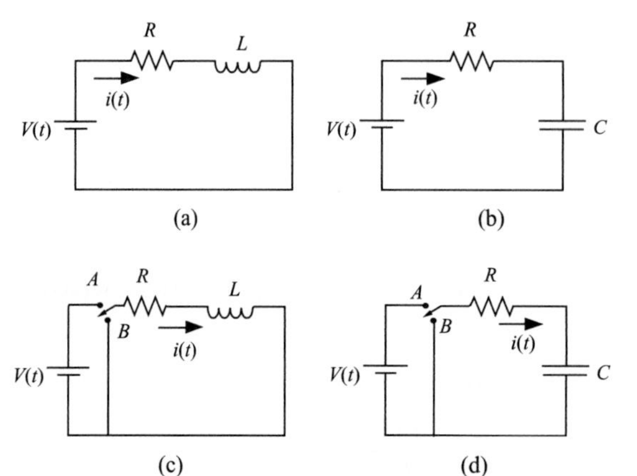


**FIGURE 5.5:** (a) RL circuit, (b) RC circuit, (c) RL circuit with a switch, and (d) RC circuit with a switch.

where the real part of admittance ($Re(Y(jw))$) is called the *conductive part* or the *conductance* and the imaginary part ($Im(Y(jw))$) is called the *susceptive part* or *susceptance*.

Similarly, let us consider the RC circuit [Fig. 5.5 (b)]. The voltage and current relationship of the circuit is

$$V(t) = R \cdot i(t) + \frac{1}{C} \int_0^t i(\tau)\, d\tau \tag{5.43}$$

$$V(jw) = \left(R + \frac{1}{Cjw}\right) \cdot i(jw) \tag{5.44}$$

$$\frac{V(jw)}{i(jw)} = \frac{1 + RC\ jw}{Cjw} \tag{5.45}$$

The impedance of the RC circuit is defined as

$$Z(jw) = \frac{V(jw)}{i(jw)} = \frac{1 + RC\ jw}{Cjw} \tag{5.46}$$

Notice that if we were interested in the relationship between the input voltage and the voltage across the capacitor ($V_C(t)$) as the output voltage (i.e., the RC circuit is used as a filter between input and output voltages), the transfer function between them would be

$$V_C(t) = \frac{1}{C} \int_0^t i(\tau)\, d\tau \tag{5.47}$$

$$V_C(jw) = \frac{1}{C \cdot jw} \cdot i(jw) \tag{5.48}$$

$$i(jw) = C \cdot jw \cdot V_C(jw) \tag{5.49}$$

and

$$\frac{V_C(jw)}{V(jw)} = \frac{1}{1 + RC\ jw} \tag{5.50}$$

which defines a low-pass filter between the input voltage and output voltages.

It is rather easy to show that the impedance of a resistor, capacitor, and inductor is as follows (Fig. 5.6),

$$Z_R(jw) = R \tag{5.51}$$

$$Z_C(jw) = \frac{1}{Cjw}; \quad \text{also called capacitive reactance} \tag{5.52}$$

$$Z_L(jw) = jw\, L; \quad \text{also called inductive reactance} \tag{5.53}$$

Impedance between two points in a circuit is the Fourier transform of the voltage (output) and current (input) ratio. It is also called the transfer function between voltage and current.

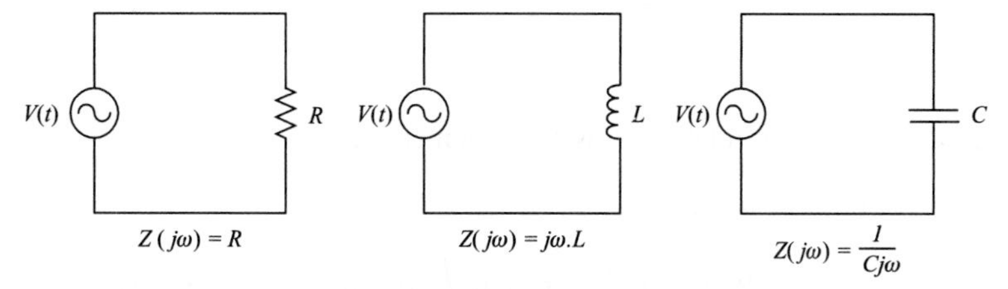


**FIGURE 5.6:** Concept of impedance: generalized resistance as impedance.

Impedance is a complex quantity. It has magnitude and phase, and is a function of frequency. When we refer to large or small impedance, we generally refer to its magnitude.

***Example*** Consider the RL and RC circuits shown in Fig. 5.5 (c,d), respectively. Switch in each circuit is connected to the supply voltage (A side) and B side at specified time instants. Assume the following circuit parameters: $L = 1000\,\text{mH} = 1000 \times 10^{-3}\,\text{H} = 1\,\text{H}$, $C = 0.01\ \mu\text{F} = 0.01 \times 10^{-6}\ \text{F}$, $R = 10\ \text{k}\Omega$. Let the supply voltage be $V_s(t) = 24$ VDC. Assume in each circuit, initial conditions on the current is zero and initial charge in the capacitor is zero. Starting time is $t_0 = 0.0$ sec. At time $t_1 = 100\ \mu$sec the switch is connected to the supply, and at time $t_2 = 500\ \mu$sec, the switch is disconnected from the supply and connected to the B side of the circuit. Plot the voltage across each component and current as function of time for the time period of $t_0 = 0.0$ sec to $t_f = 1000\ \mu$sec.

For the RL circuit, when it is connected to the supply,

$$V_s(t) = L \cdot \frac{di(t)}{dt} + R \cdot i(t); \quad t_1 \leq t \leq t_2 \tag{5.54}$$

and when it is disconnected from the supply and connected to the B side,

$$0 = L \cdot \frac{di(t)}{dt} + R \cdot i(t); \quad t_2 \leq t \leq t_f \tag{5.55}$$

with the initial conditions of current being the last value during the previous state. Another way of treating this problem is to consider the first equation with a pulse input voltage,

$$V_s(t) = L \cdot \frac{di(t)}{dt} + R \cdot i(t); \quad 0 \leq t \leq t_f \tag{5.56}$$

where $V_s(t)$ is a pulse signal, that is, the circuit is connected to 24 VDC (switch is on A side) and 0.0 VDC (switch is on B side),

$$V_s(t) = 24 \cdot (1(t - t_1) - 1(t - t_2)) \tag{5.57}$$

where $1(t)$ represents the unit step function, $1(t - t_1)$ represents the unit step function shifted in time by $t_1$.

The solution of the above differential equations can be shown to be as follows (see the last section in Appendix B for details),

$$i(t) = 0.0\,; \quad t_0 \leq t \leq t_1 \tag{5.58}$$

$$i(t) = 24/(10 \times 10^3) \cdot \left(1 - e^{-(t-t_1)/(L/R)}\right)\ \text{A} \tag{5.59}$$

$$= 2.4 \times \left(1 - e^{-(t-0.0001)/0.0001}\right)\ \text{mA}\,; \quad t_1 \leq t \leq t_2 \tag{5.60}$$

$$i(t) = i(t_2) \cdot \left(e^{-(t-t_2)/(L/R)}\right)\ A = 2.356 \cdot e^{-(t-0.0005)/0.0001}\ \text{mA}\,; \quad t_2 \leq t \leq t_f \tag{5.61}$$

Then the voltage across the resistor and inductor in each time period can be found by

$$V_R(t) = R \cdot i(t) \tag{5.62}$$

$$V_L(t) = L \cdot \frac{di(t)}{dt} = V_s(t) - R \cdot i(t) \tag{5.63}$$

For RC circuit, the circuit relationships are, when connected to supply,

$$V_s(t) = R \cdot i(t) + V_c(t) \tag{5.64}$$

$$= R \cdot i(t) + \frac{1}{C}(Q(t_1) + \int_{t_1}^{t} i(\tau)\,d\tau); \quad t_1 \leq t \leq t_2 \tag{5.65}$$

and when it is disconnected from the supply and connected to the B side,

$$0.0 = R \cdot i(t) + \frac{1}{C}\left(Q(t_2) + \int_{t_2}^{t} i(\tau)\,d\tau\right); \quad t_2 \le t \le t_f \tag{5.66}$$

The voltage across the capacitor at any given time ($t$) is the initial voltage due to initial charge at a given time ($t_0$) plus the net change in the charge (integral of the current from the initial time ($t_0$) to present time ($t$))

$$V_c(t) = \frac{1}{C}\left(Q(t_0) + \int_{t_0}^{t} i(\tau)d\tau\right) \tag{5.67}$$

Again, the above two different state of the RC circuit can be represented with the first equation with a pulse input voltage representation,

$$V_s(t) = R \cdot i(t) + \frac{1}{C}\left(\int_{t_1}^{t} i(\tau)d\tau\right); \quad t_1 \le t \le t_f \tag{5.68}$$

where we used the fact that $Q(t_1) = 0.0$ since initially the capacitor is assumed uncharged and

$$V_s(t) = 24 \cdot (1(t - t_1) - 1(t - t_2)) \tag{5.69}$$

The above differential equation can be solved using Laplace transforms method for $i(t)$, then $V_c(t)$ and $V_R(t)$ can be obtained from the current–voltage relationship across capacitor and resistor components. (See the last section in Appendix B for detailed solution with Laplace transforms and using Simulink.)

$$i(t) = 0.0; \quad t_0 \le t \le t_1 \tag{5.70}$$

$$i(t) = 2.4 \cdot e^{-(t-0.0001)/0.0001}\text{mA}; \quad t_1 \le t \le t_2 \tag{5.71}$$

$$i(t) = -2.356 \cdot e^{-(t-0.0001)/0.0001}\text{mA}; \quad t_2 \le t \le t_f \tag{5.72}$$

$$V_c(t) = 0.0; \quad t_0 \le t \le t_1 \tag{5.73}$$

$$V_c(t) = 24 \cdot (1 - e^{-(t-t_1)/(RC)})\ V = 24 \cdot (1 - e^{(t-0.0001)/0.0001})\ V; \quad t_1 \le t \le t_2 \tag{5.74}$$

$$V_c(t) = V_c(t_2) \cdot (e^{-(t-t_2)/(RC)})\ V = 23.56 \cdot e^{-(t-0.0005)/0.0001}\ V; \quad t_2 \le t \le t_f \tag{5.75}$$

It is important to recognize that both circuits are first order filters and that the response speed is dominated by the time constant of the circuit, $\tau = L/R = 100\,\mu$sec for the RL circuit and $\tau = RC = 100\,\mu$sec for the RC circuit.

### 5.4.2 Amplifier: Gain, Input Impedance, and Output Impedance

Electronic circuits consist of connections between components where the output of one component is connected to the input of another component. The input and output impedances of connected components are important and can significantly affect the transmitted signal. For instance, op-amps are used to amplify and filter its input signal before passing it to the next component.

An ideal amplifier, in its simplest form, amplifies its input signal (Fig. 5.7) and presents the result as its output signal. It does not change the original signal shape due to the fact that the signal is connected to the amplifier. Consider an operational amplifier (op-amp) connected to an input source (i.e., a sensor signal) and an output load. Ideally, the op-amp has a gain ($K_{amp}$), input impedance ($Z_{in}$ or $R_{in}$), and output impedance ($Z_{out}$ or $R_{out}$). An ideal op-amp has infinite input impedance and zero-output impedance. In reality, it has a

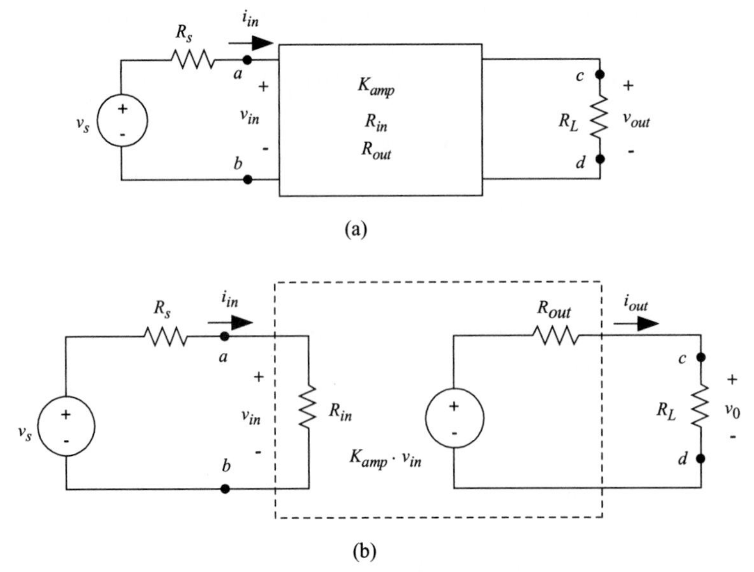


**FIGURE 5.7:** Input and output loading effects as a result of a connection of an op-amp to an input signal source and output load: (a) input and output connection of the op-amp device, and (b) model of the op-amp with input impedance, output impedance, and open-circuit voltage gain.

very large input impedance and very small output impedance. The $Z_{in}$ is the generalized version of $R_{in}$, and $Z_{out}$ is the generalized version of $R_{out}$.

The open-circuit voltage gain, $K_{amp}$, is assumed constant for low frequency signals relative to the bandwidth of the amplifier,

$$V_{out,open} = K_{amp} \cdot V_{in} \tag{5.76}$$

where the $V_{in}$ is the voltage across the input terminals of the amplifier, and $V_{out,open}$ is the voltage across the output terminals of the amplifier when it is in open-loop condition.

Let us recall the definition of the *impedance* and the fact that it is a generalization of resistance. For the amplifier model shown in Fig. 5.7, the input and output impedances are defined as

$$Z_{in}(jw) = \frac{V_{in}(jw)}{i_{in}(jw)} \tag{5.77}$$

$$Z_{out}(jw) = \frac{V_{out}(jw)}{i_{out}(jw)} \tag{5.78}$$

The input impedance is the generalized resistance seen by the input signal source. The output impedance is the generalized resistance of a voltage source which is same as the output resistance of its equivalent Thevenin's circuit. The gain, input, and output impendances are three parameters that effectively define the performance characteristics of an amplifier.

### 5.4.3 Input and Output Loading Errors

Let us consider an amplifier connected between a voltage source and a load (Fig. 5.7). The input voltage is affected by the input impedance of the amplifier,

$$V_{in} = \frac{R_{in}}{R_{in} + R_s} v_s \tag{5.79}$$

and the output voltage is effected by the output impedance,

$$V_{out} = \frac{R_L}{R_L + R_{out}} K_{amp} V_{in} \tag{5.80}$$

Notice that ideally, when $R_{in} \to \infty$ and $R_{out} \to 0$, $V_{in} = v_s$ and $V_{out} = K_{amp}\, v_s$, where the input and output loading effects are zero. In reality, the finite input and output impedances introduces the *loading errors* to the signal amplification. It is minimized when the input impedance is very large relative to the source impedance, and output impedance is very small relative to the load impedance. The net gain of the amplifier in a real amplifier is

$$v_o = \left(\frac{R_L}{R_L + R_{out}}\right)\left(\frac{R_{in}}{R_{in} + R_s}\right) K_{amp}\, v_s \tag{5.81}$$

where the portion of the amplifier gain

$$\frac{R_L}{R_L + R_{out}} \frac{R_{in}}{R_{in} + R_s} \tag{5.82}$$

is due to the input and output loading effects. In order to minimize the effect of input loading errors,

$$R_{in} \gg R_s \tag{5.83}$$

Similarly, in order to minimize the output loading errors,

$$R_{out} \ll R_L \tag{5.84}$$

Furthermore, more accurate analysis of the amplifier would take the frequency dependance of the input and output impedances into account ($Z_{in}(jw)$ instead of $R_{in}$, $Z_{out}(jw)$ instead of $R_{out}$).

**Example** Let us consider a sensor, a signal amplifier, and a measurement device—all connected in series as shown in Fig. 5.7. The sensor provides a voltage proportional to the measured physical variable. Let us call that voltage $v_s(t) = 10$ V at steady-state condition. For simplicity let us assume that the gain of the amplifier is unity, $K_{amp} = 1.0$. Determine the voltage measured at the measuring device (i.e., digital voltmeter or oscilloscope) for the following two conditions:

1. $R_s = 100\ \Omega$, $R_L = 100\ \Omega$, $R_{in} = 100\ \Omega$, $R_{out} = 100\ \Omega$
2. $R_s = 100\ \Omega$, $R_L = 100\ \Omega$, $R_{in} = 1{,}000{,}000\ \Omega$, $R_{out} = 1\ \Omega$

For the first case, the measured voltage is

$$v_{out} = \left(\frac{R_L}{R_L + R_o}\right)\left(\frac{R_{in}}{R_{in} + R_s}\right) K_{amp}\, v_s \tag{5.85}$$

$$= \left(\frac{100}{100 + 100}\right)\left(\frac{100}{100 + 100}\right) 1\, v_s \tag{5.86}$$

$$= 0.25 \cdot v_s \tag{5.87}$$

$$= 2.5 \text{ V} \tag{5.88}$$

which shows an error of 75% of the correct value of the voltage.

For the second case the measured voltage is

$$v_{out} = \left(\frac{R_L}{R_L + R_{out}}\right)\left(\frac{R_{in}}{R_{in} + R_s}\right) K_{amp}\, v_s \tag{5.89}$$

$$= \left(\frac{100}{100+1}\right)\left(\frac{1{,}000{,}000}{1{,}000{,}000+100}\right) \cdot 1 \cdot v_s \quad (5.90)$$

$$= \left(\frac{100}{101}\right)\left(\frac{1{,}000{,}000}{1{,}000{,}100}\right) \cdot v_s \quad (5.91)$$

$$= 0.990 \cdot v_s \quad (5.92)$$

$$= 9.90 \text{ V} \quad (5.93)$$

which shows an error of 1% of the correct value of the voltage. Notice that there are two components to the error in the gain due to input and output loading, one due to the input loading ($K_1$) and and the other due to the output loading ($K_2$),

$$v_{out} = K_2 \cdot K_1 \cdot K_{amp}\, v_s \quad (5.94)$$

where

$$K_1 = \left(\frac{R_{in}}{R_{in} + R_s}\right) \quad (5.95)$$

$$= \frac{1{,}000{,}000}{1{,}000{,}100} = 0.9999 \quad (5.96)$$

$$K_2 = \left(\frac{R_L}{R_L + R_{out}}\right) \quad (5.97)$$

$$= \frac{100}{101} = 0.990 \quad (5.98)$$

This example illustrates that in order to minimize the effect of loading errors, components should have very large input impedance (ideally infinite), and very small output impendace (ideally zero).

## 5.5 SEMICONDUCTOR ELECTRONIC DEVICES

Electronic systems include components made using *semiconductor* materials, such as diodes, silicon controlled rectifiers (SCRs), transistors, and integrated circuits (ICs). Pictures of some common semiconductor devices are shown in Fig. 5.8. Because current flow is accomplished through the flow of electrons in the solid crystal structure of semiconductor materials, components made from semiconductor materials are also called *solid-state* devices. The difference between an electrical and electronic system lies on whether solid-state components (diodes, transistors, ICs) are used in the circuit or not.

### 5.5.1 Semiconductor Materials

Semiconductor materials comprise the group IV elements in the periodic table. Most commonly used semiconductor materials are silicon (Si) and germanium (Ge). Semiconductor materials have electrical conductivity property that is somewhere in between the conductors and insulators (nonconductors), hence the name *semiconductors*. A silicon atom has 14 electrons, and four of these electrons revolve in the outermost orbit around the nucleous [Fig. 5.9 (a)]. In its pure form, silicon is not much of use as a semiconductor. Crystalline structure of silicon is similar to diamond structure of carbon, which is a very stable structure [Fig. 5.9 (b)]. A three-dimensional view of the crystal structure shows that each silicon

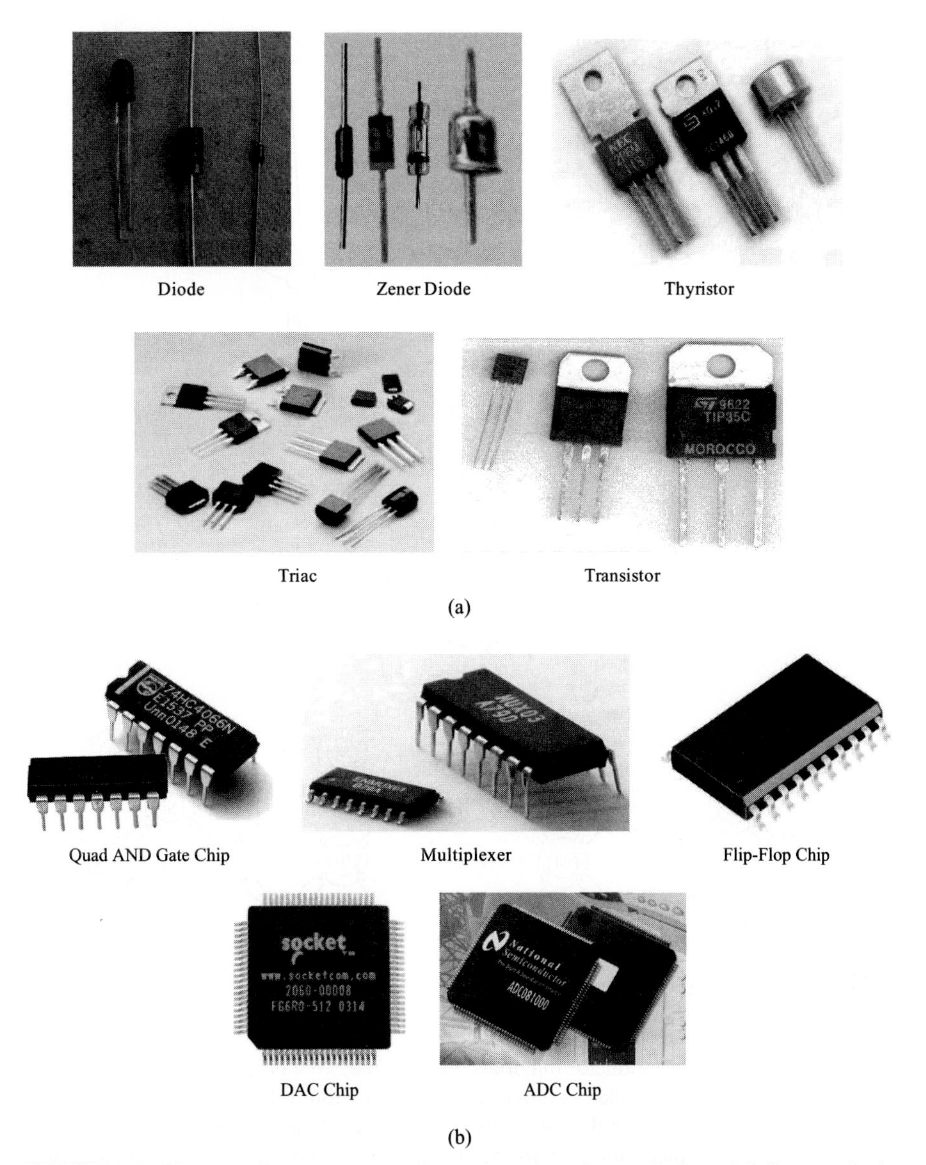


**FIGURE 5.8:** Pictures of some commonly used semiconductor devices: (a) discrete devices (diodes, transistors) and (b) integrated circuit chips (ICs).

atom is surrounded by four other atoms, and they each share one electron in each connection [Fig. 5.9 (a,b,c)].

Impurities from group III and group V elements are added to pure silicon. The resulting silicon is also called *doped silicon*. When a material from group III of periodic table [i.e., boron (B)] is added to the silicon, there will be one missing electron for each added impurity atom in the crystal structure. One of the four connections around the group III atom will be missing an electron [Fig. 5.9 (d)]. This effective positive charge, also called a *hole*, is loosely connected to the atom. Its ability to move around the crystal structure gives the material the ability to conduct electricity. Since the net added electrical charge is positive, such semiconductors are called *p-type semiconductors* [Fig. 5.9 (d)]. Similarly, if the added material is from group V of the periodic table [i.e., phospore (P), arsenic (As)], there will be an extra electron around the added impurity atom. This provides the conductor property to

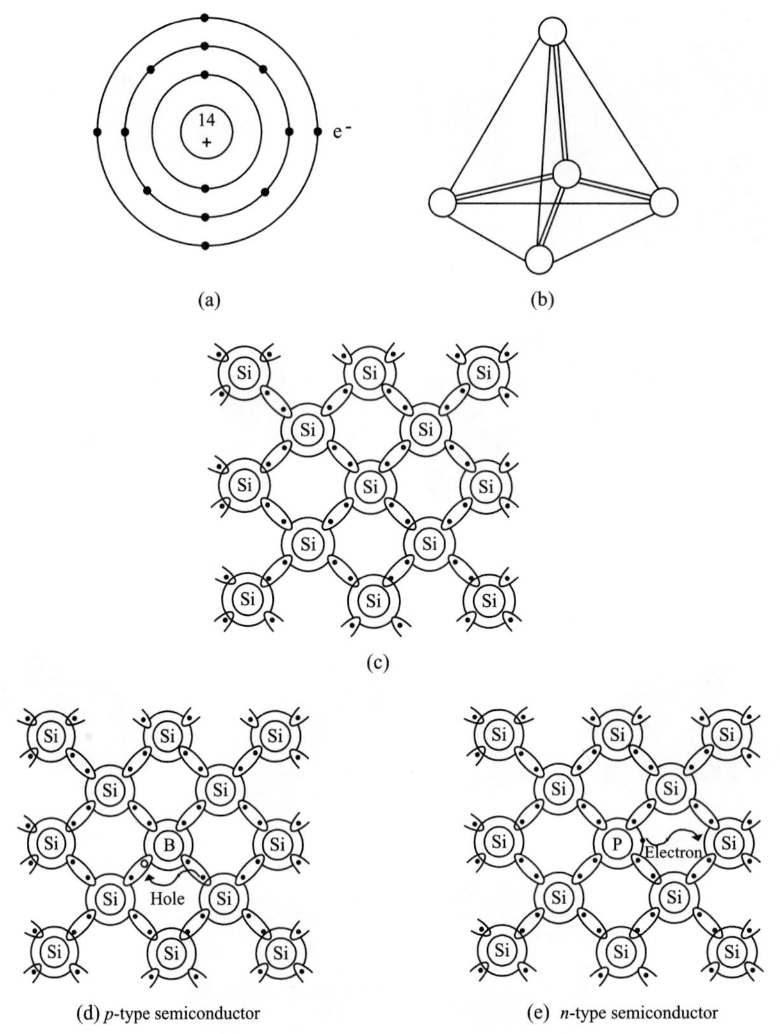


**FIGURE 5.9:** Semiconductor materials: (a) silicon atom has four electrons in the outermost oribit. (b) and (c) each silicon atom is surrounded by four other atoms in the crystal structure where it is shown both in three-dimensional and two-dimensional representation. (d) *p*-type and (e) *n*-type semiconductor material (doped silicon) crystal structure (from reference [42]).

the semiconductor material. Since the net added electrical charge to the crystal structure is a negative charge due to extra electron, this type is called *n-type semiconductor* [Fig. 5.9(e)]. All semiconductor materials conduct electricity through the holes and electrons introduced by the doping material into its crystal structure.

Semiconductor materials are the basis of electronic switches which contains various junctions of *n*-type and *p*-type semiconductor materials. The conductivity property of these junctions are controlled by a base current under our control. The device acts as a conductor (closed switch) and as a nonconductor (insulator, open switch) under different operating conditions.

### 5.5.2 Diodes

A diode is a two-terminal device made of a *p-n* junction [Fig. 5.10 (a)]. It works like an electronic switch that allows current flow only in one direction. It functions like a one-directional check valve. It lets the flow in one direction when the pressure is above a threshold value and it blocks the flow in the reverse direction. It is a conductor in positive direction, and insulator in the negative direction. Once it starts conducting, the resistance of it is negligible. It is like a closed switch. The voltage and current relationship across a diode has three regions: (1) forward bias region, (2) reverse bias region, and (3) reverse breakdown region. When the voltage is larger than the forward bias voltage value ($V_F = 0.3$ V for germanium diode, $V_F = 0.6$ V to 0.7 V for silicon diode), $V_D > V_F$, the diode becomes a conductor and acts as a closed switch. Hence, the current increases very fast indicating very low resistance at this device. The actual value of the current is determined by the rest of the circuit. In this region, the diode can be considered a closed switch with negligible resistance and a small voltage drop of $V_F$. When the diode is in reverse bias region, $V_Z < V_D < V_F$, then the diode acts as an insulator or open switch. In other words, it acts like a very large resistor. When the $V_D < V_Z$, the diode operates in the breakdown region, and acts as a closed switch. This is referred to as the *avalanche effect*, and large current flows in the reverse direction. If this reverse current is larger than a rated value for the diode, the component will fail. The design parameters of a diode include:

1. $V_F$, forward bias voltage
2. $V_Z$, reverse breakdown voltage
3. $f_{sw}$, maximum switching frequency the diode can respond to

The most common applications of diodes are half-wave and full-wave rectification of AC signals to DC signals.

**Zener Diodes** Zener diode is a special type of diode which makes use of the reverse breakdown voltage ($V_Z$). It operates in the breakdown region without destroying itself. Given a well-defined reverse breakdown voltage, Zener diode is generally used as a voltage regulator under varying supply and load voltage conditions [Fig. 5.10 (b)]. Any voltage above the $V_Z$ value across the Zener diode, hence across the load, will be reduced to $V_Z$ due to very high conductance across the diode. Therefore, the current across the load and diode are

$$i_L = \frac{V_Z}{R_L}; \quad V_S \geq V_Z \tag{5.99}$$

$$i_S = \frac{V_S - V_Z}{R_S}; \quad V_s \geq V_Z \tag{5.100}$$

$$i_Z = i_S - i_L; \quad V_S \geq V_Z \tag{5.101}$$

The power dissipated on a Zener diode is

$$P_Z = i_Z \cdot V_Z = \left[ \frac{(V_S - V_Z)}{R_S} - \frac{V_Z}{R_L} \right] \cdot V_Z \tag{5.102}$$

which must not exceed the power rating of the diode. The main design parameters of a diode are the maximum power dissipation capacity ($P_{Z,max}$, i.e., 0.25 W to 50 W) and the reverse breakdown voltage ($V_Z$, i.e., 5 V to 100 V range).

**Thyristors** Thyristor (also called *silicon controller rectifier* (SCR)) is a three-terminal device and made of three junctions (junctions of PNPN) using two $p$-type and two $n$-type semiconductor materials [Fig. 5.10 (c)]. The three terminals are anode (A), cathode (K), and gate (G). The input–output relationship in steady state is the voltage ($V_{AK}$) and current ($i_{AK}$) between A and K where this relationship is a function of the gate current ($i_G$). SCR is a controllable diode. The gate is used to turn ON the SCR, but it cannot be used to turn it OFF. SCR turns OFF when the $i_{AK}$ goes to zero, then the gate can be used to turn it ON

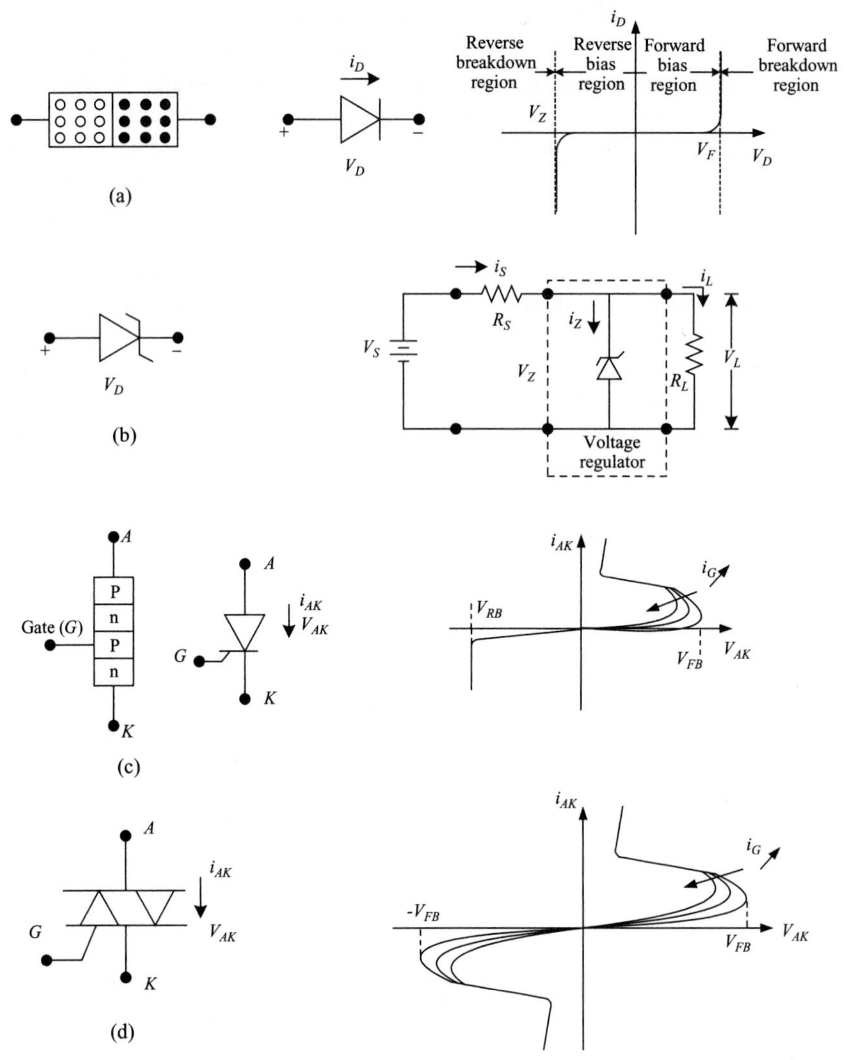


**FIGURE 5.10:** Some semiconductor devices: (a) Diode (*p-n* junction), (b) Zener diode, (c) Thyristor (silicon controller rectifier or SCR), and (d) Triac.

again. In order to turn ON the SCR, it must be forward biased ($V_{AK} \geq V_{FB}(i_G)$) plus the gate current must be above the *latching current* specification ($i_G > i_{G,min}$), the minimum amount of current required to turn ON the SCR. The forward voltage drop $V_{FB}$ is in the range of 0.5 V to 1.0 V. The current conducted by the SCR can range between 100 mA and 100 A. The power dissipated across the SCR is the voltage drop across it times the conducting current,

$$P_{SCR} = V_{drop} \cdot i_{AK} < P_{max} \tag{5.103}$$

By controlling the timing of the gate current while the SCR is forward biased, a controlled portion of a forward alternating voltage can be conducted. This is used in speed control of DC motors from AC source. If SCR is forward biased suddenly (i.e., $dV_{AK}/dt > 50V/\mu\text{sec}$), it may turn ON even though base current is not applied. To reduce such undesirable effects, a low-pass passive RC circuit (also called *snubber circuit*, R and C are in series) is used in parallel with the SCR. In addition, when a SCR is turned ON, the rate of current change, *di(t)/dt*, may be very large. In order to reduce that, the load should have inductance that is larger than a minimum required value. If the inductance is very low, the rate of current change can be too large.

**Triacs** Triac is equivalent to two SCRs connected in reverse-parallel configuration [Fig. 5.10 (d)]. While an SCR can be turned ON only in the forward-biased direction, a triac can be turned ON in both directions. Typical voltage drop across a triac is about 2 V while it is conducting. This voltage times the load current determines the energy dissipation in the form of heat by a triac. SCR and triac components are generally used to switch AC loads. The amount of power that is allowed to conduct is controlled by the timing gate voltage of SCR and triac relative to the AC voltage applied between terminals $A$ and $K$.

**Example** Consider the Zener diode application circuit shown in [Fig. 5.10 (b)]. Let us assume that $V_s = 24$ V, $R_s = 1000\ \Omega$, $V_Z = 12$ V, and $R_L$ varies between 1000 Ω and 2000 Ω. Let us determine the currents in each branch of the circuit, voltage across the load, and power dissipated across the zener diode. Kirchhoff's voltage law in the loop including the power supply and zener diode gives

$$V_s = R_s \cdot i_s + V_Z \tag{5.104}$$

$$24 = 1000 \cdot i_s + 12 \tag{5.105}$$

$$i_s = \frac{24 - 12}{1000} = 0.012 \text{ A} = 12 \text{ mA} \tag{5.106}$$

Since voltage drop across the zener diode is $V_Z = 12$ V always for $V_S > V_Z$, the same voltage potential exists across the parallel load resistor,

$$V_{R_L} = V_Z = R_L \cdot i_L \tag{5.107}$$

$$12 = R_L \cdot i_L \tag{5.108}$$

$$i_L = \frac{12}{R_L} \tag{5.109}$$

when $R_L = 1000\ \Omega$, $i_L = 12.0$ mA, and when $R_L = 2000\ \Omega$, $i_L = 6.0$ mA. Since

$$i_s = i_L + i_Z \tag{5.110}$$

the current across the diode is

$$i_Z = i_s - i_L \tag{5.111}$$

which varies as the load resistance varies. In other words, the current across the zener diode is $i_Z = 0.0$ mA when $R_L = 1000\ \Omega$ and $i_Z = 6.0$ mA when $R_L = 2000\ \Omega$. The zener diode provides a constant 12 V voltage across the load resistor. As the load resistor varies, the zener diode dumps the excess current while providing the $V_Z = 12$ V constant voltage across the two terminals. If $R_L < 1000\ \Omega$, i.e., $R_L = 500\ \Omega$, the diode does not conduct. Assume that the diode is not conducting; then, $i_Z = 0$,

$$i_s = i_L = \frac{V_s}{R_L + R_s} = \frac{24}{1500} = 0.016 \text{ A} = 16 \text{ mA} \tag{5.112}$$

$$V_L = R_L \cdot i_L = 500\ \Omega/6 \text{ mA} = 8 \text{ V} < V_Z \tag{5.113}$$

So the Zener diode acts as a component to dump the excess voltage, but does not make up

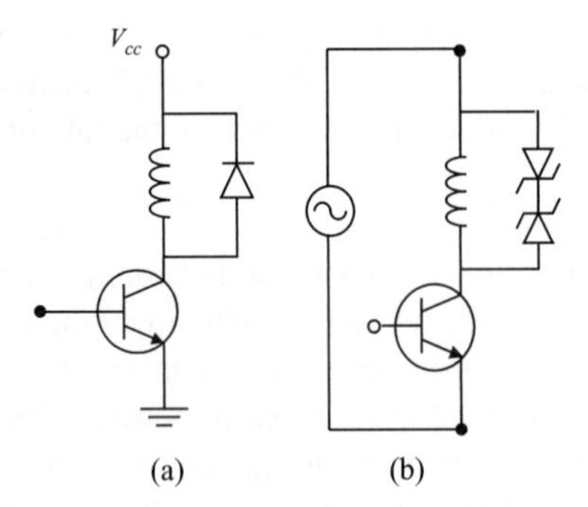


**FIGURE 5.11:** Diode application: voltage surge protection when inductive load is present in the circuit. (a) For DC circuits, and (b) for AC circuits.

for lower voltages. The maximum power dissipated across the Zener diode in this example is

$$P_Z = V_Z \cdot i_Z = 12\text{ V} \cdot 6\text{ mA} = 72\text{ mW} \tag{5.114}$$

Therefore, a Zener diode with 1/4 W power rating and 12 V breakdown voltage rating would be sufficient for this circuit.

Figure 5.11 shows the application of diodes for *voltage surge protection* due to inductive loads. Such use of diodes is called *freewheeling diodes*. When a diode is placed parallel to the transistor in motor control applications, it is called *bypass diode*. Diode limits the maximum collector voltage to $V_{CC}$ plus the diode forward bias voltage. When there is an inductive load (a coil winding of conductor, i.e., in relays, solenoids, motors), sudden switching of the current either by mechanical switches or electronic switches (transistors) results in large voltage surges due to the inductive load. Recall that for an inductor

$$V(t) = L \cdot \frac{di(t)}{dt} \tag{5.115}$$

When the current is switched suddenly, there can be large voltage surges because $di(t)/dt$ can be very large. Especially when the current is switched OFF, the surge in current does not have path to travel but shows up across the switch which can be damaged. In addition, the voltage surge may be so large as to damage the insulation of the conductor. In order to reduce the effects of the surge voltages, a diode is used across each inductive load. If the voltage supplied to inductive load is DC, a single diode is sufficient. If it is AC, then two opposite Zener diodes are used to handle the voltage surges in both directions (Fig. 5.11). In order to quickly dissipate the power trapped in the inductive load after the switch-OFF, some designs include a small resistor in series with the diodes.

Light-emitting diodes (LEDs) and light-sensitive diodes (LSDs) are diodes that give out light with intensity that is proportional to the current and pass currents proportional to the received light intensity, respectively. Furthermore, the light frequency can be modulated or pulsed (i.e., up to the 10-MHz range). LEDs and LSDs are used as *opto-couplers* in electrical circuits in order to electrically isolate two circuits and couple them optically (Fig. 5.12).

### 5.5.3 Transistors

A transistor can function in two ways: (1) ON/OFF switch, (2) proportional amplifier. It is an electronic switch (solid-state switch) that can be opened or closed completely or partially. Another way of looking at a transistor is that it is a variable resistor where the resistor value is controlled by the gate current. It is an active element which is used to modulate

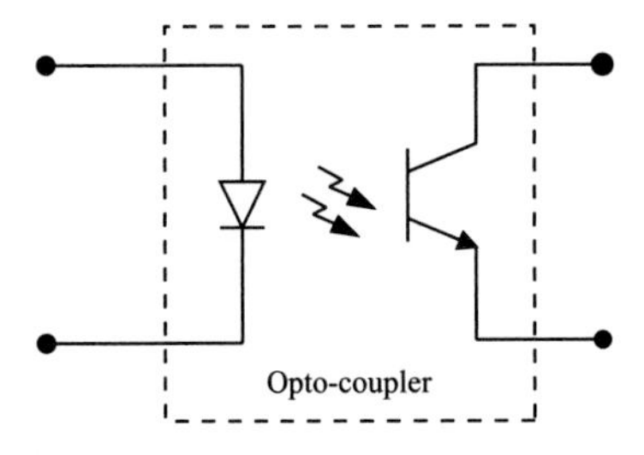


**FIGURE 5.12:** Optocoupler symbol: light-emitting diode (LED) and light-sensitive transistor (LST) pair used to electrically isolate two sides of the circuit using optical light as the coupling medium.

(control) the flow of power from source to load. A transistor is a three-terminal device made of two junctions of *p*-type and *n*-type semiconductors. If the junctions are made of *n-p-n* order, it is called *npn*-type transistor. If the junctions are made of *p-n-p* order, it is called *pnp*-type transistor. The difference between them in a circuit application is that the supply connection polarity is opposite, hence the direction of current is opposite. The three terminals are collector (C), emittor (E), and base (B) (Fig. 5.13). *One of the important features of input and output relations of a transistor is that the input strongly affects the output. However, the output variations (i.e., due to load changes) do not affect the input.* Therefore, the base of a transistor is always part of the input circuit which controls the output circuit. Unlike passive components such as resistors, capacitors, and inductors, the transistor is an active device. It requires a power supply to operate and can increase the power between input and output. Input circuit acts as a control circuit which controls a much larger power level in the output circuit using a much smaller power level at the input circuit. Transistor is the electronic analogy of a hydraulic valve.

The main types of transistors are discussed below. Bipolar junction transistors (BJT) are the most common transistors. Metal oxide semiconductor field effect transistors (MOSFET) require smaller gate current, have better efficiency, and have higher switching frequency. Insulated gate bipolar transistor (IGBT) is a more recent transistor type which attempts to combine the advantages of BJT and MOSFET. A transistor performance is characterized by the maximum base current, current between the two main terminals (collector-emitter, drain-source), forward-bias voltage, reverse breakdown voltage, maximum switching frequency, forward current gain, reverse voltage gain, input impedance, and output impedance.

***Bipolar Junction Transistors (BJT)*** There are two types of BJTs: *npn*-type and *pnp*-type, each type being a three-terminal device [Fig. 5.13 (a)]. The input–output relationship between collector (C) and emitter (E) terminals is controlled by the base (B) signal. A BJT can be used as an electronic switch (ON/OFF) or a proportional amplifier (output current $i_C$ is proportional to input current $i_B$). The voltage and current relationship across the C and E terminals of the transistor ($V_{CE}, i_{CE}$) is a function of the base current and the output circuit. There are three main design parameters of interest: current gain ($\beta$), voltage drop between base and emitter when the transistor is conducting ($V_{BE} = V_{FB}$), and the minimum voltage drop across the collector and emitter when the transistor is saturated or fully ON ($V_{CE} = V_{SAT}$). Note that $i_E = i_B + i_C$ always.

There are three main modes of operation of the BJT in steady state,

1. *Cut-off Region:* $V_{BE} < V_{FB}, i_B = 0$, hence, $i_C = 0$, $V_{CE} = V_{supply}$. A transistor acts like an OFF state switch in the cut-off region. There is no current flow between the C and E. $V_{FB}$ of a transistor can vary between 0.6 V to 0.8 V as a result of manufacturing variations.

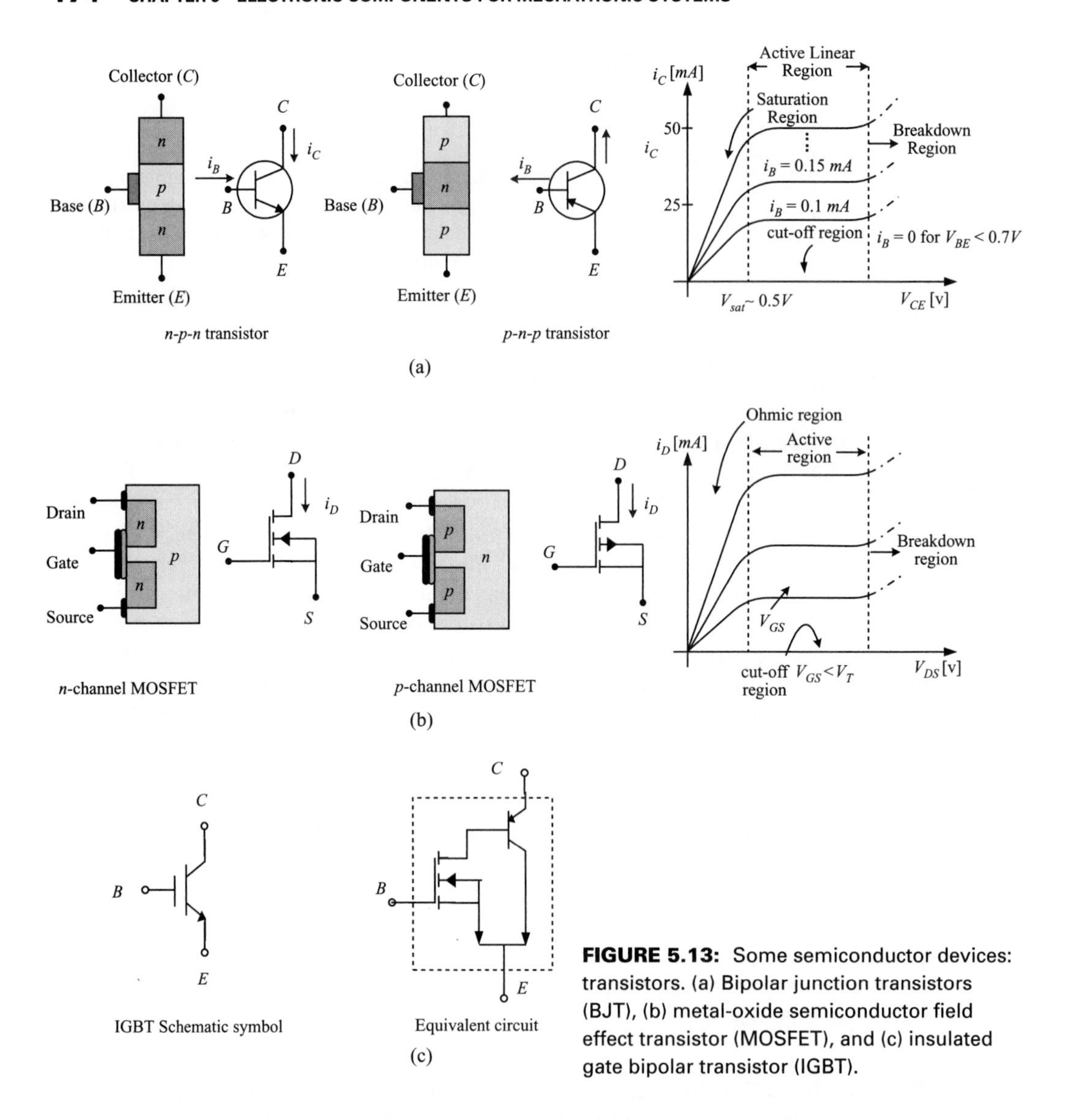


**FIGURE 5.13:** Some semiconductor devices: transistors. (a) Bipolar junction transistors (BJT), (b) metal-oxide semiconductor field effect transistor (MOSFET), and (c) insulated gate bipolar transistor (IGBT).

2. *Active Linear Region:* $V_{BE} = V_{FB}, \ i_B \neq 0; \ i_C = \beta i_B, \ V_{SAT} < V_{CE} < V_{supply}$. The typical value of $V_{SAT} \approx 0.2$ V to 0.5 V. Transistor acts like a current amplifier. The output current, $i_C$, is proportional to the base current, $i_B$, where the proportionality constant (gain) is a design parameter of the transistor, which is typically around 100 and can range from 50 to 200. Good circuit designs should not rely on this open-loop transistor gain for it varies significantly from one copy to another of the same transistor as well as function of temperature. The current $i_C$ is also function of the $V_{CE}$ slightly. Under constant base current conditions, the collector current increases slightly as the collector-to-emitter voltage increases.
3. *Saturation Region:* $V_{BE} = V_{FB}, \ i_B > i_{C,max}/\beta, \ V_{CE} \leq V_{SAT} \approx 0.2\text{ V} - 0.5\text{ V}$. In this mode the transistor operates like a closed (ON) switch between C and E terminals. The actual value of $i_C$ is determined by the rest of the output circuit, which is analogous to a completely open valve where the flow rate is determined by the supply and load pressures.

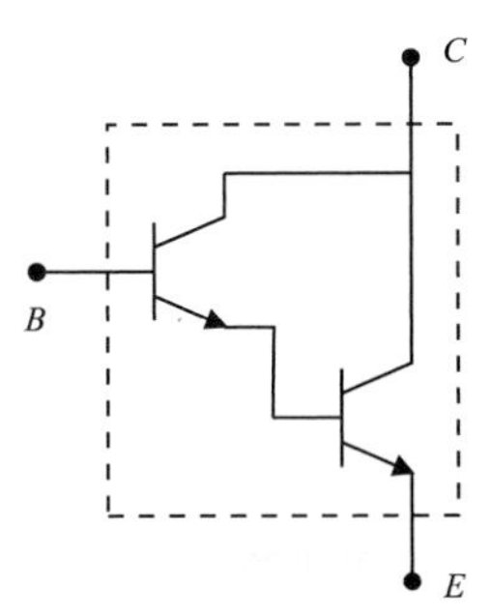


**FIGURE 5.14:** Darlington transistor symbol.

In approximate calculations, it is customary to assume $V_{FB} = V_{SAT} = 0.0$ and $i_E = i_C$ in transistor circuits.

BJT type transistors can support collector currents in the range of 100 mA to 10 A. The BJTs rated for above 500-mA current are called *power transistors* and must be mounted on a heat sink.

In order to obtain higher gain, $\beta$, from BJTs, a commonly used transistor is the Darlington transistor which gives a gain that is the multiplication of the two stages of the transistors (Fig. 5.14)

$$\beta = \beta_1 \cdot \beta_2 \tag{5.116}$$

where the gain $\beta$ of a Darlington transistor can be in the order of 500 to 20,000.

Note that the power dissipation across a BJT is

$$P_{BJT} = V_{CE} \cdot i_{CE} \tag{5.117}$$

which should be below the rated power of the transistor in a given design. The power dissipation across the transistor is a key factor to consider for two different reasons:

1. Failure of the component due to excessive heating and the associated heat dissipation issues
2. Efficiency of the component by reducing the wasted heat energy

When a transistor operates in fully ON state (in saturated region), the voltage drop across it is very small, $V_{CE} \approx 0.2$ V. Hence the power dissipation is small. Similarly, when the transistor is fully OFF (in cutoff region), even though the voltage drop across it is large, the current conducted is zero, $i_C \approx 0.0$. Hence power dissipation is again small. The observation we make here is that when the transistor is operated either in fully ON or fully OFF mode, the power dissipation is minimized and its operational efficiency is improved. Let us consider that we control the gate of the transistor with a high-frequency signal (i.e., $w_{sw} = 10$ KHz). One period of the gate signal is $t_{sw} = 100\ \mu$sec long. If we control the width of ON portion of the signal within each period that will saturate the transistor (fully ON) and the width of the signal that will turn OFF the transistor, we can control the average gain of the transistor by the average of the ON-OFF widths of the gate signal. By increasing and decreasing the ON-OFF width periods within each $t_{sw}$ period, we can vary the net average gain. Since only two states of the transistor control are needed, the gate signal only has to have two voltage levels: high level to saturate the transistor and low level to cutoff the transistor. Then the only control problem is the problem of controlling the ON-OFF widths of the switching pulses. This is called the *pulse width modulated* (PWM) control of transistors. PWM control method results in more efficient operation of the transistors. PWM control method operates a transistor in one of two regions: cut-off region (OFF state) and saturation region (ON state).

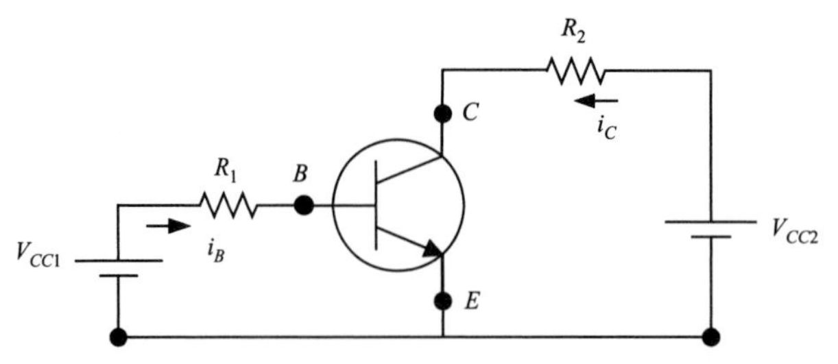


**FIGURE 5.15:** Common emitter configuration and voltage amplifier usage of a transistor.

If the transistor is operated in the *active region*, which is the region where the output current is linearly proportional to the gate current, we use the transistor in linear (or proportional) mode. In this case both the $i_C$ and $V_{CE}$ can have finite values, neither one close to zero. Hence, in linear gate signal control mode, the power dissipation is larger. In linear operating mode, the efficiency of the transistor is lower compared to the efficiency in PWM mode.

In the *common-emitter* configuration (input circuit and output circuit have the emitter as common as shown in Fig. 5.15) of a transistor, input circuit voltage-resistor combination determines the base current. Base current times the transistor gain determines the collector current. Then, the voltage drop across the C and E terminals is determined by the voltage balance in the output circuit loop given the calculated values of collector current, voltage, and resistor. If the calculated $i_C$ requires a $V_{CE}$ value that is less than $V_{CE,sat} = 0.2$ V, then the transistor saturates. Then actual $i_C$ value is determined by the output circuit.

***Metal Oxide Semiconductor Field Effect Transistors (MOSFET)*** MOSFET is also a three-terminal transistor with terminals called drain (D), source (S) and gate (G) [Fig. 5.13 (b)]. Like BJTs, there are two types of MOSFETs: *npn*-type and *pnp*-type. The input (control) variable is the gate voltage, $V_G$. The output is the drain current, $i_D$. There are three regions of operation of a MOSFET,

1. *Cut-off Region:* $V_{GS} < V_T$ and $i_G = 0$, hence, $i_D = 0$, where $V_T$ is the gate-source threshold voltage. The threshold voltage takes on different values for different types of FETs (i.e., $V_T \approx -4$ V for junction FETs, $V_T \approx -5$ V for MOSFET in depletion mode, $V_T \approx 4$ V for MOSFET in enhancement mode). There is negligible current flow through the D terminal and the connection between D and S terminals is in an open-switch state.
2. *Active Region:* $V_{GS} > V_T$, hence, $i_D \propto (V_{GS} - V_T)^2$, $V_{DS} > (V_{GS} - V_T)$. The transistor functions as a voltage controlled current amplifier, where the output current is proportional to the square of the net GS voltage. Notice that for the MOSFET to operate in this region, $V_{DS}$ must be above a certain threshold. Otherwise, MOSFET operates in the *ohmic region*.
3. *Ohmic Region:* When $V_{GS}$ is large enough ($V_{GS} > V_T$) and $V_{DS} < V_{GS} - V_T$, the $i_D$ is determined by the source circuit connected to the D terminal and the transistor behaves like a closed switch between D and S terminals. $V_{GS} \gg V_T$, then, and $i_D = V_{DS}/R_{ON}(V_{GS})$. The MOSFET acts like a nonlinear resistor where the value of the resistance is nonlinear function of the gate voltage.

It should be noted that the advantages of MOSFETs over BJTs are higher efficiency, higher switching frequency, and better thermal stability. On the other hand, MOSFETs are more sensitive to static voltage. In general, MOSFETs have been replacing BJTs in low-voltage ($< 500$ V) applications.

***Insulated Gate Bipolar Transistors (IGBT)*** IGBTs are the new alternative to BJTs in high-voltage ($>500$ V) applications where it combines the advantages of BJTs and MOSFETs. IGBT is a four-layer device with three terminals similar to BJT [Fig. 5.13 (c)]. IGBTs are widely used in motor drive applications and operated in PWM mode for high efficiency.

***Example*** In Fig. 5.15, let $V_{CC1} = 5$ V, $V_{CC2} = 25$ V, $R_1 = 100$ KΩ, $R_2 = 1$ KΩ, $\beta = 100$. The base current is (assuming $V_{BE} = 0.0$ for simplicity)

$$i_B = \frac{5\text{ V}}{100\text{ K}\Omega} = 0.05\text{ mA} \tag{5.118}$$

The collector current is

$$i_C = i_C(i_B, V_{CE}) \tag{5.119}$$

$$i_C \approx \beta \cdot i_B = 5\text{ mA} \tag{5.120}$$

Hence, the voltage balance in the output circuit dictates that

$$25\text{ V} = 1000 \cdot 5 \cdot 10^{-3} + V_{CE} \tag{5.121}$$

$$V_{CE} = 25 - 5 = 20\text{ VDC} \tag{5.122}$$

In general, the output voltage across the transistor for the common-emitter configuration shown in Fig. 5.15 is (let $V_{out} = V_{CE}$, $V_{in} = V_{CC1}$, $V_s = V_{CC2}$)

$$V_{out} = V_s - \frac{R_2}{R_1} \cdot \beta \cdot V_{in} \tag{5.123}$$

and the voltage potential across the output circuit resistor ($V_{R2}$) is

$$V_{R2} = \frac{R_2}{R_1} \cdot \beta \cdot V_{in} \tag{5.124}$$

The common-emitter configuration of the transistor as shown in Fig. 5.15 can be used as a voltage amplifier. If the output voltage is taken as the voltage across the transistor,

$$V_{out} = V_s - K_1 \cdot V_{in} \tag{5.125}$$

or if the output voltage is taken as the voltage across the resistor preceeding the collector,

$$V_{out} = K_1 \cdot V_{in} \tag{5.126}$$

***Example*** In order to illustrate the fact that a transistor can be operated like a switch (ON or OFF) as well as a proportional amplifier, let us consider the following cases of input voltage in the same Fig. 5.15. Furthermore, let us more accurately assume that the maximum voltage drop across the base and emitter is $V_{BE} = 0.7$ V, and minimum voltage drop across the collector and emitter is $V_{CE} = 0.2$ V.

1. ***Case 1:*** $V_{CC1} = 0.0$ V. Then $i_B = 0$, hence, $i_C = 0$. Therefore, the voltage measured between the collector and the emitter (let us call it the output voltage, $V_{out}$) is $V_{out} = V_{CC2} = 25$ VDC . We can call this OFF state of the transistor.
2. ***Case 2:*** $V_{CC1} = 30.7$ V (more precisely, the input voltage at the gate is large enough to generate a large base current that will saturate the transistor output circuit). Then $i_B =$

$(30.7 - 0.7)/100K = 0.3$ mA, hence, $i_C = \beta i_B = 30$ mA. Therefore, the voltage measured between the collector and the emitter (let us call it the output voltage, $V_{out}$) is $V_{out} = V_{CC2} - R_2 \cdot i_C = 25 - 1000 \cdot 30 \text{ mA} = -5$ V. Since the output voltage at the collector cannot be less than the emitter voltage plus the 0.2 V minimum voltage drop across the collector and emittor, it must be $V_{out} = 0.2$ V. Hence the actual collector current saturates at $i_C = (V_{CC2} - V_{CE})/R_2 = (25 - 0.2)/1000 = 24.8$ mA and the output voltage at the collector is $V_{out} = 0.2$ V. We can call this the ON state of the transistor.

3. ***Case 3:*** When the input voltage is above cutoff voltage ($V_{BE} = 0.7$ V) but below the saturation voltage, the transistor operates as a proportional voltage amplifier. The saturation voltage can be calculated as follows. The minimum voltage drop across C and E when the transistor is saturated is

$$V_{CE} = 0.2 \text{ V}; \quad \text{when saturated} \tag{5.127}$$

At this point, current is

$$i_C = \frac{V_{CC2} - V_{CE}}{R_2} = \frac{25 - 0.2}{1000} = 24.8 \text{ mA} \tag{5.128}$$

The base current must be

$$i_B = \frac{1}{\beta} i_C = \frac{1}{100} i_C \tag{5.129}$$

$$= \frac{V_{in,sat} - V_{BE}}{R_1} \tag{5.130}$$

$$V_{in,sat} = 0.248 \text{ mA} \cdot 100 \text{ K}\Omega + 0.7 = 25.3 \text{ V} \tag{5.131}$$

Base voltage value larger than that operates the transistor in saturation region.

$$V_{in} < V_{BE}; \quad \text{transistor is in fully OFF state} \tag{5.132}$$

$$V_{BE} < V_{in} < V_{in,sat}; \quad \text{linear region} \tag{5.133}$$

$$V_{in} > V_{in,sat}; \quad \text{saturated region (fully ON state)} \tag{5.134}$$

Cases 1 and 2 are shown as a mechanical switch analogy in Fig. 5.16. Figure 5.16(a) shows the so-called sinking connection, and Fig. 5.16(b) shows the so-called sourcing connection of the transistor to the load. The names *sinking* and *sourcing* are given from the point of view of the transistor that it either *sinks* the current from the load or *sources* current to the load.

**Example** A transistor circuit shown in Fig. 5.17 can be used to switch the power on a load which is controlled by a mechanical switch. Let's assume that $V_{BE} = 0.7$ V and $V_{CE} = 0.2$ V. If the load is the inductive type, a diode should be used in parallel with the load. When the transistor goes from ON to OFF very fast, large voltage transients develop due to inductance. These large voltage transients can be easily large enough to damage the transistor if it is larger than forward break-down voltage of the transistor. The diode limits the voltage to the supply voltage by providing a current flow path during that transient period. The current flow due to inductive voltage is then dissipated in that diode-load loop within a short period of time. The resistor ($R_2$) which connects the base to the ground is not necessary, but makes the circuit a better one by providing a ground to the base when the transistor is not turned ON (when mechanical switch is open). In the following discussion assume the components in the dotted blocks do not exist for simplicity. Let

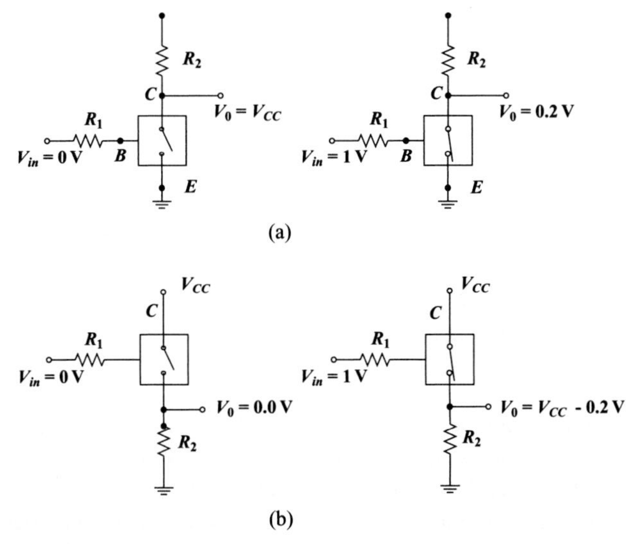


**FIGURE 5.16:** Mechanical switch analogy for a transistor: (a) current sinking connection (sinking current from the load), (b) current sourcing connection (sourcing current to the load).

$L = 0$, $R_2 = 0$. When the mechanical switch is OFF, the base voltage and current are zero, the transistor is OFF, no current flows through the load. When the mechanical switch is ON, the base current and collector current are

$$i_B = \frac{V_{AB}}{R_1} = \frac{(10 - 0.7)V}{1000\ \Omega} = 9.3 \text{ mA} \tag{5.135}$$

$$i_C = \beta \cdot i_B = 0.93 \text{ A} \tag{5.136}$$

where it is assumed that $\beta = 100$. But the maximum current that the supply can support is

$$i_{C,max} = \frac{V_{CC} - V_{CE}}{R_L} = \frac{10 - 0.2}{100} = 0.098 \text{ A} \tag{5.137}$$

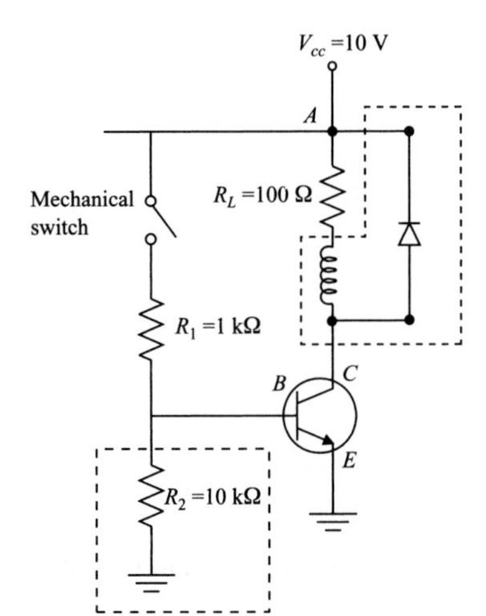


**FIGURE 5.17:** A transistor circuit used as an electronic switch. Diode protects the transistor against the voltage spikes due to inductance when the transistor switches from ON to OFF state.

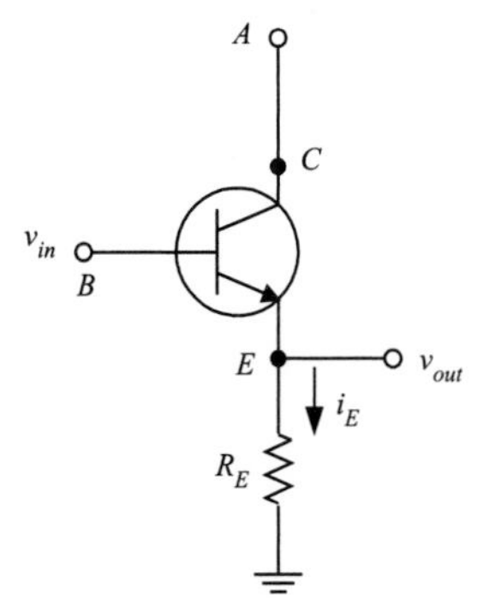


**FIGURE 5.18:** A transistor circuit as a current source.

Hence, the transistor saturates with the maximum voltage drop across the load, $V_L = 10 - 0.2$, and minimum voltage drop across the transistor, $V_{CE} = 0.2$ V. The transistor acts like a very low resistance switch when it is saturated. Voltage drop across it is $V_{CE} = 0.2$ V. Providing excess base current to saturate the transistor is a good idea and provides a safety margin to make sure the transistor is fully turned ON (in its saturation region) and hence providing the maximum voltage drop across the load. Notice that when the transistor is ON, the current drain through the resistor $R_2$ is

$$i_2 = \frac{V_{BE}}{R_2} = \frac{0.7}{10000} = 0.07 \text{ mA} \tag{5.138}$$

which is a negligible load.

***Example*** Another common use of a transistor is in a current-source circuit (Fig. 5.18). Notice that this circuit does not have any voltage amplification, but it has current amplification. Output voltage tracks the base voltage as shown below with the exception of the small voltage drop across the base and emitter. Let us assume voltage drop across the base emitter is $V_{BE} = 0.6$ V.

$$V_{out} = V_{in} - 0.6; \quad when \quad V_{in} \geq 0.6 \text{ V} \tag{5.139}$$

$$V_{out} = 0; \quad when \quad V_{in} < 0.6 \text{ V} \tag{5.140}$$

The current through the $R_E$,

$$i_E = \frac{V_{out}}{R_E} = \frac{V_{in} - 0.6}{R_E} \tag{5.141}$$

and the current through the collector,

$$i_C = \beta \cdot i_B \tag{5.142}$$

$$i_E = i_C + i_B = (\beta + 1) \cdot i_B \tag{5.143}$$

$$= \frac{(\beta + 1)}{\beta} i_C \tag{5.144}$$

and

$$i_C = \frac{\beta}{\beta + 1} i_E = \frac{\beta}{\beta + 1} \frac{V_{in} - 0.6}{R_E} \tag{5.145}$$

as long as the transistor is not saturated, ($V_C > V_E + 0.2$), the current output ($i_E$) is proportional to the base voltage ($V_{in}$).

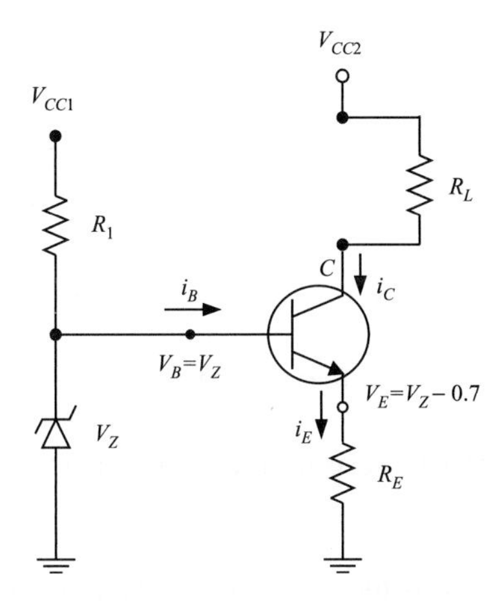


**FIGURE 5.19:** A transistor circuit as a constant current source.

Note that

$$V_{in} = V_{BE} + V_E \tag{5.146}$$

$$V_E = V_{in} - V_{BE} \geq 0.0 \tag{5.147}$$

$$V_C = V_{CE} + V_E \tag{5.148}$$

The voltage at the emitter follows the base voltage. Current through the collector and emitter is proportional to the base voltage.

***Example*** There are many applications where a load resistor varies, and despite these variations, it is desired to provide a constant current through the load.

Let us modify the circuit of the previous example as shown in Fig. 5.19. The main additions are an input circuit supply voltage $V_{CC1}$, a resistor $R_1$, and a Zener diode with breakdown voltage of $V_Z$. In the output circuit, we have a load resistor $R_L$ connected between the output supply $V_{CC2}$ and the collector terminal of the transistor.

Notice that as long as the $V_{CC1} > V_Z$, the voltage across the base and ground will be $V_B = V_Z$, and base current is

$$i_B = (V_{CC1} - V_Z)/R_1 \tag{5.149}$$

The voltage between the emitter and ground is

$$V_E = V_Z - V_{BE} = V_Z - 0.7 \tag{5.150}$$

Hence, current across the resistor $R_E$ is

$$i_E = V_E/R_E = (V_Z - 0.7)/R_E \tag{5.151}$$

The collector current which also is the load current is

$$i_C = i_E - i_B \tag{5.152}$$

Notice that $i_B$ is very small compared to $i_E$. The $i_E$ is constant as long as the $V_{CC1} > V_Z$ and the $R_E$ is constant. Therefore, the $i_C$ will remain a constant current despite the variations in the load resistance $R_L$. Notice that

$$i_C = \beta \cdot i_B = \frac{V_{CC2} - V_C}{R_L}$$

$$V_C = V_{CC2} - R_L \cdot i_C$$

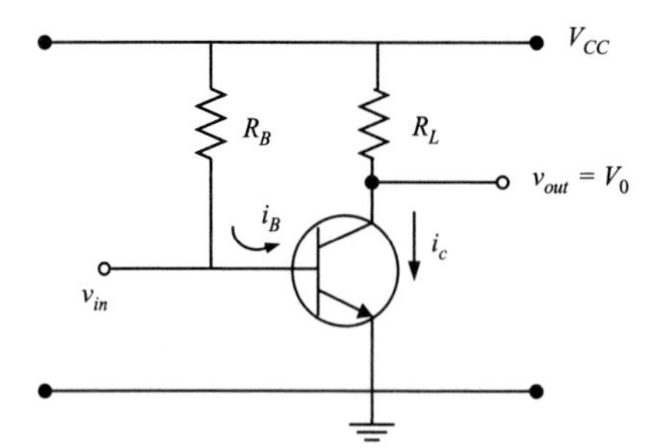


**FIGURE 5.20:** Bias voltage circuit for an amplifier.

which means $V_C$ will vary as $R_L$ varies in order to keep $i_C$ at constant value as long as $V_C \geq V_{CE} = 0.2$ V.

***Bias Voltage*** So far we have considered transistor amplifiers that amplify current or voltage and that when the input signal is zero, output signal is zero as well, i.e.,

$$V_{out} = K_v \cdot V_{in} \tag{5.153}$$

Many applications require offset voltage in the input and output voltage relationship as follows,

$$V_{out} = V_0 + K_v \cdot V_{in} \tag{5.154}$$

where $V_0$ is the bias voltage. There are many different versions of circuit designs to introduce bias voltage to the amplifier, one of which is shown in Fig. 5.20. When the input voltage ($V_{in}$) is zero, there is still a base current due to the power supply voltage ($V_{CC}$) and the resistor $R_B$. Given the supply voltage, load resistance, amplifier current gain, and the desired bias voltage, $V_{CC}$, $R_L$, $\beta$, $V_0$, determine the bias resistance ($R_B$) needed.

The output circuit relations of the transistor is

$$V_{CC} = R_L \cdot i_C + V_{CE} = R_L \cdot i_C + V_{out} \tag{5.155}$$

$$i_C = \frac{V_{CC} - V_{out}}{R_L} \tag{5.156}$$

The input circuit of the transistor has the relationship (assuming the transistor is operating in the linear active region), and $V_{in} = 0.0$,

$$i_B = \frac{1}{\beta} \cdot i_C \tag{5.157}$$

$$i_B = \frac{V_{R_B}}{R_B} \tag{5.158}$$

$$= \frac{V_{CC} - V_{BE}}{R_B} \tag{5.159}$$

$$\approx \frac{V_{CC}}{R_B} \tag{5.160}$$

Let use denote $V_{out} = V_0$ to indicate the bias voltage, since $V_{in} = 0$. From the above relations, we can determine the bias resistor needed to provide the desired bias voltage for the given supply, load, and transistor:

$$V_0 = V_{CC} - R_L \cdot i_C \tag{5.161}$$

$$= V_{CC} - R_L \cdot \beta \cdot i_B \tag{5.162}$$

$$= V_{CC} - R_L \cdot \beta \cdot \frac{V_{CC}}{R_B} \tag{5.163}$$

$$V_0 = V_{CC}\left(1 - \frac{\beta \cdot R_L}{R_B}\right) \tag{5.164}$$

The bias resistor is then

$$R_B = \beta \cdot \frac{V_{CC}}{V_{CC} - V_0} \cdot R_L \tag{5.165}$$

It should be noted again that this is only one of many possible and simplest ways of introducing bias voltage to an transistor amplifier. In this design, the bias voltage is linear function of the amplifier current gain, $\beta$, which is known to vary significantly due to manufacturing variations and temperature. In a given initial design, if it turns out that the bias voltage ($V_0$) is larger than anticipated as a result of variations in $\beta$, then the designer may need to increase or reduce $R_B$.

***Example*** Consider the bias voltage circuit for a transistor as shown in Fig. 5.20. Let the parameters of this circuit be given as follows:

$$V_{CC} = 12V, \quad \beta = 100, \quad R_L = 10\ \mathrm{K}\Omega \tag{5.166}$$

Determine the value of the bias resistor, $R_B$, so that the bias voltage is $V_0 = 6$ V.

From the above equation, the bias resistor is

$$R_B = \beta \cdot \frac{V_{CC}}{V_{CC} - V_0} \cdot R_L \tag{5.167}$$

$$= 100 \cdot 2 \cdot 10\ \mathrm{K}\Omega \tag{5.168}$$

$$= 2\ \mathrm{M}\Omega \tag{5.169}$$

After this initial design, if the circuit indicates that the $V_0$ is different than the desired value (which is so due to the variation in the transistor gain $\beta$), then the value of the resistor $R_B$ should be modified according to the above relation until the desired bias voltage is obtained.

## 5.6 OPERATIONAL AMPLIFIERS

Operational amplifiers (op-amps) were first developed in late 1940s. Today, op-amps are linear integrated circuits which are low cost, reliable, and are made in hundreds of million units per year. As the name implies, an op-amp performs an *operational amplification* function where the *operation* may be add, subtract, multiply, filter, compare, convert, etc. Op-amp is a device with very high open-loop gain (ideally infinity, in reality finite but very large). The main function of an op-amp is often defined by external feedback components. There are a number of operational amplifiers manufactured by multiple vendors under license. For instance, op-amp 741 is available under trade names LM741, NE741, $\mu$A741 by different manufacturers. LM117/LM217/LM317 is a higher performance version of 741. LM117, LM217, and LM317 are rated for military, industrial and commercial applications, respectively. The 301, LM339, and LM311 (comparator IC chip), LM317 (voltage regulator), 555 timer IC chip, and XR2240 counter/timer are widely used op-amps. The integrated circuit design incorporates multiple discrete devices (transistors, resistors, capacitors, diodes) into one compact chip which is typically in dual-inline-package (DIP) form.

The 741 op-amp has 17 BJT transistors, 12 resistors, 1 capacitor, and 4 diodes. Note that actual internal component count and design may vary from manufacturer to manufacturer. The coding standard used to identify an op-amp includes the manufacturer code, op-amp functional code, rating for commercial or military use, and mechanical package form (i.e., DIP). In order to use an op-amp in a design, the designer needs the physical size, functionality described in data sheet, and pinout information of the op-amp. With the wide availability and proven designs of integrated circuit (IC) op-amps, it is very rare that one designs circuits from discrete components (using transistors, resistors, capacitors, etc.) for analog signal processing purposes. For almost every analog signal processing need, there is an IC op-amp available in the market. Most often, the designer does not need to know all the internal circuit details of the IC op-amp. The characteristics of the IC op-amp that is needed are its input–output function (i.e., low-pass filter), gain, bandwidth, and input–output impedances. Most IC op-amp packages include multiple stages in the IC design, i.e., input stage generally includes a differential amplifier for its high–input impedance, followed by an intermediate gain stage and finally an output stage. Op-amps perform signal processing functions on low power signals which may have frequency content as high as 50 MHz. Typical output current that can be drawn from an op-amp is about 10 mA.

### 5.6.1 Basic Op-Amp

Figur 5.21 shows the DIP package and symbol for a basic op-amp. Notice that in an 8-pin DIP package, the integrated circuit chip is actually a small portion of the DIP package. A triangle is always used as a symbol of amplification in electrical circuits. The symbol of a basic op-amp shows the following five terminal, connections:

1. Power supply (bipolar) terminals ($V^+$, $V^-$, i.e., ±15 VDC, ±12 VDC, ±9 VDC, ±6 VDC).
2. Inverting (−) and noninverting (+) input terminals, each are referenced to ground, with voltages $v^-$ and $v^+$.
3. Output terminal referenced to ground, where output voltage is designated as $V_o$.

This is an open-loop op-amp. Most uses of op-amps involve adding external components to it between its terminals to implement the desired functionalities. However, key to the understanding the applications of op-amps (open loop or closed loop) is the open-loop properties of it. Here are the idealized assumptions of an open-loop op-amp. Notice that in practice, actual parameters are very close to the idealized assumptions that the performance difference between the two is negligible in most cases.

Idealized assumptions on op-amps are as follows (Fig. 5.22):

1. Input impedance of the op-amp is infinite. In reality, it is a very large number compared to source impedance.

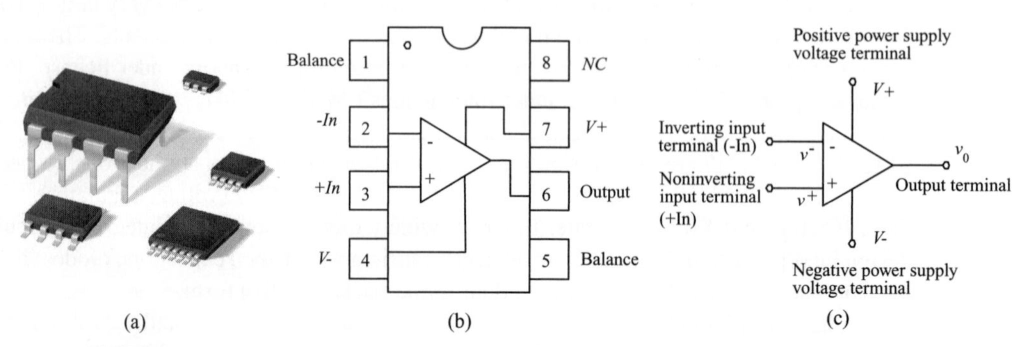


**FIGURE 5.21:** Operational amplifier: (a) op-amp DIP-packages, (b) op-amp pins, (c) op-amp symbol.

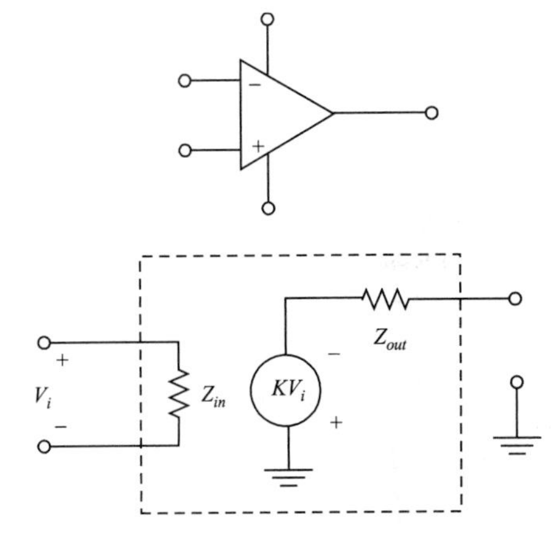


**FIGURE 5.22:** Model of an op-amp: Single-ended output op-amp. Idealized model assumes infinite input impedance and open-loop gain, and zero-output impendance, $Z_{in} = \infty$, $Z_{out} = 0$, and $K = \infty$.

2. Output impedance of an op-amp is zero. In reality, it is a small number compared to load impedance.
3. Open-loop gain of the op-amp ($K_{OL}$) is infinite. In reality, it is a very large number, i.e., $10^5$ to $10^6$.
4. Bandwidth of dynamic response is assumed to be infinite, but in reality, it is a finite large number. Most op-amps operate on signals up to 1 MHz, some special ones up to 50 MHz frequency content.

The implications of these assumptions are as follows, which are very useful in deriving the input–output relationship of any op-amp configuration.

1. Because input impedance is infinite, current flow into the op-amp from either input terminal is zero, $i^- = i^+ = 0$.
2. In addition, the differential voltage between the two input terminals is zero, $v^+ = v^-$ or $E_d = v^+ - v^- = 0$.

Note that these are approximate conclusions based on idealized assumptions. In reality, they are not exactly zero, but close. The output voltage $V_o = K_{OL} \cdot E_d$ is finite since $K_{OL}$ is a very large number and $E_d$ is a very small number. When the op-amp output is saturated, the output voltage is at most equal to the supply voltage,

$$V_o = V_{sat} = V^+ \quad \text{or} \quad V_o = -V_{sat} = -V^- \tag{5.170}$$

Typically, saturation voltage is 1 V to 2 V lower than the supply voltage. The implication of the fact that a practical op-amp has large input impedance and very small output impedance makes it ideal as a buffer between a signal source (input signal) and a load (output signal). Using an op-amp (i.e., unity gain op-amp) between a signal source and a load isolates the input signal from the effects of the load. This is a fundamental benefit and is widely used in processing sensor signals.

When a signal voltage is measured between a wire carrying the signal and the ground signal, it is called a *single-ended signal* [Fig. 5.23(a)]. When the signal voltage is measured between two wires independent of the ground, it is called *differential-ended signal* [Fig. 5.23(b)]. The actual signal information is contained in the difference between the two wires. Any noise induced on the wires during transmission would likely be equally induced on both wires. Since the differential amplifier would only pass the difference between them, and reject the common signal, the noise would be cancelled.

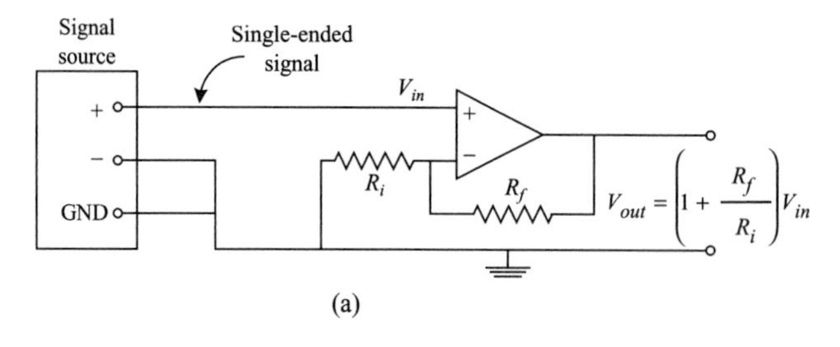


(a)

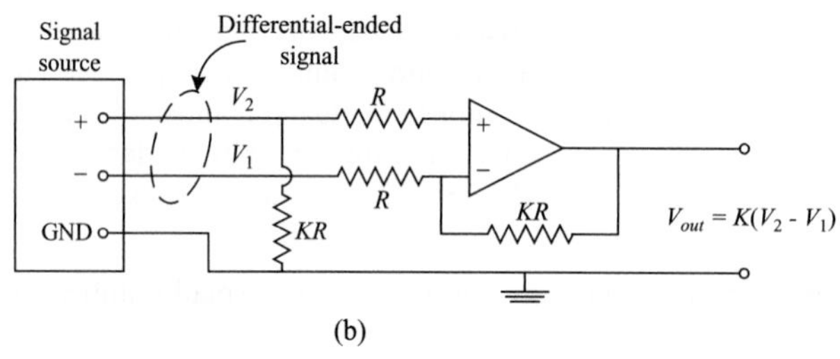


(b)

**FIGURE 5.23:** (a) Single-ended signal and op-amp where the signal return line (–) and ground are connected together, and (b) differential-ended signal and op-amp.

The performance characteristics of a differential amplifier are defined by the differential gain, common mode gain, and common-mode rejection ratio (CMMR),

$$K_{diff} = \frac{V_{out}}{V_{in1} - V_{in2}} \tag{5.171}$$

$$K_{cm} = \frac{V_{out}}{((V_{in1} + V_{in2})/2)} \tag{5.172}$$

$$CMRR = \frac{K_{diff}}{K_{cm}} \tag{5.173}$$

Notice that we only need to know two of these parameters. Most commonly referenced ones are $K_{diff}$ and $CMRR$. Ideally, the differential gain is the desired gain of the amplifier (i.e., 1.0 or 10.0) and the common mode gain is zero. That is, the noise induced on both lines that are common are rejected totally by the differential amplifier. Hence, an ideal differential amplifier would have infinite CMRR. In reality, the common-mode gain is small but finite, and CMRR is large, but finite. Typical values of CMRR are in the order of 80 dB to 120 dB,[1] which means that the amplifier amplifies the differential signal $10^4$ to $10^6$ times more than the common signal.

The parameters of a real op-amp that are of interest for designers are as follows:

1. Open-loop gain (i.e., $10^4$–$10^7$ range)
2. Bandwidth (i.e., 1-MHz range)
3. Input impedance (i.e., $Z_{in} = R_{in} = 10^6\ \Omega$ range)
4. Output impedance (i.e., $Z_{out} = R_{out} = 10^2\ \Omega$ range)
5. Common mode rejection ratio (CMRR) for differential amplifiers, (i.e., $CMRR =$ 60 dB to 120 dB range)

[1] $K_{dB} = 20\, log_{10} K$ or $K = 10^{\frac{K_{dB}}{20}}$, i.e., $K = 0.01, 0.1, 1, 10, 100$ is same as $K_{dB} = -40, -20, 0, 20, 40$, respectively.

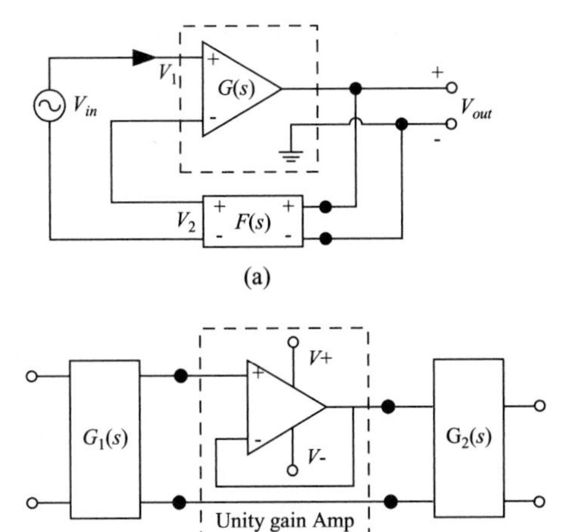


**FIGURE 5.24:** (a) Negative feedback op-amp, (b) unity gain op-amp used between two filters.

6. Power supply required (i.e., 1.5-VDC to 30-VDC range, unipolar or bipolar)
7. Maximum power dissipated internally by the op-amp (which increases approximately as a linear function of the power supply voltage)
8. Maximum input voltage and differential input voltage levels
9. Operating temperature range (commercial rating: 0°C to 70°C, industrial rating: −25°C to 85°C, military rating: −55°C to 125°C)

***Negative Feedback*** Negative feedback around a very high gain amplifier results in a circuit with precise gain characteristics that depends only on the feedback components used. Negative feedback is used to reduce the effect of variation of the open-loop gain of the amplifier on the closed-loop transfer function. An open-loop op-amp gain may vary between the $10^5$ and $10^7$ range. Consider the negative feedback op-amp shown in Fig. 5.24 (a). The transfer function between $V_{in}(s)$ and $V_{out}(s)$ can be determined as follows:

$$V_{out}(s) = G(s) \cdot V_1(s) = G(s) \cdot (V_{in}(s) - V_2(s)) \tag{5.174}$$

$$= G(s) \cdot (V_{in}(s) - F(s) \cdot V_{out}(s)) \tag{5.175}$$

$$= \frac{G(s)}{1 + G(s)F(s)} \cdot V_{in}(s) \tag{5.176}$$

Hence, the closed-loop transfer function is

$$\frac{V_{out}(s)}{V_{in}(s)} = \frac{G(s)}{1 + G(s)F(s)} \tag{5.177}$$

The key observation on the effect of negative feedback on the transfer function between the input and output signals is that when $G(s)F(s) >> 1$,

$$\frac{V_{out}(s)}{V_{in}(s)} \approx \frac{1}{F(s)} \tag{5.178}$$

which states that the closed-loop transfer function is approximately independent of the open-loop transfer function of the op-amp, hence is not affected (or the effect is negligibly small) by the variations of the open-loop transfer function of the op-amp.

Consider a special case, $F(s) = 1$, then we have a unity gain amplifier transfer function [Fig. 5.24 (b)]. Let us assume that the gain of $G(s)$ varies between $10^5$ and $10^7$, which is common, and $F(s) = 1.0$. The closed-loop transfer function gain varies as a result of the variation in $G(s)$ as follows:

$$\frac{V_{out}(s)}{V_{in}(s)} = \frac{G(s)}{1 + G(s)F(s)} \tag{5.179}$$

$$= \frac{10^5}{1 + 10^5} \sim \frac{10^7}{1 + 10^7} \tag{5.180}$$

$$= 0.999990 \sim 0.99999990 \tag{5.181}$$

which shows that the variation in the closed-loop transfer function is so small (0.0000099/1.0) even though the open-loop op-amp transfer function gain varies by a factor of 100 ($10^5$–$10^7$ range). The unity gain op-amp is often used as a buffer between a signal source and load in order to provide impedance isolation. It eliminates (isolates) the loading effect of $G_2(s)$ on the output signal of $G_1(s)$. The op-amp between $G_1(s)$ and $G_2(s)$ acts as a buffer amplifier, which has large (or infinite for practical approximations) input impedance and low (or zero for practical approximations) output impedance.

### 5.6.2 Common Op-Amp Circuits

Op-amps are used in both open-loop and closed-loop configurations. Various op-amp circuits for specific functions are discussed below in three categories: (1) open loop op-amps that are used as comparator op-amps, (2) op-amps with positive feedback, (3) op-amps with negative feedback.

***(1) Comparator Op-Amp*** The comparator functionality is to compare two signals, and turn ON (i.e., $V_o = V_{sat}$) or OFF ($V_o = -V_{sat}$) the output of the op-amp based on relative values of the two input signals (Fig. 5.25). A reference signal $V_{ref}$ is connected to

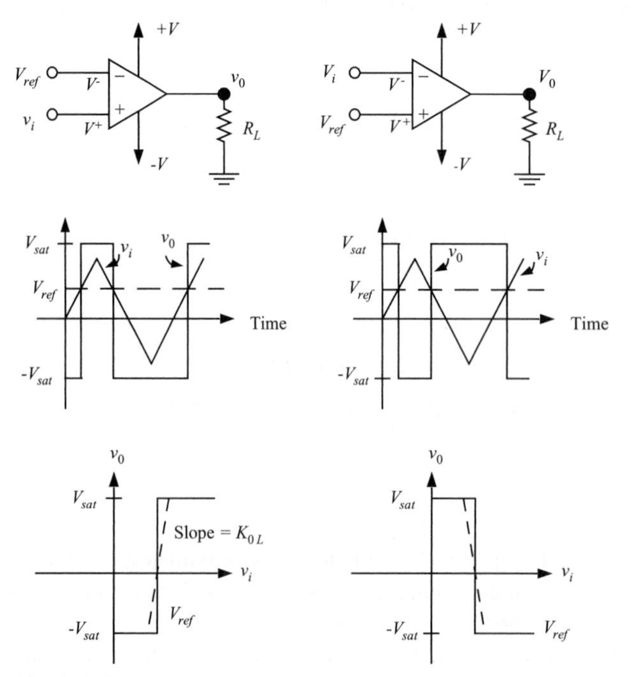


**FIGURE 5.25:** Open-loop op-amp used as a comparator: Direct polarity and reverse polarity.

the inverting (−) input terminal. The other input signal ($V_i$) is connected to the noninverting (+) input terminal. The output of the op-amp will be either $V_o = V_{sat}$ or $V_o = -V_{sat}$. The op-amp circuit determines whether input signal $V_i$ is above or below the reference signal, $V_{ref}$. By changing the connection of $V_{ref}$ and $V_i$ into (−) and (+) terminals, the output polarity of the op-amp is changed. The open-loop op-amp circuit input–output relationship as a comparator can be summarized as

$$V_o = K_{OL} \cdot (V_i - V_{ref}); \tag{5.182}$$

$$\text{if} \quad -V_{sat} < K_{OL} \cdot (V_i - V_{ref}) < V_{sat} \tag{5.183}$$

$$= V_{sat}; \quad \text{if} \quad K_{OL} \cdot (V_i - V_{ref}) > V_{sat} \tag{5.184}$$

$$= -V_{sat}; \quad \text{if} \quad K_{OL} \cdot (V_i - V_{ref}) < -V_{sat} \tag{5.185}$$

Let us consider an op-amp with the following parameters:

$$K_{OL} = 10^6, \quad V_{sat} = 12 \text{ VDC} \tag{5.186}$$

Then, when the difference between the two input terminal voltages is within $+/-12\ \mu$V, the output is proportional to the difference with a voltage output between $+/-12\ \mu$V. Otherwise, the output is either saturated at 12 V or at −12 V.

The same circuit is also used as a PWM modulator which converts the reference $V_i$ analog voltage level signal to a pulse width modulated (PWM) signal when $V_{ref}$ is a fixed frequency periodic signal (i.e., triangular or sinusoidal signal). Similarly, comparator op-amp configurations are also used as signal generator circuits. The LM311 (and LM111/LM211) family of integrated circuit op-amps are commonly used as high-speed comparators.

**Example** Consider the op-amp comparator circuits shown in Fig. 5.26. Assume that the resistances at the input terminals are

$$R_1 = 4\ \text{k}\Omega \quad R_2 = 6\ \text{k}\Omega \tag{5.187}$$

and the supply voltage $V_{C1} = 10$ VDC. Let the saturation output voltage be $V_{sat} = 13$ VDC. The input voltage $v_{in} = 9\ sin(2\pi t)$. Notice that the difference between the two circuits is the connection of reference voltage and input voltage to the op-amp terminals. Draw the output voltage as a function of time for both circuits. Notice that the reference voltage is

$$V_{ref} = \frac{R_2}{R_1 + R_2} V_{C1} = \frac{6}{4+6} \cdot 10 = 6 \text{ V} \tag{5.188}$$

Notice that this is a comparator op-amp circuit. Let us neglect the raise time transient of the output voltage response at this time scale range. Hence, we can show the change of output state like a step function. When the $(v^+ - v^-) > V_{sat}/K_{OL} \approx 0.0$, the output voltage is $V_o = V_{sat}$ and othwerwise. When the $(v^+ - v^-) < -V_{sat}/K_{OL} \approx 0.0$, the output voltage is $V_o = -V_{sat}$. For all pratical purposes, for the first connection, when $V_{in} > 6$ V, the $V_{out} = V_{sat} = 13$ V, and when $V_{in} < 6$ V, the $V_{out} = -V_{sat} = -13$ V. For the second configuration, the output polarity would be opposite the first configuration. The input–output voltage is shown in Fig. 5.26.

**Example** The circuit shown in Fig. 5.27 determines whether input voltage $V_{in}$ is inside the window defined by two reference voltages, $[V_{ref,l}, V_{ref,h}]$. The output voltage will be $V_{sat}$ if the input voltage is outside the window. It will be $-V_{sat}$ if the input is *inside* the voltage window defined by $[V_{ref,l}, V_{ref,h}]$. The LED will be ON when the input voltage is outside the window.

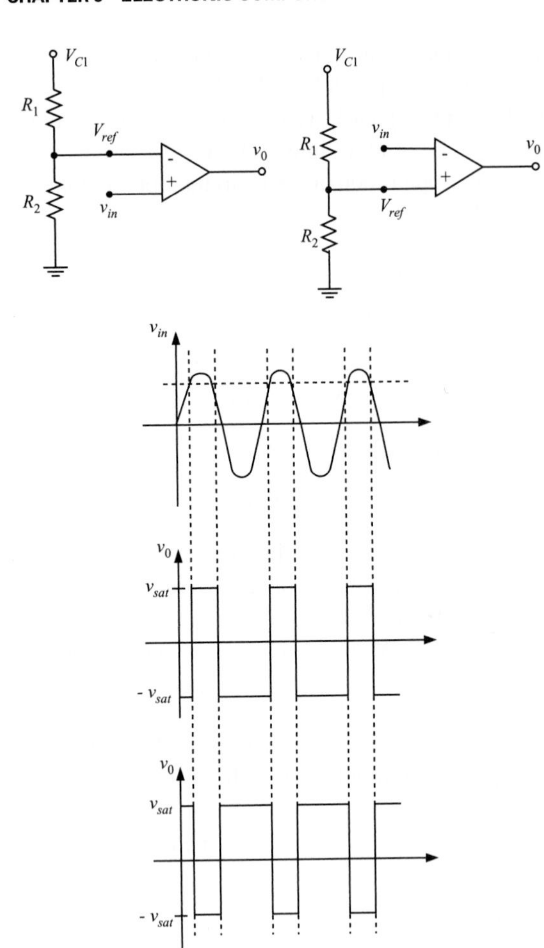


**FIGURE 5.26:** Open-loop op-amp used as a comparator: $V_{ref}$ reference voltage and $V_{in}$ input voltage (i.e., voltage from a sensor) are switched at the op-amp terminals between the two cases. The net result is the output polarity change.

***(2) Op-Amps with Positive Feedback*** Modifications to op-amp feedback loop on the noninverting terminal can provide hysteresis in the comparator to implement ON/OFF relay control with hysteresis, such as those used in home temperature control. A well-known circuit with this function is *Schmitt trigger* (inverting and noninverting types) (Fig. 5.28). The op-amp uses positive feedback which results in a hysteresis relationship between input and output. *Positive feedback is the key to the hysteresis function realization.* Such circuits, used in ON/OFF control systems with desired hysteresis, rejects undesirable noise in the signal provided the noise magnitude is smaller than the hysteresis magnitude (Fig. 5.28). A more general form of the Schmitt trigger op-amp circuit allows adjustment of both the hysteresis magnitude and center input voltage value about which the hysteresis is designed. Schmitt trigger circuits are also used in digital circuits to eliminate the ON/OFF digital state changes due to noise, i.e., LM7414 integrated circuit package incorporates six Schmitt triggers in one IC package for digital circuit applications.

Consider an inverting Schmitt trigger op-amp [Fig. 5.28 (b)]. The voltage at the (+) terminal is the same as the voltage across $R_2$,

$$V_{R_2} = \frac{R_2}{R_1 + R_2} \cdot V_o = \frac{R_2}{R_1 + R_2} \cdot V_{sat} \tag{5.189}$$

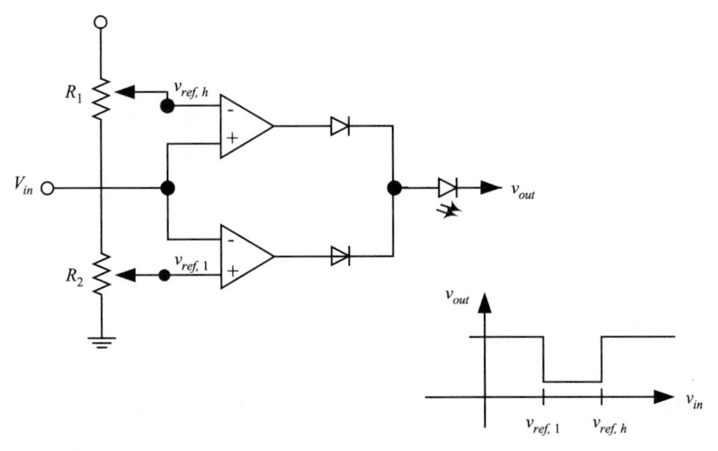


**FIGURE 5.27:** Two open-loop op-amps used as a window comparator: $V_{ref,l} \leq V_{in} \leq V_{ref,h}$.

where $V_o = V_{sat}$ or $V_o = -V_{sat}$, the output of the amplifier is essentially always saturated depending on the signals on its input terminals (+) and (−).

$$v^+ = \frac{R_2}{R_1 + R_2} \cdot V_{sat} \quad \text{during} \quad V_o = V_{sat} \tag{5.190}$$

$$v^+ = -\frac{R_2}{R_1 + R_2} \cdot V_{sat} \quad \text{during} \quad V_o = -V_{sat} \tag{5.191}$$

Let us trace the operation on the op-amp along the input–output relationship curve that has the hysteresis loop. Assume that initially we start on the curve from the left-hand side,

$$V_o = V_{sat}, \quad v^+ = \frac{R_2}{R_1 + R_2} \cdot V_{sat}, \quad v_i < 0 \tag{5.192}$$

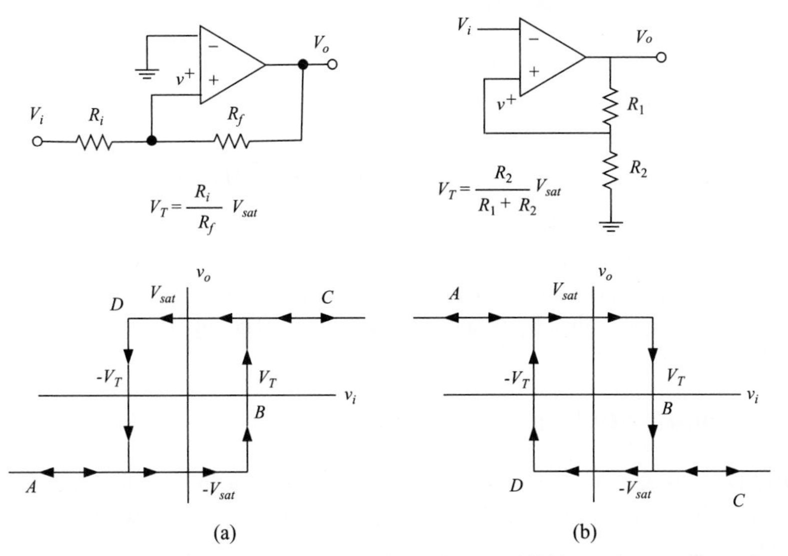


**FIGURE 5.28:** Schmitt trigger: (a) noninverting, and (b) inverting configuration. ON/OFF output with hysteresis function.

For the op-amp output to change state, $V_o = -V_{sat}$ (move to the C part of the curve), the inverting input must slightly exceed $(v_i = v^-) > (v^+)$. In other words, we move along B on the input–output curve, where $v^-$ must be larger than $v^+ = \frac{R_2}{R_1+R_2}V_{sat} > 0$. Here, we neglect the transient response details of the change in the output state of op-amp. At that point, $v^+ = \frac{R_2}{R_1+R_2}V_{sat}$ and $V_o = -V_{sat}$. In order to return to the previous state, now the inverting input must be slightly more negative than the noinverting input which is at the value of $v^+ = -\frac{R_2}{R_1+R_2}V_{sat}$. This can happen on the curve to the left side of region D. Then, the state of the op-amp output switched back to ON (D line along the hysteresis loop). The noninverting Schmitt trigger circuit works with the same principle, except the output polarity is different.

Notice that the nonlinear hysteresis function is accomplished by the feedback of the output signal to the noninverting (positive) input of the op-amp, not the negative input. Feedback to negative input is used for implementing linear functions with an op-amp.

***Example*** Consider the op-amp circuit shown in Fig. 5.28(b). Let us assume that the saturation voltage is $V_{sat} = \pm 13$ V and that the input signal $V_i(t)$ is a sinusoidal signal with magnitude 10 V and frequency 1.0 Hz (Fig. 5.29.a),

$$V_i(t) = 10\,sin(2\pi t) \tag{5.193}$$

Draw the output voltage of the op-amp for the following values of the feedback resistors,

1. $R_1 = 100$ kΩ, $R_2 = 100$ Ω
2. $R_1 = 100$ kΩ, $R_2 = 100$ kΩ

For case 1, the width of the hysteresis voltage, $V_T$, is

$$V_T = \frac{R_2}{R_1 + R_2} \cdot V_{sat} = \frac{100}{100{,}000 + 100} \cdot 13 \tag{5.194}$$

$$= 13/1001 \text{ V} \approx 13 \text{ mV} \tag{5.195}$$

The output voltage will switch between $V_{sat} = +13$ V and $-V_{sat} = -13$ V when the input signal passes the band of $-13$ mV and 13 mV around zero voltage. Due to the hysteresis, the output state switching is dependent on the previous direction of the input voltage crossing the hysteresis band. Notice that due to the scale, the hysteresis behavior is not visually detectable in the figure (Fig. 5.29.b).

For case 2, the width of the hysteresis voltage, $V_T$, is

$$V_T = \frac{R_2}{R_1 + R_2} \cdot V_{sat} = \frac{100{,}000}{100{,}000 + 100{,}000} \cdot 13 \tag{5.196}$$

$$= 6.5 \text{ V} \tag{5.197}$$

In this case, the output voltage switches between the $V_{sat} = +13$ V and $-V_{sat} = -13$ V when the input signal goes through the hysteresis band of $+6.5$ V and $-6.5$ V. The hysteresis magnitude is large and visually observed on the figure. The input signal and output signals for both of the cases above are shown in Fig. 5.29.c.

### *(3) Op-Amps with Negative Feedback*

***Inverting Op-Amp*** The functionality is to amplify the input voltage to output voltage with a negative gain. Neglecting the transient delay of response between input and output voltages,

$$V_o(t) = K_{CL} \cdot V_i(t) \tag{5.198}$$

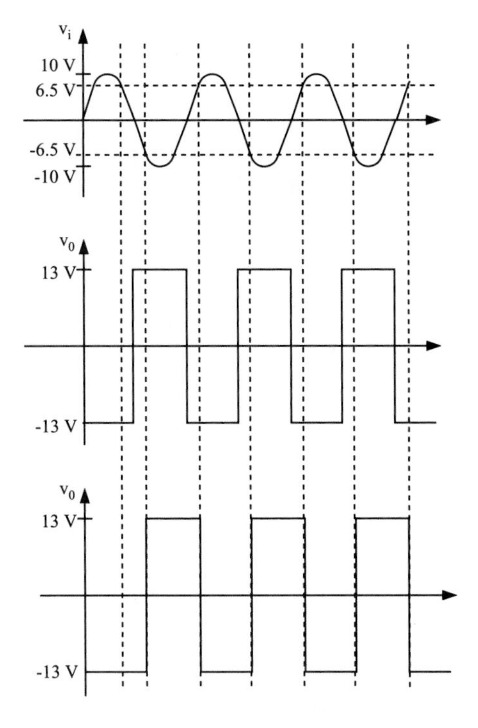


**FIGURE 5.29:** Schmitt trigger (inverting configuration) op-amp circuit used to convert a periodic input signal to a square wave output signal.

Inverting op-amp [Fig. 5.30 (a)] connects the (+) input terminal to ground, and input signal is connected to the (−) input terminal. There are two resistors around the op-amp: $R_i$ and $R_f$. Let us show the relationship between $V_i$ and $V_o$ using the ideal op-amp assumptions. Recall that ideal op-amp assumptions state, $i^+ = i^- = 0,\ E_d = v^+ - v^- = 0,\ i_f = i_{in}$. Notice that since the noninverting terminal of the op-amp is grounded,

$$v^+ = v^- = 0 \tag{5.199}$$

Then,

$$i_{in} = V_i / R_i \tag{5.200}$$

$$i_f = i_{in} \tag{5.201}$$

$$V_f = R_f \cdot i_f = R_f \cdot V_i / R_i \tag{5.202}$$

Hence, since the output voltage will have opposite polarity to $V_f$,

$$V_o = -V_f = -\frac{R_f}{R_i} \cdot V_i \tag{5.203}$$

$$V_o = K_{CL} \cdot V_i \tag{5.204}$$

where the gain of the inverting op-amp

$$K_{CL} = -\frac{R_f}{R_i} \tag{5.205}$$

***Noninverting Op-Amp*** Noninverting amplifier simply amplifies an input voltage to output voltage with a positive gain. This is accomplished by the feedback connections shown in Fig. 5.30(b). Following the same ideal op-amp assumptions ($v+ = v^-, i^+ = i^- = 0$), the input–output relationship (neglecting transient response differences) can be derived as

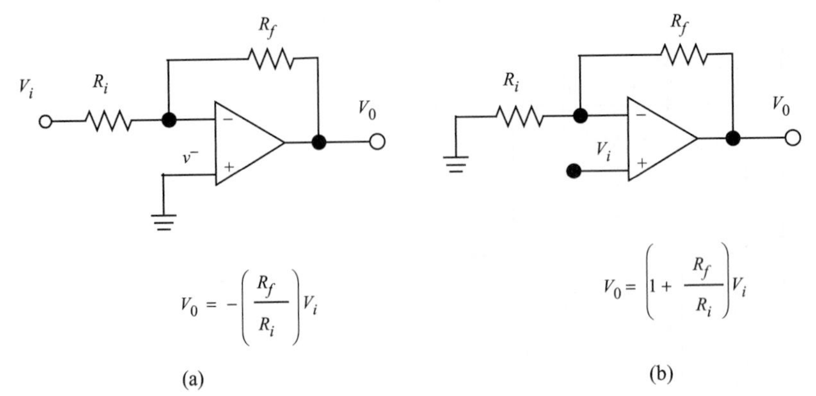


**FIGURE 5.30:** Basic feedback (closed-loop) configuration of op-amps: (a) inverting, and (b) noninverting configuration as a gain amplifier.

follows:

$$v^+ = v^- = V_i \tag{5.206}$$

$$i_{in} = V_i/R_i \tag{5.207}$$

$$i_f = i_{in} \tag{5.208}$$

$$V_0 = (R_i + R_f) \cdot i_f \tag{5.209}$$

Since this is a noninverting amplifier,

$$V_o = \frac{R_i + R_f}{R_i} \cdot V_i \tag{5.210}$$

$$= K_{CL} \cdot V_i \tag{5.211}$$

where the gain of the noninverting op-amp is

$$K_{CL} = 1 + \frac{R_f}{R_i} \tag{5.212}$$

which is always larger than one.

**Example** A special case of the noninverting op-amp is obtained when there is no resistors in the configuration. Effectively, this is same as $R_f = 0$, and $R_i = \infty$. Hence, the gain of the amplifier is unity. Such op-amp configuration is called the *voltage follower op-amp* or *buffer op-amp* and used to isolate the source and load [Fig. 5.24 (b)]. The voltage gain is unity, but the current gain is larger than unity in order to isolate the source from the load.

**Example** Let us consider a noninverting op-amp with saturation output voltage of $V_{sat} =$ 13 V, $R_i = 10\ \text{k}\Omega$, and $R_f = 10\ \text{k}\Omega$. The input voltage and output voltage relationship can be easily determined by

$$V_o = \frac{R_i + R_f}{R_i} \cdot V_i \tag{5.213}$$

$$= 2.0 \cdot V_i \tag{5.214}$$

As a result, the input voltage range is limited to $\pm 6.5$ V range. Beyond that, the output voltage saturates at 13 V.

Notice that a typical op-amp can handle input voltages up to the supply voltage values or a few volts less. However, the feedback resistor values can be such that the output of the linear amplifier saturates much earlier than the maximum input voltage the op-amp can accept. For further discussion, consider the same example with $R_i = 1\,\text{k}\Omega$ and $R_f = 99\,\text{k}\Omega$. Then, the nominal gain of the noninverting op-amp is 100.0. The output voltage would saturate when the input voltage is outside the range of $\pm 0.13$ V.

***Differential Input Op-Amp*** The desired function is to determine the difference between two signals and possibly multiply the difference with a gain,

$$V_o = K \cdot (V_1 - V_2) \tag{5.215}$$

which is used in closed-loop control circuits as the summing junctions, i.e., find the difference between a command signal and sensor signal. Figure 5.31(a) shows a differential input op-amp circuit. In its general form, the input–output relationship can be obtained using the superposition principle. The output is the sum of the outputs due to the inverting input and the noninverting input. The superposition principle can be used in the derivation: (1) connect $V_2$ to ground and solve for $v_o' = K_1 \cdot V_1$, and (2) connect $V_1$ to ground and solve for $v_o'' = K_2 \cdot V_2$. Then add them together to get $V_o = v_o' + v_o''$. The output due to input at its noninverting terminal is [Fig. 5.30 (b)]:

$$v^+ = \frac{R_2}{R_1 + R_2} V_1 \tag{5.216}$$

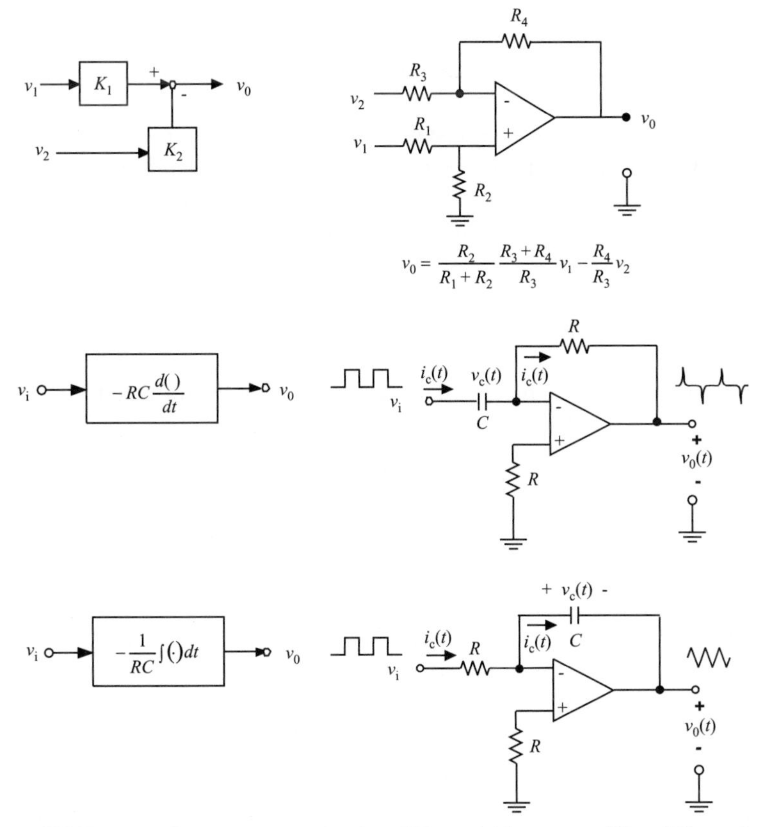


**FIGURE 5.31:** Some op-amp circuits: differential input amplifier, differentiator, integrator.

$$v_o' = \frac{R_3 + R_4}{R_3} v^+ \tag{5.217}$$

$$= \frac{R_3 + R_4}{R_3} \frac{R_2}{R_1 + R_2} \cdot V_1 \tag{5.218}$$

And the output due to input at its inverting terminal is [Fig. 5.30(a)]:

$$v_o'' = -\frac{R_4}{R_3} \cdot V_2 \tag{5.219}$$

The total output is

$$V_o = v_o' + v_o'' \tag{5.220}$$

$$V_o = \left(\frac{R_2}{R_1 + R_2}\right)\left(\frac{R_3 + R_4}{R_3}\right) \cdot V_1 - \left(\frac{R_4}{R_3}\right) \cdot V_2 \tag{5.221}$$

Note that when $R_1 = R_2 = R_3 = R_4$, the input–output relationship is

$$V_o = V_1 - V_2 \tag{5.222}$$

Similarly, when $R_1 = R_3 = R$ and $R_2 = R_4 = K \cdot R$,

$$V_o = K \cdot (V_1 - V_2) \tag{5.223}$$

One of the main usages of differential op-amps is in amplifying noise-sensitive signals. As discussed in Fig. 5.23, single-ended signals are referenced with respect to ground. Any noise induced on the signal wire coming into the op-amp would be amplified. This is particularly problematic when the noise signal is comparable to the actual signal magnitude. In such cases, it is best to transmit the signal voltage in differential-ended format; that is, using two wires and the signal information is the voltage difference between the two wires. If any noise is induced during the transmission, it would be induced on both lines and the difference between them would still be unaffected by noise. Amplification of differential-ended signals is one of the most common applications of differential op-amps. Later we will review an improved version of this op-amp, that is, the instrumentation amplifier.

***Example*** Consider the differential op-amp circuit shown in Fig. 5.31. Calculate the values of the resistors in order to obtain the following input–output voltage relationship, $v_o = v_1 - 2v_2$.

The output voltage and input voltage relationship for the differential op-amp is

$$v_o = \frac{R_2}{R_1 + R_2} \frac{R_3 + R_4}{R_3} \cdot v_1 - \frac{R_4}{R_3} \cdot v_2 \tag{5.224}$$

Since we want $v_2$ to have gain of 2, then

$$\frac{R_4}{R_3} = 2 \tag{5.225}$$

$$R_4 = 2 \cdot R_3 \tag{5.226}$$

Since we want $v_1$ to have a gain of 1, then

$$\frac{R_2}{R_1 + R_2} \frac{R_3 + R_4}{R_3} = 1 \tag{5.227}$$

$$R_1 = 2 \cdot R_2 \tag{5.228}$$

Let $R_2 = R_3 = 10\ \text{k}\Omega$, then $R_1 = R_4 = 20\ \text{k}\Omega$.

***Derivative Op-Amp*** The desired function is to take the derivative of the input voltage signal and provide that as output voltage signal,

$$V_o(t) = K\frac{d}{dt}(V_i(t)) \tag{5.229}$$

Figure 5.31 shows an op-amp circuit for differentiation. Using the ideal op-amp assumptions, the input–output relationship is derived as follows:

$$i_c = C \cdot \frac{dV_i(t)}{dt} \tag{5.230}$$

$$i_f = i_c \tag{5.231}$$

$$V_f = R \cdot i_f \tag{5.232}$$

$$V_o = -V_f \tag{5.233}$$

Hence,

$$V_o = (-RC) \cdot \frac{dV_i(t)}{dt} \tag{5.234}$$

***Integrating Op-Amp*** If we change the locations of the resistor and capacitor in the derivative op-amp, we obtain an integrating op-amp circuit (Fig. 5.31). The desired function is

$$V_o(t) = K \int (V_i(\tau)\, d\tau) + V_o(0) \tag{5.235}$$

where $V_o(0)$ is the initial voltage. The derivation of the I/O relationship is straightforward,

$$i_c = V_i(t)/R \tag{5.236}$$

$$i_f = i_c \tag{5.237}$$

$$V_f(t) = \frac{1}{C}\int_0^t i_f(\tau)\, d\tau \tag{5.238}$$

$$V_o(t) = -V_f(t) \tag{5.239}$$

$$= -\frac{1}{RC}\int_0^t V_i(\tau)\, d\tau \tag{5.240}$$

where the initial voltage values in the integrations have been neglected.

Next we present the op-amp circuits and the input–output relation for filtering operations used in signal processing and control systems. We provide op-amp circuits for the low-pass, high-pass, bandpass, and band-reject (notch) filters. It should be noted that digital implementations of filters in software provide more flexibility than the analog op-amp implementations. However, op-amp implementation is simpler because it does not require real-time software.

***Low-Pass Filter Op-Amp*** Low-pass filter passes the low-frequency content of a signal and suppresses the high-frequency content (Fig. 5.32). The break frequency at which the transition from low to high frequency occurs is defined by the filter parameters. In addition, the rate of transition and the phase lag are determined by the filter order. The frequency domain input–output voltage relationship of a low-pass op-amp filter is

$$\frac{V_O(jw)}{V_{in}(jw)} = \frac{1}{\tau_1 jw + 1} \tag{5.241}$$

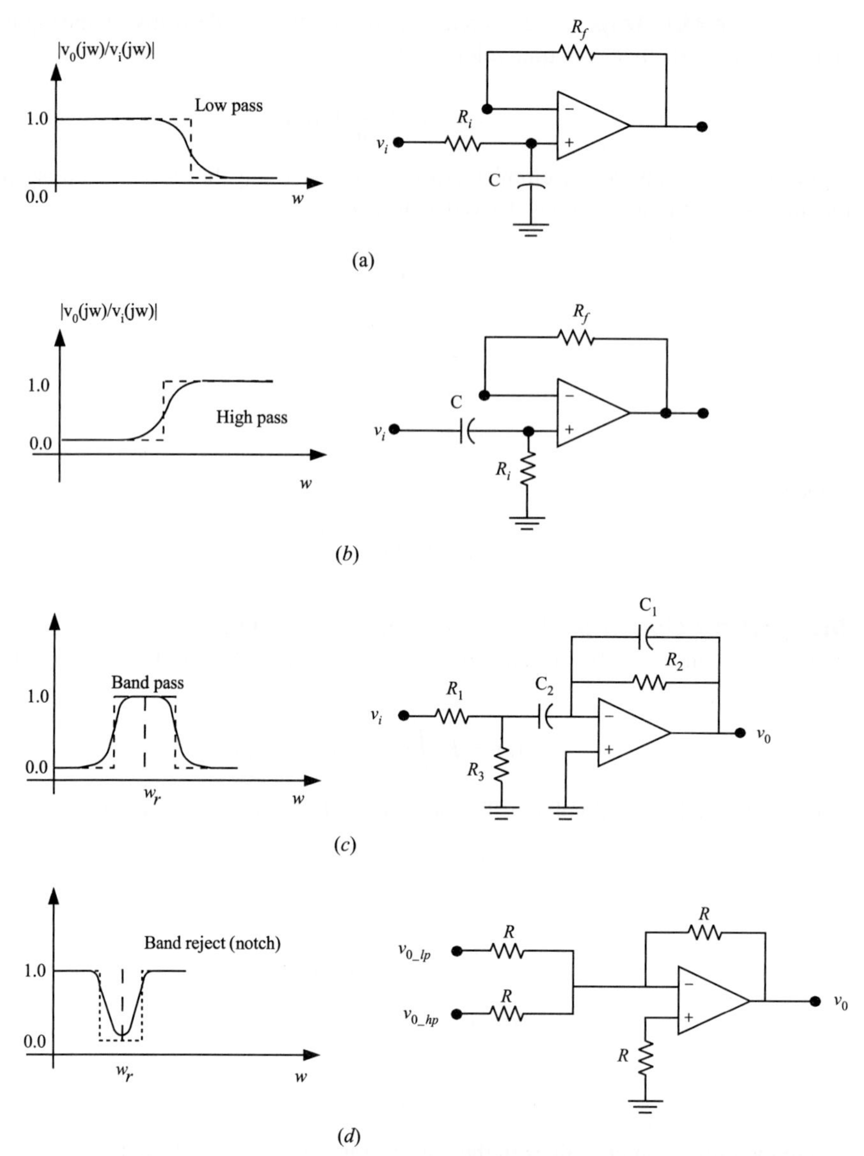


**FIGURE 5.32:** Some op-amp filter circuits: low-pass, high-pass, bandpass, and band-reject (notch) filters.

where the time contant of the first order filter $\tau_1 = RC$, where $R_i = R_f = R$ is assumed. It is important that there are other negative feedback op-amp circuits to realize the same type of filters.

***High-Pass Filter Op-Amp*** High-pass filter suppresses the low-frequency content of the input signal, and passes the high-frequency content. An op-amp implementation of a high-pass filter is shown in Fig. 5.32. The frequency domain input–output voltage relationship of a high-pass, op-amp filter is

$$\frac{V_0(jw)}{V_{in}(jw)} = \frac{j\tau, w}{1 + j\tau, w} \tag{5.242}$$

where the time contant of the first-order filter $\tau_1 = RC$, $R = R_i = R_f$ is assumed. Notice that the only difference between the low-pass and high-pass filter is the placement of the resistor and capacitor in the (+) input terminal.

***Bandpass Filter Op-Amp*** Bandpass filter passes a selected narrow band of frequencies, and suppresses the rest (Fig. 5.32). The design parameters of the filter is to select the frequency band to pass: center frequency and the width around that frequency, $w_r$ and $\Delta w_B$. Typically $C_1 = C_f = C = 0.01\ \mu$F. Then it can be shown that given the desired $w_r$, $\Delta w_B$, and $c$, the filter parameters are $R_1 = \frac{1}{\Delta W_B \cdot C}$, $R_2 = 2R_1$, $R_3 = \frac{R_2}{4\left(\frac{w_C}{\Delta w_B}\right)^2 - 2}$.

***Band-Reject Filter Op-Amp*** Band-reject filters pass all frequencies, except a selected narrow band of frequencies (Fig. 5.32). This is also called the *notch filter*, which basically does the opposite of the bandpass filter. The design parameters of a notch filter are to select the center frequency and the width of the frequency around the center frequency that will be suppressed, $w_r$ and $\Delta w_B$. A band reject filter can also be built using a low-pass filter and a high-pass filter in parallel, then use a summing op-amp to form the combined effect in frequency domain. The crossover frequencies of low-pass and high-pass filters must be appropriately chosen to obtain the desired notch frequency and its width.

***Instrumentation Op-Amp*** Instrumentation op-amp is used to amplify small sensor signals in noisy environments (Fig. 5.33). It is a modified version of the differential op-amp with improved performance characteristics (i.e., higher input impedance, easy adjustment of gain). The instrumentation amplifier has higher CMRR (common mode rejection ratio) which is an advantage for noise environments, and the gain of the op-amp is adjustable with a single resistor ($R_x$) for both input terminals. The fact that neither input is grounded at the instrumentation amplifier terminals (note that the instrumentation amplifier is a differential amplifier), the ground loops are eliminated.

***Current-to-Voltage Converter and Voltage-to-Current Converter*** In electronics circuits, we often need to convert a current signal to a proportional voltage and convert a voltage signal to a proportional current signal. The first case is used typically in sensor signal transmission where a current source (sensor) indicates the value of a measured variable. At the controller end, we may need to convert the current to a proportional voltage signal. Figure 5.34(a) shows a modified version of an inverting op-amp used as a

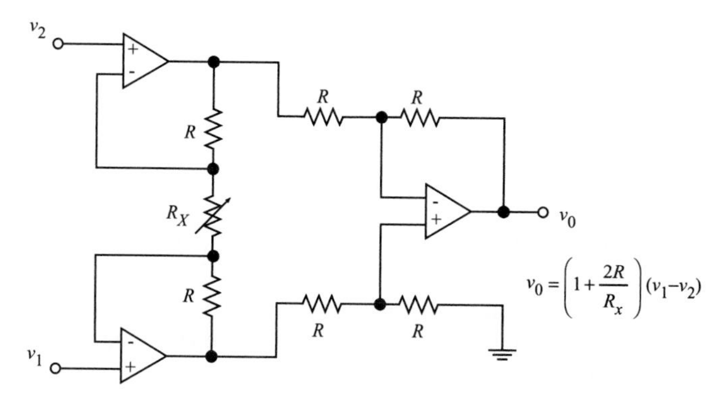


**FIGURE 5.33:** Instrumentation op-amp: modified version of differential op-amp with improved characteristics to amplify sensor signals in noisy environment.

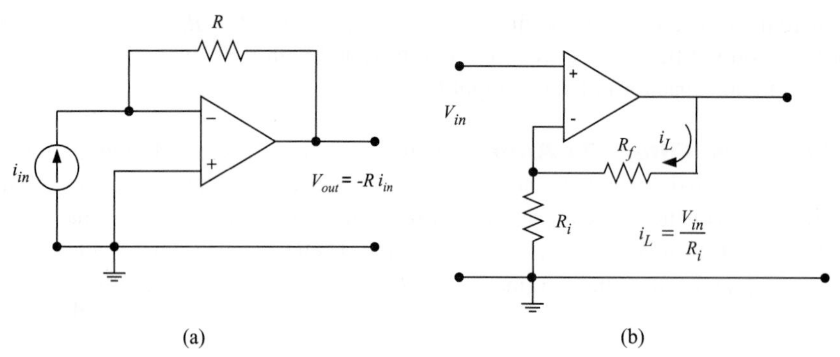


**FIGURE 5.34:** (a) Current-to-voltage converter op-amp circuit. (b) Voltage-to-current converter op-amp circuit.

current-to-voltage converter. Notice that since the current flow between two terminals of the op-amp is zero, the source current must flow through the resistor $R$. Furthermore, since the terminals of the op-amp is connected to the ground, the output voltage must be negative relative to the ground when the current is positive. Hence, the input current-to-output voltage-conversion relationship is

$$V_{out} = -R \cdot i_{in} \tag{5.243}$$

Figure 5.34(b) shows a circuit which performs the opposite function: converts voltage to current. Op-amp is used as a current source. This is a noninverting op-amp. The only difference here is that the circuit output that we are interested in is the current through the load resistor, $R_L$. Such a function is used to drive small DC motors and solenoids. For instance, the input voltage is the command signal from a controller which is proportional to the desired torque. In DC motors, torque is proportional to the current passing through its windings. Therefore, in order to obtain the desired torque output from the motor, we must provide a current value proportional to the commanded voltage signal. In the figure, the load (i.e., motor winding) is shown as having resistance $R_f$ (or can be impedance $Z_f$ for more general case of electrical load) and is placed in the feedback loop of the op-amp. Since this is a noninverting op-amp,

$$V_{out} = \frac{R_i + R_f}{R_i} \cdot V_{in} \tag{5.244}$$

and the output current over the load is

$$i_L = \frac{V_{out}}{R_i + R_f} = \frac{1}{R_i} V_{in} \tag{5.245}$$

Notice that the load current is independent of the load resistance ($R_f$) and its variations. For linear operating region without saturation,

$$V_{out} = (R_i + R_f) \cdot i_L < V_{sat} \tag{5.246}$$

When the output load requires current levels above 0.5 mAmp, the output current should be amplified using power transistors (i.e., BJT, MOSFET, or IGBTs).

Figure 5.35 shows simple examples of the use of these two op-amp configurations. In the first case, a solar cell is used to generate a current proportional to the light it receives. The input current is converted to a proportional output voltage by the op-amp. Hence, the voltage output is proportional to the light received. This circuit can be used as an analog light intensity sensor. The other circuit input voltage is manually adjusted, and the output

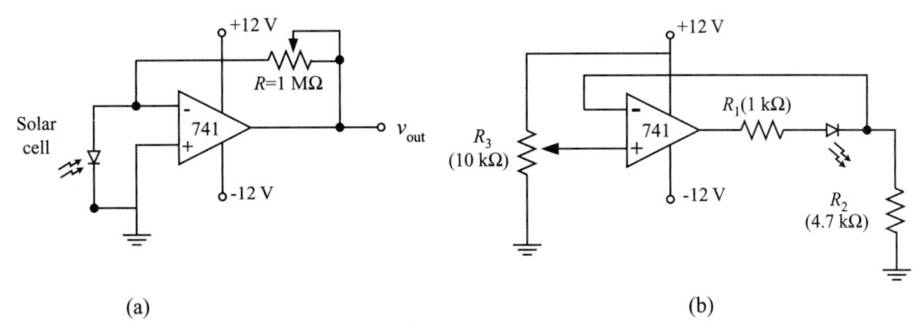


**FIGURE 5.35:** Example op-amp circuits: (a) current-to-voltage converter op-amp circuit used to provide output voltage that is proportional to the current generated by the solar cell, (b) voltage-to-current converter op-amp circuit where the intensity of LED light is proportional to the current, which is proportional to the input voltage pickup by the adjustable resistor $R_3$.

current is proportional to the voltage presented by the resistor $R_3$. The output current is proportional to this voltage, where the proportionality constant is $R_2$. Hence, the intensity of light emitted by the LED is proportional to the input voltage. The 741 op-amp can be replaced by other similar op-amps.

## 5.7 DIGITAL ELECTRONIC DEVICES

Logic ON state is represented by 1, and logic OFF state is represented by 0. Two of the most popular digital device types are *transistor-transistor logic* (TTL) and *complementary metal oxide silicon* (CMOS). Logic device families differ from each other in terms of their power consumption and the speed of operation. When mixed series of logic ICs are used in a circuit the most important factors to consider are the current loading and current driving capacity of the device. The design should make sure that each device can drive the gates it is connected to and does not present overload current to the other devices.

TTL nominal voltage level for logic 1 is 5 VDC, and for logic 0 is 0 VDC. Supply voltage must be in the range of 4.75 V to 5.25 V. Per-gate power consumption of a TTL device is in the order of few mAmp. The output current sinking capacity is about 30 mA. CMOS device supply voltage can be in the 3- to 18-VDC range. Per-gate power consumption of a CMOS device is 80% less than that of a TTL device. Digital logic devices are realized by complex networks of transistors on an integrated circuit (IC). The basic device number for the TTL NAND gate is 7400 (military version is 5400 for extended temperature range) and has four NAND gates. The variations of 7400 series take the following forms: 74L00 for low power but slower devices, 74H00 for high power and high speed, and more recent version is the 74LS00 series for low power but high-speed performance. The TTL series IC chips are numbered in 74LSxxx form where the xxx code is assigned based on the cronological introduction of the device into the market. CMOS logic family requires a less-stringent power supply voltage, consumes less power but is sensitive to static voltages.

When an IC chip is used in a circuit and some of the pins are not used, they should be connected to common or pulled up to a high-voltage state. Open pins are not a recommended design practice because they tend to fluctuate and give bad logic state.

### 5.7.1 Logic Devices

AND, OR, XOR, NOT (inverter), NAND, and NOR are the most common logic gates in digital electronics. AND gate has two inputs and one output. The output logic is 1 if both

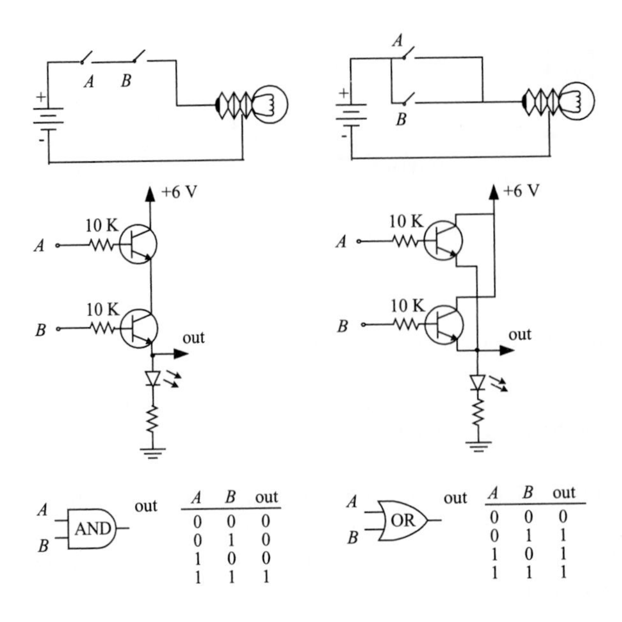


**FIGURE 5.36:** Logic gates: AND and OR gates—concept, transistor implementation, IC symbol.

inputs are 1; otherwise, it is logic 0 state. AND logic among more than two inputs can be implemented by cascading many AND gates in one chip. NOT gate inverts the logic state of input. The output is always at the opposite logic state to that of the input. Combining AND, OR, and NOT gates, we form NAND and NOR gates. The gate symbols and logic diagrams are illustrated in Fig. 5.36 and Fig. 5.37.

Another important logic device is a three-state buffer. A three-state buffer is used to interface digital devices. The device has three ports: input, output, and enable. When the buffer is enabled, the input is connected to the output. Output can be in one of two states: ON or OFF. When the buffer is disabled, it is in third state called the *high-impedance* state during which the output port is in open-circuit condition and not connected to the input port.

### 5.7.2 Decoders

Decoders are used as device selection components in a computer bus system. When a computer places the address of a device on the bus, the decoder attached to each device checks that and gives either an ON or OFF output signal if it is the addressed device. So, only one decoder should give logic 1 output in response to a unique address in the bus. By combining AND and NOT gates, it is straightforward to design address decoders. In general an 8-bit general-purpose decoder is set to respond to a specific address by setting eight-dip switches ON/OFF, which define the address of the device the decoder connected to (Fig. 5.38).

### 5.7.3 Multiplexer

Multiplexer is used to connect one of multiple input lines to an output line. Typical application of a multiplexer is with analog-to-digital converters (ADC). For instance, there can be four or eight analog signal channels connected to one ADC. Under program control, each channel is connected to the ADC for conversion. Such ADC converters are referred to as

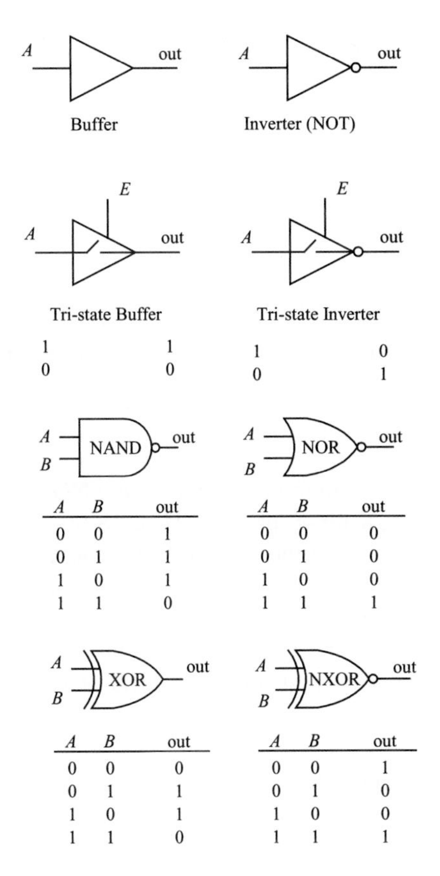


**FIGURE 5.37:** Logic gates: Buffer, NOT (inverter), tri-state buffer and inverter, NAND, NOR, XOR, Not XOR gates.

four-channel, multiplexed ADC or eight-channel multiplexed ADC. The desired channel is selected by providing the binary code on the multiplexer control lines; i.e., two lines are needed for a four-channel multiplexer, three-lines are needed for a eight-channel multiplexer (Fig. 5.39). A four-channel multiplexer circuit is shown in Fig. 5.40. Depending on the code presented in channel select lines A and B (00, 01, 10, 11), one of the four channels (Ch1, Ch2, Ch3, Ch4) is connected to the output under program control.

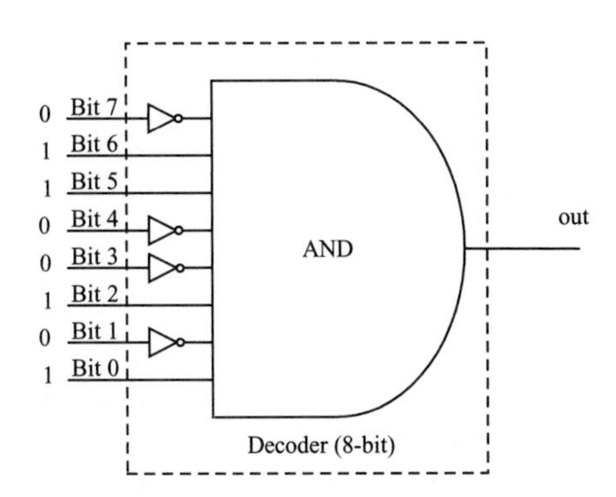


**FIGURE 5.38:** An 8-bit address decoder circuit inplemented with NOT and AND gates.

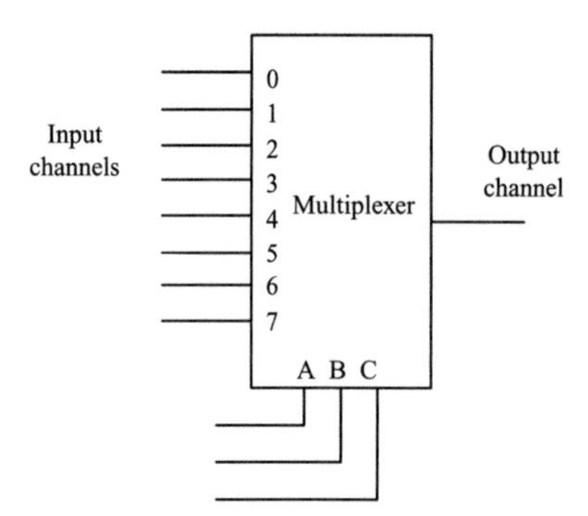


**FIGURE 5.39:** Multiplexer circuit and its function.

### 5.7.4 Flip-Flops

Flip-flop circuits are implemented using combination of AND, OR, and NOT gates. The most common flip-flop types are D, RS, and JK flip-flops (Fig. 5.41). RS flip-flop is commonly used as a debouncer for single-pole, double-throw (SPDT) mechanical switches. When a SPDT mechanical switch is closed and opened (ON/OFF), it makes multiple ON-OFF contacts in the order of milliseconds time period. In human terms, this may look like a single OFF/ON transition, but in digital electronics time scale, the many transitions of OFF/ON are detected. RS flip-flop is used in *debouncing* the mechanical switch input (Fig. 5.44).

***D-Type Flip-Flop*** The logic level present in D-input is transferred to output Q and latched when E-input is ON or transition of state on the E-channel occurs. When the

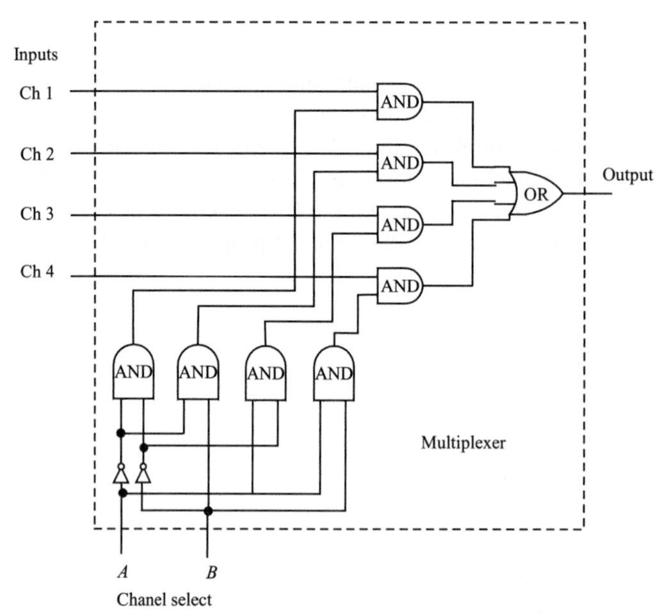


**FIGURE 5.40:** A multiplexer circuit for selecting one of four digital input channels. For instance, digital code on channel select lines AB selects the following channels: 00 select channel 1, 01 selects channel 2, 10 selects channel 3, and 11 selects channel 4.

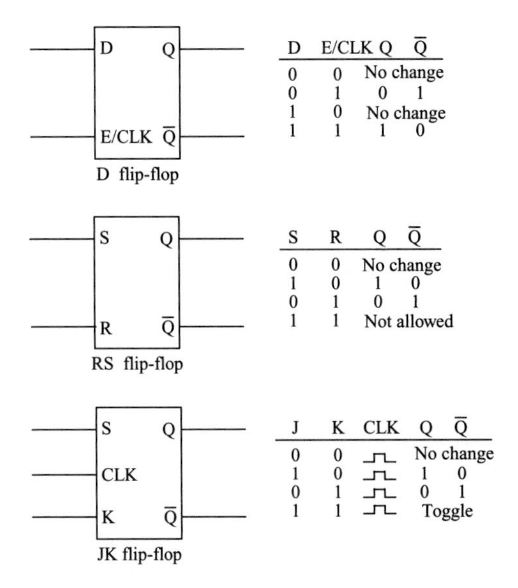

D
Q
E/CLK
Q̄
D flip-flop
D E/CLK Q Q̄
0 0 No change
0 1 0 1
1 0 No change
1 1 1 0
S
Q
R
Q̄
RS flip-flop
S R Q Q̄
0 0 No change
1 0 1 0
0 1 0 1
1 1 Not allowed
S
Q
CLK
K
Q̄
JK flip-flop
J K CLK Q Q̄
0 0 No change
1 0 1 0
0 1 0 1
1 1 Toggle


**FIGURE 5.41:** Flip-flops: D-type, RS-type, and JK-type.

E-input is low, the D-input is ignored and Q maintains its previous state. D-latch is used to strobe and buffer (latch) an input signal state. D-type flip-flops are used in groups to form $n$-bit data buffers (Fig. 5.42). For output operations from a digital computer to D/A converter or discrete output lines, the data from the computer bus are latched and maintained (until updated by the computer) using D-flip-flops.

***RS-Type Flip-Flop*** Reset-set flip-flop has two inputs (R and S) and two outputs (Q, $\bar{Q}$) where the two outputs are opposite each other. The logic between the input and output terminals is as follows: When S = 1 and R = 0, Q is latched to the 1. When S = 0 and R = 1, Q is reset (Q = 0) and latched at that state. When both S and R are 0, then the outputs

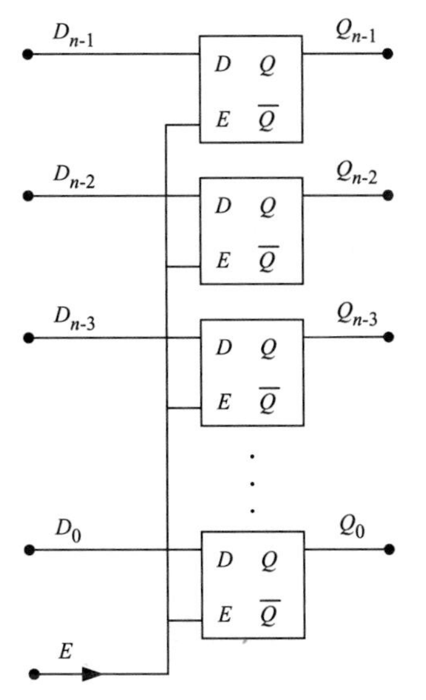


**FIGURE 5.42:** $n$-bit data latch buffer using $n$-set of D-type flip-flops for data input and output between a digital computer bus and I/O devices. Data are placed in the data bus ($D_0 \ldots D_{n-1}$), and then E-line is pulsed to latch the data in the D-flip-flops. After that, the changes in the data bus do not affect the output of the D-flip-flops (latched data) until a new pulse in the E-line.

do not change. The state of R and S both being 1 is not allowed. If that occurs, both outputs go to 0. Basically, the output Q is set and reset by S and R pulses.

***JK-Type Flip-Flop*** JK flip-flop is similar to RS flip-flop with only one difference: If both inputs are 1 simultaneously, the outputs of JK reverse their states. The I/O transfer is triggered either by a predifined edge transition signal of a clock or input sampled in high state of clock signal and transferred to output at the trailing edge of the clock signal. JK flip-flops are used in counters. For instance, four JK-flip-flops can be used to count up to 16.

## 5.8 DIGITAL AND ANALOG I/O AND THEIR COMPUTER INTERFACE

Figure 5.43 shows the interface between a generic computer bus and a parallel input and a parallel output device. For the purpose of the present discussion, let us consider the digital computer as having a CPU and a bus. Here we assume that the CPU includes the microprocessor's clock, the CPU, and the memory [random access memory (RAM) and

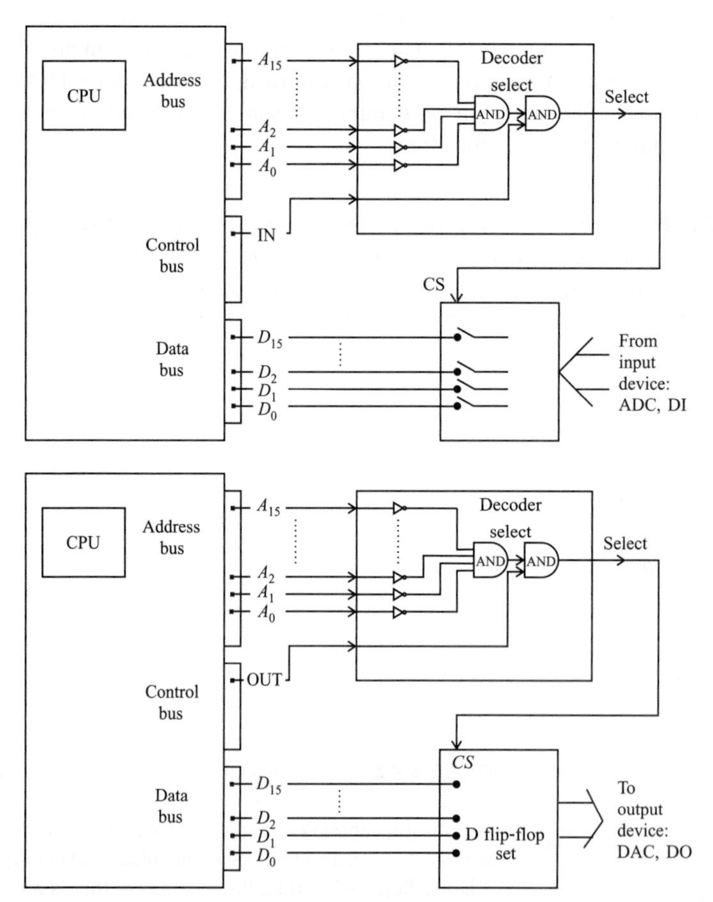


**FIGURE 5.43:** Interface circuit between a digital computer and parallel data input and output device. For instance, parallel data input may be the register of an ADC, parallel output may be the register of a DAC.

read-only memory (ROM)]. The bus is made of three main groups of digital lines: (1) *address bus* used to address devices and memory by the CPU, (2) *control bus* lines used to indicate whether the operation is a read or a write as well as used for interface handshaking signals between CPU and I/O devices, and (3) *data bus* which carries the data between the CPU and I/O devices (or memory). Let us assume that the CPU executes programmed instructions to perform logic and I/O with the external devices.

The address decoder circuit of each I/O device uniquely specifies the address of the I/O device on the bus. The data lines of the device are connected to the data bus of the computer. Every device that the CPU communicates with (reads or writes) must be addressed by the CPU first. A particular device is selected on the address bus via the decoder of the device whose address has been placed on the address bus by the CPU. The address number to which the decoder responds by turning its output ON is set by jumpers or DIP switches which define whether or not each address bus line has an inverter or direct connection.

Control bus of the computer is used to strobe the I/O device to process the data bus information. When a read (input) or write (output) operation is performed by the CPU, the following sequence of events is generated by the CPU at the machine instruction level:

1. CPU places the address of the I/O device on the address bus. Only one device decoder will provide an active output in response to a unique address in the address bus.
2. CPU places the data on the data bus for write operation.
3. CPU turns on the OUT signal of the control bus to tell the I/O device that the data are ready.
4. When the I/O device is given enough time to read (or signals the CPU that it is done via a handshake line in the control bus), the OUT signal is dropped, and CPU goes on with other operations.

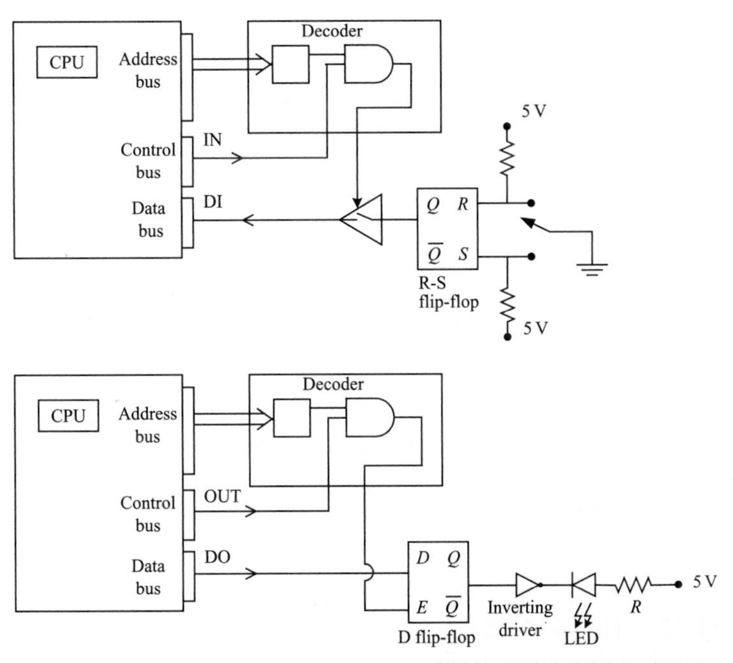


**FIGURE 5.44:** Interface circuit between a digital computer and discrete input and discrete output devices. RS-flip-flop is used to eliminate the debouncing problem of the switch. The D-flip-flop is used so that the data are maintained at the output of the flip-flop even when this device is no longer addressed.

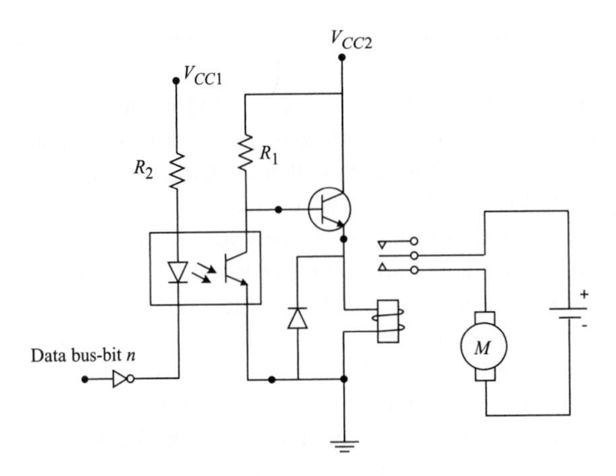


**FIGURE 5.45:** ON/OFF control of a relay through an opto-coupler and transistor. Data bus bit *n* output in this circuit can come from a circuit as shown in Fig. 5.44.

For read (input) operation, the sequence of steps 2 and 3 are changed:

1. CPU places the address of the I/O device on the address bus. Only one device decoder will provide an active output in response to a unique address in the address bus.
2. CPU turns on the IN signal of the control bus to tell the I/O device it is ready to read the data.
3. CPU reads the data on the data bus.
4. When the CPU is done (transfer the data to their registers from the data bus), it drops the IN signal, and CPU goes on with other operations.

In process control applications, the output device is a set of D-type flip-flops connected to discrete output lines or D/A converter, and the input device is a set of R-S flip-flops connected to discrete input lines or A/D converter (Fig. 5.44).

***Example*** Figure 5.45 shows the interface between a digital computer data output line and a relay. The digital data line is set high or low under software control. This controls the opto-coupler, which in turn turns ON/OFF the transistor. The transistor powers the control circuit to the relay coil. Once the relay coil is energized, its contact conducts the current in the output circuit which may turn ON/OFF a device, i.e., light or a motor. Notice that in this example, since no flip-flop is used, the data line must be dedicated for the control of this relay.

***Example*** Figure 5.46 shows various interfaces between a digital computer data output line and an ON/OFF output device through a transistor. In case 1, an output LED light is turned ON/OFF by the computer data bus via a transistor. In cases 2 and 3, a relay and a SCR is controlled (turned ON/OFF), respectively. By controlling the timing of the SCR gate signal, a proportional control is approximately achieved. Diode in parallel with the inductive component relay protects the transistor against voltage spikes when the transistor goes from ON to OFF state.

## 5.9 D/A AND A/D CONVERTERS AND THEIR COMPUTER INTERFACE

A D/A converter converts a digital number to an anlog signal, i.e., generally a voltage level. A/D converter does the opposite. It converts an analog signal to a digital number. The D/A and A/D converters are essential components in interfacing digital world to analog world.

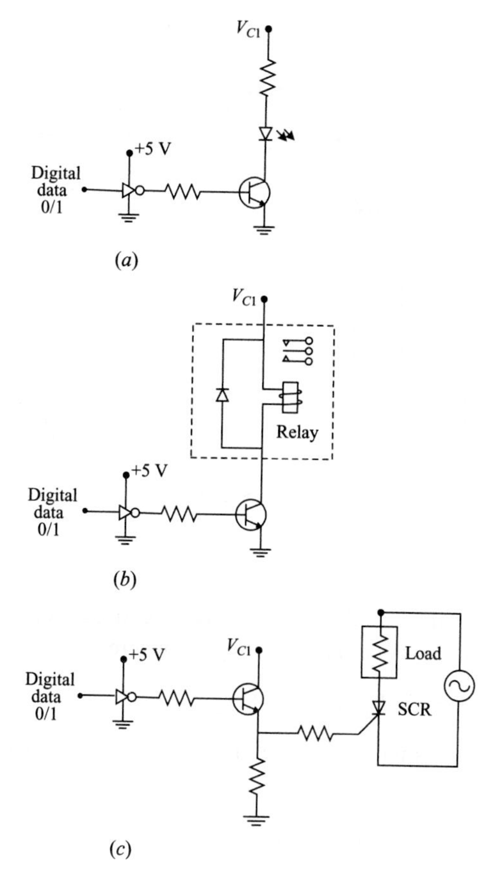


**FIGURE 5.46:** ON/OFF control using a digital computer data bus line: (a) LED (ON/OFF) light output, (b) relay ON/OFF control, (c) SCR control in an AC circuit. In all cases, a transistor is used as an electronic switch between the computer and the output circuit.

The processes of converting signals from analog form to digital form involve two operations:

1. Sample the signal
2. Quantize the signal to the resolution the A/D can represent with $n$-bits

Sampling is the process in which a finite number samples of a continuous signal is taken and converted into a discrete number sequence. The samples are the only information available regarding the signal in the computer. The sampling process and its implications will be discussed at length in a later chapter.

Let us consider how a D/A converter converts a digital number to an analog signal (Fig. 5.47). The voltage potential across the resistor bank at the output of the D flip-flop memory device is $V_h$ and held constant, i.e., 10 VDC. The resistor value $R_f$ is selectable to usually one of four different values. This is used to change the output range of the D/A converter. If a bit is ON (D0-D7, which is the same as Q0-Q7), there will be a current flowing through the corresponding resistor. If the bit is OFF, there will not be any current flowing through that resistor. Therefore, since the sum of current at point A of the op-amp should be zero, the output voltage for a given digital number $N$ sent to the D/A converter can be calculated as follows:

$$V_0/R_f = V_h(b_0/R + b_1/(R/2) + \cdots + b_7/(R/2^7))$$

$$V_0 = V_h(R_f/R)N$$

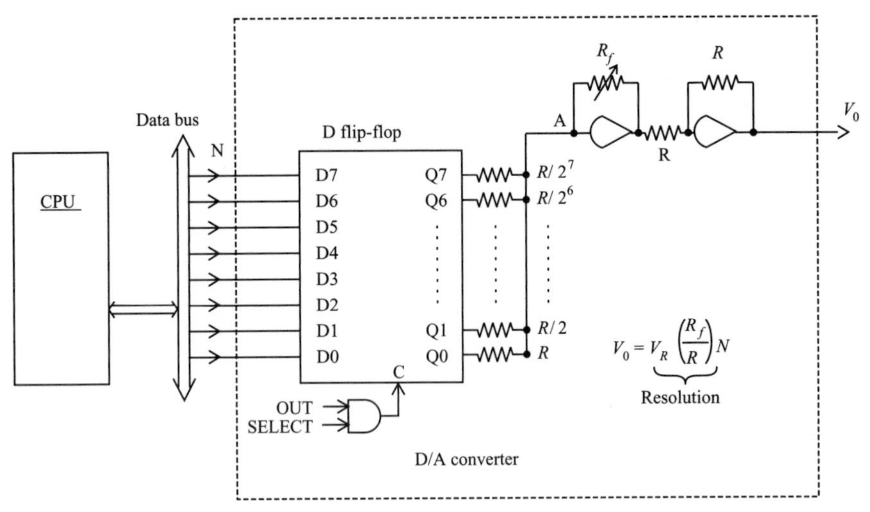


**FIGURE 5.47:** Operating principles of a D/A converter.

where $b_i$ is zero (0) or one (1) representing the value of bit $i$, N is the corresponding number sent to the D/A converter. The range it can cover is zero to $V_{range}$, where

$$V_{range} = V_h(R_f/R) * (2^n) \tag{5.247}$$

The resolution of a D/A converter (the smallest voltage change that can be made at the output of the D/A converter) is

$$\Delta V = V_h(R_f/R) = V_{range}/2^n \tag{5.248}$$

In that range, the D/A can provide $2^n$ different levels of voltage.

The basic A/D converter operation uses D/A converter plus additional circuit (Fig. 5.48). Analog signal is passed through an anti-aliasing filter. This filter may or may not be an integral part of the A/D converter. Then the signal is sampled and held at the sampled level (Fig. 5.49). While it is being held, a D/A converter cycles through a sequence of numbers

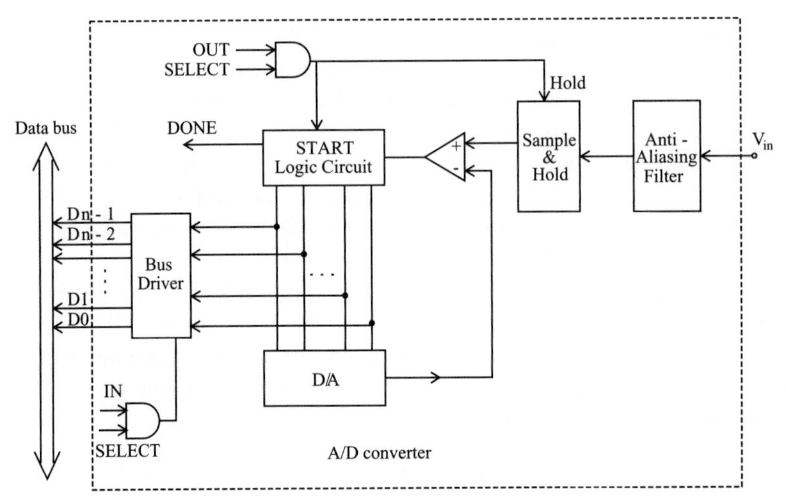


**FIGURE 5.48:** Operating principles of an A/D converter.

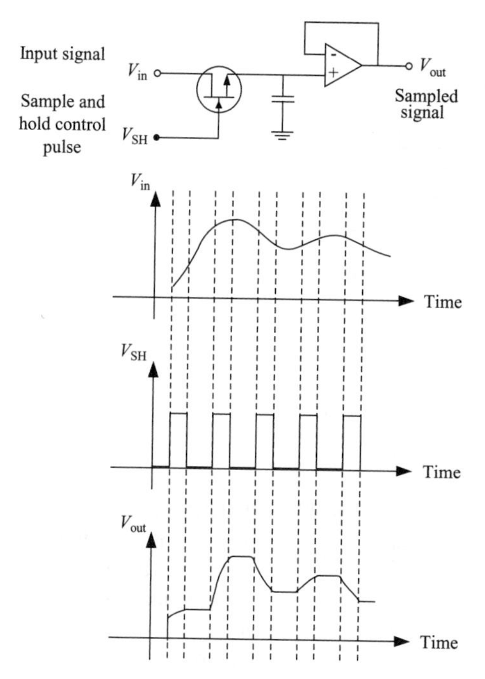


**FIGURE 5.49:** Sample-and-hold circuit using a buffer op-amp, JFET type transistor switch, and a capacitor.

based on a search algorithm to match the D/A output signal to the sampled signal. When the two are equal, the comparator will indicate that the correct signal level is determined by the D/A. The digital representation of the sampled analog signal is same as the digital number used by the D/A to generate the equal signal. The conversion is accurate to the resolution of the A/D converter. Then the CPU will be signaled about the fact that the A/D conversion is complete. The CPU would then read the data on the D/A portion of the A/D through the data bus. Notice that this type of A/D conversion is an interative process and takes longer to complete than the D/A conversion process. There are other types of A/D converters such as the flash A/D converter that make the conversion faster without using an D/A converter in the circuit, but are also more expensive.

The sample-and-hold circuit is a fundamental circuit for analog-to-digital signal conversion. For the conversion, we must sample the signal in periodic intervals (which must be fast enough to capture the frequency details of the signal) and hold the sampled value constant while the A/D converter tries to determine its digital equivalent. A basic sample and hold circuit includes a voltage follower op-amp (buffer), a JFET type transistor switch to turn ON and OFF the connection of the input voltage, and a capacitor at the op-amp input terminal to charge and hold the signal constant. The sampling process is done typically in fixed frequency, i.e., $f_s = 1$ kHz. The input transistor switch is turned ON every $T_s = 1/f_s$ sec, for a very short period of time that is long enough to charge the capacitor to the input voltage level present, i.e., $T_{on} = 0.05 \cdot T_s$. Then the transistor switch is turned OFF and the capacitor holds the last (sampled) value of the input voltage constant. The $T_{on}$ time must be long enough so that the circuit output voltage reaches the input voltage value within a desired accuracy. Typical values of $T_{on}$ are in the order of micro seconds. It should be pointed out that the sample and hold circuit shown in Fig. 5.49 is only a basic concept circuit. Actual sample and hold circuits involve two or more op-amps and are a bit more complicated. One op-amp is used to buffer the input, and the other op-amp is used to buffer the sampled output.

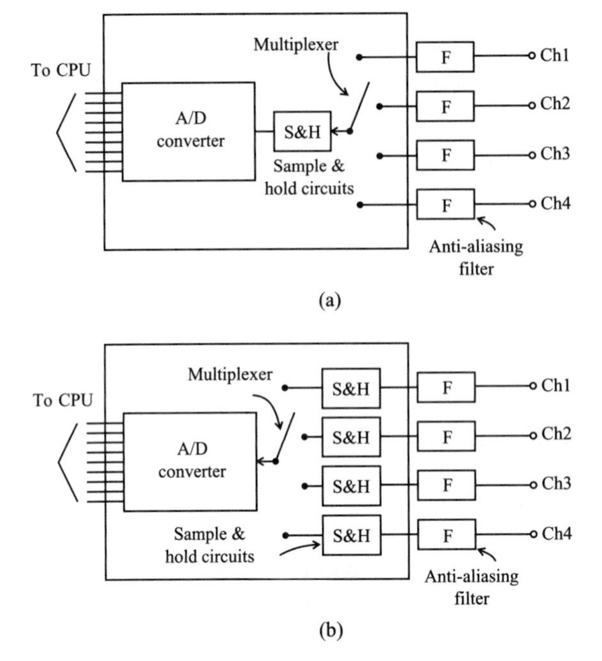


**FIGURE 5.50:** Multiplexed multi-channel analog signal and sampling via a single A/D converter: (a) multiplexed sample and hold circuit, (b) simultaneous sample and hold circuit.

Quite often an A/D converter is used to sample and convert signals from multiple channels. Multiplexer has many input channels and one output channel. It selectively connects one of the input channels to the output channel. The selection is made by a digital code in a set of digital lines in the interface. Figure 5.50 shows two different implementations. In case (a), there is only one sample and hold circuit. Therefore, the signals converted from different channels are not taken at the same time instant. For instance, the A/D converter selects channel 1, samples-hold-converts. This takes finite time, $T_{conv}$. Then it selects-samples-holds-converts channel 2. This takes another $T_{conv}$ time. Hence, the channel 3 will be sampled 2 $T_{conv}$ time later than the channel 1. In case (b), there are individual sample and hold circuits for each channel. All the channels are sampled at the same time, and signals are held at that value. Then the A/D conversion goes through each channel using the multiplexer. This is called the *simultaneous sample and hold circuit.*

***Quantization Error of A/D and D/A Converters: Resolution*** An $n$-bit device can represent $2^n$ different states. The resolution of the device is one part in $2^n-1$. In physical signal terms, the resolution is 1 part in whole range divided by $2^n - 1$. Therefore, the resolution of an $n$-bit A/D or D/A converter is

$$V_{range}/(2^n - 1)$$

Quantization is the result of the fact that the A/D, D/A, and also the CPU are finite word and finite resolution devices (Fig. 5.51). Let us consider that we are sampling a signal in the 0–7-V-DC range. Assume that we have a 3-bit A/D converter. The A/D converter can represent only 0, 1, 2, 3, 4, 5, 6, 7 VDC signals exactly. If the analog signal were 4.6 V, it can be represented as either 4 or 5 by the A/D conversion. If truncation method is used in the quantization, it will be converted to 4. If round-off method is used, it will be converted to 5. The maximum value of the quantization error is the resolution of the A/D or D/A converter.

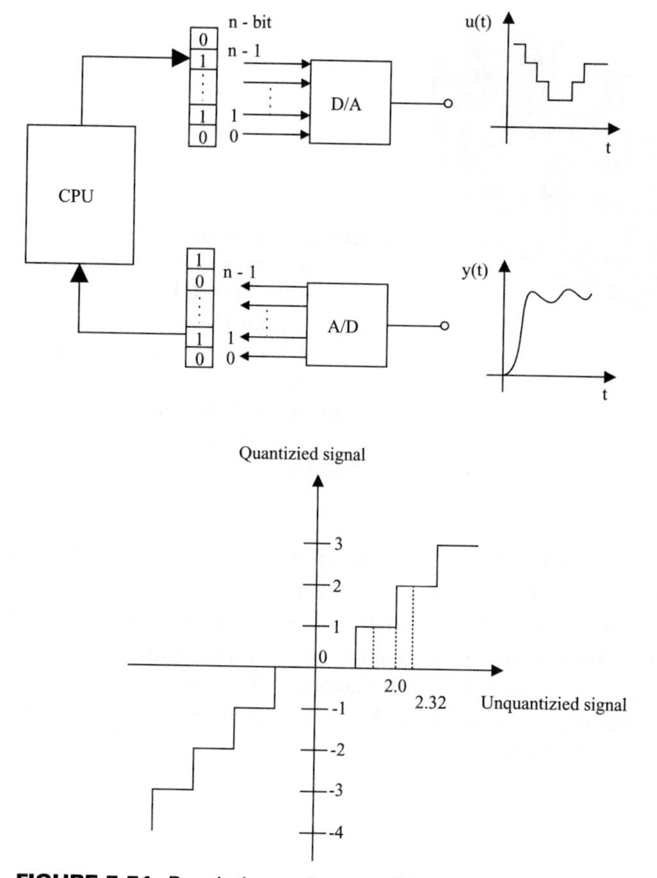


**FIGURE 5.51:** Resolution and range of A/D and D/A converters.

For instance, an 8-bit A/D converter can represent 256 different states. If an analog sensor with a signal range of 0–10 VDC is connected to it, the smallest change the A/D can detect in the signal is 10/255 VDC. Any change in the signal less than 10/255 VDC will not be detected by the computer. Similiarly, if the A/D converter is 12 bit, the smallest change detectable in the signal is equal to the signal range divided by $2^{12} - 1 = 4095$. This is called the resolution of the A/D converter. If it is 16-bit A/D converter, its resolution is one part in $2^{16} - 1 = 65{,}535$.

The D/A converter converts the digital numbers into analog signals. This is also referred to as the signal reconstruction stage. Discrete values of control action are presented to the D/A converter once per sampling period. The interpolation performed by the D/A converter for the values of the signal during the period between the updates is called the reconstruction approximation. The most commonly used form of D/A converter is the zero-order-hold type where the current value of the signal is held constant until a new value is sent. The resolution of the D/A converter is the smallest change it can send out. This is one part in the whole range it can cover, 1 part in $2^n - 1$, where $n$ is the number of bits the D/A converter has. If the D/A converter has an output range of $-10$ VDC to 10 VDC, lets us call $R = 10 - (-10) = 20$ VDC. If the D/A converter is 8-bit, it can have $N = 2^8$ different states. Therefore, the smallest change it can make in the output is

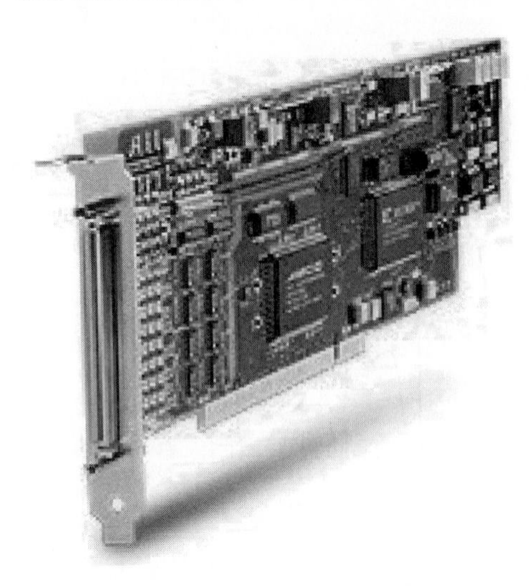

**FIGURE 5.52:** A commercially available data-acquistion board for a PC bus (Model-KPCI-1801HC by Keithley). I/O capabilities are 12-bit 32-channel differential ended or 64-channel single-ended analog input (multiplexed A/D converter), 12-bit 2-channel D/A converter, 4-channel digital input, 8-channel digital output, maximum sample rate is 333 kHz.

$R/N = R/(2^n) = 20/(255)$. Clearly, as the number of bits of the A/D and D/A gets larger (i.e., 16 bit), the resolution and the resultant quantization error become less significant.

Figure 5.52 shows a commercially available data-acquisition card for PCs (PCI or ISA bus). The board has 12-bit resolution, 32-channel differential ended or 64-channel single-ended, multiplexed A/D converter, 12-bit 2-channel D/A converter, four-channel digital input, and eight-channel digital output lines. Maximum sampling rate that is supported by the board is 333 kHz (333 samples/sec).

## 5.10 PROBLEMS

**1.** Consider the RC and RL circuits shown in Fig. 5.5 c–d. Let $V_s(t) = 12$ VDC, $R = 100$ kΩ, $L = 100$ mH, $C = 0.1\ \mu$F. Let us assume that initialy the current in the circuit and charge capacitor is zero. Simulate the current and voltage across each component for the case that the switch is connected to the supply voltage at time zero, and flipped over instantly to B position at time $t = 250\ \mu$sec. Solve this problem using Simulink and present the results in five plots including the state of the switch as function of time, plus $i(t)$, $V_L(t)$, $V_R(t)$, $V_C(t)$. Experiment with the system response by varying the circuit component parameters $R, C, L$. How does increasing $R$ affect the time constant of the system in RC and RL circuits?

**2.** Consider a voltage amplfier using a bipolar junction transistor as shown in Fig. 5.15. Let the resistances $R_1 = R_2 = 10$ kΩ, and supply voltage $V_{cc2} = 24$ VDC. Assume that the current gain of the transistor is 100, and the base-to-emitter voltage drop when it is conducting is 0.7 VDC. Calculate and tabulate the following results for these cases of input voltage, $V_{in} = V_{cc1} = 0.0, 0.5, 0.7, 0.75, 0.80, 0.85, 0.9, 1.0$. The results to be calculated are base current ($i_b$), collector current ($i_c$), emitter current ($i_e$), and output voltage ($V_o$) measured between the collector and emitter.

**3.** Design an op-amp circuit that will change the offset and slope of an input voltage and provide output voltage. Mathematically, the desired relationship between the input and output voltages of the op-amp circuit is

$$V_{out} = K_1 \cdot (V_{in} - V_{offset}) \tag{5.249}$$

For numerical calculations, assume $V_{in}$ range is 2.0 V to 3.0 V. Desired output voltage range is 0.0 V to 10 V. An example application where such a circuit may be useful is shown in Fig. 5.53.

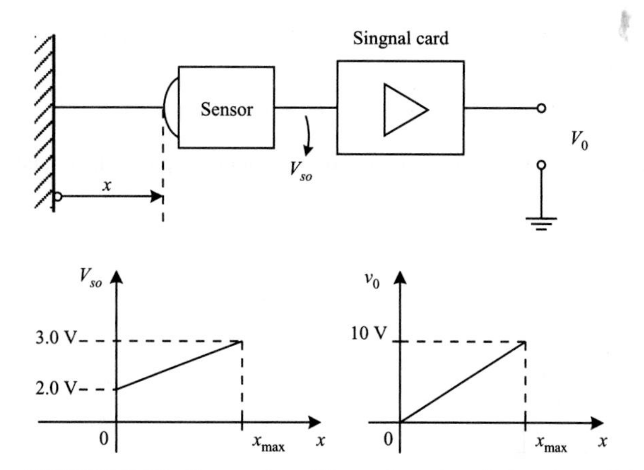


**FIGURE 5.53:** Sensor head (transducer) and signal conditioner to adjust offset voltage and the gain on the signal generated by the sensor.

**4.** Consider the op-amp circuit shown in Fig. 5.54. Derive the relationship between the input voltages $V_{in1}$, $V_{in2}$ and the output voltage $V_{out}$ as function of $R_1$, $R_2$, $R_3$, $R_4$, $R_5$. Select values for the resistors so that the following relationship is obtained,

$$V_{out} = 5.0 \cdot (V_{in2} - 3.0 \cdot V_{in1}) \tag{5.250}$$

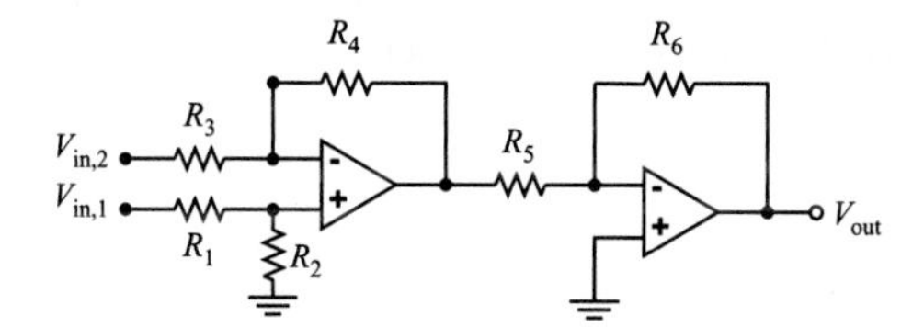


**FIGURE 5.54:** Op-amp circuit for problem 4.

**5.** Determine the minimum voltage at the base of the transistor ($V_{in}$) that will saturate the transistor and hence provide the maximum glow from the LED (Fig. 5.55). Assume the following for the LED and transistor: forward bias voltage of the LED is $V_F = 2.5$ V, and the minimum voltage drop between collector and emittor of transistor is $V_{CE} = 0.3$ V (i.e., when the transistor fully saturated), and gain of the transistor $\beta = 100$.

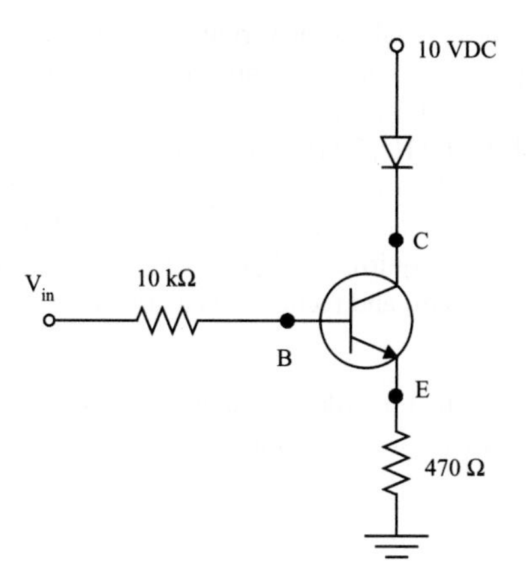


**FIGURE 5.55:** Figure for problem 5.

**6.** Design an op-amp circuit to implement a PD (proportional plus derivative) control where the proportional gain and the derivative gain are both adjustable. The circuit should have two input voltages ($V_{i1}(t)$, $V_{i2}(t)$) and one output voltage ($V_o(t)$). The desired mathematical function to be realized by the op-amp circuit is

$$V_o(t) = K_p \cdot (V_{i1}(t) - V_{i2}(t)) + K_d \cdot d/dt(V_{i1}(t) - V_{i2}(t)) \tag{5.251}$$

**7.** Consider the op-amp circuits shown in Fig. 5.56. Derive the input–output voltage relationships.

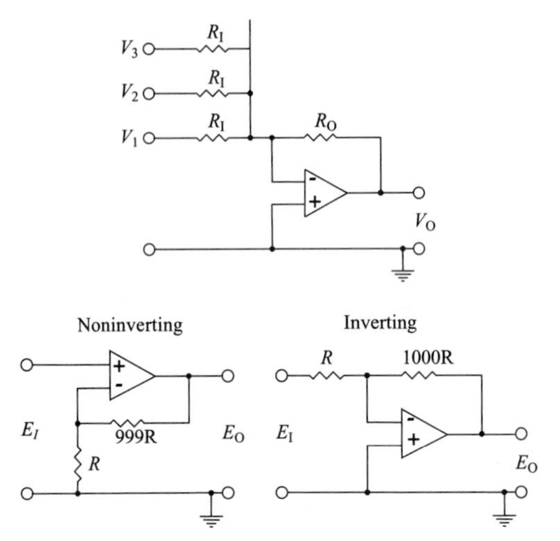


**FIGURE 5.56:** Figure for problem 7.

**8.** Consider the op-amp circuits shown in Fig 5.58. Determine the output voltage when the input voltage is $V_i = 0.1$ V, and the current that flows to the inverting input.

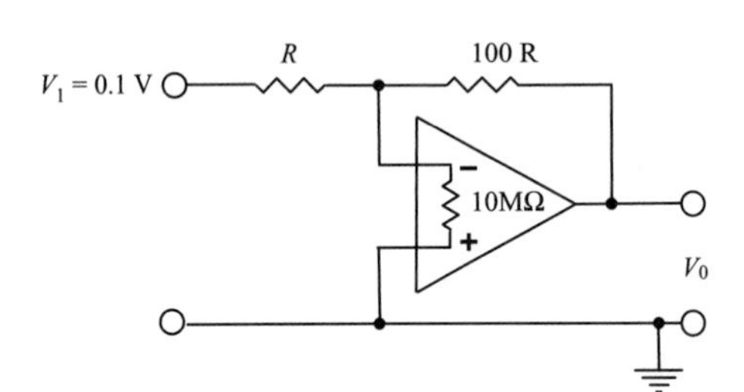


**FIGURE 5.57:** Figure for problem 8.

**9.** Consider a data-acquisition system that samples signals from various sensors. Let us assume the following: There are four sensors to be sampled. The voltage output from each sensor is in the following ranges for sensors 1 through 4: ±10 VDC, ±1 VDC, 0–5 VDC, and 0–2 VDC. Expected maximum frequency content in each signal is 1 kHZ, 100 Hz, 20 Hz, and 5 Hz for sensor 1 through 4, respectively. The error introduced due to sampling should not be more than ±0.01% (1 part in 10,000) of the maximum value of the signal. Determine the specifications for a four-channel ADC (analog-to-digital converter) for this application that will meet the requirments. Specifiy the minimum sampling rate for each channel (according to the sampling theorem) and recommend a practical sampling rate for each channel.

**10.** Consider the circuit shown in Fig. 5.3(d). Show that in order to maximize the power drawn by the load from the supply (the dotted part in the figure), the load resistance must be equal to the supply resistance, $R_l = R_i$.

# CHAPTER 6

# SENSORS

## 6.1 INTRODUCTION TO MEASUREMENT DEVICES

Measurements of variables are needed for monitoring and control purposes. Typical variables that need to be measured in a data-acquisition and control system are,

1. Position, velocity, acceleration
2. Force, torque, strain, pressure
3. Temperature
4. Flow rate
5. Humidity

Figure 6.1 shows the basic concept of a measurement device. The measurement device is called the *sensor*. We will discuss different types of sensors to measure the above listed variables. A sensor is placed in the environment where a variable is to be measured. The sensor is exposed to the effect of the measured variable. There are three basic phenomenon in effect in any sensor operation,

1. The change (or the absolute value) in the measured physical variable (i.e., pressure, temperature, displacement) is translated into a change in the property (resistance, capacitance, magnetic coupling) of the sensor. This is called the *transduction*. The change of the measured variable is converted to an equivalent property change in the sensor.
2. The change in the property of the sensor is translated into a low-power-level electrical signal in the form of voltage or current.
3. This low-power sensor signal is amplified, conditioned, and transmitted to an intelligent device for processing, i.e., display or use in a closed-loop control algorithm.

Sensor types vary in the transduction stage in measuring a physical variable. In response to the physical variable, a sensor may be designed to change its resistances, capacitance, inductance, induced current, or induced voltage.

In any measurement system, accuracy is a major specification. Let us clarify the terminology used regarding accuracy. Figure 6.2 shows the meaning of *accuracy, repeatablity*, and *resolution*. Resolution refers to the smallest change in the measured variable that can be detected by the sensor. Accuracy refers to the difference between the actual value and the measured value. Accuracy of a measurement can be determined only if there is another way of more accurately measuring the variable so that the sensor measurement can be compared with it. In other words, accuracy of a measurement can be determined only if we know the true value of the variable or a more accurate measurement of the variable. Repeatability refers to the average error in between consecutive measurements of the same value.

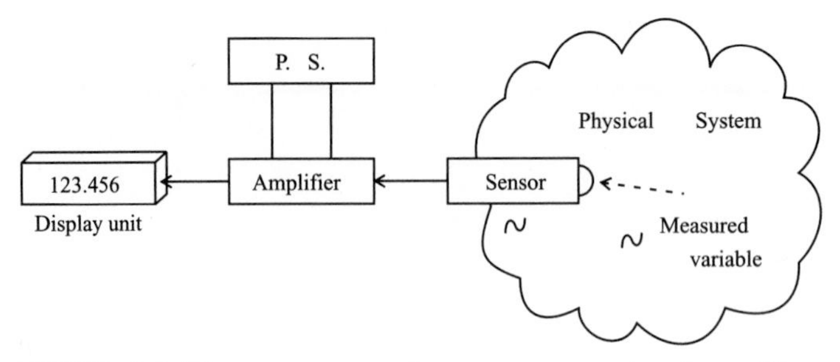


**FIGURE 6.1:** The components of a sensor: sensor head, amplifier, power supply, display, or processing unit.

The same definitions apply to the accuracy of a control system as well. In a measurement system, repeatability can be at best as good as the resolution. Resolution (smallest change the sensor can detect on the measured variable) is the property of the sensor. Repeatability (the variation in the measurement of the same variable value among different measurement samples) is the property of the sensor in a particular application environment. Hence, the repeatability is determined both by the sensor and the way it is integrated into a measurement application.

Let us focus on the input–output behavior of a generic sensor as shown in Fig. 6.3. A sensor has a dynamic response bandwidth as well as steady-state (static) input–output characteristic. The dynamic response of a sensor can be represented by its frequency response or by its bandwith specification. The bandwidth of the sensor determines the maximum frequency of the physical signal that the sensor can measure. For accurate dynamic signal measurements, the sensor bandwidth must be at least one order of magnitude (x10) larger than the maximum frequency content of the measured variable.

A sensor can be considered as a filter with a certain bandwidth and static input–output characteristics. Let us focus on the static input–output relation of a generic sensor. An ideal sensor would have a linear relationship between the sensed physical variable (input) and the output signal. This linear relationship is a function of the *transduction* and amplification stage. Typical nonideal characteristics of a sensor include (Fig. 6.4):

1. Gain changes
2. Offset (bias or zero-shift) changes
3. Saturation
4. Hysterisis
5. Deadband
6. Drift in time

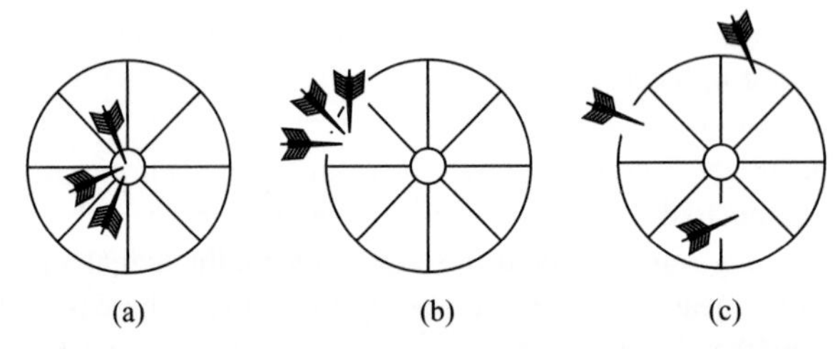


**FIGURE 6.2:** The definitions of accuracy and repeatability: (a) accurate, (b) repeatable, but not accurate, (c) not repeatable, not accurate. Resolution is the smallest positional change the arrow can be placed on the target (imagine that the target has many small closely spaced holes and the arrow can only go into one of these holes).

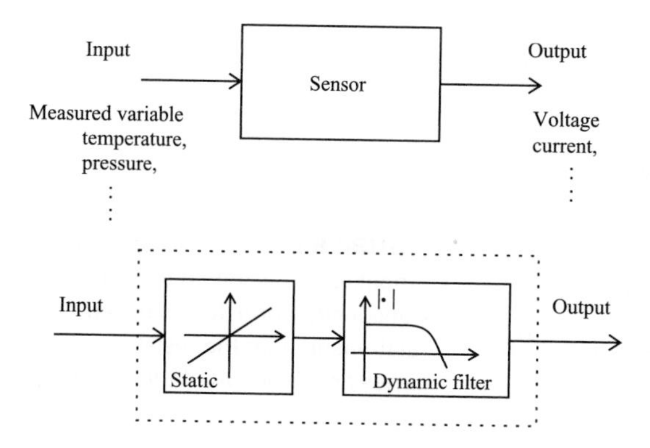


**FIGURE 6.3:** Input–output model of a sensor: steady-state (static) input and output relationship plus the dynamic filtering effect.

The static input–output relationship of a sensor can be identified by changing the physical variable in known increments, then waiting long enough for the sensor output to reach its steady-state value before recording it, and repeat this process until the whole measurement range is covered. The result can be plotted to represent the static input–output characteristic of the particular sensor. If this nonideal, input–output behavior is repeatable, then a digital signal processor can incorporate the information into the sensor signal processing algorithm in order to extract the correct measurement despite the nonlinear behavior. Repeatability of the nonlinearities is the key requirement for accurate signal processing of the sensor signals. If the nonlinearities are known to be repeatable, then they can be compensated in software in order to obtain accurate measurement.

In general, a sensor needs to be *calibrated* to customize it for an application. If the sensor exhibits drift in time, then it must also be calibrated periodically. Sensor calibration refers to adjustments in the sensor amplifier to compensate for the above variations so that the input (measured physical variable) and output (sensor output signal) relationship stays the same. The calibration process involves adjustments to compensate for variations in gain, offset, saturation, hysterisis, deadband, and drift in time.

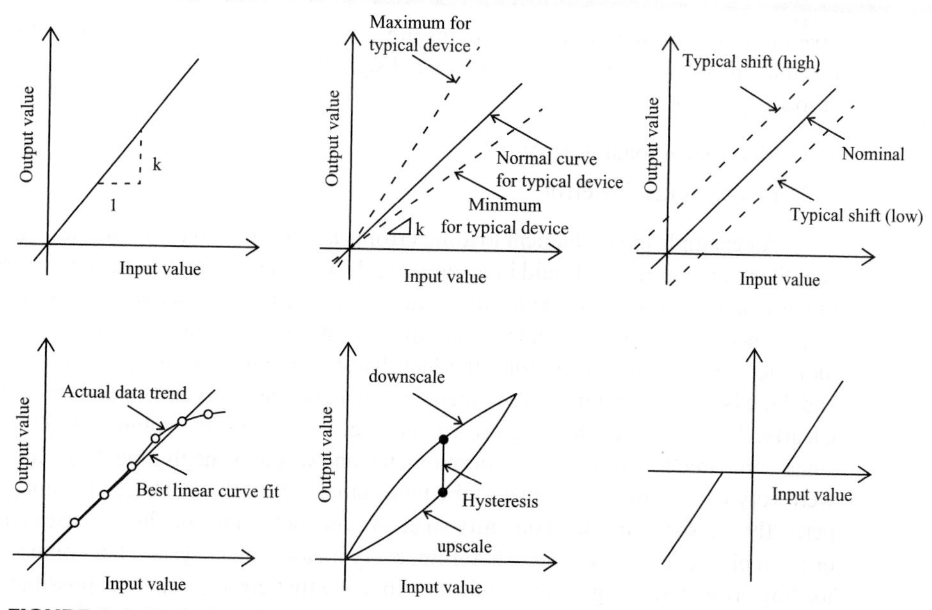


**FIGURE 6.4:** Typical nonlinear variations of static input–output relationship from the ideal behavior of a sensor.

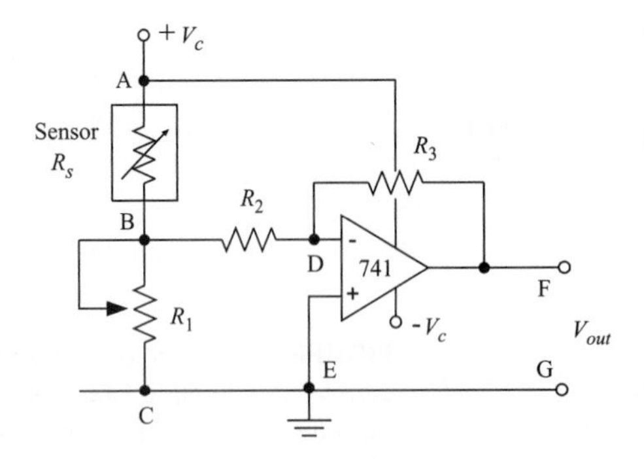


**FIGURE 6.5:** Resistive-type sensor signal amplification using an op-amp. The sensor transduction principle is based on the physical relationship that the resistance of the sensor varies as a function of the measured variable.

Figure 6.5 shows a circuit for a resistance type sensor and its signal amplification using an op-amp. The sensor transduction is based on the change of resistance as a function of measured variable (i.e., temperature, strain, pressure). The resistance change is converted to voltage change, which is typically a small value. Then it is amplified by an inverting type op-amp to bring the sensor signal (voltage) to a practical level. Notice that the resistor $R_1$ is used to calibrate the sensor for offset (bias) adjustments. The $R_1$ and $R_s$ act as the voltage dividers. The op-amp is in inverting configuration where the resistors $R_2$ and $R_3$ determine the gain of the amplifier. It can be easily shown that the output voltage from the circuit as function of the sensor resistance is

$$V_{out} = -\frac{R_3}{R_2} \cdot \frac{R_1}{R_s(x) + R_1} \cdot V_c \tag{6.1}$$

where $x$ represents the measured variable, $R_s(x)$ is the resistance of the sensor which varies as function of the measured variable.

## 6.2 MEASUREMENT DEVICE LOADING ERRORS

Loading errors in measurement systems are errors introduced to the measurement of the variable due to the sensor and its associated signal processing circuit. There are two types of loading errors:

1. Mechanical loading error
2. Electrical loading error

An example of mechanical loading error is as follows. Consider that we want to measure the temperature of a liquid in a container. If we insert a mercury-in-glass thermometer in the container, there will be a finite amount of heat transfer between the liquid and the thermometer. The measurement will stabilize when both temperatures of the liquid and thermometer are the same. Therefore, the fact that the thermometer is inserted into the liquid and there is a heat transfer between them has changed the original temperature of the liquid. Clearly, this mechanical loading error would be very large if the volume of the liquid were small compared to the size of the thermometer, and would be negligible if the liquid volume were very large compared to that of the thermometer. Strictly speaking, it is not possible to perfectly measure a physical quantity since the very act of introducing a sensor changes the original physical environment. Therefore, every measurement system has some mechanical loading error. The design question is how to make that error as small as possible.

The electrical loading error issue exists in electrical circuits used in measurement systems. Consider the voltage measurement across the resistor in the following figure (Fig. 6.6).

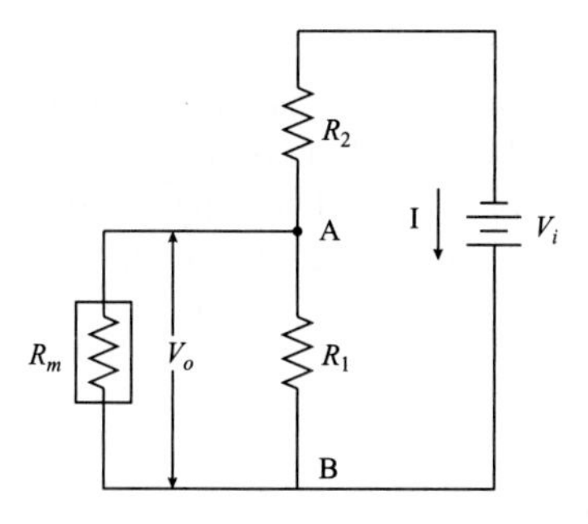


**FIGURE 6.6:** Electrical loading error in measurement systems and sensors.

Assume that the measurement device has internal resitance of $R_m$. Without the measurement device, the ideal value of the voltage we want to measure is

$$V_o^* = \frac{R_1}{R_1 + R_2} \cdot V_i \tag{6.2}$$

Once the measurement device is connected to the circuit at points A and B, it changes the electrical circuit. The equivalent resistance between point A and B is

$$R_1^* = \frac{R_1 R_m}{R_1 + R_m} \tag{6.3}$$

and the measured voltage across points A and B is

$$V_o = \frac{R_1^*}{R_1^* + R_2} \cdot V_i \tag{6.4}$$

which is

$$V_o = V_i \frac{R_1 R_m/(R_1 + R_m)}{R_1 R_m/(R_1 + R_m) + R_2} \tag{6.5}$$

Notice that as $R_m \to \infty$, $V_o = V_o^*$. However, if $R_m$ is close to the value of $R_1$, the $V_o/V_o^*$ deviates from unity, i.e., $R_m = R_1$

$$V_o = V_i \frac{R_1}{R_1 + 2R_2} \tag{6.6}$$

In most measurement systems, the relationship between the $R_m$ and $R_1$ is such that

$$R_m = 10^3 \cdot R_1 \tag{6.7}$$

Consider for simplicity that $R_1 = R_2$. The voltage measured ($V_o$) and the ideal voltage should have been measured if $R_m$ was infinity ($V_o^*$),

$$V_o = V_i \frac{1000}{2001} \tag{6.8}$$

and the ideal voltage

$$V_o^* = V_i \frac{1000}{2000} \tag{6.9}$$

The voltage measurement error percentage due to the electrical loading in this case is

$$e_v = \frac{(1000/2001) - (1/2)}{(1/2)} \cdot 100 = -0.0499\% \tag{6.10}$$

For cases in which the measurement device input impedance is about $10^3$ times the resistance of the equivalent two-port device whose voltage is measured, the voltage measurement error introduced due to the electrical loading effect is negligible.

Therefore, in order to minimize the effect of electrical loading errors due to the circuits used for the measurement, the measurement device should have large input resistance (input impedance). The larger the input impedance, the smaller the electrical loading error in the voltage measurement.

## 6.3 WHEATSTONE BRIDGE CIRCUIT

The Wheatstone bridge circuit is used to convert the change in resistance into voltage output. It is a standard circuit used as part of sensor signal conditioners (Fig. 6.7). Wheatstone bridge has a power supply voltage, $V_i$, and four resistances arranged in bridge circuit, $R_1$, $R_2$, $R_3$, $R_4$. Usually, one of the resistance branches is the resistance of the sensor. The sensor resistance changes as a function of the measured variable, i.e., RTD resistance as a function of temperature and strain-gauge resistance as a function of strain.

Let us consider the case in which the bridge is balanced, the voltage differential between points B and C is zero ($V_{BC} = 0$), and the current passing through the measurement device (i.e., galvonometer, digital voltmeter, or ADC circuit for a data-acquisition system) is zero ($i_m = 0$). Since $V_{BC} = 0$,

$$i_1 R_1 - i_3 R_3 = 0 \tag{6.11}$$

$$i_2 R_2 - i_4 R_4 = 0 \tag{6.12}$$

and $i_m = 0$,

$$i_1 = i_2 \tag{6.13}$$

$$i_3 = i_4 \tag{6.14}$$

Solving these equations, the following relations must hold for the bridge to be balanced ($V_{BC} = 0$ and $i_m = 0$),

$$\frac{R_1}{R_2} = \frac{R_3}{R_4} \tag{6.15}$$

Now, let us examine two different methods for using the Wheatstone bridge circuit in order to measure the variation in resistance in the form of proportional output voltage.

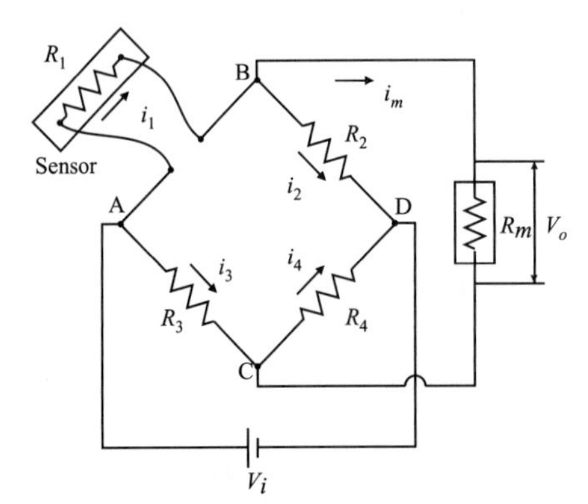


**FIGURE 6.7:** Wheatstone bridge circuit which is very often used in many different types of sensor applications as part of the signal conditioner and amplifier circuit of the sensor.

### 6.3.1 Null Method

$R_1$ represents the sensor whose resistance changes as a function of the measured variable. $R_2$ is a calibrated adjustable resistor (either manually or automatically adjusted). $R_3$ and $R_4$ are resistances fixed and known for the circuit. Initially the circuit is calibrated so that the balanced bridge condition is maintained and no voltage is observed across points B and C. Assume that the sensor resistance, $R_1$, changes as a function of the measured variable. Then, we can adjust $R_2$ in order to maintain the balanced bridge condition, $V_{BC} = 0$,

$$\frac{R_1}{R_2} = \frac{R_3}{R_4} \tag{6.16}$$

Then, when the $V_{BC} = 0$, the resistance of the sensor ($R_1$) can be determined since $R_3$, $R_4$ are fixed and known, and $R_2$ can be read from the adjustable resistor. Notice that this method of determining the sensor resistance is not sensitive to the changes in the supply voltage, $V_i$. But it is suitable only for measuring steady-state or slowly varying resistance changes, hence, slowly varying changes in the measured variable (i.e., temperature, strain, pressure).

### 6.3.2 Deflection Method

In order to measure time-varying and transient signals, the deflection method should be used. In this case, three legs of the resistor bridge have fixed resistances, $R_2$, $R_3$, $R_4$. The $R_1$ is the resistance of the sensor. As the sensor resistance changes, nonzero output voltage $V_{BC}$ is measured. For our first derivation, let us assume that the voltage measurement device has infinite input resistance, $R_m \to \infty$, so that no current flows through it despite a finite voltage potential across points B and C, $V_{BC} \neq 0$, $i_m = 0$,

$$V_o = i_1 R_1 - i_3 R_3 \tag{6.17}$$

Since $i_1 = i_2$ and $i_3 = i_4$,

$$i_1 = V_i/(R_1 + R_2) \tag{6.18}$$

$$i_3 = V_i/(R_3 + R_4) \tag{6.19}$$

Then,

$$V_o = V_i \left( \frac{R_1}{R_1 + R_2} - \frac{R_3}{R_3 + R_4} \right) \tag{6.20}$$

In most Wheatstone bridge circuit applications with sensors, the bridge is balanced at a reference condition, $V_o = 0$, and the initial values of the resistance arms are the same, $R_1 = R_2 = R_3 = R_4 = R_o$. Let

$$R_1 = R_o + \Delta R \tag{6.21}$$

where the $\Delta R$ is the variation from the calibrated nominal resistance. If we substitute these relations, it can be shown that

$$V_o/V_i = \frac{\Delta R/R_o}{4 + 2\Delta R/R_o} \tag{6.22}$$

In general, $\Delta R/R_o << 1$, and the above relation can be approximated,

$$V_o/V_i = \frac{\Delta R/R_o}{4} \tag{6.23}$$

Another convenient way of expressing this relationship is

$$V_o = \frac{V_i}{4R_o} \Delta R \tag{6.24}$$

Notice that if the sensor signal conditioner has an ADC converter and an embedded digital processor, the above approximation is not necessary. The more complicated relationship can be used in the software for a more accurate estimation of the measured variable using equation 6.20.

Now let us relax the assumption on the measurement device resistance. Let us assume that it is not infinite, but a large finite value. Therefore, there will be a finite, though small, amount of current passing through the measurement device, $i_m \neq 0$. The actual output voltage measurement can be found as follows. The reference voltage, $V_i$, is equal to the voltage drop on each arm of the bridge circuit,

$$V_i = i_1 R_1 + i_2 R_2 \tag{6.25}$$

$$= i_3 R_3 + i_4 R_4 \tag{6.26}$$

Since we no longer assume that $i_m$ is zero, then

$$i_1 = i_2 - i_m \tag{6.27}$$

$$i_3 = i_4 - i_m \tag{6.28}$$

If we consider the voltage drop along the closed path $R_1$, $R_m$, $R_3$, which must be equal to zero,

$$i_1 R_1 + i_m R_m - i_3 R_3 = 0 \tag{6.29}$$

and for the closed path circuit formed by $R_2$, $R_m$, $R_4$,

$$i_2 R_2 - i_m R_m - i_4 R_4 = 0 \tag{6.30}$$

We can solve for $i_m$, and noting that $V_o = i_m \cdot R_m$, and that for most Wheatstone bridge circuits, $R_1 = R_2 = R_3 = R_4 = R_0$ nominally, it can be shown that

$$V_o = V_i \frac{(\Delta R / R_o)}{4(1 + (R_o / R_m))} \tag{6.31}$$

Notice that if the measurement device input impedance, $R_m$, is very large, the measured voltage is practically same as the case when we assumed $R_m \to \infty$,

$$V_o = \frac{V_i}{4R_o} \Delta R \tag{6.32}$$

Note that the output voltage measurement device can be an anolog voltmeter, a digital voltmeter, or an analog-to-digital converter (ADC) of a data-acquisition circuit. If the measurement device input resistance is sufficiently large relative to the resistances used in the Wheatstone bridge, the output voltage measurement will be very close to the ideal case.

***Example*** Consider that an RTD type temperature sensor is used to measure the temperature of a location. The two terminals of the sensor are connected to the $R_1$ position of a Wheatstone bridge circuit. The sensor temperature-resistance relationship is as follows:

$$R = R_o(1 + \alpha(T - T_o)) \tag{6.33}$$

where from the sensor calibration data it is known that $\alpha = 0.004°\text{C}^{-1}$, $T_o = 0°\text{C}$ reference temperature, and $R_o = 200\ \Omega$ at temperature $T_o$. Assume that $V_i = 10$ VDC, and $R_2 = R_3 = R_4 = 200\ \Omega$. What is the temperature when the $V_o = 0.5$ VDC?

Let us assume that the input resistance of the voltage measurement device is infinity,

$$V_o = V_i \cdot \frac{\Delta R/R_o}{4} \tag{6.34}$$

Find $\Delta R$, then $R = R_o + \Delta R$, calculate $T$ from

$$R = R_o(1 + \alpha(T - T_o)) \tag{6.35}$$

The resulting numbers are $\Delta R = 40\ \Omega$, $T = 50°\mathrm{C}$.

Let us consider that input resistance of the output voltage measuring device is $R_m = 1\ \mathrm{M}\Omega$, instead of infinite. The resulting measurement would indicate the following temperature:

$$V_o = V_i \cdot \frac{\Delta R/R_o}{4(1 + R_o/R_m)} \tag{6.36}$$

which gives

$$\Delta R = 40 \cdot 1.0002\ \Omega \tag{6.37}$$

and the more accurate tempareture measurement is

$$T = 40 \cdot 1.0002/0.8 \tag{6.38}$$

$$= 50 \cdot 1.0002 \tag{6.39}$$

$$= 50.01°\mathrm{C} \tag{6.40}$$

Note that if the input resistance of the measurement device is small compared to the nominal resistance in the Wheatstone bridge, measurement error will be much larger. For instance, if $R_m = 1000\ \Omega$, the resulting temperature measurement would be

$$T = 50 \cdot 1.2 = 60°\mathrm{C} \tag{6.41}$$

which has a 20% error in measurement due to the low-input impedance of the measurement device relative to the impedance of the circuit.

## 6.4 POSITION SENSORS

There are two kinds of *length* measurements of interest: (1) absolute position (the distance between two points), and (2) incremental position (the change in the position). If a sensor can measure the position of an object on power up relative to a reference (the distance of the object from a reference point on power up), we call it an *absolute position sensor*. If the sensor cannot tell the distance of the object from a reference on power up, but can keep track of the change in position from that point on, we call it an *incremental position sensor*. Examples of absolute position sensors include a calibrated potentiometer, absolute optical encoder, linear variable differential transformer, resolver, and capacitive gap sensor. Examples of incremental position sensors include incremental optical encoder and laser interferometer. Most of the position sensors have rotary and translational (linear) position sensor versions.

### 6.4.1 Potentiometer

Potentiometer relates change in the position (linear or rotary) into change in resistance (Figs. 6.8 and 6.9). The resistance change is converted to a proportional voltage change in the electrical circuit portion of the sensor. Hence, the relationship between the measured

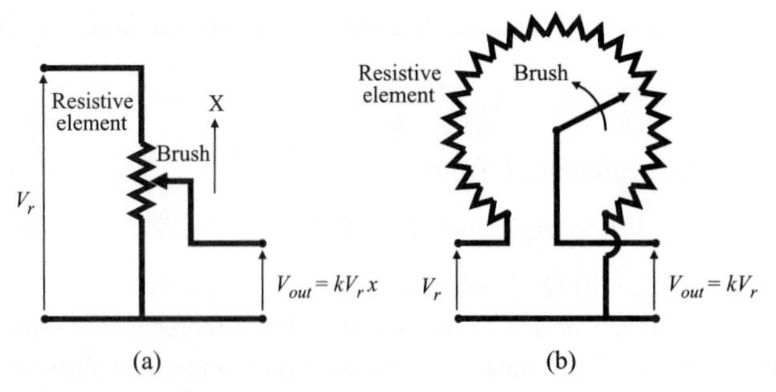


**FIGURE 6.8:** Linear and a rotary potentiometer for position measurement.

physical variable, translational displacement $x$ or rotary displacement $\theta$, and the output voltage for an ideal potentiometer is

$$V_{out} = k \cdot V_r \cdot x \tag{6.42}$$

or

$$V_{out} = k \cdot V_r \cdot \theta \tag{6.43}$$

The sensitivity, $k \cdot V_r$, of the potentiometer in equations (6.42) and (6.43) is a function of the winding resistance and physical shape of the winding. The range and resolution of the potentiometer are designed into the sensor as a balanced compromise—the higher the resolution, the smaller the range of the potentiometer. Due to the brush-resistor contact, the accuracy is limited. As the contact arm moves over the resistor winding, the output voltage changes in small, discrete steps, which defines the resolution of the potentiometer. For very long length measurements where the distance may not be a straight line, i.e., 5.0 $m$ curve, a spring-loaded multiturn rotary potentiometer arm is connected to a string. Then the string moves with the measured curve distance. As the string is pulled, the potentiometer arm moves around the multiturn resistor. The output voltage is then proportional to the

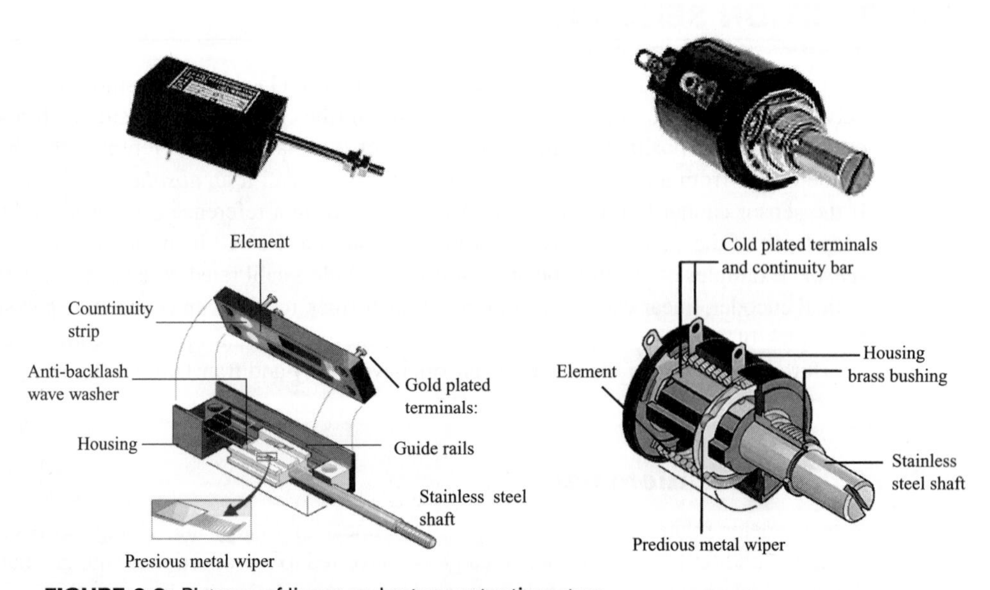


**FIGURE 6.9:** Pictures of linear and rotary potentiometers.

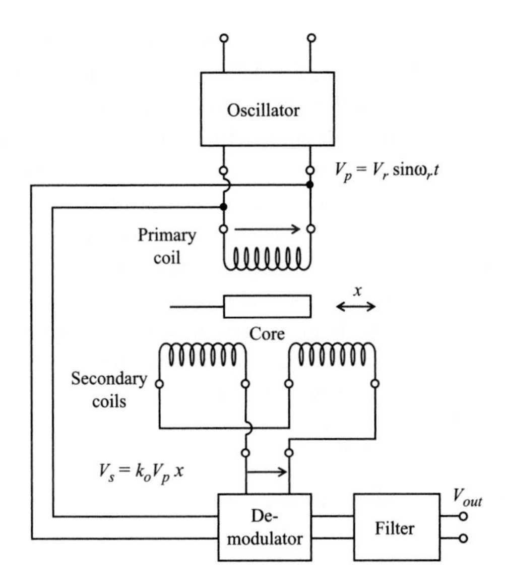


**FIGURE 6.10:** Linear variable differential transformer (LVDT) and its operating principles. An oscillator circuit generates the excitation signal for the primary winding. The demodulator circuit removes the high-freqeuncy signal content and obtains the magnitude of the induced voltage which is related to the core position.

length of the pulled string. The total distance measured can be any shape. Potentiometers are considered low-cost, low-accuracy, limited-range, simple, reliable, absolute-position sensors. Typical resistance of the potentiometer is around 1 KΩ per inch. Since there is a supply voltage, there will be a finite power dissipation over the potentiometer. However, it is a small amount of power and less than 1 W/in.

### 6.4.2 LVDT, Resolver, and Syncro

Linear variable differential transformer (LVDT), resolver, and syncro are sensors which operate based on the *transformer principle*. The key to their operating principle is that the change in the position of the rotor element changes the electromagnetic coupling (magnetic flux linkage) between the two windings (Figs. 6.10 and 6.11), primary and secondary windings. The rotor element is made of a material which has high magnetic conductivity (permeability). As a result, the induced voltage between the two windings changes in relation to the position. Hence, we have a well-defined relationship between the induced voltage and the position. In LVDTs, both windings are stationary, and a rotor core made of a material with high-magnetic permeability couples the two winding electromagnetically. In rotary LVDTs (resolvers and syncros), the primary winding is located on the rotor, and the secondary winding on the stator. Either the rotor winding or the stator windings can be excited externally by a known voltage, and the induced voltage on the other winding is measured which is related to the position. The operating principles of a LVDT, resolver, and syncro are shown in Figs. 6.10–6.15. Notice that the syncro is just a three-phase stator version of the resolver.

**FIGURE 6.11:** Pictures of LVDTs and a resolver.

LVDT is an absolute position sensor. On power up, the sensor can tell the position of the magnetic core relative to the neutral position. LVDT's primary winding is excited by a sinusoidal voltage signal. The induced voltage on the secondary windings has the same frequency except that the magnitude of the voltage is a function of the position of the magnetic core. In other words, the displacement modulates the magnitude of the induced voltage. As the core displacement increases from the center, the magnitude of the voltage differential between the two stator windings increases. The core material must have a large magnetic permeability compated to air, i.e., iron-nickel alloy. A nonmagnetic stainless steel rod is used to connect the core to the part whose displacement is to be measured. The sign (direction) of the magnitude of the voltage differential is determined by relating the induced voltage phase to the reference voltage phase. It is a function of the direction of the magnetic core displacement from neutral position. The primary winding is excited by

$$V_p(t) = V_r \cdot sin(w_r t) \tag{6.44}$$

and the voltage differential between the secondary windings is

$$V_s(t) = k_0 \cdot V_p(t) \cdot x \tag{6.45}$$

$$= k_0 \cdot V_r \cdot sin(w_r t) \cdot x \tag{6.46}$$

Once the $V_s(t)$ is demodulated in frequency, the output signal is presented as a DC voltage,

$$V_{out}(t) = k_1 \cdot V_r \cdot x(t) \tag{6.47}$$

which is proportional to the core displacement.

LVDTs can be used for high-resolution position measurement (i.e., 1/10,000 in. resolution) but with a relatively small range (up to 10 in. range). Excitation frequency of the primary coil is in the range of 50 Hz to 25 kHz. The bandwidth of the sensor is about 1/10 of the excitation frequency (Fig. 6.11).

The resolver (Fig. 6.12) and syncro (Fig. 6.13) sensors operate on the same principle as the LVDT. Resolvers compete with encoders in position measurement applications. In general, resolvers have better mechanical ruggedness, but lower bandwidth than the encoders. Resolvers have two stator windings (90° out of phase), whereas syncros have three stator windings (120° out of phase). Let us examine the operation of the resolver

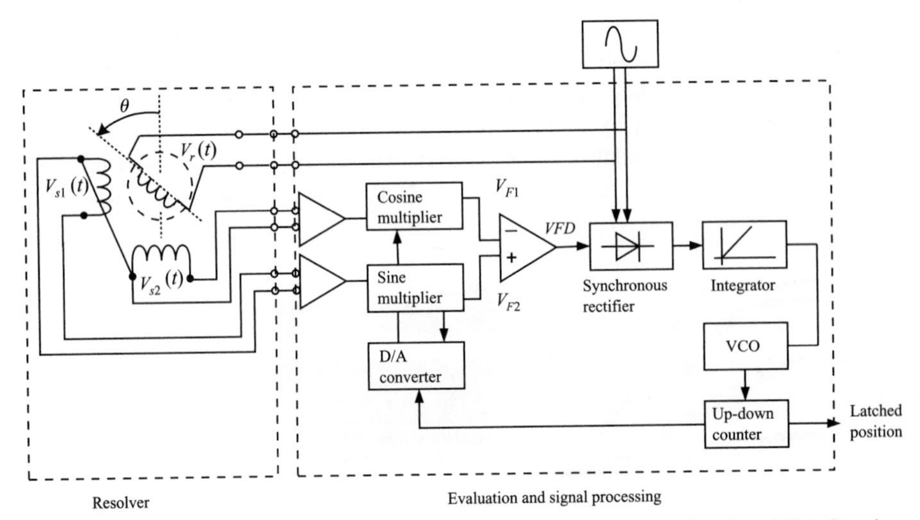


**FIGURE 6.12:** Resolver and its operating principle: resolver sensor head and the signal processing circuit (also called the resolver-to-digital converter, RTDC).

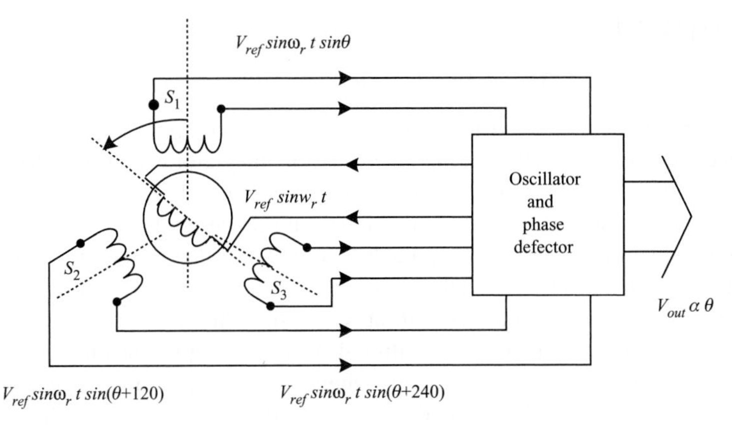


**FIGURE 6.13:** Syncro and its operating principle.

where we will consider that the rotor winding is excited externally by a sinusoidal voltage signal. The induced voltage in the stator windings is measured. The magnitude of induced voltages is a function of the angular position of the rotor. The induced voltages at the two stator windigs are 90° out of phase with each other since they are mechanically placed at 90° phase angle. Therefore, the resolver is an absolute rotary position sensor with one revolution range. The angular position changes beyond one revolution are kept track of through digital counting circuits. The rotor excitation voltage is (generated by an oscillator circuit)

$$V_r = V_{ref} \cdot sin(w_r t) \tag{6.48}$$

The induced voltage on the stator windings are (Fig. 6.14)

$$V_{s1} = k_0 \cdot V_r \cdot sin(\theta) = k_0 \cdot V_{ref} \cdot sin(w_r t) \cdot sin(\theta) \tag{6.49}$$

$$V_{s2} = k_0 \cdot V_r \cdot cos(\theta) = k_0 \cdot V_{ref} \cdot sin(w_r t) \cdot cos(\theta) \tag{6.50}$$

Next, the $V_{s1}$, $V_{s2}$ are multiplied by sin and cos functions in the RTDC circuit,

$$V_{f1} = k_0 \cdot V_r \cdot sin(\theta) \cdot cos(\alpha) \tag{6.51}$$

$$V_{f2} = k_0 \cdot V_r \cdot cos(\theta) \cdot sin(\alpha) \tag{6.52}$$

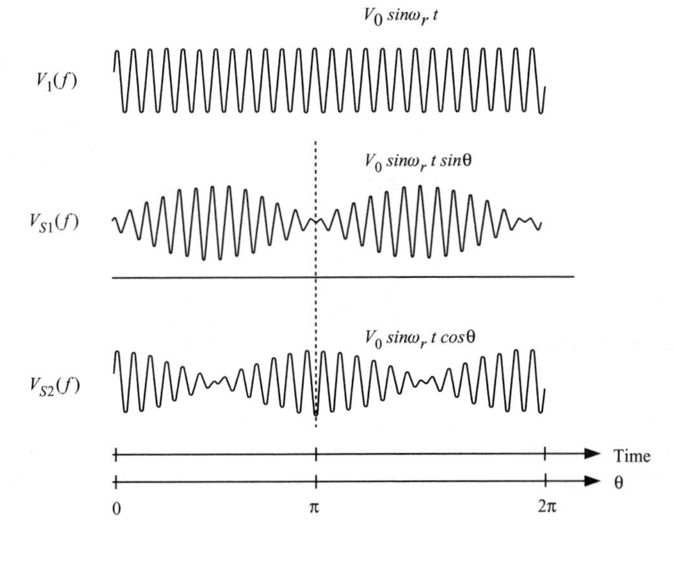


**FIGURE 6.14:** Excitation voltage for the rotor winding and the induced stator voltages. The induced voltages are shown as a function of time for a constant rotor speed over a period of one revolution.

Then the output of the error amplifier is

$$\Delta V_f = k_0 \cdot V_{ref} \cdot sin(w_r t) \cdot sin(\theta - \alpha) \tag{6.53}$$

Then the signal is demodulated to remove the $sin(w_r t)$ component by the synchronous rectifier and low-pass filter. Then the signal is fed to an integrator which provides input to the voltage-controlled oscillator (VCO). The output of the VCO is fed to an up-down counter to keep track of the position. This results in making the up-down counter to increase or decrease $\alpha$ in a direction so that $sin(\theta - \alpha)$ approaches zero. The value of the up-down counter is converted to angle $\alpha$ by a DAC as analog signal to feed the sin and cos multiplication circuit. The iteration on $\alpha$ continues until $\alpha = \theta$, at which time the rotor position information is latched from the up-down counter as a digital data. The algorithm finds the angle $\theta$ by iteratively changing $\alpha$. The iteration stops when $\Delta V_f$ is equal to zero, at which time it means that $\theta = \alpha$. Functionally, the end result is that the output position value presented is proportional to the rotor angle

$$V_{out} = k_1 \cdot \theta \tag{6.54}$$

This circuit can be implemented with integrated circuits AD2S99 programmable oscillator chip and AD2S90 resolver-to-digital converter (RTDC) chip by Analog Devices (Fig. 6.15). The AD2S99 chip is a programmable sinusoidal oscillator. The input supply voltage is

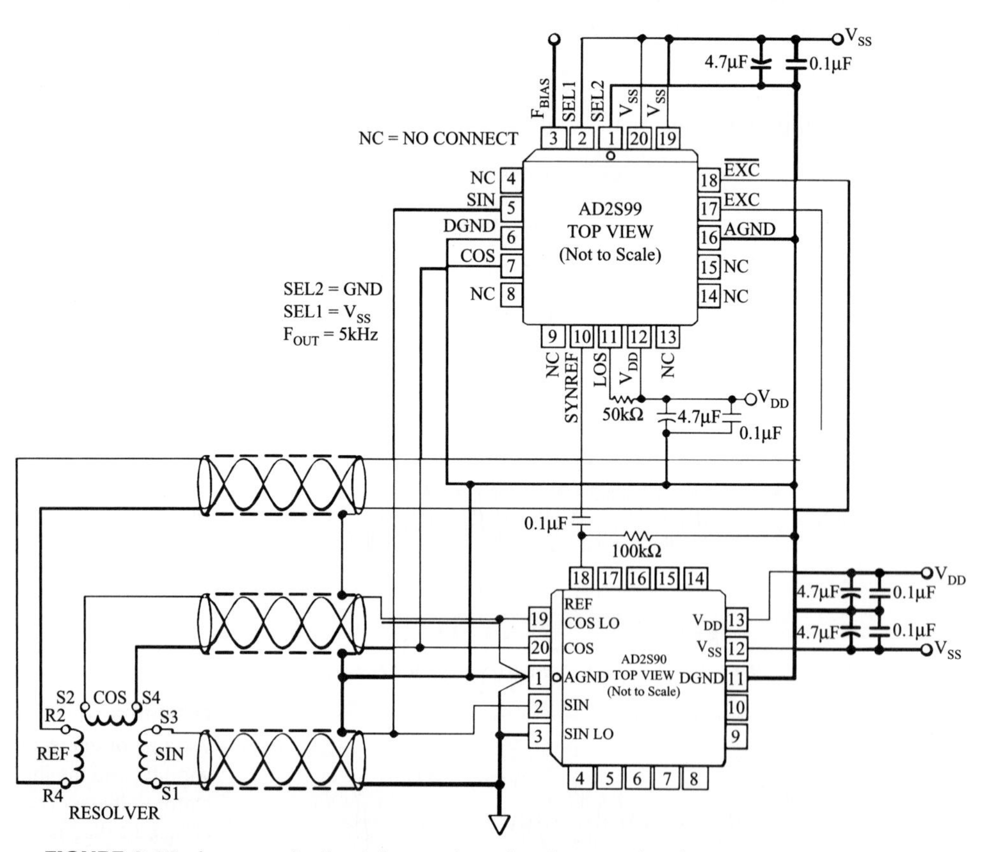


**FIGURE 6.15:** An example circuit for resolver signal processing. It can be used for a resolver which has one rotor and two stator windings. An AD2S99 chip provides the reference oscillator signal, and the AD2S90 chip processes the resolver signals for conversion to digital data.

+/− 5 VDC. It provides the sinusoidal excitation voltage for the primary winding. The excitation frequency is programmable to be one 2 kHz, 5 kHz, 10 kHz, or 20 kHz. It accepts the induced voltage signals from the secondary winding pair at the SIN and COS pins. In addition, it provides a synchronous square wave reference output signal that is phase locked with the SIN and COS signals. This signal is used by the AD2S90 chip for converting the resolver signal to digital signal. AD2S90 is the main chip that performs the conversion of the resolver analog signals to digital form. It provides the resulting digital position signal in two different formats: a 12-bit serial digital code and an incremental encoder equivalent A and B channel signals. In motor control applications, many amplifiers accept the resolver signal and use it for current commutation. In addition, they output encoder equivalent (encoder emulation) signals which can be used by a closed-loop position controller. It emulates a 1024 line per revolution incremental encoder A and B channels. When the A and B channel signals are decoded with $\times 4$, it results in 4096 counts/rev resolution which is equivalent to a 12-bit resolution. For LVDT signal processing, an AD2S93 chip is used in place of AD2S90 chip.

There are other types of signal processing circuits to extract the $\theta$ angle information from the resolver phase voltages. For instance, both stator voltages can be sampled at the same frequency $(w_r/(2\pi)$ Hz) where the start of the sampling is synchronized to the $V_r(t)$ by 90° phase angle in order to sample the maximum magnitudes. In other words, by sampling $V_{s1}$, $V_{s2}$ at the same frequency as the $w_r$, we achieve demodulation by sampling. The sampled signals are

$$V_{s1}^{adc} = V_{mag} \cdot sin(\theta) \tag{6.55}$$

$$V_{s2}^{adc} = V_{mag} \cdot cos(\theta) \tag{6.56}$$

where $V_{mag}$ is the sampled value of the $k_0 \cdot V_{ref} \cdot sin\,(w_r t)$ portion of the signal. Then we can compute the *Arctan* of the two signals to obtain the angle information,

$$\theta = Arctan\left(\frac{V_{s1}^{adc}}{V_{s2}^{adc}}\right) \tag{6.57}$$

This method uses two channels of A/D converter and a digital computational algorithm for the $Arctan(\cdot)$ function. Such a circuit for an RTDC (resolver-to-digital converter) can be implemented using an ADMC401 chip (AD converter) and AD2S99 oscillator chip (both by Analog Devices Inc.).

The operating principle of the syncro is almost identical to the resolver. The only difference is that there are three stator phases with 120° of mechanical phase angle. Hence, the induced voltages in the three stator phases will be 120° apart:

$$V_{s1} = k_0 \cdot V_{ref} \cdot sin\,(w_r t) \cdot sin\,(\theta) \tag{6.58}$$

$$V_{s2} = k_0 \cdot V_{ref} \cdot sin\,(w_r t) \cdot sin\,(\theta + 120) \tag{6.59}$$

$$V_{s3} = k_0 \cdot V_{ref} \cdot sin\,(w_r t) \cdot sin\,(\theta + 240) \tag{6.60}$$

and these signals can be processed to extract the angular position information with circuits similar to the ones used for resolvers.

Commercial LVDT and resolver sensors are packaged such that the input and output voltages to the sensor are DC voltages. The input circuit includes a modulator to generate an AC excitation signal from the DC input. The output circuit includes a demodulator which generates a DC output voltage from the AC voltage output of the secondary windings. For instance, LVDT Model -240 (by Trans-Tek, Inc.) has a position measurement range in 0.05 in. to 3.0 in., 24-VDC-input voltage, 300-Hz bandwidth, and internally generated carrier frequency of 13-kHz.

### 6.4.3 Encoders

There are two main groups of encoders: (1) absolute encoders and (2) incremental encoders. Absolute encoders can measure the position of an object relative to a reference position at any time. The output signal of the absolute encoder presents the absolute position in a digital code format. An incremental encoder can measure the change in position, not the absolute position. Therefore, the incremental encoder cannot tell the position relative to a known reference. If absolute position information is needed from incremental encoder measurement, the device must perform a so-called "home-ing" motion sequence in order to establish its reference position after the power up. From that point on, the absolute position can be tracked by digital counting. An absolute encoder does not need a counting circuit if the total position change is within the range covered by the absolute encoder.

Encoders can also be classified based on the type of position they measure: translational and rotary. Figure 6.16 shows the pictures of linear (translational) and rotary encoders. Encoders are the most widely used position sensors in electric motor control applications such as brush type DC motors, brushless DC motors, stepper motors, and induction motors. It is estimated that over 70% of all motor control applications with a position sensor use encoders as position sensors.

Operating principles of linear and rotary encoders are identical. In one case, there is a rotary disk; in the other case there is a linear scale. Figure 6.17 shows the components of a rotary and a linear incremental and absolute encoders. The main difference between the rotary and linear encoders is the glass scale with the printed pattern to interrupt the light as it moves.

An encoder has the following components (Figs. 6.18 and 6.19):

1. A disk or linear scale with light and dark paterns printed on it
2. A light source (LED possibly with a focusing lens)
3. Two or more photodetectors
4. A stationary mask

In case of an incremental encoder, the disk patern is a uniform black and opaque printed pattern around the disk (Fig. 6.18). As the disk rotates and angle changes, the disk pattern interrupts and passes the light. If the disk is metal (which may be necessary in applications

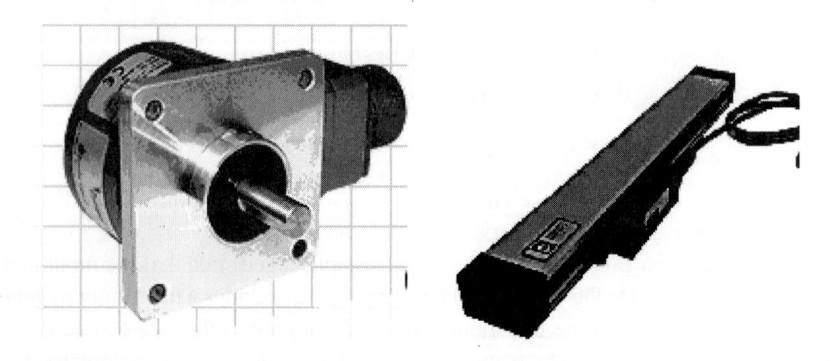

**FIGURE 6.16:** Pictures of a rotary and a linear encoder.

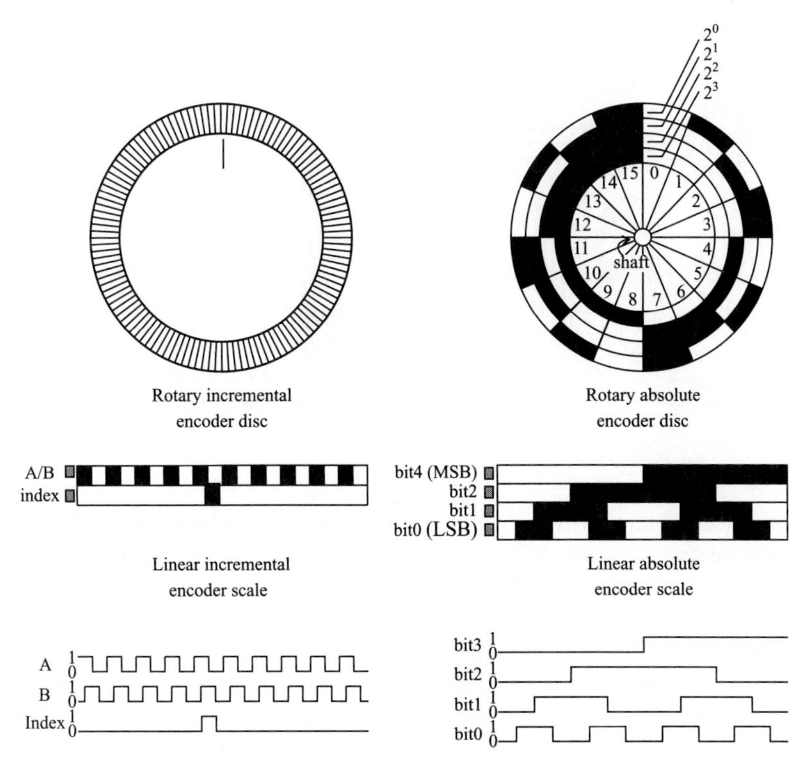


**FIGURE 6.17:** Disks of incremental (rotary and linear) and absolute encoders and the typical encoder output signals.

where the environmental conditions are extreme in terms of vibrations, shock forces, and temperature), the same principle is used as reflected light instead of pass-through light in counting the number of incremental position changes. The photodetector output turns ON and OFF every time the disk pattern passes over the LED light (Fig. 6.18). Therefore, the change in angular position can be measured by counting the number of state changes of the photodetector output. Let us assume that there are 1000 lines over the disk. Therefore,

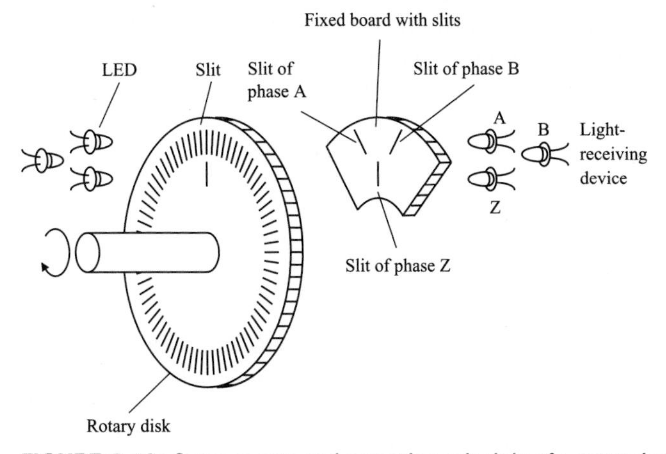


**FIGURE 6.18:** Components and operating principle of a rotary incremental encoder: rotary disk with slits, LEDs, phototransistors (light-receiving devices), and mask.

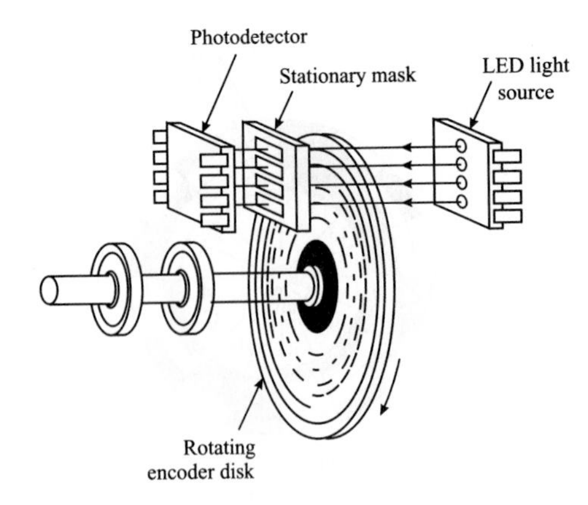


**FIGURE 6.19:** Components and operating principle of a rotary absolute encoder: rotary disk with absolute position coding in gray scale, LED set, phototransistor set, mask.

the photodetector output will change state 1000 times per revolution of the disk. Each pulse of the photodetector means $360°/1000°$ of angular position change in the shaft. If only one photodetector is used, the change in position can be detected, but not the direction of the change. By using two photodetectors (usually called A and B channels of the encoder output), which are displaced from each other by 1/2 the size of a single grading on the disk (or an integer number plus 1/2 the size of a single grading), the direction of the motion can be determined. If the disk direction of rotation changes, the phase between the A and B channel changes from $+90°$ to $-90°$. In addition to the dual photodetectors A and B, an incremental encoder also has a third photodetector which turns ON (or OFF) for one pulse period per revolution. This is accomplished by a single slit on the disk. Using this channel (usually called C or Z channel), the absolute position of the disk can be established by a home motion sequence. On power up, the angular position of the shaft relative to a zero reference position is unknown. The encoder can be rotated until the C channel is turned on. This position can be used as the zero reference position for keeping track of the absolute position.

Finally, each output channel of the encoder (A, B, C) may have a complementary channels ($\bar{A}$, $\bar{B}$, $\bar{C}$) which are used as protection against noise. Figure 6.20 illustrates how the complementary encoder output channels can be used to eliminate position measurement errors due to noise. Notice that for this approach to work, the same noise signal is assumed to be present on each channel. In short, the complementary channels improve the noise immunity of the encoder, but are not a solution for all possible noise conditions.

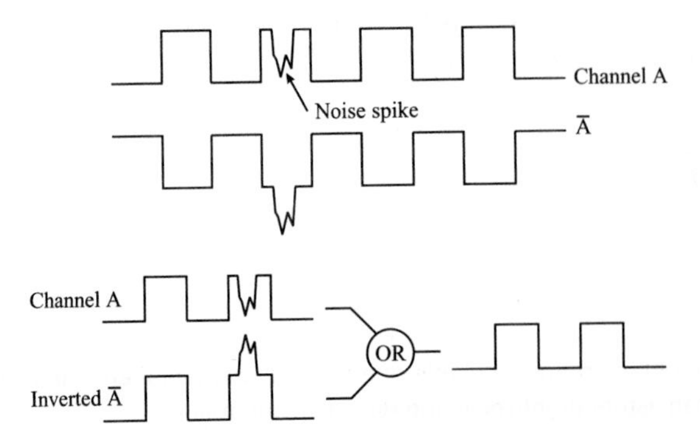


**FIGURE 6.20:** Use of complementary channel signals $\bar{A}$, $\bar{B}$ in conjuction with A and B channels to cancel noise, hence increase noise immunity of the sensor.

The linear incremental encoder works in the same principle, except that instead of a rotary disk it has a linear scale. Furthermore, the linear scale is stationary and the light assembly moves.

Absolute encoder differs from the incremental encoder in the printed pattern on the disk and the light assembly (LEDs and photodetectors). Figure 6.19 shows the components of an absolute encoder. Notice the coded pattern on the disk. Each discrete position of the disk corresponds to a unique state of photodetectors. Therefore, at any given time (i.e., after power up), the absolute position on the shaft can be determined uniquely. The encoder can tell us the absolute position within one revolution. Multiturn absolute encoders are also available. The resolution of the absolute encoder is determined by the number of photodetectors. Each photodetector output represents a bit on the digitally coded position information. If the absolute encoder has eight photodetectors (8-bit), the smallest position change that can be detected is $360°/(2^8) = 360/256°$. If the encoder is 12 bit, the resolution is $360°/(2^{12}) = 360/4096°$. For each position, the absolute encoder outputs a *unique code*. The coding of an absolute encoder is not necessarily a binary code. Gray code is known to have better noise immunity and is less likely to provide wrong reading compared with binary code.

An encoder performance is specified by:

1. *Resolution*: Number of counts per revolution. This is the number of lines on the disk or linear scale for incremental encoder. For an absolute encoder with $N$-bit resolution ($N$ set of LED and photodetector pairs), there are $2^N$ counts per revolution.
2. *Maximum speed*: The maximum speed of the encoder can be limited for electrical and mechanical reasons. The maximum state change capacity of the photodetectors determines the maximum speed limit set by electrical capability of the encoder. The frequency output of the encoder (resolution times the maximum speed) must be below that value. The maximum speed is also limited by the mechanical bearings of the encoder.
3. *Encoder output channels available*: A, B, C, $\bar{A}$, $\bar{B}$, $\bar{C}$ for incremental encoders.
4. *Electrical output signal type*: TTL, open collector, or differential line driver type. Differential line driver type is used for long cable and noisy environments.
5. *Mechanical limits*: Maximum radial and axial loads, sealing for dusk and fluid, vibration and temperature limits.

**Count Multiplication and Interpolation** By detecting the phase between A and B channels of an incremental encoder, the number of counts per line can be effectively multiplied by 1, 2, or 4. Since A and B channels are 90° out of phase, with one cycle of line change, we can actually obtain up to four counts by counting the transition in each channel (Fig. 6.21). The quadrature count multiplication ($\times 4$) is standard in the industry. For a quadrature incremental encoder the number of counts per revolution is four times the number of lines per revolution.

If further position resolution improvement is needed beyond the $\times 4$ quadrature decoding of what is physically coded to disk, then sinusiodal output photodetectors must be used. The output of the so-called sinusodial encoders are not discrete ON/OFF levels, but rather sinusoidal signals as a function of the angular position within one cycle of disk lines (Fig. 6.21). Typical signal voltage level is about 1.0 VDC peak to peak. By sampling the level of the signal finely, the effective position resolution of the encoder can be increased, i.e., by $\times 1024$ using a 10-bit sampling circuit. This type of interpolation requires a specifically designed interpolation circuit. Notice that the repeatability of the mechanical dimension of the printed pattern on the disk is ultimately the smallest resolution that can be achieved. The

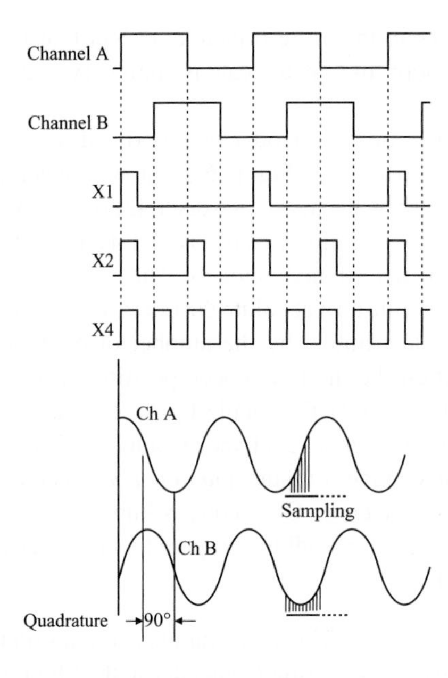


**FIGURE 6.21:** Digital processing of an incremental encoder signal channels A and B in order to increase resolution: quadrature decoding to increase resolution by ×4, and digital oversampling of sinusoidal output signal from the encoder.

electronic interpolation improves the resolution by predicting the intermediate positions through oversampling of a sinusoidal signal. The interpolation-based resolution improvement can be easily lost by noise in the sinusoidal output signal. An example of sinusoidal output linear incremental encoder is Series LR/LS (by Dynapar Corp). The sinusoidal output of the linear encoder can be sampled and interpolated to 250-nm resolution using a 10-bit sampling and interpolation circuit.

***Example*** Consider an incremental encoder with 2500-lines/rev resolution and a decoder which uses ×4 decoding logic. Then the resolution of the encoder is 10000 counts/rev. Assume that the photodetectors in the decoder circuit can handle A and B channel signals up to 1-MHz frequency.

1. Determine the maximum speed the encoder and decoder circuit can handle.
2. If the encoder resolution was 25000 lines/rev and the same decoder is used, what is the maximum speed that can be handled by the decoder circuit.

Since the decoder can handle 1-MHz A and B channel signals, the maximum speed that can be measured without data overrun is

$$w_{max} = \frac{1 \cdot 10^6 \text{ pulse/sec}}{2500 \text{ pulse/rev}} = 400 \text{ rev/sec} = 24000 \text{ rpm} \tag{6.61}$$

The angular position measurement resolution is 360/10,000°, which is the smallest detected change in angular position.

If the encoder resolution is increased by a factor of 10, then the position measurement resolution is increased by a factor of 10. However, the maximum speed the decoder can handle is reduced by the same factor,

$$w_{max} = \frac{1 \cdot 10^6 \text{ pulse/sec}}{25000 \text{ pulse/rev}} = 40 \text{ rev/sec} = 2400 \text{ rpm} \tag{6.62}$$

while the position measurement resolution is improved to 360/100,000°. Therefore, if we need to increase the position measurement resolution by increasing encoder resolution, we reduce the maximum speed measurement capacity if we use the same decoder circuit. If

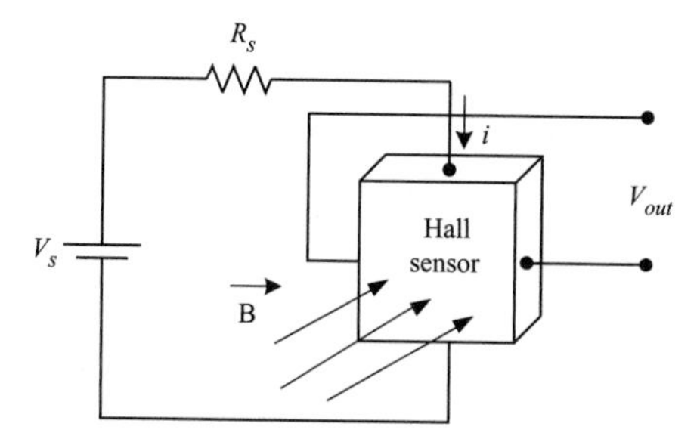


**FIGURE 6.22:** Principle of Hall effect and its usage in sensor design.

we want to increase position measurement resolution without losing the maximum speed measurement capacity, the maximum frequency capacity of the decoder circuit must be increased by the same factor. Decoder circuits which can handle encoder signal frequency up to 2 MHz are common. Decoders up to 50-MHz input frequency capacity are also available.

### 6.4.4 Hall Effect Sensors

The Hall effect (named after Edward Hall, 1879) is the phenomenon whereby semiconductor and conductor materials develop an induced voltage potential when the material is in a magnetic field and it has a current passing through it. The relationship between the induced voltage, the current, and magnetic field strength is a vector relationship (Fig. 6.22). When a sheet of semiconductor (i.e., gallium-arsenide, GaAs, indium-antimony, InSb) or conductor material has a current passing through it, and placed in a magnetic field, a voltage is induced in the direction perpendicular to both current and magnetic field vector. If the direction of the magnetic field or the current changes, so does the direction of the induced voltage. The induced voltage magnitude is also a function of the material type. It is very small for conductors, but large enough for semiconductors such that this effect is widely used in sensor designs.

$$V_{out} = V_{out}(B, i, material) \tag{6.63}$$

In a sensor application, the current ($i$) is fixed by the sensor power supply and resistor ($R_s$). The Hall effect sensor requires an external power supply for the current $i$ and magnetic field density ($B$) for it to work. The material of the sensor is also fixed. Then the output voltage varies as function of magnetic field strength for a given Hall sensor. The measured variable then must be arranged such that it changes the magnetic field strength over the Hall sensor. For instance, the measured variable may be the position of a magnet relative to the sensor. Similarly, the magnet may be part of the sensor, and a ferrous metal object entering the field changes the magnetic permeability (magnetic conductivity) around the sensor, hence change the magnetic field density ($B$). As a result the output voltage is changed (i.e., Fig. 6.34). This principle is also used in presence sensors which has only two-state ON/OFF output. Typical voltage levels produced by the Hall transducer are in the millivolts level; hence they must be amplified by an op-amp for processing in a measurement and control system. The op-amp may amplify the Hall effect sensor voltage linearly to measure analog distance or can convert it to ON/OFF output signal for presence sensing. Since the sensing element of the Hall sensor can be a semiconductor, both the sensing element and digital signal processing circuit can be integrated into a single silicon chip. For instance, AD22151 by Analog Devices, Inc. is one such integrated Hall effect sensor. It operates from a 5-VDC power supply and can operate in an environment with a temperature range of −40°C to 150°C. Such integrated Hall effect sensors also include a built-in temperature sensor so that the variations in the gain of the sensing head as temperature varies can be compensated.

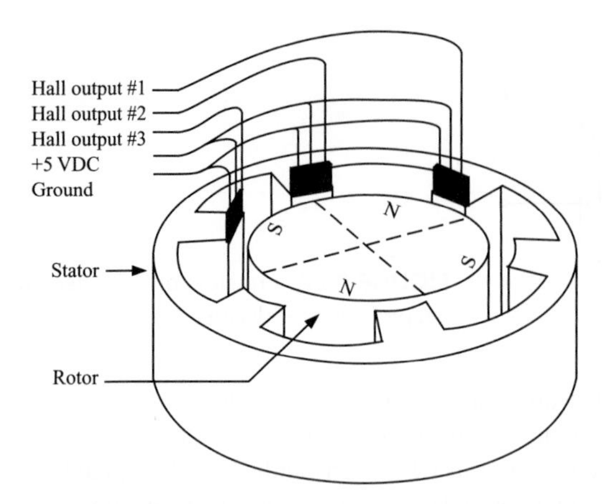


**FIGURE 6.23:** Three Hall effect sensors used to sense commutation positions in a brushless DC motor.

Figure 6.23 shows an application of Hall effect sensors in position sensing of the rotor of a brushless DC motor. Brushless DC motors need current commutation, that is, the shaping of the desired current in each phase as function of rotor position, for proper operation. In trapezoidal current commutated brushless DC motors, we need to know six ranges of rotor position in order to properly commutate the current. A set of three Hall effect sensors, each operating in ON/OFF mode, is used to provide rotor position information for current commutation. The sensor heads and permanent magnets are arranged such that at any given position, one or two of the sensors are ON.

## 6.4.5 Capacitive Gap Sensors

A capacitive gap sensor measures the distance between the front face of the sensor and a target object. It is a noncontact sensor. The target material should have high relative permittivity.[1] Metals or plastics with carbon are good target materials for sensing. The capacitance between the sensor and the target material is related approximately (Fig. 6.24),

$$C = \frac{\epsilon \cdot A}{x} \tag{6.65}$$

where $\epsilon = \kappa \cdot \epsilon_0$, and $\kappa$ is the dielectric constant of the medium separating the sensor and target material, $A$ is the area of the plates, $x$ is the distance to be measured. Therefore, there is a well-defined relationship between the distance (gap) and the capacitance. The effective dielectric constant $\kappa$ between the sensor head and targer material is a function of the target material type (conductive, i.e., iron, aluminum or nonconductive nonmetal, i.e., plastic)). Notice that since the net capacitance is function of the target material, the effective sensing distance varies with different target material types. The excitation circuit for the sensor works to maintain a constant electric field magnitude between the sensor head and the target object. As the capacitance changes, the required current to do so changes proportionally. Hence, there is a relationship between the measurable current effort and

[1] The force, $F$, between two charged particles, $q_1$ and $q_2$, which are separated from each other by distance $r$ in free space, is

$$F = k_e \frac{q_1 \cdot q_2}{r^2} \tag{6.64}$$

where $k_e = 1/(4\pi \cdot \epsilon_o) = 8.9875 \cdot 10^9 [Nm^2/C^2]$ is called the *Coulomb constant* and $\epsilon_o$ is the *permittivity of free space*.

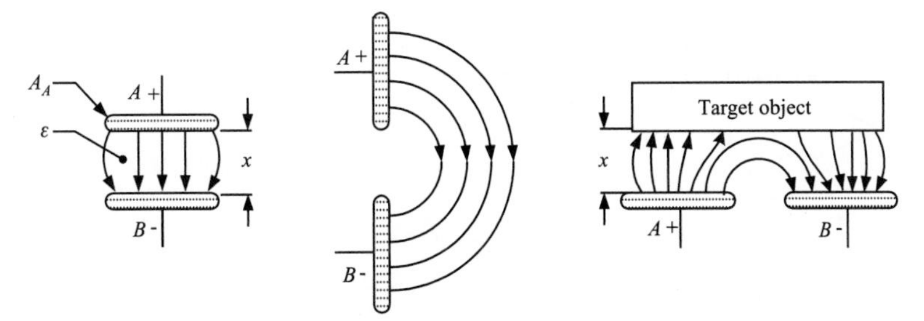


**FIGURE 6.24:** Operating principle of capacitive gap sensor: the presence of target object changes the effective capacitance.

capacitance. The excitation circuit which maintains the constant capacitance by controlling the current uses a high-frequency modulation signal (i.e., 20-KHz modulation frequency).

The resolution of the capacitive gap sensor is typically in micrometer range, but it can be made as low as a few nanometers for special applications. The range is limited to about 10 mm. The frequency response of a specific capacitive sensors can vary as function of the sensed gap distance. The bandwidth of a given sensor varies as a function of the gap measured as percentage of its total range. The smaller the gap distance measured relative to the sensor range, the higher the bandwidth of the sensor. For instance, a noncontact capacitive gap sensor can have 1000-Hz bandwidth for measuring gap distances within 10% of its range, while the same sensor may have about 100-Hz bandwidth while measuring gap distances around 80% of its range.

Capacitive presence sensors provide only two-state ON/OFF output, and sense the change in the oscillation circuit signal amplitude. When a target object enters the field-sensing distance of the sensor, the capacitance increases and the magnitude of oscillations increases. A detection and output circuit then controls the ON/OFF state of a transistor.

The capacitive gap sensor can also be used to sense the presence, density, and thickness of nonconducting objects. The nonconductor materials (such as epoxy, PVC, glass) which has different dielectric constant than air can be detected because the presence of such a material in front of the probe instead of air results in change in the capacitance.

### 6.4.6 Magnetostriction Position Sensors

Magnetostriction linear position sensors are widely used in hydraulic cylinders. Figure 6.25 shows the basic operating principle of the sensor. A permanent magnet moves with the object whose position is to be measured. The sensor head sends a current pulse along a wire which is housed inside a waveguide. The interaction between the two magnetic fields—the magnetic field of the permanent magnet and the electromagnetic field of the current pulse—produces a torsional strain pulse on the waveguide. The torsional strain pulse travels at about 9000-ft/sec speed. The time it takes for the strain pulse to arrive at the sensor head is proportional to the distance of the permanent magnet from the sensor head. Therefore, by measuring the time period between the current pulse sent out and the strain pulse reflected back, the distance can be measured:

$$x = V \cdot \Delta t \tag{6.66}$$

where $V$ travel speed of the torsional strain pulse is known, $\Delta t$ is measured, and hence the position, $x$, can be determined as the measured distance.

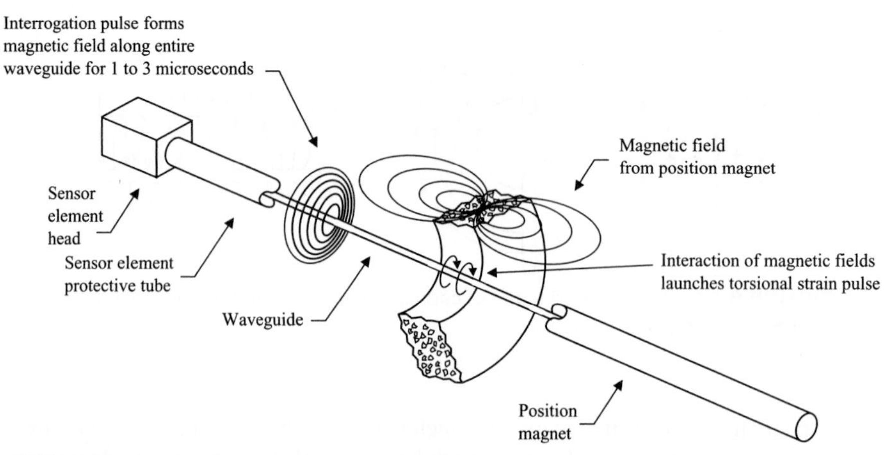


**FIGURE 6.25:** Operating principle of magnetostriction absolute position sensor—an industry standard sensor used for cylinder position measurement in electrohydraulic motion control systems.

Typical resolution is in the range of 2 $\mu$m, and range can be in the order of 10.0 m. The bandwidth of the sensor is typically in 50-Hz to 200-Hz range and is limited by the torsional strain pulse travel speed and the range (length) of the particular sensor.

## 6.4.7 Sonic Distance Sensors

Sound is transmitted through the propagation of pressure in the air. The speed of sound in the air is nominally 331 m/sec at 0°C dry air (343 m/sec at 20°C dry air). Two of the important characteristics of sound waves are *frequency* and *intensity*. The human ear can hear the frequencies in the range of 20 Hz–20 kHz. The frequencies above this range are called the ultrasonic frequencies.

Sonic distance sensors measure the distance of an object by measuring the time period between the sent ultrasonic pulse and the echoed pulse (Fig. 6.26). The sensor head sends ultrasonic sound pulses at high frequency (i.e., 200 kHz), and measures the time period between the sent pulses and the echoed pulses. For instance, it may send a short pulse of 200-kHz frequency and fixed intensity every 10 msec, at which the time period of the sent pulse is only a few milliseconds. Before a new pulse is sent out, the time instant of the reflected sound pulse is measured. This process continues periodically, i.e., every 10 msec in this example. Knowing the speed of sound, a digital signal processor embedded in the sensor can calculate the distance of the object,

$$x = V_{sound} \cdot \Delta t \tag{6.67}$$

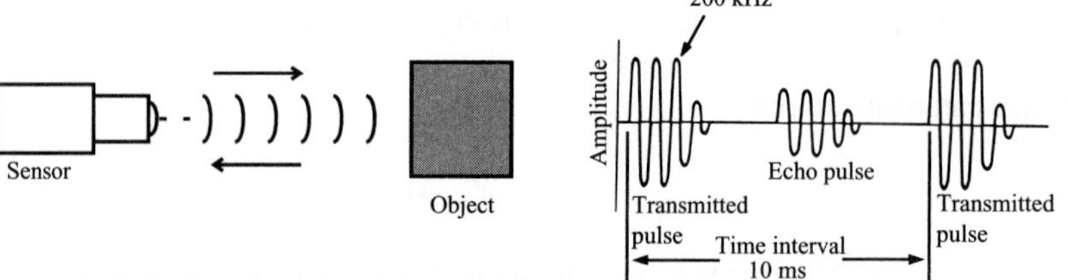


**FIGURE 6.26:** Operating principle of a sonic distance sensor.

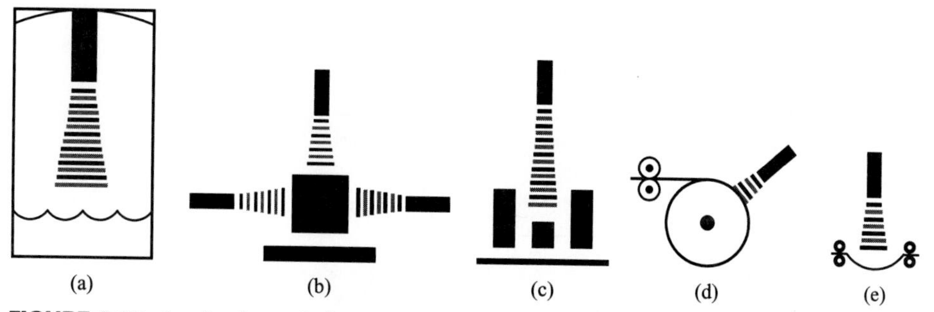


**FIGURE 6.27:** Applications of of a sonic distance sensor in various manufacturing processes: (a) liquid level sensing, (b) two-dimension sensing (i.e., box height and width), (c) one-dimension sensing (i.e., box height), (d) roll diameter sensing, and (e) web slack sensing.

where $V_{sound}$ is known, $\Delta t$ is measured, $x$ is calculated. The range of the sensor can be up to 20 m with minimum range of 5 cm. Notice that the sonic distance sensor is not appropriate for very short distance measurements, i.e., under 5 cm. The frequency response of the sensor (distance measurement update rate) varies with the distance measured; in general, it can be around 100 Hz. Clearly, sonic sensor cannot be used in high-bandwidth servo positioning applications which require 1.0 msec or faster position loop update period. Typical applications of sonic distance sensor include the roll diameter measurement, web loop length measurement, liquid level measurement, and box presence measurement on conveyors (Fig. 6.27).

## 6.4.8 Photoelectric Distance and Presence Sensors

Sensors which measure the distance or presence of an object from a reference point using "light" as the transduction mechanism are collectively called "photoelectric sensors" or "light sensors." The light sensors which provide only two discrete outputs (ON/OFF) are called the *photoelectric presence sensors* (Fig. 6.28). The sensor emitter sends a light beam. The receiver (phototransistor) turns its output either ON or OFF depending on the received light. The light threshold that separates the ON and OFF states is adjustable by the op-amps in the electronic circuit of the sensor. The light sensors are immune from the ambient light variations due to the modulated nature of the emitted light. The light-emitting diodes (LED) are tuned to emit light at high frequency. Figure 6.29 shows the frequency spectrum of light

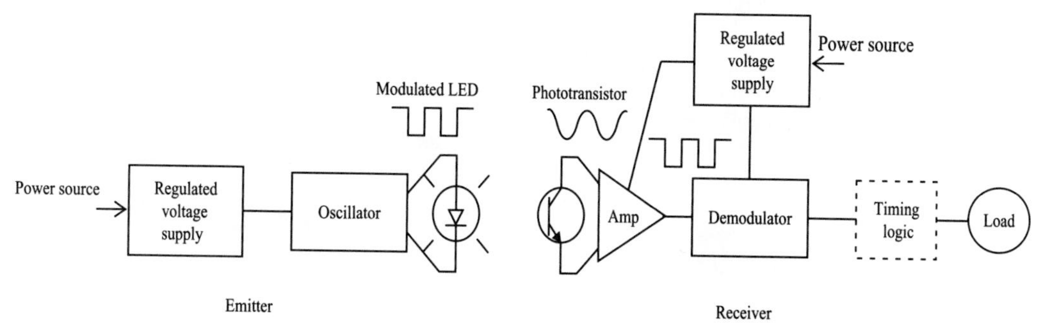


**FIGURE 6.28:** Components and operating principle of a through-beam light presence sensor. Reflective and diffusive versions of this type of sensor can also be used for distance sensing.

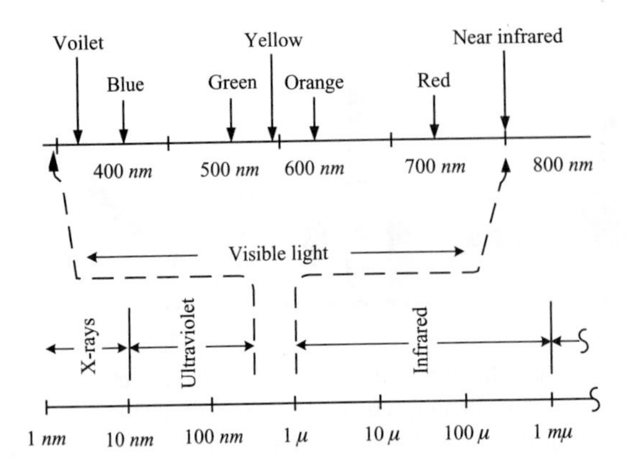


**FIGURE 6.29:** Frequency spectrum of light.

waves. The phototransistors (receivers) are also tuned to respond to that light frequency and reject the other frequencies, much like tuning a radio channel. As a result, the sensor output is very robust against the ambient light conditions. If the receiver is across from the emitter, it is called *through light sensor* (or through beam photoelectric sensor). If the receiver and emitter are on the same sensor head, and the receiver turns ON/OFF its output based on the amount of reflected light, it is called *reflective light sensor* (or reflective beam photoelectric sensor).

An analog (proportional) distance sensor using light is a modified version of the reflective type presence sensor. The receiver output is not just either ON or OFF state, but a proportional voltage output based on the reflected light strength (Fig. 6.30). The transduction principle is as follows: the emitter sends out a light signal. It is reflected from either the object or from a reflective background surface. The reflected light strength is proportional to the distance of the object from the sensor head. The sensor output voltage is proportional to the reflected light, hence to the distance of the object.

The light sensors have the largest range and resolution ratio. The bandwidth of the sensor is very high (in the order of a few kHz) and is not sensitive to the distance measured due to the high speed of light. They are also very small in physical size and easy to install.

Detecting the presence/absence of seal rubber pieces of electrolytic capacitors

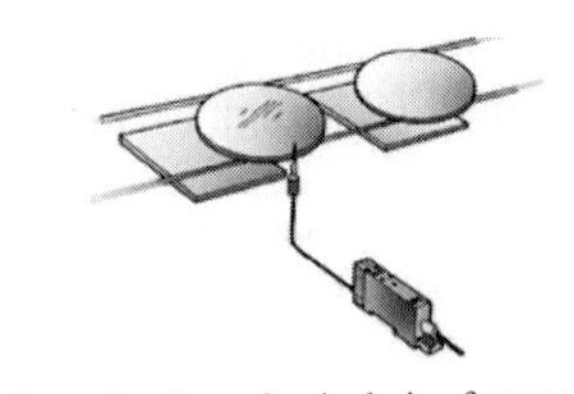

Detecting the wafers in drying furnace

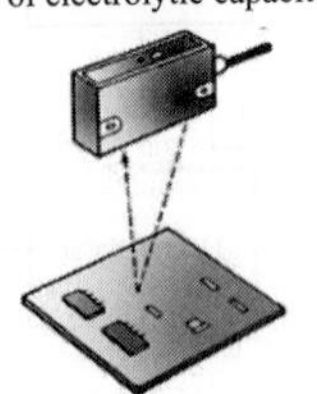

Detecting the passing of PCBs in a reflow furnace

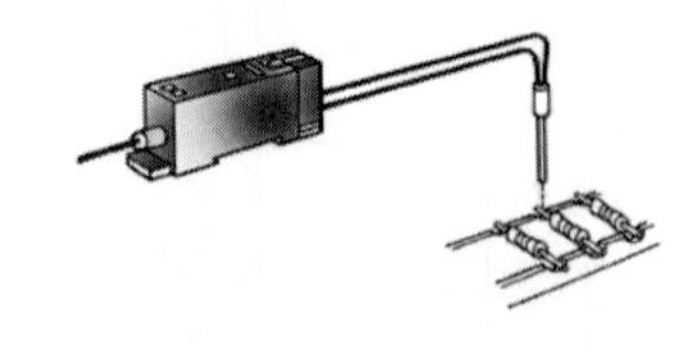

Detecting the leadwires of resistors

**FIGURE 6.30:** Applications of various types of light sensors: Presense sensors with ON/OFF output (through-beam, reflective, diffusive), and distance sensor.

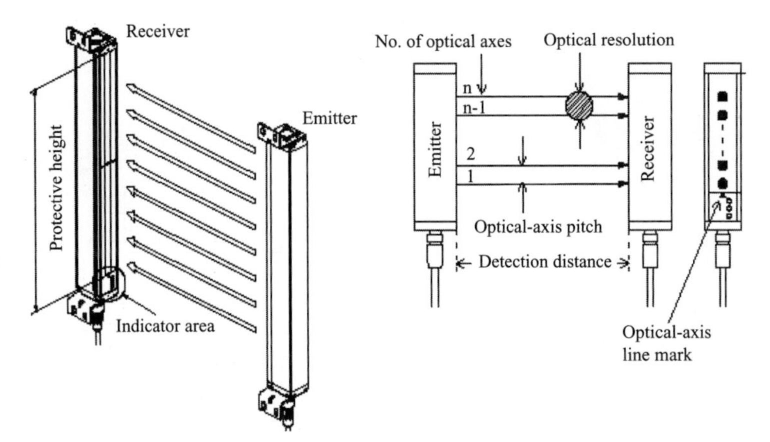


**FIGURE 6.31:** A through-beam emitter-receiver array light sensor used as a safety light curtain.

Photoelectric sensors are widely used in industry in the form of "safety light curtains" (Fig. 6.31). A set of emitter-receiver pairs forms a linear line of light beams. When any portion of the light beam is interrupted by an object passing through, the sensor output turns ON. Typically, the sensor output is connected to drive a circuit braker to shut down the power to the machine. Notice that the light curtain is formed by a finite number (i.e., 10–100 range) of emitter-receiver pairs. Therefore, the object must be at least as thick as the light curtain resolution to be detectable. Typical resolution of the light curtain is around 25 mm to 30 mm. Small objects can potentially pass through the light curtain without interrupting the light.

## 6.4.9 Presence Sensors: ON/OFF Sensors

A special class of the position-related sensors are the sensors which sense the presence of an object with a sensing range and provides one of two discrete outputs: ON or OFF. Such sensors are collectively called *presence sensors* or *ON/OFF sensors*.

We have previously discussed the light-based sensors which can be used to detect the presence of an object within the viewing field of the sensor. There are three different types of light-based (photoelectric) presence sensors: through-beam, reflective, diffusive (Fig. 6.32). The sensor has a frequency-tuned emitter and a receiver head. Depending on the receiver light and the adjustable threshold on the receiver head, the output of the sensor is turned ON or OFF.

Inductive and capacitive proximity sensors are two of the most common noncontact presence sensors used in industry. Their operating principles are shown in Fig. 6.33 and 6.24. Capacitive proximity sensor is a two state output (ON or OFF) version of a capacitive gap sensor. The sensing range of these sensors is in the 1-cm to 10-cm range. Typical

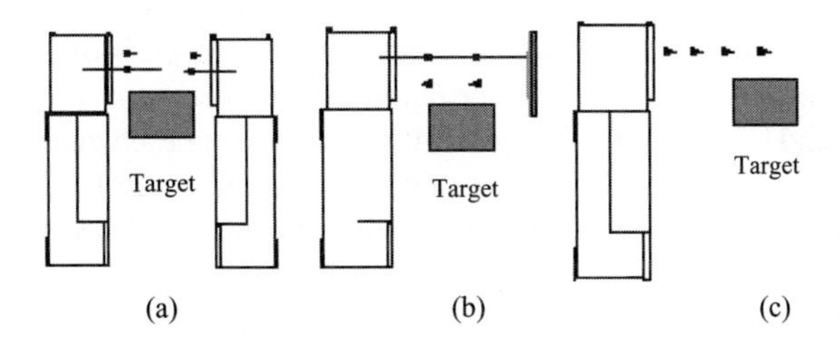


**FIGURE 6.32:** Operating principles of photoelectric presence sensors (a) through-beam type, (b) reflective type, and (c) diffusive type.

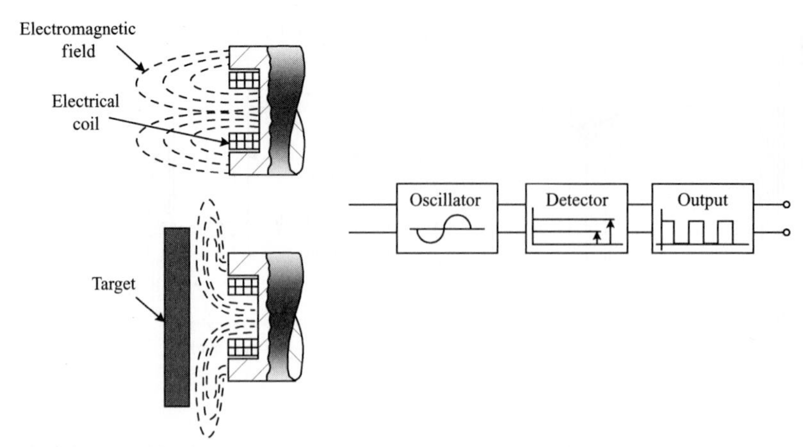


**FIGURE 6.33:** Operating principles of an inductive proximity sensor.

maximum switching frequency of an inductive presence sensor is about 1 msec (1 kHz), and the switching frequency of a capacitive presence sensor is about 10 msec (100 Hz). Notice that while the inductive proximity sensors sense only the metal targets, capacitive proximity sensor can sense nonmetalic targets as well.

An *inductive sensor* has the following main components: sensor head (ferrous core and conductor coil winding around it), oscillating current supply circuit, detection and output circuit. The oscillating supply circuit establishes an oscillating current, hence an oscillating electromagnetic field, around the sensor head. When a metallic object enters the field of the sensor, it changes the electromagnetic field density around the sensor since the effective magnetic permeability of the surrounding environment changes. The oscillating electromagnetic field induces eddy currents on the target metallic object. The eddy current losses draw energy from the supply circuit of the sensor, hence reducing the magnitude of oscillations. The detection circuit measures the drop in the oscillation current magnitude and switches (turn ON/OFF) the output circuit transistor. Inductive sensors operate with electromagnetic fields. Inductive sensors can detect metal objects. For a given inductive sensor, the detection range is higher for ferrous metals (i.e., iron, stainless steel) than the detection range for nonferrous metals (i.e., aluminum, copper). *Capacitive sensors* operate with electrostatic fields. The objects sensed by capacitive sensors must have an effect in changing the electrostatic field through the effective change in capacitance. This is typically accomplished by changing the dielectric constant around the sensor, hence changing the effective capacitance. Notice that, as a sensed part enters the field of a proximity sensor (inductive or capacitive type), the electric field around the sensor head is gradually changed. The change in the field (electromagnetic field for inductive type, electrostatic field for capacitive type) strength occurs in proportion to the position of the sensed object. As a result a threshold value in the sensed change is used to decide present (ON) or not-present (OFF) decision of the sensor.

An inductive proximity sensor and a gear set is often used as position and velocity sensor for applications which require low resolution but very rugged position and speed sensor (Fig. 6.34). As each gear tooth passes by the proximity sensor, the output of the sensor changes state between ON and OFF. This basically is equivalent to a single-channel encoder. Using a single-channel proximity sensor, the change of direction cannot be detected. Therefore, this type of sensor is appropriate for applications in which the rotational direction of the shaft is only one direction (i.e., engine output shaft rotation direction is always in the same direction). The speed is determined by the frequency of the pulses from the proximity sensor.

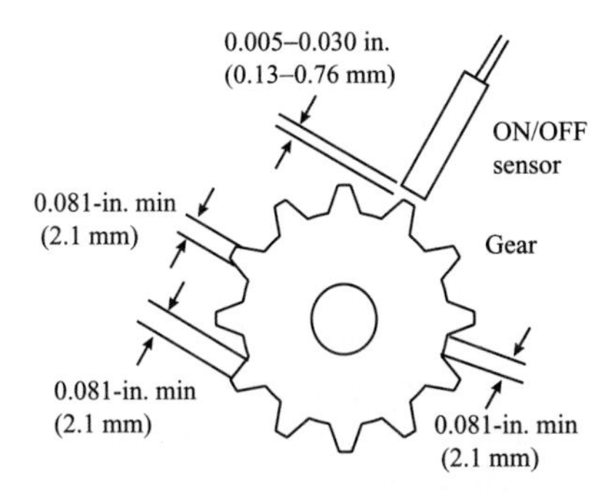


**FIGURE 6.34:** Operating principles of a proximity sensor (inductive, capacitive, hall effect types) and a gear on a shaft to measure rotary position and speed.

## 6.5 VELOCITY SENSORS

### 6.5.1 Tachometers

Construction of a tachometer is identical to the construction of a brush type DC motor, except that it is smaller because the tachometer is used for measurement purposes, not for the purpose of converting electrical power to mechanical power like an electric motor actuator. A tachometer involves a rotor winding, a permanent magnet stator, and commutator-brush assembly (Fig. 6.35, also see the chapter on electric actuators).

Tachometer is a passive analog sensor which provides an output voltage proportional to the velocity of a shaft. There is no need for external reference or excitation voltage. Let us consider the dynamic model of the electrical behavior of a brush-type DC motor. Let the resistance and inductance of the rotor winding be $R$ and $L$, respectively. The back EMF constant of the motor (which will be used as tachometer) is $K_{vw}$. The dynamic relationship between the terminal voltage, $V_t$, current, $i$, in the rotor winding, and the angular speed of the rotor, $w$, is

$$V_t(t) = L \cdot \frac{d}{dt} i(t) + R \cdot i(t) + K_{vw} \cdot \dot{\theta}(t) \tag{6.68}$$

where the voltage due to back electromotive force is

$$V_{bemf} = K_{vw} \cdot \dot{\theta}(t) \tag{6.69}$$

By design, tachometers have $L$ and $R$ parameters that result in small current ($i$). In steady state, the output voltage is proportional to the shaft velocity. The proportionality constant is a parameter determined by the size of the tachometer, type of winding, and permanent magnets used in its construction. Ideally, the gain of the tachometer should be constant, but in reality it varies with temperature ($T$) and rotor position ($\theta$) due to finite number of commutators. Hence, the sensor output voltage and angular speed relation can

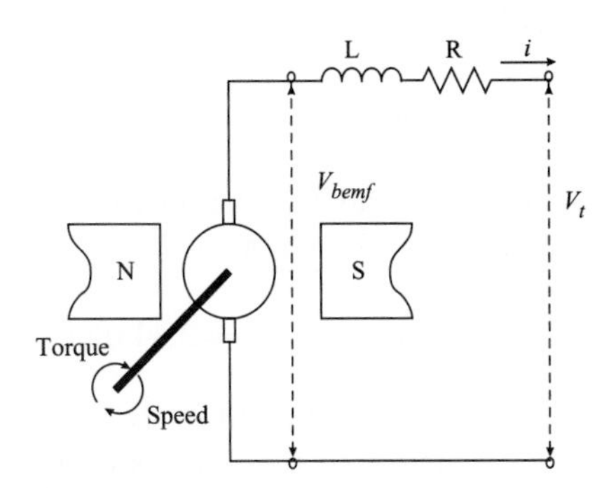


**FIGURE 6.35:** Operating principle of a tachometer for speed sensing.

be expressed as

$$V_{out}(t) = K_{vw} \cdot \dot{\theta}(t) \quad (6.70)$$

where

$$K_{vw} = K_{vw}(T, \theta) \quad (6.71)$$

The value of the gain at rated temperature and nominal rotor position is called the $K_{vwo}$. The gain changes as function of temperature. In addition, due to the finite number of commutators, the gain has a periodic ripple as a function of rotor position. The frequency of the ripple is equal to the number of commutators. The ripple due to commutation is reflected on the output voltage which is generally less than 0.1% of the maximum output voltage.

The parameters that specify the performance of a tachometer are:

1. Speed-to-voltage gain, $K_{vw}$ [V/rpm]
2. Maximum speed, $w_{max}$, limited either by the bearings or the magnetic field saturation
3. Inertia of the rotor
4. Maximum expected ripple voltage and frequency

Typical dynamic time constant of a tachometer is in the range of 10 $\mu$sec–100 $\mu$sec.

For very high speed applications, the winding and magnets are designed such that $K_{vw}$[V/krpm] is a low value, whereas for very sensitive low-speed applications, it is designed such that the $K_{vw}$ is a high value.

***Example*** Consider a tachometer with a gain of 2 V/1000 rpm = 2 V/krpm. It is interfaced to a data-acquisition system through a analog-to-digital converter (ADC) which has 12-bit resolution and ±10-V input range. The sensor specifications state that the ripple voltage due to commutators on the tachometer is 0.25% of the maximum voltage output.

1. Determine the maximum speed that the sensor and data-acquisition system can measure
2. What are the measurement errors due to the ripple voltage and due to ADC resolution?
3. If the ADC was 8 bit, which error source is more significant—ripple or ADC resolution?

Because the input to ADC saturates at 10 V, the maximum output from tachometer should be 10 V. Hence, the maximum speed that can be measured is

$$w_{max} = \frac{10 \text{ V}}{2 \text{ V/krpm}} = 5 \text{ krpm} = 5000 \text{ rpm} \quad (6.72)$$

The measurement error due to ripple voltage is

$$E_r = \frac{0.25}{100} \cdot 10 \text{ V} = 0.025 \text{ V} \quad \text{or} \quad \frac{0.25}{100} \cdot 5000 \text{ rpm} = 12.5 \text{ rpm} \quad (6.73)$$

The measurement error due to the ADC resolution is one part in $2^{12}$ because ADC has 12-bit resolution over the full range of measurement (±10 V = 20-V range),

$$E_{ADC} = \frac{20 \text{ V}}{2^{12}} = \frac{20}{4096} = 0.00488 \text{ V} \quad \text{or} \quad \frac{1}{4096} \cdot 10{,}000 \text{ rpm} = 2.44 \text{ rpm} \quad (6.74)$$

If the ADC is 12-bit resolution, the measurement error is dominated by the ripple voltage. Measurement error due to ADC resolution is negligible. If the ADC is 8 bit, then the measurement error due to the ADC resolution is

$$E_{ADC} = \frac{20 \text{ V}}{2^8} = \frac{20}{256} = 0.078 \text{ V} \quad \text{or} \quad \frac{1}{256} \cdot 10{,}000 \text{ rpm} = 39.0625 \text{ rpm} \tag{6.75}$$

In this case, the measurement error due to ADC resolution is larger than the error due to ripple voltage of the sensor.

### 6.5.2 Digital Derivation of Velocity from Position Signal

In most motion control systems, velocity is a derived information from position measurements. Accuracy of the derived (estimated) velocity depends on the resolution of the position sensor. The velocity can be derived from position measurement signal either by analog differentiation using an op-amp circuit or digital differentiation using the sampled values of the position signal.

The ideal op-amp differentiator has the following input signal to output signal relationship:

$$V_{out}(t) \approx -K \frac{d}{dt}(V_{in}(t)) \tag{6.76}$$

where $V_{in}(t)$ represents the position sensor signal and $V_{out}(t)$ represents the estimated velocity signal.

Clearly, digital derivation is more flexible because it allows more specific filtering and differentiation approximations:

$$V = \Delta X / T_{sampling} \tag{6.77}$$

An interesting method of velocity estimation in low-speed applications using encoders is to determine the time period between two consecutive pulse transitions in order to increase velocity estimation accuracy. In very low velocity applications, the number of pulse changes in encoder count within a given samping period may be as low as one or two counts. If one of the count change occurs at the edge of the sampling period, the velocity estimation can be different by 50% whether that count occurred right before or right after the sampling period ended. Notice that typically, $T_{sampling} = 1$ msec, and $\Delta X$ is one or two counts in low-speed applications. One count variation in the position changes depending on when that happens (right before the $T_{sampling}$ period expired or right after it); the estimated velocity can vary by 100%.

Instead of keeping a fixed time period and counting the number of pulses during that period to estimate the velocity, we measure *the time period between two consecutive pulses using a high-resolution timer*, i.e., using a $\mu$sec resolution timer/counter. Then the estimated velocity is

$$V = 1[count] / T_{period} \tag{6.78}$$

Since the time period measurement is very accurate, the velocity estimation is more accurate than the first case. Let us assume that we have an encoder with a resolution of 10,000 [cnt/rev] (including a $\times 4$ quadrature count multiplication), and that it is connected to a shaft which rotates at 6.0 rev/min. Assume that the sampling rate is $T_{sampling} = 1.0$ msec. This results in 60,000 counts/min or equivalently 1 cnt/msec.

Let us assume that due to small variations in the velocity, there were two count change within one sampling period instead of the normal one pulse, one at the beginning of the sampling period, and one at the 900 $\mu$sec mark. The velocity estimation using the fixed sampling period approach would give

$$V = 2\ \text{cnt}/1.0\ \text{msec} \tag{6.79}$$

$$= 2\ \text{cnt}/1.0\ \text{msec}\ (60{,}000\ \text{msec}/1\ \text{min})\ (1\ \text{rev}/10{,}000\ \text{cnt}) \tag{6.80}$$

$$= 12\ \text{rev}/\text{min}. \tag{6.81}$$

If the second pulse transition did not happen at the 900-$\mu$sec mark, but beyond the 1.0-msec sampling period, i.e., at 1.001 msec, the estimated velocity would have been

$$V = 1\ \text{cnt}/1.0\ \text{msec} \tag{6.82}$$

$$= 1\ \text{cnt}/1.0\ \text{msec}\ (60{,}000\ \text{msec}/1\ \text{min})\ (1\ \text{rev}/10{,}000\ \text{cnt}) \tag{6.83}$$

$$= 6\ \text{rev}/\text{min}. \tag{6.84}$$

If the velocity was estimated by *the time period measurement method*, then the instantenous velocity would be accurately estimated. Let us assume that we measure the time period between two consecutive pulses using a 1-$\mu$sec resolution timer/counter. Let us assume that in one case the measured time period between two pulses is 900 $\mu$sec and in another case it is 901 $\mu$sec due to timer/counter resolution. The estimated velocity from both of these cases are

$$V_1 = 1\ \text{cnt}/0.9\ \text{msec} = 6/0.9\ \text{rpm} = 6.6667\ \text{rpm} \tag{6.85}$$

and the second estimated velocity is

$$V_2 = 1\ \text{cnt}/0.901\ \text{msec} = 6/0.901\ \text{rpm} = 6.6593\ \text{rpm} \tag{6.86}$$

Clearly, using the time period measurement between two consecutive pulses using a 1 $\mu$sec resolution timer/counter results in 0.11% error in velocity measurement. Whereas, counting the number of pulses during a fixed sampling period can result in velocity measurement errors as high as 100%.

## 6.6 ACCELERATION SENSORS

Three different types of acceleration sensors, also called *accelerometers*, each based on a different *transduction principle*, are discussed as follows:

1. Inertial-motion-based accelerometers where the sensor consists of a small mass-damper-spring in an enclosure and mounted on the surface of an object whose acceleration is to be measured. The displacement ($x$) of the sensor inertia is proportional to the magnitude of acceleration ($\ddot{X}$) of the object provided that the frequency of acceleration is well within the natural frequency (bandwidth) of the sensor.

$$\ddot{X} \rightarrow x \rightarrow V_{out} \tag{6.87}$$

2. Piezoelectric-based accelerometers provide a charge ($q$) proportional to the inertial force as a result of the acceleration. Piezoelectric materials provide a charge proportional to the strain which is proportional to the inertial force.

$$\ddot{X} \rightarrow F \rightarrow q \rightarrow V_{out} \tag{6.88}$$

3. Strain gauges can be used for acceleration measurement if the sensor can transduct a strain ($\epsilon$) proportional to the acceleration. Once that is accomplished, the change

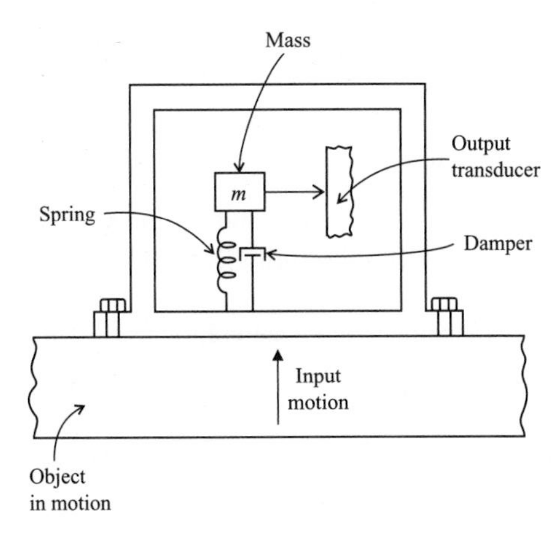


**FIGURE 6.36:** Operating principle of an inertial accelerometer.

in the strain is measured by the change in the strain-gauge resistance and hence as output voltage from a Wheatstone bridge and op-amp circuit.

$$\ddot{X} \rightarrow F \rightarrow \epsilon \rightarrow R \rightarrow V_{out} \tag{6.89}$$

### 6.6.1 Inertial Accelerometers

An inertial accelerometer is basically a small mass-spring-damper system with high natural frequency. Consider the figure shown in Fig. 6.36, which is the concept of an inertial accelerometer connected to a body whose acceleration is to be measured. Figure 6.37 shows the pictures of a number of accelerometers. The dynamic relations between the relative displacement of the sensor inertia with respect to its enclosure, $x$, and the acceleration of the body, $\ddot{x}_{base}$, are

$$m \cdot (\ddot{x}(t) + \ddot{x}_{base}(t)) + c \cdot \dot{x}(t) + k \cdot x(t) = 0 \tag{6.90}$$

$$\ddot{x}(t) + (c/m) \cdot \dot{x}(t) + (k/m) \cdot x(t) = -\ddot{x}_{base}(t) \tag{6.91}$$

Notice that if the accelerometer parameters $m, c, k$ are chosen such that the motion of the accelerometer is critically damped, then the displacement of the accelerometer relative to its enclosure, $x(t)$, is proportional to the acceleration of the base in steady state. The speed

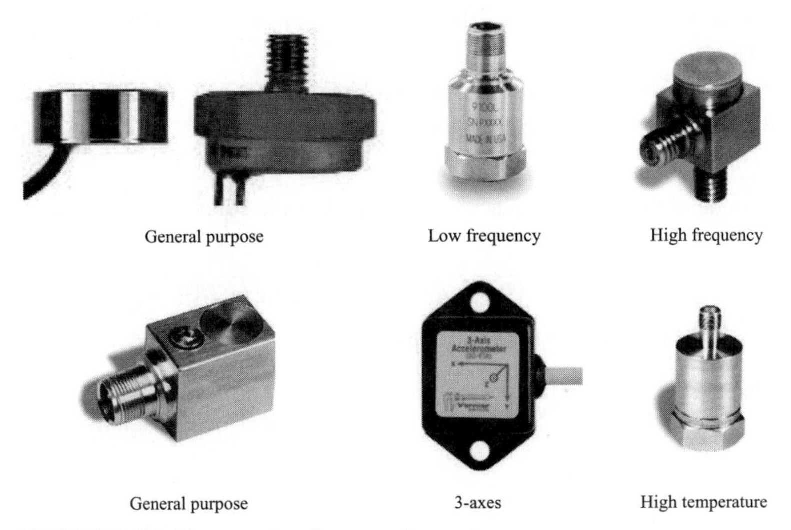


**FIGURE 6.37:** Pictures of various accelerometers.

of response is determined by the *c/m* and *k/m* ratios. Let

$$c/m = 2\xi w_n \tag{6.92}$$

$$k/m = w_n^2 \tag{6.93}$$

The values of $m, c, k$ are chosen such that $\xi = 0.7$–$1.0$ range, and $w_n$ can be chosen up to a few hundred Hz. The smaller the mass and the stiffer the spring constant of the sensor, the higher will be its bandwidth.

Let us consider a sinusoidal base displacement and acceleration as function of time:

$$x_{base}(t) = A\ sin\,(wt) \tag{6.94}$$

The resulting acceleration of the base as a result of constant magnitude sinusoidal displacement of the base is

$$\ddot{x}_{base}(t) = -A\ w^2\ sin\,(wt) \tag{6.95}$$

The steady-state response of the displacement of the sensor inertia with respect to its housing, $x(t)$, is the steady-state solution for

$$\ddot{x}(t) + (c/m) \cdot \dot{x}(t) + (k/m) \cdot x(t) = -\ddot{x}_{base}(t) \tag{6.96}$$

$$\ddot{x}(t) + (c/m) \cdot \dot{x}(t) + (k/m) \cdot x(t) = A\ w^2\ sin\,(wt) \tag{6.97}$$

In steady state,

$$x_{ss}(t) = \frac{A\ (w/w_n)^2\ sin\,(wt - \phi)}{\{[1 - (w/w_n)^2]^2 + [2\ \xi(w/w_n)]^2\}^{1/2}} \tag{6.98}$$

where

$$\phi = tan^{-1} \frac{2\xi(w/w_n)}{1 - (w/w_n)^2} \tag{6.99}$$

The steady-state displacement of the sensor has the following properties:

1. Sensor displacement has the same frequency as the acceleration of the base.
2. There is a phase shift between the sensor displacement and base acceleration and that phase angle is function of acceleration frequency $(w)$ as well the sensor parameters $(\xi, w_n)$.
3. The magnitude of the sensor displacement is proportional to the acceleration magnitude. However, the proportionality constant is a function of acceleration frequency $(w)$ as well the sensor parameters $(\xi, w_n)$.

The magnitude of the sensor displacement, which is a sinusoidal function, is

$$|x_{ss}(t)| = \frac{A\ (w/w_n)^2}{\{[1 - (w/w_n)^2]^2 + [2\ \xi(w/w_n)]^2\}^{1/2}} \tag{6.100}$$

There are two different purposes of this type of sensor:

1. If the sensor is intended to be used to measure the acceleration of the base, then we are interested in the ratio of the magnitude of the response to the magnitude of excitation:

$$\left|\frac{x_{ss}(t)}{A\ w^2}\right| = \frac{(1/w_n)^2}{\{[1 - (w/w_n)^2]^2 + [2\ \xi(w/w_n)]^2\}^{1/2}} \tag{6.101}$$

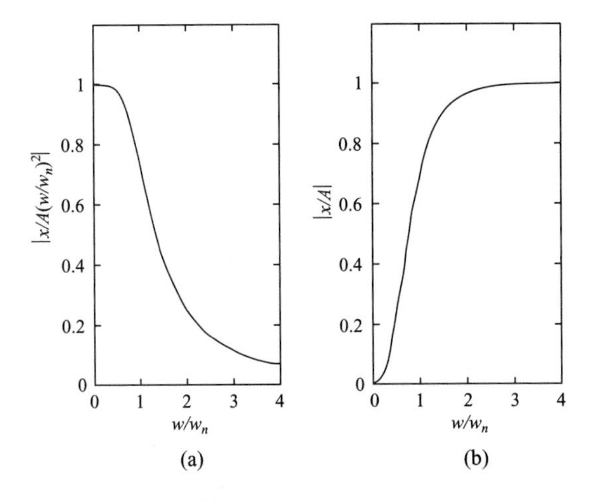


**FIGURE 6.38:** Inertial accelerometer output and input magnitude ratios: (a) sensor output to base acceleration magnitude ratio, and (b) sensor output to base displacement ratio.

2. If the sensor is intended to be used as a seismic instrument (i.e., for earthquake measurements) and to measure the displacement of the base, then we are interested in the ratio of

$$\left|\frac{x_{ss}(t)}{A}\right| = \frac{(w/w_n)^2}{\{[1-(w/w_n)^2]^2 + [2\,\xi(w/w_n)]^2\}^{1/2}} \tag{6.102}$$

The magnitude ratio of the sensor displacement to the magnitude of the acceleration for case (1) as function of frequency is shown in Fig. 6.38. In order to have a proportional relationship (constant magnitude ratio) between the sensor output and the measured quantity (acceleration in this case) within the frequency range of interest, it is necessary that [Fig. 6.38 (a)]

$$w \ll w_n; \quad w \; in \; [0, w_{max}] \tag{6.103}$$

where $w_{max}$ is the maximum frequency content of the acceleration which we expect to measure.

On the other hand, if the sensor is used to measure the displacement of the base, as in seismic applications, in order to have a proportional relationship (constant magnitude ratio) between sensor output and measured quantity (base displacement in this case), it is necessary that [Fig. 6.38 (b)]

$$w \gg w_n; \quad w_{min} < w < w_{max} \tag{6.104}$$

This relationship states that an inertial accelerometer used to measure the seismic displacement should be used in a frequency range that is larger than a certain minimum frequency, $w_{min}$, and that frequency should be at least as large as or more than the natural frequency of the sensor,

$$w_n < w_{min} < w \tag{6.105}$$

In other words, the natural frequency of an accelerometer used for seismic displacement measurement should be very small. Since the $w_n = \sqrt{k/m}$, the mass of the seismic accelerometer should be very large and the spring constant should be small. That explains the reason for the large size of seismic accelerometers compared to normal accelerometers.

A given inertial acceleration sensor can measure accelerations with frequency content below its natural frequency, and displacements with frequency content above its natural frequency.

An inertial accelerometer cannot accurately measure seismic displacements that have frequency lower than the natural frequency of the sensor. Likewise, an inertial accelerometer cannot measure accelerations that have frequency content closer to or larger than the natural frequency of the sensor. Inertial accelerometers used for seismic displacement measurement must have as small a natural frequency as possible. Inertial accelerometers used for acceleration measurement must have as large a natural frequency as possible.

### 6.6.2 Piezoelectric Accelerometers

Some materials [such as natural quartz crystal, silicon dioxide, barrium titanite, lead zirconate titanate (PZT)], called *piezo crystals*, produce a charge in response to force (or deformation) applied to it. This is called the *direct piezoelectric* effect. The same materials also have the reverse phenomenon; that is, they produce force in response to an applied charge. This is called the *reverse piezoelectric* effect. The Greek word "piezo" means "to squeeze" or "to pressure." The piezoelectric effect was first obsereved by Pierre and Jacque Curie in the 1880s. Quartz has excellent temperature stability and shows almost no decay in its piezoelectric properties over time. PZTs are polarized by applying very high DC voltages at high temperatures. PZTs show a natural decay in their piezoelectric properties in time; hence they may need to be periodically calibrated or repolarized. The manufacturing processes of piezoelectric ceramics involve the following steps:

1. Mix the powders of raw material and heat to high temperarture in order to react the constituents into the compound form.
2. Ground the resulting compond into very fine particles.
3. Using binder material, form desired shapes of ceramic parts, i.e., bars, cylinders, disks.
4. Heat it to remove the binder material.
5. This is followed by a surface grinding process in order to provide a relatively smooth surface.
6. Apply electrodes to the desired surfaces and heat it to form the ceramic to electrode binding.
7. In order to induce piezoelectric properties, heat the parts in dielectric oil bath under a very high electric field.

This step completes the polarization, hence the generation of piezoelectric property, of the ceramic. The piezoelectric properties of the produced ceramic starts to decay within the first few hours and days after the polarization. This aging process slows down significantly and is a logarithmic function of time. Notice that excessive heat, voltage, or stress can depolarize a piezoelectric material.

Piezoelectric accelerometers work on the principle that the acceleration times the sensor inertia will apply a force on the sensor. Mechanically, the piezo element acts as a very precise and stiff spring in the sensor design. As a result of the piezoelectric phenomena, the charge output, hence the output voltage, from the sensor is proportional to the inertial force

$$\ddot{x} \rightarrow F \rightarrow q \rightarrow V_{out} \tag{6.106}$$

$$q = C \cdot V_{out} \tag{6.107}$$

where $q$ is the charge produced by the piezoelectric material, $C$ is the effective capacitance, and the $V_{out}$ is the produced output voltage.

A calibrated piezoelectric acceleration sensor has the following input–output relationship:

$$V_{out} = K \cdot \ddot{x} \tag{6.108}$$

Notice that although the piezo element has a finite stiffness and acts as a stiff spring, the actual deformation of the piezo element is very small due to its large stiffness. Therefore, it can be considered almost as if there is no deformation. Piezoelectric-based acceleration sensors have a range as high as $1000 \cdot g$, with sensor bandwidth upto 100 kHz. The limiting factor in the frequency response of the sensor is the charge amplifier bandwidth, which is much slower than the sensor element. The bandwidth of the sensor element plus the charge amplifier can be up to a few-kHz range. Piezoelectric based sensors are most appropriate for measuring signals that are time varying. They are not appropriate for measuring static or low-frequency signals because the charge due to the load will slowly discharge.

The design principles of piezoelectric-based sensors for pressure, force, strain, and acceleration measurements are very similar. The external force (in case of force sensor it is the force sensed; in case of pressure sensor it is the pressure times the surface area of the diaphragm; in the case of acceleration sensor it is the inertial force, $m_{sensor} \cdot \ddot{x}$) induces a strain on the piezoelement of the sensor. The output charge is proportional to the strain (Fig. 6.39).

The transduction principle of piezoelectric sensors is that the charge is proportional to the measured variable (i.e., acceleration, stress, pressure, force). Transmitting the signal from the transducer element and amplifying it at a different location can introduce significant signal errors due to the capacitance of the transmission cable and noise (Fig. 6.40). For instance, the voltage at transducer output and at the input of op-amp (charge amplifier) is

$$V_{AB} = \frac{q}{C_1} \tag{6.109}$$

$$V_{EF} = \frac{q}{C_1 + C_2 + C_3} \tag{6.110}$$

Let us consider for a moment that the op-amp is not connected and that the $C_2$ and $C_3$ are the capacitances of the cable and the read-out instrument. The sensor output voltage at EF is function of the cable length and the read-out instrument capacitance. Therefore, such a system must be calibrated for fixed cable length. If a different cable length is used with the sensor, it must be recalibrated. When the $V_{EF}$ is amplified by a charge amplifer, the voltage

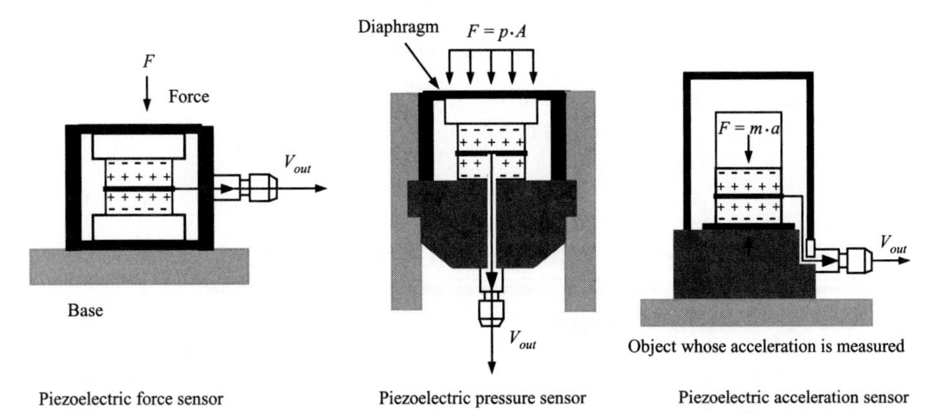


**FIGURE 6.39:** Piezoelectric principle and its usage in force, pressure, and acceleration sensors.

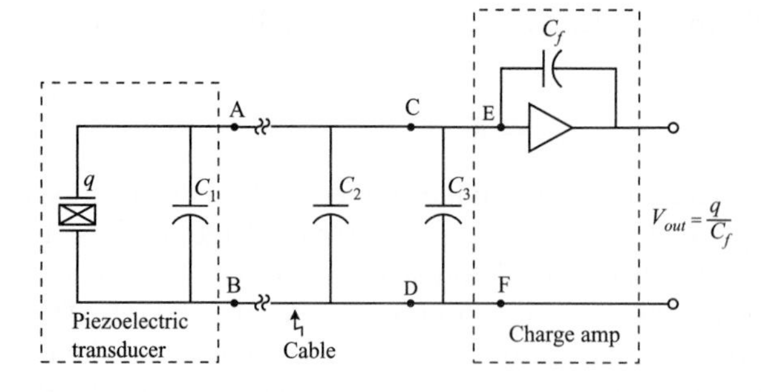


**FIGURE 6.40:** Piezoelectric sensor operating principle: piezoelectric transducer produces charge as function of measured variable, signal transmitted over cable and amplified by a charge amplifier. Cable capacitance affects the calibration.

output is

$$V_{out} = \frac{q}{C_f} \tag{6.111}$$

where the problem associated with the signal dependency on the capacitances of piezo element, cable, and read-out instrument is eliminated. The output voltage depends on the charge and the feedback capacitance. However, the problem with the charge amplifier is that the electrical noise at the output of the charge amplifer is directly proportional to the ratio of $(C_1 + C_2 + C_3)$ to the $C_f$. Therefore, cable length is still important in the case of charge amplifier. Either the signal conversion from charge to voltage and amplification should be performed at the transducer location or the sensor calibration must be performed using a well-defined coaxial cable length if the amplification is done at a separate location.

As an example, the accelerometer Model-339B01 by PCB Piezotronics has the following characteristics: voltage sensitivity $K = 100$ [mV/g], frequency range up to 2 kHz, amplitude range up to 50 g with a resolution of 0.002 g.

### 6.6.3 Strain-Gauge-Based Accelerometers

The operating principle of a strain-gage-based accelerometer is very similar to an inertial accelerometer. The only difference is that the spring function is provided by a cantilever flexible beam [Figs. 6.41–6.44 (a)]. In addition, the strain in the cantilever beam is measured, instead of the displacement. The strain is proportional to the inertial force, hence the acceleration. The voltage output from the sensor, proportional to strain, is obtained from the standard Wheatstone bridge circuit.

$$\ddot{x} \rightarrow F \rightarrow \epsilon \rightarrow R \rightarrow V_{out} \tag{6.112}$$

$$V_{out}(t) = K \cdot \ddot{x}(t) \tag{6.113}$$

The range of the accelerations this type of sensor can measure is very similar to those of piezoelectric type accelerometers, i.e., up to 1000 g. The bandwidth of the sensor can be as high as a few kHz.

## 6.7 STRAIN, FORCE, AND TORQUE SENSORS

### 6.7.1 Strain Gauges

The most common strain measurement sensor is a *strain gauge*. The strain-gauge transduction principle is based on the relationship between the change in length and its resulting change in the resistance of a conductor. Typical strain-gauge material used is constantan (55% copper and 45% nickel). A fine wire of strain-gauge material is given a directional

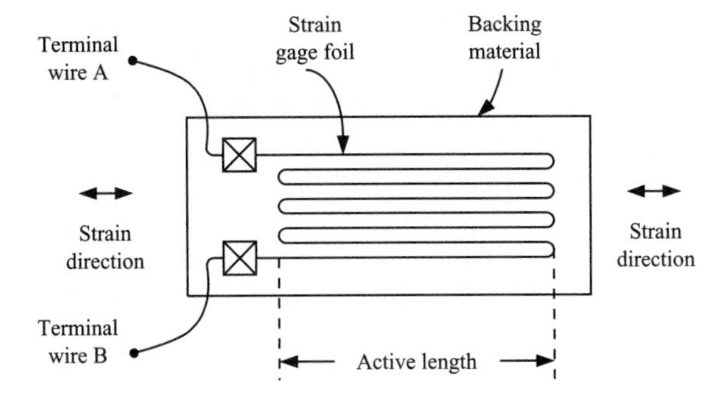


**FIGURE 6.41:** A typical strain gauge for strain measurement.

shape and bonded to the part surface using adhesive bonding materials. The resistance and strain relationship is

$$\frac{\Delta R}{R} = G\frac{\Delta L}{L} \tag{6.114}$$

where $G$ is called the *gauge factor* of the sensor. In order to increase the resulting resistance change under a given strain conditions, we need to pack more length, $L$, into a sensor size. That is the reason for the many rounds of conductor wire in one direction in the construction of a strain gauge (Figs. 6.41 and 6.42). Ideally, strain is the property at a point on a material and is the ratio of change in the length to the total length in that direction:

$$\epsilon = \frac{\Delta L}{L} \tag{6.115}$$

The change in the strain-gauge resistance is converted to a proportional voltage using a Wheatstone bridge:

$$V_{out} = K_1 \cdot \frac{\Delta R}{R} \tag{6.116}$$

$$= K_1 \cdot G \cdot \frac{\Delta L}{L} \tag{6.117}$$

$$= K_2 \cdot \frac{\Delta L}{L} \tag{6.118}$$

$$= K_2 \cdot \epsilon \tag{6.119}$$

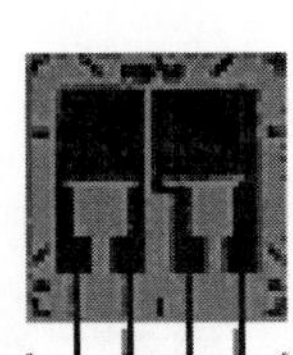
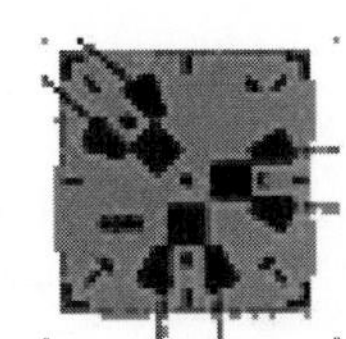
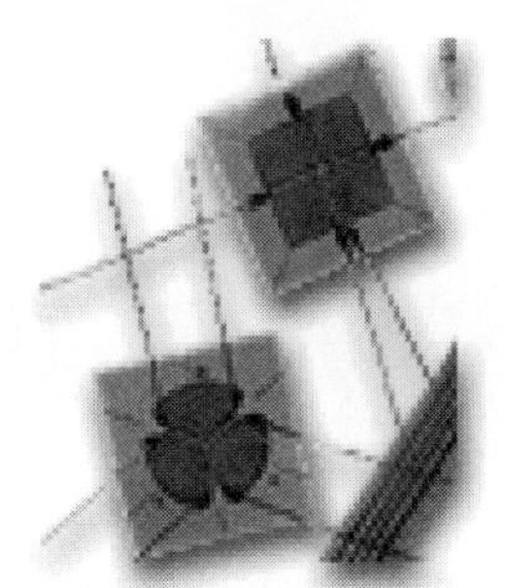

**FIGURE 6.42:** Pictures of typical strain gauges for strain measurement.

However, a strain gauge has a finite dimension. It is bounded to the surface of the part over a finite area. Therefore, the measured strain is the average strain over the area occupied by the strain gauge. The bonding of the strain gauge to the workpiece is very important for two reasons: (1) it needs to provide a uniform mechanical linkage between the surface and sensor in order to make the measurement correctly, and (2) the bonding material electrically isolates the sensor and the part. The strain gauges themselves have very high bandwidth. It is possible to build strain-gauge sensors that have almost 1-MHz bandwidth. Using silicon crystal materials, strain-gauge size can be miniaturized while having very large sensor bandwidth.

### 6.7.2 Force and Torque Sensors

Force and torque sensors operate on the same principle. There are three main types of force and torque sensors:

1. Spring displacement based force/torque sensors
2. Strain-gauge based force/torque sensors
3. Piezoelectric based force sensors

Let us consider a weighing station as a force sensor application. If we have a calibrated spring with a known spring coefficient, $K_{spring}$, the load force (for torque measurement, we would use an equivalent torsional spring) can be measured as displacement,

$$F = K_{spring} \cdot x \tag{6.120}$$

Using any of the position sensors, the displacement $x$ can be converted to a proportional voltage, and hence we obtain the force (or torque) measurement,

$$V_{out} = K_1 \cdot x \tag{6.121}$$

$$= K_2 \cdot F \tag{6.122}$$

The strain-gauge-based force and torque sensors measure the force (or torque) based on the measured strain. Again, the sensor relies on an elastic sensing component. Typical force/torque sensors are called *load cells*. A load cell has built-in elastic mechanical components on which strain gauges are mounted (Fig. 6.43). As the load cell experiences the force (or torque), it deforms a small amount which induces a strain on the sensing element.

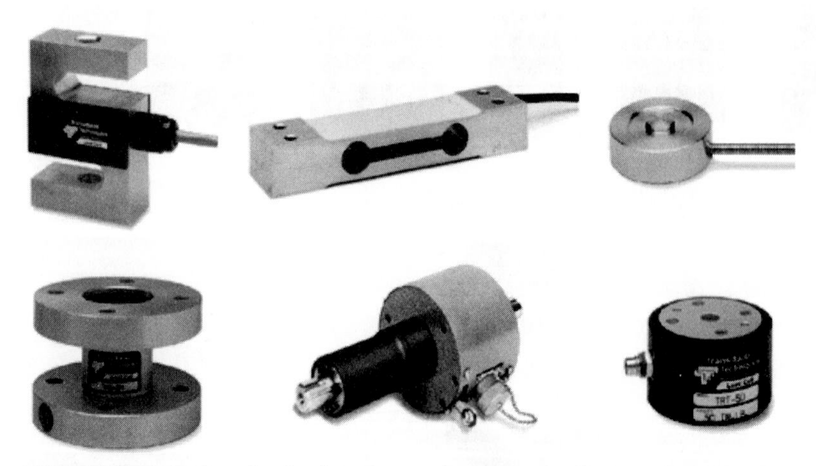

**FIGURE 6.43:** Various load cells using strain gauges for force and torque measurement.

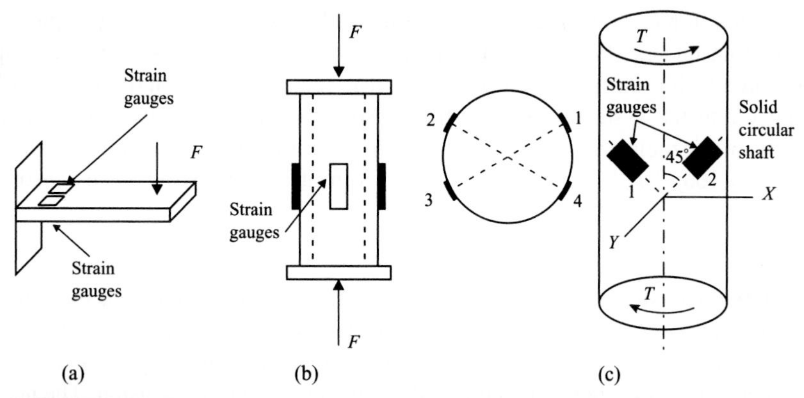


**FIGURE 6.44:** Force and torque measurement on a shaft or beam using strain gauges directly mounted on the part.

The strain is measured by the strain gauge. As a result, since the force/torque to strain relationship is linear by the design of the load cell, the measurement is proportional to the force/torque.

The other alternative strain-gauge-based force/torque sensor is to mount the strain gauges directly over the shaft on which we want to measure the force/torque. In some applications, it is not possible to install a load cell. Force or torque on a shaft must be measured without significantly changing the mechanical design. Figure 6.44 shows such force and torque sensing using four strain-gauge pairs on a shaft. Notice that most force and torque sensors use symmetrically bonded strain gauges to reduce the effect of temperature variations and drift of the strain-gauge output. There are two inherent assumptions in the stain-gauge-based force/torque sensing:

1. The strain in the material is small enough so that the deformation is in elastic range, and that

$$\epsilon = \frac{1}{E}\sigma = \frac{1}{E}\frac{F}{A} \tag{6.123}$$

2. The strain gauge is subjected to the same strain. The strain-resistance change relationship is

$$\epsilon = \frac{1}{G}\frac{\Delta R}{R} \tag{6.124}$$

where the change in the resistance is converted to an output voltage by a Wheatstone bridge circuit,

$$\frac{\Delta R}{R} = \frac{4}{V_i} \cdot V_o \tag{6.125}$$

As a result, the relationship between the strain-gauge output voltage and force is

$$F = \left(\frac{4 \cdot E \cdot A}{V_i \cdot G}\right) \cdot V_o \tag{6.126}$$

Notice that the strain-gauge output voltage to force calibration requires the information about the material ($E$) on which the strain gauge is bonded, cross-sectional area of the part ($A$), sensor gauge factor ($G$), and Wheatstone bridge circuit reference voltage ($V_i$).

Finally, force can be measured using piezoelectric sensors, similar to piezoeelectric pressure transducers (Fig. 6.40). The piezoelectric sensor creates a charge proporional to the force acting on it. The advantage of it is that it does not introduce an additional flexibility into the system as part of the sensor. The only elasticity it introduces is the elasticity of piezoelectric quartz cystal which has modulus of elasticity in the range of 100 GPa $= 100 \cdot 10^9$ [N/m$^2$]. Typical bandwidth of a piezo-based force sensor is about 10 kHz.

***Example*** Consider the force measurement using a strain gauge on a shaft under compression [Fig. 6.44 (b)]. Let us consider that the shaft material is steel. Elastic Young's modulus $E = 2 \cdot 10^8$ [kN/m$^2$], and the cross-sectional area of the shaft is $A = 10.0$ cm$^2$. We have a strain gauge bonded on the shaft in the direction of the tension. The nominal resistance of the strain gauge is $R_0 = 600\ \Omega$, and the gauge factor is $G = 2.0$. The other three legs of the Wheatstone bridge also have constant resistances of $R_2 = R_3 = R_4 = 600\ \Omega$. The reference voltage for the Wheatstone bridge is 10.0 VDC. If the output voltage measured $V_{out} = 2.0$ mV, what is the force?

Notice that the stress-strain relationship is, assuming the deformation is within the elastic range,

$$\sigma = \frac{F}{A} \tag{6.127}$$

$$\epsilon = \frac{1}{E}\sigma \tag{6.128}$$

$$\frac{\Delta R}{R} = G \cdot \epsilon \tag{6.129}$$

$$V_{out} = \frac{V_i}{4 \cdot R_o}\Delta R \tag{6.130}$$

Hence,

$$V_{out} = \frac{V_i \cdot G}{4 \cdot E \cdot A} \cdot F \tag{6.131}$$

The force corresponding to the $V_{out} = 2.0$ mV DC is $F = 80{,}000$ [N]. It is also interesting to determine the change in the strain-gauge resistance for this force measurement:

$$V_{out} = \frac{V_i}{4 \cdot R_o}\Delta R \tag{6.132}$$

$$2 \cdot 10^{-3} = \frac{10}{4 \cdot 600}\Delta R \tag{6.133}$$

$$\Delta R = 0.480\ \Omega \tag{6.134}$$

Notice that the percentage change in the resistance of the strain gauge is

$$\frac{\Delta R}{R} \cdot 100 = \frac{0.480}{600} \cdot 100 = 0.08\% \tag{6.135}$$

Because the gauge factor $G = 2$, the strain (change in the length of the shaft) is

$$\epsilon = \frac{\Delta l}{l} \tag{6.136}$$

$$= \frac{1}{G}\frac{\Delta R}{R} \tag{6.137}$$

$$= \frac{1}{2} \cdot 0.08\% \tag{6.138}$$

$$\frac{\Delta l}{l} = 0.04\% \tag{6.139}$$

$$\Delta l = \frac{0.04}{100} \cdot l \tag{6.140}$$

## 6.8 PRESSURE SENSORS

Absolute pressure is measured relative to perfect vacuum where the pressure is zero. The local atmospheric pressure is the pressure due to the weight of the air of the atmosphere at that particular location (Fig. 6.45). Therefore, the local pressure varies from one location to another as a function of the height of the location from sea level. Average atmospheric absolute pressure due to the weight of the air of the atmosphere is 14.7 lb/in$^2$ = 14.7 psi. This means that the weight acting over an area of 1 in$^2$ due to the weight of the air in the atmosphere is 14.7 lb. Therefore, the absolute atmospheric pressure at sea level is nominally 14.7 lb/in$^2$ (14.7 psia).

*Pascal's law* states that pressure in a contained fluid is transmitted equally in all directions. Using this physical principle, a *barometer* is used to measure absolute pressure. The pressure acting on the fluid due to the atmosphere at the surface is balanced by the pressure due to the fluid (i.e., mercury) weight in the tube (Fig. 6.46). Using this measurement method, it can be observed that the atmospheric pressure is equivalent to the pressure applied by a 29.92 in. (760 mm) column height of mercury. A column of mercury with 29.92 in.

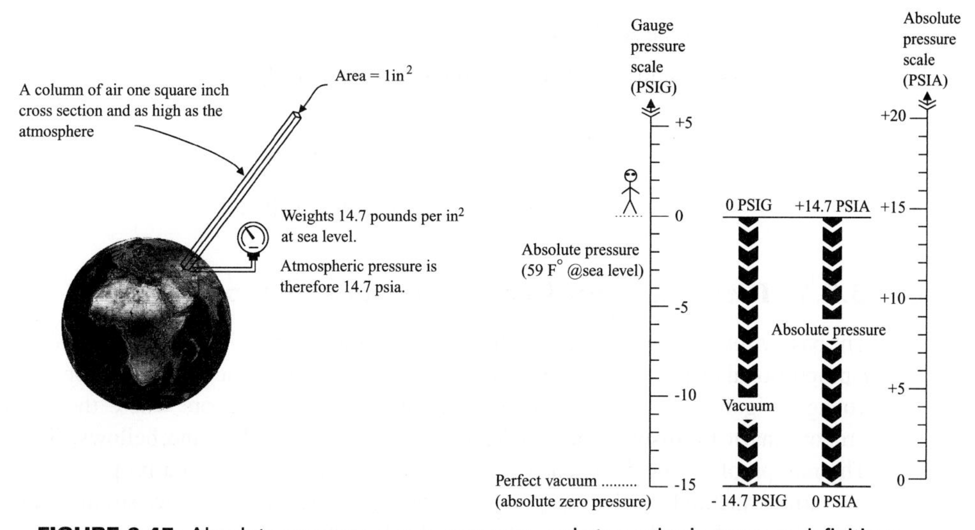


**FIGURE 6.45:** Absolute pressure, gauge pressure, and atmospheric pressure definitions.

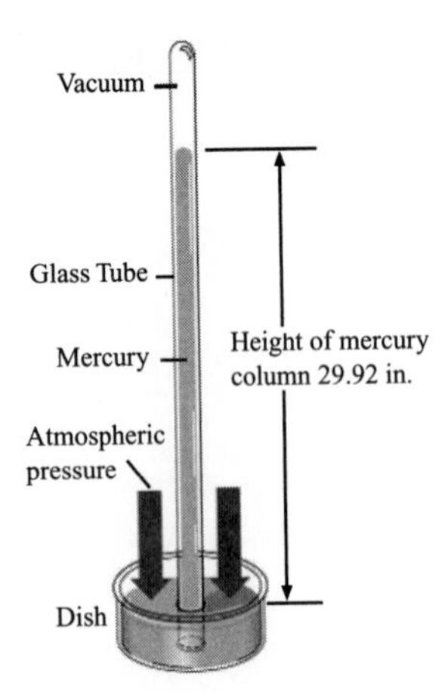


**FIGURE 6.46:** Barometer to measure the local atmospheric absolute pressure.

height and 1 $in^2$ cross section has 14.7 lb of weight. If water is used in the barometer instead of mercury, the height of the water in the tube to balance the atmospheric pressure would be 33.95 in., which produces the same 14.7 $lb/in^2$ pressure because the density of water is lower than mercury. Notice that the top of the tube in the barometer must be vacuum. In order to establish a vacuum at the top of the tube, the tube filled with mercury is placed upside-down in the container filled with mercury. The height of the mercury will drop until its height is 29.92 in., which generates a pressure of 14.7 psi at the surface level of the container.

The pressure sensed by most pressure sensors are the *relative pressure* with respect to the local atmospheric pressure. However, a sensor can be calibrated to measure the absolute pressure as well (Fig. 6.45). If a sensor measures pressure relative to the vacuum pressure, it is referred to as the absolute pressure and the units denoted as [psia]. If a sensor measures pressure relative to the local atmospheric pressure, it is referred to as the *relative* or *gage pressure*, and the unit used to indicate that is [psig]. The notation [psi] refers to [psig] by standard convention in notation.

Notice that the atmospheric absolute pressure varies from the nominal value [14.7 $lb/in^2$ (14.7 psia)] due to variations in:

1. Height from the sea level
2. Temperature and the resulting variations in air density

$$p_{atm} = p_{atm,0} + \Delta p(h, T) \tag{6.141}$$

$$= 14.7 + \Delta p(h, T)[\text{psia}] \tag{6.142}$$

## 6.8.1 Displacement-Based Pressure Sensors

The basic *transduction* principle in this type of pressure sensor is to convert the pressure into a proportional displacement, and then convert this displacement to a proportional electrical voltage. Figure 6.47 shows various concepts of pressure sensors where the pressure is proportional to the displacement of the sensing element (Bourdon tune, bellows, diaphragm). The motion of the flexible sensing element can be translated into a proportional voltage by various methods including position sensing, capacitance change, strain change, and piezoelectric effects. For instance, the Bourdon tube-based pressure sensor can be connected

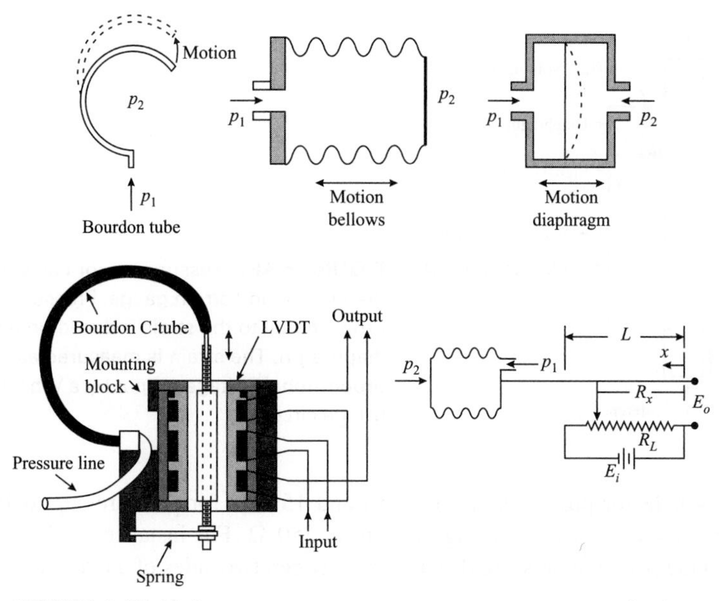


**FIGURE 6.47:** Various pressure sensor concepts: pressure to displacement transduction and then displacement measurement in order to obtain pressure measurement.

to an LVDT or a linear potentiometer to get a voltage signal proportional to the pressure:

$$\Delta x = k_p \cdot \Delta P \tag{6.143}$$

$$V_{out} = k_{vx} \cdot \Delta x \tag{6.144}$$

$$= (k_p \cdot k_{vx}) \cdot \Delta P \tag{6.145}$$

$$= k_{pv} \cdot \Delta P \tag{6.146}$$

The pressure sensors shown in Fig. 6.47 measure the relative pressure between $p_1$ and $p_2$; that is, the pressure difference between them, $\Delta p = p_1 - p_2$. If either $p_1$ or $p_2$ is the vacuum pressure (absolute zero pressure), then the sensor measures the absolute pressure.

## 6.8.2 Strain-Gauge-Based Pressure Sensor

The pressure-induced deformation of the diaphragm is measured by strain gauges on it. The strain on the diaphragm is proportional to the pressure. The resistance of the strain gauge changes in proportion to its strain. Using a Wheatstone bridge circuit, a proportional output voltage is obtained from the strain gauge. The relationship between pressure and strain-gauge voltage output is (Fig. 6.48)

$$\Delta x = k_p \cdot \Delta P \tag{6.147}$$

$$\epsilon = k_1 \cdot \Delta x \tag{6.148}$$

$$V_{out} = k_2 \cdot \epsilon \tag{6.149}$$

$$= k \cdot \Delta P \tag{6.150}$$

An example of a strain-gauge-based differential pressure sensor is the Model P2100 series by Schaevitz Inc. The sensor has two pressure input ports and can measure pressure differential between the two ports up to 5000 psi. The mechanical natural frequency of

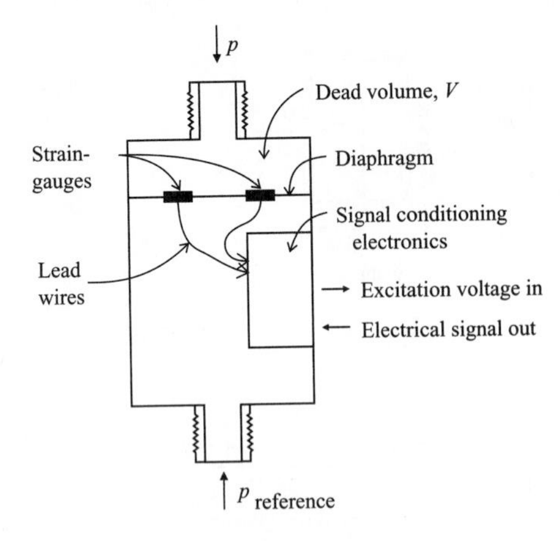


**FIGURE 6.48:** Pressure sensor using a diaphragm and strain gauge. Pressure is proportional to the strain induced on the diaphragm. The strain is measured as a proportional voltage output by a Wheatstone bridge circuit.

the sensor is 4 kHz for pressures up to 75 psi and 15 kHz for pressures up to 1000 psi. The nominal resistance of the sensing element is 350 Ω. For instance, such a sensor is appropriate to measure the pressure differential between two sides of a hydraulic cylinder.

### 6.8.3 Piezoelectric-Based Pressure Sensor

Piezoelectric-based pressure sensors are the most versatile pressure sensor types. The pressure of the diaphragm is converted to a force acting on the piezoelectric element (Fig. 6.39). The piezoelectric element will generate a voltage proportional to the force acting on it which is proportional to the pressure. The piezo-based pressure sensors can have bandwidth in the order of kHz range.

The charge produced by the piezoelectric effect as a result of pressure is

$$q = K_{qp} \cdot p \tag{6.151}$$

and the output voltage is

$$V_{out} = \frac{q}{C} \tag{6.152}$$

Then

$$V_{out} = \frac{K_{qp}}{C} \cdot p \tag{6.153}$$

$$= K_1 \cdot p \tag{6.154}$$

### 6.8.4 Capacitance-Based Pressure Sensor

The diaphragm pressure sensing concept can also be used to change the capacitance between two charged plates inside the sensor. The displacement of the diaphragm results in a proportional change in capacitance. Using an operational amplifier, reference voltage, and reference capacitor, the change in the capacitance of the sensor can be converted to a voltage output signal proportionally (Fig. 6.49).

$$x = K_1 \cdot \Delta P \tag{6.155}$$

$$C = \frac{K_2 \cdot A}{x} \tag{6.156}$$

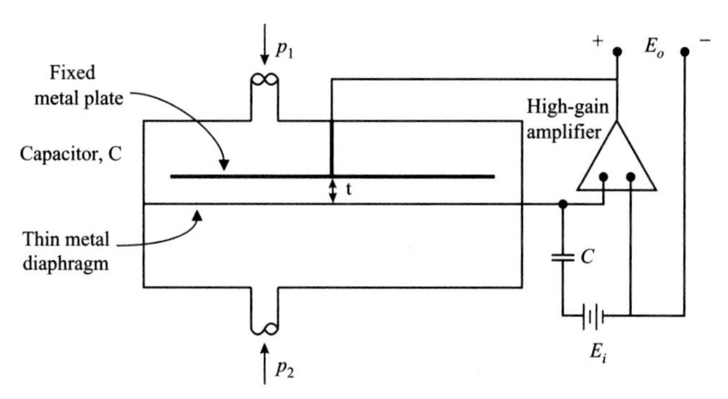


**FIGURE 6.49:** Capacitive pressure sensor and its operating principle.

Using the operational amplifier,

$$V_{out} = V_r \cdot (C_r/C) \tag{6.157}$$

$$= \frac{V_r \cdot C_r}{K_2 \cdot A} \cdot x \tag{6.158}$$

$$= \frac{V_r \cdot C_r \cdot K_1}{K_2 \cdot A} \cdot \Delta P \tag{6.159}$$

The signal flow relationship for the sensor operation is as follows:

$$\Delta P \rightarrow x \rightarrow C \rightarrow V_{out} \tag{6.160}$$

where the pressure differential results in change in the distance between two plates of the capacitive sensor which in turn changes the capacitance of the sensor.

## 6.9 TEMPERATURE SENSORS

Three classes of temperature sensors are discussed below:

1. Sensors which change physical dimension as a function of temperature
2. Sensors which change resistance as a function of temperature (RTD and thermistors)
3. Sensors which work based on thermoelectric phenomena (thermocouples)

Pictures of various RTD and thermocouple type temperature sensors are shown in Fig. 6.50.

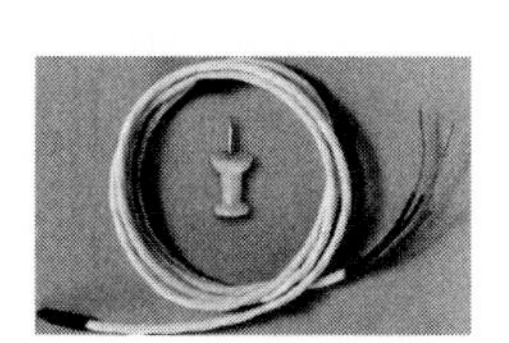
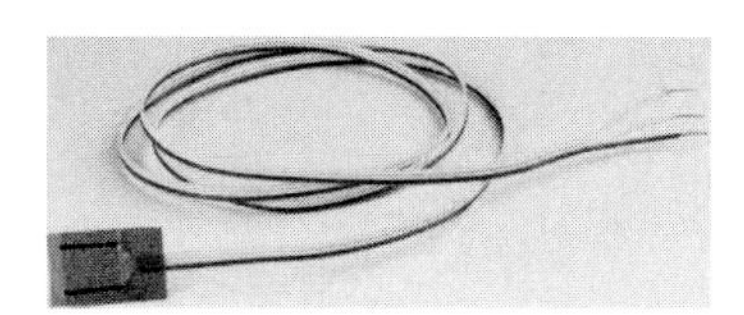
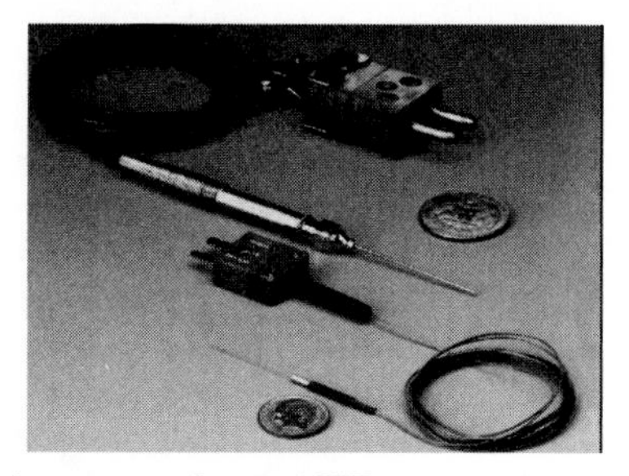

**FIGURE 6.50:** Pictures of various temperature sensors: thermocouples and RTDs.

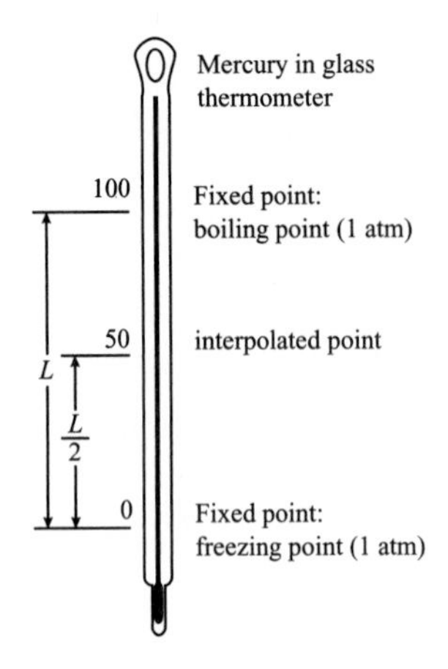


**FIGURE 6.51:** Mercury-in-glass thermometer for manual temperature measurement. It uses the transduction principle that the volume of the mercury expands as a linear function of temperature.

## 6.9.1 Temperature Sensors Based on Dimensional Change

Temperature is an indicator of the molecular motion of matter. Most metals and liquids change their dimension as a function of temperature. In particular, mercury is used in glass thermometers to measure temperature because its volume increases proportionally with the temperature. Then the glass tube can be scaled to indicate the measured temperature (Fig. 6.51). It has typical accuracy of about $+/-0.5°C$. Similarly, bimetallic solid materials change their dimension as a function of temperature. As a result, it can be used as a temperature sensor by converting the change in the dimension of the bimetalic component into a voltage.

## 6.9.2 Temperature Sensors Based on Resistance

***RTD Temperature Sensors*** An RTD (resistance temperature detector) temperature sensor operates on the transduction principle that the resistance of the RTD material changes with the temperature. Then the resistance change can be converted to a proportional voltage using a Wheatstone bridge circuit.

A good approximation to the resistance and temperature relationship for most RTD materials is

$$R = R_o(1 + \alpha(T - T_o)) \tag{6.161}$$

where $\alpha$ is the sensitivity of the material resistance to temperature variation. The construction of a typical RTD sensor is shown in Fig. 6.52. The sensitivity constants $\alpha$ for various materials are shown in Table 6.1.

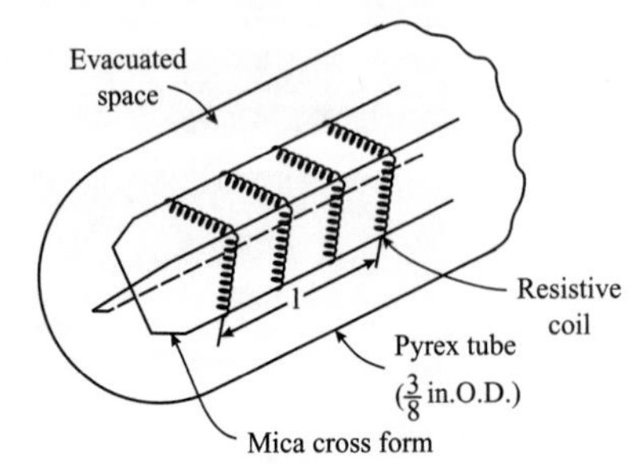


**FIGURE 6.52:** Resistance temperature detector (RTD) sensor for temperature measurement.

**TABLE 6.1: RTD Temperature-Sensor Materials and the Resistance-Temperature Sensitivity Coefficient**

| Material | $\alpha$ |
|---|---|
| Aluminum | 0.00429 |
| Copper | 0.0043 |
| Gold | 0.004 |
| Platinum | 0.003927 |
| Tungsten | 0.0048 |

RTDs may be used to measure the cryogenic temperature to approximately 700°C temperature range. Platinum is the most common material used in RTD sensors. The main advantages of RTD sensors are that the resistance-temperature relationship is fairly linear over a wide temperature range and that the measurment accuracy can be as small as ±0.005°C. Furthermore, drift of the sensor over time is very small, typically in the range of less than 0.1°C/year. As a result, RTDs do not require frequent calibration. RTD is a passive device. It has a resistance where the resistance changes linearly with temperature. In order to measure the change in resistance, the RTD must be supplied by a current source and measure the voltage across it. One good way of doing this is to use the RTD in a Wheatstone bridge circuit. The dynamic response of the RTD sensor is relatively slow compared to other temperature sensors. RTDs cannot be used to measure high-frequency transient temperature variations.

***Thermistor Temperature Sensors*** Thermistor sensors are based on semiconductor materials where the resistance of the sensing element reduces exponentially with the temperature. Typical resistance and temperature relationship for a thermistor is approximately

$$R = R_o \cdot e^{\beta(1/T - 1/T_o)} \tag{6.162}$$

where $\beta$ is also a function of temperature and a property of the semiconductor material. The variation in the resistance of a thermistor for a given temperature change is much larger than the variation in resistance of a RTD sensor. This type of sensor is used for their high sensitivity, high bandwidth, and ruggedness compared to RTDs. However, the manufacturing variations in thermistors can be large from one sensor to another. Therefore, they cannot be used as direct replacement to one another. Each sensor must be properly calibrated before replacement.

## 6.9.3 Thermocouples

Thermocouples are perhaps the most popular, easy to use, and inexpensive temperature sensors. A thermocouple has two electrical conductors made of different metals. The two conductors are connected as shown in Fig. 6.53. The key requirement is that the connections between the two conductors at both ends must form a good electrical connection. The

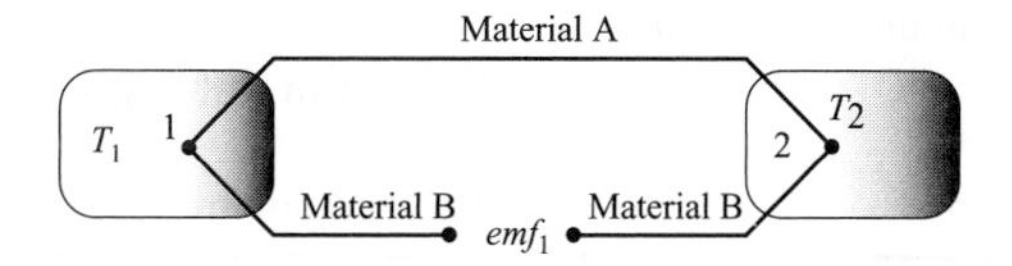


**FIGURE 6.53:** Thermocouple temperature sensor and its operating principle.

fundamental *thermoelectric phenomenon* is that there is a voltage differential developed between the open circuit end of the conductor proportional to the temperature of the one of the junction relative to the temperature of the other junction. The thermoelectric phenomena is a result of the flow of both heat and electricity over a conductor. This is called the *Seebeck effect*, named after Thomas J. Seebeck who first observed this phenomenon in 1821. The voltage differential measured at the output of the thermocouple is approximately proportional to the temperature differential between the two points ($V_{out}$ in Fig. 6.53):

$$V_{out} \approx K \cdot (T_1 - T_2) \tag{6.163}$$

Notice that the proportionality constant is a function of the thermocouple materials. The thermocouple materials refer to the material types used for conductors A and B. Furthermore, it is not exactly constant, but rather varies with temperature.

The proportionality constant is a very good approximation for many types of thermocouples over large temperature ranges. This makes the thermocouples very attractive sensors due to their linearity over large temperature ranges. The voltage output of the thermocouple is in millivolt (mV) range and must be amplified by an op-amp circuit before it is used by a data-aquisition system.

Thermocouple measures the temperature difference between its two junctions. In order to measure the temperature of one of the junctions, the temperature of the other junction must be known. Therefore, a reference temperature is required for the operation of the thermocouple. This reference can be provided by either *ice-water* or by built-in electronic reference temperature. The measurement error in most thermocouples is around $+/-1$ to 2°C. Different thermocouple material pairs are designated with a standard letter to simplify the references to them (Table 6.2).

In most cases, the output of the thermocouple is processed by a digital computer system. The reference temperature is provided by a thermistor-based sensor as part of the thermocouple interface circuit of the data-aquisition board (DAQ). Multiple thermocouples can be connected in series to sum the sensor generated signal or in parallel to measure the average temperature over a finite area. Computer interface cards for thermocouple signal processing makes use of the standard thermocouple tables for the voltage to temperature conversion for each specific type of thermocouple, instead of using linear approximation to the voltage–temperature relationship. Such standard reference tables are generated by organizations such as the National Institute of Standards and Technology (NIST) for different types of thermocouples.

**TABLE 6.2: Thermocouple Types and Their Applications**

| Type | Material A | Material B | Applications |
|---|---|---|---|
| E | Chromel (90% nickel, 10% chromium) | Constantan (55% copper, 45% nickel) | Highest sensitivity, <1000°C |
| J | Iron | Constantan | Non-oxidizing environment, <700°C |
| K | Chromel (90% nickel, 10% chromium) | Alumel (94% nickel, 3% manganese, 2% aluminum, 1% silicon) | <1400°C |
| S | Platinum and 10% rhodium | Platinum | Long-term stability, <1700°C |
| T | Copper | Constantan | Vacuum environment, <400°C |

## 6.10 FLOW RATE SENSORS

There are four main groups of sensors to measure the flow rate of a fluid (liquid or gas) passing through a cross-sectional area:

1. Mechanical flow rate sensors
2. Differential pressure measurement based flow rate sensors
3. Thermal flow rate sensors
4. Mass flow rate sensors

### 6.10.1 Mechanical Flow Rate Sensors

There are three major types of mechanical flow rate sensors: (1) positive displacement flow rate sensors, (2) turbine flow meter, and (3) drag flow meter. Their operating principle is based on the volume displaced by the fluid flow and drag between the fluid and the sensing element.

***Positive Displacement Flow Meters*** Positive displacement flow meters work on the same principle as the positive displacement hydraulic pumps and motors (see the chapter on Electrohydraulic Systems). The positive displacement pumps (and motors) are so named because they displace a well-defined fluid volume per revolution. For instance, the gear pump or piston pump sweeps a fixed amount of volume per revolution. This is called the *displacement* of the pump in units of $D = volume/revolution$. If the rotational speed of the pump is known, $w_{shaft}$, then the amount of fluid flow rate ($Q$) that passes through the pump or motor is determined by

$$Q = D \cdot w_{shaft} \tag{6.164}$$

The same principle can be used as a sensor in *positive displacement flow meters (PDFM)*. The most popular PDFM is gear type (Fig. 6.54). The flow meter usage has more similarity to hydraulic motor usage than hydraulic pump usage. The flow force drives the flow meter. The flow meter is designed such that the hydraulic energy spent in driving the meter is minimal. Since the *displacement* of the flow meter ($D$) is known for a given pump, if we

**FIGURE 6.54:** Positive displacement type flow meters: gear and lobe type positive displacement flow meters are shown.

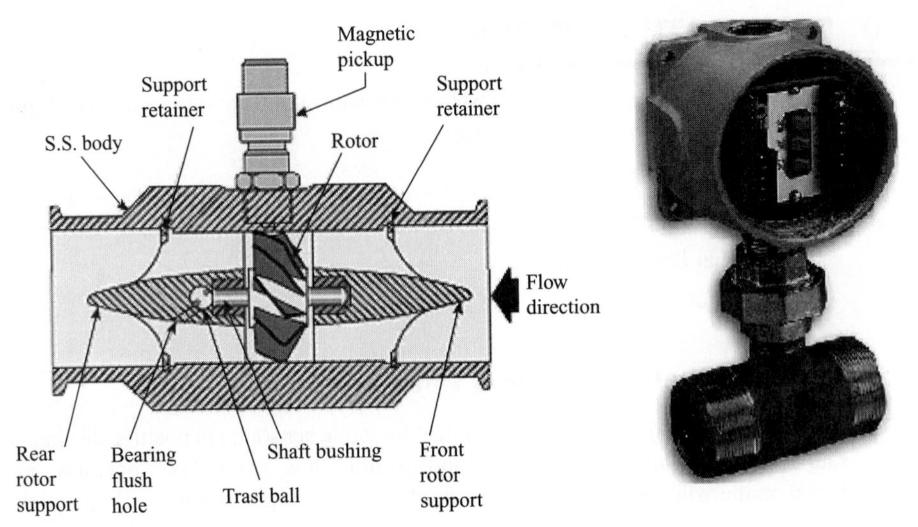


**FIGURE 6.55:** Turbine type flow rate sensor.

measure the angular speed of the shaft, then the flow rate can be calculated using the same equation given above.

The analogy between the tachometer (speed sensor) and DC brush-type electric motor is similar to the analogy between the positive displacement type flow meters and hydraulic motors. Both motor and sensor versions work on the same principle, except that in sensor applications the objective of the design is not to convert energy from one form to another (electric to mechanical or hydraulic to mechanical), but rather measure speed or flow rate.

***Turbine Flow Meters*** The turbine flow meter (Fig. 6.55) has a turbine on a shaft placed inline with the direction of the flow. As a result of the drag between the fluid flow and the turbine, the turbine rotates about its shaft.

$$Flow\ rate \rightarrow Turbine\ speed \rightarrow Speed\ sensor \rightarrow V_{out} \qquad (6.165)$$

The speed of the turbine is proportional to the flow speed (hence, the flow rate) of the fluid. The linear proportionality constant is an approximate relationship and holds well at high flow rates. Such flow meters are not suitable for measuring low flow rates. The speed of the turbine is measured and converted to output voltage using any one of the speed sensors.

***Drag Flow Meters and Vortex Flow Meters*** The drag-based flow meters insert a sensing object into the flow so as to pickup drag force from the flow. The drag force is measured by a strain-gauge-based force sensor (Fig. 6.56). It turns out that the drag force is proportional to the square of the speed:

$$F_{drag} = \frac{C_d\ A\ \rho\ u^2}{2} \qquad (6.166)$$

where $C_d$ is the drag coefficient calibrated for the specific sensor, $A$ is the drag surface of the sensing object, $\rho$ is the fluid density, and $u$ is the speed of flow. The transduction principle and sensor output relationship is as follows:

$$Flow\ rate \rightarrow Drag\ force \rightarrow Strain \rightarrow V_{out} \qquad (6.167)$$

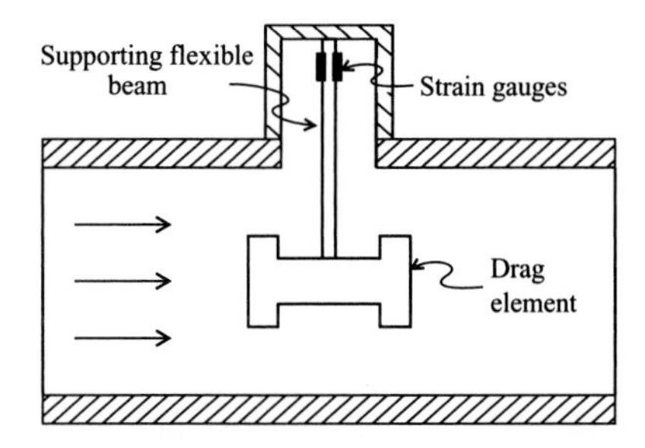


**FIGURE 6.56:** Strain-gauge-based drag measurement type flow rate sensor.

A vortex flow meter uses an inserted object into the flow field where the object sheds vorticies as the flow passes over its surface (Fig. 6.57). It turns out that the frequency of the vortex shedding is proportional to the speed of the flow, hence flow rate. Then a transducer which counts the vortex frequency can provide a proportional output voltage. As a result, the sensor output voltage can be calibrated to represent the flow rate.

### 6.10.2 Differential Pressure Flow Rate Sensors

Flow rate sensors based on differential pressure measurement make use of the Bernoulli's equation, which is a relationship between the pressure and speed of fluid flow at two different points. It is estimated that over 50% of all flow rate measuring devices are based on differential pressure type sensors. Assume that the height of the fluid does not change relative to a reference plane between the two points. Then the pressure and speed of the fluid at two separate cross sections are related by

$$p_1 + \frac{\rho\, u_1^2}{2} = p_2 + \frac{\rho\, u_2^2}{2} \tag{6.168}$$

***Pitot Tube*** The pitot tube is a differential pressure measurement sensor which makes use of the Bernoulli equation for a special case (Figs. 6.58, 6.59). The pressure is measured at two points. At one of the points, the speed of the fluid is zero, $u_2 = 0$.

Then measuring the differential pressure $p_2 - p_1$ allows us to calculate the fluid velocity along its flow stream. Once the fluid velociy is known, assuming that it is the average fluid speed, the flow rate can be calculated using the cross-sectional area information.

$$u_1 = \frac{2}{\rho}(p_2 - p_1)^{1/2} \tag{6.169}$$

Note that the Bernoulli's equation is valid for compressible fluid flow up to a Mach number of about 0.2. For higher Mach numbers, the relationship can be modified and sensor calibrated to accurately obtain the flow speed as a function of differential pressure and Mach number.

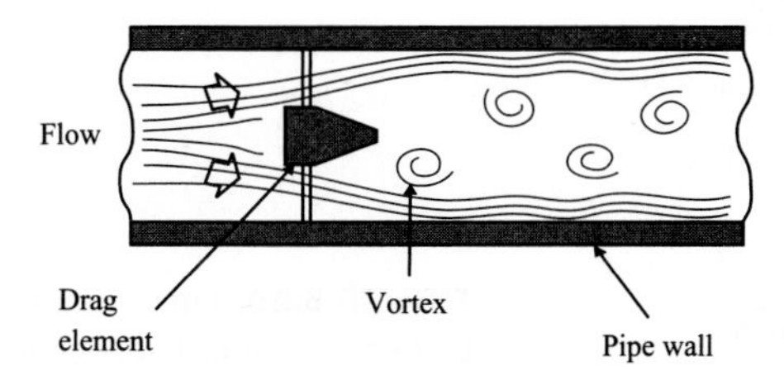


**FIGURE 6.57:** Vortex frequency counting type flow rate sensor.

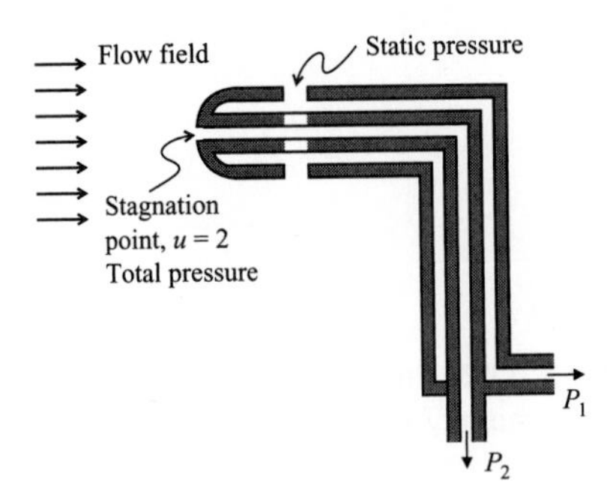


**FIGURE 6.58:** Pitot tube for flow rate sensing via differential pressure measurement.

As flow passes around the Pitot tube, vorticies are shed as a result of the tube surface and fluid interaction. The frequency of the vorticies is a function of the flow rate and the pitot tube diameter. A given Pitot tube can experience a certain vortex shedding frequency around a certain flow rate which may result in exciting the natural frequency of the Pitot tube. Each Pitot tube sensor specification includes the range of flow rate to avoid in order to make sure that it is not excited around its natural frequency.

***Obstruction Orifices*** Another method of flow rate measurement is to insert a standard profile obstruction orifice on the pipe where the flow rate measurement is desired (Fig. 6.60). The pressure differential at the input and output side of the standard obstruction profile is measured and related to the flow rate. There are many different types of standardized obstruction orifices. The flow rate is related to the differential pressure, cross-sectional area, and the geometeric shape of the the standard obstruction orifice,

$$Q = f(p_1, p_2, A, \textit{geometry of obstruction}) \tag{6.170}$$

where $A$ is the cross-sectional area. Different obstruction orifice shapes and sizes are calibrated using a higher accuracy flow rate sensor in order to define the above relationship for each specific orifice only as a function of pressure differential. Hence, for a given obstruction (geometric shape and size is defined), the flow rate as function of pressure differential

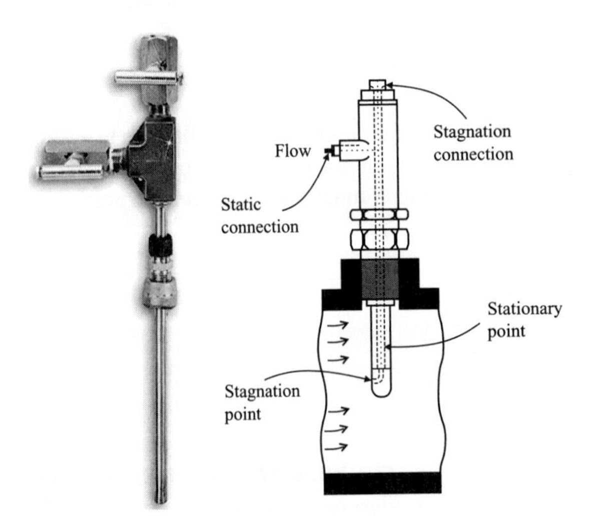


**FIGURE 6.59:** Pitot tube sensor picture and typical application.

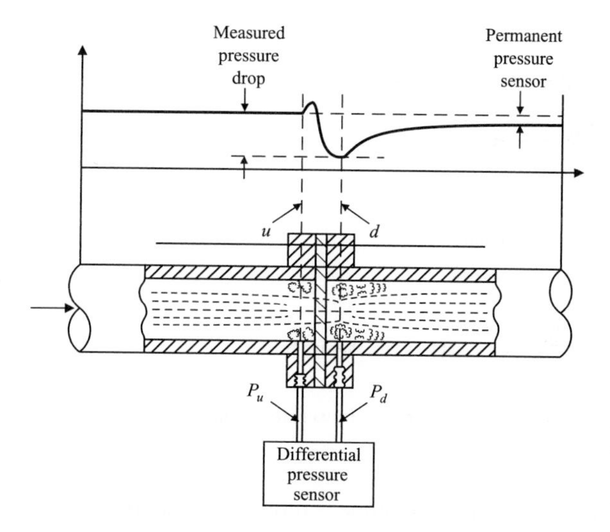


**FIGURE 6.60:** Standard obstruction orifices to measure flow rate via differential pressure measurements.

is a calibrated data table,

$$Q = f(\Delta p) \tag{6.171}$$

### 6.10.3 Thermal Flow Rate Sensors: Hot Wire Anemometer

The most well-known thermal measurement-based flow rate sensor is *hot wire anemometer*. The basic transduction principle is as follows: There is a heat transfer between any two objects with different temperatures. The rate of heat transfer is proportional to the temperature difference between them. In the case of a flow rate sensor, the two objects are the sensor head and the fluid around it (Fig. 6.61). The effective heat transfer coefficient between the sensor and the fluid is dependent on the speed of the flow. This relationship is

$$\dot{H} = (T_w - T_f)\,(K_o + K_1 u^{1/2}) \tag{6.172}$$

where $\dot{H}$ is the heat transfer rate, $T_w$ is the temperature of the tungsten wire used by the sensor, $T_f$ is the temperature of the fluid, $u$ is the fluid flow speed, and $K_o$ and $K_1$ are sensor calibration constants.

This relationship is used in the hot wire anemometer. A tiny probe with a tunsgten wire (length in the range of 1–10 mm and diameter in the range of 1–15 $\mu$m) is placed in

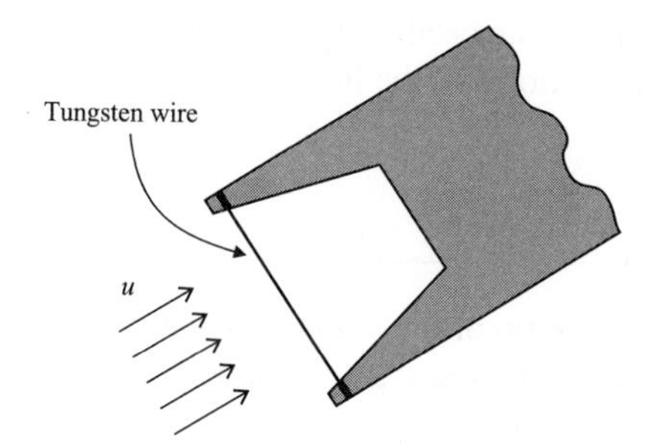


**FIGURE 6.61:** Hot wire anemometer for flow rate measurement using the thermal heat transfer principle.

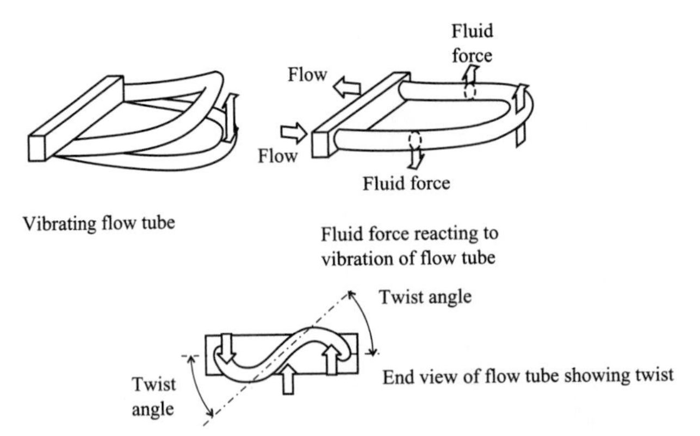


**FIGURE 6.62:** Operating principle of coriolis mass flow rate sensor.

the flow field. The resistance of the tungsten probe is proportional to its temperature.

$$R_w = R_w(T_w) \tag{6.173}$$

As current is passed through the tungsten wire, heat is transferred from the wire to the fluid.

$$\dot{H} = R_w \cdot i^2 \tag{6.174}$$

The heat transfer rate depends both on the temperature difference and fluid speed. The tungsten wire current is controlled in such a way that its temperature (hence, resistance) is held constant. The amount of heat transferred can be estimated from the current and resistance measurements on the sensor. Assuming that the fluid temperature is also constant (or measured separately by a temperature sensor), we can calculate the flow speed.

### 6.10.4 Mass Flow Rate Sensors: Coriolis Flow Meters

The coriolis flow meter measures the mass flow rate as opposed to volume flow rate. Therefore, it is not sensitive to temperature, pressure, or viscosity variation in the fluid flow. The sensor includes a U-shaped tube, a magnetically excited base which excites the U-shaped tube at around 80 Hz (Fig. 6.62). The interaction between the inertial force of the incoming fluid in one arm of the U-shaped tube and the vibration of the tube creates a force in perpendicular direction to the direction of flow and the direction of the tube vibration. The forces acting on the two sides of the U-shaped tube are in opposite directions, which creates a twist torque around the tube. The twist torque, hence the twist angle of the tube, is proportional to the mass flow rate of the fluid.

The frequency of the twist is the same as the frequency of the tube's base oscillation. The output of the twisting motion is measured as an oscillating angular displacement. The magnitude of the twisting oscillations is proportional to the flow rate.

## 6.11 HUMIDITY SENSORS

*Relative humidity* is defined as the percentage ratio of the amount of water vapor in moist air, versus the amount of water vapor in saturated air at a given temperature and pressure. Relative humidity is strongly affected by temperature.

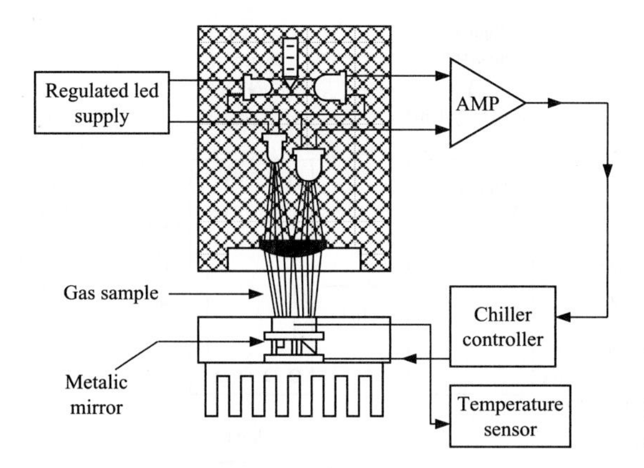


**FIGURE 6.63:** Humidity sensor: chilled mirror hygrometer.

The main types of humidity sensors use the capacitance, resistive, and optical reflection principles in the transduction stage. Capacitive humidity sensors use polymer material which changes capacitance as a function of humidity. The relationship is fairly linear. The sensor is designed as parallel plates with porous electrodes on a substrate. The electrodes are coated with a dielectric polymer material that absorbs water vapor from the environment with changes in humidity. The resulting change in dielectric constant causes a variation in capacitance. The variation in capacitance is converted to voltage by an approporiate op-amp circuit to provide a proportional voltage output.

Resistance-based humidity sensors use materials on an electrode whose resistance change as a function of humidity. In general, the resistance–humidity relationship is an exponential relationship; hence, it requires digital signal processing so that the voltage output is proportional to the measured humidity. Capacitive humidity sensors are more rugged and have less dependency on the temperature than the resistive humidity sensors.

The chilled mirror hygrometer (CMH) is one of the most accurate humidity measurement sensors. It measures humidity by dew point method (Fig. 6.63). The operating principle is based on the measurement of the reflected light from a condensation layer which forms over a cooled mirror. A metalic mirror with good thermal conductivity is chilled by a thermoelectric cooler to a temperature so that the water on the mirror surface is in equilibrium with the water vapor pressure in the gas sample above the mirror surface. When the mirror is chilled to the point that dew begins to form and the equilibrium is maintained, a beam of light is directed at the mirror surface, and the photodetectors measures the reflected light. The reflected light is scattered as a result of the dew droplets on the mirror surface. In order to maintain a constant reflected light, the photodetector output is used to control the thermoelectric heat pump to maintain the mirror at dew point temperature. Then the measured temperature is related to the humidity of the gas sample. The resolution of the CMH humidity sensor is about one part in 100 of its measurement range.

## 6.12 VISION SYSTEMS

Vision systems, also called computer vision or machine vision, are general purpose sensors. They are called the "smart sensors" in industry because what is sensed by a vision system totally depends on the image processing software. A typical sensor is used to measure a variable, i.e., temperature, pressure, length. A vision system can be used to measure shape, orientation, area, defects, differences between parts, etc. The vision technology during the

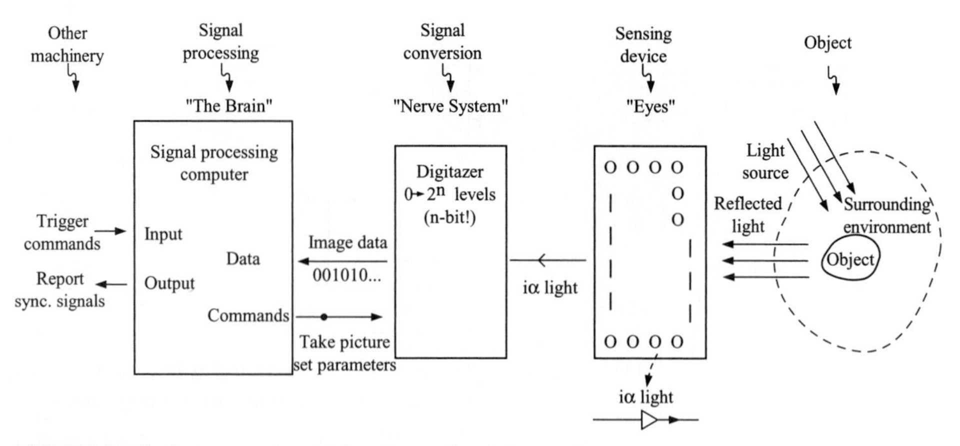


**FIGURE 6.64:** Components and functions of a vision system.

past 10 years has improved significantly in that they are rather standard smart sensing components in most factory automation systems for part inspection and location detection. Lower cost makes them increasingly attractive for use in automated processes.

There are three main components of a vision system (Figs. 6.64 and 6.65):

1. Vision camera: It is the sensor head, made of photosensitive device array, and the analog-to-digital conversion (ADC) circuit to convert the analog signal of electrical charges on the sensor head to digital form.
2. Image-processing computer and software.
3. Lighting system.

The basic principle of operation of a vision system is shown in Fig. 6.64. A vision system forms an image by measuring the reflected light from objects in its field of view. The rays of light from a source (i.e., ambient light or structured light) strike the objects in a field of view of the camera. The part of the reflected lights from the objects reaches the sensor head. The sensor head is made of an array of photosensitive, solid-state devices such

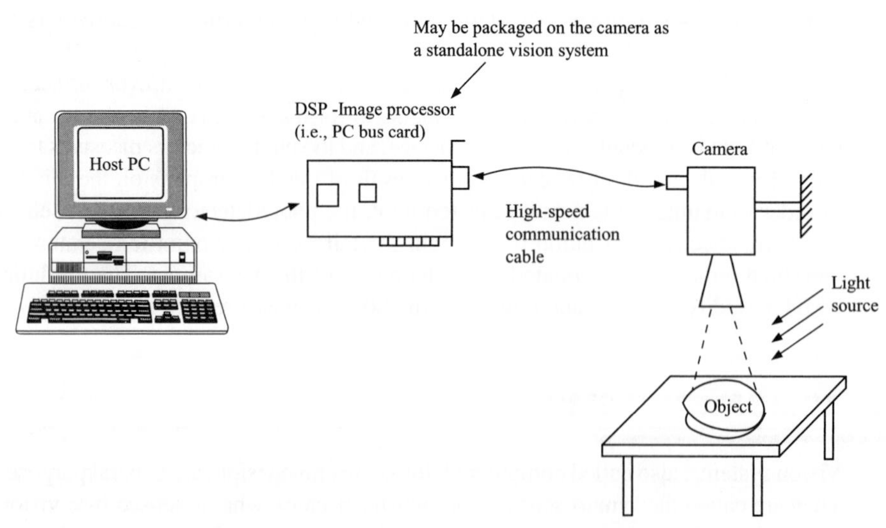


**FIGURE 6.65:** Different hardware packages of vision systems: sensor and DSP at the same physical location; or sensor head and DSP are at different physical location, and digital data are transferred from sensor head to the DSP using a high-speed communication interface.

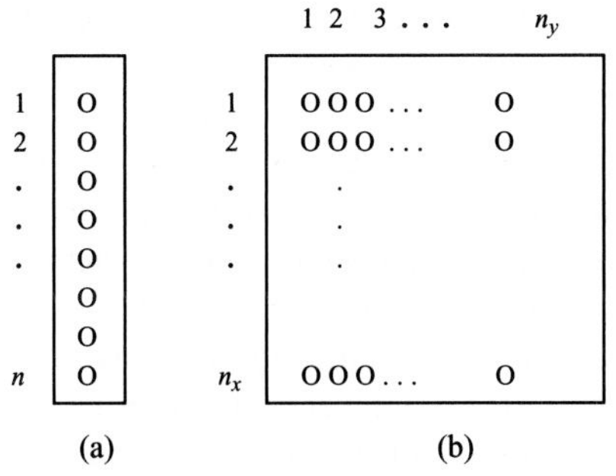

**FIGURE 6.66:** Vision sensor head types: (a) line-scan camera where the sensor array is arranged along a line; and (b) two-dimensional camera where the sensor array is arranged over a rectangular area.

as photodiodes or charge-coupled devices (CCD) where the output voltage at each element is proportional to the time integral of the light intensity received. The sensor array is a finite number of CCD elements in a line (i.e., 512 elements, 1024 elements) for the *line-scan cameras* or a finite array of two-dimensional distribution (i.e., 512 × 512, 640 × 640, 1024 × 1024) as shown in Fig. 6.66. A field of view in real-world coordinates with dimensions $[x_f, y_f]$ is mapped to the $[n_x, n_y]$ discrete sensor elements. Each sensor element is called a *pixel*. The spatial resolution of the camera, that is, the smallest length dimension it can sense in $x$ and $y$ directions, is determined by the number of pixels in the sensor array and the field of view that the camera is focused on:

$$\Delta x_f = \frac{x_f}{n_x} \tag{6.175}$$

$$\Delta y_f = \frac{y_f}{n_y} \tag{6.176}$$

where $\Delta x_f, \Delta y_f$ are the smallest dimensions in $x$ and $y$ directions the vision system can measure. Clearly, the larger the number of pixels, the better the resolution of the vision system. A camera with variable focus lens can be focused to a different field of views by adjusting the lens focus without changing the distance between the camera and field of view, hence changing the spatial resolution and range of the vision system.

Light source is a very important, often neglected, part of a successful vision system design. The vision system gathers images using the reflected light from its field of view. The reflected light is highly dependent on the source of the light. There are four major lighting methods used in vision systems:

1. Back lighting which is very suitable in edge and boundary detection applications
2. Camera-mounted lighting which is uniformly directed on the field of view and used in surface inspection applications
3. Oblique lighting which is used in inspection of the surface gloss type applications
4. Co-axial lighting which is used to inspect relatively small objects, i.e., threads in holes on small objects

An image at each individual pixel is sampled by an analog-to-digital converted (ADC). The smallest resolution the ADC can have is 1 bit. That is the image at the pixel would be considered either white or black. This is called a *binary image*. If the ADC converter has 2 bits per pixel, then the image in each pixel can be represented in one of four different levels of gray or color. Similarly, an 8-bit sampling of pixel signal results in $2^8 = 256$ different levels of gray (*gray scale image*) or colors in the image. As the sampling resolution of pixel data increases, the gray scale or color resolution of the vision system increases. In gray scale cameras, each pixel has one CCD element whose analog voltage output is proportional to the gray scale level. In color sensors, each pixel has three different CCD elements for three main colors (red, blue, green). By combining different ratios of the three major colors, different colors are obtained.

Unlike a digital camera used to take pictures where the images are viewed later, the images acquired by a computer system must be processed at periodic intervals in an automation environment. For instance, a robotic controller needs to know whether a part has a defect or not before it passes away from its reach on a conveyor. The available processing time is in the order of milliseconds and even shorter in some applications such as visual serving applications. Therefore, the amount of processing necessary to evaluate an image should be minimized. Let us consider the events involved in an image acquisition and processing.

1. A control signal initiates the exposure of the sensor head array (camera) for a period of time called *exposure time*. During this time, each array collects the reflected light and generates an output voltage. This time depends on the available external light and camera settings such as aperture.
2. Then the image in the sensor array is locked and converted to digital signal (A to D conversion).
3. The digital data are transferred from the sensor head to the signal processing computer.
4. Image processing software evaluates the data and extracts measurement information.

Notice that as the number of pixels in the camera increases, the computational load and the processing time increases since the A/D conversion, data transfer, and processing all increase with increasing number of pixels. Typical frame update rate in commercial two-dimensional vision systems is at least 30 frames/sec. Line-scan cameras can easily have a frame update rate around 1000 frames/sec.

The effectiveness of a vision system is largely determined by its software capabilities. That is, the kind of information it can extract from the image, how reliably it can do it, and how fast it can do it? Standard image processing software functions include the following capabilities:

1. Thresholding an image: Once an image is acquired in digital form, a threshold value of color or gray scale can be selected, and all pixel values below value (white value) that are set to one fixed low value, and all pixel values above that value are set to a high value (i.e., black value). This turns the image into a binary image. Various detection algorithms can be run fast on such an image, such as the edge detection algorithm.
2. Edge detection of an object: An edge is detected when a sharp change occurs from one pixel to another in the gray scale value of the image. Once such a transition is detected between two pixels in a search direction on the image array, then all the pixels around the transition pixel are searched to determine the edge boundary.
3. Color or gray scale distribution (also called histogram of image): It is a plot of the gray scale distribution of the image, that is, how many pixels ($y$-axis) has a given gray scale ($x$-axis).
4. Connectivity of object (detect discontinuities).
5. Image comparison with a reference image in memory (also called template matching), i.e., the image system may store a set of "good part" images. A real-time image is compared to the stored images to determine whether or not it matches one of the template images.
6. Position and orientation of an object relative to another reference.

7. Dimensions (length, area) of an image: Once the boundaries of a part or parts in an image are detemined, the dimensions and area information can be easily calculated.
8. Character recognition, i.e., recognize letters and numerals.
9. Geometric image transformations, i.e., mathematical operations on the matrix data of the image to move, rotate, or stretch the image.

***Example*** Consider a vision system with 1024 × 1024-pixel resolution. The camera processes 60 frames per second. The resolution of the the A/D converter system is 8 bit. What is the amount of data bytes processed per second? If the camera is focused into a surface with 10 cm × 10 cm dimensions, what is the spatial resolution in measurement?

Each pixel holds 1-byte data since the ADC converter has 8-bit resolution. The number of bytes processed per second is equal to the number of data bytes per frame times the number of frames per second:

$$N = 1024 \times 1024 \times 60 \tag{6.177}$$

$$= 62{,}914{,}560 \text{ bytes/sec} \tag{6.178}$$

$$\approx 63 \text{ MB/sec} \tag{6.179}$$

Clearly, the amount of data processed per second is very large in a high-resolution camera. The smallest distances in $x$ and $y$ directions the system can measure are

$$\Delta x_f = \frac{10 \text{ cm}}{1024} = 0.00976 \text{ cm} = 0.0976 \text{ mm} \tag{6.180}$$

$$\Delta y_f = \frac{10 \text{ cm}}{1024} = 0.00976 \text{ cm} = 0.0976 \text{ mm} \tag{6.181}$$

which indicates that the vision system can measure dimensions in $x$ and $y$ coordinates with about 0.1 mm accuracy.

## 6.13 PROBLEMS

**1.** Describe the operating principles and derive the excitation voltage, induced voltage, and output voltage relationship for a type of rotary variable differential transformer called syncro (Fig. 6.15). Suggest a signal conditioner circuit design using integrated circuits (ICs) to support the excitation voltage requirement and process the output signal of the syncro to provide an analog voltage or digital output signal that is proportional to the angular position of the rotor.

**2.** Consider a translational (linear) positioning stage driven by a servo motor (Fig. 3.3). The travel range of the table is 50 cm. The required positioning accuracy of the stage is 0.1 $\mu$m.

**(a)** Provide a list of possible position sensors for this application, discuss their relative advantages and disadvantages. What type of sensor is most appropriate for this application?

**(b)** Assume that a motor-based rotary incremental encoder is used for position sensing. What should be the resolution of the encoder, if the ball-screw pitch is 0.5 rev/mm? If the decoder circuit for the encoder can handle encoder signal frequency up to 1 MHz, what is the implication of this on the motion limits?

**(c)** Assume that the stage has a tool and that we need to detect the contact between the tool and a workpiece. After detecting the contact, we need to further move the stage in the same direction by 0.1 mm. What kind of sensor(s) would be approporiate for this purpose?

**(d)** Discuss the advantages and disadvantages of placing the position sensor on the motor shaft (rotary displacement, angle, sensor attached to motor) or on the tool (linear displacement sensor attached to the tool).

**3.** Consider a rotary motion axis and its speed control (i.e., speed control of a spindle of a CNC machine tool or speed control of a precision conveyor). Assume that both high-speed ($w_{max} = 3600$ rpm) and low-speed $w_{min} = 1$ rpm speed controls are required with 0.01 rpm regulation accuracy.

**(a)** Determine specifications for a tachometer to meet the speed control requirements.
**(b)** Select an optical incremental encoder and derive speed information from the position pulses digitally. What is the speed estimation accuracy at minimum speed if the sampling period is 1.0 msec? Show how the speed estimation can be improved with a *time period* measurement technique. What is the maximum pulse frequency of the selected encoder at the maximum speed?

**4.** Consider the following measurement problems: (1) seismic displacement measurement for earthquake monitoring, (2) three-dimensional (horizontal, vertical, and lateral) acceleration of a car body during travel, and (3) measurement of the same during a crash test. Discuss the expected range of frequency content of the signals in each application and select appropriate sensors for the measurement.

**5.** Consider the stress and force measurement on a rectangular beam. Assume that a horizontal force is applied at the tip of the beam, and the other end is clamped to a stationary base (Fig. 6.67). A strain gauge is bonded on the surface at the midpoint along the length of the beam in the direction of the deformation.

**(a)** What other information is needed in order to measure the strain, stress, and external force applied on the beam?
**(b)** If the maximum force expected is 1.0 kNt, select a strain gauge for the measurement with a Wheatstone bridge circuit. Assume the material of the beam is steel and cross-sectional dimensions (in the direction perpendicular to the force) are 10 cm × 5 cm.
**(c)** Can the same measurement system be used to measure a vertical force? If so, what other information is needed about the beam?

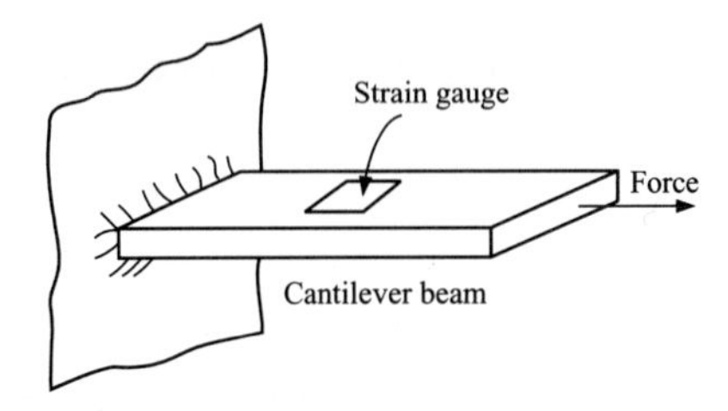


**FIGURE 6.67:** Strain, stress, and force measurement on a deformable beam under external load.

**6. (a)** Consider a strain gauge with nominal resistance of 120 Ω. It is glued on the surface of a structure. The structure deforms nominally 0.001 strain level ($\epsilon = \Delta l / l = 0.001$). Assume that the gauge factor of the strain gauge is 2.0. Determine the change in the nominal resistance of the strain gauge under this condition.
**(b)** Design a circuit to measure the strain as proportional voltage output to a data-acquisition board which has 0 to 5 V input range. When the strain level is 0.001, the output voltage should be 5.0 V.

**7.** Consider the hydraulic motion control system shown in Fig. 7.2. It is necessary to measure the pressure difference between pump output and the valve output. In other words,

$$\Delta p = max(p_P - p_A, p_P - p_B) \tag{6.182}$$

Which is typically used to control the pump displacement to provide a constant pressure drop across the valve (called *load sensing* pump control method). It is anticipated that the maximum pump output pressure is 15 MPa and the pressures at points A and B will be lower than that. Select a set of pressure transducers which will provide the desired measurement range and 1% accuracy over a frequency range of 0 Hz–1 KHz?

**8.** **(a)** What are the basic temperature-measurement principles and main differences between the following temperature sensors: (1) RTD, (2) thermistor, and (3) thermocouple?

**(b)** What type of temperature sensor would you recommend for the following cases and why? (1) Temperature measurement of a small volume electronic part with highly transient temperature condition. (2) Temperature measurement of a liquid in a large container.

**9.** Consider a PC-based temperature-measurement system. The PC has a data-acquistion card that has 12-bit resolution and can multiplex 16 channels of analog signals (16-channel, multiplexed ADC with 12-bit resolution). The ADC card has a gain of 10 which amplifies the thermocouple voltage before the ADC sampling. The range of input voltage to the board is 0–100 mV, which is amplified to 0–1.0-V range before the ADC sampling and conversion. Assume that all 16 channels use J-type thermocouples and that the PC card has a reference junction compensator which provides a reference temperature equivalent to 0°C. Draw a functional block diagram of this data-acquisition system. What is the voltage input if the nominal temperature is 100°C? What is the voltage input if the nominal temperature is 200°C? What is the estimated measurement error in units of °C, assuming that the error in reference junction compensator is 0.5°C and the accuracy of the thermocouples is ±0.25°C. In calculating the measurement error, use the worst-case scenerio in which the measurement errors are additive ($e_{sum} = \sum_{i=0}^{n} e_i$). Given the digital value of the voltage conversion by ADC, what is needed in software to relate the voltage to temperature data? Discuss if that would be different for various types of thermocouples (i.e., each channel may have a different thermocouple type, J-type, R-type, B-type, T-type).

**10.** Faradays law of induction states that an electromotive force (EMF) voltage is induced on a conductor due to changing magnetic flux and that the induced voltage opposes the change in the magnetic flux. Application of this forms the basis of many electric motors and electric sensors (see Section 8.2). A back EMF is induced in the conductors moving in a magnetic field established by the field magnets (permanent magnets or electromagnets (Fig. 8.9(d))). As the conductor moves in the magnetic field, there will be force acting on the charges in it, just as there is a force acting on a charge moving in a magnetic field. Let us consider a constant magnetic field $B$, a conductor with length $l$ moving in perpendicular direction to magnetic field vector, and its velocity $\dot{x}$ (Fig. 8.9(d)). The induced voltage because of this motion is,

$$V_{emf} = B \cdot \dot{x} \cdot l$$

This same principle can be used in a flow rate sensor, where the magnetic field $B$ is induced by a permanent or electromagnet of the sensor, $\dot{x}$ is the speed of the flow across the cross sectional area $A$. The induced voltage across the moving fluid in the direction perpendicular to both directions of $B$ and $\dot{x}$ is proportional to $B \cdot \dot{x}$ for a given sensor size. Then, the flow rate can be calculated from $Q = A \cdot \dot{x}$. The fluid must be electrically conductive for this principle to work.

**(a)** Sketch the design of a magnetic flow rate sensor based on this principle.

**(b)** Find a magnetic flow rate sensor on the web and identify its operating characteristics (i.e., minimum conductivity of the fluid, maximum flow rate it can measure, accuracy and bandwidth).

CHAPTER 7

# ELECTROHYDRAULIC MOTION CONTROL SYSTEMS

## 7.1 INTRODUCTION

In hydraulic systems, the power transmission medium is pressurized hydraulic fluid. If the control of the pressurized fluid flow is done by electrical means, then they are called electrohydraulic (EH) systems. If the control of the fluid is done by a combination of mechanical and hydraulic mechanisms, then they are called hydromechanical systems.

The concept of a hydraulic system is shown in Fig. 7.1. The basic components of an electrohydraulic motion control system is shown in Fig. 7.2. A closed-loop controlled version of a single-axis EH control system is shown in Fig. 7.3 where sensors at the valve, cylinder, and application tool may be included as part of a closed-loop digital control system. Hydraulic systems deal with the *supply and control* of fluid *pressure, flow rate, and flow direction.*

The mechanical power source (usually an internal combustion engine in mobile equipment or an electrical induction motor in industrial applications), hydraulic pump, and reservoir form the so-called *hydraulic power supply unit.* The tank acts as the hydraulic fluid supply and reservoir for the return lines. The "pump" converts the mechanical power received from the mechanical power source into the hydraulic power in the form of pressurized fluid at its outlet port. The pump is the heart of a hydraulic system. If the pump fails, no hydraulic power can be transmitted to the final actuators. In industrial applications, an electric motor is typically used to provide mechanical power to the pump, whereas the internal combustion engine is the mechanical power source to drive the pump in mobile equipment applications. Figure 7.4 shows two common examples of hydraulic power units for industrial applications. The JIC style is one of the most common hydraulic power units. The main benefit of the design is its simplicity and easy access to main components for servicing. The main drawback of it is the fact that the pump inlet port is at a higher elevation than the fluid in reservoir. This may lead to a cavitation problem at the pump inlet port. The overhead design eliminated the cavitation problem to a large extent. In addition, the pump may be submerged in the reservoir so that the circulation of fluid in the reservoir also helps cool the pump. However, the submerged style makes servicing the pump more difficult. There are other types of hydraulic power units such as L-shaped, vertically mounted motor on reservoir, and T-shaped reservoir.

The *valve* is the key control element. A valve is used to *control* (also called *modulate* or *throttle* or *meter*) the fluid flow. The valve is the metering device. The flow can be regulated in terms of its rate, direction and pressure. The type of the valve used and the "control objective" that drives the valve vary from application to application. The most common form is a proportional directional flow control valve. The valve is controlled by a solenoid or torque motor where the current in the windings generates a proportional spool

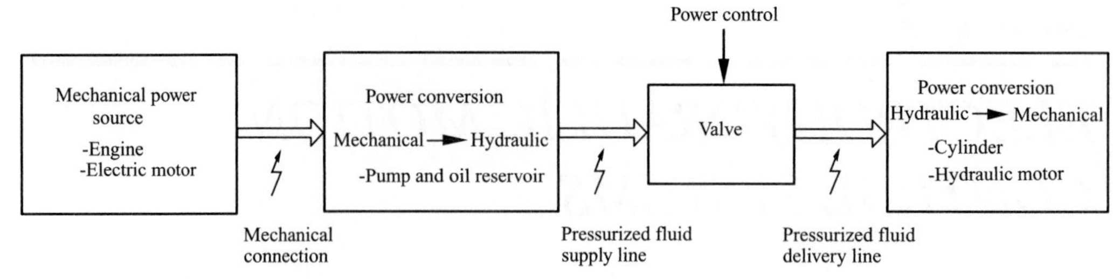


**FIGURE 7.1:** Concept of a hydraulic power system: the main functional components needed are (1) mechanical power source device, (2) mechanical-to-hydraulic power conversion device, (3) hydraulic power control device, and (4) hydraulic-to-mechanical power conversion device.

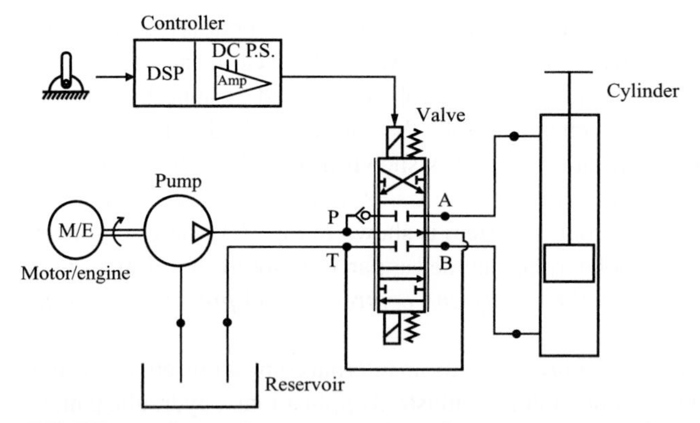


**FIGURE 7.2:** Symbolic representation of the components of an electrohydraulic motion system for one axis motion: hydraulic power unit (pump and reservoir), valve, actuator, and controller. Since there are no sensors used for the control decision, this is an open loop control system. It is also called *operator in the loop* control system.

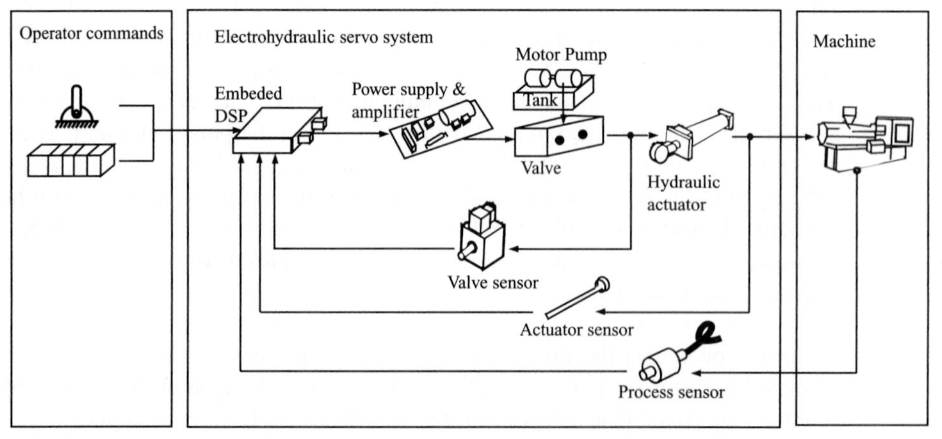


**FIGURE 7.3:** Basic components of a closed-loop controlled electrohydraulic motion system for one axis motion (pump, valve, cylinder, controller, and sensors).

(a)

(b)

**FIGURE 7.4:** Hydraulic power units for industrial applications which include electric motor, pump, reservoir, filteration, and heat exhanger. (a) JIC-style hydraulic power unit where the pump is on top of the reservoir and must have good suction in the inlet pipe in order to avoid cavitation in the pump input port. (b) Overhead reservoir where the lowest level of oil is at a higher elevation than the pump. This design is less likely to have cavitation problems. The pump can also be submerged in the reservoir which helps the cooling of the pump, but it is more difficult to service.

displacement. The spool displacement opens the metering ports; hence it controls the flow rate. The valve amplifier (considered as part of the controller block in Fig. 7.2) provides the amplified version of the low power command signal from the controller.

The actuator, power delivery component, of the EH motion system can be either a linear cylinder or a rotary hydraulic motor. For steady-state considerations, the speed of the actuator is proportional to the flow rate.

Applications of hydraulic motion control systems include:

1. Mobile equipment such as construction equipment that generates its power from an internal combustion engine and delivers power to work tools via pressurized hydraulic fluid using pump, valve, and cylinder/motor components.
2. Industrial factory automation applications:
   (a) Presses (punch presses, transfer presses)
   (b) Injection molding machines
   (c) Sheet metal thickness control drives in steel mills
   (d) Winches and hoists
3. Civil and military aircraft flight control systems and naval ships
   (a) Primary flight control systems that involve the motion of wing surfaces, rudder, and elevator
   (b) Secondary flight control systems that involves spoilers and trim surfaces
   (c) Engine fuel rate delivery control systems
4. Military ground vehicles
   (a) Turret motion control
   (b) Hydrostatic traction drive systems

The control of an EH motion system may involve the control of either or both the valve and the pump. Furthermore, the control logic may be based on "open-loop" or "closed-loop" concepts. If a human operator is in the loop, a fixed displacement pump with variable input speed (i.e., variable engine speed) controlled by an operator pedal, and a valve actuated proportional to the operator lever input (open-loop control) are sufficient. The operator would adjust the pump input speed with a pedal, hence controlling the hydraulic supply power. At the same time, the operator controls the valve shifting with a lever based on his or her observations, hence modulating the delivery to the actuator. In automated systems, closed-loop control of valve and pump may be necessary. Design of an electrohydraulic motion control system involves the following steps,

1. ***Specifications:*** The first step in any design is to specify the requirements that the system must meet, i.e., performance, operating modes, fail-safe operation.
2. ***System concept design:*** Given the specifications, an appropriate system design concept must be developed, i.e., it is an open-circuit hydraulic or closed-circuit hydraulic system, open-loop or closed-loop controlled.
3. ***Component sizing and selection:*** The proper component types and sizes must be selected. Component sizing requires power calculations to make sure the size of components are properly matched and that they have the capacity in terms of power, pressure, flow rate to meet the performance requirements.
4. ***Control algorithm design:*** In computer controlled EH systems, there is always a controller hardware and real-time control software involved in control of the system.
5. ***Modeling and simulation:*** If necessary, the EH system hardware and control algorithm may be modeled and simulated off-line on a computer to predict the performance of the overall system.
6. ***Hardware test:*** Finally, a prototype system must be built and operation of the EH system, under the control of the designed control algorithm and hardware, must be tested. If the desired performance objectives are not met, the design process is iterated for refinements.

Measured variables in an EH system for control purposes may include the following:

1. Load position and speed measured at the actuator (cylinder) and/or load end
2. Load pressure measured as pressure differential at the actuator ports or at the outlet ports (A & B) of the valve
3. Load force or torque
4. Spool displacement
5. Flow rate
6. Pump output pressure
7. Pump displacement
8. Pump speed

These measured quantities may be used in real time as part of a closed-loop control algorithm. Intermediate variables can be derived from the measured variables and regulated. For instance, if flow rate and load pressure are measured, the output power can be calculated. Then it could be used to implement a closed-loop control algorithm on the desired output power. Hence, the EH control system would operate so as to track a commanded output power level.

Figure 7.5 shows possible safety-related additions to a typical hydraulic motion system. In general, every hydraulic circuit should include safety relief valves to protect the

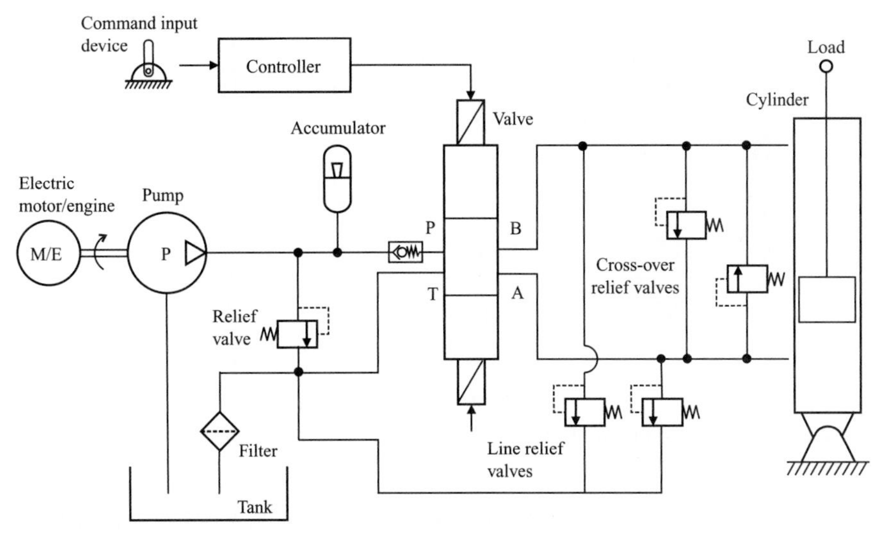


**FIGURE 7.5:** Safety valves typically added to a hydraulic motion system: pump side relief valve and load side relief valves (line-to-tank relief valves, cross-over relief valves), and check valves to limit the flow in a line to one direction only (i.e., flow is allowed from P to A or B ports, but not back to P port from A or B ports).

lines against execessive pressures at both (1) the pump side and (2) the actuator (line) side. Typical pressure safety valves designed into the system are as follows:

- Pump pressure relief valve which limits the maximum output pressure of the pump.
- Line relief valve which opens the line to tank if the line pressure exceeds a set limit, and/or cross-over relief valves which open one side of the actuator to the other side if the pressure differential between the two sides of the actuator exceeds a certain level.

***Accumulators*** The *accumulator* serves as a pressurized fluid storage component. It can help the control system in two ways in the transient response,

- Maintain the hydraulic line pressure in case of a sudden drop on the hydraulic line pressure due to sudden increase in demand or decrease in the the pump output.
- It can provide a damping effect and shock absorber function in case of large pressure spikes, i.e., as a result of sudden load changes.

On the other hand, an accumulator also reduces the open-loop natural frequency of the EH system as a result of reduced stiffness of fluid (smaller effective bulk modulus due to elasticity of the accumulator and increased fluid volume). As a result, the maximum bandwidth the control system can achieve is lower. There are three main types of accumulators: weight loaded, spring loaded, and hydropneumatic types. Hydropneumatic-type accumulators use dry nitrogen as the compressed gas. There are three major designs of hydropneumatic accumulators which are categorized in terms of the way hydraulic pressure and pressure storage components interact with each other: (a) piston type, (b) diaphram type, and (c) bladder type (Fig. 7.6). An accumulator is rated with the hydraulic fluid volume it can store (also called *working volume*), maximum and minimum operating pressures, precharge pressure, and maximum shock pressure it can tolerate. The compression ratio of an accumulator is the ratio between maximum and minimum operating pressures.

An accumulator can be sized either based on a required discharge volume or to limit the maximum pressure due to shocks. For a given discharge volume ($V_{disch}$), the required

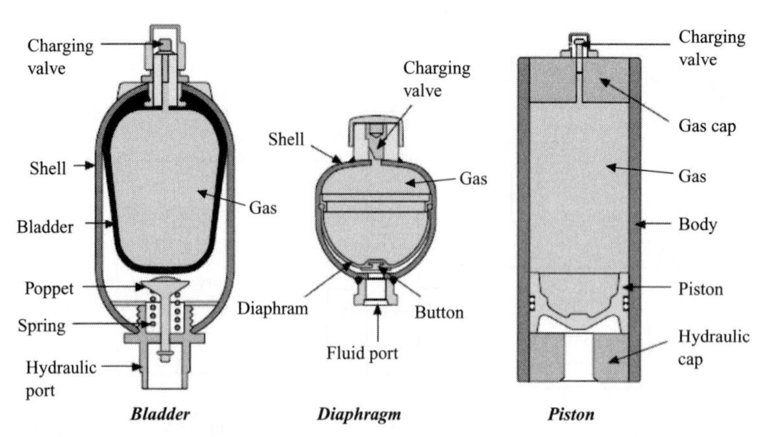


**FIGURE 7.6:** Three different types of hydropneumatic accumulators: bladder type, diaphram type, and piston type [35].

accumulator size (volume of the accumulator, $V_{acc}$) can be determined as follows:

$$V_{acc} = V_{disch} \cdot \frac{p_{min}}{p_{pre}} \Bigg/ \left[ 1 - \left( \frac{p_{min}}{p_{max}} \right)^{1/n} \right] \tag{7.1}$$

where $p_{pre}$, $p_{min}$, $p_{max}$ are precharge pressure of accumulator, minimum and maximum line pressure, $n$ is an empirical number between the 1.2 and 2.0 range and is determined by the discharge time period of the accumulator and maximum line pressure.

The pre-charge pressure is about 100 psi less than the minimum line pressure requirement,

$$p_{pre} = p_{min} - 100 \text{ [psi]} \tag{7.2}$$

After the accumulator is precharged, under the line pressure (which is presumably between minimum and maximum pressure, hence larger than the precharge pressure) charges the accumulator. Fluid flows into the accumulator until the pressure rises to the maximum pressure setting at which point the relief valve opens and discharges the accumulator. When the stored pressure in the accumulator is larger than the line pressure, i.e., sudden pressure drop in the line, the fluid flows from the accumulator to the line. This continues until the accumulator pressure reaches the minimum pressure setting, at which point the poppet valve would be closed. When the line pressure exceeds the accumulator pressure, fluid again flows into the accumulator and charges it. After the precharging, the pressure in the accumulator is limited to be between minimum and maximum pressure settings,

$$p_{min} \leq p_{acc} \leq p_{max} \tag{7.3}$$

Whenever $p_{acc} \leq p_{line}$, accumulator is charged, and whenever $p_{acc} \geq p_{line}$, accumulator is discharged. When $p_{acc} = p_{min} \geq p_{line}$, the poppet valve closes (or in the case of a piston type accumulator, end of travel is reached). When $p_{acc} = p_{max} < p_{line}$, a system relief valve opens to protect the accumulator.

Typically, the accumulator volume is about three, four, or more times the discharge volume. In general, the higher the dynamic response requirements, the larger the accumulator size relative to the discharge volume. Full discharge volume of a typical accumulator can be discharged within a few hundred milliseconds. For instance, consider a piston type accumulator with a discharge volume of 20 liters, and that it can discharge at 3600 liters/min at 3000-psi pressure differential between the accumulator and the line pressure. Therefore, the accumulator would discharge at 60-lps rate and full volume would be discharged within about 1/3 sec. Bladder- and diaphram-type accumulators have a little faster response time

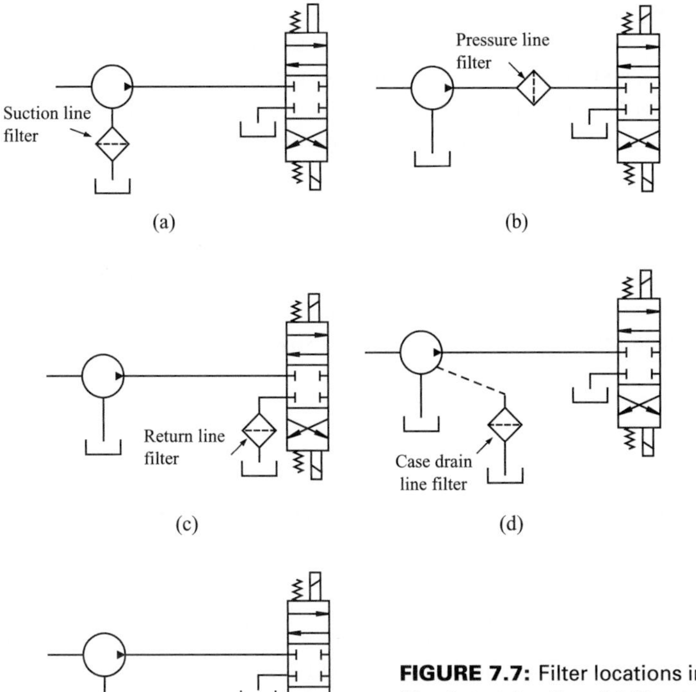


**FIGURE 7.7:** Filter locations in a hydraulic circuit: (a) filter in suction line, (b) filter in pressure line at the pump output, (c) filter in return line to the tank, (d) filter in case drain line, and (e) kidney loop filteration. There is generally a bypass check valve in parallel with the filter in order to provide a flow path in case the filter is clogged.

than the piston type due to less friction between moving parts. Hydraulic circuits incorporate valves to control the charging and discharging of the accumulator.

***Filteration*** *Filteration* removes the solid particles from the hydraulic fluid. Filters may be located at the suction line, pressure line, return line, or drain line (Fig. 7.7). In addition, there could be a dedicated loop for continuous filteration of the fluid in the reservoir. This is called the *kidney loop filteration*. If the pump is the most sensitive element in the circuit against dirt, the filter is placed at the inlet port of the pump between the reservoir and the pump. In this case it is important to maintain a pressure differential across the filter in order to avoid cavitation at the pump inlet port. If the valve is the most sensitive component (i.e., servo valve), then the filter is placed between the pump outlet port and the valve. Notice that such a filter must be rated to handle the maximum pressure and flow rate it is exposed to.

A filter has replaceable elements to remove particles above a rated size, i.e., 3 microns. As a safety measure, each filter usually has parallel bypass valve (pressure switch). In case the filter becomes clugged due to excessive dirt accumulation before a scheduled filtering element replacement, the pressure build-up in the clugged filter will open the bypass valve, and allow the flow to continue albeit it may not be filtered. Unfiltered fluid and the resulting dirt are perhaps the most common causes of component failures in high-performance hydraulic servo systems where the servo components have precision manufacturing tolerances. The three most important specifications for a filter are:

1. Minimum size of the particles it can filter (i.e., 10 microns, 5 microns).
2. Maximum pressure it can withstand (i.e., in supply line or in return line).
3. Maximum flow rate it can support.

***Viscosity of Hydraulic Fluid*** Viscosity is the resistance of a fluid to flow. It is a measure of how thick the oil is and how hard it is to move. There are two definitions of viscosity: *kinematic* and *dynamics* (or absolute viscosity). Dynamic viscosity, $\mu$, is defined as the ratio between the force necessary to move two solid pieces relative to each other at a certain velocity when they are separated by a viscous fluid,

$$F = \mu \frac{A}{\delta} \dot{x} \tag{7.4}$$

where $A$ is the surface area filled by the fluid between two pieces, $\delta$ is the fluid film thickness, and $\dot{x}$ is the relative speed between to pieces (also it is the shear rate of fluid). Kinematic viscousity is defined relative to the dynamic viscousity as

$$\upsilon = \frac{\mu}{\rho} \tag{7.5}$$

where $\rho$ is the mass density of the fluid. Kinematic viscosity of a given fluid is measured as the time period it takes for a standard volume of fluid to flow through a standard orifice. The absolute viscosity is obtained by multiplying the kinematic viscosity with density of the fluid. There are many different standards for measuring the kinematic viscosity. The Saybolt Universal Seconds (SUS) measure is the most commonly used standard. SUS viscosity measure of a fluid indicates the amount of time (in seconds) it takes for a standard volume of fluid (60 milliliters) to flow through a standard orifice, called Saybolt Viscosimeter, at a standard temperature (i.e., 40°C or 100°C). Other standards used for viscosity index are SAE (i.e., SAE-20, SAE-30W) and ISO standards. In general, the SUS viscosity of a hydraulic fluid should be within the 45–4000 SUS.

The hydraulic fluid viscosity varies as function of temperature, especially during the startup phase of operation. Therefore, the oil temperature must be kept within a certain range. Air-to-oil or water-to-oil heat exchangers are used to control the temperature of the hydraulic fluid. The *heat exchanger* transfers heat between two separate fluids (i.e., hydraulic fluid and water, or hydraulic fluid and air). If the heat is transferred from the hydraulic fluid to the other fluid (i.e., to cold water), it functions as a cooler. If the heat is transferred from the other fluid (i.e., from hot water) to the hydraulic fluid, it is used as a heater. In most applications, the hydraulic fluid needs to be cooled. However, in very cold environment temperatures, it is necessary to heat the hydraulic fluid instead of cooling it. A typical heat exhanger uses a temperature sensor in the hydraulic fluid reservior in order to control the temperature in closed-loop control mode. A regulator (analog or digital) controls a valve (which modulates the flow rate of the second fluid) based on the temperature sensor signal and the desired temperature value. In general, the viscosity of oil increases (oil gets thickker and less fluidic which resists flow more) as temperature decreases. As the oil viscosity increases, the flow rate of pumps reduces due to the fact that oil is less fluidic and the suction action of the pump cannot pull-in as much oil.

***Some Hydraulic Circuit Concepts*** Figure 7.8(a) shows a simple hydraulic system where the hydraulic power is provided by the lever motion. In this case, the lever is moved by an operator. As the lever moves up, due to the pressure differences, the check valve 1 opens, and check valve 2 closes. Hence, the lever sucks the fluid from the reservoir. In the down stroke, check valve 1 closes and check valve 2 opens. The pressure between the lever side is cylinder side is almost equal. Hence, the force is multiplied by the area ratio, while the speed of linear travel of the load and lever is divided by the same ratio. This is the result of conservation of power.

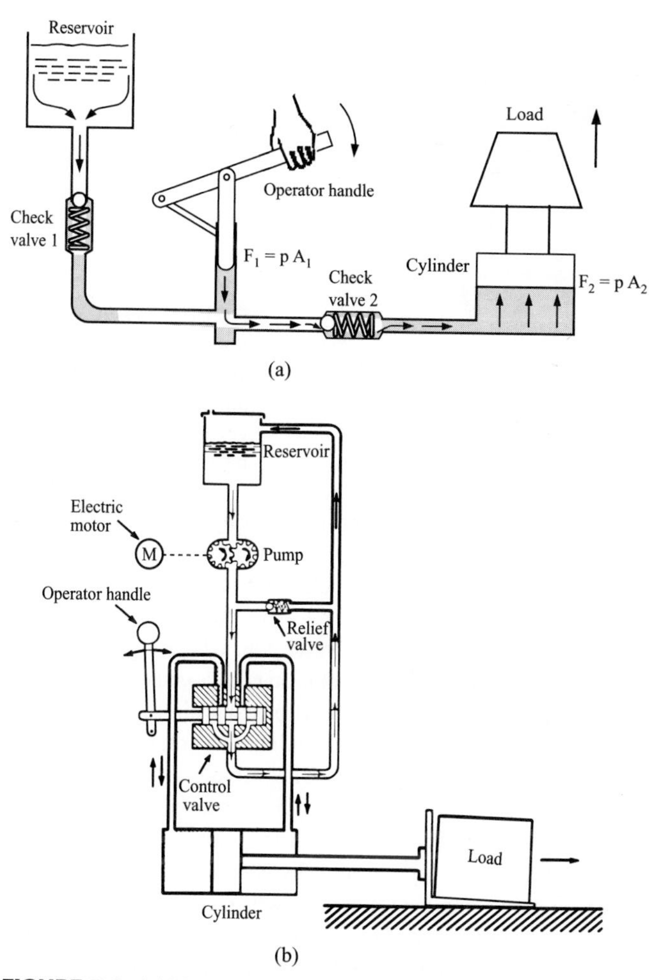


**FIGURE 7.8:** (a) Manually powered hydraulic cylinder. The pumping power and flow metering are controlled by the manual power provided to the lever arm. The flow rate is determined by the travel of the pumping action and rate of cycles per unit time. (b) Electric motor powered hydraulic system. Pump is fixed displacement type gear pump type. Flow rate is controlled manually by the valve lever.

Figure 7.8(b) shows a simple hydraulic system where the source of hydraulic power is a fixed displacement gear type pump. Input mechanical power to the pump is provided by an electric motor. An operator controls the flow to the cylinder by a manually actuated valve. Notice that the valve is *open-center* type which means that the flow circulates from pump port to tank port when the valve is in neutral position.

Figure 7.9 shows the components of a hydraulic circuit for the motion of a cylinder of an excavator. The power source is the internal combustion engine, and the hydraulic pump is mechanically geared to the engine crankshaft for mechanical power source. The directional valve is a manually controlled type.

Similarly, Fig. 7.10 shows the components of an electrohydraulic circuit for the motion of a rotary hydraulic motor for a service truck. The hydraulic motor receives power from the pump, which receives its power from the engine, and its speed is controlled by a directional flow control valve and operator input switch. Notice that hydraulic motor speed is closed-loop controlled using a tachometer speed sensor as the feedback signal.

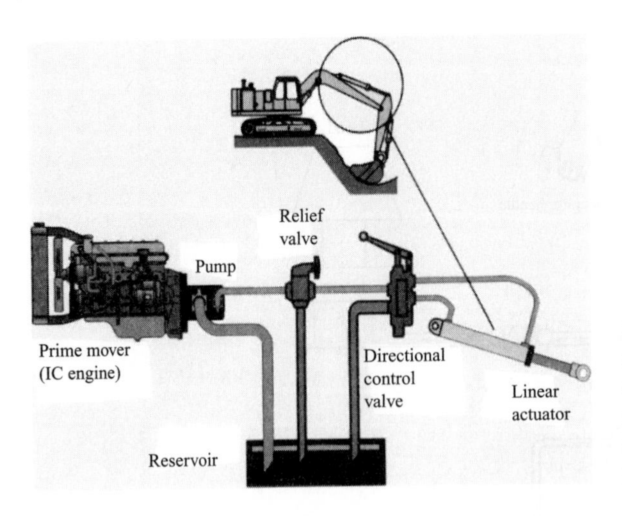


**FIGURE 7.9:** A manually operated hydraulic circuit for the cylinder motion of an excavator linkage. The main directional valve is actuated via a mechanical pedal by an operator.

A multifunction (multidegrees of freedom motion) hydraulic power based motion system is shown in Fig. 7.11. In general one pump is used (large enough to support all functions), and a valve-actuator pair is used for each function. This figure shows a *parallel connection* of each hydraulic valve-actuator pair to the pump and tank lines. In some applications, multiple circuits may have *series connection*, where a preceeding circuit has priority over the following circuits. Such implementation is necessary when it is not practical to provide a large enough pump to support full load of each circuit and when one of the circuit is more important than the others. One such implementation is used in the implement (bucket) hydraulic circuit for wheel-type loaders. The valve-cylinder pair for lift and tilt motion is connected in series (Fig. 7.12). The pump is not large enough to support both functions at their maximum flow capacity. The tilt circuit is connected ahead of the lift circuit. As a result, if tilt valve is shifted to its maximum displacement, the tilt cylinder gets the maximum flow, and the remaining small amount of flow is passed to the lift circuit through the tilt valve. Such a hydraulic circuit is said to have tilt function priority over lift function.

All of the electrohydraulic circuit examples given above circulates the flow from reservoir to pump, to valve, then to actuators and back to reservoir. Such hydraulic systems are called *open-circuit* systems. There is also the *closed-circuit* hydraulic systems, such as those used in hydrostatic transmissions. A hydraulic circuit is called closed circuit if the main flow in the power line circulates in a closed line with small leakage and replenishment

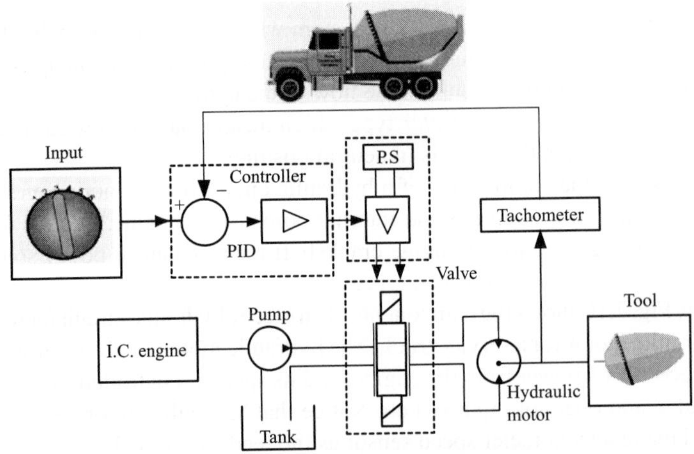


**FIGURE 7.10:** An electrohydraulic closed-loop speed control system example of a service truck.

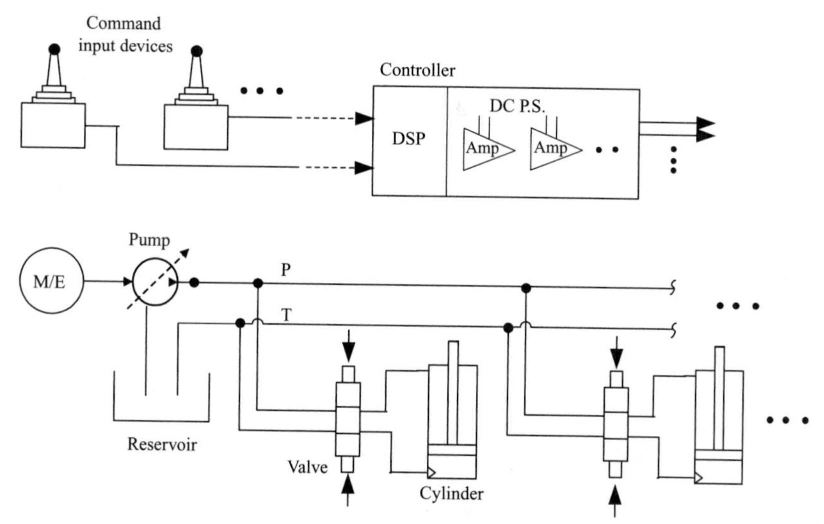


**FIGURE 7.11:** Multidegrees of freedom (multifunction) hydraulic circuit. This example shows a parallel connection of two hydraulic actuator-valve circuits to supply lines. The valves are generally closed center-type in this configuration. The pump may be variable or fixed displacement type (most likely variable displacement type).

of fluid from the reservoir. Figure 7.13 shows the hydraulic circuit and its components for a typical closed-circuit transmission. There are three major fluid flow circuits: the main flow loop between pump and motor, charge loop which supplies replenishment flow to the main loop as well as supply pressurized flow power used to actuate the pump displacement, and finally the drain loop which sends the leakage flow through the components to the reservoir. The pump control types are discussed in more detail later. Suffice it to say that the pump displacement can be controlled to regulate various variables, i.e., pump output pressure, flow rate. The power necessary to actuate the main pump displacement control system

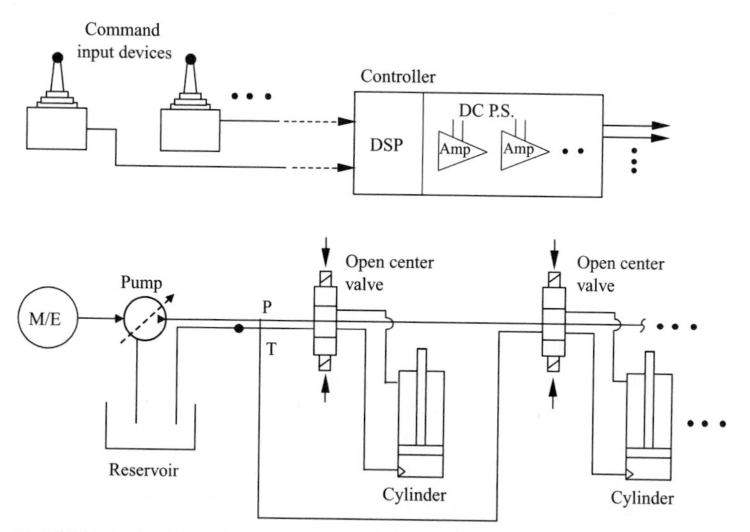


**FIGURE 7.12:** Multidegrees of freedom (multifunction) hydraulic circuit. This example shows a series connection of two hydraulic actuator-valve circuits to supply lines. Valves are necessarily the open-center type in order to supply the circuits in the serial connection when an earlier valve is in neutral position. The valves closer to the pump have priority over the ones that are connected later in the series circuit. The pump may be a variable or fixed displacement type.

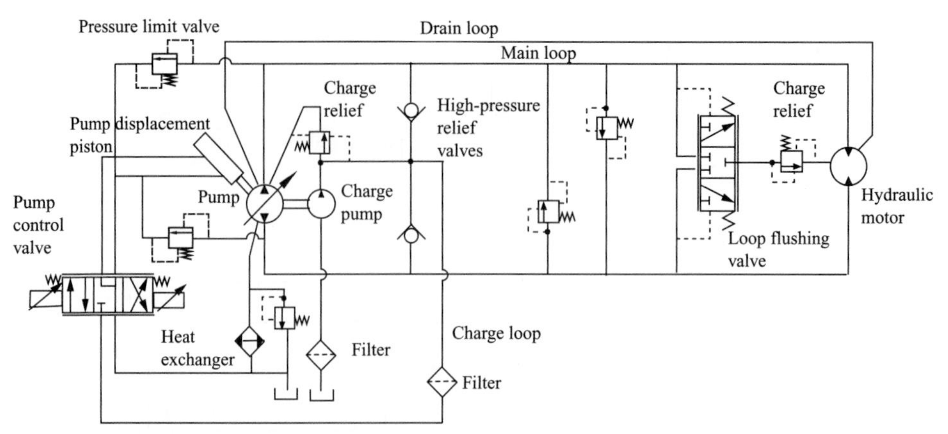


**FIGURE 7.13:** A typical closed-circuit hydrostatic transmission circuit. One of the key features of hydrostatic circuits is that there is no metering valve. Flow rate (speed) is controlled by variable displacement pumps or motors.

is provided by the charge pump. The control of pump displacement is performed by the control valve which is actuated based on the control objective. In closed-circuit hydraulic circuits, ideally there is no leakage or need for replenishment fluid. In reality, it is desirable to have a certain amount of leakage. It is through the leakage and the replenishment circuits that fluid under high pressure is removed from the main loop, cooled by heat exhanger, cleaned by filters and reinjected back into the main loop. Therefore, the leakage and charge pump replenishment is an essential part of all practical closed-circuit hydraulic systems. In fact, leakage rate is planned into the circuit design and regulated at a planned rate by a *loop flushing valve*. A typical system also includes various pressure relief valves between different sides of hydraulic lines.

The pump is a variable displacement type bidirectional pump. By changing the direction of fluid flow, the direction of speed of the hydraulic motor is changed. By controlling the pump, the flow rate, hence the speed of the hydraulic motor, is controlled. Similarly, if the pump is controlled to maintain a certain pressure, the torque output at the hydraulic motor is controlled.

A hydraulic system can be categorized in terms of the power source used to control the hydraulic valve. That is the actuation power used to move the spool. There are four major valve actuation methods (Fig. 7.14),

1. Mechanically direct actuated valve
2. Electrically direct actuated valve
3. Mechanical plus pilot hydraulics actuated valve
4. Electrical plus pilot hydraulics actuated valve

The focus in the classification is the source of actuation power for shifting the valve spool. In mechanically controlled valves, the operator pushes a lever which is connected to the main spool via a mechanical linkage. This is used in relatively small size mobile equipment applications. The same concept is extended to large-size machine applications by using an intermediate pilot pressure circuit to actuate the main valve. The motion of the pilot valve is still controlled by direct mechanical motion of the lever. The pilot valve is essentially a proportional pressure reducing valve. It has a constant pilot pressure supply port and a tank port. The output pilot pressure is proportional to the lever displacement at a value between the tank port pressure and pilot supply port pressure. The main valve spool is shifted in proportion to the output pilot pressure.

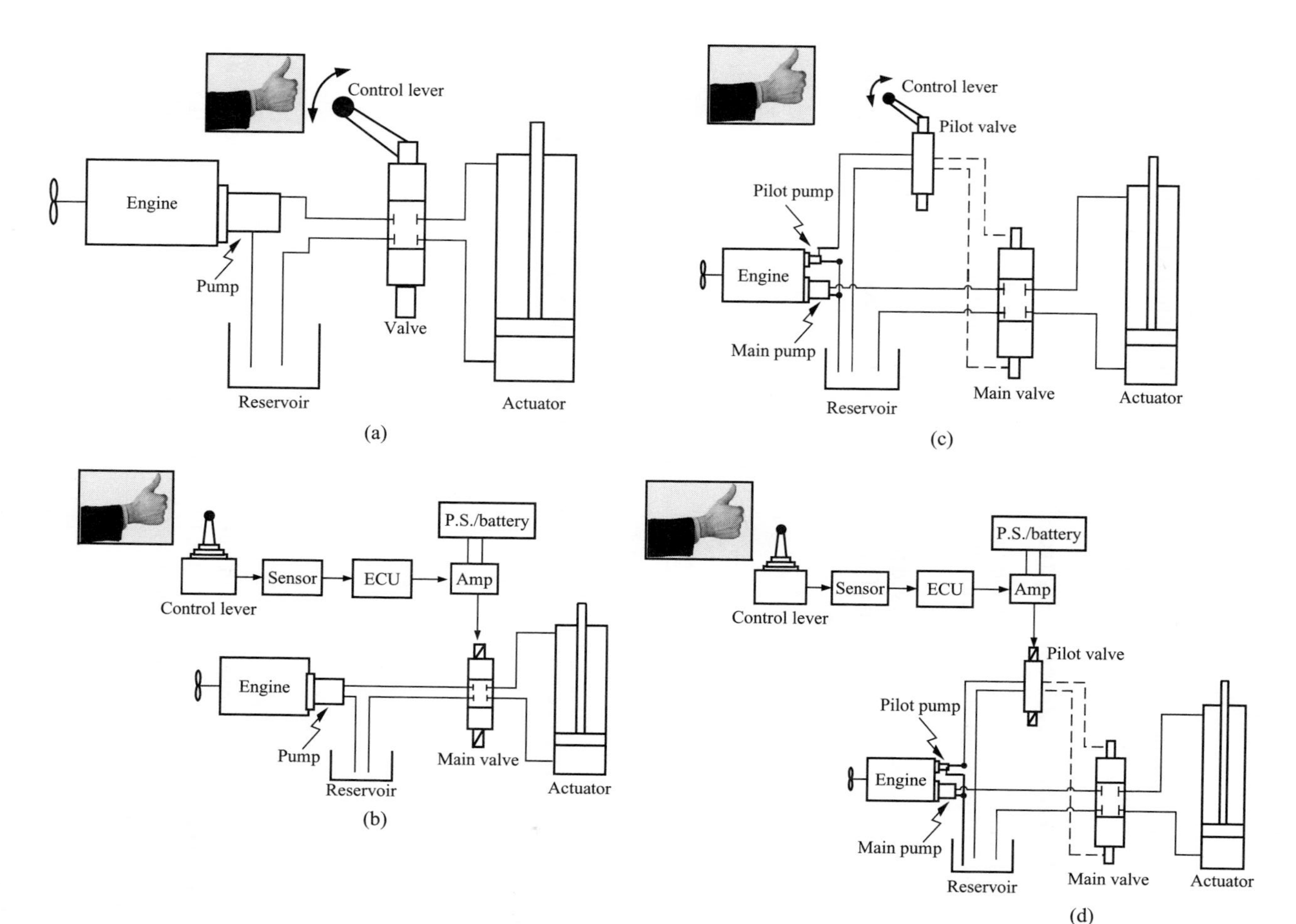


**FIGURE 7.14:** Hydraulic valve control methods: (a) mechanically direct actuated valve (single-stage valve), (b) electrically direct actuated valve (single-stage valve), (c) mechanical plus pilot hydraulics actuated valve (two-stage valve), and (d) electrical plus pilot hydraulics actuated valve (two-stage valve).

Finally, if the spool displacement power comes from the electromagnetic force which is generated by an electric current in the solenoid winding, then it is an electrically controlled valve. Such hydraulic systems are called electrohydraulic (EH) systems. The EH valves may be single stage for small power applications or multistage for large power applications. In multistage EH valves, the electrical current in the solenoid moves the first-stage valve spool (pilot spool), and then the motion of the pilot spool is amplified via a pilot output pressure line to move the main valve spool. In short, pilot-actuated multistage valves may be controlled by the mechanical connection to the control lever (mechanically controlled multistage hydraulic valve [Fig. 7.14(b)]) or by an electrical actuator (electrically controlled (EH) multistage valve [Fig. 7.14(d)]).

### 7.1.1 Fundamental Physical Principles

In this section we briefly review the fundamental principles of the physics that govern hydraulic circuits. Namely, we discuss *Pascal's law* and *Bernoulli's equation.*

***Pascal's Law*** Pascal observed that a confined fluid transmit pressure at all directions in equal magnitude. Implication of this observation is illustrated in Fig. 7.15. The pressure on both sides of the hydraulic circuit is equal to each other.

$$p_1 = p_2 \tag{7.6}$$

$$\frac{F_1}{A_1} = \frac{F_2}{A_2} \tag{7.7}$$

$$F_2 = \frac{A_2}{A_1} \cdot F_1 \tag{7.8}$$

By using a smaller weight in side one, we can move a much larger weight in side two. The circuit acts like a force multiplier. Conservation of energy principle also requires that

$$Energy\ in = Energy\ out \tag{7.9}$$

$$F_1 \cdot \Delta x_1 = F_2 \cdot \Delta x_2 \tag{7.10}$$

$$F_1 \cdot \Delta x_1 = F_1 \cdot \frac{A_2}{A_1} \cdot \Delta x_2 \tag{7.11}$$

$$A_1 \cdot \Delta x_1 = A_2 \cdot \Delta x_2 \tag{7.12}$$

$$\Delta x_2 = \frac{A_1}{A_2} \cdot \Delta x_1 \tag{7.13}$$

which confirms the conservation of fluid volume. As an analogy, this hydraulic circuit acts like a gear reducer. That is, neglecting the loss of energy due to friction, the energy is conserved (input energy equal to output energy) and the force is increased while displacement is reduced.

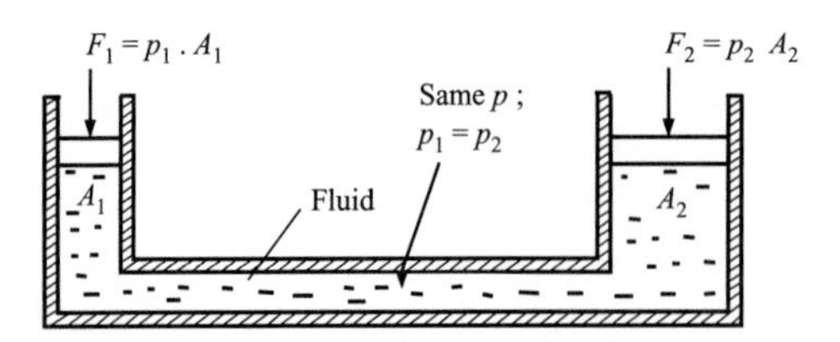


**FIGURE 7.15:** Fundamental principles of fluid flow in hydraulic circuits: Pascal's law.

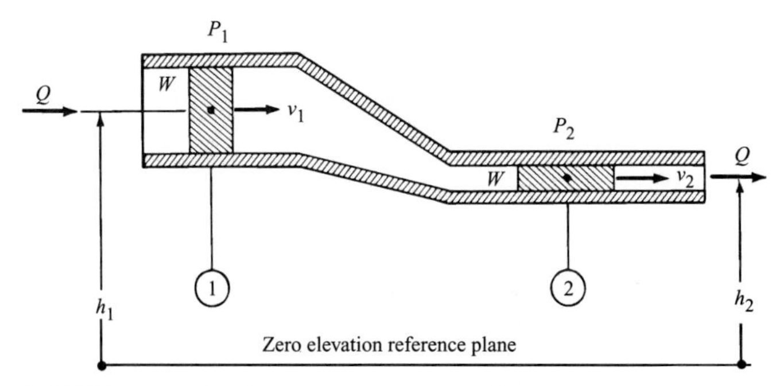


**FIGURE 7.16:** Fundamental principles of fluid flow in hydraulic circuits: conservation of mass and energy (Bernoulli's equation).

***Conservation of Mass: Continuity Equation*** The principle of convervation of mass in hydraulic circuits is used between any two cross-sectional points (Fig. 7.16). Assume that the mass of the fluid in the volume between the two points does not change. Then, the net incoming fluid mass per unit time is equal to the net outgoing fluid mass per unit time,

$$\dot{m}_{in} = \dot{m}_{out} \tag{7.14}$$

$$Q_1 \cdot \rho_1 = Q_2 \cdot \rho_2 \tag{7.15}$$

In hydraulic circuits, the compressibility of the fluid is commonly neglected. If the compressibility of the fluid is negligible, then $\rho_1 = \rho_2$. If so, the continuity equation for a noncompressible fluid flow between two cross sections is

$$Q_1 = Q_2 \tag{7.16}$$

***Conservation of Energy: Bernoulli's Equation*** Consider any two cross-sectional points in a hydraulic circuit Fig. 7.16. It can be shown that the conservation of energy principle leads to the Bernoulli's equations. If there is no energy added or taken from the flow and frictional losses are neglected, the Bernoulli's equation can be expressed in its standard form as

$$h_1 + \frac{p_1}{\gamma_1} + \frac{v_1^2}{2g} = h_2 + \frac{p_2}{\gamma_2} + \frac{v_2^2}{2g} \tag{7.17}$$

where $h_1, h_2$ are the nominal heights of the fluid from a reference point, $p_1, p_2$ are the pressures, $v_1, v_2$ are the velocities, $\gamma_1, \gamma_2$ are the weight densities at points 1 and 2, and $g$ is the gravitational acceleration ($9.81 \text{ m/s}^2 = 386 \text{ in/s}^2$). If the fluid compressibility is neglected, then $\gamma_1 = \gamma_2$.

If there is energy added between the two points (i.e., by a pump), $E_p$, or energy taken (i.e., by an motor), $E_m$, and if we take energy losses due to friction into account as $E_l$, the Bernoulli's equation can be modified as

$$h_1 + \frac{p_1}{\gamma_1} + \frac{v_1^2}{2g} + E_p - E_m - E_l = h_2 + \frac{p_2}{\gamma_2} + \frac{v_2^2}{2g} \tag{7.18}$$

where the added or removed energy terms ($E_p, E_m, E_l$) are energy units per unit weight of the fluid. Energy unit is equal to force times the displacement. Energy per unit weight is obtained by dividing that unit by weight ($weight = m \cdot g$) which has force units. Hence, the $E_p, E_m, E_l$ have effective units of length. The energy stored in the fluid has three components: kinetic energy due to its speed, potential energy stored due to its pressure, and potential energy stored due to gravitational energy due to the elevation of the fluid. Quite

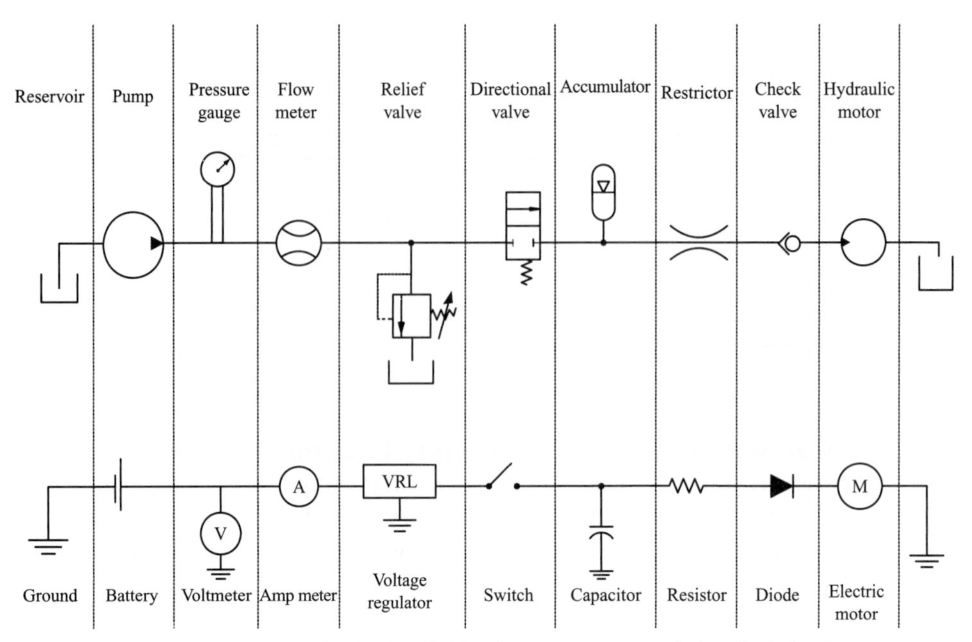


**FIGURE 7.17:** Analogy between hydraulic circuit components and electrical circuit components.

often, the change in energy due to the change in height between two points is negligible and the $h_1$ and $h_2$ terms are dropped from the above equations in hydraulic circuit analysis. It should be noted that the Bernoulli's equation is simply the expression of the principle of energy conservation for fluid systems.

### 7.1.2 Analogy Between Hydraulic and Electrical Components

The analogy between the components used in a hydraulic circuit and an electrical circuit is shown in Fig. 7.17. Notice the fact that for majority of electrical circuit components, there is a corresponding hydraulic component for the same functionality. The electrical circuits have voltage, current, resistors, capacitors, inductors, and diodes. The hydraulic circuits have pressure, flow, orifice restriction, accumulator, and check valves. However, it should be noted that there are three main differences between hydraulic and electrical circuits,

1. Flow–pressure $(P, Q)$ relationship is nonlinear, whereas current–voltage $(i, V)$ relationship is linear,

$$Q = K(x_s)\sqrt{\Delta P} \tag{7.19}$$

$$i = (1/R)\,V \tag{7.20}$$

   where $x_s$ is the displacement of the valve spool, $K(x_s)$ represents the effective orifice area and discharge coefficient function as function of $x_s$, $\Delta P$ is the pressure differential across the valve, $R$ is electrical resistance, and $V$ is voltage potential.

2. The medium (fluid or electrons) of power transmission is compressible in hydraulic systems, and its properties vary among different fluids.

3. Voltage is a relative quantity, there is no absolute zero voltage. However, there is absolute zero pressure, which is the vacuum condition.

The capacitor ($C$) and accumulator ($C_H$) analogy is as follows:

$$V = \frac{1}{C}\int i(t)\cdot dt \qquad p = \frac{1}{C_H}\int Q(t)\cdot dt \tag{7.21}$$

where $C_H$ is determined by the volume and stiffness of the accumulator, which has the units of *Length*$^5$/*Force* (i.e., $m^5$/Nt as a result of *volume*/*pressure* $= [m^3]/[\text{Nt}/m^2]$). The capacitance of a hydraulic circuit is defined as the ratio of volumetric change to pressure change.

$$\frac{dp(t)}{dt} = \frac{1}{C_H}\cdot Q(t) \tag{7.22}$$

$$C_H = \frac{Q(t)}{dp(t)/dt} \tag{7.23}$$

$$= \frac{(dV/dt)}{(dp/dt)} \tag{7.24}$$

$$= \frac{dV}{dp} \tag{7.25}$$

For a given volume of fluid, $V$, and bulk modulus , $\beta$, capacitance is

$$C_H = \frac{V}{\beta} \tag{7.26}$$

The bulk modulus $\beta$ represent the compressibility of the fluid volume,

$$\beta = -\frac{dp}{dV/V} \tag{7.27}$$

where the negative sign indicates that the volume gets smaller as the pressure increases. The bulk modulus has the same units as pressure. It is an indication of the stiffness of the fluid. The higher the bulk modulus of the fluid, the less compressible it is. It is also a function of nominal pressure and temperature. Nominal value of bulk modulus for hydraulic fluids is around $\beta = 250{,}000$ psi.

Resistance ($R$) to current and hydraulic resistance ($R_H$) to fluid flow are similar. The $R_H$ is a function of orifice area opening and its geometry. In the above flow-pressure relationship, the $K(x_s)$ is equivalent inverse resistance to flow and function of the orifice opening ($x_s$).

The inertia of moving fluid in a pipe acts like an inductor. Consider a pipe with cross-sectional area $A$, length $l$, mass density of fluid $\rho$, pressures at the two ends $p_1$, $p_2$, and flow rate $Q$ through it. Assuming the following:

1. The friction in the pipe is neglected, and
2. Compressibility of the fluid in the pipe is neglected,

the motion of the fluid mass in the pipe can be described as

$$(p_1 - p_2)\cdot A = m\cdot \ddot{x} \tag{7.28}$$

$$= (\rho\cdot l\cdot A)\cdot \frac{\dot{Q}}{A} \tag{7.29}$$

where $m = \rho \cdot V = \rho \cdot l \cdot A$, and $Q = \dot{x} \cdot A$. The pressure and flow relationship is

$$p_1 - p_2 = \left(\frac{\rho \cdot l}{A}\right) \cdot \dot{Q} \tag{7.30}$$

$$p_1 - p_2 = L \cdot \dot{Q} \tag{7.31}$$

where the hydraulic inductance is defined as $L = \rho \cdot l/A$. Notice the analogy between pressure, flow rate, and hydraulic inductance versus the voltage, current, and self-inductance,

$$\Delta p(t) = L \cdot \frac{dQ(t)}{dt} \tag{7.32}$$

$$\Delta V(t) = L \cdot \frac{di(t)}{dt} \tag{7.33}$$

Similarly, the diode and check valve analogy is that they allow flow of electricity and fluid in one direction, and block in the opposite direction,

$$i_o = i_i \quad V_i \geq V_o \tag{7.34}$$

$$= 0.0 \quad V_i < V_o \tag{7.35}$$

$$Q_o = Q_i \quad p_i \geq p_o \tag{7.36}$$

$$= 0.0 \quad p_i < p_o \tag{7.37}$$

where $i_i$, $i_o$ are input and output current, $V_i$, $V_o$ are input and output voltage, respectively. Similar notation applies for $Q_o$, $Q_i$, $p_i$, $p_o$ for flow rate and pressure.

***Example*** Consider a hydraulic fluid with bulk modulus of $\beta = 250{,}000$ psi, a nominal volume of $V_0 = 100$ in$^3$. If the fluid is compressed from atmospheric pressure level to 2500 psi level, find the change in the fluid volume.

Since

$$\beta = -\frac{dp}{dV/V_0} \tag{7.38}$$

then

$$dV = -\frac{V_0 \cdot dp}{\beta} \tag{7.39}$$

$$= -\frac{100 \cdot 2500}{250{,}000} \tag{7.40}$$

$$= -1 \text{ in}^3 \tag{7.41}$$

$$= 1\% V_0 \tag{7.42}$$

which shows that a typical hydraulic fluid will contract about 1% in volume under a 2500 psi increase in pressure. The negative sign indicates that the volumetric change is a decrease.

***Example***[1] Consider the hydraulic circuit shown in Fig. 7.18. The dimensions of the pipe are shown in the figure, $d = 20$ mm, $l = 10$ m. Assume that there is a constant flow rate in steady state, $Q_0 = 120$ liters/min $= 2$ liters/s $= 0.002$ m$^3$/s. The mass density of fluid is $\rho = 1000$ kg/m$^3$. Then, the valve closes suddenly, over a period of $\Delta t = 10$ msec, which

[1] Courtesy of Dr. Daniele Vecchiato.

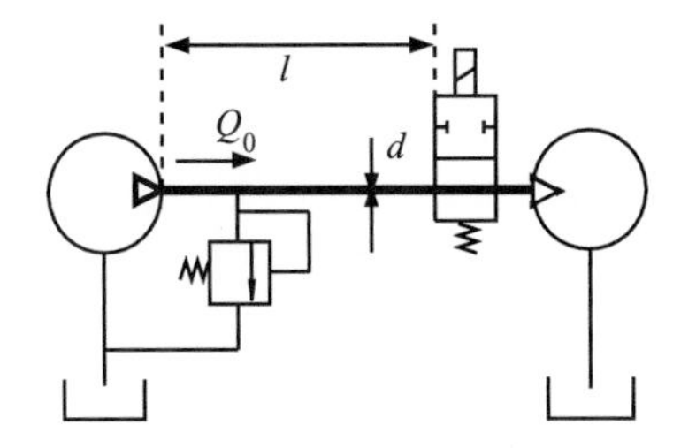


**FIGURE 7.18:** A hydraulic circuit where sudden closure of the valve results in pressure spikes due to the inertia of the moving fluid.

results in pressure spike. This phenomenon is also known as *water hammering*. Assume that the relief valve does not open.

The change in pressure due to the inertial deceleration of the fluid due to sudden closure of the valve is determined by

$$\Delta p = p_1 - p_2 = L\dot{Q} \tag{7.43}$$

$$= \frac{\rho \cdot l}{A} \frac{Q_0}{\Delta t} \tag{7.44}$$

$$= \frac{1000 \cdot 10}{\pi (0.02)^2/4} \frac{0.002}{0.01} \tag{7.45}$$

$$= \frac{0.2}{\pi} 10^8 \ [\text{Nt/m}^2] \tag{7.46}$$

$$= \frac{20}{\pi} \ [\text{MPa}] = 6.36 \ [\text{MPa}] \tag{7.47}$$

Notice that, in this example we neglected the pressure change due to the compressibility of the fluid. A long pipe (inductor or resistor equivalent) and accumulator (capacitor equivalent) pair can be used as RLC filter in hydraulic circuits to damp out pressure oscillations in pipes.

### 7.1.3 Energy Loss and Pressure Drop in Hydraulic Circuits

Any time there is a pressure drop and flow in a hydraulic circuit, there is energy loss. The lost energy is converted to heat at the loss point. Hydraulic power is equal to pressure difference times the flow rate between two points. Hence, energy differential between two points over a period of time is pressure differential times the total flow between the two points. In a typical hydraulic circuit, hydraulic fluid leaves the pump at a high pressure. There are pressure drops at the transmission pipes, valves, and actuators [Fig. 7.19 (a)]. This means energy is lost during the transmission of fluid from pump to the load. In order to maximize the utilization of hydraulic power, hence increase the efficiency of the hydraulic system, pressure drop between the source (pump) and load should be minimized.

$$Power_{12} = \Delta P_{12} \cdot Q_{12} \tag{7.48}$$

$$Energy_{12} = \int_{t_1}^{t_2} \Delta P_{12}(t) \cdot Q_{12}(t) \cdot dt \tag{7.49}$$

Pressure drop along hydraulic pipes is function of the following parameters:

1. Viscosity of the fluid, which is highly function of temperature
2. Pipe diameter
3. Pipe length

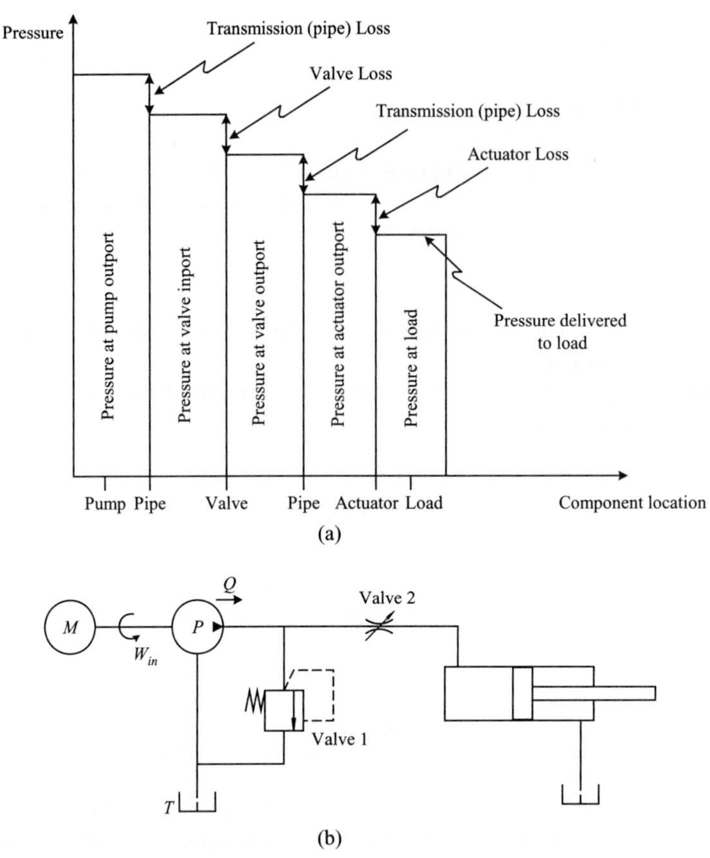


**FIGURE 7.19:** Hydraulic pressure and power loss in a hydraulic circuit: (a) pressure drop across different components in a hydraulic circuit, (b) an example circuit where pressure drop in each valve results in power loss in the form of heat.

4. Number of turns and bend in the pipe circuit
5. Flow rate

Manufacturers provide empirical data tables for pressure drop estimation as a function of above parameters.

***Example*** Consider the example hydraulic circuit shown in Fig. 7.19(b). Assume that the pump input speed ($w_{in}$) is constant, and hence the output flow is constant, $Q_p =$ 120 liters/min = 2 liters/sec. The pressure relief valve (valve 1) is set to a constant pressure value, $p_{relief} = 20$ MPa $= 20 \times 10^6$ N/m$^2$. Assume that the flow metering valve (valve 2) is sized to handle a maximum of 1/2 of total pump flow, $Q_{v,max} = 0.5 \cdot Q_p$. Let us assume that the cylinder head-end cross-sectional area is $A_{he} = 0.01$ m$^2$ and the load force is $F_l = 100{,}000$ N. Determine the total heat loss rate at the two valves under this operating conditions in steady state.

The heat loss rate is the pressure drop times the flow rate across each valve. Since main flow valve (valve 2) can handle only half of the pump flow, the other half has to go through the pressure relief valve (valve 1).

$$Q_{v1} = 0.5 \cdot Q_p = 1.0 \text{ liter/sec} = 10^{-3} \text{ m}^3\text{/sec} \tag{7.50}$$

$$Q_{v2} = Q_{v1} \tag{7.51}$$

The pressure drop across valve 1 is the same as the pump outlet pressure since the output port of the valve 1 is connected to the tank. The pressure drop across the valve 2 is the input pressure ($p_s$) minus the load pressure, $p_l$.

$$\Delta p_{v1} = p_s - p_t = p_s = 20 \times 10^6 \text{ N/m}^2 \tag{7.52}$$

$$\Delta p_{v2} = p_s - p_l = 20 \times 10^6 - \frac{10{,}0000}{0.01} = 10 \times 10^6 \text{ N/m}^2 \tag{7.53}$$

The total heat loss rate is

$$P_{loss} = p_{v1} \cdot Q_{v1} + p_{v2} \cdot Q_{v2} \tag{7.54}$$

$$= 20 \times 10^6 \cdot 1.0 \times 10^{-3} \text{ N/m}^2 \cdot \text{m}^3/\text{sec} + 10 \times 10^6 \cdot 1.0 \times 10^{-3} \text{ N/m}^2 \cdot \text{m}^3/\text{sec} \tag{7.55}$$

$$= 20{,}000 \text{ W} + 10{,}000 \text{ W} \tag{7.56}$$

$$= 30{,}000 \text{ W} \tag{7.57}$$

which indicates that in this case the heat loss over the relief valve is twice as much as the heat loss over the main flow valve. Since the total power output of pump is

$$P_{pump} = p_s \cdot Q_p = 20 \times 10^6 \cdot 2.0 \times 10^{-3} \text{ N/m}^2 \cdot \text{m}^3/\text{sec} \tag{7.58}$$

$$= 40{,}000 \text{ W} \tag{7.59}$$

the efficiency of the overall hydraulic circuit is 25%. Only 25% of the hydraulic power generated by the pump is delivered to the load.

$$\begin{aligned} P_{mech} &= \text{F} \cdot \dot{\text{X}} \\ &= 100 \text{ KN} \cdot (10^{-3} \text{ m}^3/\text{s}/0.01 \text{ m}^2) \\ &= 10{,}000 \text{ Nm/s} = 10{,}000 \text{ W} = 10 \text{ KW} \end{aligned}$$

The rest is wasted at various pressure drop components. This example also illustrates one of the drawbacks of fixed displacement pumps. It has low efficiency. Since pump outputs flow at a constant rate as function of input shaft speed regardless of the hydraulic flow demand, unused flow must be dumped (wasted) over the relief valve.

Figure 7.20 shows the list of ANSI/ISO standard symbols used in describing hydraulic components in circuits. Readers should get familiar with the symbols in order to comfortably interpret hydraulic circuits.

## 7.2 HYDRAULIC PUMPS

The functional block diagram and operating principle of a pump is shown in Fig. 7.21(a–d). The pump is the device used to convert mechanical power to hydraulic power. A positive displacement pump concept is shown in Fig. 7.21(b–d). During the in-stoke $p_3 < p_1$, oil is sucked-in from the "tank." During the out-stroke $p_3 \geq p_2$, oil is pushed out to the load. Notice that $p_3$, pump output pressure, is determined by the load pressure $p_3 \approx p_2 = p_{load} + p_{spring}$. If there is no load resistance, then the pump cannot build up pressure. In a hydraulic system including pump-valve-cylinder-tank, the pressure difference between the pump outlet port and tank is determined by the pressure drop on the valve and the load pressure created by the cylinder and load interaction. In this concept figure, the check valves control the direction of the flow. The line relief valve [Fig. 7.21(d)] limits the maximum allowed line pressure as protection and returns the flow back to the tank if the line pressure tries to exceed a set limit. Hence, the relief valve ensures that the line pressure stays less than or equal to maximum relief pressure set, $p_3 \leq p_{max}$. Notice that during the in-stroke, the volume is expanding. Similarly, during the out-stroke, volume is

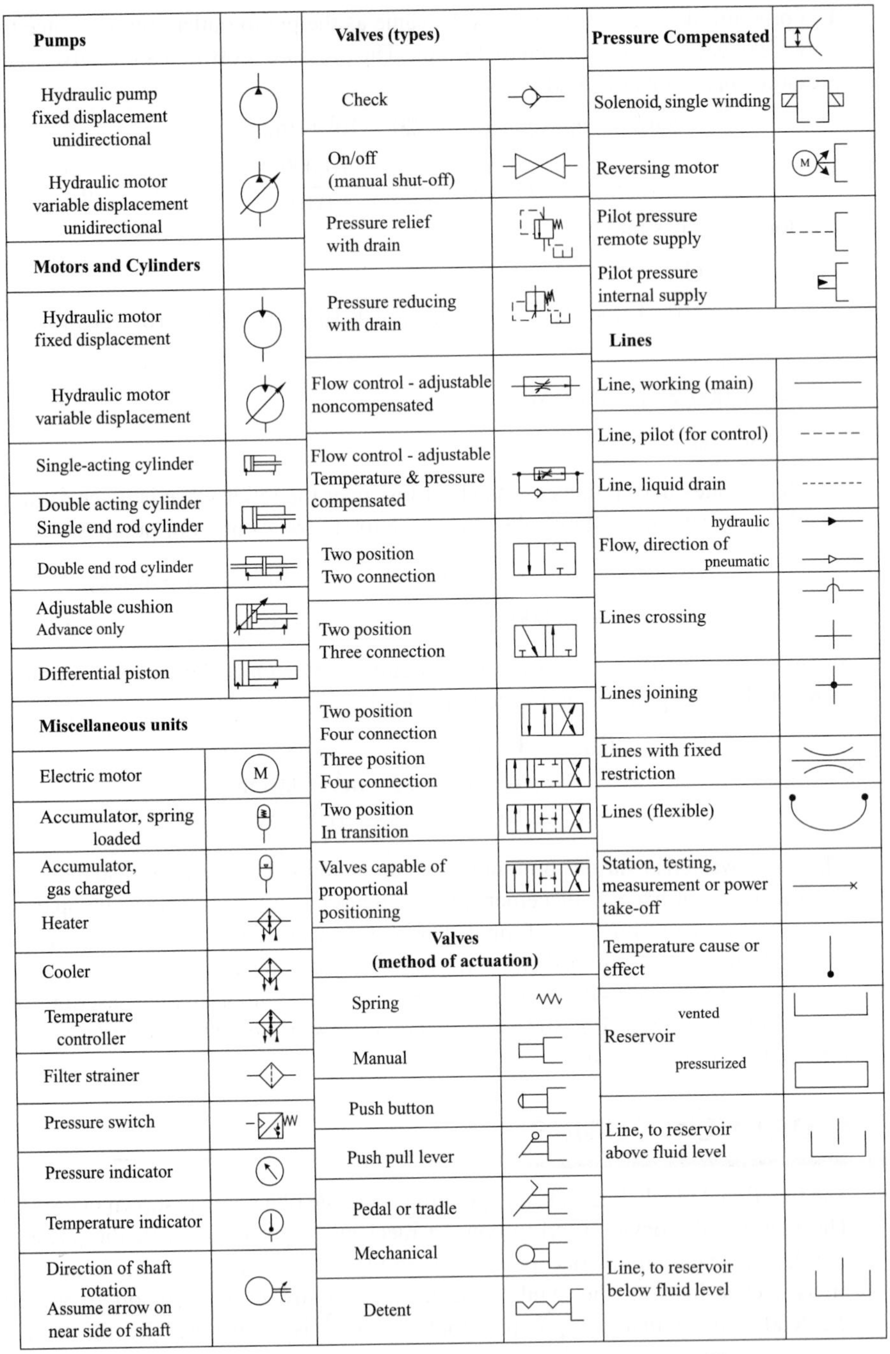

| Pumps | |
|---|---|
| Hydraulic pump fixed displacement unidirectional | |
| Hydraulic motor variable displacement unidirectional | |
| **Motors and Cylinders** | |
| Hydraulic motor fixed displacement | |
| Hydraulic motor variable displacement | |
| Single-acting cylinder | |
| Double acting cylinder Single end rod cylinder | |
| Double end rod cylinder | |
| Adjustable cushion Advance only | |
| Differential piston | |
| **Miscellaneous units** | |
| Electric motor | M |
| Accumulator, spring loaded | |
| Accumulator, gas charged | |
| Heater | |
| Cooler | |
| Temperature controller | |
| Filter strainer | |
| Pressure switch | |
| Pressure indicator | |
| Temperature indicator | |
| Direction of shaft rotation Assume arrow on near side of shaft | |

| Valves (types) | |
|---|---|
| Check | |
| On/off (manual shut-off) | |
| Pressure relief with drain | |
| Pressure reducing with drain | |
| Flow control - adjustable noncompensated | |
| Flow control - adjustable Temperature & pressure compensated | |
| Two position Two connection | |
| Two position Three connection | |
| Two position Four connection | |
| Three position Four connection | |
| Two position In transition | |
| Valves capable of proportional positioning | |
| **Valves (method of actuation)** | |
| Spring | |
| Manual | |
| Push button | |
| Push pull lever | |
| Pedal or tradle | |
| Mechanical | |
| Detent | |

| Pressure Compensated | |
|---|---|
| Solenoid, single winding | |
| Reversing motor | M |
| Pilot pressure remote supply | |
| Pilot pressure internal supply | |
| **Lines** | |
| Line, working (main) | |
| Line, pilot (for control) | |
| Line, liquid drain | |
| Flow, direction of: hydraulic / pneumatic | |
| Lines crossing | |
| Lines joining | |
| Lines with fixed restriction | |
| Lines (flexible) | |
| Station, testing, measurement or power take-off | |
| Temperature cause or effect | |
| Reservoir: vented / pressurized | |
| Line, to reservoir above fluid level | |
| Line, to reservoir below fluid level | |

**FIGURE 7.20:** ANSI/ISO standard symbols for hydraulic components. These component symbols are used in the schematic circuit design of hydraulic systems.

contracting. This phenemenon provides the suction and pumping action for the pump. The volume spanned by the piston per stroke of the pump is called the "pump displacement per stroke." If that volume is fixed (not adjustable) for a given pump, the pump is called a *fixed displacement pump*. If that volume is adjustable, the pump is called a *variable displacement pump*. If the pump is the fixed displacement type, flow volume pumped per stroke is fixed.

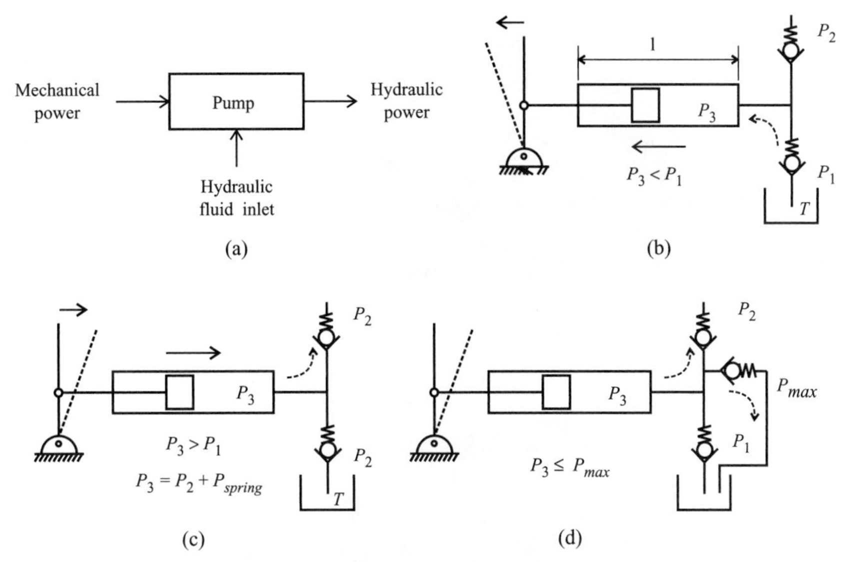


**FIGURE 7.21:** Concept of a positive displacement pump: (a) pump input–output block diagram, (b) in-stroke (suction), (c) out-stroke (pumping), and (d) pressure-limiting valve.

Therefore, the only way to control the flow rate is to control the rate of the stroke (mechanical input speed). If the pump is variable displacement type, the flow rate can be controlled by either input speed and/or by adjusting the displacement of the pump per stroke.

## 7.2.1 Types of Positive Displacement Pumps

Different types of positive displacement pumps are briefly discussed below (Fig. 7.22) and the differences among them are pointed out.

***(i) Gear Pumps*** Gear pumps work by the principle of carrying fluid between the tooth spaces of two meshed gears (Fig. 7.23). One of the two shafts is the drive shaft, called drive gear, and the other, driven by the first, is called driven or idler gear. The pumping chambers are formed by the spaces between the gears, enclosed by the pump housing and the side plates. The gear pump is always unbalanced by design because the high-pressure side pushes the gear toward the low-pressure side. The relation between the driven and driving shaft causes unbalance that will have to be taken up by the bearings. Gear pumps are always of fixed displacement type. Despite these drawbacks, gear pumps are popular because of their simplicity, low cost, and robustness. The unmeshing of gears around the inlet port creates

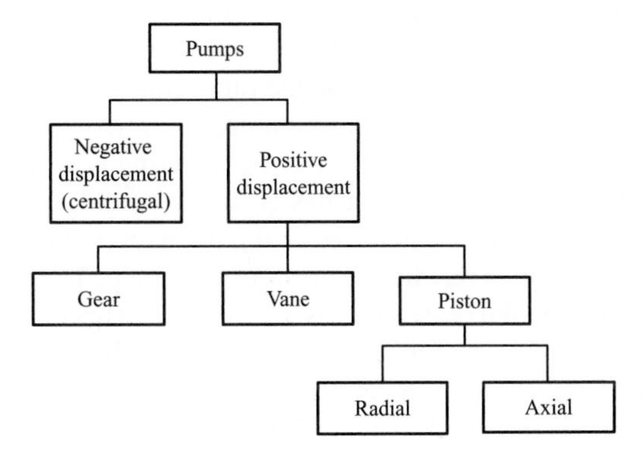


**FIGURE 7.22:** Hydraulic pump categories: positive displacement pump types—gear, vane, and piston (radial and axial).

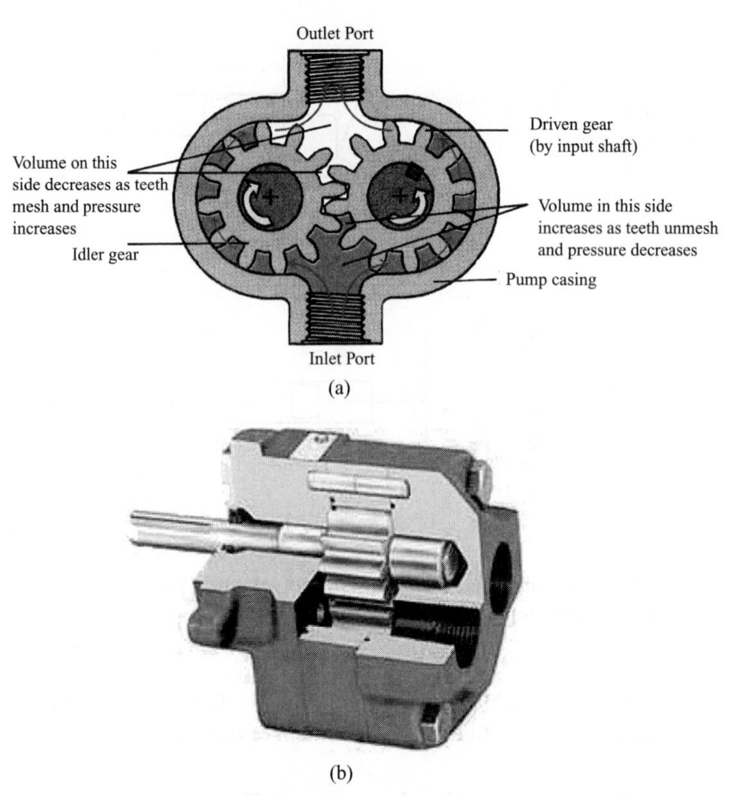


**FIGURE 7.23:** Gear pump cross-sectional views: (a) two-dimensional cross-sectional view; (b) three-dimensional cutaway view.

a low pressure and results in the suction action. Similarly, the squeezing action of the gears on the fluid between the teeth and pump housing as well as when the meshing between teeth begins around the outlet port results in the increased pressure at the outlet port. Ideally, the flow rate and pressure relationship is constant. The flow rate is determined by the input shaft speed of the pump and its fixed displacement. The pressure is determined by the load as long as it is not too high to the point of stalling the input shaft or breaking the pump housing. However, as the pressure increases, the leakage increases. Hence, the flow rate versus pressure curve drops down a little as output pressure increases.

***(ii) Vane Pumps*** Vane pumps are known to operate more quietly than gear pumps and used to have a cost benefit over piston pumps. More recent development has made piston pumps both quieter and cheaper so they now totally dominate the market for mobile hydraulics. A variable displacement vane pump is shown in Fig. 7.24. It works on the principle that oil is taken in on the side with an increasing volume and the vanes connected to the rotor then push the oil into the part of the pump with decreasing volume, therefore increasing the pressure of the oil before it reaches the outlet port. The so-called cam ring is eccentrically mounted compared to the rotor of the pump, making the design unbalanced. Changing the position of the cam ring in relation to the rotor and hence changing the ratio between the volumes of the inlet and outlet sides change the displacement of the pump. The cam ring rests on a control piston on one side, and the other side has a bias piston. By changing the pressure in the control piston (Fig. 7.24), the displacement of the pump is changed. Control piston is used to adjust the pump displacement. The maximum displacement is set by the displacement volume adjustment screw that limits the travel of the bias piston.

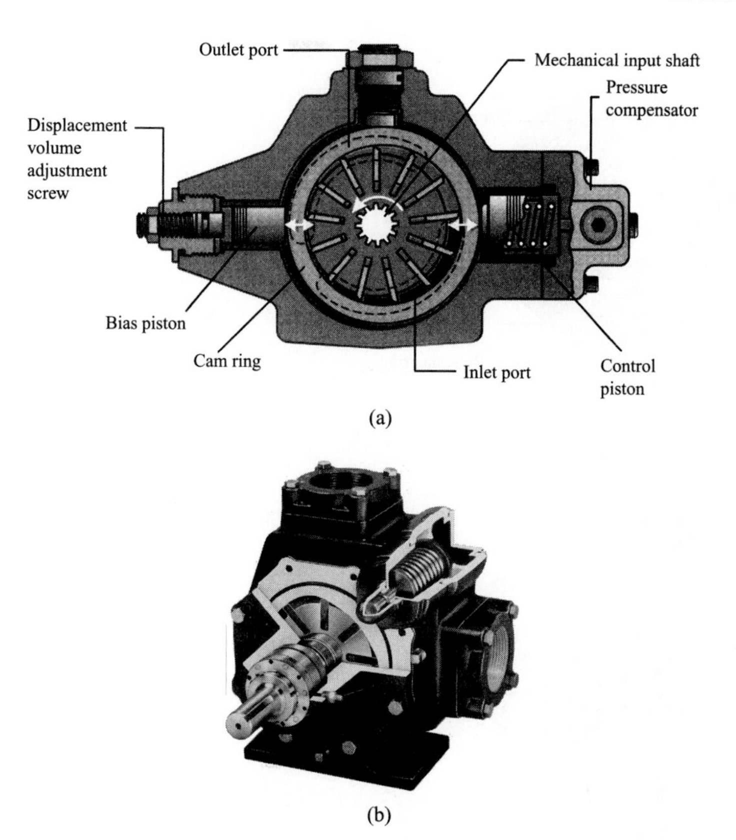


**FIGURE 7.24:** Vane pump: (a) cross section, (b) three-dimensional cutaway picture.

***(iii) Radial Piston Pumps*** A radial piston pump (Fig. 7.25) is built on a rotating shaft with the cylinder block rotating on the outside. As the piston follows the outer housing of the pump on slippers, the offset from the central position creates the pumping motion. The pistons maintain contact with the outer ring at all times. The input shaft rotation axis is perpendicular to the plane of axes along which the pistons slide. The porting in the pivot pin allows intake at low pressure into the cylinder (from the inner diameter side) as the cylinder block passes by and pressurizes outflow as the cylinder block passes the outlet port (through the inner diameter side). The number of pistons, diameter of the pistons, and the length of their stroke determine pump displacement. The piston and cylinder sleeves are manufactured and custom fitted to high tolerances for each pump. Radial piston pumps are built so that more than one port discharges fluid at the same time and more than one port are open to intake at the same time, allowing the transition between low and high pressure to be smooth. This design attempts to reduce cyclic flow rate and pressure pulsation in the output flow as a function of rotor angular position per revolution. Typical radial piston pumps provide output pressure up to about 350 bar, and flow rates of 200 lpm.

***(iv) Axial Piston Pumps*** Axial piston pumps are the most widely used pump types in mobile hydraulic applications. There are two major types:

- Inline piston pump (Fig. 7.26)
- Bent-axis axial piston pump (Fig. 7.27)

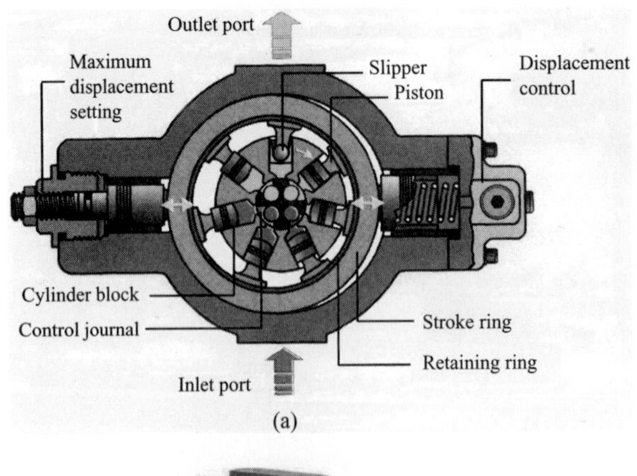

**FIGURE 7.25:** Radial piston pump.

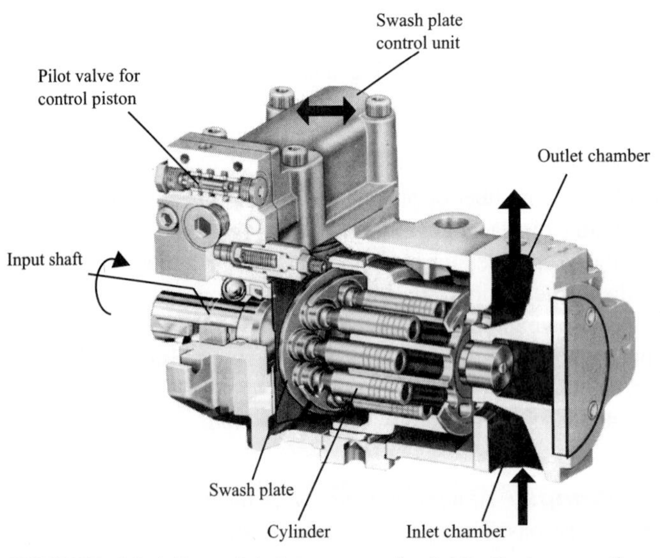


**FIGURE 7.26:** Inline axial piston pump (variable displacement) cross-sectional view.

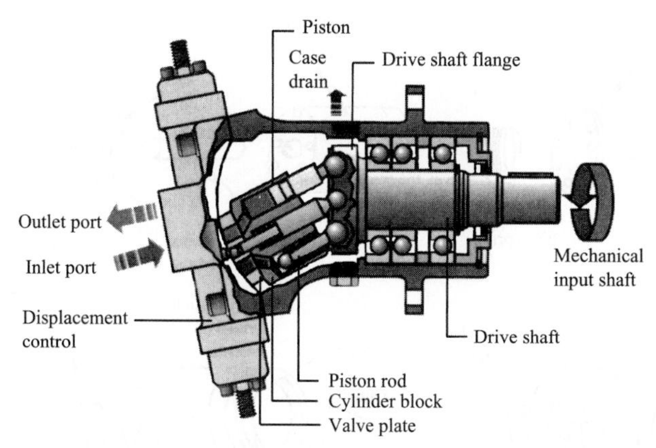


**FIGURE 7.27:** Bent-axis axial piston pump (variable displacement).

The only difference between them is the orientation of the rotation axis of the pistons with respect to the input shaft. In case of the inline piston pump, mechanical input shaft and piston rotation axes are inline, whereas in the case of bent-axis type, they are not inline. Varying the angle of the swash plate in the pump changes the displacement. Typical range of swash plate angle rotation is about 15°. The displacement of the pump is controlled by controlling the swash plate angle between its limits, i.e., between 0° and 15°. The oil is pushed into the pump through the intake side of the port plate, sometimes called kidney plate because of the shape of the ports. The cylinders are filled when they pass the inside area and the oil is pushed out on the other side of the kidney plate. When the cylinders travel along, the swash plate pushes the trapped oil toward the kidney plate and therefore raises the pressure. When the swash plate is perpendicular to the rotation axis, the displacement is zero. Then the pump is in its stand-by position and does not provide any flow. More accuratetly, at stand-by the pump provides flow only to compensate for leakage flow. It is this capability (adjustable displacement) that counts for energy-saving potential of variable displacement pumps by providing adjustable flow rate based on demand.

Axial piston pumps can be fixed displacement or variable displacement, unidirectional or bidirectional, and may have over-center control (Fig. 7.28). In the fixed displacement pump, the swash plate angle is constant. In the variable displacement pump, the swash plate angle can be varied by a control mechanism. Unidirectional pump is intended to be driven by input shaft in either clockwise or counterclockwise direction. If the pump has the *over-center* control capability, which is the ability to change the swash plate angle about its neutral position in either direction, the pump output flow direction can be changed by the over-center control even though input shaft speed direction is the same. A bidirectional pump can be driven by input shaft in either direction, and the flow direction changes with the input speed direction. Analogous operating principle applies for hydraulic motors. In case of motor, the output is the shaft rotation and input is the fluid flow. Unidirectional motor is intended to receive flow in one direction and rotate in one direction only. Bidirectional motors can receive flow in either direction; hence, can provide output shaft rotation in either direction. Over-center control allows the motor output speed direction to be changed even though the input flow direction does not change.

### 7.2.2 Pump Performance

A variable displacement pump provides the most flexibility in return for additional control complexity. In this case, the input speed of the pump can be left to be determined by other

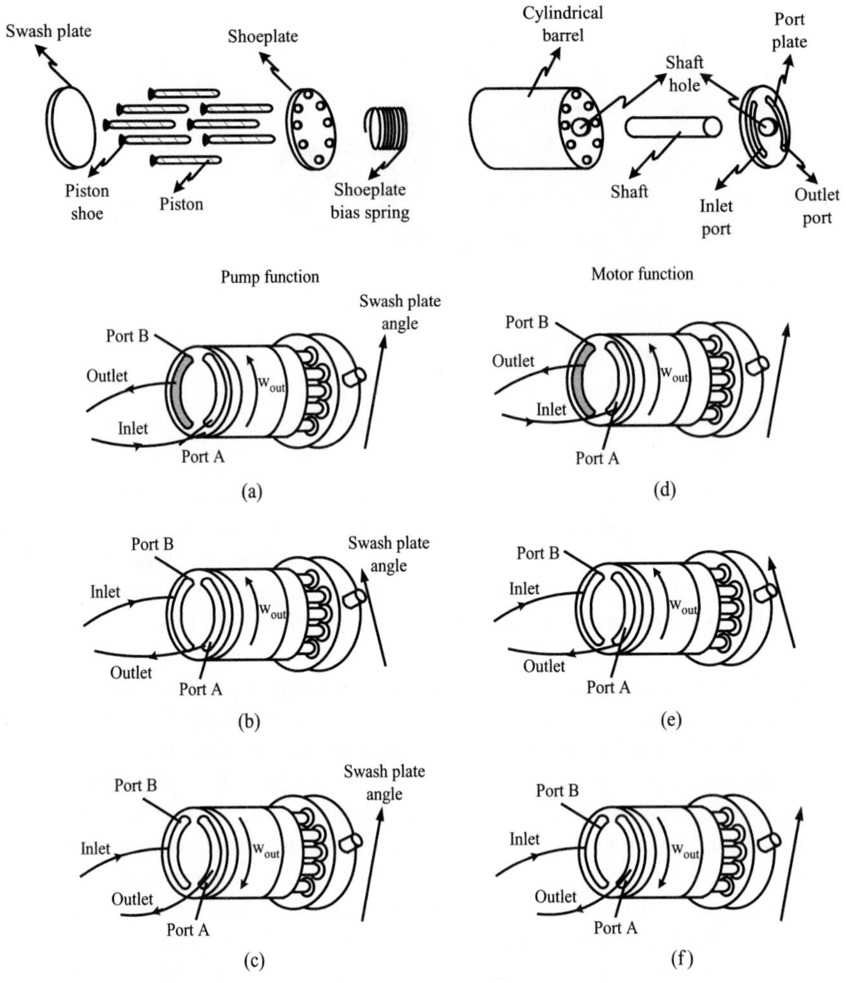


**FIGURE 7.28:** Axial piston pump and motor: components and operating principles. (a) unidirectional pump function, (a) and (b) over-center controlled pump functionality, (a) and (c) bidirectional pump functionality, (d) unidirectional motor function, (d) and (e) over-center controlled motor functionality, and (d) and (f) bidirectional motor functionality.

conditions (i.e., operator may control engine speed for other considerations, i.e., vehicle travel speed), and the desired pump output is controlled by manipulating the swash-plate angle. The swash-plate actuation logic may be based on regulating flow rate, pressure, or other derived variables of interest (Fig. 7.29). Figure 7.30 shows the typical steady-state performance characteristics of a pump in terms of its size and efficiency. The pressure, flow, and power capacity of the pump are directly related to the pump size.

The primary performance measures of a pump are specified by the following parameters:

1. Displacement [in$^3$/rev] (or [cc] per rev which is short for [cm$^3$/rev], 1 liter = $10^3$ cm$^3$) which defines the size of the pump, $D_p(\theta)$
2. Rated flow, $Q_r$, at rated speed and rated pressure
3. Rated pressure, $p_r$

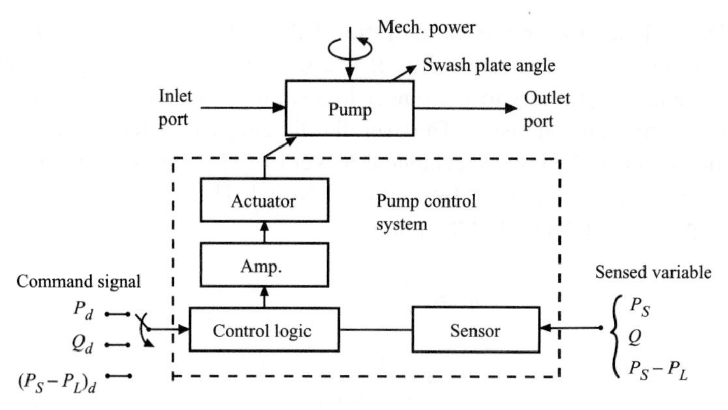


**FIGURE 7.29:** Control system block diagram for a variable displacement pump. The pump can be controlled to regulate the output pressure, flow rate, pressure difference between pump output and load pressures, or other derived variable. The control element is the swash plate angle positioning mechanism of the pump.

4. Rated power (derived quantity from rated pressure and rated flow)
5. Dynamic response bandwidth of the pump displacement control system if the pump is the variable displacement type

The secondary performance specifications include:

1. Efficiency: volumetric, mechanical, and overall
2. Maximum speed
3. Weight
4. Noise level
5. Cost

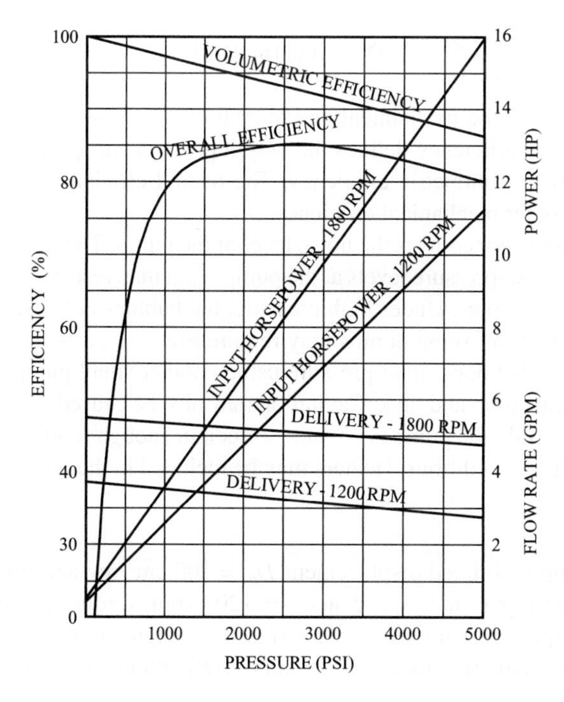


**FIGURE 7.30:** Steady-state performance characteristics of a pump: flow rate, power, and efficiency as a function of its operating pressure range at different input shaft speed.

The volumetric efficiency ($\eta_v$) refers to the ratio of the volume of the output fluid flow to the volumetric displacement of the pump. It is a measure of the leakage in the pump. Mechanical efficiency ($\eta_m$) refers to the power losses due to factors other than leakage, such as friction and mechanical losses. The overall efficiency ($\eta_o$) refers to the ratio of the output hydraulic power to the input mechanical power of the pump. Let the units of each term be as follows, $D_p$ [m$^3$/rev], $w_{shaft}$ [rev/sec], $p$ [N/m$^2$]. These efficiency measures are functions of the pressure and speed of the pump,

$$\eta_v = \frac{V_{out}}{V_{disp}} \tag{7.60}$$

$$= \frac{Q_{out}}{D_p \cdot w_{shaft}} \tag{7.61}$$

and

$$\eta_m = \frac{Power^*_{out}}{Power_{in}} \tag{7.62}$$

$$= \frac{p_{out} \cdot D_p \cdot w_{shaft}}{T \cdot w_{shaft} \cdot (2\pi)} \tag{7.63}$$

$$= \frac{p_{out} \cdot D_p}{2\pi T}; \quad \text{for } D_p \text{ in units of [m}^3\text{/rev]} \tag{7.64}$$

and

$$\eta_o = \frac{Power_{out}}{Power_{in}} = \frac{p_{out} \cdot Q_{out}}{T \cdot w_{shaft}} \tag{7.65}$$

$$= \eta_v \cdot \eta_m \tag{7.66}$$

Notice that if the unit of $D_p$ is in [m$^3$/rad], then mechanical efficiency would be defined as

$$\eta_m = \frac{p_{out} \cdot D_p}{T}; \quad \text{if } D_p \text{ unit is [m}^3\text{/rad]} \tag{7.67}$$

The difference in the definition of the mechanical and overall efficiency is the definition of output power. The mechanical efficiency defines the output power as the power if there were zero leakage, that is, 100% volumetric efficiency. The overall efficiency takes into account both the leakage and other mechanical efficiency.

One of the main operational concerns is the noise level of the pump. The noise due to cavitation can be very serious. Low-pressure levels at the pump input may result in allowing air bubbles to enter the pump inlet port. Under high pressure, the bubbles collapse, which leads to high noise levels. Therefore, some pumps may require higher than normal levels of inlet port pressure, i.e., so-called boost inlet pressure using another small pump (called charge pump) between the main pump and tank. Cavitation may also be caused by too high hydraulic fluid viscosity; as a result, the pump is not able to suck in enough fluid. This may happen especially in cold start-up conditions. Heaters may be included in the pump inlet to regulate the fluid viscosity.

***Example*** Consider a pump with fixed displacement $D_p = 100$ cm$^3$/rev and following nominal operating conditions: input shaft speed $w_{shaft} = 1200$ rpm, torque at the input shaft $T = 250$ Nt.m, the output pressure $p_{out} = 12$ MPa, and output flow rate $Q_{out} =$ 1750 cm$^3$/sec. Determine the volumetric, mechanical, and overall efficiencies of the pump at this operating condition.

$$\eta_v = \frac{Q_{out}}{D_p \cdot w_{shaft}} \tag{7.68}$$

$$= \frac{1750 \text{ cm}^3/\text{s}}{100 \text{ cm}^3 \cdot 20 \text{ rev/sec}} = 87.5\% \tag{7.69}$$

$$\eta_m = \frac{p_{out} \cdot D_p}{2\pi \cdot T} \tag{7.70}$$

$$= \frac{12 \times 10^6 \text{ N/m}^2 \cdot 100 \times 10^{-6} \text{ m}^3/\text{rev}}{2 \cdot \pi \cdot 250} = 76.4\% \tag{7.71}$$

$$\eta_o = \eta_v \cdot \eta_m \tag{7.72}$$

$$= 0.875 \cdot 0.764 \cdot 100\% = 66.85\% \tag{7.73}$$

Notice that if any of the efficiency calculations results in being larger than 100%, it is an indication that there is an error in the given information or measured variables because the efficiencies cannot be larger than 100%. In order to confirm the results, let us calculate the input mechanical power to the pump and output hydraulic power from the pump.

$$Power_{in} = T \cdot w_{shaft} \tag{7.74}$$

$$= 250 \text{ N.m} \cdot 20 \text{ rev/sec} \cdot 2\pi \text{ rad/rev} \tag{7.75}$$

$$= 31.41 \text{ kN.m/sec} = 31.41 \text{ kW} \tag{7.76}$$

$$Power_{out} = p_{out} \cdot Q_{out} \tag{7.77}$$

$$= 12 \cdot 10^6 \text{ N/m}^2 \cdot (1750 \text{ cm}^3/\text{sec}) \cdot (1 \text{ m}^3/10^6 \text{ cm}^3) \tag{7.78}$$

$$= 12 \cdot 1750 \text{ N.m/sec} \tag{7.79}$$

$$= 21 \cdot 10^3 \text{ N.m/sec} = 21 \text{ kW} \tag{7.80}$$

$$\frac{P_{out}}{P_{in}} = \frac{21}{31.41} \tag{7.81}$$

$$= 66.85\% \tag{7.82}$$

$$= \eta_v \cdot \eta_m \tag{7.83}$$

$$= \eta_o \tag{7.84}$$

***Example*** Consider a hydrostatic transmission shown in Fig. 7.31. Assume that the pressure drop across the connecting hydraulic line is 200 psi. The pump variables and motor variables are shown in the table on the next page, along with the volumetric and mechanical efficiencies.

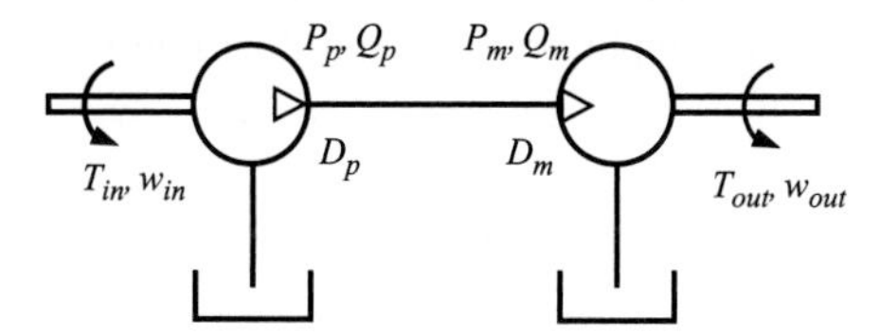


**FIGURE 7.31:** Hydraulic circuit with a hydraulic pump and a hydraulic motor.

| Pump | Motor |
|---|---|
| $D_p = 10$ in$^3$/rev | $D_m = 40$ in$^3$/rev |
| $\eta_v = 0.9$ | $\eta_v = 0.9$ |
| $\eta_m = 0.85$ | $\eta_m = 0.85$ |
| $w_p = 1200$ rpm | $w_m = ?$ |
| $p_p = 2000$ psi | $p_m = ?$ |
| $Q_p = ?$ | $Q_m = ?$ |
| $T_{in} = T_p = ?$ | $T_{out} = T_m = ?$ |
| $PumpPower_{in} = ?$ | $MotorPower_{out} = ?$ |

Let us first focus on the pump and its input power source. The net flow rate from the pump at the given input speed is

$$Q_p = \eta_v \cdot D_p \cdot w_p \tag{7.85}$$

$$= 0.9 \cdot 10 \cdot 1200 \text{ in}^3\text{/rev} \cdot \text{rev/min} \tag{7.86}$$

$$= 10800 \text{ in}^3\text{/min} \tag{7.87}$$

The input torque required to drive the pump is

$$\eta_0 = \eta_v \cdot \eta_m = \frac{PumpPower_{out}}{PumpPower_{in}} \tag{7.88}$$

$$= \frac{p_p \cdot Q_p}{T_p \cdot w_p} \tag{7.89}$$

$$T_p = \frac{1}{\eta_0} \frac{p_p \cdot Q_p}{w_p} \tag{7.90}$$

$$= \frac{1}{0.9 \cdot 0.85} \frac{2000 \text{ lb/in}^2 \cdot 10800 \text{ in}^3\text{/min} \cdot \frac{1 \text{ min}}{60 \text{ sec}}}{20 \cdot 2\pi \text{ rad/sec}} \tag{7.91}$$

$$= 3744.8 \text{ lb} \cdot \text{in} \tag{7.92}$$

and the input power that must be supplied to the pump is

$$PumpPower_{in} = T_p \cdot w_p \tag{7.93}$$

$$= 3744.8 \cdot 20 \cdot 2\pi \text{ rad/sec} \cdot \frac{1 \text{ HP}}{6600 \text{ lb.in/sec}} \tag{7.94}$$

$$= 71.3 \text{ HP} \tag{7.95}$$

Now let us focus on the motor. The pump output pressure and flow are connected to the motor. We assumed that there is 200-psi pressure drop in the hydraulic line and no leakage in the transmission line. Hence, the input pressure and flow rate to the motor are

$$p_m = p_p - 200 \text{ psi} = 1800 \text{ psi} \tag{7.96}$$

$$Q_m = Q_p = 10800 \text{ in}^3\text{/min} \tag{7.97}$$

The volumetric efficiency of the motor indicates that less than 100% of the input flow is converted to displacement,

$$\eta_v = \frac{Q_a}{Q_m} \tag{7.98}$$

$$Q_a = \eta_v \cdot Q_m = 0.9 \cdot 10800 = 9720 \text{ in}^3/\text{min} \tag{7.99}$$

and we can determine the output speed of the motor,

$$Q_a = \eta_v \cdot Q_m = D_m \cdot w_m \tag{7.100}$$

$$w_m = \frac{Q_a}{D_m} = 243 \text{ rpm} \tag{7.101}$$

The output power of the motor is determined from the ratio of its input power and output power, given the overall efficiency of the motor, (notice that efficiency definitions for hydraulic motors are opposite of those of hydraulic pumps since they perform opposite functions. Motor converts hydraulic power (input) to mechanical power (output), pump does the opposite).

$$\eta_o = \eta_v \cdot \eta_m = \frac{T_m \cdot w_m}{p_m \cdot Q_m} \tag{7.102}$$

$$T_m = \frac{\eta_o \cdot p_m \cdot Q_m}{w_m} \tag{7.103}$$

$$= \frac{0.9 \cdot 0.85 \cdot 1800 \text{ lb/in}^2 \cdot 10800 \text{ in}^3/\text{min}}{243 \text{ rev/min} \cdot 2\pi \text{ rad/rev}} \tag{7.104}$$

$$= 61200/(2\pi) \text{ lb} \cdot \text{in} = 9740 \text{ lb} \cdot \text{in} \tag{7.105}$$

Hence, the mechanical power delivered at the output shaft of the motor is

$$Motor\ Power_{out} = T_m \cdot w_m = \frac{(61200/(2\pi)) \cdot 243 \cdot 2\pi}{6600} = 37.55 \text{ HP} \tag{7.106}$$

### 7.2.3 Pump Control

The pump control element is the actuation mechanism that controls the angular position of the swash plate (Fig. 7.26). The mechanism which controls the swash plate angle is called the *compensator*. A pump control system has the following components:

1. A single or a pair of small control cylinders (also called *displacement pistons*) used to move the swash plate angle.
2. A proportional valve which meters flow to the pump swash plate actuation cylinder, and is controlled by electrical or hydromechanical means for a certain control objective (i.e., constant pressure output from the pump, load-sensing pump).
3. Sensor or sensors (either hydraulic pressure sensing lines or electrical sensors) used to implement the control objectives i.e., if pump is controlled to provide a constant output pressure, only sensor signal needed is the pump output pressure. If the pump is controlled to provide a constant pressure drop across a valve (load-sensing control), then both pump output pressure and valve output port pressure sensors are needed.

The swash plate angle is limited to a range

$$\theta_{min} \leq \theta \leq \theta_{max} \tag{7.107}$$

where $\theta_{min}$ is the minimum and $\theta_{max}$ is the maximum angular displacement of the swash plate. If the pump has two output ports, it is called a *bidirectional pump*. Output flow is directed to one or the other port by the swash plate angle control. Such a pump control is referred to as the *over-center* pump control (Fig. 7.28). If the swash plate displacement is on one side of the center, output flow is in one direction. If the swash plate displacement is on the other side of the center, then output flow is in the opposite direction. Hence the flow direction can be changed by pump control alone in a closed-circuit hydraulic systems such as hydrostatic transmission applications [30a].

In most pumps, the maximum and minimum swash plate angles can be mechanically adjusted by a set screw. In hydromechanically controlled pumps, the pressure feedback signals are provided by hydraulic lines with orifices. The command signal (i.e., desired output pressure) is implemented by an adjustable spring and screw combination. The actuator which moves the swash plate is called the control cylinder or control piston. In some pumps, the actuator provides power to move the swash plate in both directions under the control of the proportional valve, whereas in others, the actuator provides power for one direction and the power for the other direction is provided by a preloaded spring. In most cases, at start-up the preloaded spring will move the swashplate to maximum displacement. As the output pressure of the pump builds up, the compensator mechanism (proportional valve and control piston) provides control power to reduce the swash plate angle.

Pump control, that is the control of the pump displacement ($D_p(\theta)$) which is a function of the swash plate angle, may be based on different objectives such as

1. Pressure compensating and limiting
2. Flow compensated
3. Load sensing
4. Positive flow control (matched flow supply and demand)
5. Torque limiting
6. Power limiting

Among different pump control methods, pressure compensated valve control and the load-sensing valve control are the two most common methods.

**(1)** *Pressure compensated pump control:* Controls the pump displacement such that the actual pump output pressure ($P_s$) regulated about a desired pump output pressure ($P_{cmd}$),

$$\theta_{cmd} = \theta_{offset} + K \cdot (P_{cmd} - P_s) \tag{7.108}$$

where $\theta_{cmd}$ is the commanded (desired) swash plate angle, $\theta_{offset}$ is the offset value of the swash plate angle, and $K$ is the proportional gain for the pressure regulator. For simplicity, we illustrate a proportional control logic here. Clearly, more advanced control logic can be implemented in the relationship between the desired variables, measured variables, and the output of the controller. This type of pump control is called a *pressure-compensated pump control* and provides constant pressure output. The flow rate will be determined by whatever is needed to maintain the desired pressure, up to the maximum flow capacity [Fig. 7.32 (a)]. The commanded pressure does not have to be constant and typically can be set between a minimum and a maximum value for a given pump. As the flow rate increases, the regulated pressure tends to reduce a little in hydromechanically controlled pumps. But this effect can be eliminated by software control algorithms in digitally controlled pumps. When the external load is so large that it cannot be moved, the pump under pressure-compensated control will provide the desired regulated output pressure at almost zero flow rate. This is called *dead head condition.*

Pressure-limiting control is established by feeding back the pressure in the outlet to a check valve. When a preset pressure limit value is reached, the check valve opens to

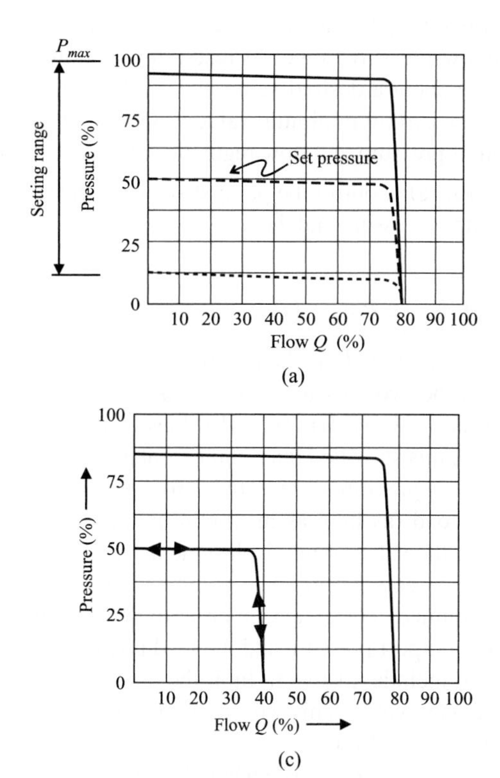


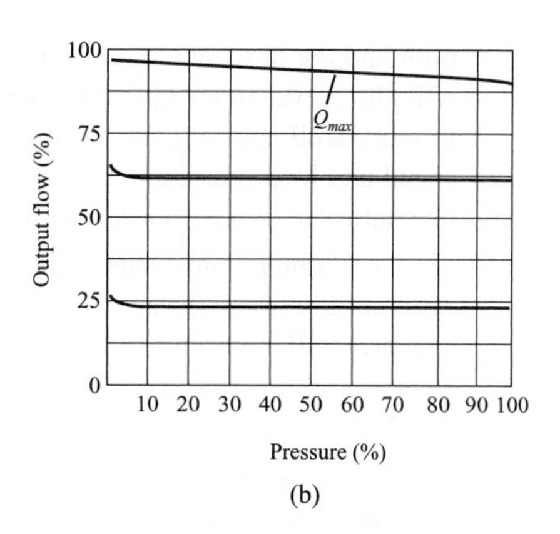


**FIGURE 7.32:** Performance characteristics of a pump under various control methods: (a) pressure regulated, (b) flow regulated, and (c) flow regulated with pressure limit or pressure regulated with flow limit.

destroke the pump swash plate to lower the output pressure. Pressure limiting is installed, as a safety to make sure that the pump does not output a pressure higher than the maximum allowed pressure for the system.

**(2)** *Flow-compensated pump control:* Control the pump displacement such that the actual pump output flow rate ($Q_p$) is regulated about a desired pump output flow rate ($Q_{cmd}$),

$$\theta_{cmd} = \theta_{offset} + K \cdot (Q_{cmd} - Q_p) \tag{7.109}$$

The flow-pressure curves for a flow-compensated pump control look as shown in [Fig. 7.32 (b)]. In order to maintain a stable output flow, the mechanical input speed of the pump may also have to be above a minimum value. In hydromechanical implementation of this control logic, we use a simple three-way regulator valve with two pressure feedbacks on the sides. There is a fixed orifice at the output port of the pump and by regulating a constant pressure drop across it, we regulate flow rate. Hence, the two pressure feedbacks come from the two ports of the fixed orifice. The three-port flow compensator valve has one port connected to the pump outlet, one port to tank, and the third one to the cylinder which actuates the swash plate of the pump. Depending on the pressure balance, the flow compensator valve throttles the cylinder port to the pump output port or to the tank port. The desired pressure differential, hence the desired flow rate, is set by the spring constant of the compensating valve. The performance curve illustrated in Fig. 7.32 (b) is for a hydromechanical implementation of the flow compensator control of the pump. The compensator valve works to maintain a constant pressure drop across a fixed orifice at the pump output port. Such a control mechanism works only if there is a *margin* pressure across the compensator. If the load pressure is very low, then the regulation quality of the flow compensator is very poor and the flow rate tends to increase at very low load pressure levels. This is the reason why the curves move upward around low-pressure range. As result, for the flow regulation to work well and regulate the flow rate about a desired value, the mechanical shaft input speed and load pressure must be above a certain minimum value. Likewise, the pump can be controlled based on a combination of pressure and flow regulation [Fig. 7.32 (c)].

**(3)** *Load-compensated (load-sensing):* Control the pump displacement such that the difference between the actual pump output pressure and the load pressure is regulated about a desired differential pressure,

$$\theta_{cmd} = \theta_{offset} + K \cdot (\Delta P_{cmd} - (P_s - P_L)) \tag{7.110}$$

where $\Delta P_{cmd}$ is the desired pressure differential, $P_s$ is the pump outlet pressure, and $P_L$ is the load pressure, which is the larger of the two pressures sensed from the two ports of the actuator (cylinder or hydraulic motor ports A and B). Typical values of $\Delta P_{cmd}$ are in the 10-bar to 30-bar range. Notice that since this so-called load-compensated (or load-sensing) pump control method maintains a constant pressure differential across the valve, the speed of the cylinders is independent of the load (as long as maximum pump output pressure limit is not reached) and proportional to the valve spool displacement. This can be seen by examining the flow equations for a valve (eqn. 7.156). The pump supplies the necessary flow (up to its saturation value at maximum displacement) in order to maintain the desired pressure differential between the pump outport pressure and the load pressure. Most load-sensing pump control mechanisms also include a valve to limit the pump output pressure to a maximum value set by a spring loaded valve. This valve output flow is zero until the pump pressure exceeds the maximum pressure limit.

Load-sensing control is used to reduce energy waste when the pump is not needed to operate at full system pressure and flow rate. It requires the load pressure sensing and pump output pressure sensing. Based on these two signals, the swash plate angle is controlled by a valve and piston mechanism. Figure 7.33 shows a load-sensing pump control system with completely hydromechanical components. This system uses a completely hydromechanical sensing, feedback, and actuation mechanism to perform a closed-loop control function on

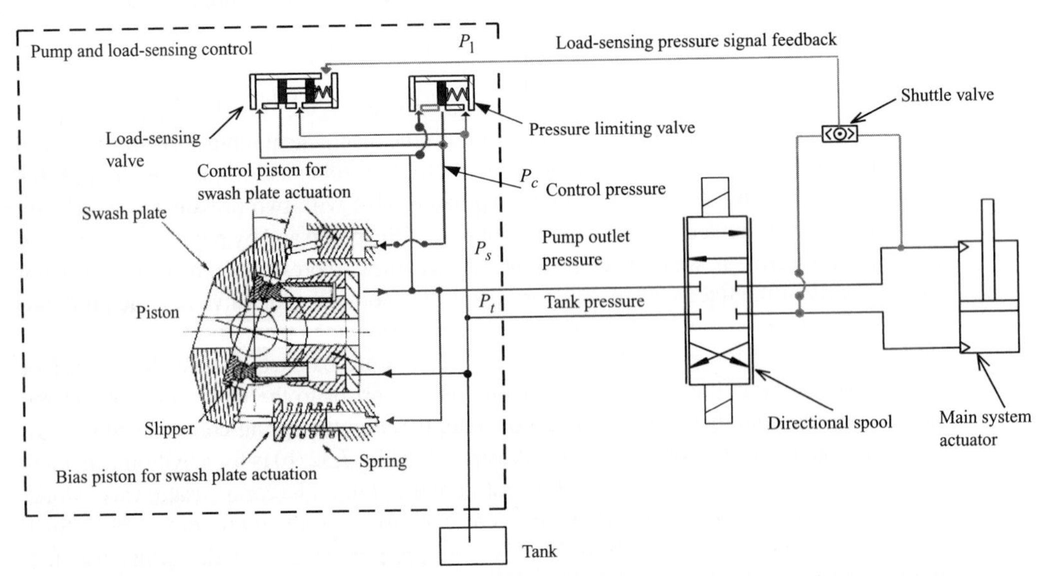


**FIGURE 7.33:** Principle of pressure-limiting and load-sensing control using a hydromechanical feedback on an axial piston pump; control element is the swash plate angle.

the swash plate angle in order to regulate a constant pressure drop across the valve. There are no electrical sensors or digital controls involved. The load-sensing valve controls the flow rate to the control piston in proportion to the difference between the pump outlet pressure ($P_s$) and the load pressure ($P_L$). As the pressure difference $P_s - P_L$ gets larger (i.e., $P_L$ is lower), the flow from load-sensing valve to control piston increases. The movement of the control piston to left reduces the swash plate angle, hence, the pump output flow rate and the pump outlet pressure. This reduces the $P_s$. If the pressure difference $P_s - P_L$ gets smaller (i.e., $P_L$ is larger), the flow rate from the load-sensing valve gets smaller, hence the control piston moves right to increase the swash plate angle. This in turn increases the pump output (pressure and flow) to increase the $P_s$.

Maintaining a constant pressure drop across the valve for all load conditions up to the maximum pump outlet pressure limit makes the flow rate through the valve proportional to its spool displacement regardless of the load pressure as long as maximum pump outlet pressure is not reached. This gives good controllability because the flow rate through the valve, hence the cylinder (actuator) speed will be proportional to the displacement of the directional spool regardless of the load. Load sensing is usually mounted together with pressure-limiting devices. The swash plate is controlled so that the pressure difference between the pump output pressure and load is maintained at a constant value. Hence, this type of control increases or decreases the pump output pressure depending on the load, up to a maximum limit set by the pressure-limiting valve.

The combination of the load-sensing valve and pressure-limiting valve is referred to as *the load sensing valve* (also called the load-sensing compensator valve; see Fig. 7.33). The load-sensing compensator has four ports: (1) pump output pressure input, (2) load pressure input, (3) tank port, and (4) output port to swash plate control piston. The pressure-limiting valve overrides the load-sensing valve control only when the pump outlet pressure reaches a preset limit set by the spring in the pressure limiting valve. When $p_s > p_{limit}$ (i.e., directional valve is in metering position and the load is very large), the flow from the pressure-limiting valve to control piston will de-stroke the swash plate and reduce the output pressure. Furthermore, if the cylinder had reached its travel limit and the directional valve is not in neutral position, pump output pressure will reach its limit but it will also de-stroke to almost zero flow. This is called the *high-pressure stand-by condition.* Otherwise, the load-sensing valve controls the swash plate angle and the pressure limit valve does not make a change in the swash plate angle control as long as pump output pressure is less than the maximum limit. The pump output pressure is constantly fed to the bias piston and to the valves for the load sensing and pressure limiting. The bias piston puts the pump at maximum displacement until flow is restricted by the actuator or the directional valve of the system that will create a pressure at the discharge port of the pump. The pressure downstream the directional valve is fed back via a shuttle valve that always gives the highest pressure of the two sides of the actuator.

Although not shown in the figure, the load-sensing shuttle valve is enabled by a check valve only when the directional control valve is not in the neutral closed center position. When in the neutral position, the cylinder is stationary and there is no connection between pump and cylinder ports. Therefore, the pump does not need to be concerned with supporting the load. As a result, the signal from the shuttle valve is blocked and the holding pressure of the load is not fed back to the pump control. In neutral position of directional control valve, the pump will go to its *low-pressure stand-by condition.*

Figure 7.34 shows the cross-sectional drawing of a variable displacement axial piston type pump along with its two valves (load-sensing valve and pressure-limiting valve). The pump and load-sensing control valve pair are generally packaged together. For the variable displacement pump to function, the load-pressure feedback line must be connected to the proper port associated with the load.

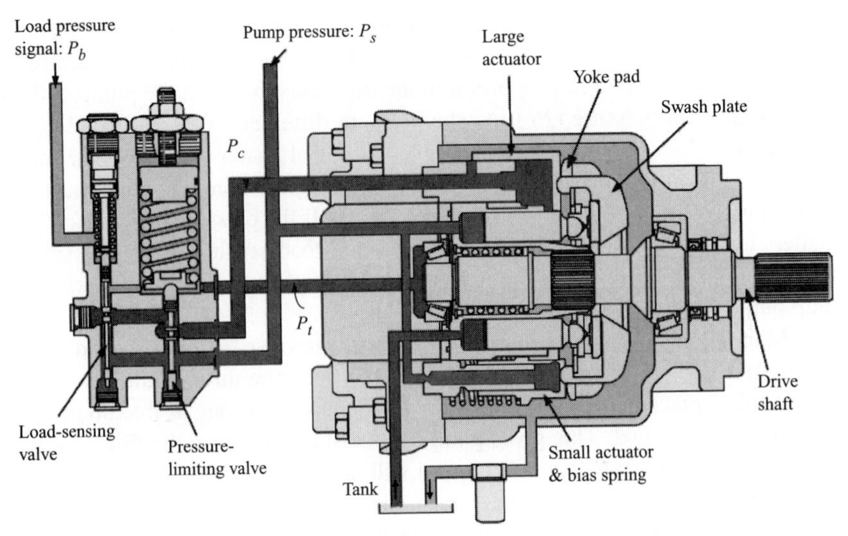


**FIGURE 7.34:** Cross-sectional component drawing of a variable displacement axial piston type pump and its load-sensing control system components (load-sensing valve and pressure-limiting valve).

Figure 7.35 shows a two-axis hydraulic circuit with a variable displacement axial piston pump which has load-sensing hydromechanical controls. Notice that the load-sensing control valve on top of the pump has three ports: (1) load pressure sensing port, (2) pump output pressure sensing port, and (3) output port to control the swash plate angle of the pump. The load pressure is the maximum of two pressures sensed between the cylinder axis and rotary gear motor. Flow to each actuator is regulated by a manually operated proportional flow control valve. Two flow control valves are typically built on a single frame valve block with internal porting for P and T ports between them. Internal porting may connect P and T lines to each valve either in parallel or series connection. P and T connections are now shown in the figure. Each valve has built-in ports that sense the maximum pressure at its output ports. Using two resolver valves, maximum pressure is fed back to the pump control valve. In this configuration, the smallest load pressure signal that can be fed back to the control valve is the tank pressure, which is shown as one of the inputs to the resolver valve next to the valve that controls the cylinder.

**(4)** *Positive flow control (PFC) of pump:* Control the pump displacement such that it matches the flow demand of the line [53a, 53c]. Let $Q_p$ be the flow demand by the line, and $w_{eng}$ is the input shaft speed of the pump. The necessary pump displacement to provide the desired flow rate can be calculated from the pump performance characteristics,

$$Q_P = w_{eng} \cdot D_P(\theta) \tag{7.111}$$

$$\theta = D_P^{-1}(Q_P/w_{eng}) \tag{7.112}$$

It is desirable to implement the PFC algorithm without a flow rate sensor. Therefore, let us generate the desired pump displacement based on the predicted flow demand and using the pump map,

$$Q_{Pcmd} = Q_P(x_{s1}, x_{s2}, \ldots) \tag{7.113}$$

where $(x_{s1}, x_{s2}, \ldots)$ are the spool displacements of multiple valves supplied by the pump. The desired pump displacement is determined as (an offset value is added to account for initial nominal operating point),

$$\theta_{cmd} = \theta_{offset} + D_p^{-1}(Q_{Pcmd}/w_{eng}) \tag{7.114}$$

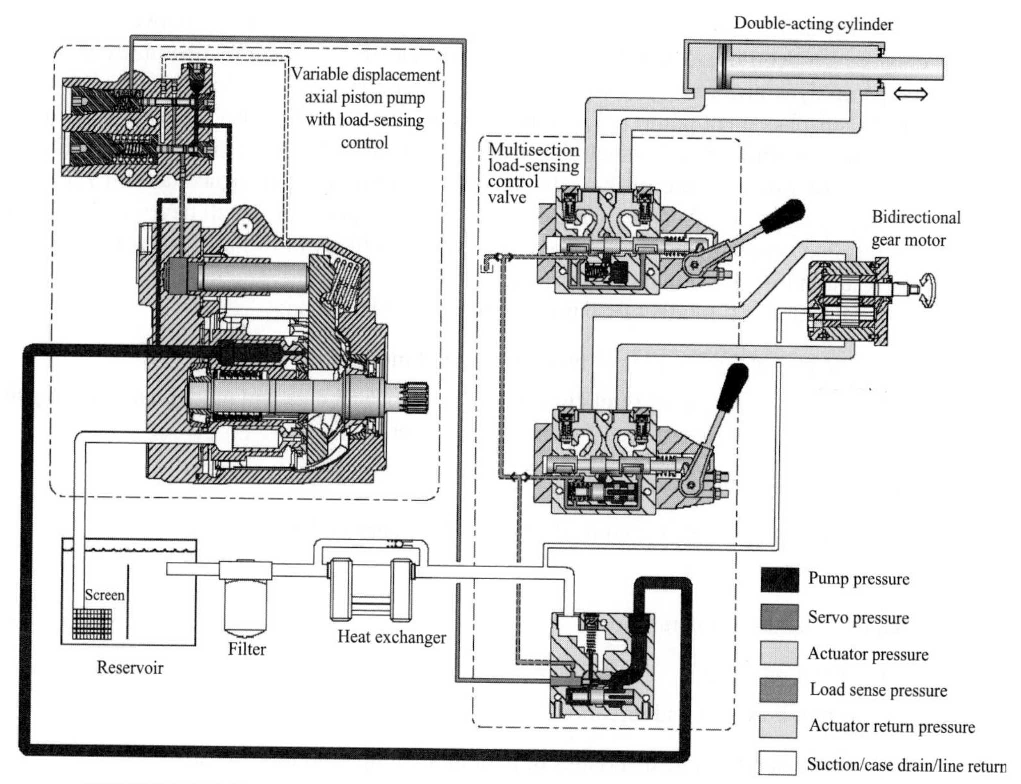


**FIGURE 7.35:** Two-axis hydraulic motion control system components: pump is a variable displacement axial piston pump with a load-sensing control (implemented hydromechanically), a linear cylinder, a bidirectional gear motor, and two manually controlled proportional valves. Pump and tank port connections to two proportional valves are not shown in the figure for simplicity (*courtesy of Sauer-Danfoss*).

Notice that in order to implement the PFC method, we need the pump map function $D_P^{-1}(Q_P, w_{end})$ for the specific valve or valves supplied by the pump. Accuracy of this pump map is important because mismatch between the flow demand and flow supply can result in serious performance degredations. The price paid for the increased energy efficiency benefit of the PFC closed-center EH systems is the increased complexity of the control algorithm.

Torque limiting and power limiting controls are implemented in pumps to prevent them from stalling when both high pressure and high flow rate occur in the system at the same time. In a mobile application the stalling will occur when the pump requires more power then the diesel engine can output, and this will eventually bring the diesel engine to a dead stop. Torque limiting destrokes the pump to where the diesel engine have enough power to drive the pump. This will also make possible for the designer to use a smaller size engine in the system and therefore save weight and energy.

***Transient Response of the Pump*** If the delay associated with the dynamic response of pump displacement controller is to be taken into account, instead of assuming that the pump displacement is equal to the commanded pump displacement, then a first- or second-order filter dynamics should be included in the pump model,

$$\theta = \frac{1}{(\tau_{p1}s + 1)(\tau_{p1}s + 1)}\theta_{cmd}(s) \tag{7.115}$$

where $\tau_{p1}$ and $\tau_{p2}$ are the time constants of the dynamic relationship between the pump displacement command and the actual pump displacement. This effective time constant of the pump dynamics is very critical in closed-center EH systems (closed-center valve and variable displacement pump EH systems). The reason is the fact that the bandwidth of the main flow control valve is much faster than the bandwidth of the pump control. During any kind of valve closure, if the valve reaches the null position (hence, almost zero flow demand) much faster than the pump can de-stroke, the pump flow will have no place to go and result in very large pressure spikes. This will most likely blow the pressure relief valves and result in low-performance operation.

Mathematical model of a pump can be derived based on:

1. The physical principles of fluid and inertial motion or
2. The input–output (I/O) relationship using empirical data and model it as a static gain (possibly nonlinear) plus dynamic filter effects

Input variables of the pump are:

1. Swash-plate angle (or equivalent control element variable)
2. Input shaft speed

Output variables of interest are:

1. Outlet pressure
2. Outlet flow rate and direction

Some of the nondeal characteristics of hydraulic pump (and motors) are:

1. Variation of displacement as function of rotor position within one revolution. Since there is a finite number of fluid cavities (cylinder-piston pairs in an piston pump), the displacement has ripple as function of the rotor position and number of the piston-cylinder pair (this is in principle the same as the commutation ripple in a brush-type DC motor).
2. Every hydraulic pump, valve, motor, and cylinder has leakage, and it increases with the pressure.

## 7.3 HYDRAULIC ACTUATORS: HYDRAULIC CYLINDER AND ROTARY MOTOR

Translational cylinder and rotary hydraulic motor are the power delivering actuators in translational and rotary motion systems, respectively. The basic functionality of the actuator is to convert the hydraulic fluid power to mechanical power which is the opposite of the pump function (Fig. 7.36). Unidirectional pumps and motors are optimized to work in one direction in terms of reduced noise and increased efficiency. Bidirectional hydraulic pumps and motors have symmetric performance in either direction. In general, a pump can operate

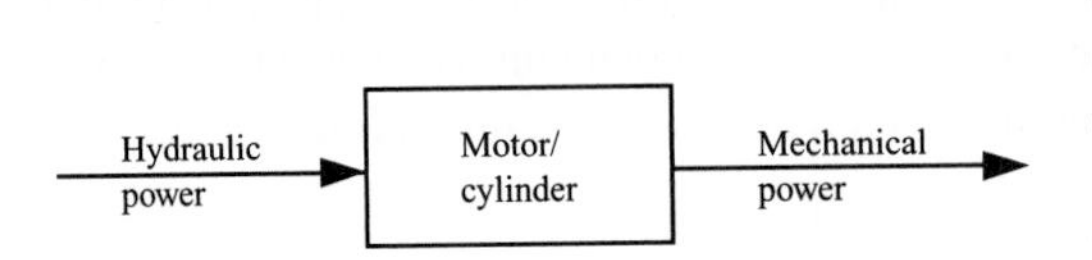


**FIGURE 7.36:** Hydraulic actuator functionality: convert hydraulic power to mechanical power. The hydraulic cylinder converts hydraulic power to translational motion power, and the hydraulic motor converts it to rotational motion power.

both in pumping or motoring mode. Similarly, a hydraulic motor can operate in motoring or pumping mode. However, there are exceptions. Some pump designs incorporate check valves in their design which makes it impossible for them to operate in motoring mode. Similarly, some motor designs are such that the hydraulic motor cannot operate in pumping mode even under *overrunning (regenerative) load* conditions (Fig. 7.28).

An *over-center* motor means that when the swash plate is moved in opposite direction relative to its neutral position, the direction of the output shaft speed of the motor changes even though the input–output hydraulic ports stay the same. An *over-center* pump means when the swash plate is moved in opposite direction relative to its neutral position, the direction of the hydraulic fluid flow changes (input port becomes output port, and output port becomes input port), while the direction of the mechanical input shaft speed stays the same.

Given the displacement ($D_m$, fixed or variable) of a rotary hydraulic motor, the motor output speed ($w$) is determined by the flow rate ($Q$) input to it,

$$w = Q/D_m \tag{7.116}$$

Similarly, for linear cylinders, the same relationship holds by analogy,

$$V = Q/A_c \tag{7.117}$$

where $w$ is the speed of motor, $V$ is the speed of cylinder, $D_m$ is the displacement of the motor (volume/rev), and $A_c$ is the cross-sectional area of the cylinder. If we neglect the power conversion inefficiencies of the actuator, the hydraulic power delivered must be equal to mechanical power at the output shaft, for rotary motor,

$$Q \cdot \Delta P_L = w \cdot T \tag{7.118}$$

and for cylinder,

$$Q \cdot \Delta P_L = V \cdot F \tag{7.119}$$

where $\Delta P_L$ is the load pressure differential acting on the actuator (cylinder or motor) between its two ports (A and B), $T$ is the torque output, $F$ is the force output. Hence, the developed torque/force, in order to support a load pressure $\Delta P_L$, is

$$T = \Delta P_L \cdot D_m \tag{7.120}$$

$$F = \Delta P_L \cdot A_c \tag{7.121}$$

If the rotary pump and motor are variable displacement type, and an input–output model is desired from the commanded displacement to the actual displacement, a first- or second-order filter dynamics can be used between the $D_m$ and $D_{cmd}$,

$$D_m(s) = \frac{1}{(\tau_{m1} \cdot s + 1)(\tau_{m2} \cdot s + 1)} D_{cmd}(s) \tag{7.122}$$

There are applications which require extremely high pressures with small flow rate which cannot be directly provided by a pump. In these cases, *pressure intensifiers* are used. The basic principle of the pressure intensifier is that it is a hydraulic power transmission unit, like a mechanical gear. Neglecting the friction and heat loss effects, the input power and output power are equal. The only function it performs is that it increases the pressure while reducing the flow rate. It is the analog of a mechanical gear reducer (increases the output torque, reduces the output speed). Figure 7.37 shows a pressure intensifier in a hydraulic

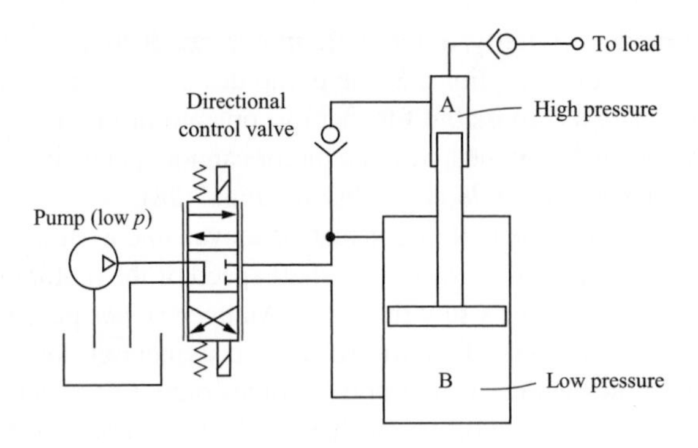


**FIGURE 7.37:** Pressure intensifier circuit.

circuit. The ideal power transmission between B and A pressure chambers means

$$Power_B = Power_A \tag{7.123}$$

$$F_B \cdot V_{cyl} = F_A \cdot V_{cyl} \tag{7.124}$$

$$p_B \cdot A_B \cdot V_{cyl} = p_A \cdot A_A \cdot V_{cyl} \tag{7.125}$$

$$p_B \cdot A_B = p_A \cdot A_A \tag{7.126}$$

Notice that the intensified pressure in the forward cycle of the cylinder is equal to the area ratios between the cylinder head-end and the intensifier ram (i.e., 10),

$$\frac{p_A}{p_B} = \frac{A_B}{A_A} \tag{7.127}$$

During the forward stroke, the pressure in chamber A is amplified by the area ratios defined by the above equation. During the retraction stroke, the intensifier and rod-end of the cylinder is filled with hydraulic fluid and is not considered a work cycle.

***Example*** Consider a pump that supplies $Q_p = 60$ liter/min constant flow to a double-acting cylinder. The cylinder bore diameter is $d_1 = 6$ cm and rod diameter $d_2 = 3$ cm. Assume the rod is extended through both sides of the cylinder. The load connected to the cylinder rod is $F = 10000$ Nt. The flow is directed between the pump and the cylinder by a four-way proportional control valve. Neglect the pressure drop and losses at the valve. Deterimine the pressure in the cylinder, velocity, and power delivered by the cylinder during extension cycle. Assuming 80% overall pump efficiency, determine the power necessary to drive the pump.

We need to determine $\Delta p_l$, $V$, and $P$ for both extension and retraction cycle. Let us determine the areas of the cylinder.

$$A_c = \frac{\pi\left(d_1^2 - d_2^2\right)}{4} = \frac{\pi(6^2 - 3^2)}{4}\ \text{cm}^2 = 19.63\ \text{cm}^2 \tag{7.128}$$

$$= 19.63 \times 10^{-4}\ \text{m}^2 \tag{7.129}$$

The pressure during extension stroke that must be present as differential pressure between two sides of the cylinder in order to support the load force is

$$F = \Delta p_l \cdot A_c \tag{7.130}$$

$$\Delta p_l = \frac{F}{A_c} = \frac{10000}{19.63 \times 10^{-4}} \tag{7.131}$$

$$= 5.09 \times 10^6\ \text{N/m}^2 = 5.09\ \text{MPa} \tag{7.132}$$

The linear velocity is determined by the conservation of flow,

$$Q = A_c \cdot V \tag{7.133}$$

$$V = \frac{Q}{A_c} = \frac{60 \text{ liter/min } 10^{-3} \text{ m}^3/\text{liter} \cdot 1 \text{ min/60 sec}}{19.63 \times 10^{-4} \text{ m}^2} \tag{7.134}$$

$$= 0.509 \text{ m/s} \tag{7.135}$$

The power delivered to the load by the cylinder is

$$Power_m = F \cdot V = \Delta p_L \cdot Q \tag{7.136}$$

$$= 10000 \text{ Nt} \cdot 0.509 \text{ m/s} = 5090 \text{ Watt} = 5.09 \text{ kW} \tag{7.137}$$

and the necessary pump power rating is

$$Power_p = \frac{1}{\eta_o} \cdot Power_m \tag{7.138}$$

$$= \frac{1}{0.8} \cdot 5.09 \text{ kW} = 6.36 \text{ kW} \tag{7.139}$$

***Example*** Consider a double-acting cylinder and its hydraulic lines connecting to its ports (Fig. 7.38). Let us assume that the maximum linear speed of fluid in the transmission lines is to be limited to 15 ft/sec in order to reduce flow turbulance and resulting flow resistance. Calculate the forward and reverse speed of the cylinder that can be achieved for different values of the line diameter, rod diameter, and cylinder head end diameter (see Table 7.1). Assume the rod diameter is half of the cylinder inner diameter. Consider the line diameters for $d_{line} = 0.5$ in., 0.75 in., 1.0 in., 1.5 in., and $d_{he} = 4.0$ in., 6.0 in., 8.0 in.

From flow continuity,

$$Q = A_{line} \cdot V_{line} = A_{he} \cdot V_{fwd} \quad \text{in forward motion} \tag{7.140}$$

$$= A_{re} \cdot V_{rev} \quad \text{in reverse motion} \tag{7.141}$$

where we limit $V_{line} = 15$ ft/sec. Then,

$$V_{fwd} = \frac{A_{line}}{A_{he}} \cdot V_{line} \tag{7.142}$$

$$V_{rev} = \frac{A_{line}}{A_{re}} \cdot V_{line} = \frac{A_{he}}{A_{re}} \cdot V_{fwd} \tag{7.143}$$

If we focus on the case where the cylinder inner diameter is twice the rod diameter (Table 7.1),

$$V_{fwd} = \frac{\pi {d_{line}}^2/4}{\pi {d_{he}}^2/4} \cdot V_{line} \tag{7.144}$$

$$V_{rev} = \frac{{d_{he}}^2}{{d_{he}}^2 - {d_{re}}^2} \cdot V_{fwd} \tag{7.145}$$

$$= \frac{4}{3} \cdot V_{fwd} \tag{7.146}$$

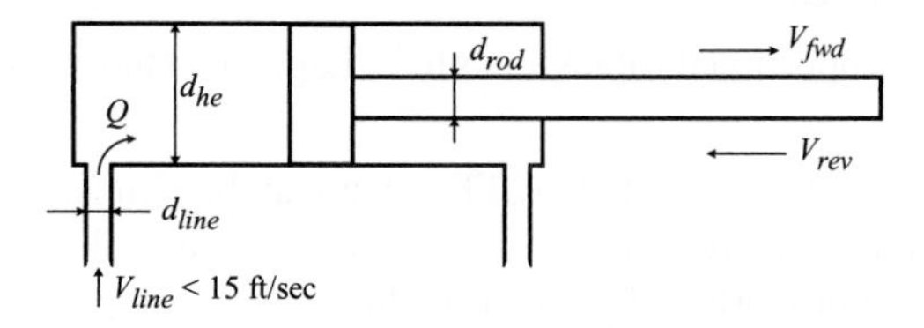


**FIGURE 7.38:** Maximum cylinder speed in forward and reverse directions, given the maximum linear speed of fluid flow (15 ft/sec), and dimensions of the cylinder and connecting lines.

**TABLE 7.1: Cylinder speed (forward and reverse direction in units of in/sec) for different values of line pipe diameter and cylinder diameter. It is assumed that the rod diameter is half of cylinder inner diameter. The linear line speed of fluid is limited to 15 ft/sec.**

| | $d_{line}$ | | | |
|---|---|---|---|---|
| $d_{he}$ | 0.5 | 0.75 | 1.0 | 1.5 |
| 4.0 | (3.75, 5.0) | (8.44, 11.25) | (15.0, 20.0) | (33.75, 45.0) |
| 6.0 | (1.67, 2.23) | (3.75, 5.0) | (6.67, 8.89) | (15.0, 20.0) |
| 8.0 | (0.94, 1.25) | (2.11, 2.81) | (3.75, 5.0) | (8.44, 11.25) |

## 7.4 HYDRAULIC VALVES

The valve is the main control component in a hydraulic circuit. It is the *metering* component for the fluid flow. The metering of the fluid is done by moving a spool to adjust the orifice area. There are two main valve output variables of interest which can be controlled by the spool movement:

1. Flow (rate and direction)
2. Pressure

If the movement of the spool or poppet is determined in order to maintain a desired pressure, the valve is called a *pressure control valve*. If the movement of the spool is determined in order to maintain a desired flow rate, it is called a *flow control valve*. If the flow direction is changed between three or more ports, they are called *directional flow control valves*.

If the spool position of the valve is controlled only to two discrete positions, they are called *ON/OFF valves* (Fig. 7.39). If the spool position of the valve can be controlled to be anywhere between fully open and fully closed positions (i.e., in proportion to a signal), then they are called *proportional valves*. The ON/OFF type valves are used in applications which require discrete positions, i.e., open the door and close the door. Here, we are interested primarily in proportional and servo control valves.

A valve is sized based on three major considerations:

1. *Flow rating*: Flow rate it can support at a certain pressure drop across the valve port (i.e., maximum flow at 1000-psi pressure drop for servo valves, at 150 psi for proportional valves)
2. *Pressure rating*: Rated pressure drop across the valve and the maximum supply port pressure
3. *Speed of response*: Bandwidth of the valve from current signal to spool displacement

The main differences between ON/OFF valves and proportional valves are as follows:

1. *Solenoid design*: In proportional valves, the gain of the current–force relationship is fairly constant in the travel range of the solenoid, whereas in ON/OFF valves, the main thing is to generate the maximum force and the linearity of the current–force gain relationship is not that critical.
2. Centering spring constant in proportional valves tends to be larger than the comparable size ON/OFF valves.
3. While the valve body of a proportional and ON/OFF valve may be almost identical, the spool designs are different. The proportional valve spools are carefully designed with desired orifice profiles to proportionally meter the flow.

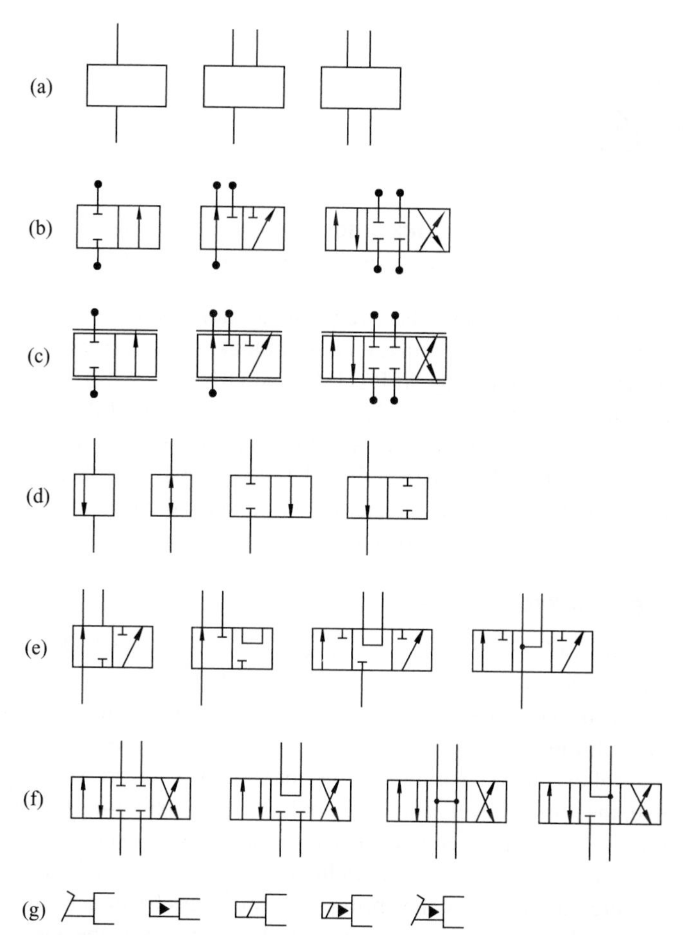


**FIGURE 7.39:** Valve categories: (a) number of ports (also called *ways*): two-port (or two-way), three-port, four-port valves, (b) two-, three-, and four-port valves with two-, three-, and four-position configurations, (c) same valves with proportional flow metering capability (proportional valves), (d) normally closed, normally open conditions, that is the flow condition of the valve, when it is in default condition, (e) three-port valve with various discrete positions, (f) various flow conditions when a four-port valve is in the center position (closed center, tandem center, open center, float center), (g) actuation method of the valve: manual, pilot, solenoid, or the combinations of them (solenoid plus pilot, manual plus pilot).

Most proportional valve geometries are designed so that the flow is approximately proportional to the spool displacement under constant pressure drop conditions.

Valves can be categorized according to the following:

1. Number of ports, i.e., two, three, four ports
2. Number of discrete positions of the spool, i.e., one, two, three, or four discrete positions or ability to position the spool proportionally within its travel range
3. Flow condition of the valve at center position, i.e., open center, closed center
4. Actuation method of valve, i.e., manual, electric actuator, pilot actuated, or combination of them
5. Metering element type: sliding spool, rotary spool, poppet (or ball)
6. Main control purpose of valve: pressure, flow (direction only, direction and flow rate which is referred to as proportional)

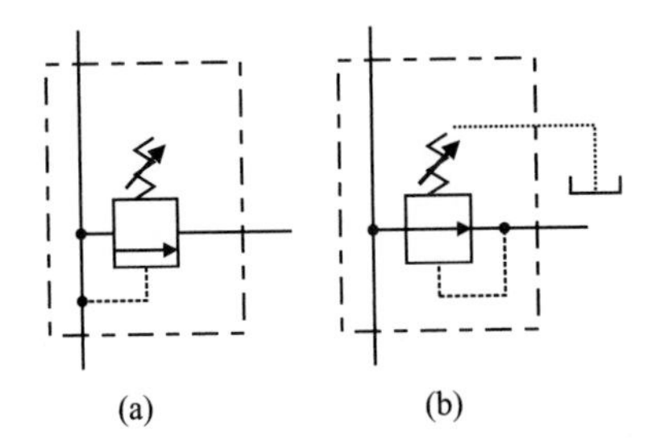


**FIGURE 7.40:** Pressure control valve symbols: (a) relief valve, (b) pressure-reducing valve.

7. Mounting method of the valve into the circuit: stand-alone body, multifunction integrated valve block, subplate mounting, manifold block, stacked (sandwich) block. Mounting standards are specified by NFPA, ISO, and DIN [36–38]. NFPA standards are referred with letter D, and DIN standards are referred by NG, ISO standards are referred by CETOP. Equivalent standard specifications are DO3/NG6/CETOP 03, D05/NG10/CETOP 05, D07/NG16/CETOP 07, D08/NG25/CETOP 08, D010/NG32/CETOP 10. The mounting standards specify the mechanical dimensions of the ports, their locations as well as screw holes on the mounting plate.

### 7.4.1 Pressure Control Valves

Pressure control valves involve two or three-port connections. The spool which controls the orifice area is actuated based on a sensed pressure. Relief valve, pressure reducing valve, counterbalance valve, sequence valve, brake valve are examples of pressure control valves (Fig. 7.40). *Relief valve* limits the maximum pressure of the valve by venting the flow to tank. In a mechanical relief valve, a spring sets the maximum input pressure. In an EH controlled relief valve, the current level of the solenoid sets the maximum pressure; hence an EH controlled relief valve can be used to control a variable line pressure. There are two major types of pressure relief valves: direct and indirect acting (Fig. 7.41). In direct acting relief valve, the relief pressure is set by the spring. The line pressure directly acts on the area of the poppet spool. When the line pressure exceeds the set pressure, the poppet moves up and opens the line to the tank.

$$p_{line} \cdot A_{poppet} \leq k_{spring} \cdot x_{spring}; \quad valve\ closed \tag{7.147}$$

$$p_{line} \cdot A_{poppet} \approx k_{spring} \cdot x_{spring}; \quad valve\ open \tag{7.148}$$

$$p_{line} \approx p_{spring} = (k_{spring} \cdot x_{spring})/A_{poppet}; \tag{7.149}$$

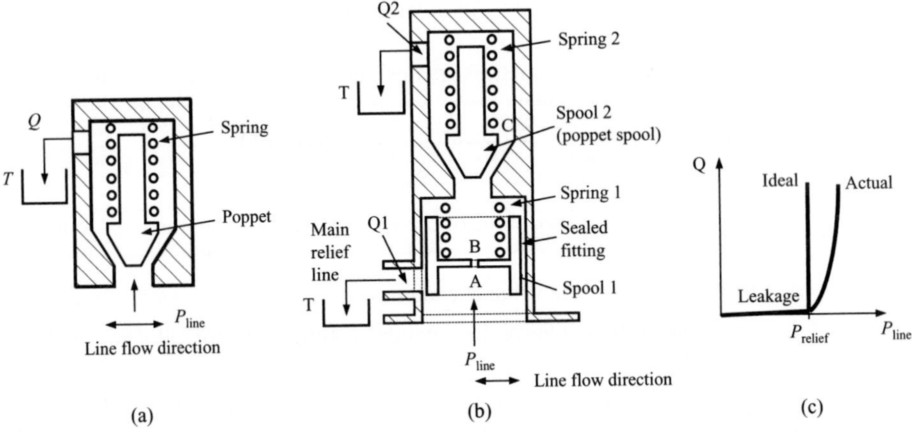


**FIGURE 7.41:** Pressure relief valve: (a) direct acting, (b) indirect acting, (c) line pressure, set relief pressure, and flow rate to tank relationship in steady state (ideal case and realistic case shown).

The indirect acting relief valve uses an intermediate orifice and spring between the poppet valve section and the line pressure. The function of the orifice in the indirect pressure relief valve is as follows: When the set pressure of poppet stage is exceeded (spring 2 in Fig. 7.41b), the poppet spool opens and flow crosses the orifice. The role of the orifice is to effectively create a pressure drop between A and B locations, hence allowing the flow to overcome preload of spring 1 to open the main relief line. When the poppet valve opens, the line pressure is relieved to tank not through the poppet valve orifice, but a separate orifice on the side. Poppet section has small (or almost zero) leakage flow to tank. Compared to the direct acting relief valve, the indirect acting valve operation is less sensitive to flow rate, is more stable, but has a little slower response due to the orifice. It is not unusual that a direct acting relief valve can operate in a high-frequency open and close motion due to pressure spikes in the line, hence result in valve failures in a matter of minutes in large power applications.

Poppet valves are commonly used as pressure regulator or pressure relief valves (Fig. 7.42). Main components of a poppet valve are: poppet (spherical ball or conical shaped), seat, valve body, and poppet actuator which may include a spring. Poppet valves are easy to manufacture, have lower leakage, and are insensitive to clogging by dirt particles compared to other types of spool valves. As a result, poppet valves are often used in load-holding applications to minimize leakage and hence movement of the load [63]. Depending on the flow rate supported by the valve, it may be actuated directly by a solenoid for small flow rate valves or via a secondary pilot stage amplification of actuation force for large flow rate valves. Poppet valves are used as screw-in cartridge valves with manifold blocks. A poppet valve performance is rated in terms of its maximum operating pressure ($p_{max}$), rated flow through the valve at a given pressure drop across it ($Q_r$ at $\Delta p_v$). In recent years, poppet valves have been developed as proportional flow control valves. However, since the valve actuation force directly acts against the dynamic flow forces, the dynamically stable control of the valve is a difficult task. The cross-sectional shape of the poppet and its seat make a significant difference in flow forces and proportional flow metering ability of a poppet valve.

*Pressure-reducing valve* maintains an output pressure which is lower than the input pressure by venting the excess flow to tank or limiting the orifice area based on desired and actual output pressure sensing. Figure 7.43 shows the basic design, symbol, and steady-state input–output relation of the valve. When the output pressure is less than the output pressure set by the spring, the spool does not move. Output pressure is very close to the

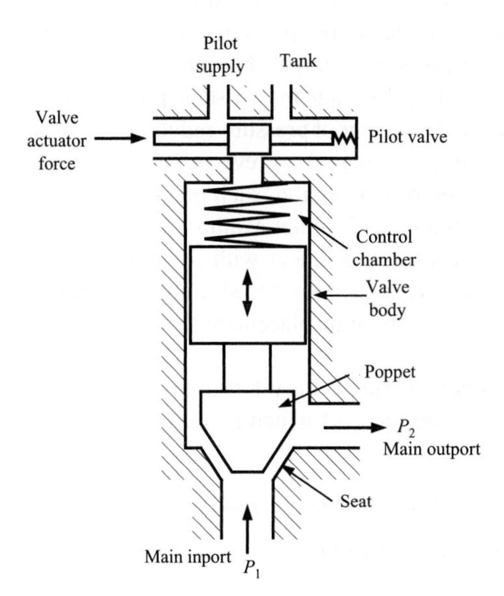


**FIGURE 7.42:** Poppet valve construction: valve body and orifice seat, poppet, control chamber with a mechanism to control the pressure or direct actuation of the poppet.

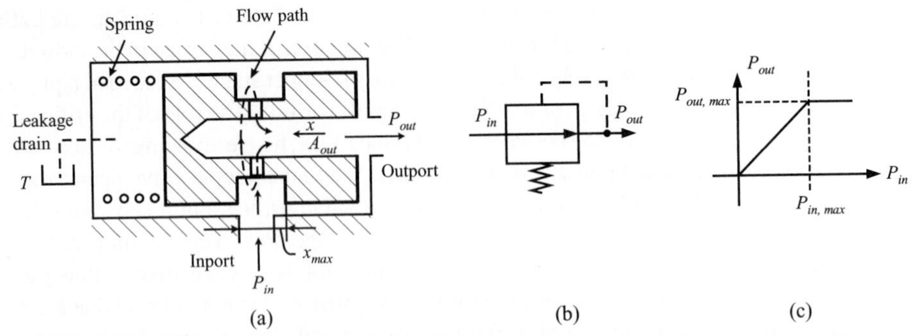


**FIGURE 7.43:** Pressure-reducing valve: (a) valve design, (b) valve symbol, (c) steady-state input and output pressure relationship. Notice that only the output pressure is fed back to affect the movement of the spool and hence the metering orifice. Input pressure ideally does not affect the force on the spool as a result of the way the flow is routed around the spool.

input pressure.

$$p_{out} \cdot A_{out} < F_{spring,0} \tag{7.150}$$

As the output pressure increases due to increase in the input pressure or load, the spool force due to the pressure feedback will increase and start to move the spool against the spring until the spring force balances the output pressure feedback force,

$$p_{out} \cdot A_{out} \approx k_{spring} \cdot x_{spring} = F_{spring} \tag{7.151}$$

As the spool moves, it restricts the flow orifice between the input and output port. Eventually, when the output pressure is high enough to further move the spool to close off the orifice between input and output ports, the output pressure is limited to the maximum value set by the spring and orifice design of the valve. The valve then regulates within the vicinity of this point and maintains a constant output pressure which is set by the spring as long as the input pressure is larger than the set (desired output) pressure.

A modified version of the same valve concept (pressure-reducing valve) is used as an adjustable pressure-reducing valve in lever or solenoid operated *pilot valves*. Figure 7.44 shows a modification of the pressure-reducing valve. The base of the spring is moved by a lever (or an electric actuator in case of an EH type pressure-reducing valve), hence changing the set spring pressure. Therefore, the output pressure will be proportional to the lever displacement as long as the input pressure is larger than the set pressure. A pair of such pressure-reducing valves is used as the pilot stage in two-stage proportional valves. In mechanically actuated versions, a lever moves a pair of pressure-reducing valves. In forward motion of the lever, one of the pilot valve is actuated, in reverse motion the second pilot valve is actuated. Similarly, in the EH-controlled version, each pressure-reducing valve is actuated by a solenoid based on lever command. The net result of such a valve is that the output pressure is proportional to the lever displacement with its maximum value being the input pilot pressure. This output pressure is then used to shift the main spool against a centering spring in order to generate a main spool displacement that is proportional to the lever displacement.

In pilot valve applications, the input pressure is the pilot supply pressure ($p_{pilot,s}$), and output pressure is the pilot control pressure ($p_{pilot,c}$) which is used to shift the spool of a larger flow control valve,

$$p_{pilot,c} = [k_{spring}(x_{spring})]/A_{out} \leq p_{pilot,s} \tag{7.152}$$

$$\approx K \cdot x_{spring} \leq p_{pilot,s} \tag{7.153}$$

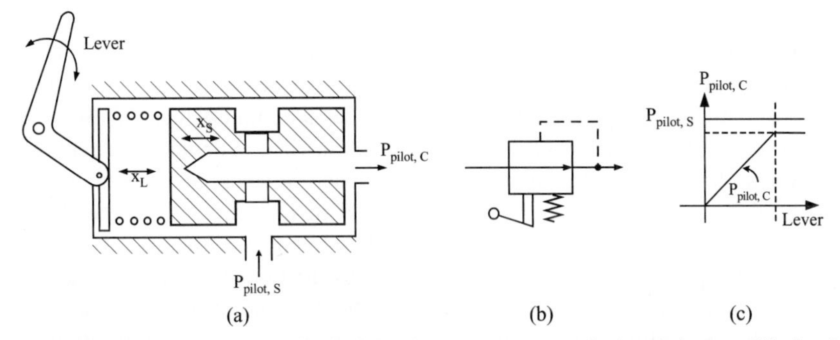


**FIGURE 7.44:** Lever-controlled pilot valve: a pressure-reducing valve modified to be used as a pilot valve. The output pressure is always smaller or equal to input pressure. The output pressure is proportional to the lever displacement, $P_{pilot,c} \propto x_L$, while $P_{pilot,c} \leq P_{pilot,s}$ (a) valve design concept, (b) symbol, (c) inport and outport pressures and control lever relationship. The lever movement sets the desired output pressure. The feedback from the output pressure to the spool moves to balance that and regulate to the set pressure.

A different version of pilot valve has two in-ports: pilot pressure supply and tank ports. The output pilot control pressure is regulated by the lever motion to be between the tank pressure and pilot supply pressure,

$$p_{tank} \leq p_{pilot,c} \approx K \cdot x_{spring} \leq p_{pilot,s} \tag{7.154}$$

where $x_{spring} = x_l + x_s$ in Fig. 7.44.

*Unloading valve* is a pressure relief valve where the pilot pressure which activates the valve to open comes from a line past the valve. For instance, a fixed displacement pump may need to charge an accumulator which maintains a line pressure. When the accumulator pressure exceeds the pump output pressure, the pilot pressure opens the relief valve, the check valve closes, and the pump flow is sent to the tank without having to work against the load, hence the name *unloading* valve [Fig. 7.45(a)].

*Sequence valve* is a pressure relief valve used in a circuit to make sure that a hydraulic line does not open until a certain pressure requirement at another location is met [Fig. 7.45(b)]. For instance, a pressure relief valve can be used to open or close a hydraulic line between two points based on the pressure feedback from another third location in the circuit.

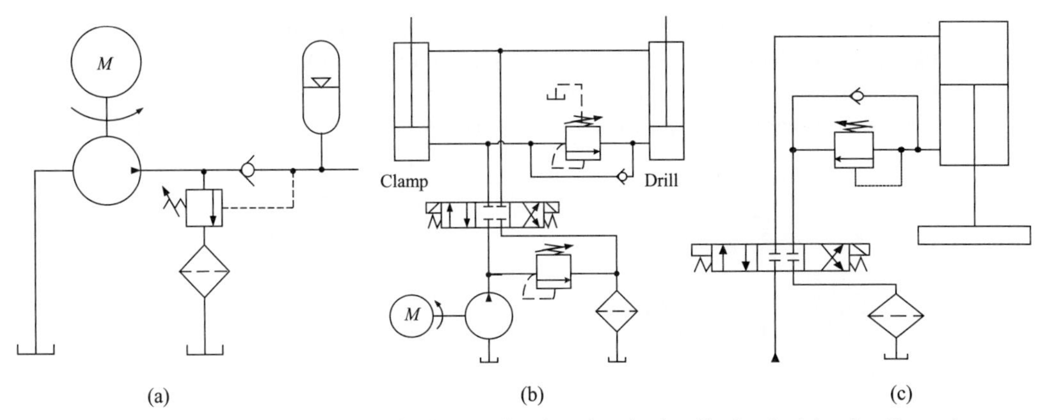


**FIGURE 7.45:** Pressure relief valve applications in a hydraulic circuit: (a) unloading valve, (b) sequence valve, (c) counterbalance valve.

Until the pressure in the third location reaches a preset value, the line between locations one and two is blocked. Such a valve may be used between a single-directional control valve and two cylinders: clamp and drill cylinders. The objective is that the drill cylinder should not move until the clamp cylinder pressure reaches a certain value. This is accomplished by a pressure relief type sequence valve between the drill cylinder and directional control valve output line. The control signal for the sequence valve comes from the pressure in the clamp cylinder. When the directional valve opens, the flow first goes to the clamp cylinder because the sequence valve makes the flow path to the drill cylinder have more resistance. When the clamp cylinder extends and pressure in the line increases, then the sequence valve opens and guides the flow to the drill cylinder. It should be noted that the sequence of motion between the two cylinders is based on pressure sensing, not based on position. Therefore, if the pressure increases in the clamp cylinder for some reason before the actual clamping operation is done, the drill cylinder will still actuate.

*Counterbalance valve* (also called load-holding valve) is a pressure relief valve. It does not open until a desired pressure differential is met in order to avoid sudden movements [Fig. 7.45 (c)]. They are used in load lowering applications such as lift trucks and hydraulic presses. The valve can be used to control the back pressure between cylinder and tank port so that the speed of the cylinder does not move uncontrollably due to large load, but is controlled by the valve on the cylinder to tank line.

The *shuttle* or *resolver* valve is commonly used as a selector valve in hydraulic circuits (Fig. 7.46). The valve has two input ports and one output port. The output port has the larger of the two input port pressures,

$$p_o = max(p_{i1}, p_{i2}) \tag{7.155}$$

### 7.4.2 Example: Multifunction Hydraulic Circuit with Poppet Valves

Figure 7.47 (also see Fig. 7.11) shows the hydraulic circuit schematics of the bucket control for the wheel loader model L120 by Volvo [90]. The load-sensing signal (LS) is used in the same way to control the pump displacement. That is, the maximum load pressure signal is resolved by use of a series of shuttle valves, and the resulting pressure is used to control the displacement of the pump. The maximum load signal is determined by comparing the load signals from the bucket hydraulics and steering hydraulic system (steering hydraulic system is not shown in the figure). That way, the pump is able to support the circuit with the most demanding load.

The lever command and pilot valves (one pair of pressure reducing valve (PMV1, PMV2) for shifting the spool of each main flow control valve) are very similar. The tilt and lift valves are closed-center, proportional flow control valves (MV1, MV2). The lift valve also has a so-called *float position* which is used when the bucket is riding on the ground

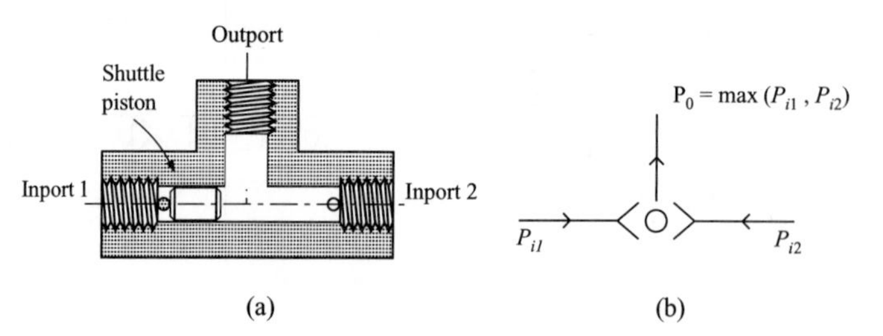


**FIGURE 7.46:** Shuttle valve is used as a maximum function. It selects the larger of the two pressures at its input ports and feeds it to its output port. a) cross-section, b) function.

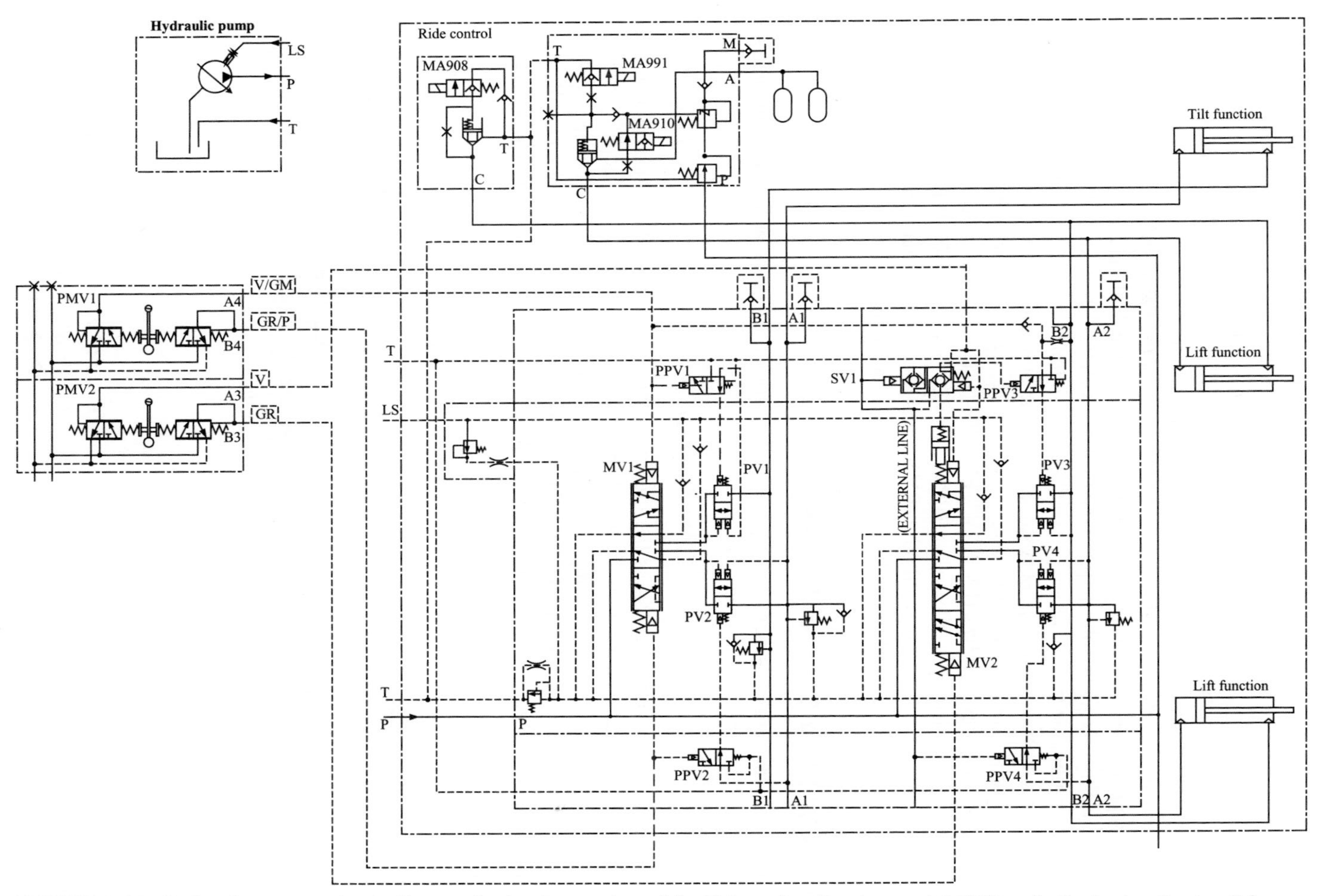


**FIGURE 7.47:** Hydraulic system circuit diagram for bucket motion of the Volvo wheel loader model 120E: only the hydraulic circuit for bucket motion is shown. The pump is the variable displacement axial piston type. Main flow control valves are connected to the hydraulic supply in parallel configuration and are closed-center, nonpressure compensated type. Poppet valves are used to seal the load on each end of cylinders for load holding valve function.

surface and desired to track the natural contour of the ground. At the float position, both sides of the cylinder are connected to the tank. This way, the cylinder follows the natural contour of the ground surface. The shuttle valve (SV1) is used to actuate the main valve spool to the float position when the command signal is larger than a certain set value. Also shown in the figure is the *ride control* or *boom suspension* circuit which reduces the oscillations transmitted to the machine frame between road surface and lift cylinder.

The flow between each service port (A and B) of the main flow control valve and the cylinder ports (head-end and rod-end) passes through a poppet valve. There are two poppet valves per valve–cylinder pair, one for each of the valve–cylinder port connection. The poppet valve provides a leakage free sealing on a line. Unlike a spool-type valve where some leakage is inevitable, poppet valves are excellent for leakage free sealing. In this configuration, the poppet valves are used as load-holding valves, i.e., when the main flow valve is in neutral position, poppet valve blocks the flow and seals the line. Hence, the load does not drift in position due to leakage problems. Hence, for lift and tilt circuits combined, there are four poppet valves (PV1, PV2, PV3, PV4). The spool displacement of the poppet valve is controlled by a pilot valve (PPV1, PPV2, PPV3, PPV4). When a pilot control pressure from the lever operated pilot (PMV1) valve acts on the main spool (MV1), it also acts on the pilot valve (PPV1) which then actuates the poppet valve.

### 7.4.3 Flow Control Valves

Flow rate through an orifice or restriction is a function of both the area of opening and the pressure differential across the orifice,

$$Q = K \cdot A(x_s) \cdot \sqrt{\Delta p} \tag{7.156}$$

where $Q$ is the flow rate, $\Delta p$ is the pressure drop across the valve, $A(x_s)$ is the valve orifice opening area as function of spool displacement $x_s$, and $K$ is proportionality constant (discharge coefficient). Noncompensated flow control valves set the orifice area only by moving a spool or needle based on a command signal. Figure 7.48(a) shows a needle valve used as a flow control valve where the needle position is manually adjusted. The orifice area

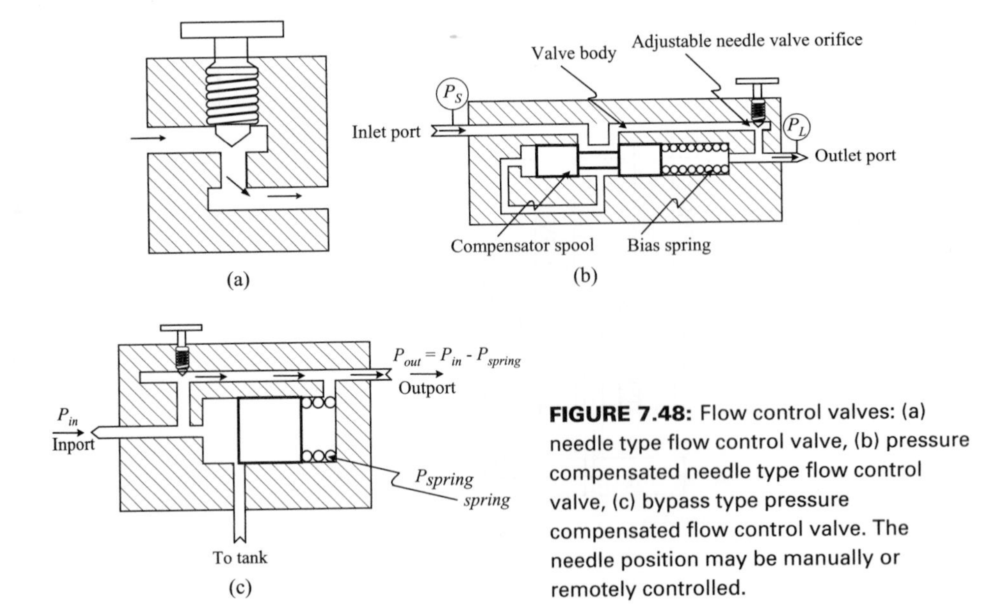


**FIGURE 7.48:** Flow control valves: (a) needle type flow control valve, (b) pressure compensated needle type flow control valve, (c) bypass type pressure compensated flow control valve. The needle position may be manually or remotely controlled.

is approximately proportional to the needle position. If the input or output pressure changes, the flow rate changes for a set needle position in accordance with the above orifice equation.

If it is desired that the flow rate should not change with pressure variations, the standard flow control valve can be modified with a pressure compensator spool and orifice. Such a valve is called *pressure compensated flow control valve*. There are two types of pressure compensated flow control valves: restrictor type and bypass type [Fig. 7.48 (b,c)].

There are two spool and two orifice areas in a pressure compensated flow control valve: One pair is the needle-orifice pair which sets the nominal orifice opening. Another pair modulates the second orifice opening based on input–output pressure feedback signals in order to maintain a constant pressure drop across the needle-orifice area. As a result, a constant flow rate is maintained at a constant setting of the needle even though input and output pressures may vary (since the second spool would compensate for it) as long as valve operating conditions do not reach saturation. This type is called *restrictor type* pressure compensated flow control valve because flow is regulated against pressure variations by adding restriction in the flow line [Fig. 7.48 (b)]. The valve regulates the pressure drop (tries to maintain it at a constant value) across the needle orifice.

Another type is the *bypass* type where an orifice opening bypasses excess flow to tank port as a function of pressure feedback signal. The output pressure is maintained at the load pressure plus the spring due to pressure, i.e., $p_{out} = p_{in} - p_{spring}$ [Fig. 7.48(c)].

Notice that the desired flow rate is set by the main orifice opening which is shown in the figures as being controlled by a manually moved needle-screw. This mechanism can also be proportionally controlled by an electric actuator, such as a proportional solenoid, which is the case in many pressure compensated EH flow control valves (see [67], www.HydraForce.com).

In a *pressure compensated valve*, a compensator spool moves in proportion to the difference between two pressure feedback signals. It is also called an *adjustable pressure compensated flow control valve*. In the *pre-compensator valve configuration*, one pressure is fed back from the output of the compensator valve and the other pressure is fed back from the output of the flow control valve (Fig. 7.49). In the *post-compensator valve configuration*, one pressure feedback is from the input pressure to the compensator valve, and the other pressure feedback is from the load pressure (Fig. 7.50).

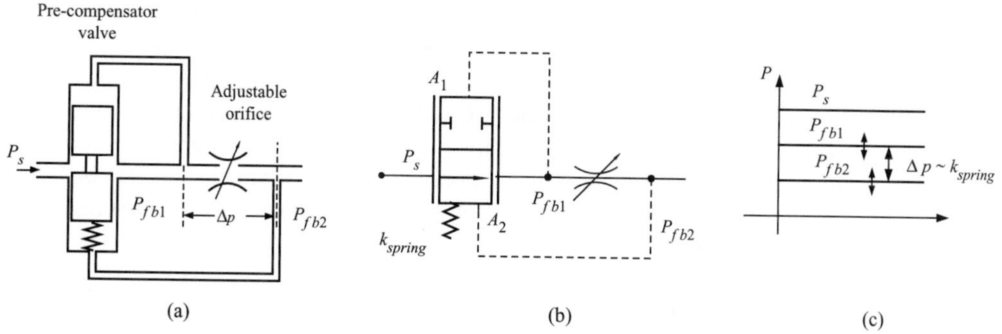


**FIGURE 7.49:** Pre-compensator valve configuration. Compensator valve is used to maintain a constant pressure drop across an orifice. This is accomplished by adding hydraulic resistance (pressure drop) in a circuit. The pre-compensator valve uses two pressure feedback signals for its control: compensator output port pressure signal and the output of the flow control valve following the compensator valve (i.e., maximum of two pressures of the cylinder if the flow control valve is connected to a cylinder). The pressure drop regulated across the main flow control valve is proportional to the spring constant of the pre-compensator valve. (a) component cross-sectional circuit diagram, (b) symbolic circuit diagram, (c) pressure relationships.

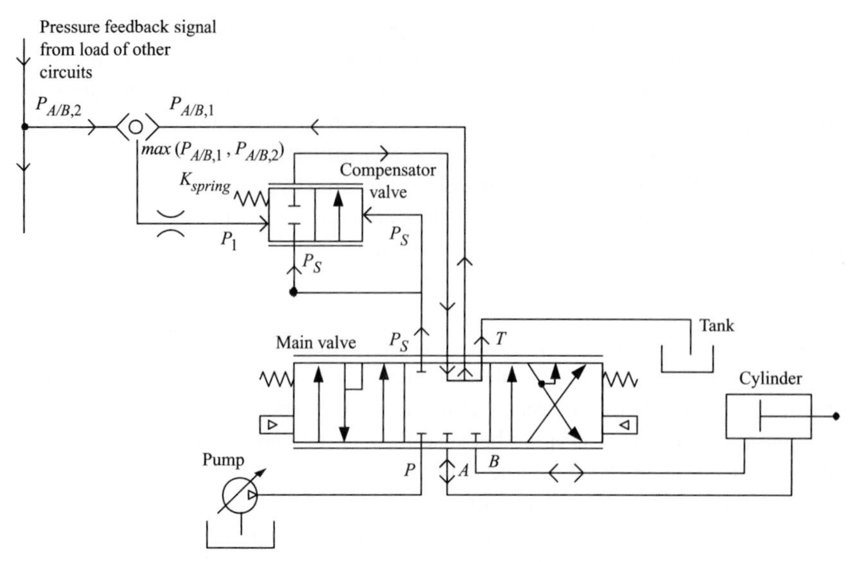


**FIGURE 7.50:** Post-compensator valve configuration. The two pressure feedback signal sources are the input pressure signal to the compensator valve and the load pressure. In multicircuit configurations, the load signal is the maximum of all load signals among circuits.

The objective of the compensator valve is to maintain a constant pressure drop across the main flow control valve. Hence, when the main flow control valve spool is at a certain position, flow rate across it would not be affected by the pressure variations at the pump supply or load side as long as the pressure variations do not reach saturation levels. The function of a compensator valve is to add restriction (pressure drop) in a hydraulic circuit. It is the analogy for variable resistance in series with another resistance in electrical circuits. A directional flow control valve in series with a pressure compensator valve is called a *pressure compensated directional flow control valve* or just *pressure compensated directional valve.* Pressure compensated valves are used in:

1. Circuits where load is variable and the flow rate should not vary as function of the load as long as valve is not saturated (fully open or fully closed).
2. Multiple parallel hydraulic circuits where all circuits share the same pump supply pressure, but the load pressure in each cylinder may be different. If a compensator valve is not used, most of the flow would go to the valve which has the lowest load, and other circuits would get very little flow. In order to provide equal flow on demand to each circuit regardless of the load pressure in other circuits, a post compensator valve is used in each circuit to add additional restriction (hence, pressure drop, just as a larger load would do) in the circuit so that all circuits effectively see the same load pressure [43a].

The pressure relationship in Fig. 7.49 is as follows:

$$p_s \geq p_{fb1} \geq p_{fb2} \tag{7.157}$$

that is, the outlet pressure of compensator valve is smaller or equal to the input pressure, and similarly, the output pressure of the flow control valve is smaller or equal to the pressure at its input. The $p_{fb2}$ pressure is determined by the load conditions downstream. The task of the compensator pressure is to maintain a constant pressure differential between the input–output ports of the main flow control valve, labeled as adjustable orifice in Fig. 7.49,

$$\Delta p_{set} = p_{fb1} - p_{fp2} \tag{7.158}$$

by controlling the intermediate pressure $p_{fb1}$ through the movement of compensator valve spool displacement ($\Delta x_{cs}$),

$$\Delta x_{cs} = \frac{1}{K_{spring}}(p_{fb1} \cdot A_1 - p_{fb2} \cdot A_2) \quad (7.159)$$

and this can be accomplished provided that the $p_s$ is large enough because $p_s \geq p_{fb1}$ is required. If the load pressure ($p_{fb2}$) is so large that the pressure differential between $p_s$ and $p_{fb2}$ is less than the desired pressure diffential, then the compensator valve modulation saturates and it tries to do its best by making $p_{fb1}$ as close as possible to $p_s$ (minimize the restriction). Hence, when supply pressure is constant and the load pressure increases to a level so large that the desired pressure differential between two ports is not possible, the compensator valve fully opens, trying to minimize the pressure drop,

$$p_{fb1} \approx p_s \quad (7.160)$$

$$\Delta P_v \approx p_s - p_{fb2} \leq \Delta P_{set} \quad (7.161)$$

This is the condition in which pump pressure supply has reached its maximum (saturation) level and load pressure is very high.

The post-compensator configuration (Fig. 7.50) uses two pressure feedback to its spool: input port pressure feedback to one side and output (i.e., maximum load) pressure feedback to the other side.

The spool movement of compensator valve is controlled by the following relationship (Fig. 7.50):

$$\Delta x_{cs} = \frac{1}{K_{spring}}(p_s \cdot A_s - p_l \cdot A_l) \quad (7.162)$$

and the compensator valve spool movement is proportional to the pressure difference between the supply and load pressure. The pressure $p_l$ is the load pressure feedback signal from either A or B service port of the cylinder. Let us consider that $p_s$ is constant. The load pressure $p_l$ drops due to smaller load. In order to maintain the same desired pressure drop, compensator valve spool displacement ($x_s$) will increase which in turn will increase the effective load pressure. Likewise, if the load pressure increases, compensator spool displacement will decrease to maintain constant pressure drop across the main flow control valve. In short, the compensator valve adds restriction (pressure drop) between pump line and the output line (A or B) as a function of the difference between the pump pressure and the load pressure.

In multiple circuits, a set of shuttle valves are used to select the maximum load pressure among all circuits. The maximum load pressure is needed for two purposes: (1) to control the pump displacement so that pump is able to support the circuit with most load (worst case) and (2) to control multiple compensator valves and try to equalize the flow distribution among circuits. There are two major types of pressure compensation in multifunction circuits: pre-compensator type and post-compensator type (Figs. 7.51 and 7.52).

Figure 7.51 shows a multifunction hydraulic circuit supplied by a variable displacement pump. Each circuit has a pre-compensated valve along with the proportional directional flow control valve. Each compensator valve uses two pressure feedback to regulate pressure differential across the main valve ahead of it. The pressure feedback signal for each compensator valve is local to that circuit: output pressure of the compensator and load pressure of the cylinder (maximum pressure of cylinder sides A and B are selected by a shuttle valve). Finally, the pump displacement is controlled by the maximum load pressure among all circuits so that pump runs at a displacement that can support the largest load among all circuits.

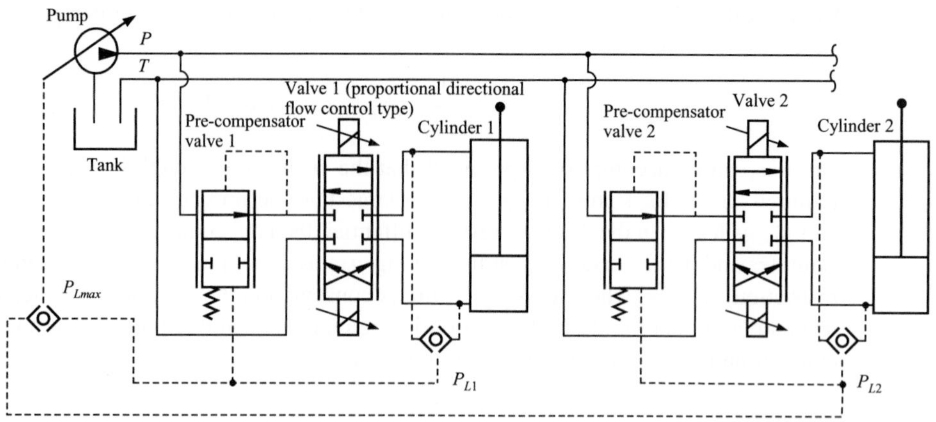


**FIGURE 7.51:** Pre-compensator valve configuration in a multifunction hydraulic circuit. The pump is the variable displacement type. Each function is connected to the supply line in parallel. Each compensator valve uses two pressure feedback signals for its control: compensator output port pressure signal and its load pressure signal (i.e., maximum of two pressures of the cylinder it is controlling). Each compensator uses its own circuit load pressure signal as feedback. In multifunction circuits, the pump is controlled by the maximum of all load pressure signals among multiple circuits. When the pump saturates (maximum flow capacity is reached), the flow will go to the circuit with the lowest load pressure, and the circuits with higher load pressures will get lower or no flow.

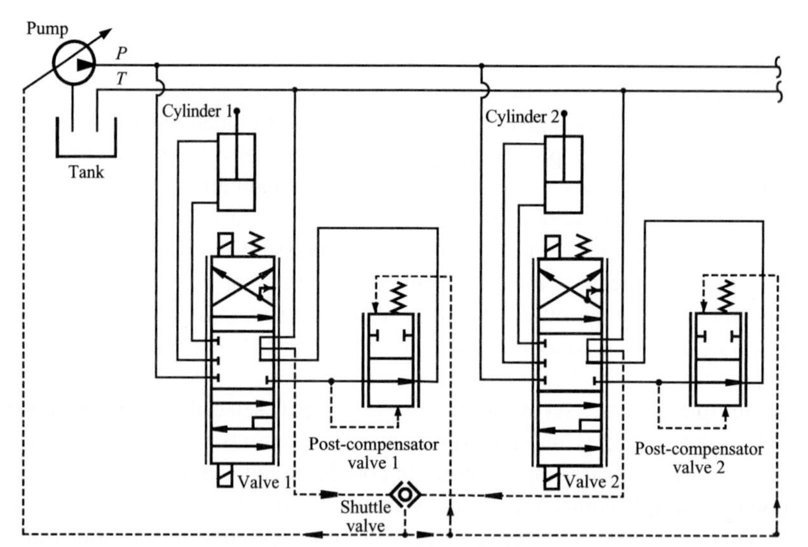


**FIGURE 7.52:** Post-compensator valve configuration in a multifunction hydraulic circuit. The pump is variable displacement type. Each function is connected to the supply line in parallel. Each post-pressure compensator valve uses two pressure feedback signals for its control: the input port pressure to it and the maximum of all the load pressures among multiple circuits. Therefore, one of the two pressure feedback signals to a post-compensator valve may originate from a different circuit in multicircuit applications. The same pressure signal is used for the pump control. Post-pressure compensator valves are used to equalize the flow distribution among multiple circuits. When the pump is saturated, all circuits get proportionally less flow (unlike the pre-compensated case). For instance, hydraulic implement lift and tilt circuits of the John Deere wheel loader model 644H (Fig. 7.53) use the same type of post-compensators along with each proportional directional flow control valve.

Figure 7.52 shows a similar multicircuit. The only difference is that the location of the compensator valves: post-compensator configuration. In each circuit, the sources of the two pressure feedback to the compensator valve are not all local: One of the pressure feedback is the pump pressure line which is the input line pressure to the post-compensator, and the second load pressure feedback is the maximum load pressure among all circuits. Hence, in the post-compensated configuration, the load pressure feedback (the second pressure feedback) to each compensator valve is determined by the rest of the circuits. It is always the maximum of all load pressures. The same maximum load pressure signal is used to control the pump displacement.

Under nonsaturated conditions, the performances of pre-compensated and post-compensated configuration of multicircuit hydraulic systems are very similar. When the pump capacity is saturated, the post-compensator configuration backs off the flow demand from each line proportionally and tries to maintain even flow delivery based on demand to each circuit, whereas in pre-compensated multicircuits, when the pump saturates, the circuit which has the largest load pressure gets less flow and may eventually get zero flow. In three or more function circuits, progressively the higher load pressure circuits get less flow and only the lowest pressure circuit gets most of the flow. The orifice in the pressure feedback line from the load has increased dampening effect (in addition to the added hydraulic resistance) so that high-frequency pressure spikes are filtered from the feedback signal to the compensator valve [43a].

The pressure compensated flow control valves works on the basic operating principle of maintaining constant flow despite pressure variations. Examining equation 7.156 shows that for a given spool displacement, the flow rate will change as function of pressure differential across the valve which may vary as a function of load and supply pressure variations. The basic principle is to change the valve opening $A(x_s)$ by shifting the spool as a function of pressure feedback. As a result, the effective spool displacement is not only function of the current but also the pressure drop. Hence, if the pressure drop increases, feedback moves the spool to reduce the orifice area. Similarly, if the pressure drop reduces, feedback moves to increase the orifice area. The end result is to maintain the almost constant flow for a constant current signal under varying load conditions. The feedback mechanism to shift the spool as a function of pressure drop can be implemented with hydromechanical or electronic means. Normally, a pressure compensated flow control valve is implemented with two valves in series: One valve is the proportional directional flow control valve, and the other valve is the pressure compensator valve. Examples of such implementations are shown in Figs. 7.50–7.52. It should be noted that if two pressure sensors were available at the input and output ports of the proportional valve, the solenoid current can be controlled to perform the function of the second valve, while at the same time metering the flow. In other words, the control algorithm that decides on the solenoid current to shift the spool can take not only the desired command signal but also the two pressure signals into account. Hence, we can implement a pressure compensated flow control valve using a solenoid controlled proportional valve, two pressure sensors, and digital control algorithm.

### 7.4.4 Example: A Multifunction Hydraulic Circuit Using Post-Pressure Compensated Proportional Valves

Figure 7.53 (also see Fig. 7.11) shows the hydraulic circuit schematics of the bucket control for the wheel loader model 644H by John Deere [91]. This circuit serves as a good example of post-pressure compensated proportional flow control valves. The figure shows only the hydraulic system relevant to control the bucket (lift, tilt, and auxiliary functions). The rest

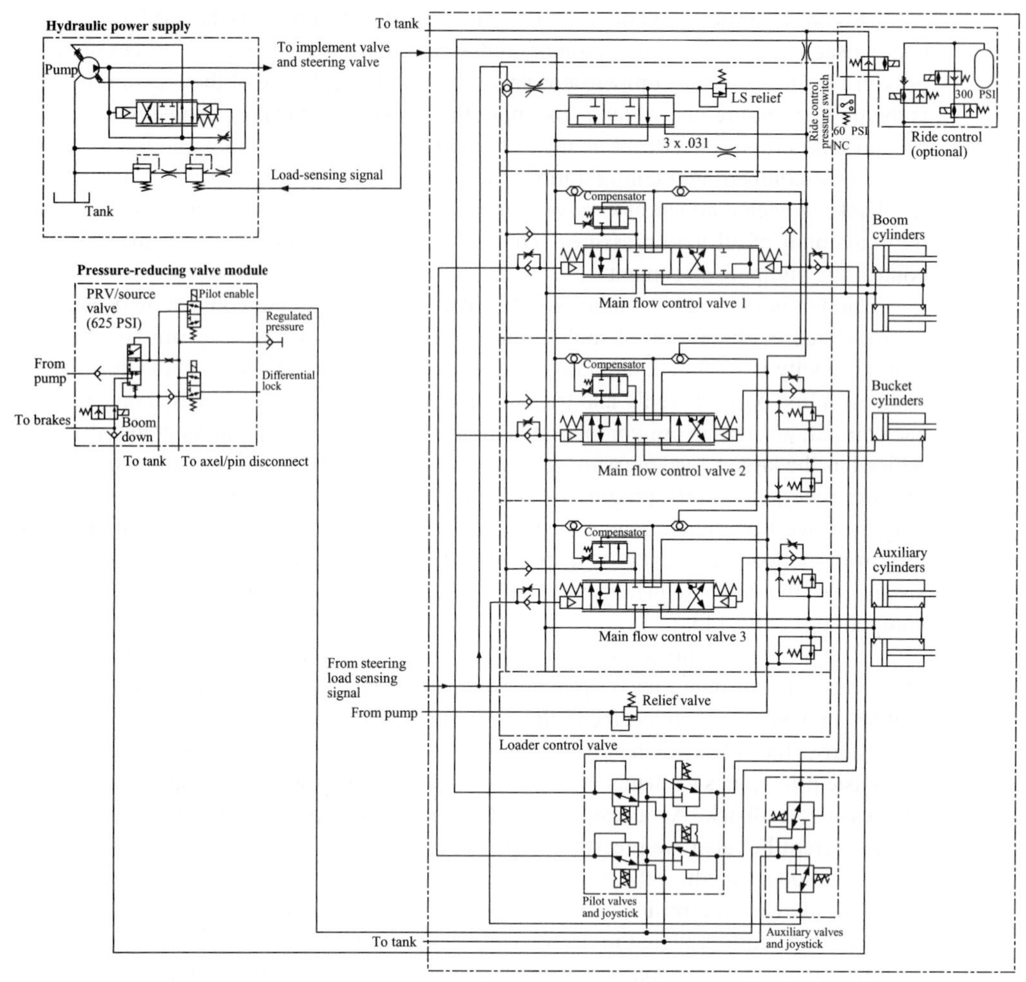


**FIGURE 7.53:** Hydraulic system circuit diagram for bucket motion of the John Deere wheel loader model 644H: only the hydraulic circuit for bucket motion is shown. Hydraulic circuits for steering, brake, and cooling fan are not shown. Pump is the variable displacement axial piston type. Main flow control valves are connected to hydraulic supply in parallel configuration. They are closed-center types and post-compensated for pressure drop.

of the hydraulic circuit for the machine (steering, brakes, cooling fan drive) is not shown. The key features of this hydraulic circuit are:

1. Pump is variable displacement type and controlled by a load-sensing hydraulic mechanism.
2. Valve–cylinder pairs for each function are closed-center types and are connected to the hydraulic supply lines (P and T) in parallel configuration.
3. Each proportional directional main flow control valve has a post-compensator valve in order to maintain constant pressure drop across each main valve.

The hydraulic circuits (steering, implement hydraulics, and brake hydraulics) are supplied by a single variable displacement axial piston pump. The pump is controlled by a load-sensing hydraulic circuit. Since the same pump supports multiple hydraulic circuits which may have different loads, it is necessary to sense all loads and use the highest load signal in control of the pump. Therefore, the pump displacement is determined by the maximum load pressure signal among steering circuits (not shown in the figure), boom circuit, bucket, and auxiliary circuits. The selection of the maximum load pressure is made by a series of shuttle valves.

The valve–cylinder pair for each function is connected in parallel configuration to the pump and tank lines. Furthermore, each directional flow control valve (proportional type) is accompanied by a post-compensator valve. As discussed in Fig. 7.52, the post-compensator valve works to maintain a constant pressure drop across the main valve so that the function speed is proportional to the lever command regardless of the load as long as the load is not so large as to saturate the pump. The load pressure feedback is the maximum load pressure among all circuit load pressures. This selection is made by a series of shuttle valves. Notice that due to the load-sensing signal needed for the compensator, the main valve has three additional lines (in addition to the P, T, A, B main lines): Two of them are for the input and output of the compensator valve connection, and one of them is the load pressure sensing line for feedback control of the compensator. In addition, there are two pilot pressure ports, one for each side of the spool, which comes from the pilot valves. The check valve and fixed orifice in parallel with it between the pilot control valve signal and the main valve spool on both sides have the effect of dampening the pressure oscillations on the pilot control line and result in smoother operation of the main valve.

The pilot pressure supply for the main valve control of implement hydraulic is derived from the output of the main pump pressure using a pressure-reducing valve. The output of the pressure-reducing valve is a constant pilot supply pressure. The pilot valves allow the operator to control the pilot output pressure to the main valves. For each function (lift, tilt, auxilary), there are two pilot valves. Each valve has two inports (pilot pressure supply port and tank pressure port) and one output port (output pilot pressure used to control the main valve). The output pilot pressure is approximately proportional to the mechanical movement of the lever controlled by the operator. As the operator moves the lever, i.e., to lift the bucket, the pilot valve connected to the lift lever sends pilot pressure (proportional to the lever displacement) to the main valve spool's control port. The main valve spool is spring centered, and its displacement is proportional to the pilot pressure differential between the two sides (one side is at the tank pressure, the other side is modulated by the pilot valve output). Hence, if we assume a constant pressure drop across the main valve, the cylinder speed will be proportional to the main spool displacement which is in turn proportional to the lever displacement. In approximate terms, the cylinder speed is proportional to the lever displacement.

### 7.4.5 Directional Flow Control Valves: Proportional and Servo Valves

Directional flow control valves are categorized based on the following design characteristics (ISO 6404 standard):

1. Number of external ports: two-port, three-port, four-port. Number of ports refers to the plumbing connections to the valve which can be 2, 3, 4, or more. A four-port valve connects the pump and tank (P, T) ports to two load ports (A, B) (Fig. 7.39).
2. Number of discrete or continously adjustable spool position: ON/OFF two-position, ON-OFF three position, proportional valves.
3. Neutral spool position flow characteristics: open center where one or more ports are connected to tank (i.e., P to T, A to T, B to T, A and B to T, P and A to T, P and B to T) or closed center where all ports are blocked. Closed-center valves are generally

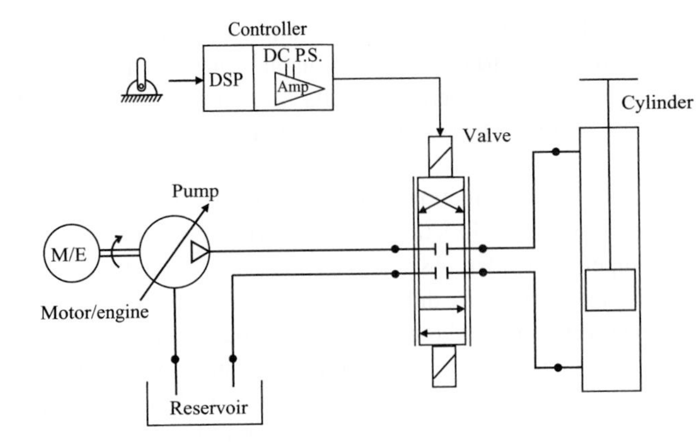


**FIGURE 7.54:** Closed-center EH motion system for one-axis motion: valve is closed-center type, pump is variable displacement type.

used with variable displacement pumps, whereas open-center valves are used with fixed displacement pumps (Figs. 7.54 and 7.2).

4. Number of actuation stage: single-stage, two-stage, and three-stage valve spool actuation.
5. Actuation method: referring to the first stage actuation whether by mechanical lever (manual), electrical actuator (i.e., solenoid, torque motor, linear force motor) or by another modulated pilot pressure source.

The "servo" valve and "proportional" valve names both refer to controlling the spool displacement of the valve *proportional* to the command signal (i.e., solenoid current). A servo valve has custom matched spool and sleeve, machined to higher tolerances than the spool for a proportional valve. A proportional valve has spool and valve body. The fitting clearance between spool-sleeve of servo valves is in the range of 2–5 microns, whereas the clearance between the spool and body of proportional valve is in the range of 8–15 microns or more. For all practical purposes, the only differences between these two types of valves are that:

- Servo valves use feedback from the position of main spool to decide on the control signal (i.e., the feedback may be implemented by completely hydromechanical means or by electrical means), whereas proportional valves do not use main spool position feedback. However, it should be pointed out that more and more proportional valves have started to use spool position feedback in recent years. As a result, the differentiation between servo and proportional valves based on spool position feedback started to disappear.
- Servo valves generally refer to higher bandwidth valves.
- Servo valves provide better gain control around null position of the spool. As a result, servo valves can provide higher accuracy positioning control. The average deadband in servo valves is in 1–3% of maximum displacement, whereas it can be as high as 30 to 35% in proportional valves.

Otherwise, their construction and control are very similar.

The performance specifications of a proportional directional EH valve include:

1. Flow rating at a certain standard pressure drop, i.e., 180 liters/min at a standard pressure drop across the valve, when valve is fully open. The standard pressure drop for flow rating of the servo valves is 1000 psi and of proportional valves is 150 psi.

2. Maximum and minimum operating pressures, $p_{max}$, $p_{min}$
3. Maximum tank line pressure, $p_{t,max}$
4. Number of stages (single-stage, double-stage, triple-stage)
5. Control signal type: manual, pilot pressure, or electrical signal controlled
6. Pilot pressure range ($p_{pl,max}$, $p_{pl,min}$) and pilot flow ($Q_{pl}$) required if pilot pressure is used
7. Open-center or closed-center type spool design
8. Linearity of current–flow relationship at a constant pressure drop (Fig. 7.68)
9. Symmetry of current-flow relationship between plus and minus side of current
10. Nominal deadband
11. Nominal hysterisis
12. Maximum current from the amplifier stage if solenoid operated (and PWM frequency and dither frequency if used)
13. Bandwidth of the valve at a specified supply pressure ($w_v$ [Hz]) or rise time for a step command change in flow at a certain percentage (i.e., 25%) of maximum flow rate
14. Operating temperature range

The specifications for the electronic driver (power supply and amplifier circuit) include:

1. Input power supply voltage (i.e., 24-VDC nominal, or ±15 VDC) and current rating to the drive power supply section
2. Control signal type and range (i.e., +/ − 10 VDC, ±10 mA, or 4 to 20 mA, analog or PWM signal)
3. Recommended *dither* signal frequency and magnitude (i.e., less than 10% of rated signal at a higher frequency than the valve bandwidth) in order to reduce the effect of friction between the spool and valve body
4. Feedback sensor signal type (if any for a driver which uses local valve spool position feedback to control the spool position)

For flow rates under 20 to 50 gal/min (gpm), a single-stage direct actuated valve is generally sufficient. For flow rate ranges between 50 and 500 gpm, a two-stage valve is used. For flow rates over 500 gpm, typically a three-stage valve is used. In a single-stage valve, the electrical current sent to the solenoid (or linear torque motor), which creates electromagnetic force, directly forces the main spool and moves it (Fig. 7.55). Multistage valves (two- or three-stage) that use pilot pressure to amplify the electrical actuator signal to shift the main spool use pilot pressure typically at 50% or more of the main supply

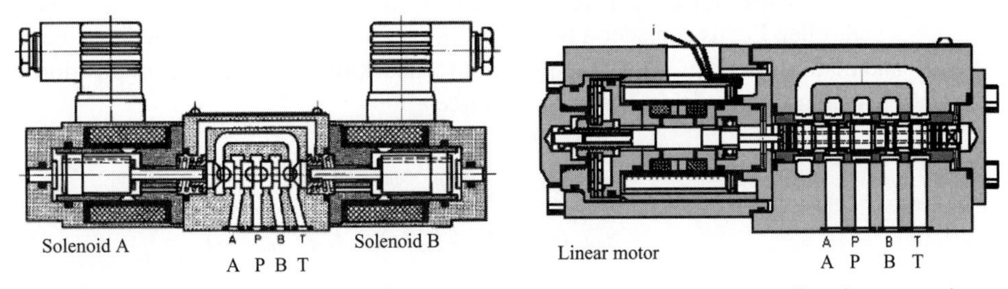


**FIGURE 7.55:** Single-stage EH valve-solenoid (or linear motor) armature directly moves the valve spool.

pressure. This provides a very large amplification gain between the first-stage electrical actuator signal and second-stage pilot force. This gain cannot be matched with any direct drive electric motor technology currently. The large pilot stage gain generated shifting power also makes the valve less sensitive to main stage contamination problems because the large shifting pressure of pilot stage is very likely to able to force through contamination problems. However, the contamination problem is more likely to occur at the pilot stage of the valve because the orifices at the pilot stage are much smaller than those in main stage.

There are three main types of two-stage proportional and servo valve designs:

1. Spool-spool (double-spool) design
2. Double-nozzle flapper design
3. Jet pipe design

In a two-stage valve, the electrical current moves an intermediate spool, which then amplifies power using the pilot pressure line to move the second main spool (Fig. 7.56). In multistage valve cases, the valve is so large that the electrically generated force by the solenoid (or electric actuator) is not large enough to move the main spool directly.

The same concept applies to the three-stage valves (Fig. 7.57) where there are three spools (or poppets) in the valve, first two acting as the amplifiers to move the final third stage main valve spool. The great majority of valves used in mobile equipment applications are two-stage valves. From a control system perspective, the functionality of single-stage and multistage valves is the same: Input current is translated proportionally into the main spool displacement (with some dynamic delay and filtering effects, of course) which is then results in proportional to the flow rate under constant pressure drop across the valve.

Notice that the second-stage spool position (main spool) of a two-stage valve is integral of the first-stage spool position (pilot spool) if there is no feedback from the main spool position to the pilot spool position. Two-stage spool valve is basically a single-stage direct acting valve connected to a second valve spool. The second valve can be viewed as a small cylinder connected to a single-stage valve. Therefore, the current-main spool displacement relationship is not proportional, but an integral relationship.

$$x_{main}(s) = \frac{1}{s} K_{mp} \cdot x_{pilot}(s) \tag{7.163}$$

$$= \frac{1}{s} K_{mi} \cdot i_{sol}(s) \tag{7.164}$$

*In order to make the current-main spool displacement a proportional relationship, like the case in a single-stage valve, there has to be a "feedback mechanism" on the main spool position.* The pilot spool position is determined by the solenoid force and the feedback from the main spool position. The feedback from the main spool to pilot spool may be in the form of a mechanical linkage, spring, pressure, or in the form of an electronic sensor signal to a controller. Let us consider a two-stage valve which uses pressure feedback between pilot and main stage (Fig. 7.56). The pressure feedback from the main spool position acts like a balancing spring against the solenoid force

$$x_{main}(s) = \frac{1}{s} K_{mp} \cdot x_{pilot}(s) \tag{7.165}$$

$$x_{pilot}(s) = K_{pi} \cdot i_{sol}(s) - K_{pf} \cdot x_{main}(s) \tag{7.166}$$

$$x_{main}(s) = \frac{K_{mp} \cdot K_{pi}}{s + (K_{pf} \cdot K_{mp})} \cdot i_{sol}(s) \tag{7.167}$$

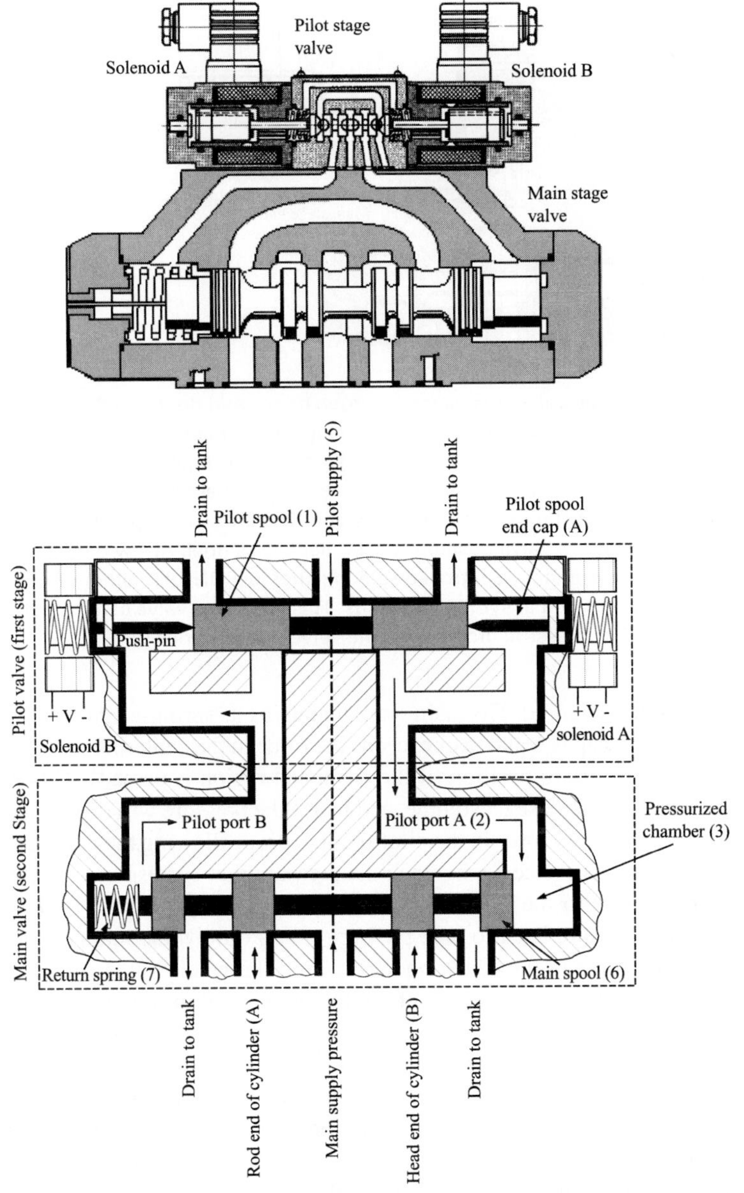


**FIGURE 7.56:** Two-stage valve: first-stage spool is for pilot stage, second-stage spool is for the main stage.

This feedback mechanism can be implemented either by hydromechanical means or by using spool position sensor and closed-loop control algorithms.

Let us discuss how this feedback is physically accomplished with hydromechanical means (Fig. 7.56). When solenoid B is activated, its corresponding force pushes directly against the pilot spool to open the pilot port A to pilot pressure supply. The pressure built up in the chamber on the right side of the main spool is also directly fed back to the end cap (A) of the pilot spool. When the force in the end cap equals or exceeds the force from the solenoid in B, the spool shifts back and closes the opening to pilot supply. The solenoid-feedback combination will now hold a pressure in port A that is proportional to

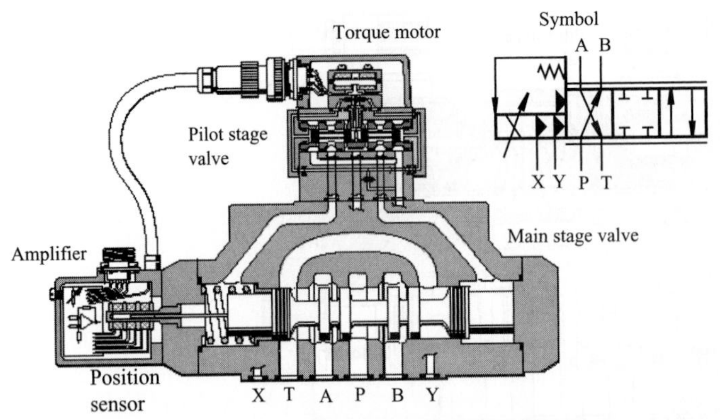


**FIGURE 7.57:** Three-stage valve: first stage is torque motor and double-nozzle flapper, second-stage spool is the pilot valve, and the third stage is the main valve. Second stage amplifies the actuation force of the first stage to move the third stage (main spool). The X and Y letters in the valve symbol indicates that the valve has two pilot stages. Hence, it is a three-stage valve, two pilot stages plus the main stage. The arrow on top of the solenoid signal around the spring indicates a valve position sensor feedback signal. Arrow over the solenoid indicates that it is a proportionally controlled solenoid, not just ON/OFF, although this representation is often omitted in symbol drawings. This particular valve can support flow rates up to 1500 l/min at 10 bar pressure drop across the valve (Moog series D663).

the input current to the solenoid. The main spool maintains its spring-centered position until the pressure in the end gap is equal to or exceeds the force in the centering spring. The main spool will shift to a metering position (if the deadband region is passed). This movement will cause the port A to increase its volume, and hence the pressure will drop. This pressure drop will cause the feedback force in the pilot spool to drop, allowing the pilot spool to shift open to supply pressure to once again reach the pressure that corresponds to the solenoid input. When the current to the solenoid is cut, the pilot spool will shift back to open to tank and lower the pressure back to tank pressure. Accordingly, the main spool will shift back because of the force in the centering spring until the whole valve is back to neutral position. In summary, during transient motion, the pilot spool moves in proportion to the solenoid current, then the built-up pressure in pilot port moves the main spool. The same pilot pressure is feedback to the pilot spool end caps. In steady state, the main spool is at a position where the spring force balances the built-up pilot pressure. The built-up pilot pressure is proportional to the solenoid current. The pilot spool returns to neutral position (closed) after the transients when a constant current is applied and main spool reaches the proportional position. Without the pressure feedback to the pilot spool end caps, under constant current, the pilot spool would shift a constant amount, and the main spool position would keep increasing as integral of the pilot spool, hence that of solenoid current. The key to making the current and main spool position proportional in steady state is the pressure feedback between the main spool and pilot spool end caps.

There are also two-stage proportional valves where the main spool is shifted by two pressure-reducing valves, one on each side (Fig. 7.58 [92]). Such valves are widely used in construction equipment applications (Figs. 7.47 and 7.85). A pair of pressure-reducing valves acts as a pilot valve for the main stage. The output pressure of the pilot valve is proportional to the lever displacement or solenoid current. A proportional main spool displacement is developed as a result of the balance between the the pilot pressure output of the pressure-reducing valve and the centering spring of the main spool. The input–output relations in two-stage proportional valves where the first stage is a pair of pressure-reducing

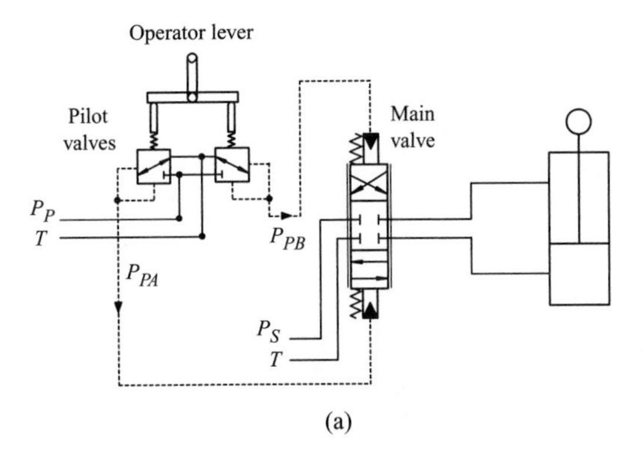


(a)

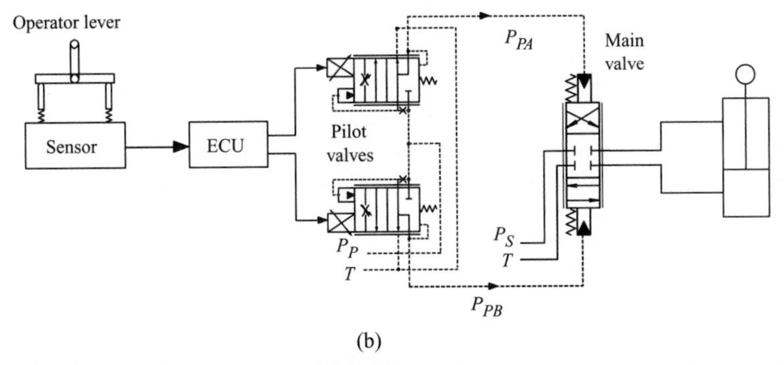


(b)

**FIGURE 7.58:** Two-stage proportional valve with two pressure-reducing valves at the pilot stage. Pilot stage output is a pressure proportional to the input command (mechanical or electrical). The main spool displacement is proportional to the pilot output pressure (control pressure) as a result of the centering springs. (a) Mechanically actuated pilot valves (for instance, this type of valves is used on Komatsu wheel loader model WA450-5L for its bucket control system), (b) electrically actuated pilot valves.

valves are as follows:

$$p_{pc} = \frac{K_{pi}}{(\tau_{pi}s + 1)} \cdot i_{sol} \tag{7.168}$$

$$x_{main} = \frac{1}{K_{spring}} A_{main} \cdot p_{pc} \tag{7.169}$$

$$= \frac{K_{pi} A_{main}}{K_{spring}(\tau_{pi}s + 1)} \cdot i_{sol}(s) \tag{7.170}$$

where $p_{pc}$ is the pilot pressure output form the pressure-reducing pilot valves to the main spool end caps, $i_{sol}$ is the solenoid current, $x_{main}$ is the main spool displacement, $K_{spring}$ is the centering spring constant of the main spool, $A_{main}$ is the cross-sectional area of the main spool at the end caps where the pilot control pressure acts, and $\tau_{pi}$ is the time constant of the transient response between current and displacement. Notice that each pilot control valve (pressure-reducing pilot valve) may be controlled electrically by a solenoid (Fig. 7.58b) or manually by a mechanical lever (Fig. 7.58a). In mechanically actuated pilot valve version, the pressure-reducing pilot valve pair is mounted right under the operator control lever. The output pilot control pressure line is then routed to the main valve which is likely to be closer to the cylinder. In the electrically controlled version (EH version), the main valve and the pilot valve along with solenoid pair can all be located at the same location and control signals

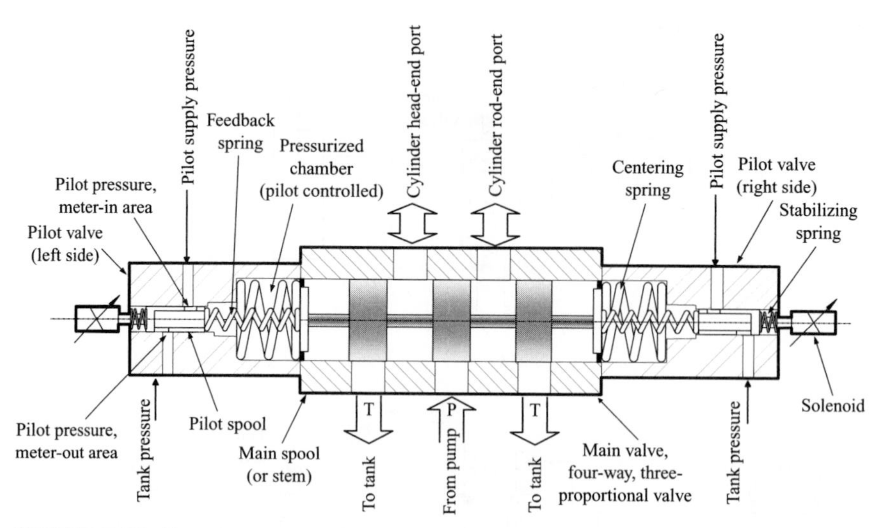


**FIGURE 7.59:** Two-stage proportional valve with two pressure-reducing valves at the pilot stage in line with the main spool. Pilot stage output is a pressure proportional to the solenoid current. The main spool displacement is proportional to the pilot control pressure as a result of the centering spring. The feedback between the main spool and pilot spools (one on each side of main spool) has mechanical springs in addition to the hydraulic feedback.

from the operator lever sent to the solenoids electrically. Therefore, the EH version has fewer transient delays associated with the pilot control pressure transmission.

A variation of the two-stage proportional valve with pressure-reducing pilot valve pair is shown in Fig. 7.59. The differences between the two versions (Fig. 7.58 and Fig. 7.59) are as follows:

1. Pilot valve pair (pressure-reducing valve pair) is in line with the main spool, and there is a mechanical spring feedback between the pilot stage spools (one on each side of main spool) in addition to the hydraulic pressure feedback.
2. Main spool is actuated by reducing the pilot control pressure in one side as opposed to increasing it in the previous version. In order to move the main spool to the left, the solenoid on the left side is energized and pilot pressure is dropped to a lower value proportionally on that side. Hence, the main spool moves to the left until the centering spring on the left side balances the forces on both sides. When the current to the solenoid is reduced, the pilot control pressure increases and pushes back the main spool to neutral position. The operation for shifting the main spool to the right follows a similar relationship.
3. Centering springs follow the main spool to the pilot side but not into the main spool body side. As a result, when the main spool shifts to the left, it works against only the centering spring on the left side. When it shifts to the right side, it works against only the centering spring on the right side.

The double-nozzle flapper design is the most common two-stage servo valve used in high-performance applications. This is the most accurate, but expensive, type of servo valve. Stage 1 is the torque motor plus the double-nozzle flapper mechanism, stage 2 is the main valve spool. Figure 7.60 shows the cross-sectional picture and operating principle of a two-stage double-nozzle flapper type valve. Torque motor is a limited rotation permanent magnet electric motor with a coil winding. Direction of the current in the winding determines the direction of the torque generated. The permanent magnet strength and air gap between

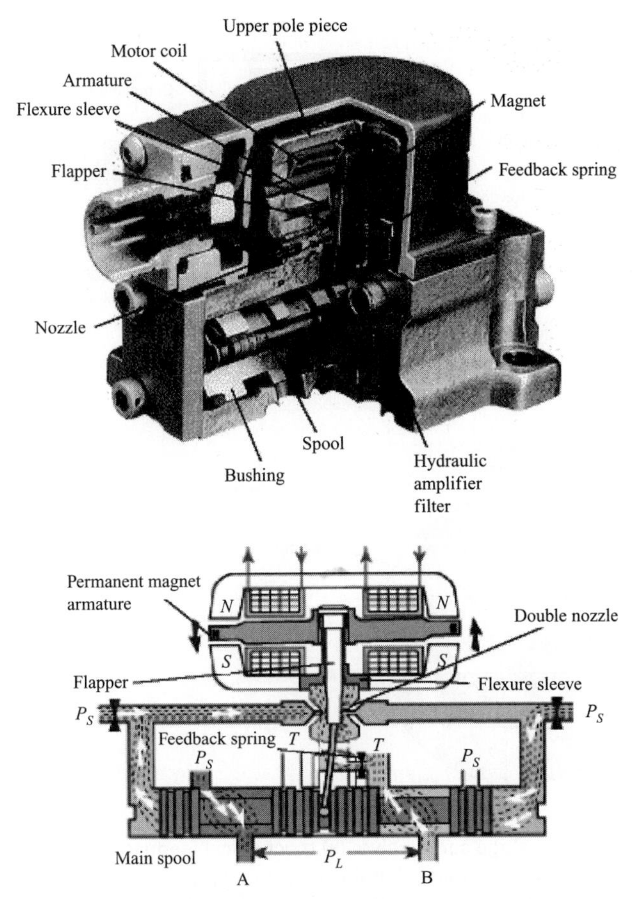


**FIGURE 7.60:** Double-nozzle flapper type two-stage servo valve—the most popular two-stage servo valve type.

the armature and the winding determine the current to torque gain. Hence, the current magnitude determines the torque magnitude. Armature and flexure tube are mounted around a low friction pivot point about which they rotate as the motor generates torque. When current is applied to the torque motor winding, the permanent magnet armature rotates in the direction based on the current direction, and by an amount proportional to the current magnitude. The flexure sleeve allows the armature-flapper assembly to tilt. As a result, the nozzle at the tip of the flapper is changed and hence different pressure differential is created between the two sides. The pressure on the side of the nozzle where the flapper is closer gets larger. This pressure differential moves the main spool and along with it the feedback spring until the pressure is balanced on both sides of the flapper at the double-nozzle point. At this time, the main spool is positioned in proportion to the input current. The input current is translated proportionally into the main spool displacement. Notice that due to the small nozzles at the double-nozzle and flapper interface, the filteration of the fluid at the pilot stage is very important in order to avoid contamination-related failures of the valve operation at the pilot stage.

The jet pipe valve design is very similar to the double-nozzle flapper design, except that the pilot stage amplification mechanism is different (Fig. 7.61). Stage 1 is the torque motor plus the jet pipe nozzle and two receiver nozzles. Stage 2 is the main spool. The pilot pressure is fed through the jet pipe nozzle which directs a very fine stream of fluid to two receivers. When current in torque motor is zero (null position), the jet pipe directs this very

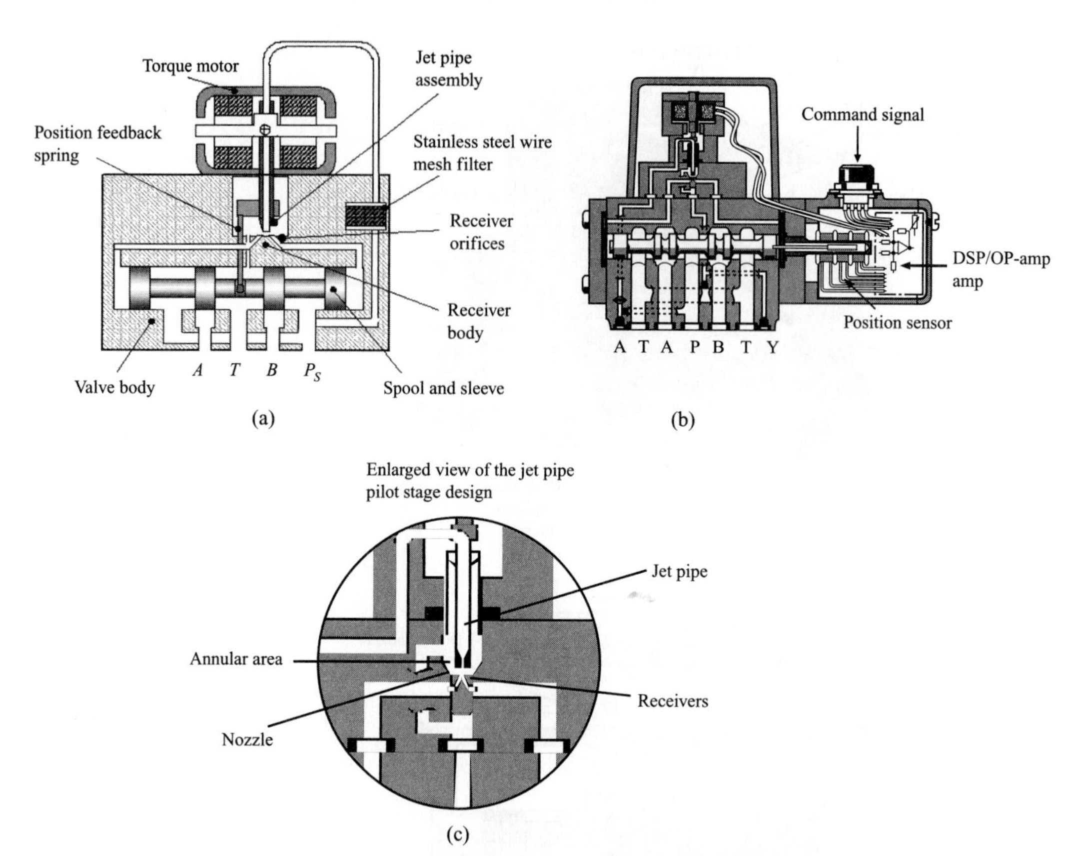


**FIGURE 7.61:** Jet pipe type two-stage servo valve: (a) mechanical feedback of the main stage spool position, (b) electrical sensor feedback of main stage spool position, (c) enlarged view of the jet pipe and receiver nozzles. In steady state the main spool displacement is proportional to the current of the torque motor.

fine stream of "jet" flow equally to both receivers and there is a balance between the two control ports holding the main spool in null position. When current is applied, the torque motor deflects the armature and the jet pipe proportionally. As a result, the jet pipe directs a different amount of fluid stream to two receivers, hence it creates a pressure differential in the control ports (two sides of the main spool). This pressure differential moves the main spool until the pressure differential in the control ports are stabilized. The feedback from the main spool movement to the jet pipe is provided either mechanically by a flexure spring between the nozzle and the main spool position or electrically by a spool position sensor (typically an LVDT) and closed-loop spool position control algorithm used to determine the torque motor current (Fig. 7.61).

In general, jet pipe design has lower bandwidth, requires lower manufacturing tolerances, and is able to tolerate more contamination compared with the double-nozzle flapper servo valve. The jet pipe pilot stage can support larger pilot force than the double-nozzle flapper valves, hence it tends to be used in large valves more often. In high flow rate applications (i.e., 1500 liters/min), a double-nozzle flapper or jet-pipe valve is used as a pilot stage of a three-stage servo valve. In modern three-stage proportional or servo valves, the position loop on the main spool (third stage) is generally closed using an electrical position sensor. Notice that in both types of servo valves, the pressure for the hydraulic amplification stage can be provided either from a separate pilot pressure supply source or can be derived

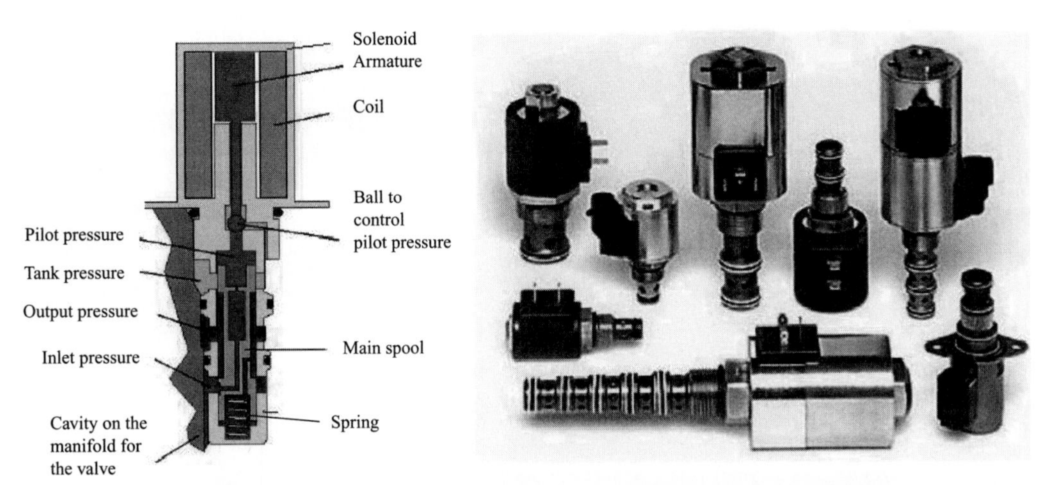


**FIGURE 7.62:** Examples of cartridge valves. The cross-sectional figure shows a cartridge valve with a sliding spool type metering element.

from the main supply line pressure using a pressure-reducing valve or orifice. Filtration of the fluid especially in the hydraulic amplifier stage (pilot stage) is very important since the nozzle and receivers have very small diameter dimensions and are prone to contamination. Flow orifices in the main stage spool are rather large and less prone to contamination. Therefore, filteration of the fluid is most important for the pilot stage.

Single-stage direct drive valves (DDV) currently support up to about 150 liters/min (lpm) flow rate (i.e., Moog D633 through D636 series). Two-stage jet pipe servo valves support flow rates up to about 550 lpm at 70 bar pressure drop across the valve (i.e., Moog D661 through D664 series). A three-stage valve where the first two stage is either a double-nozzle flapper or jet pipe type servo valve can support up to 1500 lpm flow rate (i.e., Moog D665 and D792 series) [66].

Cartridge valves are designed to be assembled on a manifold. Manifold can be made of a single cartridge valve (single function manifold) or for multiple valves (multifunction manifold). A manifold block may typically hold multiple cartridges, and other types of valves (Fig. 7.62 and Fig. 7.63). Cavity sizes (diameter, depth, tread) on the manifolds are standardized so that cartridge valves from different manufacturers can be used interchangeably.

Cartridge valves can be categorized in terms of different criteria as follows:

1. Mechanical connection to the manifold:
   - **(a)** Screw-in type which is installed by screwing valve threads into manifold cavity threads.
   - **(b)** Slip-in type which is installed in the manifold by a bolted cover to manifold. Screw-in type cartridge valves support flow rates up to about 150 lpm, and slip-in types support flow rates above 150 lpm. Slip-in types have the advantage over the screw-in types in that they do not squeeze the ports and hence achieve better repeatability in the assembly. There are seven standard slip-in cartridge valve sizes (specified by ISO 7368 and DIN 24342) where the nominal valve port diameter is 16 mm, 25 mm, 32 mm, 40 mm, 50 mm, 63 mm, and 100 mm, supporting flow rates in the range of 200 lpm to 7000 lpm.

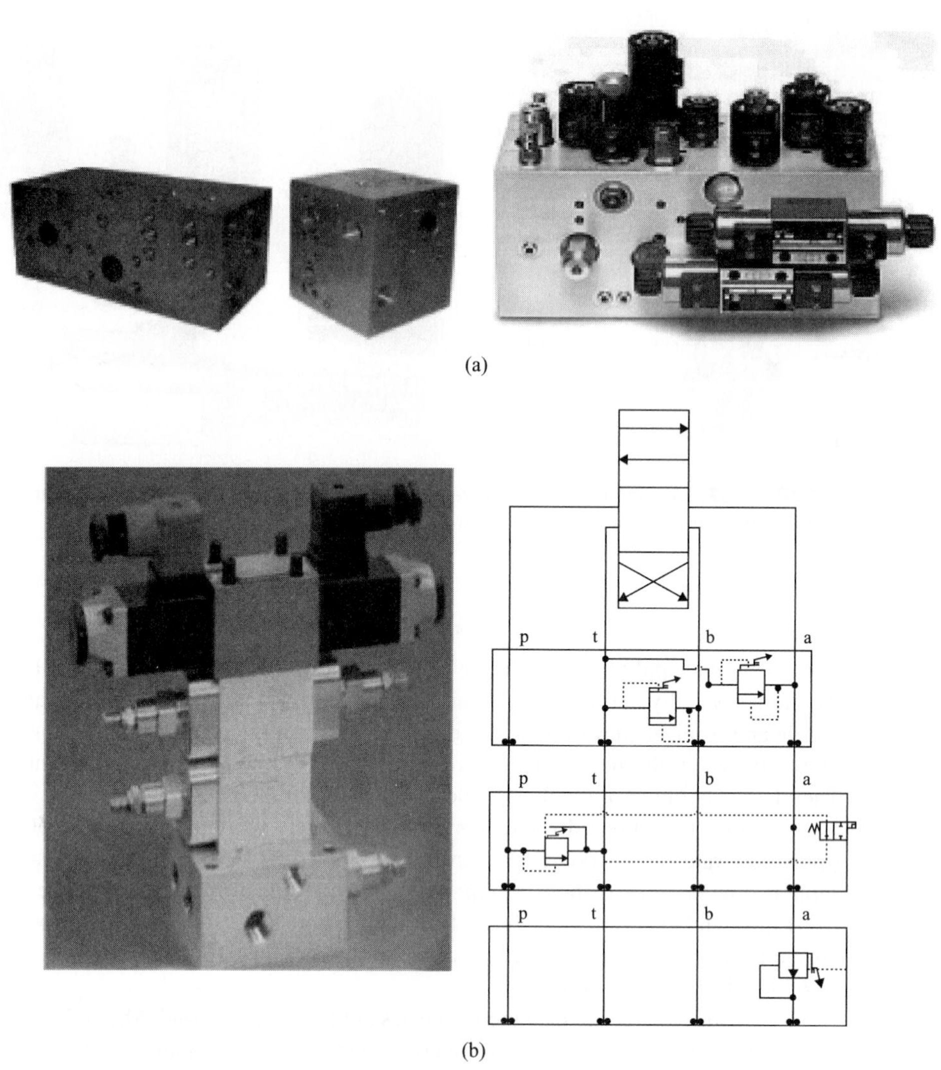


**FIGURE 7.63:** (a) Manifold block for valve mounting and internal piping: examples of manifold blocks, and a manifold block with valves mounted on it. (b) Sandwich-style mounting blocks: typically used to integrate a directional or proportional valve function with additional functions such as relief valve, check valve, and pressure-reducing valve functions.

**2.** Metering component:
   **(a)** *Spool type*: flow metering element can be a spool similar to standard spool valve-body assembly. Spool type cartridge valves can be two-way, three-way, four-way, or more type.
   **(b)** *Poppet type*: The flow is controlled by a poppet and its seat. Poppet type cartridge valves are typically two-way valves. Cartridge valves use O-rings on the stationary component of the valve body in order to seal the valve ports from each other and minimize leakage. O-rings also help increase the damping effect on the valve, but add hysteresis to the valve input current-flow characteristics.

**3.** Valve actuation methods:
   **(a)** Directly actuated by a solenoid with ON/OFF or proportional control.

**(b)** Two-stage actuation with pilot pressure in final actuation stage being proportional to the first-stage command signal. The command signal may be generated by manual controls or solenoid.
**(c)** Three-stage actuation for very large flow rate applications.

In general, manifold base plus a set of cartridge valves result in integrated hydraulic valve systems with low leakage and reliable operation. Cartridge valves can perform the directional flow and pressure control valve functions (i.e., proportional flow metering, pressure relief, check valve, unloading valve).

*Cartridge valves* for small flow rates can be directly actuated by a solenoid. Cartridge valves for large flow rates typically use a two-stage actuation mechanism which includes a first-stage electrical actuator and a second-stage pilot amplifier. For instance, let us consider a proportional spool type cartridge valve (Fig. 7.62). The valve has three ports: inlet port (main input pressure, i.e., pump supply), tank port, and output port. The output port pressure is regulated to be somewhere between the inlet and tank port pressures. It is essentially a pressure-reducing valve. The outlet pressure is regulated to a value between the two limit pressures (inlet and tank pressures) by the pilot stage presssure. Pilot stage supply pressure is obtained by a pressure-reducing valve section from the main inlet pressure. The actual pilot pressure in the pilot control chamber is regulated by the poppet ball which is controlled by the solenoid. Therefore, the solenoid current regulates a pilot chamber pressure proportional to it which in turn positions the main stage spool proportional to the pilot pressure with the balancing force from the spring. The output pressure is proportional to the main spool position, hence to the solenoid current.

For very large cartridge valves, valve actuator may take the form of a double-nozzle flapper or jet pipe valve as the pilot stage. Such a valve can support very high flow rates (i.e., up to 9600 lpm by Moog DSHR series). However, the addition of a pilot stage servo valve along with poppet position feedback makes the cost of the valve significantly higher than the cost of proportional spool-type valves. Screw-in cartridge valves provide check, relief, flow, pressure, and direction control functionalities for two, three, four, or more port configurations.

The recent design of the *direct drive valve* concept uses electrical actuators with higher power density to directly shift the main spool of a two-stage valve. The valve then becomes a single stage with the flow capacity of a two-stage valve. The electrical actuator is either a rotary motor (hybrid permanent magnet stepper motor or bushless DC motor coupled with a rotary-to-linear motion conversion mechanism, i.e., helical cam, ball-screw) or a direct drive linear electric motor (i.e., linear force motor, linear brushless DC, linear stepper motor). The main benefit of direct drive valve is that it eliminates the pilot pressure stage between the electrical actuator (solenoid in current designs) and the main spool motion. Due to power density and cost, current applications are still limited to valves with low flow rates. As the power density of electric actuators (permanent magnets and power amplifiers) gets larger and cost gets lower, the application of direct drive valve technology is likely to increase in the coming years.

### 7.4.6 Mounting of Valves in a Hydraulic Circuit

In hydraulic motion systems, a set of valve functions can be machined into a single *manifold block* (Fig. 7.63). Manifold block, a set of cartridge valves as well as other noncartridge valves provide a very reliable and leakage free design for hydraulic valve groups. The manifold block provides the hydraulic connections from P and T ports to different valves internally. The hydraulic plumbing functions between P and T and service ports (A1, B1, A2, B2, etc.) are machined into the block. For instance, in a multifunction circuit, parallel or series hydraulic connection of valve ports to P and T lines can be machined into the manifold. For each function, there is a valve (typically a screw-in cartridge valve) that controls (meters) the hydraulic flow between the ports. Technically, a single P port and T port is sufficient to the whole manifold block. There are as many A and B ports as there are the number of valve functions used. Furthermore, the manifold block is machined to accommodate open-center or closed-center functions by selectively plugging certain flow orifices on the manifold. Check valves, relief valves, and filter connections can also be built into the manifold design. The main advantages of the manifold block approach to multivalve hydraulic circuits instead of individual valves are:

- Manifold block modularizes and simplifies hydraulic plumbing in installation and maintenance.
- Reduces leakage and can support higher pressure circuits.
- Compact size.

The manifold block port locations and sizes are standardized by ISO-4401 standard (i.e., ISO-4401-03, -05, -06, -07, -08, -10 specify different number of ports, size, and locations for external connections). These are also referred to as CETOP-03, ..., CETOP-10 standards. For pressures up to 3000 psi, aluminum manifolds are recommended, and for higher pressures (5000 psi) cast iron manifolds are recommended. In addition to manifold blocks, valves are also made with *stackable standard mounting plates* which makes connecting supply and tank lines between valves easier.

Other mounting methods for hydraulic plumbing are *subplates, inline bar manifolds, mounting plates, valve adaptors,* and *sandwich-style mounting plates*. In particular, sandwich style mounting is typically used to integrate a directional or proportional valve and a number of relief, check, and pressure-reducing functions in one stack. Sandwich-type mounting plates also have DIN, ISO, and NFPA standard interface dimensions.

### 7.4.7 Performance Characteristics of Proportional and Servo Valves

The spool and orifice geometry around the null position is an important factor in proportional and servo valve performance. The spool may be machined so that at null position it overlaps, zero-laps, or underlaps the flow orifices. The zero-lapped spool is the ideal spool, but is difficult to accomplish due to tight manufacturing tolerances. Overlapped spool results in a mechanical deadband between the current and flow relationship. The underlapped spool provides a large gain around the null position (Fig. 7.64).

The null position is defined as the position of the spool at which the pressure versus spool position curve ($\Delta P_L, x_s$ or $\Delta P_L, i$) goes through zero value at pressure axis (Fig. 7.64e). The null position test is conducted on a valve by fully blocking the ports between P-T and A-B (Fig. 7.65). The null position of a valve usually adjusted mechanically under zero current condition. The mechanical adjustment of the null position may be provided by an adjustable screw on the valve which is used to move the spool around neutral position by a small amount. This is part of the valve calibration procedure.

Null position performance is highly dependent on the machining tolerances of the spool lands, sleeve and valve orifices, pressure, and temperature. Even high accuracy servo valves exhibit variation in their flow gain as function of input current (under constant pressure drop) in the null position vicinity. Recall that the flow is function of spool position and pressure,

$$Q = Q(x_{spool}, \Delta P_v) \tag{7.171}$$

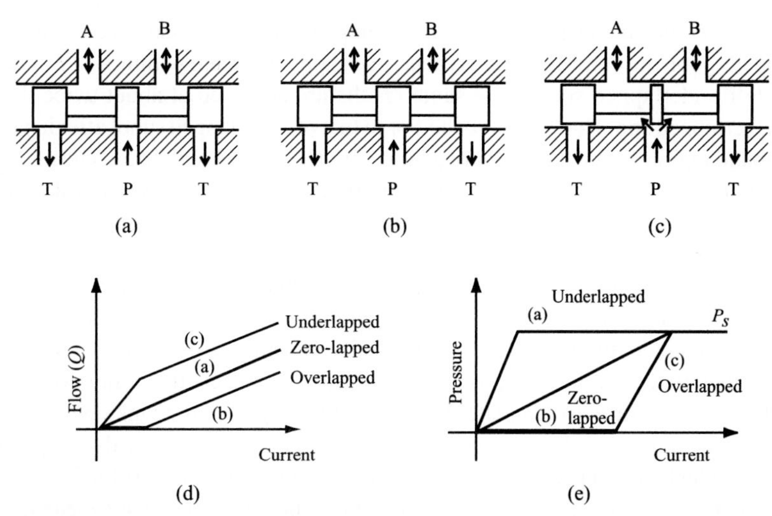


**FIGURE 7.64:** Spool and sleeve orifice geometry at null position: underlapped, critically lapped, overlapped. The resulting flow versus current and pressure versus current gains are also shown.

The valve characteristics around the null position are represented by *flow gain, pressure gain, flow-pressure gain* (also called the *leakage coefficient*). The flow gain

$$K_q = \frac{\partial Q}{\partial x_{spool}} \tag{7.172}$$

can vary between 50 and 200% of the nominal gain within the ±2.5% of maximum current value. The *pressure gain* of a valve is defined as the rate of change of output pressure as a function of solenoid current when output ports (A, B) are blocked (Fig. 7.65),

$$K_p = \frac{\partial P_L}{\partial x_{spool}} \tag{7.173}$$

The *flow-pressure gain* is defined as

$$K_{pq} = \frac{\partial Q_L}{\partial P_{spool}} = \frac{K_q}{K_p} \tag{7.174}$$

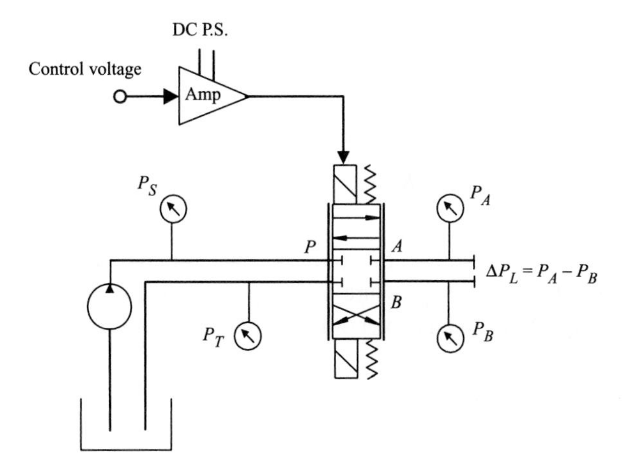


**FIGURE 7.65:** Test setup for measuring spool null position characteristics: pressure gain when valve output ports are blocked.

The output pressure under this condition reaches the supply pressure very quickly. For a zero-lapped spool, the output pressure reaches the supply pressure within 3–4% of maximum current. For underlapped spools, the same results are achieved at about twice the current values. Overlapped spool has similar behavior to that of zero-lapped spool after it passes over the deadband region where the pressure gain is zero (Figs. 7.64 and 7.65). The pressure gain of a valve around its null position is very important in servo position control applications because it strongly affects the stiffness of the positioning loop against external forces.

It is important to note that the null spool position is affected by the variations in supply pressure and temperature. The variation in null position as function of supply pressure and temperature is called *pressure null shift* and *temperature null shift*, respectively.

Flow characteristic of a valve in the vicinity of null position is an important performance characteristic. At the null position, if the valve connects the pump port to the tank port and hence continues to circulate the fluid between pump and tank, even though load does not require any, it is called the *open-center* valve (Fig. 7.2). If the valve blocks the flow from pump to tank at null position, it is called the *closed-center* valve (Fig. 7.54). An actual closed-center valve still exhibits some open-center characteristics due to leakage. Notice that a open-center valve can be used with a fixed displacement pump (Fig. 7.2). The control task is simpler, but the system is not energy efficient because it continuously circulates flow in the system even if the load does not require it. A closed-center valve requires a variable displacement pump so that when the valve is at null position, the pump is de-stroked to stop the flow or provide just enough flow to make up for the leakage (Fig. 7.54). It is energy efficient because it provides pressurized fluid flow on demand and shuts it off when there is no demand. The control task is a little more complicated since both valve and pump must be controlled and coordinated to avoid large pressure spikes (i.e., closing the valve to null position but keeping the pump running at high displacement will result in pressure spikes in the system and will most likely blow relief valves). Because of this, some degree of open-center characteristics is built into the valve, circulating a small percentage of its rated flow at null position in order to simplifiy coordinated control of valve and pump and increase system fault tolerance against the control timing errors between valve and pump.

The output flow, $Q$, from a valve is a function of orifice area ($A_{valve}$) and pressure differential across the port ($\Delta P_{valve}$). The orifice area is a function of the spool displacement ($x_{spool}$) and the geometric design of the spool-orifice geometry. The spool displacement is proporional to the solenoid current ($i_{sol}$). Hence,

$$Q = Q(A_{valve}, \Delta P_{valve}) \tag{7.175}$$

where

$$A_{valve} = A_{valve}(x_{spool}) \tag{7.176}$$

$$x_{spool} = K_{ix} \cdot i_{sol} \tag{7.177}$$

Let us consider a proportional valve, with a flow rating of $Q_{nl}$ at maximum current, $i_{max}$, and no-load conditions (all of the supply pressure is dropped across the valve). Then the flow rate at any pressure drop ($\Delta P_{valve}$) and solenoid current ($i_{sol}$) can be expressed as

$$Q/Q_{nl} = (i_{sol}/i_{max})\sqrt{\Delta P_{valve}/P_s} \tag{7.178}$$

$$= (i_{sol}/i_{max})\sqrt{1 - (\Delta P_l/P_s)} \tag{7.179}$$

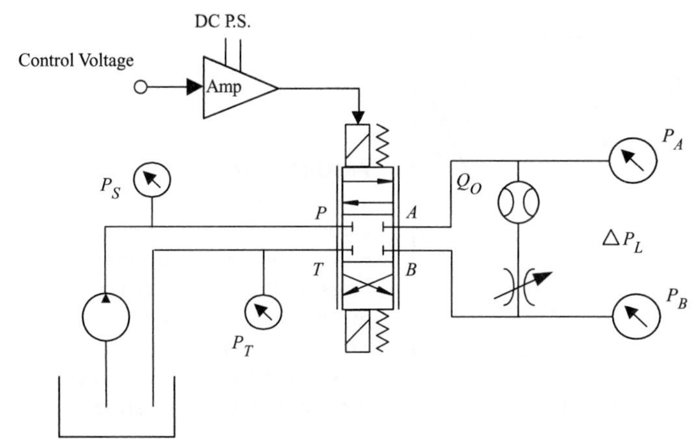


**FIGURE 7.66:** Test set-up for measuring the flow, load pressure, and solenoid current relationship. In order to measure flow-current relationship (approximate performance number to represent this relationship is flow gain, $K_p$), set the load pressure to a standard value. A standard way of measuring flow-current relationship for servo and proportional valves: set $P_s - P_T$ to a standard value (i.e., 1000 psi for servo valves, 150 psi for proportional valves), connect port A to port B directly (hydraulic short circuit, $\Delta P_L = 0$), hence pressure drop across the valve $\Delta P_v = P_s - P_T - \Delta P_l = P_s - P_T$. If $P_s - P_T$ cannot be set to a standard desired value, then connect port A and port B with an adjustable restriction ($\Delta P_l \neq 0$) so that the pressure drop across the valve is a standard value.

where the pressure drop across the valve is the difference between the pump and tank pressure (net supply pressure) minus the load pressure (Fig. 7.66),

$$\Delta P_{valve} = P_s - P_t - \Delta P_l \quad (7.180)$$

$$= P_s - P_t - |P_A - P_B| \quad (7.181)$$

$$= (P_s - P_A) + (P_B - P_t) \quad or \quad (7.182)$$

$$= (P_s - P_B) + (P_A - P_t) \quad (7.183)$$

Let us assume that the supply pressure is constant. When the load pressure is constant, the flow rate is linearly proportional to the current in the solenoid. Similarly, for a constant solenoid current (equivalently a constant spool orifice opening), the flow rate is related to the load pressure with square root relationship. As the load pressure increases, the available pressure drop across the valve decreases and hence, the flow rate decreases. When the load pressure is the same as the tank pressure, all of the supply pressure is lost across the valve and maximum flow rate is achieved, which is called the *no-load flow* ($Q_{nl}$). However, the no load flow at the output of the valve has no pressure to exert force or torque to the actuator (cylinder or rotary hydraulic motor). Therefore, in a good design, part of the supply pressure is used to support the necessary flow rate across the valve, and part of it is used to provide a load pressure to generate actuator force/torque,

$$P_s - P_t = \Delta P_{valve} + \Delta P_l \quad (7.184)$$

In practice, a valve flow rating ($Q_r$) is given at a standard pressure drop across the valve ($\Delta P_r = 1000$ psi for servo valves and 150 psi for proportional valves) when the spool is fully shifted. The flow rate of the valve ($Q$) for a different pressure drop ($\Delta P_{valve}$) across the valve when the spool is fully shifted can be obtained approximately by

$$Q = Q_r \cdot \sqrt{\Delta P_{valve}/\Delta P_r} \quad (7.185)$$

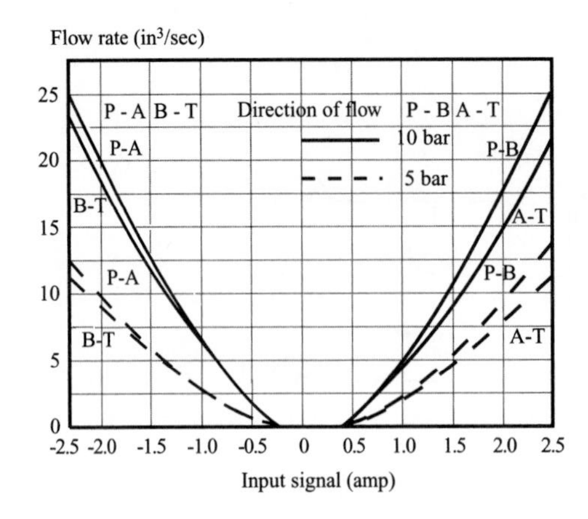


**FIGURE 7.67:** Valve flow rate versus solenoid current for different pressure drop across ports. Four-way proportional directional valve is considered for different pressure drop values. Servo valves have much smaller deadband, and flow rate-current relationship for a constant pressure drop is more linear compared to those of proportional valves.

For valve measurements, $P_s$ and $P_t$ are set to constant values, and pressure drop is set to a number of discrete values (Fig. 7.66). For each value of the load pressure ($\Delta P_L$), current ($i$) is varied from zero to maximum value ($i_{max}$), and flow rate ($Q$) is measured. The results are plotted as valve flow rate, load pressure, and current relationship as shown in Fig. 7.67 for fixed $P_s$ and $P_t$ values. In Fig. 7.67, the valve deadband is clearly observed. In reality, valve current-flow relationship also displays hysteresis characteristics in addition to deadband (Fig. 7.68).

The flow-current-pressure relationship given above neglects the leakage flow in the valve. Servo and proportional valves used in precise positioning applications generally have critically lapped (almost zero-lapped) spools, hence have measurable leakage at zero spool displacement [42]. When the current is zero and the spool is nominally at null position, as the pressure drop across the valve is increased, the leakage flow increases. Taking this fact into account, flow-current-differential pressure relationship can be expressed as [Fig. 7.69(a), $i = 0$ curve]

$$Q/Q_{nl} = (i/i_{max})\sqrt{1 - (\Delta P_L/P_s)} - K_{pq} \cdot (\Delta P_L/P_s) \tag{7.186}$$

where the last term accounts for the leakage when the spool is around the null position, $K_{pq}$ is the leakage coefficient. For different values of constant current, flow versus the load pressure curves are shown in Fig. 7.69 (a). When the load pressure equals the pump pressure, the flow rate is the leakage flow at all values of spool displacement. Let us focus on the flow-current-differential pressure relationship around the two extreme conditions:

1. No-load condition, $\Delta P_L = 0$
2. Blocked-load condition, $\Delta P_L = P_s$

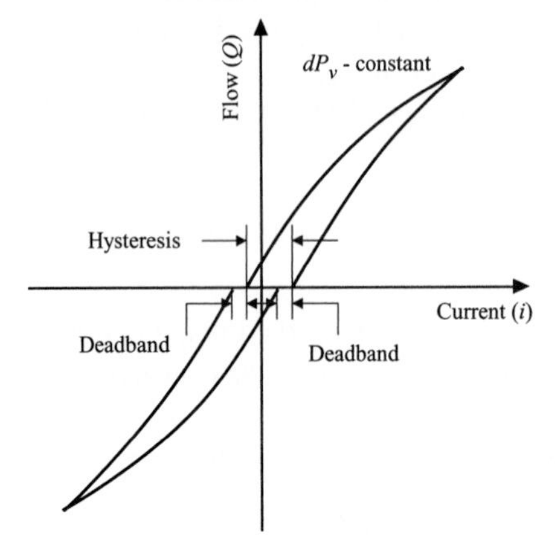


**FIGURE 7.68:** Nonideal characteristics of valve current-flow relationship under constant pressure drop condition across the valve: hysteresis, deadband, and offset.

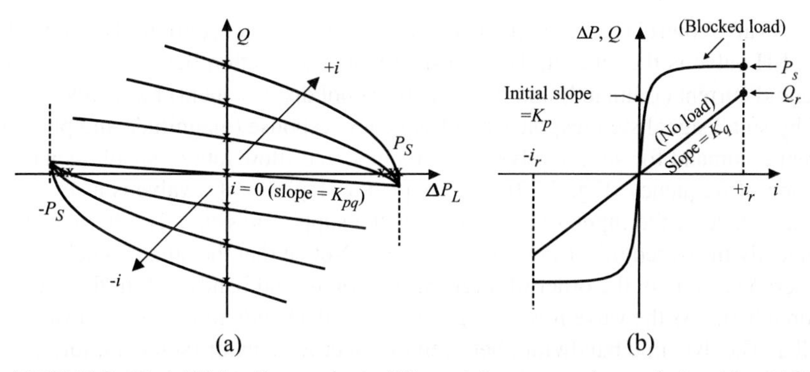


**FIGURE 7.69:** (a) Valve flow-current-differential pressure drop relationship, (b) No-load flow-current and blocked load pressure-current relationship. The figure in (b) is obtained from the data on figure in (a).

The blocked-load condition data plotted in Fig. 7.69 (b) is also conveyed in Fig. 7.69 (a). The data for the port pressure versus current [Fig. 7.69 (b)] is same as the data in Fig. 7.69 (a) for flow rate equal to zero cases ($Q = 0$) for different values of $i$ and $\Delta P_L$. These are the data points obtained along the $x$-axis intersections. The data for the no-load flow versus current on Fig. 7.69 (b) is obtained from Fig. 7.69 (a) for zero values of load pressure, $\Delta P_L = 0$. These are the data points obtained for $Q$ and $i$ along the $y$-axis intersection.

The current versus the no-load flow, and the current versus the pressure drop across the valve when load is blocked are shown in Fig. 7.69 (b). Notice that the $K_q$ is the no-load case current-to-flow gain (or flow gain), and $K_p$ is the blocked load case current-to-pressure gain (or pressure gain). The flow-pressure gain (leakage coefficient) is $K_{pq} = K_q/K_p = \frac{dQ}{dP}\big|_{\approx 0}$.

Some of the nonideal characteristics of a proportional valve is illustrated in Fig. 7.68. They are:

- Deadband due to friction between the spool, sleeve, and leakage
- Hysteresis due to the magnetic hysteresis in the electromagnetic circuit of the solenoid (or torque motor)
- Zero-position current bias due to manufacturing tolerances in the spool and feedback spring

In fact, the source of deadband and hysteresis in the valve cannot be exactly separated into friction and magnetic hysteresis. It is their combined effects that create the deadband and hysteresis in the valve input–output behavior.

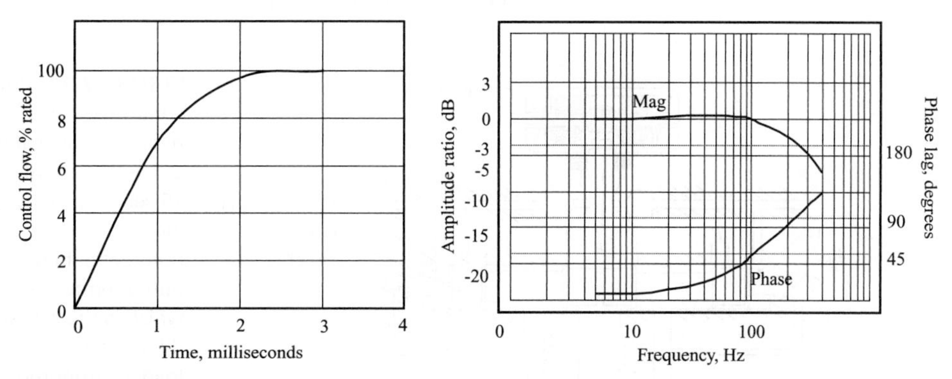


**FIGURE 7.70:** Dynamic characteristics of a typical servo valve: (a) step response, and (b) frequency response.

A valve control can be based on regulating flow, pressure, or both. The control variable of an EH valve is the current. The transient response relation between the control signal (solenoid current command) and flow rate (or spool displacement) for a valve is characterized by small signal step response and frequency response (magnitude and phase between current command for an EH valve) and output signal (flow rate or spool displacement) as function of frequency (Fig. 7.70). The frequency response of a valve is also a function of the magnitude of the input signal as well as the supply pressure. The frequency response is typically measured under no-load conditions. Notice that the small signal step response of a servo valve is in the order of a few milliseconds, and bandwidth in the order of a few hundred hertz. As the valve flow rating gets larger, the bandwidth of a given valve type gets smaller. The dynamic bandwidth between the electric actuator (solenoid, torque motor, or linear force motor) current and the main spool displacement is determined by two factors:

1. The bandwidth of the electric actuator.
2. Bandwidth of the pilot amplification stage. (In direct drive valves, there is no pilot stage.)

For a given valve, as the pilot stage supply pressure increases, the bandwidth of this stage also increases slightly.

The pilot stage may be supplied either by external pilot pump or derived internally from main line pressure. When the pilot supply is derived internally from the main line, the flow goes through a pressure-reducing valve in order to regulate a constant pilot pressure supply so that the pilot pressure does not fluctuate with the supply pressure (i.e., if the main pressure supply is a load-sensing pump, as the load varies, the main supply pressure varies). The pilot supply would vary by a small amount as the supply pressure varies. However, if the modulating bandwidth of the pressure-reducing valve is much higher than the main valve and the actuator, the transient effect of such variation in the pilot pressure is insignificant. For instance, when the pilot pressure is derived from a main line pump which is controlled using load-sensing feedback, the lowest stand-by pressure setting of the pump should be above a certain minimum value to make sure the pilot supply is properly maintained.

***Pressure Regulating Servo Valves*** A servo valve can be locally controlled to regulate the load pressure [42a]. Figures 7.71 and 7.72 show two types of valves controlled to regulate the load pressure: One is a completely mechanical pressure feedback system (Fig. 7.71), the other includes a sensor and embedded digital controller (Fig. 7.72). Under a constant solenoid current condition, the main spool is positioned in steady state such that

$$\Delta P_{12} = K_{pi} \cdot i_{sol} \tag{7.187}$$

$$\Delta P_{AB} \cdot A_{fb} = \Delta P_{12} \cdot A_{12} \tag{7.188}$$

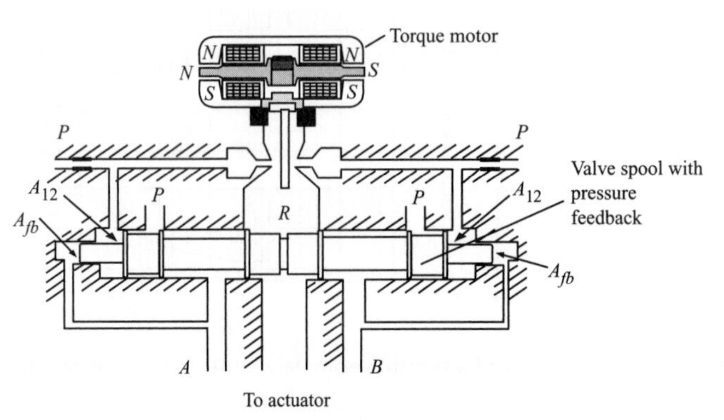


**FIGURE 7.71:** Pressure-controlled valve with hydromechanical means of feedback control (Series 15 servo valve by Moog, Inc.). The pressure differential between the output ports A and B, which is the pressure differential applied to the load, is proportional to the current applied to the torque motor.

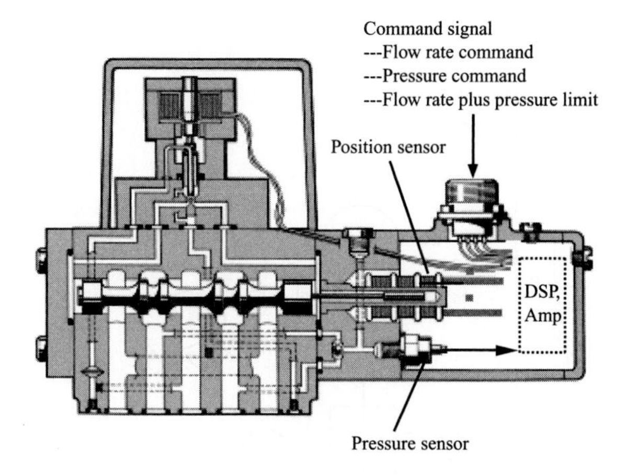


**FIGURE 7.72:** Pressure and flow controlled valve with electrical means of actuation control.

where $\Delta P_{12}$ is the pilot pressure differential acting on spool area $A_{12}$, $\Delta P_{AB}$ is the load pressure differential between the two ports A and B, and $A_{fb}$ is the cross-sectional area of the spool where the load pressure differential acts. The objective for a pressure control valve is to provide an output pressure differential that is proportional to the solenoid current, as achieved by this valve,

$$\Delta P_{AB} = \frac{A_{12}}{A_{fb}} \cdot K_{pi} \cdot i_{sol} \tag{7.189}$$

In general, the dynamic characteristics of an EH system with a pressure control valve is highly dependent on the load dynamics.

The digitally controlled version of the valve can easily be modified in software to implement a flow control plus a programmable pressure limit control logic (Figs. 7.72 and 7.73). Once the spool position and pressure sensors are available, the spool actuation of the valve can be controlled to achieve either flow or pressure regulation by simply selecting a different software mode. The trend in valve technology is to implement the mechanical control mechanisms by mechatronic control components, i.e., instead of using hydromechanical control mechanisms, use electrical sensors, actuators, and embedded digital controllers.

## 7.5 SIZING OF HYDRAULIC MOTION SYSTEM COMPONENTS

Let us consider the component sizing problem for a one-axis hydraulic motion control system shown in Fig. 7.74. The question of component sizing is to determine the size of

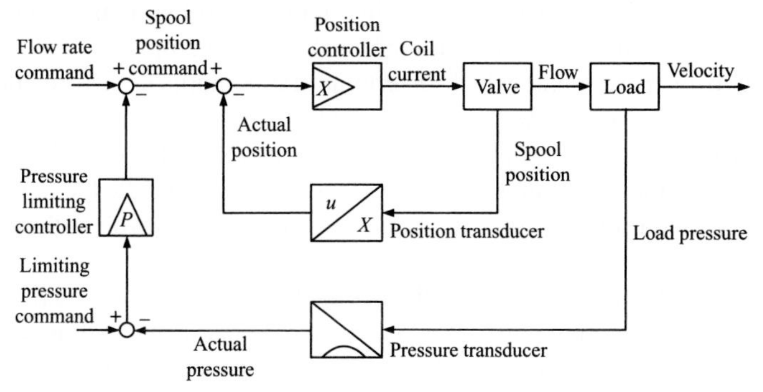


**FIGURE 7.73:** Pressure flow control algorithm block diagram.

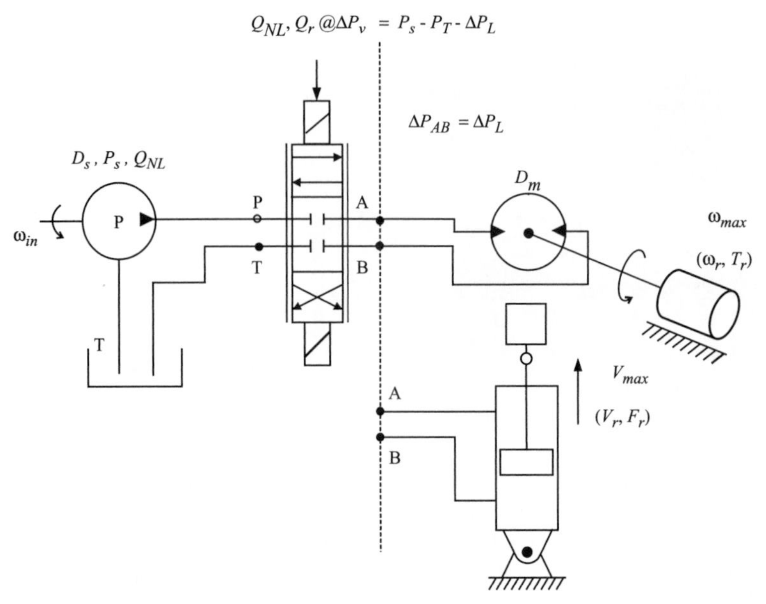


**FIGURE 7.74:** Component sizing for a one-axis EH motion system: pump, valve, cylinder (or rotary hydraulic motor) sizing for given load requirements.

the hydraulic motor or cylinder, the valve, and the pump. Oversizing of components results in lower accuracy, lower bandwidth, and higher cost. Undersizing of components results in lower than required power capacity.

As a rule of thumb, a servo valve size should be selected such that the pressure drop at the valve at maximum flow should be about 1/3 of the supply pressure. In general, such a balanced distribution of supply pressure between valve and load is possible in servo valve applications. In proportional valve applications, especially in large construction equipment, pressure drop across the valve can be much less than 1/3 of the supply pressure. In such application, precision motion control is not critical. It is desirable that more of the hydraulic power be delivered to the load and less be lost in the process of metering it. If the valve is sized too large and the pressure drop across it is not large enough, the flow will not be modulated well until the valve almost closes. As a result, resolution of the motion control will be low. If the valve is too small, it will not be able to support the desired flow rates or the pressure drop across it will be too large. Valve is the flow metering component. It is desirable to have a good resolution in metering the flow. In general, the higher the metering resolution required, the higher the pressure drop that must exist across the valve. Higher pressure drops lead to higher losses and lower efficiency. Therefore, the metering resolution and efficiency are two conflicting variables in a hydraulic control system. A good design targets an acceptable metering resolution with minimal pressure drop.

There are two main variables which determine the component size requirements: *required force* and *speed*. Another way of stating that is *required pressure* and *flow rate*. Based on these two requirements, supply (pump) and actuator (cylinder and motors) are sized. The valve, which modulates and directs the flow, is sized to handle the flow rate and pressure required. The control question in hydraulic systems always involves the control of flow rate and direction, and/or the pressure.

In general, the load speed requirement dictates the flow rate, and the load force/torque requirement dictates the operating pressure. Consider a cylinder as the hydraulic actuator. A given load force/torque requirement can be met with different pressure and cylinder cross-sectional area. For instance, in order to provide a certain force, many pressure and cylinder area combinations are possible, $F = p \cdot A$. Similarly, a given load speed requirement can be met with different combinations of flow rate and cylinder cross-sectional area, $Q = V \cdot A$. The trend in industry is to use higher operating pressures which results in smaller components. However, high-pressure circuits require more frequent maintenance and have lower life cycle.

The application requirements typically specify the the load conditions in terms of three variables: the no-load maximum speed ($w_{nl}$ for rotary actuator or $V_{nl}$ for translational actuator) and speed at rated load ($w_r$ at $T_r$ or $V_r$ at $F_r$). For rotary and linear system the specifications are, respectively,

$$\{w_{nl}, (w_r, T_r)\} \quad \text{or} \quad \{V_{nl}, (V_r, F_r)\} \tag{7.190}$$

Recall that the relationships between the hydraulic variables and mechanical variables for *the actuator* (hydraulic rotary motor or hydraulic linear cylinder) are

$$w = Q/D_m \quad \text{or} \quad V = Q/A_c \tag{7.191}$$

$$T = \Delta P_L \cdot D_m \quad \text{or} \quad F = \Delta P_L \cdot A_c \tag{7.192}$$

where $\Delta P_L$ is the pressure difference between the two sides of the actuator (two sides of the cylinder), $D_m$ is the hydraulic motor displacement (volumetric displacement per revolution), and $A_c$ is the cross-sectional area of the cylinder (assuming it is a symmetric cylinder).

When the load force (or torque) is specified, the total force output from the actuator should include the force to be exerted ($F_{ext}$) on the external load (i.e., pressing force in a press or testing application), plus the force needed to accelerate the total moving mass (cylinder piston, rod, and load inertia, $m \cdot \ddot{x}$), and finally some force needed to overcome friction, ($F_f$),

$$F = F_L = m \cdot \ddot{x} + F_f + F_{ext} \tag{7.193}$$

Estimating the friction force, $F_f$, is rather difficult. One common way to take it into account is to use a safety factor as follows:

$$F = F_L = SF \cdot (m \cdot \ddot{x} + F_{ext}) \tag{7.194}$$

where $SF$ is the safety factor (i.e., $SF = 1.2$). For rotary actuators, analogous relationships hold:

$$T = T_L = SF \cdot (J \cdot \ddot{\theta} + T_{ext}) \tag{7.195}$$

where $J$ is the total mass moment of inertia of rotary load, $\ddot{\theta}$ is angular acceleration, $T_{ext}$ is the torque to be delivered to the load. The flow-current-pressure differential relation for a proportional or servo valve which is positioned between the pump supply line and load line,

$$Q/Q_{nl} = (i/i_{max})\sqrt{1 - (\Delta P_L/P_S)} \tag{7.196}$$

$$= (i/i_{max})\sqrt{(\Delta P_v/P_S)} \tag{7.197}$$

where $\Delta P_L$ is the load pressure (i.e., in the case of a linear cylinder this is the pressure difference between the two sides of the cylinder, and in case of a rotaty hydraulic motor this is the inport and outport pressure differential). In other words, $\Delta P_L$ is the pressure differential between the A and B ports of the valve. $\Delta P_v = P_S - \Delta P_L - P_T$ is the pressure drop across the valve, $P_S$ is supply pressure from pump, $P_T$ tank return line pressure. Quite often, $P_T$ is taken as zero pressure. $Q_{nl}$ is the flow through the valve when it is fully open ($i = i_{max}$) and the pressure drop across the valve is $P_S$. When there is no load, all of the supply pressure is dropped across the valve. In order to determine the flow rate ($Q_r$) of a valve at a particular pressure drop ($P_S - \Delta P_L$) when it is fully open, we set $i = i_{max}$ in equation 7.197:

$$Q_r/Q_{nl} = \sqrt{1 - (\Delta P_L/P_S)} \tag{7.198}$$

$$= \sqrt{(\Delta P_v/P_S)} \tag{7.199}$$

A proportional valve is rated in terms of its flow rate capacity in catalogs for a pressure drop of $\Delta P_v = 150$ psi across the valve. A servo valve is rated for $\Delta P_v = 1000$ psi.

### The Component Sizing Algorithm 1

***Step 1:*** Determine the performance specifications based on load requirements [42b]:

$$\{w_{nl}, (w_r, T_r)\} \quad \text{or} \quad \{V_{nl}, (V_r, F_r)\}$$

***Step 2:*** Pick either displacement per unit stroke for the actuator ($D_m$ or $A_c$) or the supply pressure, $P_s$. Let us assume we pick the displacement ($D_m$ or $A_c$). This selection specifies the actuator size (hydraulic motor or cylinder).

***Step 3:*** Calculate $Q_{nl}$, $Q_r$, $\Delta P_L$ using equations 7.191 and 7.192 for the specifications.

$$Q_{nl} = D_m \cdot w_{nl} \qquad Q_{nl} = A_c \cdot V_{nl} \tag{7.200}$$

$$Q_r = D_m \cdot w_r \qquad Q_r = A_c \cdot V_r \tag{7.201}$$

$$\Delta P_L = T_r/D_m \qquad \Delta P_L = F_r/A_c \tag{7.202}$$

***Step 4:*** Assume $i = i_{max}$, calculate the $P_s$ from equation 7.198. Add a 10–20% safety factor to the $P_s$ as a safety factor.

$$Q_r/Q_{nl} = \sqrt{1 - \Delta P_L/P_S} \tag{7.203}$$

***Step 5:*** Given the calculated value of $Q_{nl}$, and given the operating input shaft speed of the pump $w_{in}$ which is determined by the mechanical power source device driving the pump (i.e., electric motor in industrial applications, internal combustion engine in mobile applications), calculate the pump displacement requirement, $D_p$, using equation 7.191. Select proper pump given $Q_{nl}$, $P_s$, $D_p$.

$$D_p = Q_{nl}/w_{in} \tag{7.204}$$

***Step 6:*** Given the calculated $Q_{nl}$, $P_s$, $Q_r$, calculate the rated flow for the valve, $Q_v$, at its rated pressure $\Delta P_v$ for $i = i_{max}$ using equation 7.199. Notice that rated flow for a valve is defined at a standard pressure drop across the valve (i.e., typically $\Delta P_v = 1000$ psi for servo valves, $\Delta P_v = 150$ psi for proportional valves) and at the fully shifted spool position ($i = i_{max}$). Select the proper valve size based on the calulated $Q_v$ and assumed $\Delta P_v$.

$$Q_v = Q_{nl}\sqrt{\Delta P_v/P_S} \tag{7.205}$$

In summary, the load conditions determine these specifications:

- Given: $\{w_{nl}, (w_r, T_r)\}$ [for linear actuator $\{V_{nl}, (V_r, F_r)\}$]
- As design choices, pick $D_m$ hydraulic motor displacement (for linear actuator $A_c$ cylinder cross-sectional area) and pick $\Delta P_v$ valve pressure rating
- Then calculate the rest of the sizing parameters: $Q_{nl}$, $Q_r$, $\Delta P_L$, $P_s$, $D_p$, $Q_v$, which are the size requirements for pump, valve, and motor/cylinder using equations 7.191, 7.192, 7.198, and 7.205

### The Component Sizing Algorithm 2

***Step 1:*** Determine the performance specifications based on load requirements:

$$\{w_{nl}, (w_r, T_r)\} \quad \text{or} \quad \{V_{nl}, (V_r, F_r)\}$$

***Step 2:*** Pick either displacement per unit stroke for the actuator ($D_m$ or $A_c$) or the supply pressure, $P_s$. Let us assume we pick pump pressure, $P_s$.

***Step 3:*** Pick the pressure differential to be delivered to the actuator to drive the load $\Delta P_L$, or calculate from eqn. (7.203), noting $Q_r/Q_{nl} = \frac{V_r}{V_{nl}} = \frac{w_r}{w_{nl}}$ calculate hydraulic motor displacement ($D_m$) or cylinder cross-sectional area ($A_c$).

$$D_m = T_r/\Delta P_L \qquad A_c = F_r/\Delta P_L \tag{7.206}$$

***Step 4:*** Calculate $Q_{nl}$ from equation 7.191 and $Q_r$ at $\Delta P_L$, $P_s$, and $i = i_{max}$ from equation 7.198:

$$Q_{nl} = D_m \cdot w_{nl} \qquad Q_{nl} = A_c \cdot V_{nl} \tag{7.207}$$

$$Q_r = D_m \cdot w_r \qquad Q_r = A_c \cdot V_r \tag{7.208}$$

***Step 5:*** Given the calculated value of $Q_{nl}$, and given the operating input shaft speed of the pump $w_{in}$, calculate the pump displacement requirement, $D_p$, using equation 1.191. Select proper pump given $Q_{nl}$, $P_s$, $D_p$:

$$D_p = Q_{nl}/w_{in} \tag{7.209}$$

***Step 6:*** Given the calculated $Q_{nl}$, $P_s$, calculate the rated flow for the valve, $Q_v$, at its rated pressure $\Delta P_v$. Select the proper valve size based on the calulated $Q_v$ and assumed $\Delta P_v$:

$$Q_v = Q_{nl}\sqrt{\Delta P_v/P_s} \tag{7.210}$$

In summary, the load conditions determine three specifications:

- Given $\{w_{nl}, (w_r, T_r)\}$ [for linear actuator $\{V_{nl}, (V_r, F_r)\}$].
- As design choices, pick $P_s$ hydraulic pump rated pressure output and pick or calculate $\Delta P_L$ load pressure differential.
- Then calculate the rest of the sizing parameters: $D_m$, $D_p$, $Q_{nl}$, $Q_r$, $Q_v$, which are the size requirements for pump, valve, and motor/cylinder.

Notice that the pump and motor displacement and pressure ratings are available in standard sizes, such as

$$D_p, D_m = 1.0, 2.5, 5.0, 7.5 \text{ in}^3\text{/rev} \tag{7.211}$$

$$P_s = 3000 \text{ psi}, \ldots, 5000 \text{ psi} \tag{7.212}$$

Similarly, the valves are available at standard rated flow capacity (measured at $\Delta P_v =$ 1000 psi and at maximum current $i = i_{max}$ for servo valves) such as

$$Q_v = 1.0 \text{ gpm}, 2.5 \text{ gpm}, 5.0 \text{ gpm}, 10 \text{ gpm} \tag{7.213}$$

Based on the calculated sizes and safety margins, one of the closest size for each component should be selected with appropriate safety margin.

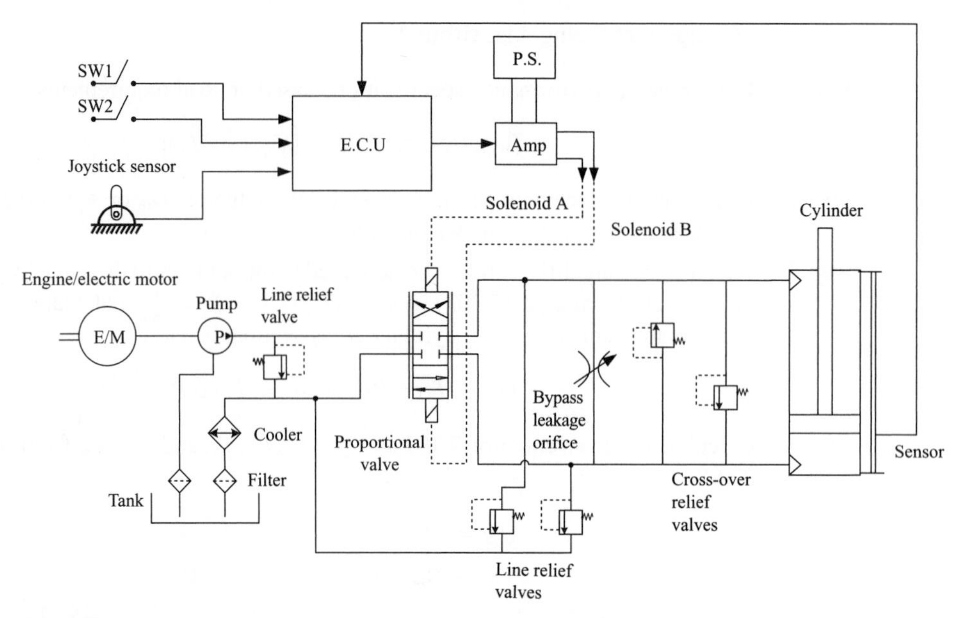


**FIGURE 7.75:** Control system components and circuit diagram for a single-axis EH control system which operates in two modes using closed-loop control: (1) speed mode and (2) force mode.

***Example*** Consider a single-axis EH motion control system shown in Fig. 7.75, which may be used for a hydraulic press. The system will be controlled in one of two possible modes:

- ***Mode 1:*** Closed-loop speed control where the commanded speed is obtained from a programmed command generator or from the displacement of the joystick. The cylinder speed is measured and a closed-loop control algorithm implemented in the electronic control unit (ECU) which controls the valve.
- ***Mode 2:*** Closed-loop force control where the force command is obtained from a programmed command generator or from the displacement of the joystick. The cylinder force is measured using a pair of pressure sensors on both ends of the cylinder.

In order for the operator to tell the ECU which mode to operate (speed or force) and from where to obtain the command signals, there should be two two-position discrete switches. Switch 1 OFF position means speed control mode, and ON position means force control mode. Switch 2 OFF position means the source of the command signal is programmed into the ECU, ON position means the joystick provides the command signal (Fig. 7.75).

Select proper components (components with appropriate technology such as variable displacement or fixed displacement pumps, ON/OFF or proportional or servo valves) and proper size (calculate component size requirements) for pump, valve, and cylinder in order to meet the following specifications:

1. Maximum speed of the cylinder under no-load conditions is $V_{nl} = 2.0$ m/sec,
2. Provide an effective output force of $F_r = 10{,}000$ Nt at the cylinder rod while moving at the speed of $V_r = 1.5$ m/sec.
3. The desired regulation accuracy of the speed control loop is 0.01% of maximum speed.
4. The desired regulation accuracy of the force control loop is 0.1% of maximum force.

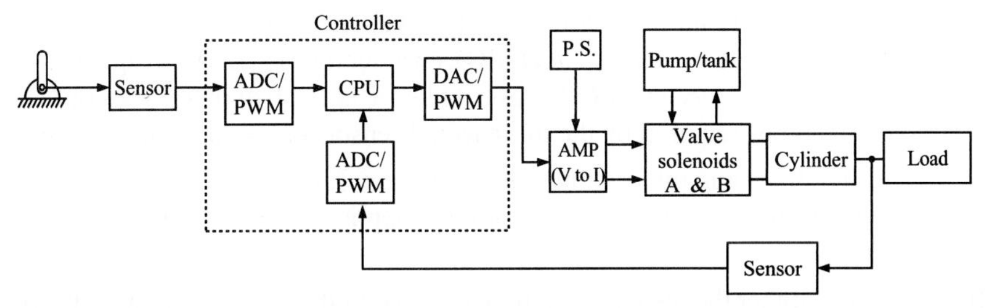


**FIGURE 7.76:** Components and block diagram of a single-axis electrohydraulic motion control system. Controller is a programmable embedded computer.

In addition, select an operator input device (joystick for motion command) with proper sensor, electronic control unit (ECU) with necessary analog–digital signal interfaces, and position and force sensor on the cylinder for closed-loop control (Fig. 7.76).

The range and resolution requirements for the measured variables determine the sensor and ECU interface requirements (DAC and ADC components). Let us assume that the maximum speed will be mapped to a maximum analog sensor feedback signal (i.e., 10 VDC), and the required accuracy is 0.01% of maximum value which is one part in 10,000. In general, measurement resolution should be at least 2 to 3 times better than the required regulation (control) accuracy. Therefore, we need a speed sensor with a data-acquisition component that will have one part in 30,000 resolution. Let us assume that we have selected a magnetostrictive position sensor (also called *Temposonic sensor*) from which speed information is digitally calculated. The output of the sensor can be scaled so that it gives 10 VDC when the speed is at 2.0 m/sec. The sensor nonlinearity should be better that 0.01% of the maximum range. In addition, the analog-to-digital converter (ADC) must be at least 15-bit resolution in order to give a sampling resolution of better than 1/30,000 parts over the whole range (Fig. 7.76).

The command signal (either in speed or force control mode) is obtained either from a real-time program or joystick. The real-time program can command motion in any accuracy desired, i.e., one part in 10,000 depending on the data type used in the control algorithm. If 2-byte signed integers are used and scaled to cover the whole range, the commanded signal resolution can be one part in $(2^{15} - 1)$. If 4-byte signed long integers are used in the servo control algorithm and numbers are scaled to cover the whole motion range, the commanded signal resolution can be one part in $(2^{31} - 1)$.

If the command signal is obtained from the joystick sensor, it is important to note that a human operator cannot command motion without shaking his or her hand better than 1% of the maximum displacement range of the joystick. Therefore, the resolution requirements of the joystick command is much smaller. Even if we provided a high-resolution sensor and ADC, the human operator cannot actually change the command in finer resolutions than about one part in 100. Therefore, an 8-bit ADC converter with an analog speed sensor that can provide 10 VDC with accuracy of 1% is sufficient. If an incremental encoder is used for the sensor, an encoder with 512 lines/rev also would have the sufficient range and resolution for command signal.

ECU should have resources in the following areas:

1. Speed of the CPU (microprocessor or DSP chip) in order to implement the control logic fast enough. Generally, this is not a problem with the current state of art.
2. Memory resources to store control code and sensory data for control purposes as well as for off-line analysis purposes (ROM or battery-backed RAM for program storage, RAM for real-time data storage and program execution).

3. I/O interface circuit: analog-to-digital converter (ADC), digital-to-analog converter (DAC), interface for discrete I/O (DIO), encoder interface, PWM signal input and output interface. Notice that ECU does not have to have all of these types of I/O interface, only the types required by the the selected sensor inputs and amplifier outputs.
4. Interrupt lines and interrupt handling software. Interrupts are key to the operation of most real-time systems, even though this application may be solved without use of interrupts.

In general, the higher the resolution (number of bits) of the ADC and DAC converter circuit, the better it is. If the programming is to be done for the servo loop by the designer, then it is preferrable to have the ECU based on a microprocessor or DSP which has a C-compiler so that we can use a high-level programming language to implement the servo control loops and use integer, long integer, or floating point data types. If we had to program the ECU in assembly language, then we have to manage the data and its size explicitly. In general, the speed, memory, and signal interface resources of the ECU are not limiting factors in the design.

Let us assume that the valve rating is specified for $\Delta P_v = 1000$ psi (6.8948 MPa $= 6.8948 \times 10^6$ Pa $= 6.8948 \times 10^6$ Nt/m$^2$-pressure drop, and that the input shaft speed at the pump is $w_{in} = 1000$ rpm. Following the component sizing algorithm, let us pick a cylinder diameter size, $d_c = 0.05$ m, which determines the cross-sectional area, $A_c$. Using equations 7.191–7.192, calculate the $Q_{nl}$, $Q_r$, and $\Delta P_L$,

$$A_c = \pi d_c^2/4 = 0.0019635 \text{ m}^2 \approx 0.002 \text{ m}^2 \tag{7.214}$$

$$Q_{nl} = V_{nl} \cdot A_c = 240 \text{ liter/min} \tag{7.215}$$

$$Q_r = V_r \cdot A_c = 180 \text{ liter/min} \tag{7.216}$$

$$\Delta P_L = F_r/A_c = 5 \times 10^6 \text{ Nt/m}^2 = 5 \text{ MPa} \tag{7.217}$$

Assume $i = i_{max}$, and using equation (7.198), calculate $P_s$ and add 10% safety margin to it. Then calculate the necessary servo valve flow rating,

$$Q_r/Q_{nl} = \sqrt{1 - (\Delta P_L/P_s)} \tag{7.218}$$

$$\Delta P_L/P_s = 1 - \left(Q_r^2/Q_{nl}^2\right) = \left(Q_{nl}^2 - Q_r^2\right)/Q_{nl}^2 \tag{7.219}$$

$$P_s = \frac{Q_{nl}^2}{Q_{nl}^2 - Q_r^2} \cdot \Delta P_L \tag{7.220}$$

$$= 11.42 \text{ MPa} = 1658 \text{ psi} \tag{7.221}$$

Select $P_s = 2000$ psi,

$$Q_v = Q_{nl}\sqrt{\frac{\Delta P_v}{P_s}} \tag{7.222}$$

$$= 240 \cdot \sqrt{1000/2000} \text{ lt/min} \tag{7.223}$$

$$= 169 \text{ lt/min} \tag{7.224}$$

where we assumed a pressure drop of 1000 psi across the valve for its flow rating. Because we know the maximum flow rate needed and the input speed to the pump, we can calculate the pump displacement, $D_p$, from equation 7.191,

$$D_p = 0.240 \text{ liter/rev} \tag{7.225}$$

In summary, the following component sizes are required: Cylinder with bore diameter $d_c = 0.05$ m (bore cross-sectional area $A_c = 0.002$ m$^2$), valve with flow rating of $Q_v = 200$ lt/min or higher at 1000 psi pressure drop, and pump with displacement $D_p = 0.240$ lt/rev and output pressure capacity of $P_s = 2000$ psi or higher value.

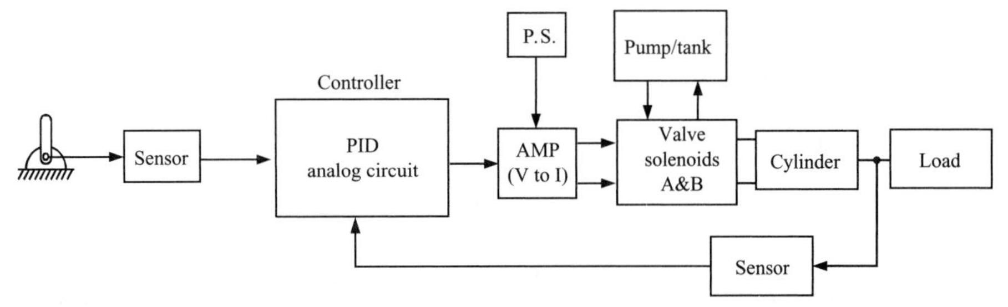


**FIGURE 7.77:** Components and block diagram of a single-axis electrohydraulic motion control system. Controller is an analog OP-AMP which implements a closed-loop PID control between the operator command, sensor signal, and valve-drive command.

Finally, an amplifier with proper power supply is needed to drive the valve. Amplifier type which has current feedback loop is preferred so that the current sent to the solenoid is maintained proportional to the signal from the ECU regardless of the variations in the power supply or solenoid resistance as function of temperature. A commercial amplifier has tunable parameters to adjust the input voltage offset, gain, and maximum output current values (i.e., Parker BD98A or EW554 series amplfiers for servo valves plus a +/−15 VDC power supply, such as Model PS15, which provides DC bus voltage of +/−15 VDC up to 1.5 amps to the amplifier using a regular 85–132 VAC single-phase line power). Similarly, if a variable displacement pump is used and the valve that controls the pump displacement is EH controlled, we would need another amplifier-power supply matched to the size of the pump control valve. Notice that if the EH motion axis were to be controlled only through the joystick inputs, and no programming was necessary, the closed-loop control algorithm could be implemented with analog OP-AMP circuits in the amplifier and the need for the ECU can be eliminated. The drive can have a PID circuit where the command signal from the joystick sensor and the feedback signal from the cylinder motion connects to the input of the OP-AMP circuit (Fig. 7.77). The error signal output of the OP-AMP is amplified as a current signal to the solenoids. However, the software programmability of the motion system is not possible in this case. One such drive is Parker Series EZ595 used together with Series BD98A amplifier, where the EZ595 implements the PID function and interfaces and BD98A implements the voltage to current amplification function.

List of components selected for this design are as follows:

1. ***Pump and reservoir:*** Oil gear model PVWH-60 axial piston pump, with pressure rating of 2000 psi, and flow rate of 60 gpm when driven by an electric motor (i.e., AC induction motor) at 1800 rpm. The reservoir should handle 150 gal (3–5 times the rated flow capacity per minute of the pump) of hydraulic fluid, fitted with filters in both return line and pump input line from the reservoir, as well as cooler and pressure relief valve.

2. ***Cylinder:*** Parker industrial cylinder (Series 2H) with 2.0-in. bore, 40-in. stroke, with threaded rod end for load connection, with base lugs mounting option at both ends.

3. ***Valve:*** MOOG servo valve model D791, rated flow up to 65 gpm at 1000 psi pressure drop across the valve when valve is fully open, equivalent first-order filter time constant is about 3.0 msec.

4. ***Valve drive (power supply and amplfier):*** MOOG power supply and amplfier circuit matched to the servo valve: MOOG model "snap track" servo valve drives.

5. ***Sensors:*** *Position sensor:* linear incremental encoder. DYNAPAR series LR linear scale, range 1.0 m, 5.0 micron (0.0002 in.) linear resolution, maximum speed 20 m/sec, frequency response 1 MHz, quadrature output channels $A, B, C, \bar{A}, \bar{B}, \bar{C}$.

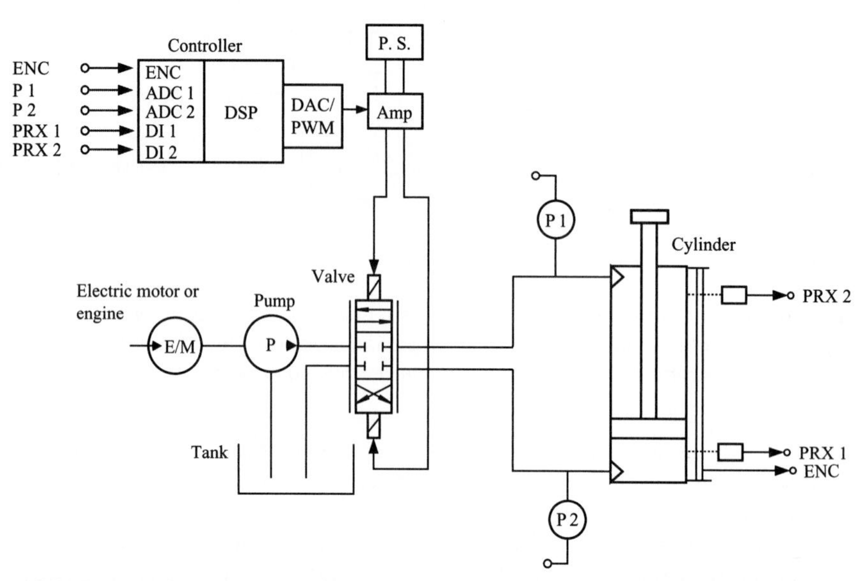


**FIGURE 7.78:** Control system components and circuit diagram for a single-axis EH control system which operates in two modes using closed-loop control: (1) speed mode and (2) force mode.

*Pressure sensor:* differential pressure sensor, Schaevitz series P2100, up to 5000 psi line pressure, and up to 3500 psi differential pressure, bandwidth in 15-KHz range. Pressure sensors can be placed on the valve manifold.

6. ***Controller:*** PC-based, data-acquisition card (National Instruments: NI-6111) with I/O interface capabilities for encoder interface, two channels of 16-bit DAC output, two channels of 12-bit ADC converter, and 8 channels of TTL I/O lines. The control logic can be implemented on the PC, and the data-acquisition card is used for I/O interface.
7. ***Operator input/output devices:*** Joystick and mode selection switches (switch 1 and 2). One degree of freedom joystick with potentiometer motion sensor and analog voltage output in the 0- to 10-VDC range—Penny and Giles model JC 150 or ITT Industries model AJ3 [82a, 82b].

***Example*** As an extension of the previous example, let us assume that we have selected two sensors for the hydraulic cylinder (Fig. 7.78):

1. A position sensor (i.e., an absolute encoder or an incremental encoder plus two discrete proximity sensors, to indicate mechanical travel limits and establish an absolute reference position: ENC1, PRX1, PRX2)
2. A pair of pressure sensors which measures the pressure at both ends of the cylinder (P1, P2)

It is desired that the actuator operate in programmed mode and ignore the joystick input. The objective is that the actuator is to approach the load at a predefined speed until a certain amount of pressure differential is sensed between P1 and P2 signals. The position range of the actuator around which that is expected is known. When a certain pressure differential is sensed, the control system is suppose to automatically switch to force (pressure) regulation mode. Maintain the desired pressure until a certain amount of time is passed or another discrete ON/OFF state sensor input is high (i.e., PRX2). Let us assume that we will use the time period option to decide how long to apply the pressure. Then reverse

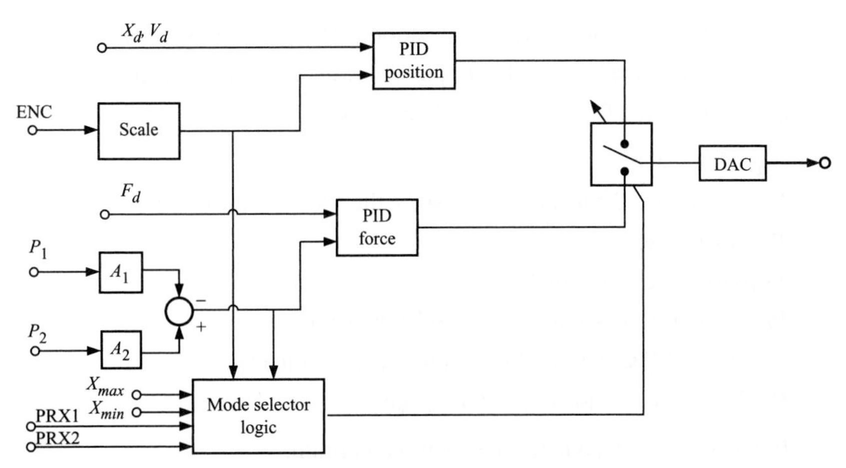


**FIGURE 7.79:** Real-time control algorithm for an EH control system using both position and force feedback, and implement a "bumpless mode transfer."

the motion, and return to the original position under a programmed speed profile. Such EH motion control systems are used in mechanical testing of products, injection molding, nip roll positioning, and press applications (Fig. 7.79).

**(a)** Modify the design of the EH system used in the previous example.

**(b)** Draw a block diagram of the control logic and closed-loop servo algorithms for both position and force servo modes. Suggest ways to make a "bumpless transfer" between the two modes.

There are two closed-loop control modes: (1) position servo mode where actual position is sensed using an encoder (ENC, PRX1, PRX2) and the desired position is programmed and (2) force servo mode where force information is derived from the measured differential pressure (P1 and P2). The desired force profile (constant or time varying) is preprogrammed, and its magnitude can be changed for each product. Therefore, we need two servo control algorithms (i.e., of PID type). Each loop has its commanded signal (desired position or desired force) and feedback signal (encoder signal or differential pressure sensor signal). At the power-up, the cylinder makes a home search motion sequence in order to find an absolute reference position (also called *home position*) so that every cycle is started from the same position. Using an absolute encoder, the controller can read the current position at power-up, and command the axis to move to the desired position for the home reference (using position servo loop). If an incremental encoder is used, then an external ON/OFF sensor (i.e., proximity or photoelectric sensor PRX1 and PRX2) can be used to establish the absolute reference.

Once the starting position is accurately established after power-up (home motion sequence is completed), the system is to operate in position servo mode, followed by a force servo mode and followed by position servo mode again to return to the starting position. Therefore, all that is needed is the logic to decide when to switch between the two modes.

Let us assume that we have the parameters for motion, and the I/O registers mapped to variables as follows:

Given: Programmable motion parameters:

$v_{d\text{-}fwd}$ desired forward motion speed

$a_{d\text{-}fwd}$ desired forward motion acceleration

$v_{d\text{-}rev}$ desired reverse motion speed

$a_{d\text{-}rev}$ desired reverse motion acceleration

$x_{min}$ minimum value of cylinder position where pressure is expected to rise

$x_{max}$ maximum value of cylinder position where pressure is expected to rise

$t_{press}$ time period for which to apply pressure

I/O signal registers:

*Enc* accumulated encoder counts since the end of home cycle

$P_1$ pressure sensor signal 1

$P_2$ pressure sensor signal 2

$Prx_1$ ON/OFF sensor used to establish home reference

$Prx_2$ ON/OFF sensor used to establish end of force cycle

*DAC* digital-to-analog converter register.

Below is a pseudocode for the control algorithm,

```
Home_Motion_Sequence() ;
while(true)
{
  1. Wait until cycle start command is received, or
     trigger signal is ON or a predefined time period
     expires,
  2. Enable position servo mode and move forward with
     desired motion profile,
  3. Monitor pressure differential and switch mode if it
     reaches above a certain value when cylinder position
     is within the defined limits,
  4. Operate in force servo mode until the predefined
     time period expired,
  5. Then switch to position servo mode and command
     reverse motion to home position.

}
void function PID_Position(x_d, v_d)
{
  static float K_p = 1.0, K_d = 0.0, K_i=0.01 ;
  static float u_i = 0.0 ;
  static Scale = 1.0 ;

   u_i = u_i + K_i * (x_d - Scale * ENC) ;
   DAC = K_p * (x_d - Scale * ENC) + u_i ;

  return ;
}
void function PID_Force(F_d)
{
  static float K_pf = 1.0, K_df = 0.0, K_if=0.01 ;
  static float u_if = 0.0 ;
  static float A1= 0.025, A2=0.0125 ;

   u_if = u_if + K_if * (F_d - (P1*A1-P2*A2) ) ;
   DAC = K_pf * (F_d - (P1*A1-P2*A2)) + u_if ;

  return ;
}
```

## 7.6 EH MOTION AXIS NATURAL FREQUENCY AND BANDWIDTH LIMIT

As a general "rule of thumb," a control system should aim to limit the bandwidth of the closed-loop system ($w_{bw}$) to less than 1/3 of the natural frequency ($w_n$) of the open-loop system,

$$w_{bw} < \frac{1}{3} w_n \tag{7.226}$$

Otherwise, too large closed-loop control gains will result in closed-loop instability or very lightly damped oscillatory response. Therefore, it is important to estimate the open-loop natural frequency of a given EH motion axis. The most significant factors affecting the natural frequency of an EH axis are the compressibility of the fluid (its spring effect) and the inertial load on the axis. The fluid spring effect is modeled by its bulk modulus and fluid volume. The lowest natural frequency is approximated as

$$w_n = \sqrt{K/M} \tag{7.227}$$

where $M$ is the total inertia moved by the axis. For varying load conditions, the worst-case inertia should be considered. The spring coefficient, $K$, of the axis is approximated by

$$K = \beta \left( \frac{A_{he}^2}{V_{he}} + \frac{A_{re}^2}{V_{re}} \right) \tag{7.228}$$

where $A$ and $V$ represent cross-sectional area and volume, and $_{he}$ and $_{re}$ subscripts refer to the head-end and rod-end of the cylinder, respectively. The fluid bulk modulus, $\beta$, is defined as

$$\beta = \frac{-\Delta P}{\Delta V/V} \tag{7.229}$$

The typical range of values for $\beta$ is 2 to $3 \cdot 10^5$ psi.

It is important to note that entrapped air in hydraulic fluid reduces the effective bulk modulus very significantly, whereas dissolved air has very little effect. Let us consider a hydraulic fluid and some entrapped air in a fluid hose. Let the bulk modulus of the liquid be $\beta_l$, of the entrapped air $\beta_g$, and of the hose (container) $\beta_c$. It can be shown [30] that the effective bulk modulus is

$$\frac{1}{\beta} = \frac{1}{\beta_l} + \frac{1}{\beta_g}\frac{V_g}{V_t} + \frac{1}{\beta_c} \tag{7.230}$$

where

$$\beta_l = 2.2 \times 10^5 \text{ psi}; \quad \text{for most hydraulic fluids} \tag{7.231}$$

$$\beta_c = \frac{\delta x}{D} \cdot E; \quad \text{for container} \tag{7.232}$$

$$\beta_g \approx 1.4 \times p; \quad \text{for gas} \tag{7.233}$$

where $\delta x$ is wall thickness of the container, $D$ is radius of container or hose diameter, $E$ is Young's modulus of the container material, $p$ is the nominal line pressure, $V_t$ is total volume, and $V_g$ is entrapped gas volume. Notice that even 1% of air entrapment results in significant loss of bulk modulus while increasing nominal pressure helps reduce that effect.

Because the effective stiffness is a function of the cylinder position since $V_{he}$ and $V_{re}$ are functions of cylinder position, the natural frequency of the axis varies as a function of the cylinder stroke. The minimum value of the natural frequency is around the middle position of the cylinder stroke. Unless advanced control algorithms are used, a closed-loop control system gains should be adjusted so that the closed-loop system bandwidth stays below 1/3 of this open-loop natural frequency. In general the natural frequency of the valve is much larger than the cylinder and load hydraulic natural frequency. Therefore, the valve natural frequency is not the limiting factor in closed-loop performance of most applications.

When the open-loop system and load dynamics have low damping, the closed-loop system bandwidth using position feedback is severely limited due to low damping ratio, i.e., to much lower values than the 1/3 $w_n$, where $w_n$ is the open-loop system natural frequency. Therefore, it is necessary to add damping into the closed-loop system in order to achieve higher closed loop bandwidth. This can be achieved in two ways:

1. Bypass leakage orifice at the valve between two sides of the actuator. This, however, has two drawbacks: (1) It wastes energy and (2) the static stiffness of the closed-loop system against load disturbance is reduced.
2. Velocity and pressure feedback in the valve control.

The standard pressure compensated flow control valve increases the effective damping in the closed-loop system by modifying the pressure-flow characteristics of the valve. The fact that the valve affects flow as function of pressure, i.e., a constant times the pressure, it indirectly affects the velocity because flow rate and actuator velocity are closely related. Hence, the feedback in the form of *velocity* or *pressure* adds damping. However, the cost is the reduced static stiffness of the closed-loop system.

***Example*** Let us consider an EH motion axis with the following parameters for the cylinder and load,

$$A_{he} = 2.0 \text{ in}^2 \tag{7.234}$$

$$A_{re} = 1.0 \text{ in}^2 \tag{7.235}$$

$$L = 20 \text{ in} \tag{7.236}$$

$$W = 1000 \text{ lb} \tag{7.237}$$

Let us consider the natural frequency at the midstroke point,

$$V_{he} = A_{he} \cdot L/2 = 20 \text{ in}^3 \tag{7.238}$$

and

$$V_{re} = A_{re} \cdot L/2 = 10 \text{ in}^3 \tag{7.239}$$

Let us approximate the bulk modulus of hydraulic fluid to be $2.5 \times 10^5$ psi. The weight of the rod and load $W = 1000$ lb, hence the mass, $M = W/g = 1000 \text{ lb}/386 \text{ in/sec}^2$. The open-loop natural frequency of the EH axis when the cylinder is at midstroke point is

$$w_n = \sqrt{K/M} \tag{7.240}$$

$$M = \frac{1000}{386} \text{ lb/[in/sec}^2] = 2.59 \text{ lb/[in/sec}^2] \tag{7.241}$$

$$K = 2.5 \cdot 10^5 \cdot \left(\frac{2^2}{20} + \frac{1^2}{10}\right) \text{ lb/in}^2 \text{ (in}^2)^2/\text{in}^3 \tag{7.242}$$

$$= 7.5 \cdot 10^4 \text{ lb/in} \tag{7.243}$$

Then,

$$w_n = \sqrt{K/M} \tag{7.244}$$

$$= \sqrt{7.5 \cdot 10^4/2.59 \text{ (lb/in)/(lb} \cdot \text{sec}^2\text{/in)}} \tag{7.245}$$

$$= 170 \text{ rad/sec} \tag{7.246}$$

$$= 27 \text{ Hz} \tag{7.247}$$

Therefore, a closed-loop controller should not attempt to reach a bandwidth higher than

$$w_{bm} < (1/3 \cdot w_n) = 9.0 \text{ Hz} \tag{7.248}$$

This particular EH motion axis cannot accurately follow cyclic small motion commands with frequency higher than 9.0 Hz.

## 7.7 LINEAR DYNAMIC MODEL OF A ONE-AXIS HYDRAULIC MOTION SYSTEM

Linear models assume that the hydraulic system operates about a nominal operating condition, i.e., valve is operating about the null position with small movements (Fig. 7.64). The fluid compressibility effects are neglected. The pump is assumed to provide a constant supply pressure and the gain of valve from current to flow rate is assumed constant. Figure 7.80 shows a block diagram of a closed-loop-controlled EH system. Figure 7.81 shows a position servo loop version of the system, where each component is represented with a linear model. The dynamics of the valve and its nonlinear flow-current-pressure relationship approximated with a linearized version, load inertia, compliances ($K_s$, $K_a$) and sensor dynamics are neglected. Furthermore, the valve-flow, current, and pressure relationship is approximated by a linear relationship between flow and current (the flow gain, $K_q$), and linear relationship between load and leakage ($K_{pq}$). The compliance of the actuator to base mounting (represented as $K_s$), and the compliance due to the load and hydraulic oil compressibility (represented as $K_a$) are neglected in the linear model analysis. Such compliances determine the open-loop natural frequency of the system which sets the upper limit for closed-loop system bandwidth.

Figure 7.82 shows a one-axis EH motion system where the load pressure is regulated by the control element instead of the load position. In the linear block diagram models, the symbols have the following physical meaning:

$K_{sa}$ is the amplifier command voltage to current gain.

$K_q$ is the valve solenoid current to flow rate gain.

$K_p$ is the pressure gain.

$K_{pq}$ is the leakage gain.

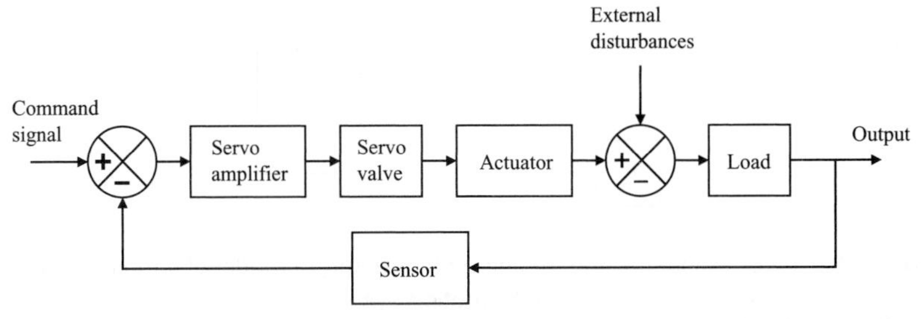


**FIGURE 7.80:** Block diagram of a closed-loop controlled, one-axis electrohydraulic motion system.

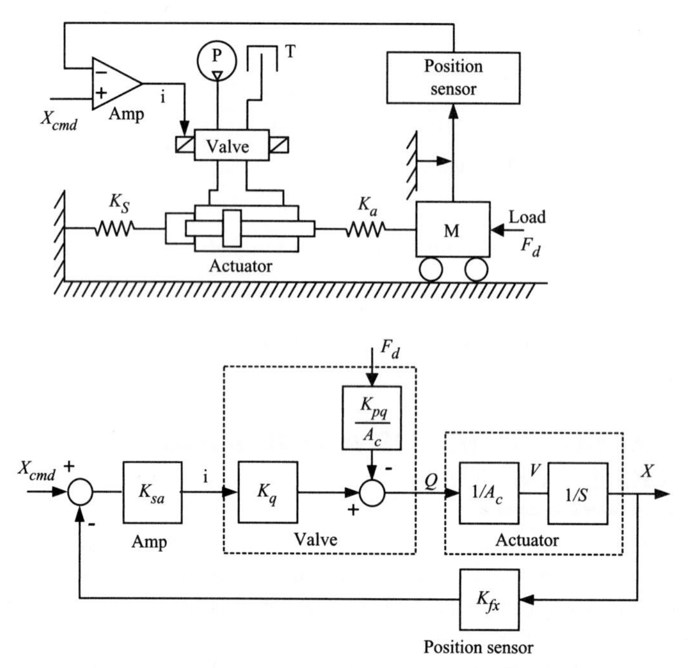


**FIGURE 7.81:** (a) Components, and (b) linear block diagram model of a position-servo controlled, one-axis electrohydraulic motion system.

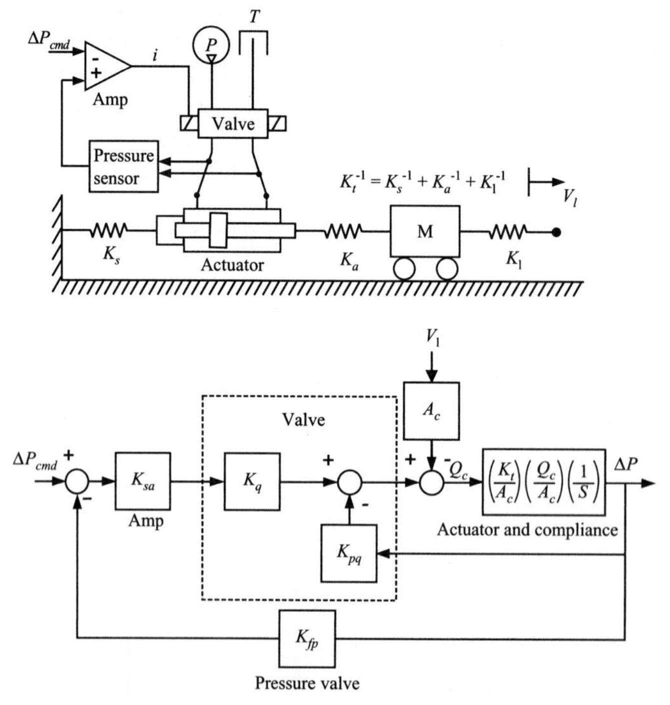


**FIGURE 7.82:** (a) Components and (b) linear block diagram model of a force (load pressure) servo controlled, one-axis electrohydraulic motion system.

$A_c$ is the cylinder cross-sectional area (assumed to be equal areas on both sides).

$K_{fx}$ is the position sensor gain.

$K_{fp}$ is the pressure sensor gain.

$Q_c$ is the compliance flow.

### 7.7.1 Position Controlled Electrohydraulic Motion Axes

The controlled output, position ($x$), is determined by the commanded position ($x_{cmd}$) signal and load force ($F_d$). Using the basic block diagram algebra, the transfer function from $x_{cmd}$ and $F_d$ to $x$ can be calculated using the linear models for each component,

$$X(s) = \frac{1}{A_c \cdot s} \cdot \left(K_q \cdot i(s) - \left(\frac{K_{pq}}{A_c}\right) \cdot F_d(s)\right) \quad (7.249)$$

$$X(s) = \frac{1}{A_c \cdot s} \quad (K_q \cdot K_{sa} \cdot (X_{cmd}(s) - K_{fx} \cdot X(s)) \quad (7.250)$$

$$- \left(\frac{K_{pq}}{A_c}\right) \cdot (F_d(s)) \quad (7.251)$$

$$(A_c \cdot s + K_{sa} \cdot K_q \cdot K_{fx}) \cdot X(s) = K_{sa} \cdot K_q \cdot X_{cmd}(s) - \left(\frac{K_{pq}}{A_c}\right) \cdot F_d(s) \quad (7.252)$$

$$X(s) = + \frac{(K_{sa} \cdot K_q / A_c)}{s + (K_{sa} \cdot K_q \cdot K_{fx}/A_c)} \cdot X_{cmd}(s) \quad (7.253)$$

$$- \frac{K_{pq}/A_c^2}{s + (K_{sa} \cdot K_q \cdot K_{fx}/A_c)} \cdot F_d(s) \quad (7.254)$$

Using the last expression for the transfer function and the final value theorem for Laplace transforms, we can determine the following key characteristics of the closed-loop position control system.

- Steady-state position following error in response to a step position command and no disturbance force condition, $X_{cmd} = X_o/s$ and $F_d(s) = 0$

$$\lim_{t\to\infty} x(t) = \lim_{s\to 0} s \cdot X(s) \quad (7.255)$$

$$= \lim_{s\to 0} s \cdot \frac{(K_{sa} \cdot K_q / A_c)}{(s + (K_{sa} \cdot K_q \cdot K_{fx}/A_c)} \cdot X_o/s \quad (7.256)$$

$$= (1/K_{fx}) X_o \quad (7.257)$$

  When a constant displacement is commanded and there is no disturbance force, the actual position will be proportional to the commanded signal by the feedback sensor gain. By scaling the commanded signal, the effective ratio between the desired and actual position can be made unity. In other words, the output position is exactly equal to the commanded position in steady state when the commanded position is a constant value.

- What is maximum speed error, $V_e$, generated if following error is $X_e(s) = X_{cmd}(s) - K_{fx} X(s)$ in response to ramp command signal (ramp position command means a step velocity command).

$$X_{cmd}(s) = V_0/s^2 \tag{7.258}$$

$$F_d(s) = 0 \tag{7.259}$$

The tracking error transfer function is

$$X_e(s) = X_{cmd}(s) - K_{fx}X(s) \tag{7.260}$$

$$= X_{cmd}(s) - \frac{K_{fx} \cdot K_{sa} \cdot K_q/A_c}{s + (K_{sa} \cdot K_q \cdot K_{fx}/A_c)} \cdot X_{cmd}(s) \tag{7.261}$$

$$= \frac{s}{(s + K_{fx} \cdot K_{sa} \cdot K_q/A_c)} \cdot V_o/s^2 \tag{7.262}$$

Using the final value theorem, it can be determined that the steady-state position following error while tracking a ramp position command (constant speed command) is

$$\lim_{t\to\infty} X_e(t) = \lim_{s\to 0} s \cdot X_e(s) \tag{7.263}$$

$$= \frac{1}{(K_{sa} \cdot K_q \cdot K_{fx}/A_c)} \cdot V_o \tag{7.264}$$

where the gain is called the loop velocity gain of the control system,

$$K_{vx} = \frac{K_{sa} \cdot K_q \cdot K_{fx}}{A_c} \tag{7.265}$$

and that if position $X(s)$ is disturbed by any external reason to a value of $X_e$, the transient response will create the following velocity while trying to correct it:

$$V_e = K_{vx} \cdot X_e \tag{7.266}$$

- Another important characteristic is the closed-loop stiffness of the positioning system. What is the steady-state change in position if there is a constant disturbance force acting on the system, $F_d(s) = F_{do}/s$, $X_{cmd}(s) = 0.0$, $K_{cl} = F_{do}/X$? Using the final value theorem,

$$X(s) = -\frac{\left(K_{pq}/A_c^2\right)}{s + (K_{sa} \cdot K_q \cdot K_{fx}/A_c)} \cdot F_d(s) \tag{7.267}$$

$$\lim_{s\to 0} s \cdot X(s) = -s \cdot \frac{\left(K_{pq}/A_c^2\right)}{s + (K_{sa} \cdot K_q \cdot K_{fx}/A_c)} \cdot F_{do}/s \tag{7.268}$$

$$x(\infty) = \left(K_{pq}/A_c^2\right)/K_{vx} \cdot F_{do} \tag{7.269}$$

$$\frac{F_{do}}{x(\infty)} = K_{cl} \tag{7.270}$$

The closed-loop, steady-state stiffness is

$$K_{cl} = \frac{K_{vx} \cdot A_c^2}{K_{pq}} \tag{7.271}$$

$$= \frac{\frac{K_{sa} \cdot K_q \cdot K_{fx}}{A_c} \cdot A_c^2}{K_{pq}} \tag{7.272}$$

$$= \frac{K_{sa} \cdot K_q \cdot K_{fx} \cdot A_c}{K_{pq}} \tag{7.273}$$

In general the leakage coefficient for servo valves is

$$K_{pq} \approx 0.02 \cdot \frac{Q_r}{P_S} \tag{7.274}$$

Hence, the stiffness of the closed-position control servo with such a valve is approximately,

$$K_{cl} = 50 \cdot K_{vx} \frac{A_c^2 \cdot P_S}{Q_r} \tag{7.275}$$

which clearly indicates the effect of leakage coefficient on the closed-loop stiffness.

It is important to recall that the natural frequency of the electrohydraulic system and its load imposes an upper limit on the closed-loop system bandwidth. The position servo control system described above has approximately the closed-loop bandwidth (closed-loop transfer function pole) equal to the $K_{vx}$, and the value of it is limited by the natural frequency of the system due to stability considerations:

$$w_{bw} = K_{vx} \leq \frac{1}{3} \cdot w_n \tag{7.276}$$

The characteristics of the valve around its null position significantly affects the steady-state positioning accuracy of the closed-loop system. Valve imperfections around the null position which were neglected in the linear block diagram analysis are: (1) deadband, (2) hysteresis, and (3) null threshold shift. These properties of the valve were discussed earlier in this chapter. The resulting positioning error as a result of null-position-related imperfections of the valve can be calculated from the block diagram. Deadband, hysteresis, and threshold values of the current all add up to a range of input current to the valve for which no position correction is made in the valve. Given a total value of spurious valve current due to null-position imperfections, $i_{err}$, the resulting steady-state positioning error can be determined as

$$X_e = \frac{i_{err}}{K_{sa} \cdot K_{fx}} \tag{7.277}$$

***Example*** Consider a single-axis electrohydraulic motion control system as shown in Fig. 7.81 and its block diagram representation in Fig. 7.80. We assume that the cylinder is rigidly connected to its base and load. Let us consider the small movements of the valve around its null position, i.e., a case that is maintaining a commanded position. Determine the cylinder positioning error for the following conditions:

1. Amplifier gain is $K_{sa} = 200\ \text{mA}/10\ \text{V} = 20$ [mA/V],
2. Valve gain around the null position operation, $K_q = 20$ [in$^3$/sec]/200 [mA] $=$ 0.1 [in$^3$/sec]/[mA],
3. Cross-sectional area of the cylinder is $A_c = 2.0$ in on both sides (rod is extended to both sides),
4. The sensor gain is $K_{fx} = 10\ \text{V}/10$ [in] $= 1$ [V/in],
5. Deadband of the valve is 10% of the maximum input current, $i_{db} = 0.1 \cdot i_{max} = 0.1 \cdot 200\ \text{mA} = 20\ \text{mA}$.

As long as the error magnitude times the gain of the amplifier is less than the deadband, there will not be any flow through the valve and no position correction will be made. Only after the error times the amplifier gain is larger than the deadband of the valve, will there be flow through the valve and correction to the position error. Hence, the error in the following range is an inherent limitation of this system, since it does not result in any flow or motion in the system,

$$X_{e,max} \cdot K_{sa} = i_{db} \tag{7.278}$$

$$X_{e,max} = \frac{i_{db}}{K_{sa}} \tag{7.279}$$

$$= \frac{20 \text{ mA}}{20 \text{ mA/V}} \tag{7.280}$$

$$= 1 \text{ V} \tag{7.281}$$

which is 10% of the total displacement range.

Any positioning error less than that value, while maintaining a commanded position (by the regulating motion of the valve around the null position), is inherently not corrected,

$$X_e \leq X_{e,max} \tag{7.282}$$

$$\leq \frac{i_{db}}{K_{sa}} \tag{7.283}$$

$$\leq 1 \text{ V} \tag{7.284}$$

which is equal to 1.0 in. in physical units since $K_{fx} = 1$ [V/in].

Notice that if the amplifier gain is increased, i.e., by 10, the positioning error due to the same valve deadband is reduced by the same factor.

$$X_e \leq \frac{i_{db}}{K_{sa}} = \frac{20 \text{ mA}}{200 \text{ mA/V}} = 0.1 \text{ V} \tag{7.285}$$

which is equal to 0.1 in. in physical units since $K_{fx} = 1$ [V/in].

In order to ensure that positioning error due to valve deadband is less than a desired value, the amplifier gain should be larger than a minimum value as follows:

$$K_{sa} \geq \frac{i_{db}}{X_{e,max}} \tag{7.286}$$

$$\geq \frac{20 \text{ mA}}{X_{e,max} \text{ V}} \tag{7.287}$$

However, the amplifier gain cannot be made arbitrarily large due to the closed-loop bandwidth limitations set by the natural frequency of the open-loop system,

$$f_1(i_{db}, e_{max}) \leq K_{sa} \leq f_2(w_{wb}) \tag{7.288}$$

$$w_{bw} = K_{vx} = \frac{K_{sa} \cdot K_q \cdot K_{fx}}{A_c} \leq \frac{1}{3} \cdot w_n \tag{7.289}$$

which indicates that the lower limit of the amplifier gain is set by a function of the deadband and desired positioning accuracy, and upper limit is set by the open-loop bandwidth of the hydraulic axis.

### 7.7.2 Load Pressure Controlled Electrohydraulic Motion Axes

Let us also consider the same EH motion system, except that this time the closed-loop control objective is to control the load pressure. The load pressure is measured as the differential pressure between the two output ports of the valve. We will assume that the pressure dynamics between the valve output ports and the actuator ports are negligible (Fig. 7.82). The commanded signal represents the desired load pressure, $\Delta P_{cmd}$. Using the basic block diagram algebra, the transfer function from commanded load pressure and the external load speed to the load pressure can be obtained as

$$\Delta P(s) = \frac{K_t}{A_c}\frac{1}{A_c}\frac{1}{s} \cdot (K_q \cdot i(s) - K_{pq} \cdot \Delta P(s) - A_c \cdot V_l(s)) \tag{7.290}$$

Notice that the internal leakage term due $K_{pq} \cdot \Delta P(s)$ is relatively small for servo valves and can be neglected in the analysis. Assuming the leakage term is neglected, the closed-loop transfer function can be expressed as

$$\Delta P(s) = + \frac{\left(K_t \cdot /A_c^2\right) \cdot K_{sa}K_q)}{\left(s + K_{fp}K_{sa}K_qK_t/A_c^2\right)} \cdot \Delta P_{cmd}(s) \tag{7.291}$$

$$- \frac{(K_t/A_c)}{\left(s + K_{fp}K_{sa}K_qK_t/A_c^2\right)} \cdot V_l(s) \tag{7.292}$$

In terms of command signal and output signal relationship, the transfer function behavior of the force servo is identical to the position servo. The closed-loop system has a first-order filter dynamic behavior. The velocity gain for the pressure loop (force servo) system is

$$K_{vp} = \frac{K_{fp}K_{sa}K_qK_t}{A_c^2} \tag{7.293}$$

In the above linear models, the only dynamics included are the integrator behavior of the actuator. The transient response of the valve from current to flow is not modeled. Experimental studies indicate that the transient response of proportional or servo valve from current to flow can be approximated by a first-order or a second-order filter depending on the accuracy of approximation needed. The transient response model of a valve is then

$$\frac{Q_o(s)}{i(s)} = \frac{K_q}{(\tau_{v1}s + 1)(\tau_{v2}s + 1)} \tag{7.294}$$

where $\tau_{v1}$ and $\tau_{v2}$ are the two time constants for the second-order model. One of them is set to zero for a first-order model.

## 7.8 NONLINEAR DYNAMIC MODEL OF A HYDRAULIC MOTION SYSTEM

Nonlinear dynamic model for a single-axis linear hydraulic motion system, where the actuator is a cylinder, is discussed below. The same identical equations apply for a rotary hydraulic motion system, where the actuator is a rotary hydraulic motor, with analogous parameter replacements. The fluid compressibility is also taken into account in the model (Figs. 7.2 and 7.81) but leakage is neglected. Let us consider the motion of the piston-rod-load assuming they are rigidly connected to each other. Using Newton's second law for the force-motion relationships of the cylinder and the load,

**(1)** Extension motion:

$$m \cdot \ddot{y} = p_A \cdot A_A - p_B \cdot A_B - F_{ext}; \quad 0 \le y \le l_{cyl} \tag{7.295}$$

and the pressure transients in the control volumes on both sides of the cylinder (from eqn. 7.27),

$$\dot{p}_A = \frac{\beta}{y \cdot A_A}(Q_{PA} - \dot{y} \cdot A_A) \tag{7.296}$$

$$\dot{p}_B = \frac{\beta}{(l_{cyl} - y) \cdot A_B}(-Q_{BT} + \dot{y} \cdot A_B) \tag{7.297}$$

$$\dot{p}_P = \frac{\beta}{V_{hose,pv}}(Q_P - Q_{PA} - Q_{PT}) \tag{7.298}$$

where

$$Q_P = w_{pump} \cdot D_p(\theta_{sw}) \tag{7.299}$$

$$Q_{PT} = C_d \cdot A_{PT}(x_s) \cdot \sqrt{(2/\rho) \cdot (p_P - p_T)} \tag{7.300}$$

$$Q_{PA} = C_d \cdot A_{PA}(x_s) \cdot \sqrt{(2/\rho) \cdot (p_P - p_A)} \tag{7.301}$$

$$Q_{BT} = C_d \cdot A_{BT}(x_s) \cdot \sqrt{(2/\rho) \cdot (p_B - p_T)} \tag{7.302}$$

**(2)** Retraction motion:

$$m \cdot \ddot{y} = p_A \cdot A_A - p_B \cdot A_B - F_{ext}; \quad 0 \le y \le l_{cyl} \tag{7.303}$$

and the pressure transients in the control volumes on both sides of the cylinder,

$$\dot{p}_A = \frac{\beta}{y \cdot A_A}(-Q_{AT} - \dot{y} \cdot A_A) \tag{7.304}$$

$$\dot{p}_B = \frac{\beta}{(l_{cyl} - y) \cdot A_B}(Q_{PB} + \dot{y} \cdot A_B) \tag{7.305}$$

$$\dot{p}_P = \frac{\beta}{V_{hose,pv}}(Q_P - Q_{PB} - Q_{PT}) \tag{7.306}$$

where

$$Q_P = w_{pump} \cdot D_p(\theta_{sw}) \tag{7.307}$$

$$Q_{PT} = C_d \cdot A_{PT}(x_s) \cdot \sqrt{(2/\rho) \cdot (p_P - p_T)} \tag{7.308}$$

$$Q_{PB} = C_d \cdot A_{PB}(x_s) \cdot \sqrt{(2/\rho) \cdot (p_P - p_B)} \tag{7.309}$$

$$Q_{AT} = C_d \cdot A_{AT}(x_s) \cdot \sqrt{(2/\rho) \cdot (p_A - p_T)} \tag{7.310}$$

where the design parameters and variables for the hydraulic system are

$m$—inertia

$A_A$, $A_B$—cylinder head-end and rod-end cross-sectional area

$\beta$—bulk modulus of hydraulic fluid due to its finite stiffness

$V_{hose,pv}$—volume of the hose between pump and valve

$l_{cyl}$—the travel range of the cylinder

$D_p$—flow rating of the pump [volume/revolution]

$C_d$—valve flow gain constant

$A_{PA}(x_s)$, $A_{PB}(x_s)$, $A_{AT}(x_s)$, $A_{BT}(x_s)$, $A_{PT}(x_s)$—valve flow areas as functions of spool displacement, between pump, cylinder, and tank

$Q_P$, $Q_{PT}$, $Q_{PA}$, $Q_{PB}$, $Q_{AT}$, $Q_{BT}$—flow rate from pump, pump-to-tank, pump-to-cylinder, and cylinder-to-tank

$x_s$—valve spool displacement

$\theta_{sw}$—swash plate angle

$\theta_{sw0}$—constant value of swash plate angle

$y$—cylinder displacement

$p_P$, $p_A$, $p_B$, $p_T$—pressure at pump, head-end, rod-end, and tank

$w_{pump}$—speed of the pump

**Relief Valves** Relief valves are used to limit the pressure in hydraulic lines, hence protect the lines against excessive pressure. The maximum pressure limit of relief valve is adjustable. Let us assume that the maximum pressure is set to $p_{max,p}$ for the pump side, and to $p_{max,l}$ for the line side. If $p_P > p_{max,p}$, $p_P = p_{max,p}$; pump side relief valve opens and limits the maximum output pressure of the pump. If either of the cylinder side line pressures exceeds the relief valve settings, $p_A, p_B > p_{max,l}$, $p_A, p_B = p_{max,l}$; cylinder side relief valve opens.

**Directional Check Valves** In many EH motion applications, it is not desirable to allow flow from cylinder back to the pump, which can happen if $p_A > p_P$ or $p_B > p_P$. This is accomplished by one directional load check valve on each line. Load check valve closes to prevent back flow from cylinder to pump. During extension, if $p_A > p_P$, then $Q_{PA} = 0.0$; during retraction, if $p_B > p_P$, then $Q_{PB} = 0.0$.

**Open-Center EH Systems** An "open-center" EH system has fixed displacement pump and open-center valve, where there is an orifice between pump and tank and the pump displacement is constant:

$$A_{PT}(x_s) \neq 0 \tag{7.311}$$

$$Q_P = w_{pump} \cdot D_p(\theta_{sw0}) \tag{7.312}$$

**Closed-Center EH Systems** A closed-center EH system has variable displacement pump and closed-center valves, where there is not any orifice directly between pump and tank ports, and the pump displacement is variable:

$$A_{PT}(x_s) = 0; \tag{7.313}$$

$$Q_P = w_{pump} \cdot D_p(\theta_{sw}) \tag{7.314}$$

## 7.9 CURRENT TRENDS IN ELECTROHYDRAULICS

The current trend in future electrohydraulic technology is to increase:

- The power/weight ratio to reduce physical size of components, hence, to reduce cost
- Software programmable components [45a]

In order to increase the power/weight ratio, the system pressure must be increased which results in smaller size components to deliver the same power. However, increased system pressure reduces the resonant frequency of the hydraulic system due to oil compressibility; hence, the control-loop system bandwidth limit is lower. As the supply pressure gets higher, it is more important to minimize cavitation and air bubbles in the hydraulic lines. Othwerwise, the system response will be significantly slower or even go unstable in the case of closed-loop control. Furthermore, cavitation leads to damage to the hydraulic components and increases noise.

Another way of reducing the size of components and the cost is to make more effective use of the components. Consider the hydraulic circuits (implement, steering, brake, cooling fan, pilot hydraulics) and the pumps used to support them in construction equipment applications. Traditional designs dedicate one or more pumps per circuit. Since all systems are not used in maximum flow demand at all times, all of the pumps are not used in their maximum capability. Furthermore, duplicate pumps are provided for safety back-up reasons for critical systems (i.e., steering). The concept of sharing pumps among multiple

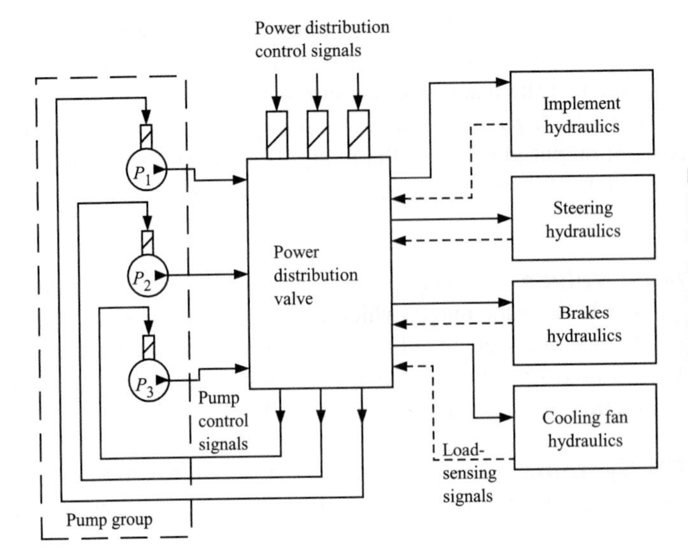


**FIGURE 7.83:** Programmable power allocation in multipump, multicircuit hydraulic systems. Instead of dedicating a pump to each circuit in hardware, the output of all pumps is brought into a controllable distribution valve. The valve directs the desired amount of flow to each subsystem based on demand.

circuits through a controllable power distribution valve has been emerging in recent years (Fig. 7.83). Instead of dedicating a pump to each circuit in hardware, the total hydraulic power of all pumps is combined at a distribution valve, and under program control, the hydraulic power is distributed to different subsystems based on demand. This approach has the promise of making better use of available component capability, reduced cost, and improved performance.

The valve is the main critical component in a hydraulic system from a control system perspective. All of the valves we discussed so far have a single spool for each stage. One-spool geometry defines the orifice areas between the four ports of the valve: pump (P), tank (T), A and B side of the cylinder. The single-variable, spool displacement, $x_{spool}$, determines the orifice areas

$$A_{PA}(x_v),\ A_{PB}(x_v),\ A_{AT}(x_v),\ A_{BT}(x_v),\ A_{PT}(x_v) \tag{7.315}$$

These geometric relationships between the spool displacement versus the orifice areas are designed and physically machined into each valve spool. Once the valve is machined, its orifice characteristics are fixed. In order to accommodate many different application specific requirements on orifice functions, many different spool geometry variations are often needed. This requires machining many variations of the basically same spool geometry for different applications. For instance, it has been estimated that one of the major construction equipment manufacturers alone machines over 1600 different valve spool geometries. It would be desirable to reduce the number of different spool geometries that must be physically machined. This idea had led to the development of the *independently metered valves* (IMV) concept. The idea is to define the orifice areas in software by actively and independently controlling each orifice area by a separate spool (Fig. 7.84). The IMV valve has up to six independently operated spools and solenoids, one for each port connection orifice area:

$$A_{PA}(x_1),\ A_{PB}(x_2),\ A_{AT}(x_3),\ A_{BT}(x_4),\ A_{PT}(x_5),\ A_{AB}(x_6) \tag{7.316}$$

where each spool position is proportional to the associated solenoid current. Therefore, the orifice area functions can be equally expressed as a function of solenoid currents.

$$A_{PA}(i_1),\ A_{PB}(i_2),\ A_{AT}(i_3),\ A_{BT}(i_4),\ A_{PT}(i_5),\ A_{AB}(i_6) \tag{7.317}$$

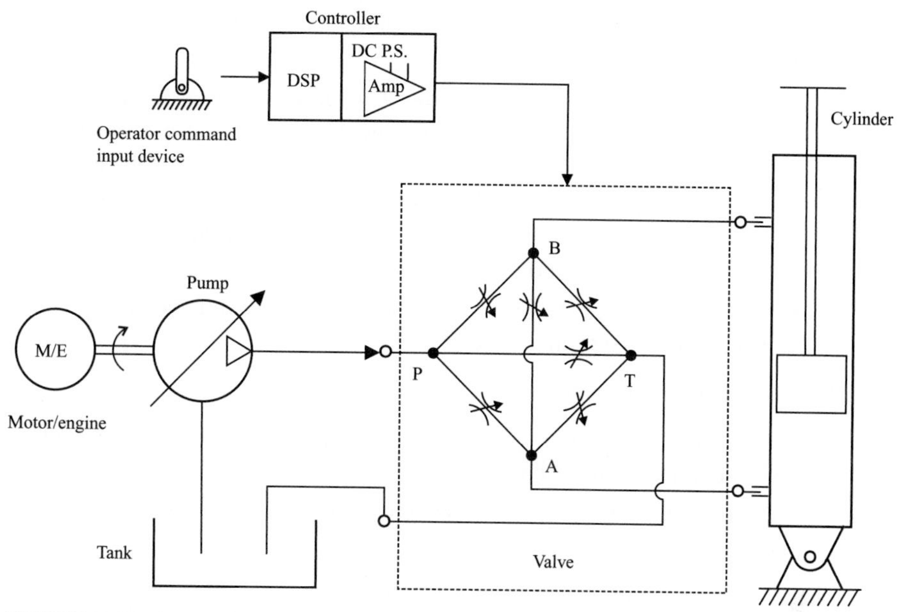


**FIGURE 7.84:** Independently metered valve (IMV) concept. Valve flow orifices are independently controlled and defined in real-time control software.

There are several patents on this concept ([53b]). The valve flow-orifice-pressure differential relationships are still the same, except that in the case of IMV, each flow rate can be independently metered,

$$Q_{PA} = C_d \cdot A_{PA}(i_1)\sqrt{(p_P - p_A)} \tag{7.318}$$

$$Q_{PB} = C_d \cdot A_{PB}(i_2)\sqrt{(p_P - p_B)} \tag{7.319}$$

$$Q_{AT} = C_d \cdot A_{AT}(i_3)\sqrt{(p_A - p_T)} \tag{7.320}$$

$$Q_{BT} = C_d \cdot A_{BT}(i_4)\sqrt{(p_B - p_T)} \tag{7.321}$$

$$Q_{PT} = C_d \cdot A_{PT}(i_5)\sqrt{(p_P - p_T)} \tag{7.322}$$

$$Q_{AB} = C_d \cdot A_{AB}(i_6)\sqrt{(p_A - p_B)} \tag{7.323}$$

Since each solenoid pair is independently controlled, each orifice area can be controlled and their relationship to each other, like the case in a single-spool valve, can be defined in software. Therefore, the same IMV valve can be controlled to behave like a valve with a different geometry by changing the control software. In other words, the effective valve geometry is defined in software. For instance, if the IMV control is to emulate a standard closed-center valve, then $i_5, i_6$ should be controlled such that

$$A_{PT}(i_5) = 0 \quad A_{AB}(i_6) = 0 \tag{7.324}$$

If open-center valve emulation is desired, then $i_6$ should be controlled such that

$$A_{AB}(i_6) = 0 \tag{7.325}$$

If regenerative power capability is desired in a closed-center valve emulation, then $i_5, i_6$ should be controlled such that

$$A_{PT}(i_5) = 0 \quad A_{AB}(i_6) \neq 0 \tag{7.326}$$

The IMV concept offers the following advantages and disadvantages. The advantages are as follows:

- Valve geometry is defined in software. Therefore, only three or four mechanical valves would be needed to cover the low-, medium-, and high-power applications. The application specific spool geometry would be defined in the software.
- Emulated valve geometry in software does not have to be the equivalent of a single mechanical valve. It can function as the equivalent of different types of mechanical valves during different operating conditions to optimize the performance.
- Through the added flexibility in valve control, regenerative energy can be used to operate the EH system more energy efficiently.

The disadvantages are as follows:

- The control task is more complicated. While a standard valve has a single controlled solenoid, IMV has six controlled solenoids.
- Because of the increased number of electrical components, the number of possible failures is higher.

The IMV concept is one of the most significant new EH technology in recent years. As embedded digital control of valves increases, the functionality of a valve component is defined in software. The mechanical design of it gets simpler, since the functionality no longer needs to be machined into the valve. However, the software aspect of the valve becomes more complicated, and without the software the component would not be functional.

## 7.10 CASE STUDIES

### 7.10.1 Case Study: Multifunction Hydraulic Circuit of a Caterpillar Wheel Loader

Figure 7.85 (also see Fig. 7.12) shows the complete hydraulic system schematics for wheel loader model 950G by Caterpillar [93]. The steering subsystem is discussed separately in detail below. The hydraulic system power is supplied by four major pumps: a variable displacement pump for the steering subsystem, and three fixed displacement pumps for implement hydraulics main pressure lines, for brake charging and pilot pressure lines, and for cooling the fan motor. Steering system also has a secondary power source as a safety backup using a battery, electric motor, and pump (the fifth pump). This design uses:

1. Mostly fixed displacement pumps (except the primary steering pump), and each pump is dedicated to one circuit (i.e., steering pump is dedicated to support steering subsystem, implement pump is dedicated to the implement subsystem).
2. Lift, tilt, and auxiliary function valve-cylinder functions are connected in series to the hydraulic supply lines (P and T) and are necessarily open-center types (see Fig. 7.12). Furthermore, the function closer to the pump has priority over the functions following it.
3. Main flow control valves do not have pressure compensation (no compensator valves); hence, the function speed will vary with load for a given command.

The valves of the implement hydraulics are the electrohydraulic (EH) type, meaning that the pilot control valves are controlled proportionally by electric current sent to a solenoid for each pilot valve. The hydraulic system is also available with mechanically actuated pilot valves instead of the EH pilot valves. Proportional flow control valves for mobile equipment

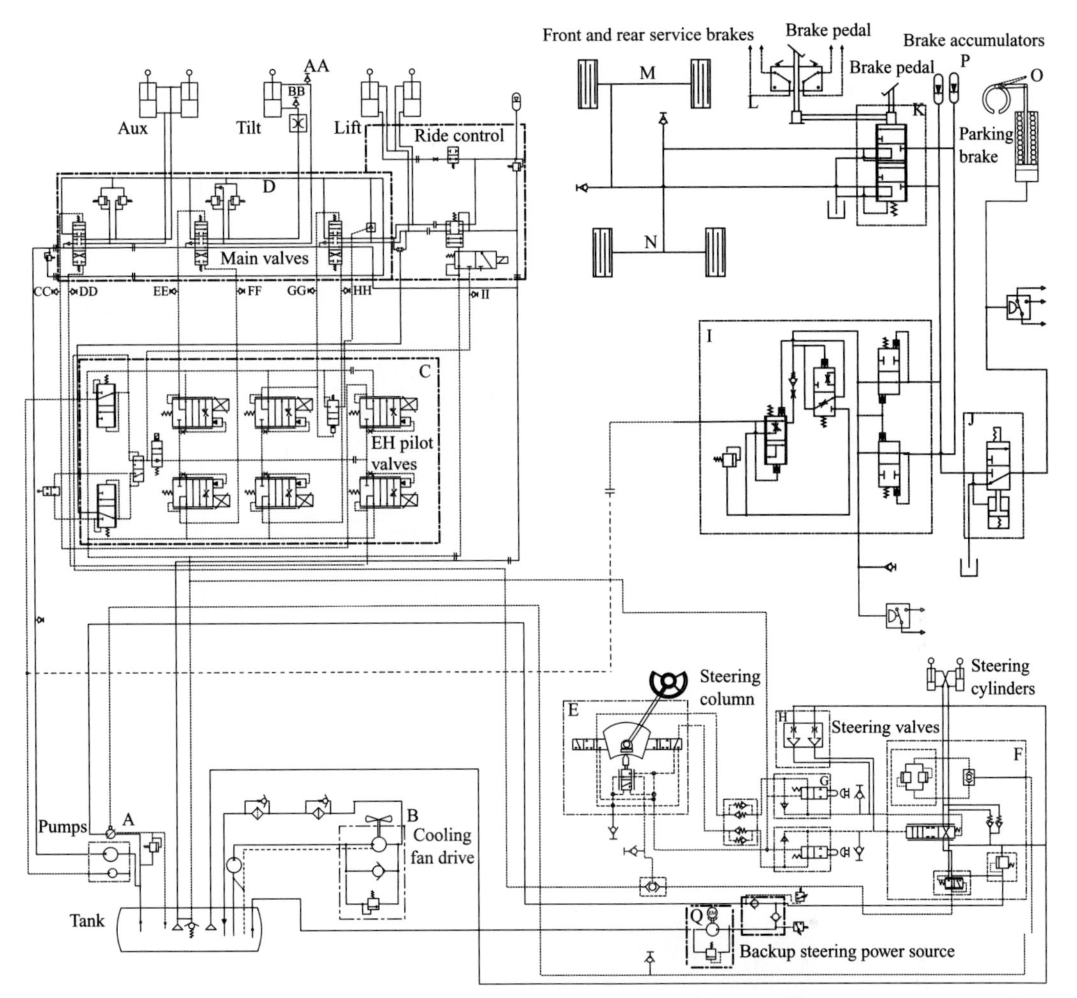


**FIGURE 7.85:** Hydraulic system circuit diagram of the Caterpillar wheel loader model 950G.

are typically fabricated in stackable blocks, i.e., a valve stack may have up to 10 valve sections for 10 different functions (Fig. 7.86).

In some cases, the pilot and main flow sections of the valves are designed as separate blocks. The advantage of this is that the main section of the valve accepts the pilot control pressure. The pilot control pressure may be contolled either by a mechanically actuated pilot valve or solenoid actuated EH valve. Hence, the main flow section of the valve can be used with either mechanically actuated pilot valves or electrically (solenoid) actuated pilot valves. It is instructive to note that the main flow control valve for multifunction hydraulic circuits, such as the the lift and tilt circuit in Fig. 7.47 (component V on the figure) is manufactured as a single-valve block with all the individual valve functions incorporated into one block design that has the external port connections (pump, tank, service ports A and B for lift and tilt, and pilot pressure control ports, Fig. 7.87).

Notice that the difference between mechanical lever controlled and electrical solenoid controlled pressure reducing pilot control valves is the source of actuation power to shift the spool of the pilot valve (see Fig. 7.58). In a mechanical lever controlled system, as the operator moves the lever, the lever motion moves a spring in the pilot valve which then moves the pilot valve spool and changes the output pilot control pressure. In solenoid controlled pilot valves, the lever displacement is measured by a sensor, and the electronic control

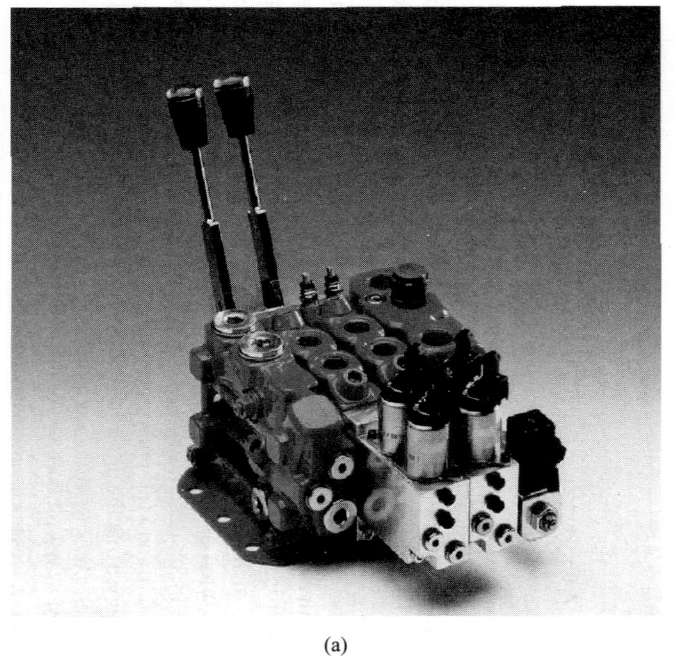

(a)

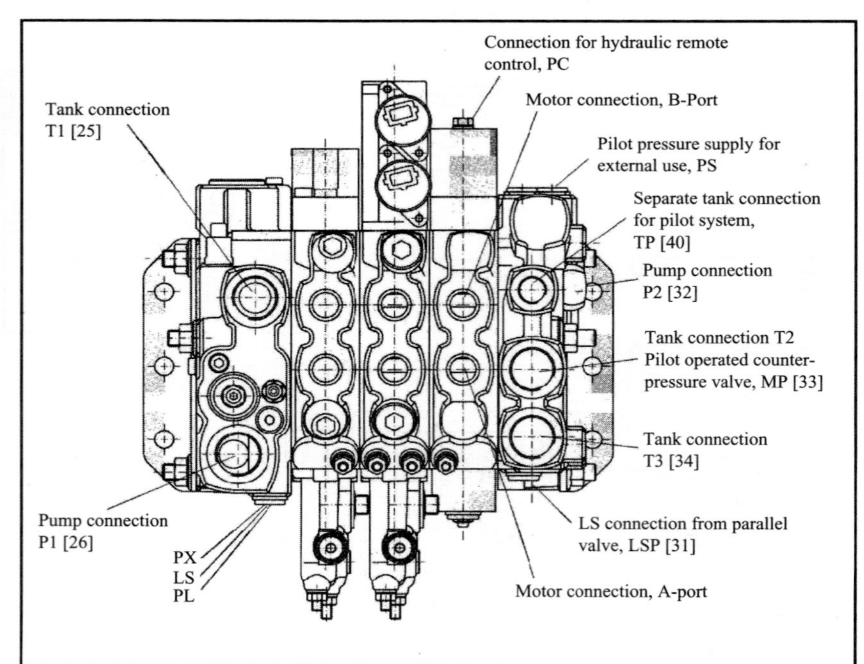


(b)

**FIGURE 7.86:** Stackable valve block for mobile applications: (a) picture, and (b) top view.

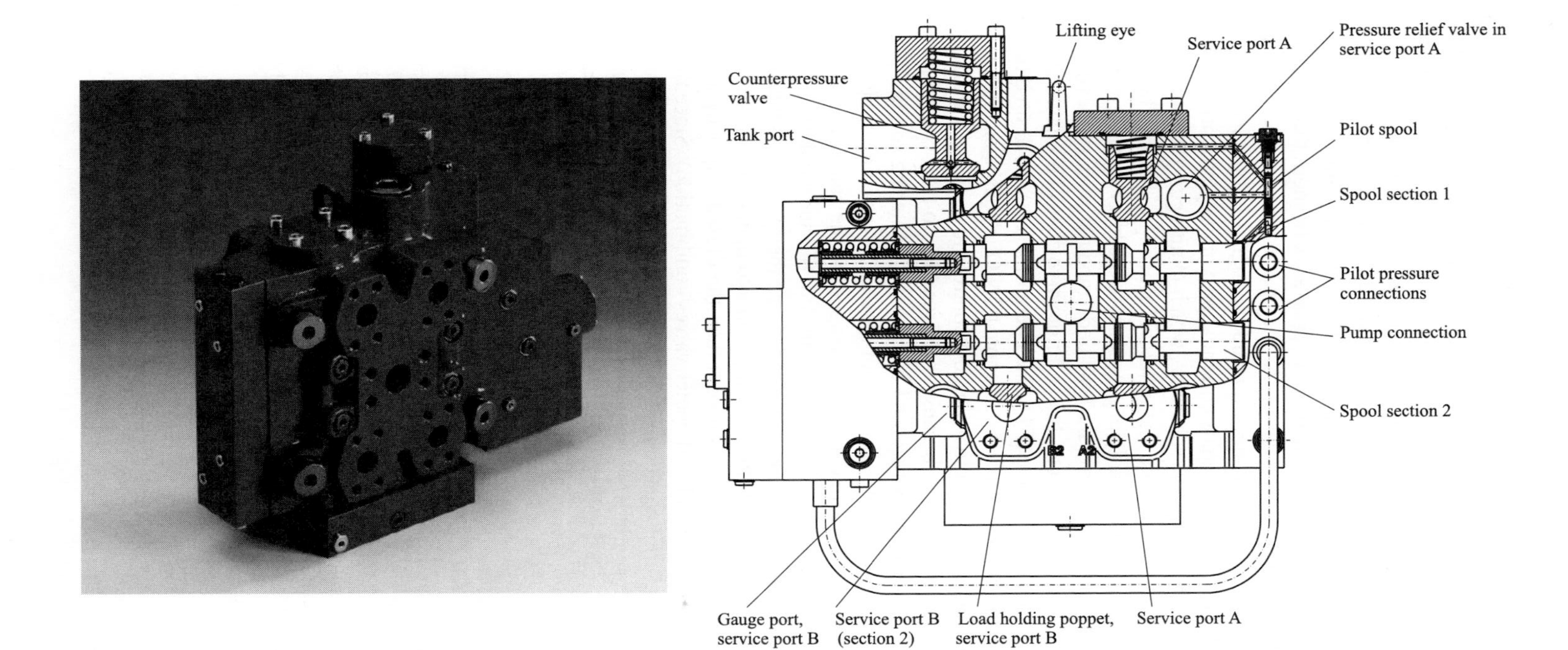


**FIGURE 7.87:** Valve block (Model M400LS by Parker Hannifin AB) used in various hydraulic circuits of mobile equipment applications. Notice that the external port connections are provided for pump (P), tank (T), service ports (A1, B1, A2, B2), and pilot pressure ports. All other hydraulic connections (i.e., control of poppet valve, line relief valves, anticavitation make-up valves) are internally machined into the valve block and do not require any plumbing.

module (ECM) samples the sensor signal and then sends a current to the pilot valve to move its spool. The rest of the power hydraulic circuit is the same.

There are two pilot valves per cylinder function, i.e., lift raise and lift lower pilot valves for the lift circuit. Each pilot valve is proportionally controlled by a solenoid in EH system. Hence, there are two solenoid outputs per function. Each pilot valve has two inports: pilot pressure supply and tank ports. The outport carries the pilot control pressure to the main flow control valve. The outport pilot control pressure is proportionally controlled by the solenoid current between the tank pressure (minimum outport pilot pressure) and the pilot pressure supply (maximum outport pilot pressure). In short, these are proportional solenoid controlled pressure-reducing valves. As the operator moves a lever (i.e., lift lever), a sensor senses the displacement of the lever and the signal is sampled by the electronic control module (ECM). The ECM sends current approximately proportional to the lever displacement signal to the pilot valve solenoids, which in turn generates a proportional pilot control pressure to main valves. Note that in wheel loader applications, the solenoid current sent to solenoid by ECM is not exactly proportional to the lever displacement, but includes a deadband around neutral position of the lever and a modulation curve beyond the deadband region that is not necessarily perfectly linear. Pilot valve control is enabled/disabled by a solenoid controlled two-position valve (E1). The operator selects the desired function by a switch in the cab, and the ECM controls the solenoid (ON/OFF) based on that switch state (ON/OFF).

The displacement of the main flow control valve, hence the flow rate, is proportional to the pilot control pressure because the force (due to pilot control pressure) acting on the sides of the valve spool is balanced by a pair of centering springs. It is important to note that the main flow control valves do not have any flow compensator valves. As a result, as the load changes, the cylinder speed changes for the same lever command. The cylinder speeds are load dependent. Furthermore, the lift, tilt, and auxilary function main flow control valves are hydraulically connected in series, and they are open-center types. The open-center hydraulic system (fixed displacement pump and open-center valves combination) are less energy efficient (wastes more energy) than the closed-center hydraulic systems (variable displacement pump and closed-center valves combination). Since the tilt valve is closer to the pump in the serial connection, the tilt circuit has priority over the lift circuit. Tilt and auxiliary function cylinders have anticavitation make-up valves connected to a tank as well as line pressure relief valves. The lift cylinder has a ride control circuit that is engaged or disengaged by the operator using a selector switch.

## 7.11 PROBLEMS

**1.** Consider a pump with fixed displacement $D_p = 50$ cm$^3$/rev and following nominal operating conditions: input shaft speed $w_{shaft} = 600$ rpm, torque at the input shaft $T = 50$ N $\cdot$ m, the output pressure $p_{out} = 5$ MPa, and output flow rate $Q_{out} = 450$ cm$^3$/sec. Determine the volumetric, mechanical, and overall efficiency of the pump at this operating condition. Consider another version of this problem statement where $p_{out} = 10$ MPa, and all other data are the same. Is there any error in this pump data? If so, discuss what might be the source of error and suggest ways to correct the data.

**2.** Consider a bidirectional hydraulic motor with fixed displacement of $D_m = 100$ cm$^3$/rev. The load torque it needs to provide at its output shaft is 100 N $\cdot$ m, and it must maintain a rotational speed of 600 rpm. Determine the necessary flow rate and differential pressure between the two ports (input and output) of the motor that must be supplied by an hydraulic circuit (i.e., a pump and valve preceeding the hydraulic motor). What is the power rating of the pump to supply this motor assuming an overall pump efficiency of 80% and neglecting the losses at the valve? Assume 100% efficiency for the hydraulic motor. (See Fig. 7.74.)

**3.** Consider the two-axes hydraulic motion system shown in Fig. 7.11. The sytem is to operate under the control of an operator. The operator commands the desired speed of each cylinder by two joysticks. There is no need for feedback sensors because this is an operator in the loop control system. Cylinder 1 needs to be able to provide a force of 5000 Nt at a rated speed of 0.5 m/s and maximum no-load speed of 1.0 m/s. Cylinder 2 needs to be able to provide 2500 Nt force at the same rated speed, and have the same maximum no-load speed as cylinder 1.

**(a)** Select components and size them for a completely hydromechanical control of the system. There should be no digital or analog computers involved. Draw the block diagram of the control system, and indicate the function of each component in the circuit. Assume that we use a fixed displacement pump. As a designer, make decisions regarding the anticipated realistic pressure that can be delivered at the cylinder ports and decide on realistic cylinder bore size.

**(b)** Select components for an embedded computer-controlled system. Discuss the differences between this design and the previous design in terms of the components and their differences.

**4.** Consider a hydraulic fluid in a hydraulic hose that has some entrapped air in it. Let the house diameter be $D = 6$ in. and the wall thickness of the hose is $\delta x = 1$ in. Let the Young's modulus of the hose material be $E = 30 \times 10^6$ psi. Determine the effective bulk modulus for zero and 1% air entrapment for operating pressures of 500 psi and 5000 psi.

**5.** Let us consider a single-axis EH motion control system with the following parameters. For the cylinder core diameter, consider the following size information: $d_{bore} = 10$ cm, and the rod diameter $d_{rod} = 5$ cm, stroke is $L = 1.0$ m. Let us approximate the bulk modulus of hydraulic fluid to be $2.5 \times 10^5$ psi $= 1.723 \times 10^9$Pa $= 1.723$ GPa. The mass carried by the cylinder (including the mass of the rod and piston and load) is $M = 200$ kg.

**(a)** Determine the lowest natural frequency of the system due to the compressibility of the fluid in the cylinder and the inertia.

**(b)** What is the maximum recommended closed-loop control system bandwidth?

**6.** Consider a single-axis electrohydraulic motion control system as shown in Fig. 7.81 and its block diagram representation in Fig. 7.80. We assume that the cylinder is rigidly connected to its base and load. Let us consider the small movements of the valve around its null position, i.e., a case that is maintaining a commanded position.

**(a)** Amplifier gain is $K_{sa} = 200$ mA/10 V $= 20$ [mA/V].

**(b)** Valve gain around the null position operation, $K_q = 20$ [in$^3$/sec]/200 [mA] $= 0.1$ [in$^3$/sec]/[mA].

**(c)** Cross-sectional area of the cylinder is $A_c = 2.0$ in$^2$ on both sides (rod is extended to both sides).

**(d)** The sensor gain is $K_{fx} = 10$ V/10 [in.] $= 1$ [V/in.].

**(e)** Deadband of the valve is 2% of the maximum input current, $i_{db} = 0.02 \cdot i_{max} = 0.02 \cdot 200$ mA $=$ 4 mA.

Assume that the total inertial load cylinder moves is $W = 1000$ lb, $m = 1000$ lb/386 in/s$^2 = 2.59$ lb$^2$/in, and the cylinder is currently at the midposition of its total travel range of $l = 10$ in. Note that the approximate bulk modulus of the hydraulic fluid is $\beta = 2.5 \times 10^5$ psi.

In terms of the amplifier gain, what is necessary to reduce the positioning error? Is there an upper limit set on the value of the amplifier gain? If so, determine that limit. Discuss the limitations imposed on the amplifier gain by the small positioning error requirement and closed-loop system stability. Determine the amplifier gain that provides the maximum practical closed-loop bandwidth while minimizing the positioning error due to the deadband of the valve.

**7.** Consider a single-axis EH motion control system shown in Fig. 7.76. The system needs to operate in two modes under the control of a programmed embedded computer:

- ***Mode 1:*** Closed-loop speed control where the commanded speed is obtained from a programmed command generator. The cylinder speed is measured and a closed-loop control algorithm implemented in the electronic control unit (ECU) which controls the valve.

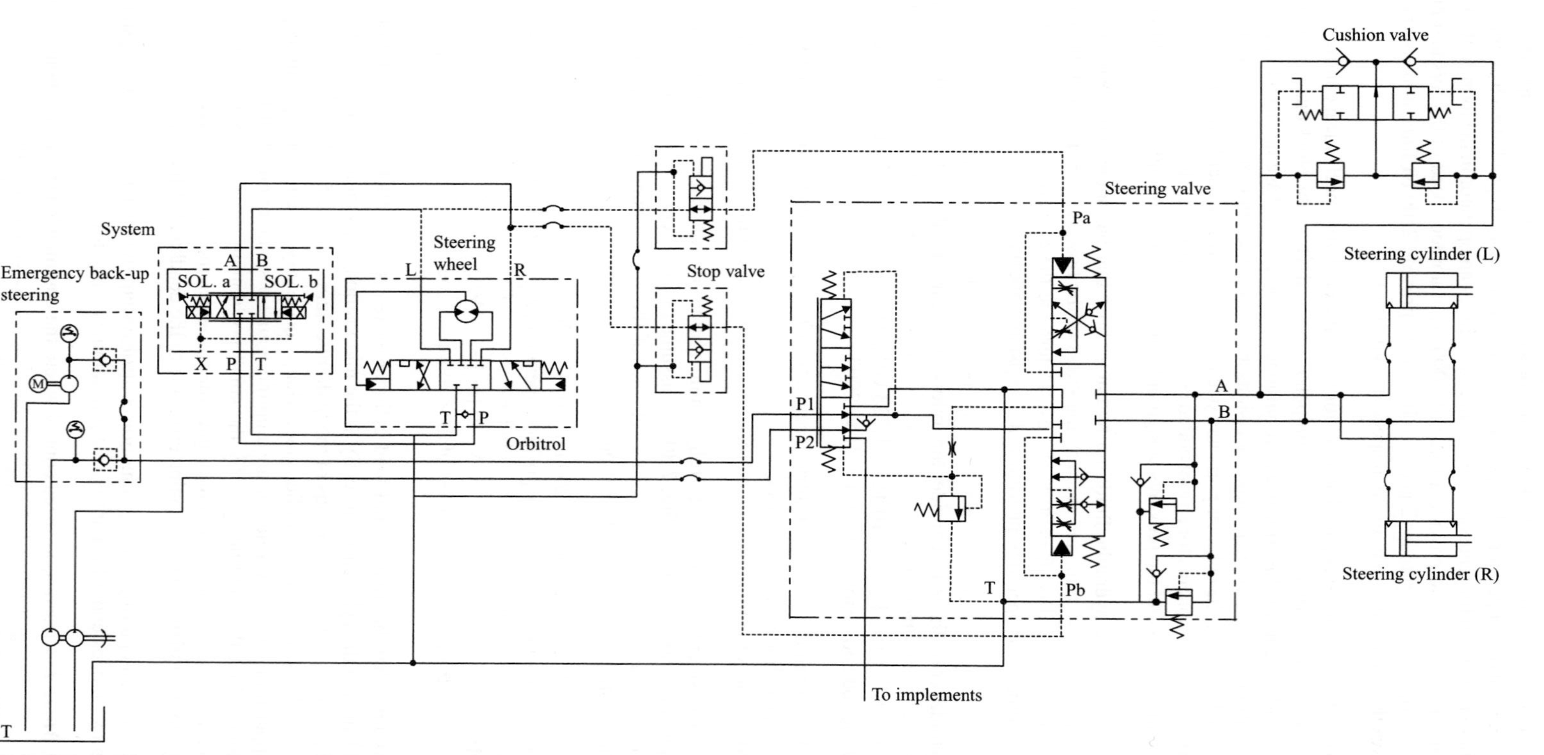


FIGURE 7.88: Hydraulic schematics for hand metering unit (HMU) steering system of wheel loader model WA450-5L by Komatsu.

- ***Mode 2:*** Closed-loop force control where the force command is obtained from a programmed command generator. The cylinder force is measured using a pair of pressure sensors on both ends of the cylinder.

A real-time control algorithm programmed in the embedded computer decides when to operate the axis motion in speed control or force control mode (Fig. 7.75).

**(a)** Draw a block diagram of the components and their interconnection in the circuit.
**(b)** Select proper components in order to meet the following specifications:
  (1) Maximum speed of the cylinder under no load conditions is $V_{nl} = 1.0$ m/sec.
  (2) Provide an effective output force of $F_r = 1000$ Nt at the cylinder rod while moving at the speed of $V_r = 0.75$ m/sec.
  (3) The desired regulation accuracy of the speed control loop is 0.1% of maximum speed.
  (4) The desired regulation accuracy of the force control loop is 1.0% of maximum force.

The components to be selected and sized include electronic control unit (ECU) with necessary analog–digital signal interfaces, position, and force sensor on the cylinder for closed-loop control (Fig. 7.76), pump, valve, and cylinder. Assume that the valve rating is specified for $\Delta P_v = 7$ MPa ($1\ Pa = 1\ \mathrm{N} \cdot \mathrm{m}^2$) pressure drop, and that the input shaft speed at the pump is $w_{in} = 1200$ rpm.

**8.** For the previous problem, write a pseudocode for the real-time control software in order to implement the following logic for the operation of this hydraulic motion control system.

**(a)** First, make a list of all inputs and outputs from the controller point of view and give symbolic names to them.
**(b)** Then write the pseudocode.

The objective is that the actuator is to approach the load at a predefined speed until a certain amount of pressure differential is sensed between P1 and P2 signals. The predefined speed is a trapezoidal speed profile, cylinder is to start from stationary position and accelerate to a top speed and run for a while at the top speed while monitoring the pressure sensors. The position range of the actuator around which that is expected is known. When a certain pressure differential is sensed, the control system is supposed to automatically switch to force (pressure) regulation mode. Maintain the desired pressure until and move a defined distance at a lower speed. Then reverse the motion, and return to the original position under another programmed speed profile.

**9.** Consider the hydraulic circuit shown in Fig. 7.88.

**(a)** Describe the operation of the circuit.
**(b)** What is the role of the emergency back-up steering block and how does it work?
**(c)** What is the role of the joystick and wheel steering, and how do they work?

CHAPTER **8**

# ELECTRIC ACTUATORS: MOTOR AND DRIVE TECHNOLOGY

## 8.1 INTRODUCTION

The term *actuator* in motion control systems refers to the component that delivers the motion (Fig. 8.1). It is the component that delivers the mechanical power which may be converted from an electric, hydraulic, or pneumatic power sources. In electric power based actuator category, the motor and the drive are two power conversion components that work together. In a motion control system, when we refer to the performance characteristics of a motor, we always have to refer to it in conjunction with the type of "drive" that the motor is used with, for the type of drive determines the behavior of the motor. The term *drive* is generically used in industry to describe the *power amplification* and the *power supply* components together.

The discussion in this chapter is limited to motor-drive technologies that can be used in high-performance motion control applications, i.e., involving closed-loop position and velocity control with high accuracy and bandwidth. Low-cost constant speed motor-drive components, which are used in mass quantities in applications such as fans and pumps, are not discussed. To this end, we will discuss the following motor-drive technologies:

1. DC motors (brush type and brush-less type) and drives
2. AC induction motors and field-oriented vector control drives
3. Step motors and drives:
   (a) Permanent magnet step motors
   (b) Hybrid step motors
   (c) Switched (variable) reluctance step motors
   along with full-step, half-step, and micro-stepping drives.

The operating principle of any electric motor involves one or more of the following three physical phenomena:

1. Opposite magnetic poles attract, and same magnetic poles repel each other.
2. Magnets attract iron and seek to move to a position to minimize the reluctance to magnetic flux.
3. Current-carrying conductors create an electromagnet and act like a current-controlled magnet.

Every motor has the following components:

1. Rotor on a shaft (moving component)
2. Stator (stationary component)
3. Housing (with end plates for rotary motors)

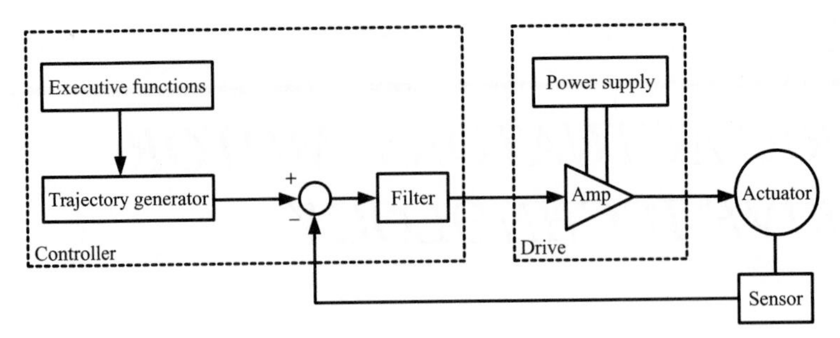


**FIGURE 8.1:** Functional blocks of a closed-loop motion control system: actuator, sensor, amplifier, commutation, and controller filters.

4. Two bearings, one for each end, to support the rotor in the housing, including some washers to allow axial play between the shaft and the housings

In addition, brush-type motors have commutator and brush assembly to direct current into the proper coil segment as a function of rotor position. Brushless motors have some type of rotor position sensor for electronic commutation of the current (i.e., Hall effect sensors or incremental encoders). Commutation means the distribution of current into appropriate coils as function of rotor position.

Traditionally AC induction motors have been used in constant speed applications, whereas DC motors have been used in variable speed applications. With the advances in solid-state power electronics and digital signal processors (DSP), an AC motor can be controlled in such a way that it behaves like a DC motor. One way of accomplishing this is the "field-oriented vector control" algorithm used in the drive for current commutation.

In the following discussion, a magnetic pole refers to a north (N) or south (S) pole, a magnetic pole pair refers to an N and an S pole. When we refer to a two-pole motor, we mean it has one N and one S pole. Likewise, a four-pole motor has two N and two S poles.

An electric motor is a power conversion device. It converts electrical power to mechanical power. Input to the motor is in the form of *voltage* and *current*, and the output is mechanical *torque* and *speed*. The key physical phenomenon in this conversion process is different for various motors.

1. In the case of DC motors, there are two magnetic fields. In brush-type DC motors, one of the magnetic fields is due to the current through the armature winding on the rotor, and the other magnetic field is due to the permanent magnets in the stator (or due to field excitation of the stator winding if electromagnets are used instead of permanent magnets). In the case of brushless DC motors, the roles of rotor and stator are swapped. When the two magnetic field vectors are perpendicular, maximum torque is generated per unit current.
2. In the case of AC motors, the first magnetic field is set up by the excitation current on the stator. This magnetic field in turn induces a voltage in the rotor conductors by Faraday's induction principle. The induced voltage at the rotor conductors results in current which in turn sets up its own magnetic field, which is the second magnetic field. The torque again is produced by the interaction of the two magnetic fields. In the case of a DC motor and an AC induction motor (with field-oriented vector control), the two magnetic fields are always maintained at a 90 degree angle in order to maximize the torque generation capability per unit current. This is accomplished by commutating the stator current (mechanically or electronically) as a function of the rotor position.
3. Stepper motors (permanent magnet (PM) type) work on basically the same principle as brushless DC motors, except that the stator winding distribution is different. A

given stator excitation state defines a stable rotor position as a result of the attraction between electromagnetic poles of the stator and permanent magnets of the rotor. The rotor moves to minimize the magnetic reluctance. At a stable rotor position of a step motor, two magnetic fields are parallel. In the case of switched reluctance motors, the rotor is not PM, but a soft, ferromagnetic material such as iron. As the electromagnetic pole state of stator changes by changing the current in stator winding phases, the rotor moves to minimize the magnetic reluctance while it is being temporarily magnetized by the stator's field.

The torque generation, that is, the electrical energy to mechanical energy conversion process, in any electric motor can be viewed as a result of the interaction of two magnetic flux density vectors: one generated by the stator ($\vec{B}_s$) and one generated by the rotor ($\vec{B}_r$). In different motor types, the way these vectors generated is different. For instance, in a permanent magnet brushless motor the magnetic flux of rotor is generated by permanent magnets and the magnetic flux of stator is generated by current in the windings. In the case of an AC induction motor, the stator magnetic flux vector is generated by the current in the stator winding, and the rotor magnetic flux vector is generated by induced voltages on the rotor conductors by the stator field and resulting current in the rotor conductors. It can be shown that the torque production in an electric motor is proportional to the strength of the two magnetic flux vectors (stator's and rotor's) and the sine of the angle between the two vectors. The proportionality constant depends on the motor size and design parameters.

$$T_m = K \cdot B_r \cdot B_s \cdot sin(\theta_{rs}) \tag{8.1}$$

where $K$ is the proportionality constant, and $\theta_{rs}$ is the angle between the $\vec{B}_s$ and $\vec{B}_r$, and $T_m$ is the torque.

Every motor requires some sort of current commutation by mechanical means as in the case of brush type DC motors or by electrical means as in the case of brushless DC motors. Current commutation means modifying the direction and magnitude of current in the windings as a function of rotor position. The goal of the commutation is to give the motor the ability to produce torque efficiently, i.e., maintain $\theta_{rs} = 90°$.

The design of an electric motor seeks to determine the following;

1. Shape of the effective magnetic reluctance of the motor by proper selection of materials and geometry of the motor
2. Distribution of coil wires, coil wire diameter, and its material (i.e., copper or aluminum)
3. Permanent magnets (number of poles, geometric dimensions, and PM material)

The engineering analysis is concerned with determining the resulting force/torque for a given motor design and coil currents. In addition, we also need to examine the flux density and flux lines in order to evaluate the overall quality of the design so that there is no excessive saturation in the flux path. These results are obtained from the solution of Maxwell's equations for magnetic fields. Modern engineering analysis software tools are based on finite element method (FEM) to solve Maxwell's equations and used for motor design (examples include Maxwell 2D/3D by Ansoft Corporation, Flux2D/3D by Magsoft Corporation, and PC-BLDC by The Speed Laboratory).

### 8.1.1 Steady-State Torque-Speed Range, Regeneration, and Power Dumping

Electric motors can act either as a motor, that is to convert electrical power to mechanical power, to drive loads, or as a generator, that is to convert mechanical power to electrical

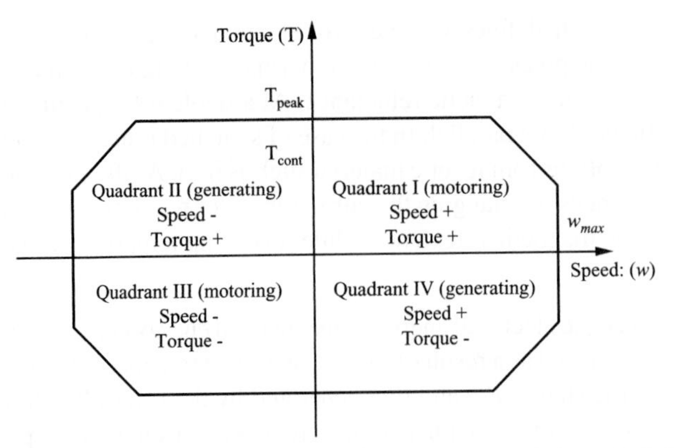


**FIGURE 8.2:** Four-quadrant region torque-speed characteristics of an electric motor.

power, when driven externally by the load. Let us consider the steady-state torque versus speed plane (Fig. 8.2). Motor-drive combinations that can operate in all four quadrants of the torque-speed plane are called *four-quadrant operation* devices and can act as motor and generator during different modes of an application. In quadrants I and III of the torque-speed plane, the mechanical power output is positive.

$$P_m = T \cdot w > 0.0; \quad \text{motoring mode} \tag{8.2}$$

When the motor speed and torque are in the same direction, the device is in *motoring mode*. In quadrants II and IV of the torque-speed plane, the mechanical power output is negative. That means the motor takes mechanical energy from the load instead of delivering mechanical energy to the load. The device is in *generator mode* or *regenerative braking mode*.

$$P_m = T \cdot w < 0.0; \quad \text{generating mode} \tag{8.3}$$

This energy can either be dissipated in the motor-drive combination, stored in a battery or capacitor set, or returned to the supply line by the drive.

During acceleration, the motor adds mechanical energy to the load. It acts as a *motor*. During deceleration, the motor takes away energy from the load. It acts like a brake or *generator*. This means that energy is put into the load inertia during acceleration, and energy is taken from the load inertia (returned to the drive) during deceleration (Fig. 8.2).

Some drives can convert the generated electric power and put it back to into the electric supply line while others dump the regenerative energy as heat through resistors. The amount of the regenerative energy depends on the load inertia, deceleration rate, time period, and load forces.

There are two different motion conditions where regenerative energy exists and satisfies the $T \cdot w < 0$ condition:

1. During deceleration of a load, where the applied torque is in opposite direction to the speed of inertia
2. In load-driven applications, i.e., in tension-controlled web handling applications, a motor may need to apply a torque to the web in opposite direction to the motion of the motor and web in order to maintain a desired tension. Another example is the case where the gravitational force provides more than needed force to move an inertia, and the actuator needs to apply force in the direction opposite to the motion in order to provide a desired speed.

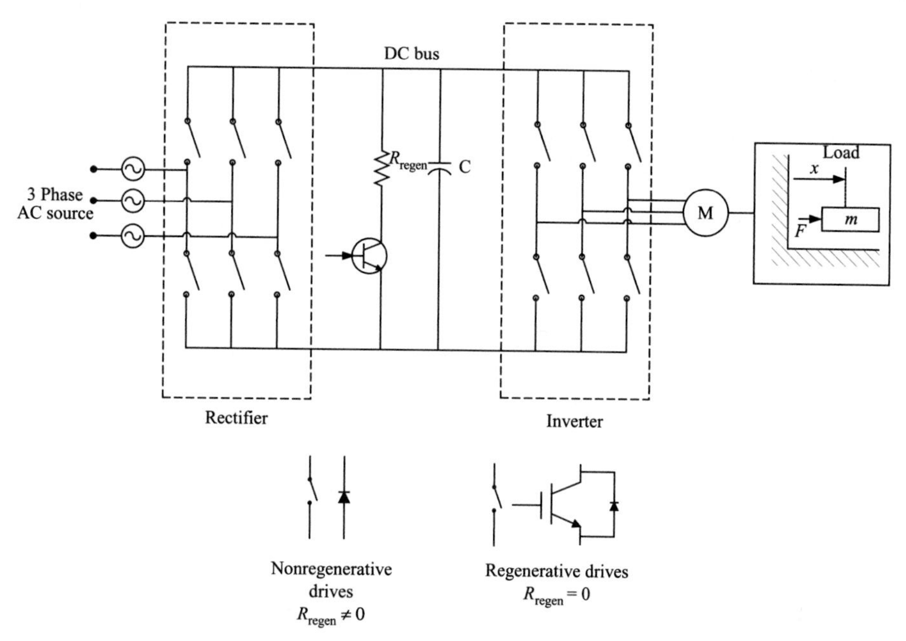


**FIGURE 8.3:** Regenerative energy in motion, its storage and dissipation.

***Example*** Consider an electric motor driven load shown in Figs. 8.3 and 8.4. Assume that the load is a translational inertia and electric motor is a perfect linear force generator. Consider an incremental motion that moves the inertia from position $x_1$ to position $x_2$ using a square force input. For simplicity, let us neglect all the losses. We will assume that the motor-drive combination converts electrical power ($P_e(t)$) to mechanical power ($P_m(t)$) with 100% efficiency in motoring mode, and mechanical power to electrical power with 100% efficiency in generator mode.

$$P_e(t) = P_m(t) \tag{8.4}$$

The force-motion relationship from Newton's second law is

$$F(t) = m \cdot \ddot{x}(t) \tag{8.5}$$

The mechanical power delivered to the inertia is

$$P_m(t) = F(t) \cdot \dot{x}(t) \tag{8.6}$$

which is supplied by the electric motor and drive combination.

Notice that when the force and speed are in the same direction, the mechanical power delivered to the inertia is positive, and the motor-drive operates in motoring mode, that is, to convert electrical energy to mechanical energy. Similarly, when the direction of force is opposite to the direction of speed, the mechanical power is negative, which means the inertia gives out energy intead of taking energy. This energy is converted to electrical energy by the motor because it acts like a generator under this condition.

$$P_m(t) = P_e(t) = F(t) \cdot \dot{x}(t) \tag{8.7}$$

$$P_m(t) = P_e(t) = F(t) \cdot \dot{x}(t) > 0 \quad \text{motoring mode} \tag{8.8}$$

$$P_m(t) = P_e(t) = F(t) \cdot \dot{x}(t) < 0 \quad \text{generating mode} \tag{8.9}$$

In motoring mode, the motor-drive provides energy to the load. In generator mode, motor-drive takes away energy from the load. This energy must be either stored, returned to line, or dissipated in resistors. One of the most common approaches in servo applications is to

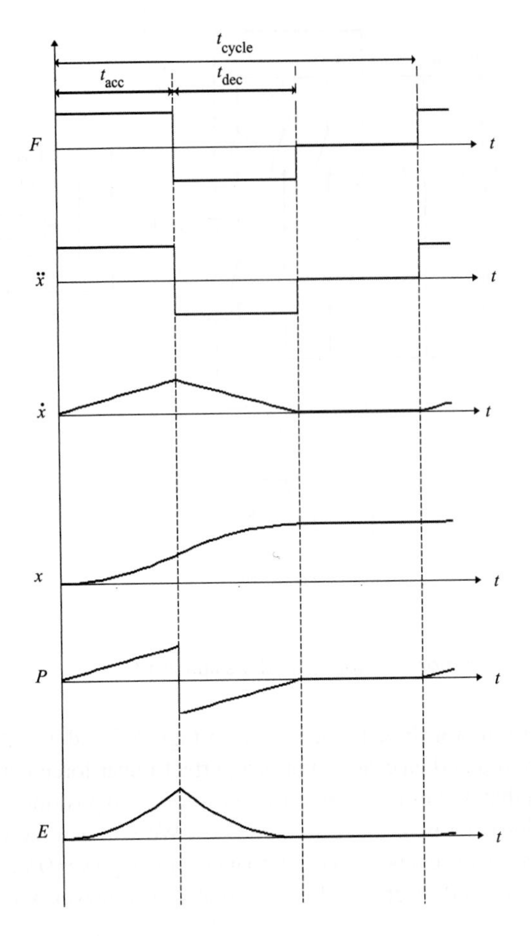


**FIGURE 8.4:** Typical output force to load and motion profile and regenerative energy opportunity for energy recovery during deceleration phase of the motion.

store a small portion of the energy in the DC bus capacitor and dissipate the rest as heat over external resistors added specifically for this purpose. In applications where this *regenerative energy* is large, external resistors are added for the purpose of dumping it.

In a given application, the amount of regenerative energy is a function of inertia, deceleration rate, and time period. In this example, the regenerative energy is

$$E_{reg}(t) = \int_0^{t_{dec}} P_m(t) \cdot dt \tag{8.10}$$

$$= \int_0^{t_{dec}} F(t) \cdot \dot{x}(t) \cdot dt \tag{8.11}$$

$$= \int_0^{x_{dec}} F(x) \cdot dx \tag{8.12}$$

$$= \int_0^{x_{dec}} m \cdot \ddot{x}(t) \cdot dx \tag{8.13}$$

$$= \int_0^{x_{dec}} m \cdot \dot{x}(t) \frac{d\dot{x}}{dx} \cdot dx \tag{8.14}$$

$$= \int_0^{x_{dec}} m \cdot \dot{x}(t) \cdot d\dot{x} \tag{8.15}$$

$$= \frac{1}{2} \cdot m \cdot \left(\dot{x}_1^2 - \dot{x}_2^2\right) \tag{8.16}$$

This energy ($E_{reg}$) must be dissipated at the regen resistors ($E_{ri}$) and partially stored in the DC bus capacitors ($E_{cap}$).

$$E_{reg} = E_{cap} + E_{ri} \tag{8.17}$$

The amount of energy that can be stored in the capacitor is

$$E_{cap} = \frac{1}{2} \cdot C \cdot \left(V_{max}^2 - V_{nom}^2\right) \tag{8.18}$$

where $C$ is the capacitance, $V_{max}$ is the maximimum DC bus voltage allowed before fault condition occurs, and $V_{nom}$ is the nominal DC bus voltage. Clearly, the capacitor can store a finite amount of energy and its size grows as the required energy storage capacity increases. Therefore, the remainder of the regenerative energy must either be returned to the supply line through a voltage regulating inverter or dissipated as heat at the regenerative resistor. Peak and continuous power rating of the regenerative resistors can be calculated from

$$P_{peak} = \frac{E_{reg} - E_{cap}}{t_{dec}} \tag{8.19}$$

$$P_{cont} = \frac{E_{reg} - E_{cap}}{t_{cycle}} \tag{8.20}$$

$$P_{peak} = R_{reg} \cdot i^2 \tag{8.21}$$

$$= \frac{V_{reg}^2}{R_{reg}} \tag{8.22}$$

$$R_{reg} \leq \frac{V_{reg}^2}{P_{peak}} \tag{8.23}$$

where $P_{peak}$, $P_{cont}$ are peak, continuous power dissipation capacity required from the regenerative resistors, $V_{reg}$ is the nominal DC bus voltage over which the regenerative circuit is active (i.e., a value between the nominal DC bus voltage and maximum DC bus voltage), and $R_{reg}$ is the resistance value of the regenetative resistors. In some applications, the regenerative power may be so small that the DC bus capacitor is large enough to store the energy without the need for dissipating it as heat over resistors. Notice that the capacitor (or battery) is sized based on maximum energy, whereas the resistor is sized based on maximum power.

In load-driven applications, i.e., tension-controlled web handling or gravity driven loads, motor continuously operates in regenerative power mode (generator mode) and the resistors must be sized based on

$$P_{peak} = max(F_{tension}(t) \cdot \dot{x}(t)) \tag{8.24}$$

$$P_{cont} = RMS(F_{tension}(t) \cdot \dot{x}(t)) \tag{8.25}$$

where the motor must continuously provide the resistive tension force while maintaining a speed in the opposite direction.

### 8.1.2 Electric Fields and Magnetic Fields

There are two types of fields in electrical systems: *electric* and *magnetic* (also called *electromagnetic*) fields. Although we are primarily interested in electromagnetic fields for the study of electric motors, we will discuss both briefly for completeness. Electric fields ($\vec{E}$) are generated by static charges. Magnetic fields ($\vec{H}$, also called electromagnetic fields) are generated by moving charges (current).

*Electric field* is a distibuted vector field in space whose strength at a location depends on the charge distribution in space. It is a function of the static location of charges and the amount of charges. By convention, electric fields start (emitted) from positive charges and end (received) in negative charges [Fig. 8.5(a)]. Capacitors are commonly used to store charges and generate electric fields. The smallest known charge is that of an electron (negative charge) and a proton (positive charge) with units of *Coulomb* [C],

$$|e^-| = |p^+| = 1.60219 \times 10^{-19} [C] \tag{8.26}$$

Electric field $\vec{E}$ at a point in space $(s)$ due to $n$ many charges $(q_1, q_2, \ldots, q_n)$ at various locations can be determined from [Fig. 8.5(b)],

$$\vec{E}(s) = k_e \sum_{i=1}^{n} \frac{q_i}{r_i^2} \cdot \vec{e}_i [\text{N/C}] \; or \; [\text{V/m}] \tag{8.27}$$

where $k_e = 8.9875 \times 10^9$ [N·m²/C²] is called the *Coulomb constant*, $r_i$ is the distance between the location of charge $(i)$ and the point in space considered $(s)$, and $\vec{e}_i$ is the unit vector between each charge location and the point $(s)$ $q_i$ is the charge at location $i$. For negative charges, the unit vector is directed toward the charge; for positive charges it is directed away from the charge. Coulomb constant is closely related to another well-known constant as follows:

$$k_e = \frac{1}{4 \cdot \pi \cdot \epsilon_o} \tag{8.28}$$

where $\epsilon_o = 8.8542 \times 10^{-12}$ [C²/N·m²] is the *permittivity of free space*. Force $(\vec{F})$ exerted on a charge $(q)$ which is in an electric field $(\vec{E})$ is [Fig. 8.5(c)]

$$\vec{F} = q \cdot \vec{E} \tag{8.29}$$

and if the charge is free to move, the resulting motion is governed by

$$\vec{F} = m_q \cdot \vec{a} \tag{8.30}$$

where the generated force results in the acceleration $(\vec{a})$ of the charge mass $(m_q)$ based on Newton's second law. This last equation is used to study the motion of charged particles in electric fields, i.e., motion of electrons in cathode ray tube (CRT), motion of charged small dropplets in ink-jet printing machines. The motion trajectory of a particle of a known mass can be controlled by controlling the force acting on it. The force can be controlled either by controlling its charge or the electric field in which it travels. Generally, the electric field is kept constant, and charge on each particle is controlled before it enters the electric field. Another electric field quantity of interest is the *electric flux*, $\Phi_E$, which is the area integral of the electric field over a closed surface. The electric flux over a closed surface is proportional to the net electric charge inside the surface [Fig. 8.5(d)],

$$\Phi_E = \oint_A \vec{E} \cdot d\vec{A} = \frac{q_{net}}{\epsilon_o} \tag{8.31}$$

where $d\vec{A}$ is a differential vector normal to the surface. The line integral of electric field between any two points is the *electric potential* difference between the two points (voltage) [Fig. 8.5(e)],

$$V_{AB} = -\int_A^B \vec{E} \cdot d\vec{s} \tag{8.32}$$

where the $d\vec{s}$ vector is differential vector that is tangent to the path traveled from $A$ to $B$.

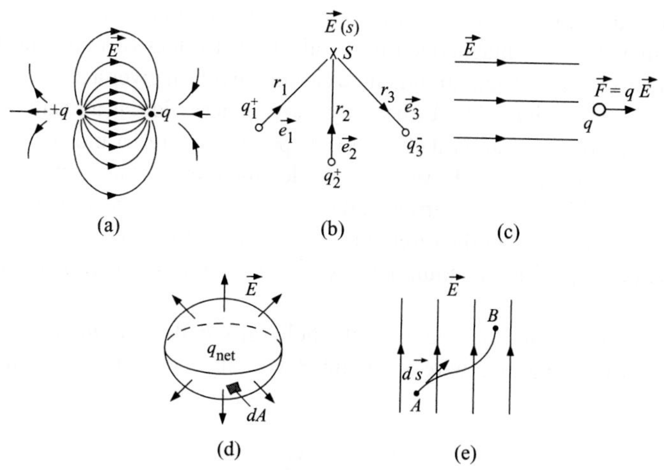


**FIGURE 8.5:** Electric fields due to stationary electric charges. (a) By convention, electric fields are assumed to eminate from positive charges and terminate at the negative charges. (b) Electric field at a point in space due to charged particles at other locations. (c) Force acting on a charged particle due to an electric field. (d) Integral of electric field over a closed surface is proportional to the net charge inside the volume spanned by the closed surface. (e) Voltage between two points in space (A and B) is equal to the line integral of electric field between the two points.

The *magnetic fields* are generated by moving charges (Fig. 8.6). There are two sources to generate and sustain a magnetic field:

1. Current (moving charge) over a conductor
2. Permanent magnetic materials

In the case of current-carrying conductors, the magnetic field generated by the current (moving charges) is called the *electromagnetic field*. In the case of permanent magnet materials, the magnetic field is generated by orbital rotation of the electrons around the nucleous and spin motion of electrons around their own axis. The net magnetic field of the material in macro scale is the result of vector sum of the magnetic fields of its electrons. The magnetic fields of the nucleous due to the charge and motion of protons do not make

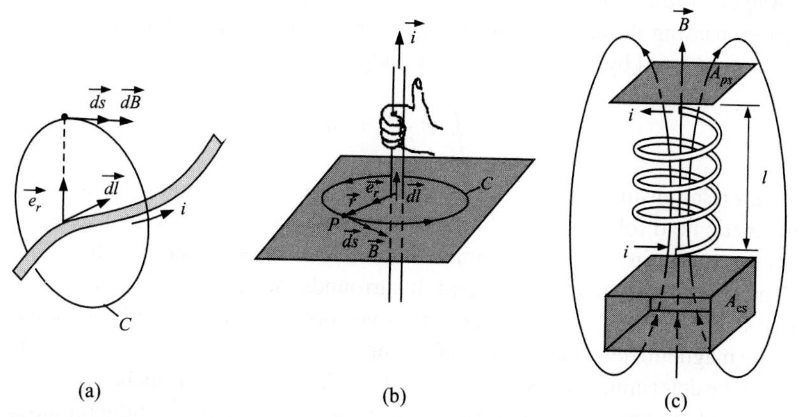


**FIGURE 8.6:** Magnetic fields due to current (moving charge): (a) magnetic field at any point P due to current over a general shape conductor; (b) magnetic field around an infinitely long, straight current carrying conductor; and (c) magnetic field inside a coil due to current.

much difference because they tend to be randomly oriented, smaller in magnitude, and cancel out each other. In a nonmagnetized material, the net effect of magnetic fields of electrons cancel out each other. Their alignment in certain direction gives a material nonzero magnetization in a specific orientation. Either way, the magnetic field is a result of moving charges. The electric field vector starts at charges (i.e., positive charges) and ends in charges (negative charges). The magnetic field vector encircles the current that generates it (Fig. 8.6). The vector relationship between current and the magnetic field it generates follows the right-hand rule. If the current is in the direction of the thumb, the magnetic field is in the direction of the fingers encircling the thumb. If the current changes direction, the magnetic field changes direction.

The *Biot-Savart law* states that the magnetic field (also called *magnetic flux density*), $\vec{B}$, generated by a current on a long wire at a point $P$ with distance $r$ from the wire is [Fig. 8.6(a) and (b)]

$$d\vec{B} = \frac{\mu\, i\, d\vec{l} \times \vec{e}_r}{4\pi\, r^2} \tag{8.33}$$

where $\mu$ is the *permeability* of the medium around the conductor ($\mu = \mu_0$ for free space), and $\vec{e}_r$ is unit vector of $\vec{r}$. This is called the *Biot-Savart law*. The permeability of free space is called $\mu_0$:

$$\mu_0 = 4\pi \cdot 10^{-7}\ [\text{Tesla} \cdot \text{m/A}] \tag{8.34}$$

Quite often, the permeability of a material ($\mu_m$) is given relative to the permeability of free space,

$$\mu_m = \mu_r \cdot \mu_0 \tag{8.35}$$

where $\mu_r$ is the relative permeability of the material with respect to free space. If the Biot–Savart law is applied to a conductor over the length of $l$, we obtain the magnetic flux density $\vec{B}$ at point $P$ at a distance $r$ from the conductor due to current $i$ [Fig. 8.6(a)],

$$\vec{B} = \int_0^l \frac{\mu\, i\, d\vec{l} \times \vec{e}_r}{4\pi\, r^2} \tag{8.36}$$

The units of $\vec{B}$ expressed in SI units is [Tesla] or [T].

*Ampere's law* states that the integral of magnetic field over a closed path is equal to the current passing through the area covered by the closed path times the permeability of the medium covered by the closed path of integration [Fig. 8.6(a) and (b)],

$$\oint_C \vec{B} \cdot d\vec{s} = \mu \cdot i \tag{8.37}$$

The vector relationship between the current, position vector of point $P$ with respect to wire and magnetic field follows the right-hand rule. It describes how electromagnetic fields are created by current in a given medium with a known magnetic permeability. The magnetic flux density is a continuous vector field. It surrounds the current that generates it based on the right-hand rule. The magnetic flux density vectors are always closed, continuous vectors.

The magnetic field due to current flow through a conductor of any shape in an electrical circuit can be determined by using either the Biot–Savart law or Ampere's law. For instance, the magnetic field generated around an infinitely long and straight conductor having current $i$ in free space at a distance $r$ from it can be calculated [Fig. 8.6(b)]

$$|\vec{B}| = B = \frac{\mu_0 \cdot i}{2\pi \cdot r} \tag{8.38}$$

Similarly, the magnetic field inside a coil of solenoid is [Fig. 8.6(c)]

$$|\vec{B}| = B = \frac{\mu_0 \cdot N \cdot i}{l} \tag{8.39}$$

where $N$ is the number of turns of the solenoids and $l$ is the length of the solenoid. It is assumed that the magnetic field distribution inside the solenoid is uniform and the medium inside the solenoid is free space. Notice that if the medium inside the coil is different than free space, i.e., steel with $\mu_m \gg \mu_0$, then the magnetic flux density developed inside the coil would be much higher.

Magnetic flux ($\Phi_B$) is defined as the integral of magnetic flux density ($\vec{B}$) over a cross-sectional area perpendicular to the flux lines [Fig. 8.6(c)],

$$\Phi_B = \int_{A_{ps}} \vec{B} \cdot d\vec{A}_{ps} \text{ [Tesla} \cdot \text{m}^2\text{] or [Weber]} \tag{8.40}$$

where $d\vec{A}_{ps}$ is differential vector normal to the surface ($A_{ps}$). The area is the effective perpendicular area to the magnetic field vector. It is important not to confuse this relationship with Gauss's law.

*Gauss's law* states that the integral of magnetic field over *a closed surface* that encloses a volume is zero (integration over the closed surface $A_{cs}$ shown in Fig. 8.6(c)),

$$\oint \vec{B} \cdot d\vec{A}_{cs} = 0 \tag{8.41}$$

where $d\vec{A}_{cs}$ is differential area over a closed surface ($A_{cs}$), not a cross-sectional perpendicular area to flux lines. This integral is over a closed surface. In other words, net magnetic flux over a closed surface are zero. The physical interpretation of this result is that magnetic fields form closed flux lines. Unlike electric fields, magnetic fields do not start in one location and end in another location. Therefore, the net flow-in and flow-out lines over a closed surface are zero [Fig. 8.6(c)].

Let us define the concept of *flux linkage*. Consider that a magnetic flux ($\Phi_B$) is generated by a coil or permanent magnet or a similar external source. If it crosses one turn of conductor wire, the magnetic flux passing through that wire is called the *flux linkage* between the existing magnetic flux and the conductor,

$$\lambda = \Phi_B; \quad \text{for one turn coil} \tag{8.42}$$

If the conductor coil had $N$ turns instead of one, the amount of flux linkage between the external magnetic flux $\Phi_B$ and the $N$ turn coil is

$$\lambda = N \cdot \Phi_B; \quad \text{for } N \text{ turn coil} \tag{8.43}$$

*Magnetic field strength* ($\vec{H}$) is related to the magnetic flux density $\vec{B}$ with the permeability of the medium,

$$\vec{B} = \mu_m \cdot \vec{H} \tag{8.44}$$

*Magnetomotive force (MMF)* is defined as

$$MMF = H \cdot l \tag{8.45}$$

where $l$ is the length of the magnetic flux path.

Reluctance of a medium to the flow of magnetic flux is analogous to the electrical resistance of a medium to the flow of current. The *reluctance* of a medium with cross-sectional area $A$ and thickness $l$ can be defined as

$$R_B = \frac{l}{\mu_m \cdot A} \tag{8.46}$$

where $\mu_m$ is the permeability of the medium, i.e., air, iron. *Permeance* of a magnetic medium is defined as the inverse of reluctance,

$$P_B = \frac{1}{R_B} \tag{8.47}$$

Reluctance is analogous to resistance, and permeance is analogous to conductivity.

The design task of shaping a magnetic circuit is the design task of defining the reluctance paths to the flow of magnetic field lines in the circuit. It is a function of material and geometry. Series and parallel reluctances add following the same rules for electrical resistance. Iron and its variations are the most commonly used materials in shaping a magnetic circuit, i.e., in design of electric actuators. The material and geometry of the magnetic circuit determine the resistance paths to the flow of magnetic flux.

For instance, the magnetic flux in a coil can be determined as follows (the magnetic field has been given above eqn. 8.39):

$$B = \frac{\mu_0 \cdot N \cdot i}{l} \tag{8.48}$$

$$= \mu_0 \cdot H \tag{8.49}$$

$$H = \frac{N \cdot i}{l} \tag{8.50}$$

$$= \frac{MMF}{l} \tag{8.51}$$

$$MMF = N \cdot i \tag{8.52}$$

The magnetic flux is defined as the integral of the magnetic flux density over a surface perpendicular to the flux density vector [Fig. 8.6 (c)]:

$$\Phi_B = B \cdot A = \frac{\mu_0 \cdot N \cdot i}{l} \cdot A \tag{8.53}$$

$$= \frac{N \cdot i}{[l/(\mu_0 \cdot A)]} \tag{8.54}$$

$$= \frac{MMF}{R_B} \tag{8.55}$$

For instance, a coil with $N$ turns and current $i$ can be modeled in an electromagnetic circuit as an *MMF* source (similar to voltage source) and magnetic reluctance $R_B$ (similar to electrical resistance) in series with the source (Fig. 8.7),

$$MMF = N \cdot i \tag{8.56}$$

$$R_B = \frac{l}{\mu \cdot A} \tag{8.57}$$

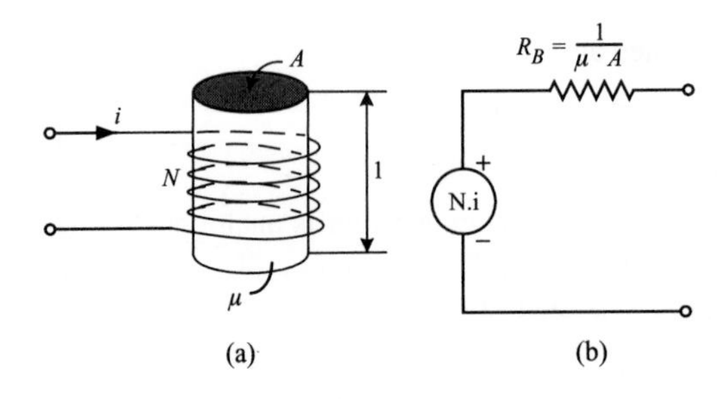


**FIGURE 8.7:** (a) A coil winding, and (b) its magnetic model. Coil is modelled as having an $MMF = N \cdot i$, magnetomotive source, and magnetic resistance (reluctance) $R_B = l/(\mu A)$ in series with the *MMF*.

Then, the magnetic flux (analogous to current) through the coil is

$$\Phi_B = \frac{MMF}{R_B} \tag{8.58}$$

Quite often, analogy between electrical circuits and magnetic circuits is used.

By analogy to electrical circuits, there are three main principles used in analyzing magnetic circuits as follows:

1. Sum of MMF drop across a closed path is zero. This is similar to Kirchhoff's law for voltages, which says the sum of voltages over a closed path is zero.

$$\sum_i MMF_i = 0; \quad \text{over a closed path} \tag{8.59}$$

2. Sum of flux at any cross section in a magnetic circuit is equal to zero (that is the sum of in-coming and out-going flux through a cross section). This is similar to Kirchhoff's law for currents which says that at a node, algebraic sum of currents is zero (sum of in-coming and out-going currents).

$$\sum_i \Phi_{Bi} = 0; \quad \text{at a cross section} \tag{8.60}$$

3. Flux and MMF are related by reluctance of the path of the magnetic medium, similar to voltage, current, and resistance relationship (Table 8.1).

$$MMF = R_B \cdot \Phi_B \tag{8.61}$$

Magnetic circuits typically have current-carrying conductors in coil form which act as the source of magnetic field, permanent magnets, iron-based material to guide the magnetic flux, and air. The geometry and material of the medium uniquely determine the reluctance distribution in space. Current-carrying coils and magnets determine the magnetic source. Interaction between the two (magnetic source and reluctance) determines the magnetic flux.

Force ($\vec{F}$) in a magnetic field ($\vec{B}$) and a moving charge ($q$) has vector relationship [Fig. 8.8(a)],

$$\vec{F} = q\vec{v} \times \vec{B} \tag{8.62}$$

where $\vec{v}$ is the speed vector of the moving charge. This relationship can be extended for a current-carrying conductor instead of a single charge. The force acting on a conductor of over length $l$ due to the current $i$ and magnetic field $\vec{B}$ interaction is [Fig. 8.8 (b)],

$$\vec{F} = l \cdot \vec{i} \times \vec{B} \tag{8.63}$$

**TABLE 8.1: Analogy Between Electrical and Electromagnetic Circuits**

| Electric Circuits | Magnetic Circuits |
|---|---|
| $V$ | $MMF$ |
| $i$ | $\Phi_B$ |
| $R$ | $R_B$ |
| $i = \frac{V}{R}$ | $\Phi_B = \frac{MMF}{R_B}$ |

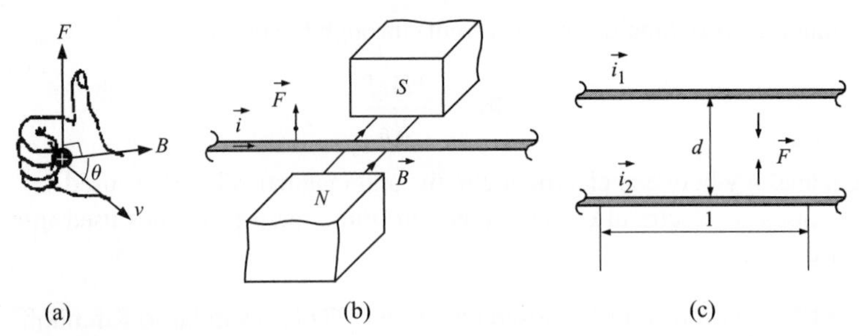


**FIGURE 8.8:** Magnetic forces: (a) magnetic force acting on a moving charge in a magnetic field, (b) magnetic force acting on a conductor with current in a magnetic field, and (c) magnetic force between two current-carrying conductors.

This relationship is convenient to derive the [Tesla] or [T] unit,

$$1 \text{ Tesla} = 1 \frac{N}{\text{C} \cdot \text{m/s}} = 1 \frac{N}{\text{A} \cdot \text{m}} \tag{8.64}$$

This is the basic physical principle for the electromechanical power conversion for *motor action*. Notice that the force is a vector function of the current and the magnetic flux density. $\vec{B}$ may be generated by a permanent magnet and/or electromagnet.

The force between two current carying parallel conductors can be described as the interaction of current in one conductor with the electromagnetic field generated by the current in the other conductor. Consider two conductors parallel to each other, carrying currents $i_1$ and $i_2$, separated from each other by a distance $d$, and have length $l$. The force acting between them is [Fig. 8.8(c)],

$$|\vec{F}| = \mu_o \cdot \left(\frac{l}{2\pi \cdot d}\right) \cdot i_1 \cdot i_2 \tag{8.65}$$

where $\mu_o$ is the permeability of space between the two conductors. The force is attractive if the two currents are in the same direction, and repulsive if they are opposite.

Similarly, there is a dual phenomenon called the *generator action*, which is a result of Faraday's law of induction. *Faraday's law of induction* states that an electromotive force (EMF) voltage is induced on a circuit due to changing magnetic flux and that the induced voltage opposes the change in the magnetic flux. We can think of this as the relationship between magnetic and electric fields: a changing magnetic field induces an electric field (induced voltage) where the induced electric field opposes the change in the magnetic field [Fig. 8.9(a)].

$$V = \int \vec{E} \cdot d\vec{s} = -\frac{d\Phi_B}{dt} \tag{8.66}$$

Note that the time rate of change in magnetic flux can be due to the change in the magnetic field source or due to the motion of a component inside a constant magnetic field which results in a change of effective reluctance or both [Fig. 8.9(a), (b), (c), (d)]. If we consider the induced voltage on a coil with $n$ turns and the magnetic flux passing through each of the turns is $\Phi_B$, then

$$V = -n \cdot \frac{d\Phi_B}{dt} = -\frac{d\lambda}{dt} \tag{8.67}$$

where $\lambda = n \cdot \Phi_B$ is called the *flux linkage*, which is the amount of flux linking the $n$ turns of coil.

In electromagnetism, the current is the source (cause) and magnetic field is the effect of it. Faraday's induction law states that the time rate of change in magnetic flux induces EMF

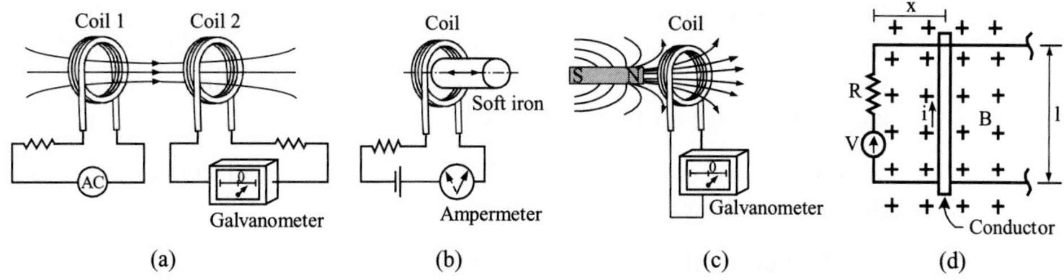


**FIGURE 8.9:** Faraday's law of induction: change in magnetic flux induces an opposing voltage in an electrical circuit affected by the field. The induced voltage is proportional to the time rate of change in the magnetic flux and in opposite direction to it. The source of the change in magnetic flux can be (a) a changing current source which generates the electromagnetic field. (b) It can be due to the reluctance change in a magnetic field. In other words, the inductance of the electrical circuit changes. In inductive circuits, the back emf can be self-inductance itself and the change in the inductance. (c) Or, the change can be due to mechanical motion. The last two cases are observed as back emf on solenoids and DC motors, respectively. The first case is observed in transformers. (d) Generator action as a result of Faraday's law of induction.

voltage on a circuit affected by it. The source of electromagnetic induction is the change in the magnetic flux. This change may be caused by the following sources (Fig. 8.9):

1. *A changing magnetic flux itself, i.e., changing magnetic flux created by a changing source current.* In the case of a transformer, the AC current in the primary winding creates a changing magnetic flux. The flux is guided by the iron core of the transformer to the secondary winding. The changing magnetic flux induces a voltage in the secondary winding [Figs. 8.9(a), (c) and 8.12].
2. *A changing magnetic flux created as a result of changing inductance.* In a circuit, even though current is constant, change in magnetic flux can be caused by the change in the geometry and permeability of the medium (change in reluctance). This results in induced back EMF. This phenomenon is at work in the case of solenoids and variable reluctance motors. In general this voltage has the form [Fig. 8.9(b)],

$$V_{bemf} = -\frac{d\Phi_B}{dt} \tag{8.68}$$

$$= -\frac{d(L(x)i(t))}{dt} \tag{8.69}$$

$$= -L(x)\frac{di(t)}{dt} - \frac{dL(x)}{dx} \cdot i(t) \cdot \dot{x}(t) \tag{8.70}$$

   where the first term is the back emf due to self-inductance of the circuit ($L(x)$), and the second term is the back emf induced due to the change in the inductance ($dL(x)/dx$).
3. *A back EMF induced in the conductors moving in a fixed magnetic field established by the field magnets [permanent magnets or electromagnets (Fig. 8.9(d))].* As the conductor moves in the magnetic field, there will be force acting on the charges in it just as there is a force acting on a charge moving in a magnetic field. Let us consider a constant magnetic field $B$, a conductor with length $l$ moving in perpendicular direction to magnetic field vector, and its current position $x$ [Fig. 8.9(d)]. The induced back emf because of this *motion* is

$$V_{bemf} = -\frac{d\Phi_B}{dt} = -\frac{d(B \cdot l \cdot x)}{dt} = -B \cdot l \cdot \dot{x} \tag{8.71}$$

In the case of a brush-type DC motor, the conductor is the winding on the rotor which moves relative to stator magnets. In the case of bruhless DC motors, the permanent magnets on the rotor moves relative to the fixed stator windings. The resulting induced back emf voltage effect is the same.

We should point out the relationship between SI units and CGS units commonly used in practice are as follows:

$$1\ [\text{A turn/m}] = 4\pi \cdot 10^{-3}\ [\text{Oerstead}] \tag{8.72}$$

$$1\ [\text{Tesla}] = 10^4\ [\text{Gauss}] \quad \text{or} \quad 1\ [\text{T}] = 10^4\ [\text{G}] \tag{8.73}$$

$$1\ [\text{Weber}] = 10^8\ [\text{Gauss} \cdot \text{cm}^2] \quad \text{or} \quad 1\ [\text{Wb}] = 10^8\ [\text{Maxwell}] \tag{8.74}$$

Consider a coil with $N$ turns and an external voltage source which forces current through it. When the circuit is first turned ON, due to the change in the magnetic flux, there will be back emf voltage opposing the change. This is a result of Faraday's law of induction. In a coil, the induced back emf that opposes the change is always proportional to the change in current. The proportionality constant is called the *self-inductance* ($L$) of the coil. Let us define *flux linkage*, $\lambda$,

$$\lambda = N \cdot \Phi_B \tag{8.75}$$

Voltage induced on a coil by a change in the magnetic flux (or flux linkage) equals the time rate of change of flux linkage

$$V = -N \cdot \frac{d\Phi_B}{dt} = -L \cdot \frac{di}{dt} = -\frac{d\lambda}{dt} \tag{8.76}$$

Therefore, for a coil with constant inductance (linear magnetic circuit, $\lambda = L \cdot i$)

$$\lambda = L \cdot i = N \cdot \Phi_B \tag{8.77}$$

$$L = \frac{N \cdot \Phi_B}{i}\ [\text{V} \cdot \text{sec/Amp}] \text{ or } [\text{Henry}] \tag{8.78}$$

A coil of conductor has self-inductance, which opposes the change in magnetic field around it. Through self-inductance, $L$, the coil generates emf voltage that is proportional to the rate of change of current in opposite direction (Fig. 8.10). If the circuit geometry and its material properties vary (i.e., in case of a solenoid, the air-gap varies), the inductance is not constant. The inductance is a function of the geometry (i.e., number of turns in a coil) and the permeability of the medium. If the core of an inductor winding has a moving iron piece and the permeability of the medium changes as the core moves, the inductance of the coil changes. Consider the self-inductance $L$ of a solenoid coil with length $l$ and total of $N$ turns, cross-sectional area of $A$. Let us assume that the medium inside the coil is air. Let us approximate that the magnetic flux inside the coil is uniform. Then

$$L = \frac{N \cdot \Phi_B}{i} \tag{8.79}$$

$$B = \mu_o \cdot (N/l) \cdot i \tag{8.80}$$

$$\Phi_B = B \cdot A = \mu_o \cdot (N/l) \cdot i \cdot A \tag{8.81}$$

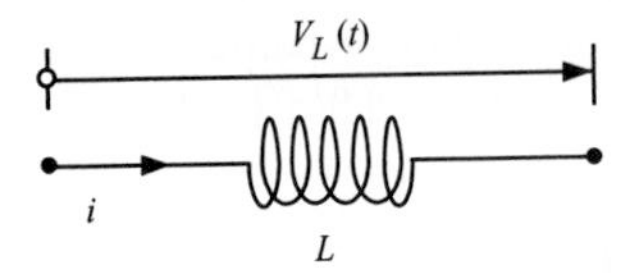


**FIGURE 8.10:** A coil of conductor and its self-inductance. Self-inductance of a coil is a function of the number of turns in the coil, its geometry, and the material properties of the medium, the coil core and its surrounding.

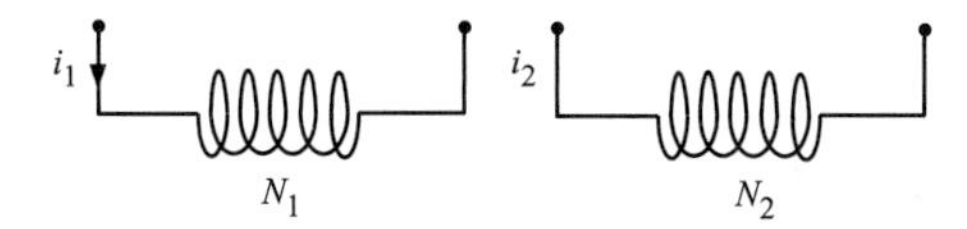


**FIGURE 8.11:** Two coils of conductors with $N_1$ and $N_2$ turns. The current $i_1$ in coil 1 induces a current $i_2$ on coil 2. That current in turn induces a current on coil 1.

$$L = \frac{N \cdot \mu_o \cdot (N/l) \cdot i \cdot A}{i} \tag{8.82}$$

$$= \frac{\mu_o \cdot N^2 \cdot A}{l} \tag{8.83}$$

$$= \frac{N^2}{R_B} \tag{8.84}$$

which shows that the inductance is function of the coil geometry and permeability of the medium. If the coil core is iron, then $\mu_o$ would be replaced by $\mu_m$ for iron which is about 1000 times higher than the permeability of air. Hence, the inductance of the coil would be higher by the same ratio.

In electric actuator design, it is desirable to have small magnetic reluctance, $R_B$, so that more flux ($\Phi_B = MMF/R_B$) is conducted per unit magnetomotive force ($MMF$). On the other hand, it is desirable to have small inductance ($L$) so that the electrical time constant of the motor is small. These are two conflicting design requirements. A particular design must find a good balance between them that is appropriate for the application.

Let us consider two coils of conductors, each having $N_1$ and $N_2$ turns [Figs. 8.11 and 8.9(a)]. Let $i_1(t)$ be the current passing through coil 1. The change in current on coil 1 will generate a changing magnetic flux. As a result, the changing magnetic flux will induce a current $i_2(t)$ on the second coil. Furthermore, the induced current $i_2(t)$ on coil 2 will in turn induce a back emf on coil 1. The proportionality constant between the induced magnetic flux and current is the mutual inductance between the two coils. Flux linkage on coil 2 due to current on coil 1,

$$\lambda_{12} = N_2 \cdot \Phi_{12} \tag{8.85}$$

$$= L_{12} \cdot i_1 \tag{8.86}$$

The voltage induced on coil 2 by the current on coil 1 through the mutual inductance,

$$V_{12} = -L_{12} \frac{di_1}{dt} \tag{8.87}$$

and the voltage induced on coil 1 due to current in coil 2 through the mutual inductance,

$$V_{21} = -L_{21} \frac{di_2}{dt} \tag{8.88}$$

It can be shown that the mutual inductances, $L_{12}$ and $L_{21}$, are equal:

$$L_{12} = L_{21} = f(\mu_m, geometry) \tag{8.89}$$

The mutual inductance is a function of the permeability of the medium between the two circuits and the geometry (shape, size, and relative orientation with respect to each other).

Let us consider the transformer shown in Fig. 8.12. A transformer has two windings, primary and secondary, and a core which magentically couples them. A transformer works based on the Faraday's induction principle; that is, voltage is induced on a conductor due to a change in the magnetic field. In case of transformers, the change in magnetic field is due to the alternating current (AC) nature of the source at the primary winding. An ideal

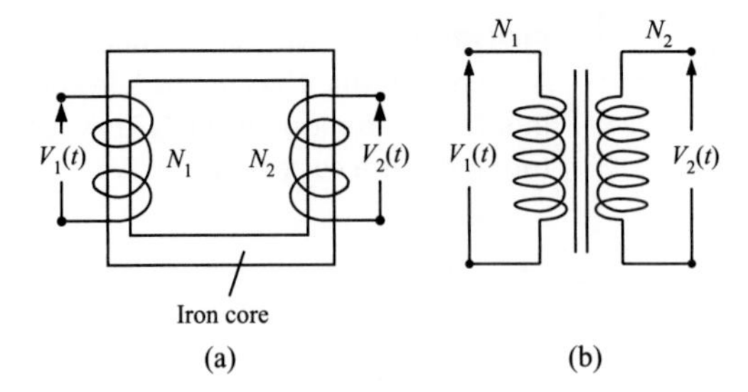


**FIGURE 8.12:** (a) An ideal transformer with primary and secondary coil windings, laminated soft iron core. (b) Circuit diagram for a transformer.

transformer can be viewed as having pure inductance, although in reality there is some resistance and capacitance of each winding.

Faraday's law states that the voltage across the primary winding is proportional to the rate of change of the magnetic flux and opposes that change,

$$v_1(t) = V_1 \cdot \sin \omega t \tag{8.90}$$

$$v_1(t) = -N_1 \frac{d\Phi_B}{dt} \tag{8.91}$$

where $v_1(t)$ is the source voltage applied to the primary winding, $N_1$ is the number of turns of the coil, and $\Phi_B$ is the magnetic flux.

Assuming no loss in the magnetic flux, the induced voltage in the secondary winding is

$$v_2(t) = -N_2 \frac{d\Phi_B}{dt} \tag{8.92}$$

$$v_2(t) = \frac{N_2}{N_1} v_1(t) \tag{8.93}$$

Notice that when $N_2 > N_1$, the transformer increases the voltage (step-up transformer), and when $N_2 < N_1$, it reduces the voltage (step-down transformer, Fig. 8.12). Notice that a transformer works on the AC voltage. The DC component of the voltage in the primary winding does not cause any change in the magnetic field, hence it does not contribute to the voltage induced in the secondary winding. Therefore, a transformer is sometimes used to *isolate (or block) the DC component* of a source voltage in signal processing applications between operational amplifiers.

***Co-Energy Concept*** Stored energy in a magnetic circuit can be defined as follows. If we consider a lossless magnetic circuit, the energy stored is time integral of power, $P = v \cdot i$,

$$dW_m = P \cdot dt = v \cdot i \cdot dt = \frac{d\lambda}{dt} \cdot i \cdot dt = i \cdot d\lambda \tag{8.94}$$

$$W_m = \int_0^{\lambda_f} i \cdot d\lambda \tag{8.95}$$

For linear magnetic systems, $\lambda = L \cdot i$

$$W_m = \int_0^{i_f} L \cdot i \cdot di = \frac{1}{2} L \cdot i_f^2 \tag{8.96}$$

The concept of *co-energy* in magnetic circuits is defined as the area on the opposite side of the $\lambda - i$ curve. The co-energy concept is very useful in determining the force and torque

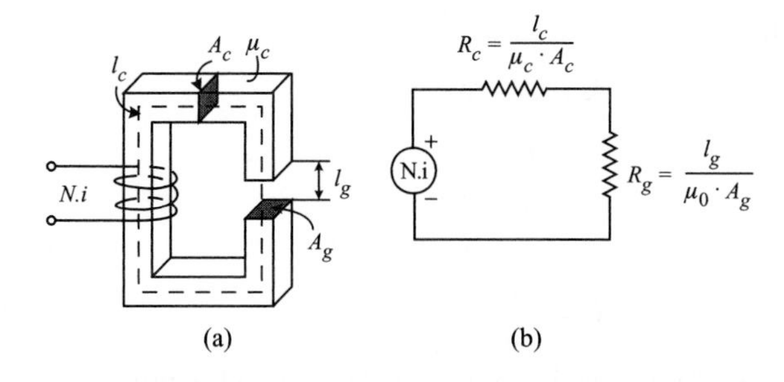


**FIGURE 8.13:** (a) An electromagnetic circuit example: coil wound over a core which has an air-gap. (b) Magnetic circuit model.

in electromagnetic actuators. Co-energy in a magnetic circuit is defined as

$$W_{co} = \int_0^{i_f} \lambda \cdot di \tag{8.97}$$

It can be shown that [21] the force, $F$, (or torque, $T$) delivered to the mechanical system from the stored magnetic energy,

$$F = -\left.\frac{\partial W_m(x)}{\partial x}\right|_{\lambda_f = constant} \tag{8.98}$$

$$T = -\left.\frac{\partial W_m(\theta)}{\partial \theta}\right|_{\lambda_f = constant} \tag{8.99}$$

The same relationship can be expressed in terms of the co-energy,

$$F = \left.\frac{\partial W_{co}(x)}{\partial x}\right|_{i_f = constant} \tag{8.100}$$

$$T = \left.\frac{\partial W_{co}(\theta)}{\partial \theta}\right|_{i_f = constant} \tag{8.101}$$

The above force and energy relationships assume that the magnetic circuit is lossless and does not have hysteresis.

***Example*** Consider an electromagnetic circuit shown in Fig. 8.13. The core of the coil winding is made of a magnetically conductive material with permeability coefficient of $\mu_c$. The cross-sectional area, the length of the core material, and the total number of turns of the solenoid are $A_c$, $l_c$, $N$, respectively. Let the air-gap distance be $l_g$. Cross-sectional area at the air-gap is $A_g$. Determine the effective reluctance and inductance of the circuit.

The reluctance of the magnetically permeable core and air-gap add in series like electrical resistance.

$$R = R_c + R_g \tag{8.102}$$

$$= \frac{l_c}{\mu_c \cdot A_c} + \frac{l_g}{\mu_0 \cdot A_g} \tag{8.103}$$

Notice that if $\mu_c \gg \mu_0$, then $R \approx R_g$. The magnetomotive force ($MMF$) generated due to the coil and current is

$$MMF = N \cdot i \tag{8.104}$$

Flux circulating in the closed path along the core and through the air-gap is

$$\Phi_B = \frac{MMF}{R} \tag{8.105}$$

The flux linkage to the coil is

$$\lambda = N \cdot \Phi_B \tag{8.106}$$

$$= L \cdot i \tag{8.107}$$

$$= N \frac{N \cdot i}{R} \tag{8.108}$$

$$L = \frac{N^2}{R} \tag{8.109}$$

It shows that self-inductance of a magnetic circuit involving a coil is proportional to the square of the number of turns and inversely proportional to the reluctance of the circuit. In actuator applications, small reluctance is desirable for generating more magnetic flux, hence force or torque. However, smaller reluctance leads to large inductance, which results in larger electrical time constant. In electric motor design applications, this conflicting design requirement must be balanced: For large force/torque, we want small reluctance; for small electrical time constant we want small inductance. However, inductance and reluctance are inversely related. As one increases, the other one decreases.

### 8.1.3 Permanent Magnetic Materials

Materials can be classified into three categories in terms of their magnetic properties:

1. Paramagnetic (aluminum, magnesium, platinum, tungsten)
2. Diamagnetic (copper, diamond, gold, lead, silver, silicon)
3. Ferromagnetic (iron, cobalt, nickel, gadolinium) materials

The difference between these materials originates from their atomic structure. The magnetic field strength, $\vec{H}$, and magnetic flux density, $\vec{B}$, relationship in a given spatial location depends on the "permeability" of the surrounding material, $\mu_m$,

$$\vec{B} = \mu_m \cdot \vec{H} \tag{8.110}$$

where $\mu_m = \mu_r \cdot \mu_0$ is called the magnetic permeability of the material which is a measure of how well a material conducts magnetic flux, $\mu_r$, and is called the relative permeability. The relationship between the magnetic field strength ($\vec{H}$) and the magnetic flux density ($\vec{B}$) is linear for paramagnetic and diamagnetic materials. Although the same relationship can be used to describe the magnetic behavior of ferromagnetic materials, the relationship is not linear and exhibits magnetic hysteresis (Fig. 8.14). *Susceptibility*, $\chi$, is defined as

$$\mu_r = 1 + \chi \tag{8.111}$$

The susceptibility is:

1. Positive but small (i.e., $10^{-4}$ to $10^{-5}$ range) for paramagnetic materials,
2. Negative but small (i.e., $-10^{-5}$ to $-10^{-10}$ range) for diamagnetic materials, and
3. Positive and several thousand times larger than 1 (i.e., in the $10^3$–$10^4$ range) for ferromagnetic materials. Furthermore, the effective $\mu_m$ in the $\vec{B}$ and $\vec{H}$ relationship is not linear, but exhibits hysteresis nonlinearity. Ferromagnetic materials are categorized into two groups based on the size of the hysteresis:
   - **(a)** Materials that exhibit small hysteresis in their B-H curves are called *soft ferromagnetic* materials [Fig. 8.14(a)],
   - **(b)** Materials that exhibit large hysteresis in their B-H curves are called *hard ferromagnetic* materials [Fig. 8.14(b)].

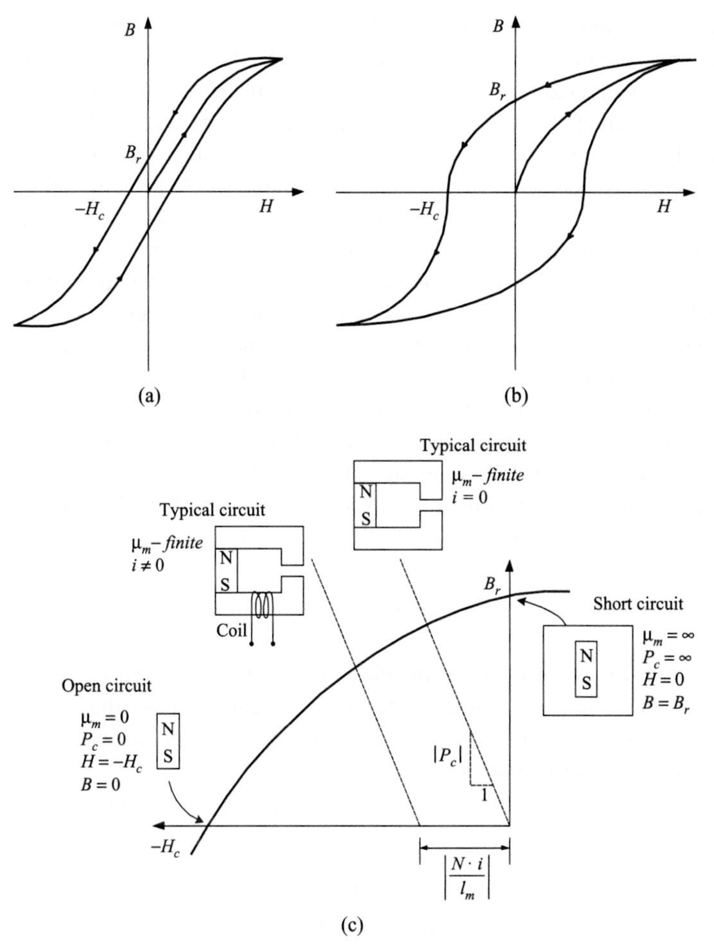


**FIGURE 8.14:** Magnetic properties of ferromagnetic materials are not linear. They exhibit nonconstant $\mu$, saturation, residual magnetism, and hysteresis. $B_r$ is called the remanence, $H_c$ is called the coercivity, $(B \cdot H)_{max}$ indicates the maximum magnetic energy that can be stored in the material per unit volume. (a) In *soft magnetic materials,* the $B_r$ and $H_c$ are small, hence they have small hysteresis. They retain a very small portion of their magnetization when the external field is removed. They are temporarily magnetized, not permanetly. (b) In *hard magnetic materials,* the $B_r$ and $H_c$ are large (with large hysteresis) which means the material retains large magnetization after external field is removed. This is the case for permanent magnetic materials. (c) Permanent magnets (PM) in electric actuator applications operate primarily in the second quadrant of the B-H curve. Rare-earth PMs (i.e., samarium-cobalt mixtures, neodymium mixtures) have almost linear characteristics in the second quadrant. The operating point on the curve is determined by the magnetic circuit and coil current.

The residual magnetization ($B_r$) left after the external magnetic field is removed, which is a result of the magnetic hysteresis, is used as a way to permanently magnetize ferromagnetic materials. This permanent magnetization property is called *remanent magnetization* or *remanence* ($B_r$) of the material, which means the remaining magnetization. As the magnitude of the residual magnetic flux density ($B_r$) increases, the capability of the material to act as a permanent magnet (PM) increases. Such materials are called "hard ferromagnetic materials" compared to the "soft ferromagnetic materials" which have small residual magnetization. The value of the the external magnetic field strength ($H_c$) necessary to remove the residual magnetism (to totally demagnetize it) is called the *coercivity* of the material. It is a measure of how hard the material must be "coersed" by external magnetic field to give up its magnetization. Notice that the area inside the hysteresis curve is the energy lost during each cycle of the magnetization between $\vec{H}$ and $\vec{B}$. This is called the *hysteresis loss*. This energy is converted to heat in the material. In electromagnetic actuator applications such as electric motors, a permanent magnet (PM) usually operates in the second quadrant of B-H curve. As long as the external field is below the $H_c$, the magnet state moves back and forth along the curve on the second quadrant. In this case, there is very small hysteresis loss [Fig. 8.14(c)]. If a PM is exposed to an external field such that its state moves below the linear region in the second quadrant of B-H curve, it will lose some of its magnetization permanently. This point beyond the linear region in the second or third quadrant of the B-H curve is called the *knee point*. If a PM operating condition reaches or passes the knee point, it permanently loses some of its magnetic strength. It will recover along the *recoil line*. The slope of the recoil line is called the *recoil permeability* of the PM. The permeability of rare-earth PM materials such as samarium-cobalt and neodymium-iron-boron (NdFeB) in the second quadrant is very close to that of air. Therefore, the *recoil permeability* defined in the second quadrant of B-H curve for a PM,

$$\mu_{rec} = \frac{1}{\mu_0} \frac{\mathrm{dB}}{\mathrm{dH}} \tag{8.112}$$

is in the range of 1.0 to 1.1 for rare-earth PMs. The lost magnetization can only be recovered by re-magnetizing the magnet. In electromechanical actuators, the electromagnetic circuit should be designed such that PM never reaches the *knee-point*, that is, the point of permanently losing some of its magnetization.

The power of a permanent magnet is measured in terms of the flux and *MMF* it can support. Since,

$$MMF = H \cdot l_m \tag{8.113}$$

$$\Phi_B = B \cdot A_m \tag{8.114}$$

where $l_m$ is the length of the magnet in the direction of magnetization, and $A_m$ is the cross section of the magnet perpendicular to the magnetization direction. In order to increase the *MMF* for a given PM with a specific B-H characteristics, it must have large thickness in the direction of magnetization ($l_m$). Similarly, in order to increase the flux, it must have large surface area that is perpendicular to its magnetization ($A_m$).

If a permanent magnet is placed in a infinitely permeable medium, no *MMF* would be lost and the magnetic field intensity coming out of the magnet would be $B = B_r$, $H = 0.0$ [Fig. 8.14(c)]. If on the other hand, the permanent magnet is placed in a medium with zero permeability, no magnetic flux can exit it. The operating point of the magnet would be $B = 0$ and $H = -H_c$. In a real application, the effective permeablity of the surrounding medium is finite. Hence, permanent magnet operates somewhere along the curve between the two extreme points in the second quadrant of the B-H curve. The nominal location of the operating point is determined by the permeance of the surrounding medium. The absolute value of the slope of the line connecting the nominal operating point to the origin is called the *permeance coefficient*, $P_c$, and the line is called the *load line*. Applied coil current shifts the net MMF (or H), and hence the operating point of the magnet along the H-axis. It is important that the applied coil current should not be large enough to force the magnet into the demagnetization zone. In permanent magnet motors, the electromagnetic circuit of the motor generally results in a load line that is $P_c = 4$–6 range. In magnetic circuits where the closed path of the flux is made of air-gap, highly permeable material, and permanent magnet, it can be shown that

$$P_c \approx \frac{l_m}{l_g} \tag{8.115}$$

where $l_m$ is the permanent magnet thickness in the direction of magnetization, and $l_g$ is the effective air-gap length.

For rare-earth permanent magnets, the B-H curve in the second quadrant can be approximated by

$$B = B_r + \mu_r \, \mu_0 \, H \tag{8.116}$$

When the magnetic circuit is such that $P_c = \infty$, then $H = 0.0$, $B = B_r$. Similarly, when the magnetic circuit is such that $P_c = 0.0$, then $H = -H_c$, $B = 0.0$. In many electric actuator applications, the circuit also includes a current carrying coil. Let the number of turns on the coil be $N$ and current be $i$ in a circuit where the magnetic field of permanent magnet and coil are in series. Then the net magnetic field $H$ is the one due to PM plus the $N \cdot i$ due to the coil. Hence the B-H relationship can be approximated by

$$B = B_r + \mu_r \, \mu_0 \left( H + \frac{N \cdot i}{l_m} \right) \tag{8.117}$$

For a nominal value of $P_c$ in a given electromagnetic circuit, the operating point of the PM moves along the B-H line in the second quadrant as function of current. It is important to design the circuit such that the current does not force the magnet to the demagnetization zone beyond the knee-point and into the third quadrant in the B-H plane.

The operating point of the magnet can be determined at any condition as follows: For a given electromagnetic circuit, $P_c$ is determined by the geometry and material of the circuit and can be calculated for any given configuration, i.e., rotor position relative to stator in an electric motor. We can calculate the $H_m$, $B_m$ point from the intersection of B-H line and load line ($P_c$ line). Then, for a given current value in the coil with $N$ turns, we can calculate the effective $H_{op} = -H_m + N \, i / l_m$, and determine the $B_{op}$ from the B-H curve. The net effect of current is to move the magnet operating point along the B-H curve by shifting the $H$ around $H_m$.

In an electromagnetic circuit, a permanent magnet can be modelled as a flux source $\Phi_r$ and a reluctance $R_m$ in parallel with it (Fig. 8.15)

$$\Phi_r = B_r \cdot A_m \tag{8.118}$$

$$R_m = \frac{l_m}{\mu_r \cdot \mu_o \cdot A_m} \tag{8.119}$$

where $l_m$ is the length along the magnetization direction, $A_m$ is the cross-sectional area perpendicular to the magnetization direction, and $\mu_r$ is the recoil permeability of the magnet material.

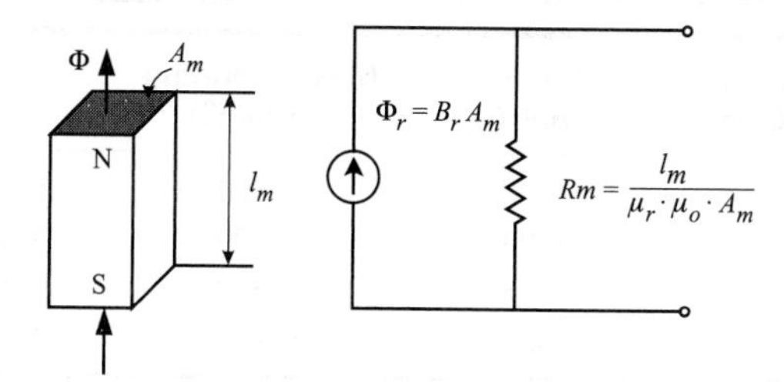


**FIGURE 8.15:** Magnetic model of a permanent magnet. The permanent magnet acts like a flux source and a parallel reluctance. The actual flux, Φ, which leaves the magnet is determined by the rest of the load circuit.

In case of soft ferromagnetic materials, the material goes through the full cycle of magnetization and demagnetization at the same frequency of the external magnetic field. For instance, the stator and rotor materials in an electric motor are made of soft ferromagnetic material. As stator current changes direction in a cyclic way, the B-H curve for the steel goes through the hysteresis loop. The energy in the hysteresis loop is lost as heat. *Hysteresis loss* is proportional to the maximum value of magnetic field intensity magnitude and its frequency. Therefore, in order to minimize the energy loss in motor and transformer cores, lamination material is chosen among the soft ferromagnetic materials. Soft ferromagentic materials have fewer hysteresis losses. But, as a result of the same property, they have small residual magnetization when the external magnetic field is removed. Therefore, we can think of them as temporarily magnetized materials. *Hard ferromagnetic* materials have large residual magnetization, which means they maintain a strong magnetic flux density even when external field is removed, hence the name permanent magnets. But, as a result of the same property, they have large hysteresis losses if they operate in full cycle of external magnetic field variation. The maximum value of the product of $B \cdot H$ on the hysteresis curve indicates the magnetic strength of the material. It is easy to confirm that the $BH$ term has energy units as follows:

$$[\text{Tesla}] \cdot [\text{A/m}] = \frac{[\text{Nt}]}{[\text{A}][\text{m}]}[\text{A/m}] = \frac{\text{Nt}}{\text{m}^2} = \frac{\text{Nt} \cdot \text{m}}{\text{m}^3} \tag{8.120}$$

$$= \frac{[\text{Joule}]}{[\text{m}^3]} \tag{8.121}$$

or in CGS units, $BH$ has units of [Gauss · Oerstead]; 1 [GOe] $= \frac{1}{4\pi} \times 10^3$ [Joule/m$^3$].

In short, remanence $B_r$, coercivity $H_c$, $(BH)_{max}$ maximum energy, and permeability $\mu_m$ are four nominal parameters which characterize the magnetic properties of a ferromagnetic material.

There are four major types of natural hard ferromagnetic materials that can be used as permanent magnets:

1. Alnico, which is an aluminum-nickel-cobalt mixture.
2. Ceramic (hard ferrite) magnetic materials, which consists of strontium, barium ferritite mixtures.
3. Samarium cobalt (samarium and cobalt mixtures, $SmCo_5$, $Sm_2Co_{17}$).
4. Neodymium (neodymium, iron and boron are the main mixture components with small amounts of other compounds). The ideal mixture is $Nd_2Fe_{14}B_1$.

Alnico and ceramic ferrite permanent magnet materials are the lowest cost types and have lower magnetic strength compared to samarium and neodymium (Table 8.2). The maximum magnetic energy of each type is shown in the next table. Today, Alnico permanent magnets (PM) are used in automotive electronics, ceramic PM materials are used in consumer

**TABLE 8.2: Comparison of Four Major Permanent Magnetic Material Types**

| Permanent Magnet Material | Max. Magnetic Energy (MGOe) | Curie Temp(°C) | Max. Operating Temp(°C) | Cost ($) |
|---|---|---|---|---|
| Alnico | 5 | ~ 1000 | ~ 500 | Low |
| Ceramic | 12 | ~ 450 | ~ 300 | Moderate |
| SmCo | 35 | ~ 800 | ~ 300 | High |
| NdFeB | 55 | ~ 350 | ~ 200 | Medium |

electronis, and samarium and neodymium are used in high-performance actuators and sensors. The cost of the PM material increaseas as the magnetic energy level increases. Notice that the enegry levels given in the table are the maximum currently achievable levels. Lower energy level versions are available at lower cost. For instance, the cost of NdFeB at 45 MGOe is twice that of NdFeB at 30 MGOe. The biggest advantage of samarium-cobalt PM material over neodymium PM material is the fact the samarium-cobalt PM material can operate at higher temperatures.

The manufacturing process for making permanent magnets from one of the above materials has the following steps (it is important to note that small variations in composition and manufacturing process make a difference in the final magnetic and mechanical properties of the magnet):

1. Mix proper amount of elements to form the magnet compound and melt it in furnace, and make ingots.
2. Crush and mill to fine powder. Mix the fine powder.
3. Place the mixed powder in die cavity, apply initial electromagnetic field to orient the magnetic directions (pre-alignement of magnetic field), and press it down to about 50% of its powder state size. This is a powder metallurgy process. The product at this state is called green.
4. Heat in furnace (i.e., vacuum chamber at 1100–1200°F for neodymium-iron-boron) will result in further shrinkage in size. Sintered PM has almost 99.9% magnetic material density and can maintain their mechanical properties under high temperatures. Bonded PM includes bonding material which reduces its density and cannot maintain its mechanical properties at high temperatures. This process may be followed by a lower temperature heat treatment around 600°C.
5. Saw and grind to the desired shape (rectangular, cylindrical) and size.
6. Coat the surface of the magnet piece if desired.
7. Magnetize each piece to magnetic saturation by an external electromagnetic field pulse (i.e., a few-milliseconds duration of external magnetic field pulse with high enough $H$ value to make the magnet reach its saturation level). Because the handling of magnetized permanent magnet pieces is difficult, it is desirable to magnetize them as late as possible in the manufacturing process for a given magnetic product application.
8. Finally, the permanent magnet batch should be treated with stabilization and calibration processes. The calibration process makes sure that the magnetic strength of each piece is within a certain tolerance of the desired specifications (i.e., $\pm 1\%$).

Typical shapes and magnetization directions of permanent magnets are shown in Fig. 8.16.

There is a critical temperature, called *Curie temperature*, for each ferromagnetic material above which the material loses its ferromagnetic properties and becomes a paramagnetic material. The Curie temperature for iron is 1043°K, for cobalt 1394°K, and for nickel 631°K. In practice, the allowed maximum operating temperature for a PM material is much lower than the Curie temperature.

The magnetic field strength of a permanent magnet gets weaker as the temperature increases and finally permanently loses its PM properties at the Curie temperature. The lost magnetic strength, while the temperature is well below Curie temperature, has two components: nonreversible and reversible parts. The nonreversible component of the magnetic property variation as function of temperature is always removed as part of the final step of the manufacturing process. The stabilization process exercises the nonreversible magnetic property; hence, the final product would have only the recoverable magnetic property against

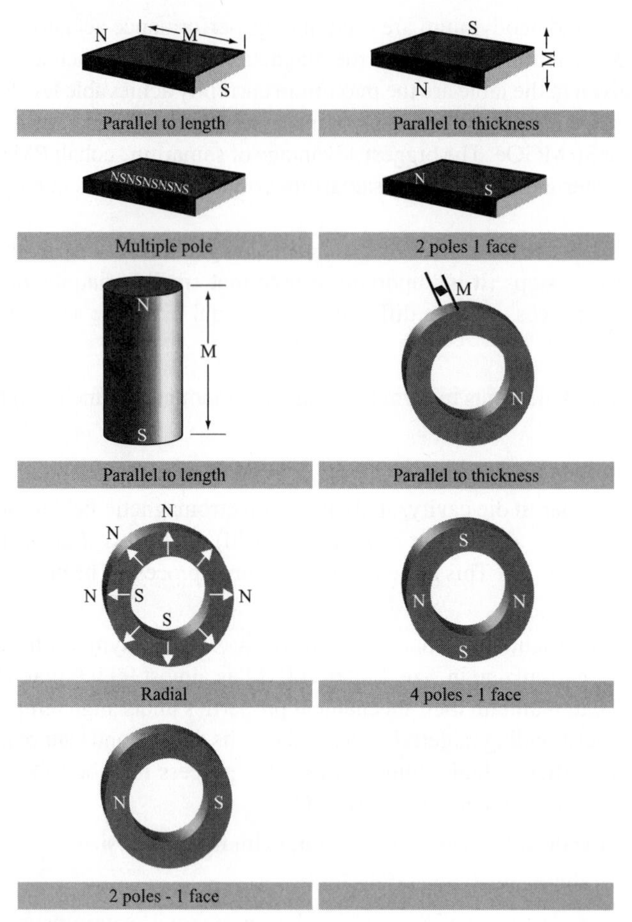


**FIGURE 8.16:** Typical shapes and magnetization directions of permanent magnets. The letter M with arrows indicates the magnetization direction of the piece.

temperature variations. It is a one-time variation in the material property. The reversible component is recovered when the temperature gets lower.

Each PM material has different susceptibility to environmental conditions. In addition to temperature, the chemical composition of the environment is most important. Alnico, ceramic, and samarium magnets are corrosion resistant, whereas neodymium magnets are very susceptible to corrosion. In the selection of a permanent magnet (PM) material, the environmental issues to consider are:

1. Oxidation and humidity level
2. Acid content
3. Salt content
4. Alkaline content
5. Radiation level

The following specifications are typically used by designers in selecting a proper permanent magnet for an application:

1. Permanent magnet material
2. Remanance, coercivity, and maximum energy, $B_r$, $H_c$, $(BH)_{max}$

3. Mechanical dimensions (shape and size)
4. Magnetization direction (radial, axial, etc.)
5. Whether PM is sintered or bonded
6. Surface coating (i.e., 2- $\mu$m to 20-$\mu$m-range thickness, aluminum, nickel, or titanium nitride coating material)

The selected PM material largely determines the temperature coefficient for the loss of magnetic field as temperature rises, and the thermal expansion coefficient of mechanical dimensions which is important in a device assembly. Neodymium magnets expand in the magnetized direction and contract in the other directions with increasing temperature.

A permanent magnet piece is usually bonded to a steel-backing material using adhesives in motor applications, i.e., in the case of electric motors on a steel rotor [3M adhesives]. The bonding material types include thermosetting epoxies, structural adhesives. There are many adhesives [3M adhesives] specifically designed for motor applications. The strength of the bond between PM and and the rotor is function of the contact surface area. Typically, the surface area is roughened in order to provide a good bonding. After the application of adhesive, the PM is clamped on the rotor and cured at high temperature. The curing temperature should be well below the demagnetization temperature of the PM, if the PM was magnetized before the bonding process. High-strength PMs should be handled with care because they have very strong magnetic attraction forces. Therefore, it is desirable to magnetize them as late in the assembly process as possible.

***Example*** Consider an electromagnetic circuit shown in Fig. 8.17 where there are two magnetic field sources: (1) the permanent magnet and (2) the coil. The coil has $N$ turns, and current is $i$. Let the permeability constant of the core be $\mu_c$. Let the air-gap, permanent magnet, core cross-sectional areas be the same, $A_m = A_c = A_g$, for simplicity. (a) Determine the inductance of the circuit, and (b) the $P_c$, magnitude of the slope of load line, assuming $\mu_c = \infty$.

**(a)** The permanent magnet (PM) is modelled as a flux source ($\Phi_r = B_r \cdot A_m$) and a reluctance ($R_m$) in parallel with it. Then we have the additional reluctances of the core and that of the air-gap in series with the magnet. The magnetic circuit is shown in Fig. 8.17(b). Total flux in the closed path along the core, air-gap, and magnet as a result of the MMF

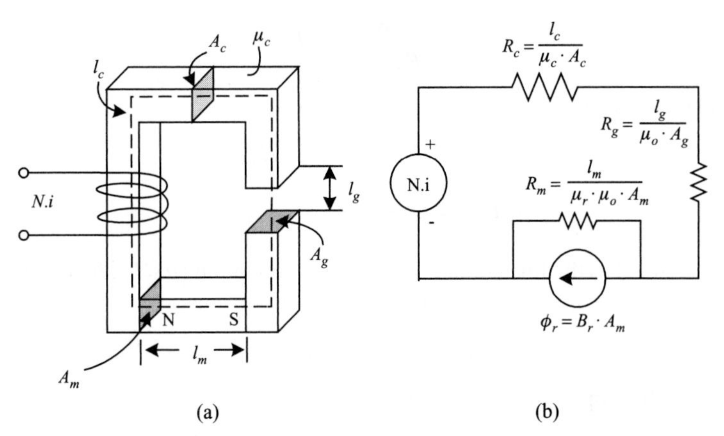


**FIGURE 8.17:** (a) An electromagnetic circuit example involving a permanent magnet, a coil wound over a core which has an air-gap. (b) Magnetic circuit model.

provided by the permanent magnet and the coil current is

$$\Phi = \Phi_m + \Phi_c \tag{8.122}$$

The flux leaving the permanent magnet $\Phi_m$ and the flux due to the coil current $\Phi_c$ are

$$\Phi_m \cdot (R_c + R_g) = (\Phi_r - \Phi_m) \cdot R_m \tag{8.123}$$

$$\Phi_m = \frac{R_m}{R_m + R_c + R_g} \cdot \Phi_r \tag{8.124}$$

$$\Phi_c = \frac{N \cdot i}{R_m + R_c + R_g} \tag{8.125}$$

$$\Phi = \frac{R_m}{R_m + R_c + R_g} \cdot \Phi_r + \frac{N \cdot i}{R_m + R_c + R_g} \tag{8.126}$$

The flux linkage is

$$\lambda = N \cdot \Phi = L \cdot i + N \cdot \Phi_m \tag{8.127}$$

$$L \cdot i = N \cdot \Phi_c \tag{8.128}$$

$$L = \frac{N^2}{R_m + R_c + R_g} \tag{8.129}$$

$$L = \frac{N^2}{R_{eqv}} \tag{8.130}$$

**(b)** Magnitude of the load line slope can be determined as follows. We need to determine the flux density leaving the permanent magnet in this circuit when $i = 0.0$. The effect of nonzero coil current on the operating point of the magnet is considered in the next example. The net flux leaving the PM,

$$\Phi_m = B_m \cdot A_m \tag{8.131}$$

$$B_m \cdot A_m = \frac{R_m}{R_m + R_c + R_g} \cdot B_r \cdot A_m \tag{8.132}$$

$$B_m = \frac{R_m}{R_m + R_c + R_g} \cdot B_r \tag{8.133}$$

For simplicity, we assumed that $\mu_c = \infty$, then $R_c = 0.0$,

$$B_m = \frac{R_m}{R_m + R_g} \cdot B_r \tag{8.134}$$

Then, $H_m$ can be found from

$$B_m = B_r + \mu_r \cdot \mu_0 \cdot H_m \tag{8.135}$$

$$\mu_0 H_m = \frac{B_m - B_r}{\mu_r} \tag{8.136}$$

The load line is defined as the line connecting the operating point $(\mu_0 H_m, B_m)$ of the magnet to the origin of the $B$ versus $\mu_0 H$ curve of the permanent magnet.

$$B_m = -P_c \cdot (\mu_0 H_m) \tag{8.137}$$

The magnitude of the slope of the load line is

$$P_c = \left| \frac{B_m}{\mu_0 H_m} \right| \tag{8.138}$$

For further insight into the load line slope, let us assume that $\mu_r = 1.0$ for the magnet. Then noting that

$$R_m = \frac{l_m}{\mu_0 A_m} \tag{8.139}$$

$$R_g = \frac{l_g}{\mu_o A_g} = \frac{l_g}{\mu_o A_c}; \quad \text{since } A_c = A_g \tag{8.140}$$

The flux leaving the permanent magnet is

$$B_m = \left(\frac{R_m}{R_m + R_g}\right) B_r \tag{8.141}$$

$$= \left(\frac{l_m}{l_m + l_g}\right) B_r \tag{8.142}$$

and the corresponding magnetic field intensity of the magnet on the B-H curve,

$$\mu_0 H_m = \frac{B_m - B_r}{\mu_r} \tag{8.143}$$

$$= \frac{\left(\frac{l_m}{l_m + l_g} - 1\right) B_r}{\mu_r} \tag{8.144}$$

$$= -\left(\frac{l_g}{l_m + l_g}\right) B_r \tag{8.145}$$

Hence, the slope of the load line (permeance coefficient of the magnetic circuit),

$$P_c = \left|\frac{B_m}{\mu_0 H_m}\right| \tag{8.146}$$

$$= \left|\frac{\frac{l_m}{l_m + l_g} B_r}{-\frac{l_g}{l_m + l_g} B_r}\right| \tag{8.147}$$

$$= \frac{l_m}{l_g} \tag{8.148}$$

***Example*** Consider the same example shown in Fig. 8.17 and determine the operating point of the permanent magnet under a given nonzero current condition at the coil. Neglect the reluctance of the core relative to the reluctance of the magnet and air-gap and let the magnet permeability be the same as that of air ($\mu_r = 1.0$). Also, let $A_m = A_g$, cross-sectional areas of magnet and air-gap be the same. Let the coil current be $i$ and the number of turns of the coil be $N$.

The total flux in the circuit through the air-gap is due to the PM and the coil,

$$\Phi = \Phi_m + \Phi_c \tag{8.149}$$

It can be shown that the flux due to the magnet (flux exiting the magnet into the closed-path circuit), $\Phi_m$, is

$$\Phi_m = \frac{R_m}{R_m + R_g} \cdot \Phi_r \tag{8.150}$$

where, reluctances are

$$R_m = \frac{l_m}{\mu_r \cdot \mu_0 \cdot A_m} = \frac{l_m}{\mu_o \cdot A_g}; \quad for \quad \mu_r = 1.0, \; A_g = A_c \tag{8.151}$$

$$R_g = \frac{l_g}{\mu_0 \cdot A_g} \tag{8.152}$$

Flux due to the coil current, $\Phi_c$, is

$$\Phi_c = \frac{N \cdot i}{R_m + R_g} \tag{8.153}$$

$$= \frac{\mu_o \, A_g}{l_m + l_g} N \cdot i \tag{8.154}$$

Total flux due to permanent magnet and coil is

$$\Phi = \Phi_m + \Phi_c \tag{8.155}$$

$$= \frac{l_m}{l_m + l_g} \Phi_r + \frac{\mu_o \, A_g}{l_m + l_g} N \cdot i \tag{8.156}$$

$$B = \frac{\Phi}{A_c}; \quad note \quad A_c = A_g \tag{8.157}$$

$$= \frac{\Phi}{A_g} \tag{8.158}$$

$$= \frac{l_m}{l_m + l_g} B_r + \frac{\mu_o}{l_m + l_g} N \cdot i \tag{8.159}$$

$$= \frac{l_m}{l_m + l_g} \left( B_r + \mu_o \frac{N \cdot i}{l_m} \right) \tag{8.160}$$

Let us determine the corresponding $\mu_o \cdot H$ point on the second quadrant of the B-H of the permanent magnet for this value of magnetic flux density,

$$B = B_r + \mu_r \mu_0 H \tag{8.161}$$

$$\mu_o H = \frac{B - B_r}{\mu_r} \approx B - B_r; \quad \text{assuming} \quad \mu_r \approx 1.0 \tag{8.162}$$

$$= \left( \frac{l_m}{l_m + l_g} - 1 \right) B_r + \frac{\mu_o}{l_m + l_g} N \cdot i \tag{8.163}$$

$$= \frac{-l_g}{l_m + l_g} B_r + \frac{\mu_o}{l_m + l_g} N \cdot i \tag{8.164}$$

Let us substract $(\mu_0 \cdot N \cdot i / l_m)$ from both sides of this relationship:

$$\mu_0 \left( H - \frac{Ni}{l_m} \right) = \frac{-l_g}{l_m + l_g} B_r + \frac{\mu_o}{l_m + l_g} N \cdot i - \mu_o \cdot \frac{N \cdot i}{l_m} \tag{8.165}$$

$$= \frac{-l_g}{l_m + l_g} B_r + \frac{l_m - (l_m + l_g)}{l_m \cdot (l_m + l_g)} \mu_0 \cdot N \cdot i \tag{8.166}$$

$$= -\frac{l_g}{l_m + l_g} \cdot \left( B_r + \mu_0 \frac{N \cdot i}{l_m} \right) \tag{8.167}$$

and the load line slope in this case is

$$P_c = \left| \frac{B}{\mu_0 \left( H - \frac{Ni}{l_m} \right)} \right| \tag{8.168}$$

$$= \left| \frac{\frac{l_m}{l_m + l_g} \left( B_r + \mu_o \frac{N \cdot i}{l_m} \right)}{-\frac{l_g}{l_m + l_g} \cdot \left( B_r + \mu_0 \frac{N \cdot i}{l_m} \right)} \right| \tag{8.169}$$

$$= \left| \frac{l_m}{l_g} \right| \tag{8.170}$$

and the new load line equation including the effect of coil current is

$$B = -P_c \cdot \mu_o \left( H - \frac{Ni}{l_m} \right) \tag{8.171}$$

where the effect of coil current is to shift the load line along the $H$ axis by $\frac{N \cdot i}{l_m}$ amount (Fig. 8.14c).

We are most concerned with the case when current is negative, which takes the magnet closer to the demagnetization zone. When current is zero, magnet operates at a nominal point determined by the permeance coefficient ($P_c$) of the magnetic circuit surrounding the magnet. The coil current effect should not be too large to force the magnet into the demagnetization zone (Fig. 8.14c).

## 8.2 SOLENOIDS

### 8.2.1 Operating Principles of Solenoids

A *solenoid* is a translational motion actuator with a rather limited motion range. Solenoids are used in fluid flow control valves and small range translational displacement actuators. A solenoid is made of a coil, a frame which is a material with high permeability to guide the magnetic flux, a plunger, a stopper (and a centering spring in most cases), and a bobin (Fig. 8.18). A *bobin* is a plastic or nonmagnetic metal on which coil is wound. It is non-magnetic so that there is no short circuit for the flux between the coil and plunger. The operating principle of solenoids is based on the tendency of the ferromagnetic plunger and coil generated magnetic flux to seek the minimum reluctance point. As a result, when the coil is energized, the plunger is pulled in toward the stopper. The higher the magnetic field strength ($n_{coil} \cdot i$, number of turns in the coil times current) and the better the magnetic permeability of the medium that guides the flux to the plunger, the higher the force generated. The plunger works on the pull principle. However, by mechanical design we can obtain pull or push motion from the solenoid (Fig. 8.18). The mechanical connection between the

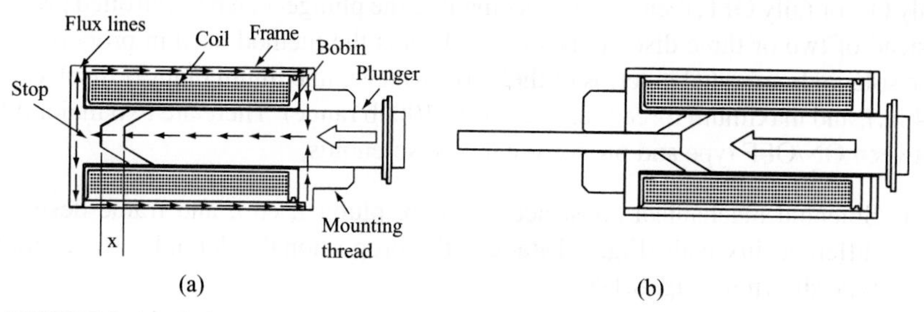


**FIGURE 8.18:** Solenoid components and its design: pull and push types.

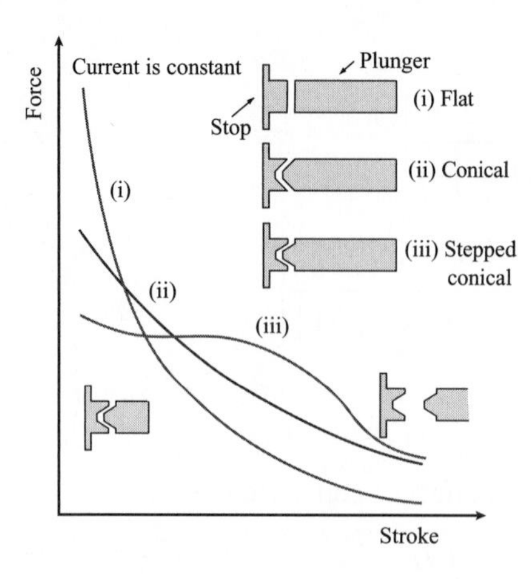


**FIGURE 8.19:** Solenoid force versus plunger displacement under constant coil current condition. As the plunger closes the air-gap between itself and the stopper, the force capacity increases. This is called the holding force, which is generally higher than the average pull-push force.

plunger, made of a ferromagnetic material, and the tool must be via a nonmagnetic material. For instance, in case of a push-type solenoid [Fig. 8.18(b)], the push-pin is made of a nonmagnetic material, i.e., stainless steel. The quality of the magnetic circuit and its ability to guide magnetic flux lines depend on the permeability of the coil frame, air-gap between the plunger and coil (fixed gap) and air-gap between the plunger and stopper (variable gap). For a given current, the force generated by the solenoid varies as a function of the air-gap between the plunger and the stopper. The smaller this gap is, the smaller the effective reluctance of the magnetic flux path, and hence, the higher the force generated. The force as a function of plunger displacement under constant current varies as shown in Fig. 8.19. Notice that the shape of the force-displacement curve for a constant current can be affected by the shape of the plunger and stopper head. For high-performance applications, in order to reduce the eddy current losses in the solenoid, iron core of the winding and the plunger may be constructed from laminated sheets of iron that are insulated.

Solenoids are referred to as *single acting* or *double acting*. A solenoid is a single-acting type device; that is, when current is applied the plunger moves in one direction in order to minimize the reluctance, regardless of the direction of the current. Force is always generated in one direction, pull or push. Double-acting solenoid can move in both directions by generating force in both directions, both pull and push directions, by using two solenoids in one package. Therefore, a double-acting solenoid is basically two solenoids with one plunger, two coils, and two stops.

A directional flow control valve may have one or two solenoids to position it in two or three discrete positions. For instance, a double-acting solenoid (two solenoids in one package) can have three positions: (1) center position when both the solenoids are deenergized, (2) left position when the left solenoid is energized, and (3) right position when the right solenoid is energized. If the current in each solenoid is controlled proportionally, instead of fully ON or fully OFF, then the displacement of the plunger can be controlled proportionally instead of two or three discrete positions. This is the method used in proportional valves. The solenoids are rated in terms of their coil voltage, maximum plunger displacement (i.e., 1/4 in.), and maximum force (i.e., 0.25 oz to 100 lb range). There are two major differences between ON/OFF type and proportional type solenoids,

1. Solenoid mechanical construction where plunger, coil, and frame design provides different flux path [Fig. 8.19(a–c)], the proportional solenoids are generally of the type shown in [Fig. 8.19(c)].
2. Current in the coil is controlled either in ON/OFF or in proportional mode.

The solenoids used in hydraulic flow control valves incorporate a tube design around the plunger. The tube performs two functions,

1. Isolate the plunger from hydraulic pressure
2. Provide proper guide for flux flow

In order to guide the flux into the plunger, the tube is made of three sections where the middle section is a nonmagnetic material. For instance, the material of three sections of the tube might be low carbon steel, brass and low carbon steel.

Coil acts as the electromagnet in all electic actuators. Current ($i$), number of turns ($n_{coil}$), and the effective permeance of the magnetic medium (core material, air-gap, etc.) determine the strength of the electromagnetic field generated by the coil. At the same time, there are mechanical size and thermal considerations. Rated current determines the minimum diameter requirement of the conductor wire. The wire diameter and the number of turns determine the mechanical size of the coil. In general, the insulation material increases the effective conductor diameter by about 10%. Different insulation materials have different temperature ratings (i.e., 105°C for formvar, 200°C for thermalex insulation material compounds commonly used industry). Once the coil wire diameter, number of turns, and mechanical size are known, the resistance of the coil is determined. Hence, the resistive heat dissipation is known. In order to make sure the temperature of the coil stays within the limits of its coil insulation rating, the thermal heat conducted from the coil should balance the resistive heat. The coil design requires the balancing of electrical capacity (current and number of turns), mechanical size, and thermal heat.

The force generated by a solenoid is a function of the current in the coil ($i$), the number of turns in the coil ($n_{coil}$), the magnetic reluctance ($R_B$, which is a function of the plunger displacement, $x$, the design shape, and material permeability, $\mu$), and temperature ($T$):

$$F_{sol} = F_{sol}(i, n_{coil}, R_B(x, \mu), T) \tag{8.172}$$

For a given solenoid, the $n_{coil}$ is fixed, and $R_B(x, \mu)$ varies with the displacement of the plunger and the air-gap between the winding coil and the plunger. The main effect of temperature is the change in the resistance of coil. This leads to change in current for a given terminal voltage. If the control system regulates the current in the coil, the effect of temperature on force other than its effect on resistance is negligible. Therefore, for a given solenoid, the generated force is a function of operating variables as follows:

$$F_{sol} = F_{sol}(i, x) \tag{8.173}$$

The basic mode of control is the control of current in the coil in order to control the force. Unlike a rotational DC motor where the current and torque relationship is constant and independent of the rotary position of the shaft, the force-current relationship of a solenoid is nonlinear and function of the rotor (plunger) position.

In general the force capacity of the solenoid is rated at 25°C. The force capacity would typically reduce to 80% value at around 100°C temperature. At a rated current, the force can either decrease or increase as function of plunger displacement depending on the type of solenoid: the force reduces as function of plunger displacement in push type and increases in pull type solenoids (Fig. 8.19). Notice that there is always a residual left in the core when the current is turned OFF (current is zero) due to the hysteresis nature of electromagnetism. Use of *annelead* steel for the core and plunger material minimizes that effect.

In basic mode of operation, a solenoid is driven by DC voltage, $V_t$ (i.e., 12 V, 24 V, 48 V) at its coil terminals, and current $i$ is developed by the resistance ratio (neglecting the inductance of coil),

$$i = \frac{V_t}{R_{coil}} \tag{8.174}$$

The physical size of the solenoid determines the maximum amount of power it can convert from electrical to mechanical power. The continuous power rating of the solenoid should not be exceeded in order to avoid overheating,

$$P = V_t \cdot i = \frac{V_t^2}{R_{coil}} < P_{rated} \tag{8.175}$$

In some applications, it is desirable to provide a larger current (the *in-rush current*) and after the initial movement, the current is reduced to a lower value, called the *holding current*. One way to reduce the in-rush current to holding current is to divide the coil winding into two resistor sections and provide an electrical contact between the two series resistor. When large in-rush current is needed, short circuit via a switch (i.e., electronic transistor switch) the second resistor section to increase current. When it is desired to reduce the current, turn OFF the switch to include the second part of the resistor in the circuit in series, hence reduce the current for the given terminal voltage. There are also solenoids made with primary and secondary windings. Such solenoids can be driven by 50-Hz or 60-Hz AC voltages (24, 120, 204 VAC). Force is related to square of current. Therefore, the direction of generated force does not oscillate with the AC current direction change. However, the magnitude oscillates at twice the frequency of the supply current frequency, i.e., if AC supply is 60 Hz, the force magnitude would oscillate at 120 Hz. In other words, the electromagnetic circuit of the solenoid acts like a "rectifier" between supply current and generated force.

## 8.2.2 DC Solenoid: Electromechanical Dynamic Model

Consider the solenoid shown in Fig. 8.18. The coil has $n_{coil}$ turns, and the voltage is controlled across the terminals of the coil, $V(t)$. The plunger moves inside in the direction of $x$. The electromechanical dynamic model of the solenoid includes three equations: (1) the electromechanical relationship which describes the voltage, current in coil, and motion of the plunger,

$$V(t) = R \cdot i(t) + \frac{d}{dt}(\lambda(x, i)) \tag{8.176}$$

where $\lambda(x, i)$ is the flux linkage. For an inductor type coil circuit, $\lambda(x, i)$ is

$$\lambda(x, i) = L(x) \cdot i(t) \tag{8.177}$$

Hence, the voltage-current-motion relationship can be expressed as

$$V(t) = R \cdot i(t) + \frac{dL(x)}{dx}\frac{dx}{dt} \cdot i(t) + L(x) \cdot \frac{di(t)}{dt} \tag{8.178}$$

$$V(t) = R \cdot i(t) + L(x)\frac{di(t)}{dt} + \left(\frac{dL(x)}{dx}\right) \cdot i(t)) \cdot \dot{x}(t) \tag{8.179}$$

The electromagnetic energy conversion mechanism generates the force as a result of the interaction between the coil generated electromagnetic field and variable reluctance of the plunger–air-gap assembly [Gamble 1996]. Let us consider the magnetic flux path in the solenoid (Fig. 8.18). Assume that the permeability of the plunger, frame, and stop is very high compared to the permeability of air-gap, $\mu_c \gg \mu_0$. We can assume the magnetic energy is only stored in the air-gap, neglecting the magnetic energy stored elsewhere. This model is similar to the magnetic circuit shown in Fig. 8.13 except that in case of solenoids, the air-gap is variable. Then the following relations hold for magnetic field strength, flux density, and flux itself:

$$H_g \cdot x = n_{coil} \cdot i \tag{8.180}$$

$$B_g = \mu_0 \cdot H_g = \mu_0 \cdot \frac{n_{coil} \cdot i}{x} \tag{8.181}$$

$$\Phi_b = B_g \cdot A_g = \mu_0 \cdot \frac{n_{coil} \cdot i}{x} \cdot A_g \tag{8.182}$$

The flux linkage and the inductance are defined as

$$\lambda(x, i) = \Phi_b \cdot n_{coil} = L(x) \cdot i(t) \tag{8.183}$$

$$= \mu_0 \cdot \frac{n_{coil}^2 \cdot i(t)}{x} \cdot A_g \tag{8.184}$$

Then, the inductance as function of plunger displacement is

$$L(x) = \mu_0 \cdot \frac{n_{coil}^2}{x} \cdot A_g \tag{8.185}$$

The force is calculated from the *co-energy equation*,

$$W_{co}(\lambda, i) = \frac{1}{2}\lambda(x, i) \cdot i = \frac{1}{2}L(x) \cdot i^2 \tag{8.186}$$

$$F(x, i) = \frac{\partial W_{co}(x, i)}{\partial x} \tag{8.187}$$

$$= \frac{1}{2}\frac{\partial L(x)}{\partial x} \cdot i^2 \tag{8.188}$$

$$F(x, i) = -\frac{1}{2}\mu_0 \cdot \frac{n_{coil}^2}{x^2} \cdot A_g \cdot i^2 \tag{8.189}$$

This is the second equation which relates current to force.

Notice that the force direction does not depend on the current, and it is proportional to the square of current and inversely proportional to the square of the air-gap. The shape of the force as function of displacement can be shaped with the design of the plunger and stopper cross sections (Fig. 8.19). Finally, the third equation is the force–inertia relationship, which defines the motion of the plunger and any load it may be driving,

$$F(t) = m_t \cdot \ddot{x}(t) + k_{spring}(x(t) - x_0) + F_{load}(t) \tag{8.190}$$

where $m_t$ is the mass of the plunger plus the load, $F_{load}$ is the load force, and $x_0$ is the preload displacement of the spring.

Notice that the electrical time constant of the solenoid, $\tau = L/R$, can become large when $L$ is large. In order to reduce the electrical time constant for fast response, one method is to increase the effective resistance by adding external resistance in series with the coil. As $R$ increases, $\tau$ decreases. However, in order to provide the rated current at increased resistance, the supply voltage level must be proportionally increased because $i = V/R$. This in turn increases the resistive loss, $P_{Ri} = R \cdot i^2$.

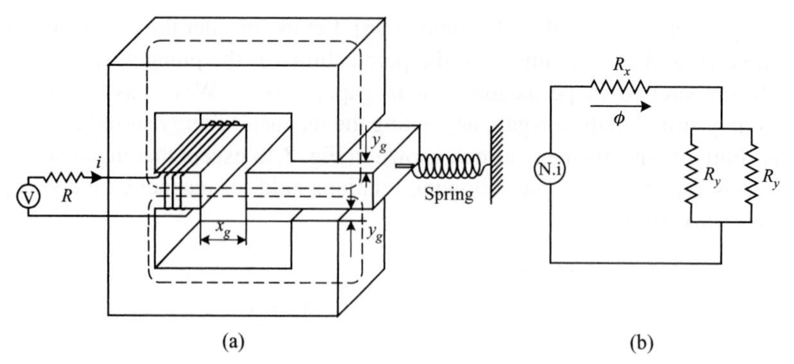


**FIGURE 8.20:** A solenoid example. The magnetic circuit is shown on the right. The force is a nonlinear function of the air-gaps and current. The air-gap $x_g$ varies with motion, and the air-gaps $y_g$ are fixed.

It is also common to provide a surge suppressor component in parallel with a solenoid in order to reduce the voltage spikes in an electrical circuit due to large inductance of the solenoid.

***Example*** Consider a solenoid type shown in Fig. 8.20. There are three air-gap paths, one $x_g$ and two $y_g$ spaces. The winding has $N$ turns, and a controlled current $i$ is supplied to the winding. Let us determine the force generated as a function of current $i$ and air-gap $x_g$. The air-gaps $y_g$ on both sides are constant. Assume that the permeability of iron core is much larger than that of the air-gaps and neglect the MMF loss along the path in the iron. Then, total MMF is stored in the air-gaps $x_g$ and $y_g$. Let $A_x$ and $A_y$ denote the cross-sectional areas at the air-gaps.

From conservation of the MMF, if we trace the flux path either on the top or bottom section, we obtain

$$H_x \cdot x_g + H_y \cdot y_g = N \cdot i \tag{8.191}$$

From symmetry,

$$\Phi_x = 2 \cdot \Phi_y \tag{8.192}$$
$$B_x \cdot A_x = 2 \cdot B_y \cdot A_y \tag{8.193}$$

Note that

$$B_x = \mu_o \cdot H_x \tag{8.194}$$
$$B_y = \mu_o \cdot H_y \tag{8.195}$$

Hence,

$$\Phi_x = 2 \cdot \Phi_y \tag{8.196}$$
$$\mu_o \cdot H_x \cdot A_x = 2 \cdot \mu_o \cdot H_y \cdot A_y \tag{8.197}$$

This results in the follow magnetic field strength relation,

$$H_y = \frac{A_x}{2A_y} H_x \tag{8.198}$$

Then,

$$H_x = \frac{N \cdot i}{\left(x_g + \frac{A_x}{2A_y} y_g\right)} \tag{8.199}$$

$$H_y = \frac{A_x}{2A_y} \frac{N \cdot i}{\left(x_g + \frac{A_x}{2A_y} y_g\right)} \tag{8.200}$$

Another way to look at this is in terms of MMFs and reluctances. The flux due to MMF and effective reluctance in the circuit,

$$\Phi = \Phi_x = \frac{1}{2}\Phi_y \tag{8.201}$$

$$= \frac{MMF}{R_{eqv}} = \frac{N \cdot i}{R_x + (R_{y,eqv})} \tag{8.202}$$

$$= \frac{\mu_o N i A_x}{\left(x_g + \frac{A_x}{2A_y} y_g\right)} \tag{8.203}$$

where the effective reluctances

$$R_x = \frac{x_g}{\mu_o \cdot A_x} \tag{8.204}$$

$$R_y = \frac{y_g}{\mu_o \cdot A_y} \tag{8.205}$$

and $R_{y,eqv}$ from parallel connection of two reluctances,

$$R_{y,eqv} = \left(\frac{1}{R_y} + \frac{1}{R_y}\right)^{-1} = R_y/2 \tag{8.206}$$

Flux linkage is a function of variable quantities $i$ and $x_g$ ($y_g$ is constant)

$$\lambda(x_g, i) = \Phi_x \cdot N = L(x_g, i) \cdot i \tag{8.207}$$

$$= B_x \cdot A_x \cdot N = \mu_o \frac{N^2 \cdot i}{\left(x_g + \frac{A_x}{2A_y} y_g\right)} A_x \tag{8.208}$$

$$L(x_g, i) = \mu_o \frac{N^2}{\left(x_g + \frac{A_x}{2A_y} y_g\right)} A_x \tag{8.209}$$

The co-energy expression as a function of flux linkage and current is

$$W_{co} = \frac{1}{2}\lambda(x_g, i) \cdot i = \frac{1}{2} L(x_g, i) \cdot i^2 \tag{8.210}$$

$$F(x_g, i) = \frac{\partial W_{co}(x_g, i)}{\partial x_g} \tag{8.211}$$

$$= -\frac{1}{2} \frac{\mu_o N^2 A_x}{\left(x_g + \frac{A_x}{2A_y} y_g\right)^2} \cdot i^2 \tag{8.212}$$

The generated force is a function of actuator geometery, $A_x, A_y, x_g, y_g$, permeability of gap, $\mu_0$, coil turn, $N$, and current, $i$.

The complete electromechanical dynamic model of this actuator can be written as

$$V(t) = L(x_g)\frac{di(t)}{dt} + R_{coil}\, i(t) + \left(\frac{dL(x_g)}{dx}\right) i(t)\frac{dx(t)}{dt} \tag{8.213}$$

$$F(x_g, i) = -\frac{1}{2}\,\frac{\mu_o\, N^2\, A_x}{\left(x_g + \frac{A_x}{2\,A_y}\, y_g\right)^2} \cdot i^2 \tag{8.214}$$

$$F(x_g, i) = m_p\frac{d^2x(t)}{dt^2} + k_{spring} \cdot x(t) \tag{8.215}$$

Notice that the coupling between the mechanical and electrical equations is through the flux linkage, $\lambda = L \cdot i$. As the plunger moves, it changes the flux linkage between the core and the plunger. Another way to look at this is that as the plunger moves, the effective magnetic reluctance distribution of the actuator geometry changes. Change in flux linkage as a result of mechanical motion results in change in the co-energy. The partial derivative of co-energy with respect to displacement is the generated force.

## 8.3 DC SERVO MOTORS AND DRIVES

DC servo motors can be divided into two general categories in terms of their commutation mechanism: (1) brush-type DC motors and (2) brushless DC motors. The brush-type DC motor has a mechanical brush pair on the motor frame and makes contact with a commutator ring assembly on the rotor in order to *commutate* current, that is, to switch current from one winding to another, as a function of rotor position so that the magnetic fields of the rotor and stator are always at a 90 degree angle relative to each other. In brush type permanent magnet DC motors, the rotor has the coil winding and the stator has the permanent magnets.

The brushless DC motor is an inside-out version of the brush-type DC motor; that is, the rotor has the permanent magnets and the stator has the winding. In order to achieve the same functionality of the brush-type motor, magnetic fields of the rotor and stator must be perpendicular to each other at all rotor positions. As the rotor rotates, the magnetic field rotates with it. In order to maintain perpendicular relationship between the rotor and stator magnetic fields, the current in the stator must be controlled as a vector quantity (both magnitude and direction) relative to the rotor position. Control of current to maintain this vector relationhips is called *commutation*. Commutation is done by solid-state power transistors based on a rotor position sensor. Notice that a rotor position sensor is necessary to operate a brushless DC motor, whereas a brush type DC motor can be operated without any position or velocity sensor as a torque source. When a motor is controlled in conjunction with a position or velocity sensor, it is considered a "servo" motor.

The field magnetics in a brush-type DC motor can be established either by permanent magnets (hence the name *permanent magnet DC motor*) or electromagnets (hence the name *field-wound DC motor*). Field-wound DC motors are used in high-power applications (i.e., 20 HP and above) where the use of permanent magnets are no longer cost effective. Permanent magnet (PM) DC motors are used in applications below 20 HP.

Coil winding (either on stator in the case of brushless DC motors, AC induction motors, stepper motors, or on the rotor in the case of brush-type DC motors) determines one of the magnetic fields essential to the operation of a motor. The coil design question

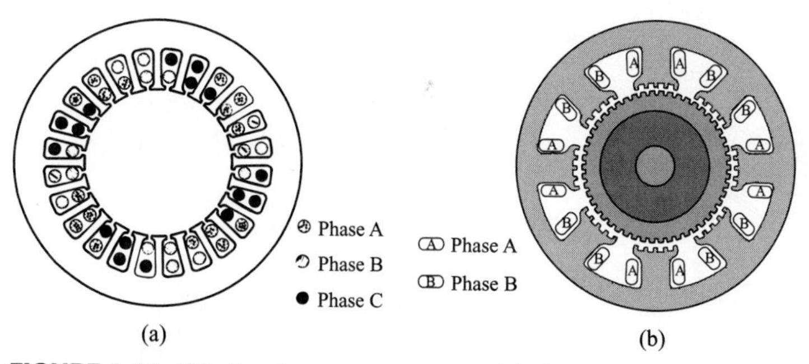


**FIGURE 8.21:** Winding types on the stator: (a) distributed winding (i.e., AC induction motors, brushless DC motors), (b) concentrated winding (i.e., stepper motors).

is the question of how to distribute the coil around the perimeter of the stator or rotor. The design parameters are [48]

1. The number of electrical phases
2. Number of coils in each phase
3. Number of turns on each coil
4. Wire diameter
5. The number of slots and how each coil is distributed over these slots

There are two types of windings in terms of the spatial distribution of a wire on the stator (Fig. 8.21):

1. Distributed winding where each phase winding is distributed over multiple slots and one phase winding has overlaps with the other windings (i.e., AC induction motors, DC brushless motors)
2. Concentrated winding where a particular winding is wound around a single pole (i.e., stepper motors)

Most common step motors have concentrated winding, whereas AC and DC motors have distributed winding. In concentrated winding, one coil is placed around a single tooth. By controlling the current direction in that particular coil, magnetic polarity (N or S) of that tooth is controlled. Hence, a desired N and S pole pattern can be generated by controlling each coil current direction and magnitude. In distributed winding, there are many variations on how to distribute the coils. The most common type is a three-phase winding, and each slot has two coil segments. The coil can be distributed to generate two pole, four pole, eight pole, etc. on the stator at any given current commutation condition. By controlling the current in each phase, both magnitude and direction of the magnetic field pattern are controlled. It is common to view the coil distribution in slots in a linear diagram by considering the unrolled version of the motor stator and rotor.

## 8.3.1 Operating Principles of DC Motors

There are three major classes of brush type permanent magnet DC (PMDC) motors:

1. Iron core armature
2. Printed-disk armature
3. Shell-armature DC motors

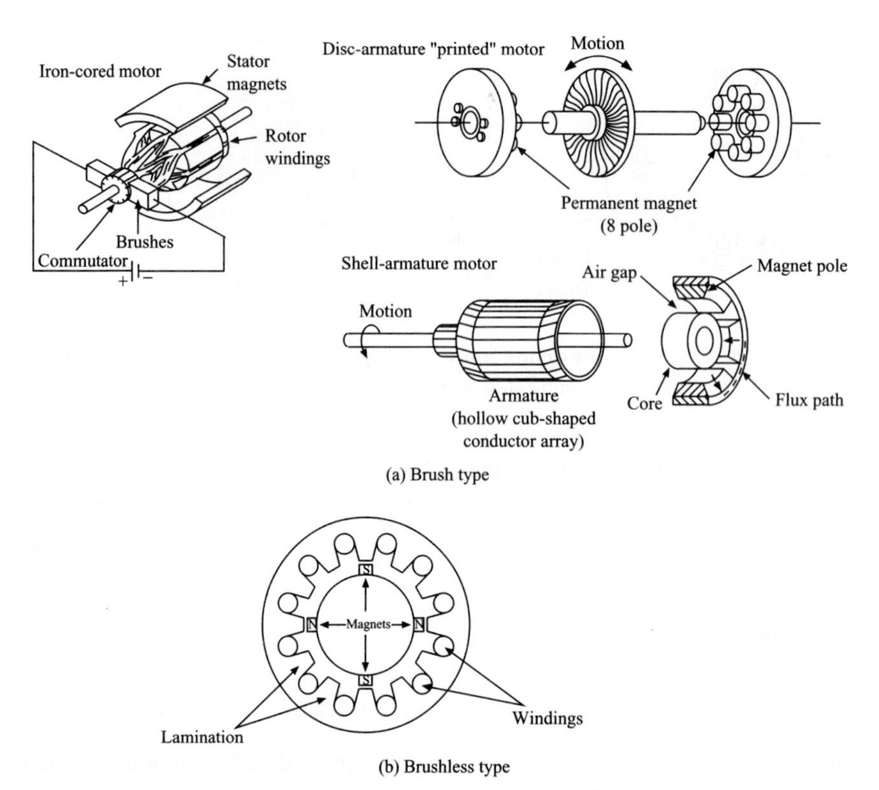


**FIGURE 8.22:** Permanent magnet DC motor types: (a) brush types (iron-core, disc-armature, shell-armature) and (b) brushless type.

Iron core armature PMDC motor is the standard DC motor where stator has the permanent magnet and rotor has the wound conductors [Figs. 8.22(a), 8.23, and 8.24]. The other two types are developed for applications which require very large torque to inertia ratio, hence the ability to accelerate and decelerate very fast (Fig. 8.22). The use of the printed-disk and shell-armature type motors have significantly reduced in recent years since comparable or better torque-to-inertia ratios can be achieved by low inertia brushless DC motors with better reliability [Figs. 8.22(b) and 8.25].

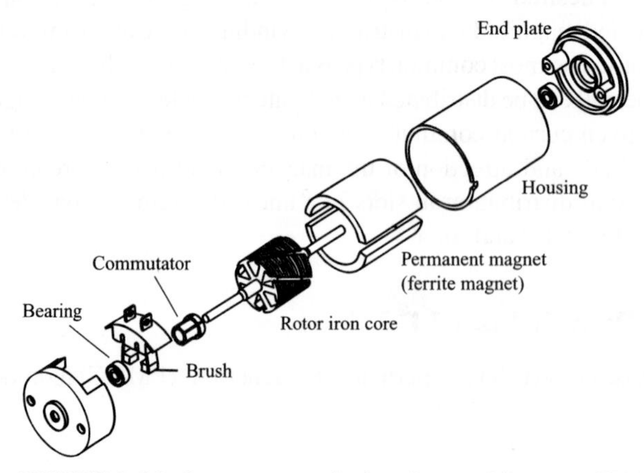


**FIGURE 8.23:** Components of a brush-type DC motor: Rotor (typically made of laminated sheets of steel), stator, commutator, brush pair, housing, end and face plates, and two bearings.

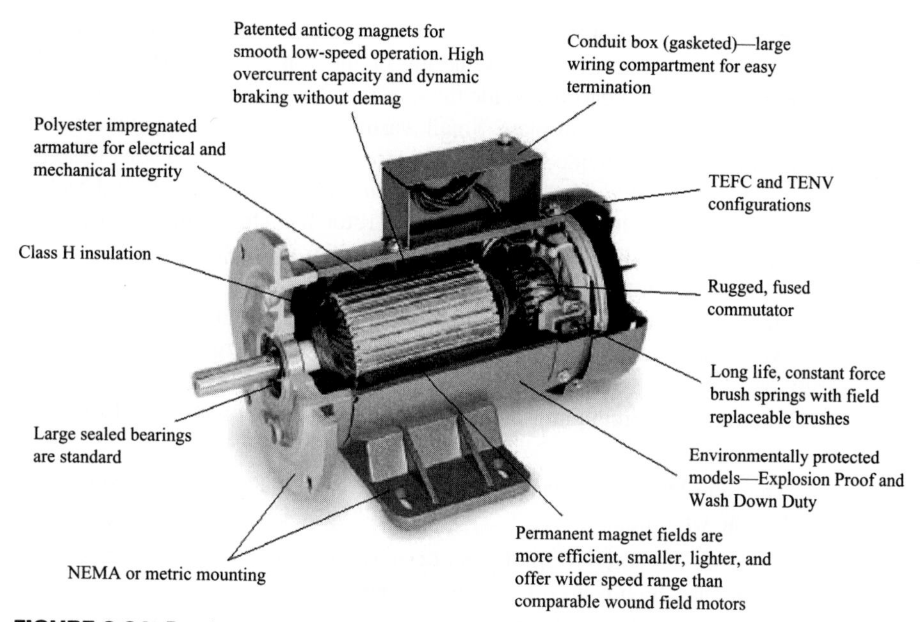


**FIGURE 8.24:** Brush-type iron core armature permanent magnet DC motor assembly.

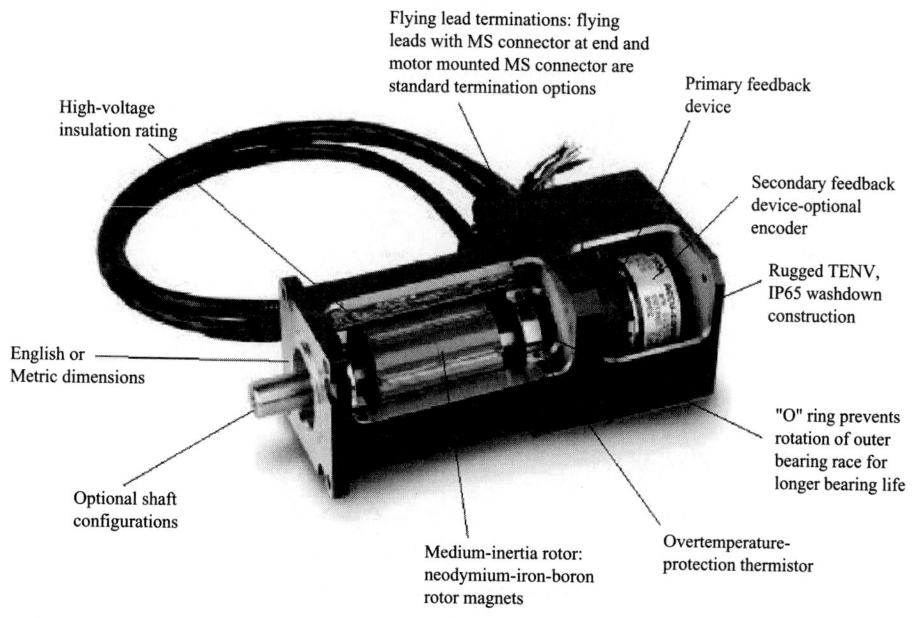


**FIGURE 8.25:** Brushless DC motor cross-sectional view showing the permanent magnet rotor, stator winding, and position sensor. Rotor has the permanent magnets glued on its periphery. In high-speed and/or high-temperature applications, a steel sleve may be fitted over the magnets to hold them in place securely. Rotor may be manufactured from laminations fitted onto the solid shaft. Stator is made of laminations and houses the windings.

Brush type DC motors have the permanent magnet as the stator (typically two-pole or four-pole configuration) and the windings on the iron core rotor (Fig. 8.23). The rotor is supported by two ball bearings inside the housing. The ends of the housing are covered by the end plate and face mount plate. Small washers are placed between the bearing and the end plates in order to provide space for the rotor and housing expansion due to temperature variations. The rotor is typically made of laminated iron sheet metal, and each lamination is insulated from each other. The laminated rotor has slots which house the windings. Core surface and windings are electrically insulated from each other by insulation material, coating on the core, or insulating thin sheet of paper. The slots may be skewed on the perimeter in order to reduce the torque ripple at the expense of lower maximum torque.

Almost identical mechanical components exist in the brushless DC motors with three exceptions:

1. There are no commutator or brushes since commutation is done electronically by the drive.
2. The rotor has the permanent magnets glued to the surface of the rotor, and the stator has the winding.
3. The rotor has some form of position sensor (i.e., hall effect sensors or encoder most common) which is used for current commutation.

In order to understand the operating principle of a permanent magnet DC (PMDC) motor, let us review the basics of electromagnetism (Fig. 8.26). A current-carrying conductor establishes a magnetic field around it. The electromagnetic field strength is proportional to the current magnitude, and the direction depends on the current direction based on the right-hand rule. The magnetic field shape can be changed by changing the physical shape of the current-carrying conductor, i.e., by forming loops of the conductor as in the case of a solenoid winding. When current passes through the winding of a solenoid, the magnetic

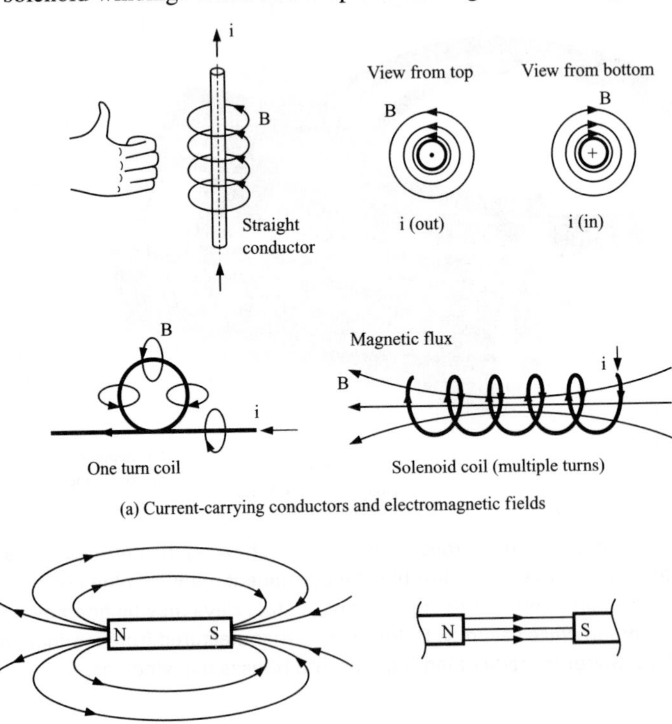


**FIGURE 8.26:** Basic principles of electromagnetism: (a) a current-carrying conductor generates a magnetic field around it; (b) the magnetic field generated by permanent magnets.

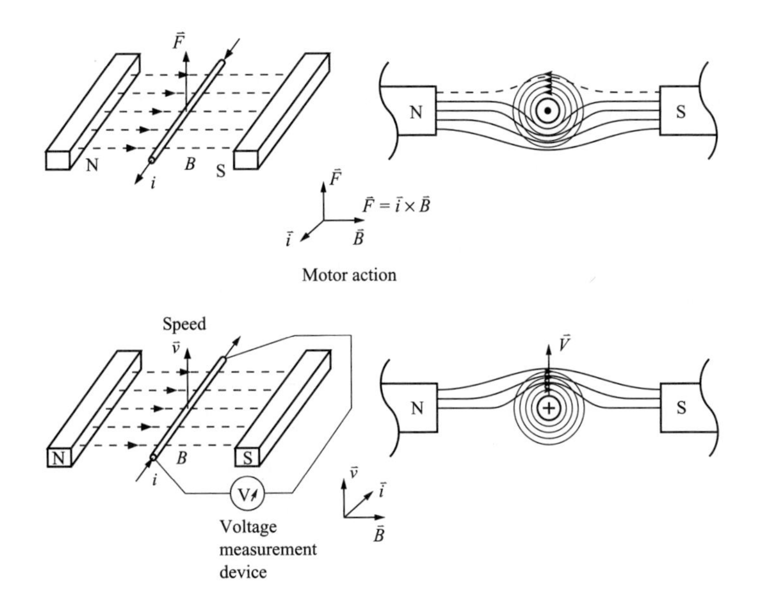


**FIGURE 8.27:** DC motor operating principles: A current-carrying conductor in a magnetic field.

field inside the coil is concentrated in one direction, which in turn temporarily magnetizes and pulls the iron core of the solenoid. This is an example of electromechanical power conversion for linear motion.

Let us consider that a current-carrying conductor is placed inside a magnetic field established by two permanent magnet poles (or the magnetic field can be established by a field-winding current in the case of field-wound DC motors). Depending on the direction of the current flow, a force is generated on the conductor as a result of the interaction between the "stator magnetic field" and the "rotor magnetic field" [Figs. 8.27(a) and 8.28].

$$\vec{F} = l\,\vec{i} \times \vec{\mathbf{B}} \tag{8.216}$$

Next, let us consider that we place a loop of conductor into the magnetic field and feed DC current to it using a pair of brushes (Fig. 8.28). Since the current directions in the two opposite sides of the conductor are in opposite direction, the exerted force on each leg of the conductor loop is in opposite directions. The force pair creates a torque on the conductor.

$$T_m = F \cdot d \tag{8.217}$$

Considering the fact that $\vec{B}$, $l$, $d$ are constant, we can deduce that

$$T_m = K_t \cdot i \tag{8.218}$$

where $K_t(B, l, d)$, the torque constant, is a function of the magnetic field strength and size of the motor. For a practical motor, the conductor loop would contain multiple turns, not just one pair. As a result, the $K_t$ constant is also a function of the number of conductor turns ($n$) or equivalently the surface area ($A_c$) over which flux density acts on the conductors, $K_t(B, l, d, n) = K_t(B, l, d, A_c)$. This is the main operating principle of a DC motor where electrical power is converted to mechanical power. This is called the *motor action*. The current in the coil is controlled by controlling the terminal voltage and is effected by the resistive, inductive, and back emf voltages. The electrical circuit relationship is

$$V_t(t) = R \cdot i(t) + \frac{d\lambda(t)}{dt} \tag{8.219}$$

$$= R \cdot i(t) + L\frac{di(t)}{dt} + K_e \cdot \dot{\theta}(t) \tag{8.220}$$

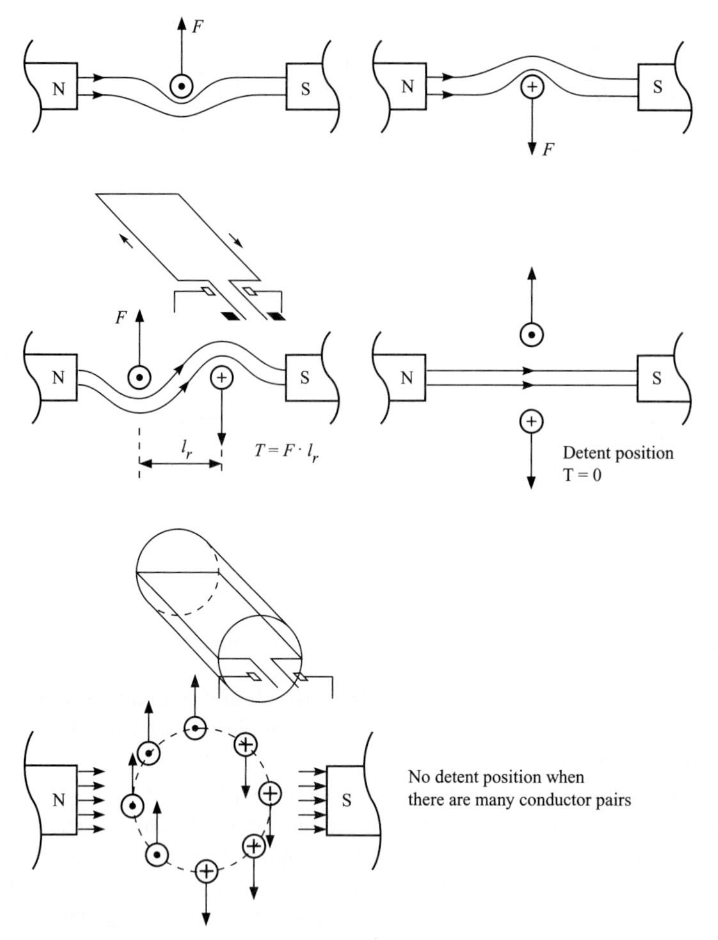


**FIGURE 8.28:** DC motor operating principles: Obtaining a continuous torque by winding a coil around a rotor.

where the $d\lambda(t)/dt$ is the induced voltage as a result of Faraday's law of induction. It has two components: the first one is due to the self-inductance of coils, and the second one is due to the generator action of the motor.

Notice that when the rotor turns 90 degrees, the moment arm between the forces is zero and no torque is generated even though each leg of the conductor has the same force. In order to provide a constant torque independent of the rotor position, for a given magnetic field strength and current, multiple rotor conductors are evenly distributed over the rotor armature. In order to switch the current direction for a continuous torque direction, a pair of brushes and commutators are used. Without the current switching commutation, the motor would only oscillate as the torque direction would oscillate between clockwise and counterclockwise direction for every 180 degree rotation. Consider that the line connecting to the brushes and the coils above and below that. At any given position, current in one half of the coil is in opposite direction to the current in the other half of the coils.

Also shown in Fig. 8.27 is the *generator action* of the same device. This is the result of the Faraday's induction principle which states that when a conductor is moved in a magnetic field, a voltage is induced across it in proportion to the speed of motion and the strength of the magnetic field. This is represented by the $K_e\dot{\theta}(t)$ term in equation 8.220. Therefore, both the motor and generator actions are at work at the same time during the operation of a DC motor.

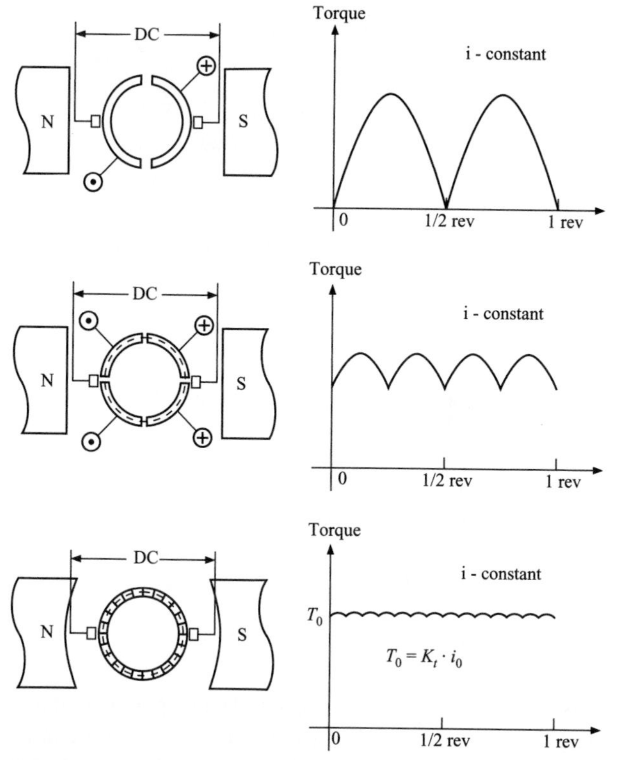


**FIGURE 8.29:** Commutation and torque variation as a function of angular position of the rotor. Torque ripple magnitude and frequency are a function of the number of commutation segments. At any given time, coils on one side of the line between the brushes (and the coils on the other side) have current in the opposite directions. Contribution of each coil to torque production under a constant current depends on its angular position relative to the magnetic field of the permanent magnets at that instant.

Figure 8.29 shows the brush and commutator arrangement and torque as a function of rotor position for different number of commutator segments. Ideally, the larger the number of commutators, the smaller the torque ripple. However, there is a practical limit on how small the brush-commutator assembly can be sectioned. If we neglect the torque ripple due to commutation resolution, torque is proportional to the armature current for a given permanent magnet field and independent of the rotor angular position. In a PMDC motor, the magnetic field strength is fixed, and current is controlled by a drive (the term *drive* is used to describe the amplifier and power supply components together as one component).

The brushless permanent magnet DC (BPMDC) motor is basically an "inside-out" version of the brush-type PMDC motor (Fig. 8.30). The rotor has the permanent magnets, and the stator has the conductor windings, usually in three electrically independent phases. The stator winding of brushless servo motors is similar to the stator winding of traditional induction motors and lends itself to the same well-established winding processes used in manufacturing induction motors.

The operating goal is the same: maintain the field (stator) and armature (rotor) magnetic fields perpendicular to each other at all times. If this can be accomplished, the electromechanical power conversion relationship and torque generation in a BPMDC motor would be identical to that of the brush-type PMDC motor. Of course, the difference is in the commutation (Fig. 8.30). In the brush-type motor, the magnetic flux generated by permanent

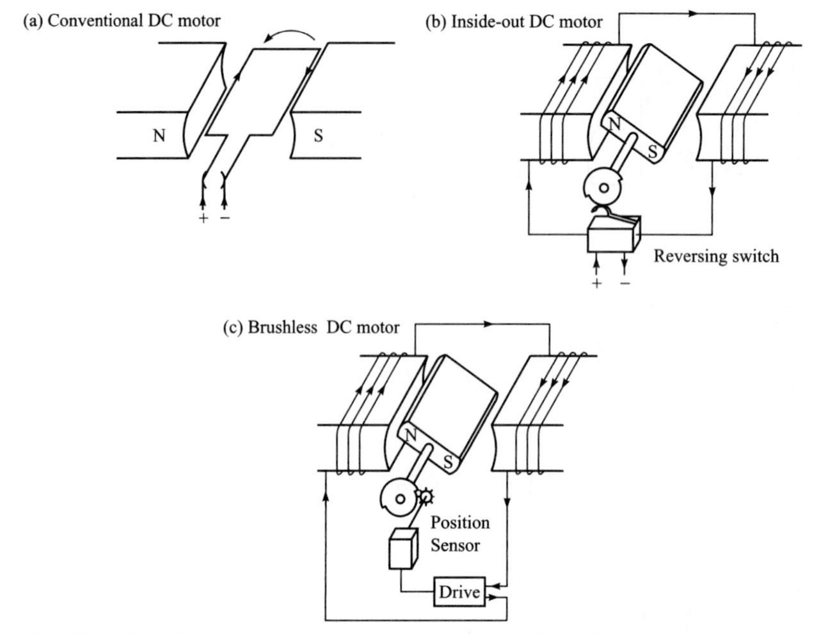


**FIGURE 8.30:** DC motor types: Brush type and brushless DC motors.

magnets (or electromagnets) of the stator is fixed in space. The magnetic field generated by armature is also maintained fixed in space by the mechanical brush-commutator assembly and perpendicular to that of the stator. In the case of brushless PMDC motor, we have the same objective. However, the field magnetics is established by the rotor and it rotates in space with the rotor. Therefore, the stator winding current has to be controlled as a function of rotor position so as to keep the stator generated magnetic field always perpendicular relative to the magnetic field of the rotor, although both rotate in space with the rotor speed. In other words, not only the magnitude of current in the stator winding, but also the vector direction of it must be controlled. This switching of the currents into the stator windings is achieved by controlling solid-state power transistors as a function of rotor position. Hence, the brushless motor requires a rotor position sensor for its power stage.

## 8.3.2 Drives for DC Brush-Type and Brushless Motors

Drive is considered as the power amplification stage of an electric motor. It is the drive that defines the performance of an electric motor. The most common type of power stage amplifier used for DC brush-type motors is an H-bridge amplifier (Fig. 8.31). The H-bridge uses four power transistors. When controlled in pairs (Q1 & Q4 and Q2 & Q3), it changes the direction of the current, hence the direction of generated torque. Notice that the pair Q1 & Q3 or the pair Q2 & Q4 should never be turned ON at the same time because it would form a short-circuit path between supply and ground. The diodes across each transistor serve the purpose of suppressing voltage spikes and provide a freewheeling path for the current to follow. Large voltage spikes occur across the transistor in the reverse direction due to the inductance of the coils. If a current flow path is not provided, the transistors may be damaged. The diodes provide the alternative current path for inductive loads and lets current pass through the coil. When the diode is ON, the current flows from negative end of the power supply to positive end due to the fact that the inductive voltage raises the negative side of the potential temporarily (transient) higher than the positive side. This current is

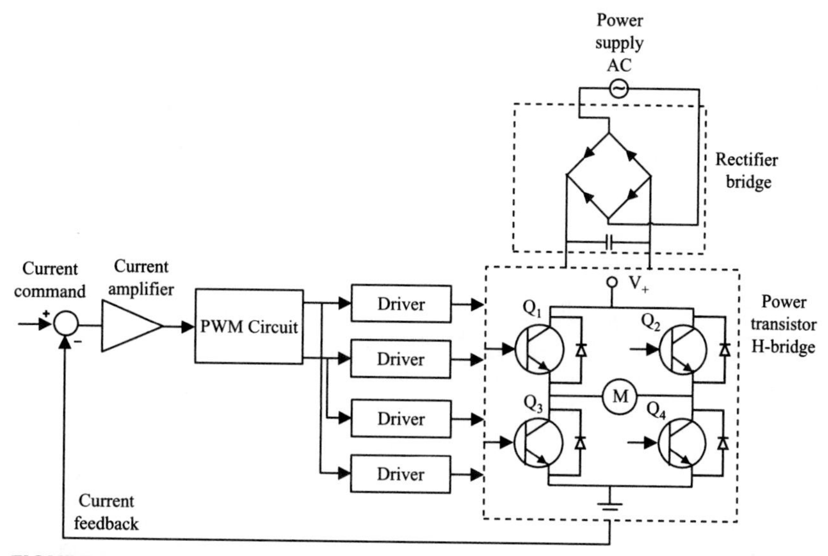


**FIGURE 8.31:** Block diagram of the brush type DC motor drive: PWM amplifier with current feedback control.

charged to a capacitor on the drive. Capacitor must be sized properly to handle this charge. The use of diodes in all power amplifiers for different motor types serves the same purpose.

Let us assume transistors $Q_1$ & $Q_4$ pair was ON and current flow from left to right thru the motor. When $Q_1$ & $Q_2$ pair are suddenly turned OFF, diodes across $Q_2$ & $Q_3$ pair provides a current flow path. The opposite occurs between the transistors $Q_2$ & $Q_3$ and diodes of $Q_1$ & $Q_4$. By controlling the current magnitude through the power transistors, the magnitude of the torque is controlled. In very small size motors (fractional horsepower), linearly operated transistor amplifiers are used. The pulse width modulation (PWM) circuit operates the transistors in all ON or all OFF mode in order to increase the efficiency. The linear amplifiers provide lower noise but are less efficient than the PWM amplifiers.

The PWM circuit converts an analog input signal (i.e., amplified error signal from the current loop which is an analog signal) to a fixed frequency but variable pulse width signal. By modulating the ON-OFF time of the pulse width at a high switching frequency, hence the name "pulse width modulation," a desired average voltage can be controlled. The PWM circuit (Fig. 8.32) uses a triangular carrier signal with high frequency (also called

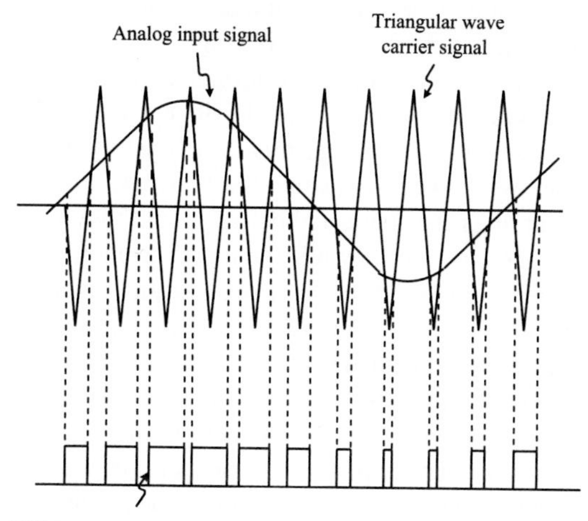


**FIGURE 8.32:** PWM circuit function: Carrier signal is a high-frequency triangular signal. The input signal is an analog signal value. The output pulse has fixed frequency, which is the carrier frequency. The ON/OFF pulse width is varied as a function of the value of the input signal relative to the carrier signal.

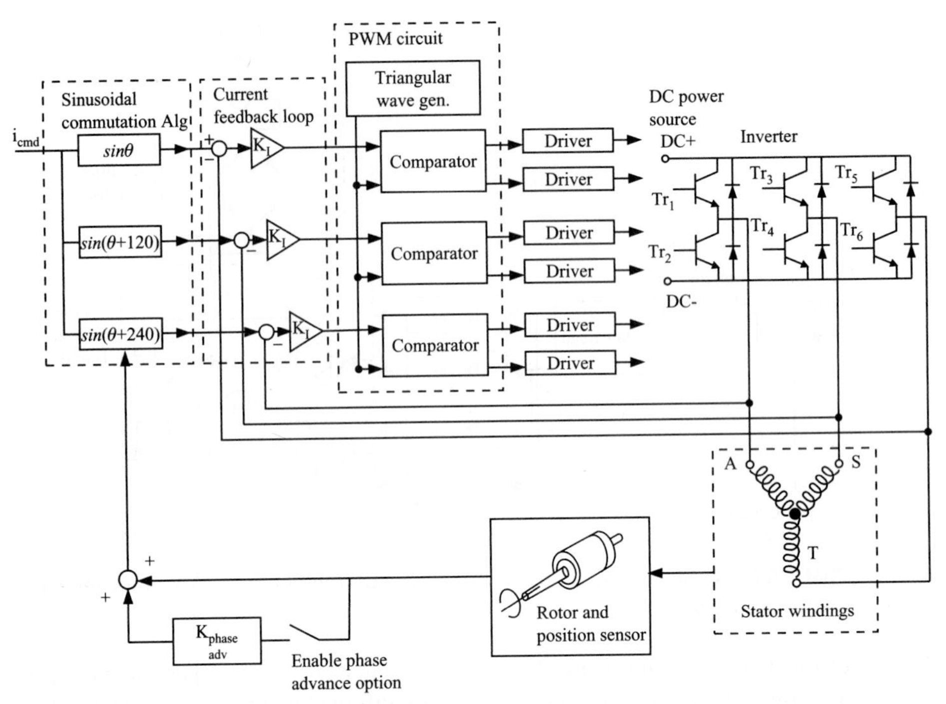


**FIGURE 8.33:** Brushless servo motor drive block diagram.

the switching frequency). When the analog input signal is larger than the carrier signal, the pulse output is ON, when it is smaller, the pulse output is OFF. PWM is another way of transmitting an analog signal which takes on a value between a minimum value and a maximum value, i.e., a value between 0 VDC and 10 VDC or −10 VDC to +10 VDC. Instead of transmitting the signal by an analog voltage level, PWM transmits the information as a percentage of the ON cycle of a fixed frequency signal. When the signal value is to have the minimum value, duty cycle (ON cycle) of the signal is set to zero percent. When the signal value is to have the maximum value, the duty cycle (ON cycle) is set to 100 percent. In other words, the analog value of the signal is conveyed as the percentage of duty cycle of the signal. The PWM switching frequency must be at least an order of magnitude (10 times or more) larger than the current loop bandwidth in order to make sure that the switching frequency does not have any significant adverse effect in the closed-loop bandwidth. The switching frequency of typical drives are in the 2- to 20-kHz range.

Figure 8.33 shows the block diagram of a current-controlled drive for a three-phase brushless DC motor. The switch set is based on the familiar H-bridge, but uses three bridge legs instead of two legs as is the case for an H-bridge drive for brush type DC motors. Since each leg of the H-bridge has two power transistors, the brushless motor drive has six power transistors. The stator windings are connected between the three bridge legs as shown in Fig. 8.33. The so-called Y-connection shown is the most common type of phase winding connection, while Δ-connection is used in rare cases. At any given time, three of the transistors are ON and three of them are OFF. Furthermore, two of the windings are connected between the DC bus voltage potential and have current passing through them in positive or negative direction, whereas the third winding terminals are both connected to the same voltage potential (either $V_{DC}$ or 0 V) and act as the balance circuit. The combination of the ON/OFF transistors determines the current pattern on the stator, hence the flux field vector generated by the stator. In order to generate the maximum torque per unit current, the

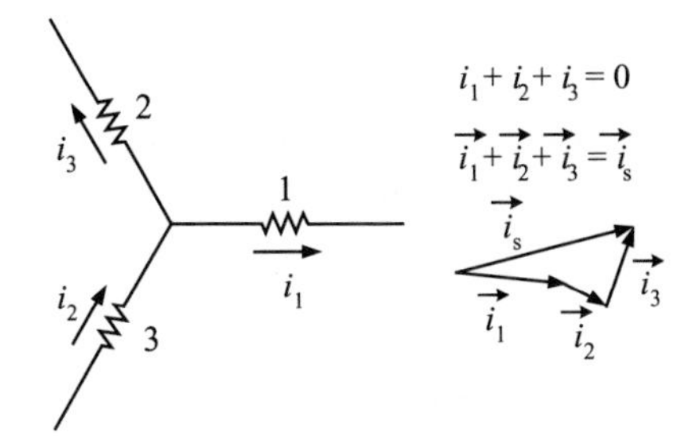


**FIGURE 8.34:** Brushless motor current vector and phase current components (3 phase Y wound).

objective is to keep the stator's magnetic field perpendicular to that of the rotor. By controlling the phase currents in the stator phase windings, we control the stators magnetic field (magnitude and direction, a vector quantity). Therefore, the torque direction and magnitude can be controlled by controlling the stator's magnetic field relative to that of the rotor.

There are two types of brushless drives based on the commutation algorithm:

1. Sinusiodal commutation
2. Trapezodial commutation

If the winding distribution and the effective magnetic circuit of the motor are such that the back EMF function is a sinusoidal function of rotor angle, such a motor should be controlled using a drive that uses sinusoidal commutation algorithm. Similarly, if the motor back EMF is trapezoidal type, the drive should be the type which uses trapezoidal current commutation. The sinusoidal commutation drive provides the best rotational uniformity at any speed or torque. The primary difference between the two types of drives is a more complex control algorithm. For best performance, the commutation method of the drive is matched to the back EMF type of the motor. The back EMF of the motor is determined primarily by its winding distribution, lamination profile, and magnets. The feedback sensor and power electronics components remain the same. In recent years, as the cost of high-performance digital signal processors has fallen, the use of the sinusoidal commutation brushless drive has exceeded that of all others.

The current in each of the three phases is controlled by 120 degrees phase shift relative to each other. The rotor position is tracked by a position sensor. The vector sum of phase currents is commutated relative to the rotor position such that two magnetic fields are always perpendicular to each other. Algebraic sum of the currents in three phases is zero, but not the vector sum (Fig. 8.34). A current feedback loop is used in the PWM circuit to regulate the current in each phase with sufficient dynamic bandwidth. Notice that the magnitude of magnetic field generated by the permanent magnets is constant and not actively controlled and it rotates with the rotor, whereas the stator field (a vector quantity which has magnitude and direction) is controlled actively by the drive current control loop.

In digital implementation, the current command, the commutation and current control algorithms can reside either in the drive or in a higher level controller. In the latter case, the drive performs only the PWM modulation and is called "the power block." In most cases, current commutation and regulation algorithms are implemented in the drive.

The stator being static by definition, the angle of each stator phase is fixed on the diagram (Fig. 8.34). The magnitude of each phase current ($i_a$, $i_b$, $i_c$) for a given total current vector ($\vec{i}_s$) is the projection of the current vector onto the phase (Fig. 8.34).

$$i_n = \vec{i}_s \cdot \vec{u}_n \tag{8.221}$$

$\vec{u}_n$ is the unit vector for the phase, $\vec{u}_a$ for phase a, $\vec{u}_b$ for phase b, and $\vec{u}_c$ for phase c.

The derivation below shows that the sinusoidal commutation algorithm for a three-phase brushless motor (which is designed to have a sinusoidal back emf) produces the same torque–current relationship as that of a brush-type DC motor. Let us assume that the brushless motor has three-phase winding and each phase has a sinudoidal back EMF as a function of rotor position. As a result, the current-to-torque gain for each individual phase has the same sinusoidal function. For each phase, they are displaced from each other by a 120 degree angle as a result of the physical distribution of the windings around the periphery of the stator.

Consider that the rotor is at angular position, $\theta$, and each phase of the stator has current values $i_a$, $i_b$, $i_c$. The torque generated by each winding is $T_a$, $T_b$, and $T_c$.

$$T_a = i_a \cdot K_T^* \cdot sin(\theta) \tag{8.222}$$

$$T_b = i_b \cdot K_T^* \cdot sin(\theta + 120°) \tag{8.223}$$

$$T_c = i_c \cdot K_T^* \cdot sin(\theta + 240°) \tag{8.224}$$

Let us control the current (with commutation and current feedback control algorithms) such that each phase is 120 degrees apart from each other and sinusoidaly modulated as a function of the rotor position.

$$i_a = i \cdot sin(\theta) \tag{8.225}$$

$$i_b = i \cdot sin(\theta + 120°) \tag{8.226}$$

$$i_c = i \cdot sin(\theta + 240°) \tag{8.227}$$

The total torque developed as a result of the contribution from each phase is

$$T_m = T_a + T_b + T_c \tag{8.228}$$

$$= K_T^* \cdot i \cdot (sin^2\theta + sin^2(\theta + 120°) + sin^2(\theta + 240°)) \tag{8.229}$$

Note the trigonometric relation,

$$(sin(\theta + 120°))^2 = (cos\,\theta\; sin120° + sin\,\theta\; cos120°)^2 \tag{8.230}$$

$$= \left(\frac{\sqrt{3}}{2} cos\,\theta - \frac{1}{2} sin\,\theta\right)^2 \tag{8.231}$$

$$= \frac{3}{4} cos^2\theta + \frac{1}{4} sin^2\theta - \frac{\sqrt{3}}{2} sin\,\theta\; cos\,\theta \tag{8.232}$$

$$(sin(\theta + 240°))^2 = \frac{3}{4} cos^2\theta + \frac{1}{4} sin^2\theta + \frac{\sqrt{3}}{2} sin\,\theta\; cos\theta \tag{8.233}$$

$$T_m = K_T^* \cdot i \cdot \frac{3}{2}(sin^2\theta + cos^2\theta) \tag{8.234}$$

Hence, the torque is a linear function of the current, independent of the rotor angular position, and the linearity constant torque gain ($K_T$) is a function of the magnetic field strength.

$$T_m = K_T \cdot i \tag{8.235}$$

where

$$K_T = K_T^* \cdot \frac{3}{2} \tag{8.236}$$

Therefore, a sinusoidally commutated brushless DC motor has the same linear relationship between current and torque as does the brush type DC motor. Most implementations make use of the fact that the algebraic sum of three phase currents is zero. Therefore, only two of the phase current commands and current feedback measurements are implemented. Third-phase information for both the command and feedback signal is obtained from the algebraic relationship (Figs. 8.33 and 8.34).

Actual back EMF of a real motor is never perfectly sinusoidal nor trapezodal. The ultimate goal in commutation is to maintain a current-torque gain that is independent of the rotor position, that is, a constant torque gain. In order to achieve that, the current commutation algorithm must be matched to the back EMF function of a particular motor. Clearly, if a trapezoidal current commutation algroithm is used with a motor which has a sinusoidal back EMF function, the current-to-torque gain will not be constant. The resulting motor is likely to have large torque ripple.

Let us consider the effect of commutation angle error on the current-torque relationship of the motor. Let $\theta_e$ be the measurement error in rotor angle. That is, $\theta_a$ is the actual motor angle, and $\theta_m$ is the measured angle that is used for commutation algorithm, $\theta_e = \theta_a - \theta_m$. It can be shown that the current–torque relationship under nonzero commutation angle error conditions is

$$T_m = K_T \cdot i \cdot cos(\theta_e) \tag{8.237}$$

Notice that when:

1. $\theta_e = 0$, we have the ideal condition.
2. $\theta_e = 90°$, then there is no torque generated by the current.
3. $-90 \leq \theta_e \leq 90$, effective torque constant is less than ideal.
4. $90 \leq \theta_e \leq 270$, effective torque constant is less than ideal and negative. If the normal position feedback and command polarity is used in the closed position control loop, the motor will run away. In other words, the closed-loop position feedback control of the motor would be unstable.

At the most inner loop, voltage fed into each phase is controlled by the PWM amplifier circuit in such a way that the current in each phase follows the commanded current. The dynamic response lag between the voltage modulation and current response in the stator is small but finite. This lag becomes an important factor at high speeds because

$$\phi_{lag} = tan^{-1}(\tau_e \cdot \omega_m) \tag{8.238}$$

where $\phi_{lag}$ is the phase lag, $\tau_e$ is electrical time constant of the current loop, and $\omega_m$ is the rotor speed. As the speed of the motor increases, the phase lag of the current control loop can become significant. As a result, effective angle between the field and armature magnetic fields will not be 90 degrees. Therefore, the motor will be producing torque at lower efficiency. Keeping this in mind, if the time constant of the current loop ($\tau_e \simeq L/R$) is known approximately or estimated in real time, the commanded current can be calculated to make not a 90-degree phase with the rotor magnetic field vector but 90 degrees plus the anticipated phase lag. In other words, anticipating the phase lag, we can feed the command signal with a phase lead to cancel out the phase lag due to current regulation loop. This is called *phase advancing* in the brushless drive commutation algorithm. This can be accomplished in real time by modifying the rotor position sensor signal as shown in Fig. 8.33.

Finally, the brushless commutation algorithm requires the absolute position measurement of the rotor within one revolution. This is needed to initialize the commutation algorithm on power-up. On power-up, incremental position sensors (i.e., incremental optical encoder) do not provide this information, whereas resolvers and absolute encoders do. Over 70% of the brushless motors are used with incremental type encoders. Therefore, on power-up a *phase finding* algorithm is needed to establish the absolute position information when an incremental position sensor is used as the position feedback device.

***Example: Drive Sizing*** Motor and drive (amplifier and power supply) sizes must be matched in a well-designed system. Let us assume that we have a motor size selected. It is characterized by its torque capacity (peak and RMS: $T_{max}$, $T_{rms}$), current-to-torque gain (which is the same as back emf gain: $K_T = K_E$), winding resistance ($R$), and maximum operating speed ($w_{max}$). Let us determine the required drive size.

The drive size determination means the determination of DC bus voltage ($V_{DC,max}$) and current ($i_{max}$, $i_{rms}$) that the drive must supply. As a result, the maximum and RMS current requirements are calculated from

$$i_{max} = T_{max}/K_T \tag{8.239}$$

$$i_{rms} = T_{rms}/K_T \tag{8.240}$$

The maximum DC bus voltage required at worst condition, that is, when providing maximum torque and running at maximum speed and neglecting the transient inductance effects,

$$V_{DC} = L \cdot \frac{di(t)}{dt} + R \cdot i(t) + K_E \cdot w(t) \tag{8.241}$$

$$V_{DC,max} \approx R \cdot i_{max} + K_E \cdot w_{max} \tag{8.242}$$

where the $L \cdot di/dt$ term in the electrical model of the motor is neglected. Then, a drive should be selected that can provide the required DC bus voltage and current requirements with some safety margin. Notice that at a given torque capacity, the *head room of supply voltage* available limits the maximum speed capacity the drive can support:

$$V_{head\text{-}room} = V_{DC,max} - R \cdot i_{max} \tag{8.243}$$

$$= K_E \cdot w_{max} \tag{8.244}$$

$$w_{max} = \frac{(V_{DC,max} - R \cdot i_{max})}{K_E} \tag{8.245}$$

Top speed is limited by the back EMF of the motor. At any given torque level $T_r$, the current required to generate that torque is $i_r = T_r/K_T$. This means that the $R \cdot i_r$ portion of the available bus voltage is used up to generate the current needed for torque. The remaining voltage $V_{DC,max} - R \cdot i_r$ is available to balance the back emf voltage. Hence, the maximum speed at a given torque output is limited by the available "head-room voltage."

***Example: Brush-Type DC Motor*** Consider a brush-type DC motor with stator coil resistance of 0.25 Ω at nominal operating temperature. The DC power supply is 24 VDC. The back EMF constant of the motor is 15 V/krpm. The voltage to the motor is turned ON and OFF by an electromechanical relay. Consider two cases: (a) motor shaft is locked and not allowed to rotate, and (b) motor speed is nominally at 1200 rpm. Calculate the steady-state current developed in the motor when the relay is turned on for both cases.

When the motor is locked and not allowed to rotate, there is not any back EMF voltage as a result of the generator action of the motor. This is the stall condition. If we neglect the transient effect of inductance, the electrical equations for the motor gives

$$V(t) = R \cdot i(t) + L\frac{di(t)}{dt} + K_E \cdot w \tag{8.246}$$

$$\approx R \cdot i(t) \tag{8.247}$$

$$i = \frac{24 \text{ V}}{0.25 \ \Omega} \tag{8.248}$$

$$= 96 \text{ A} \tag{8.249}$$

When the motor speed is nonzero, the terminal voltage minus the back EMF is available to develop current. Then,

$$V(t) = R \cdot i(t) + L\frac{di(t)}{dt} + K_E \cdot w \tag{8.250}$$

$$\approx R \cdot i(t) + K_E \cdot u \tag{8.251}$$

$$24\text{ V} = 0.25 \cdot i + 15/1000 \cdot 1200 \tag{8.252}$$

$$i = \frac{24 - 18}{0.25} \tag{8.253}$$

$$= 24\text{ A} \tag{8.254}$$

Notice that under no-load conditions ($i = 0$), the maximum speed is limited by the back EMF and it is

$$V(t) = R \cdot i(t) + L\frac{di(t)}{dt} + K_E \cdot w \tag{8.255}$$

$$24 \approx R \cdot 0 + 15/1000 \cdot w_{max} \tag{8.256}$$

$$w_{max} = \frac{24 * 1000}{15} \tag{8.257}$$

$$= 1600\,\text{rpm} \tag{8.258}$$

In SI units, $K_T = K_E$, whereas in CGS units, $K_T$ [Nm/A] $= 9.5493 \times 10^{-3}$ $K_E$ [V/krpm]. Then we can find the maximum torque developed at stall and at 1200-rpm speed.

$$T_{stall} = K_T \cdot i \tag{8.259}$$

$$= 9.5493 \times 10^{-3} \times 15 \times 96\,[\text{Nm}] \tag{8.260}$$

$$= 13.75\,[\text{Nm}] \quad at \quad stall \tag{8.261}$$

$$T_r = 9.5493 \times 10^{-3} \times 15 \times 24\,[\text{Nm}] \tag{8.262}$$

$$= 3.43\,[\text{Nm}] \quad at \quad 1200\,\text{rpm} \tag{8.263}$$

***Example: PWM Control H-Bride IC Drive for Brush-Type DC Motors—TPIC0107B*** The integrated circuit (IC) package TPIC0107B by Texas Instruments implements an H-bridge and switching control logic for brush-type DC motors (Fig. 8.35). The drive supply voltage ($V_{cc}$) must be in the 27-VDC to 36-VDC range, and it can support up to 3-A continuous bridge output current. The two terminals of the DC motor are connected between the OUT1 and OUT2 ports. DC supply voltage and ground are connected to the $V_{cc}$ and GND terminals.

Logic voltage is internally derived from the $V_{cc}$. The operation of the TPIC0107B is controlled by two input pins: DIR (IN1) and PWM (IN2). The PWM pin should be connected to the PWM output port of a microcontroller, whereas DIR pin can be connected to any digital output. The PWM switching frequency is 2 kHz. The state of OUT1 and OUT2 (H-bridge output) follows the signal in the PWM pin. The actual PWM signal needs to be formed by the PWM port of the microcontroller. For instance, current control loop must be implemented in the microcontroller. The IC is capable of sensing over-voltage, under-voltage, short circuit, over-current, under-current, and over temperature conditions, and if necessary, the IC shuts down bridge output and sets the status output pins to indicate error code (STATUS1 and STATUS2 pins).

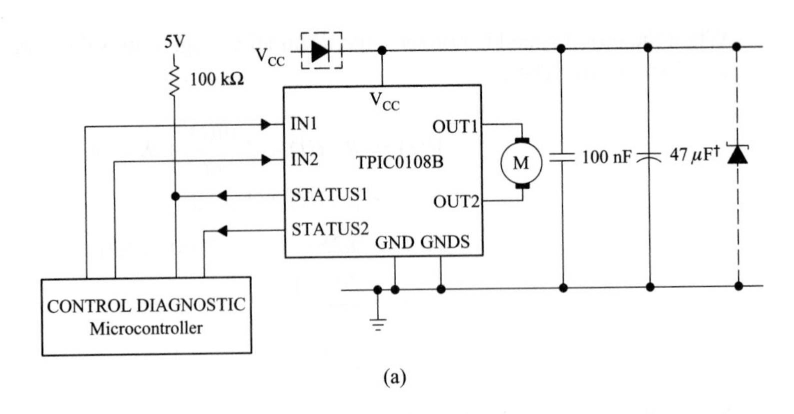


(a)

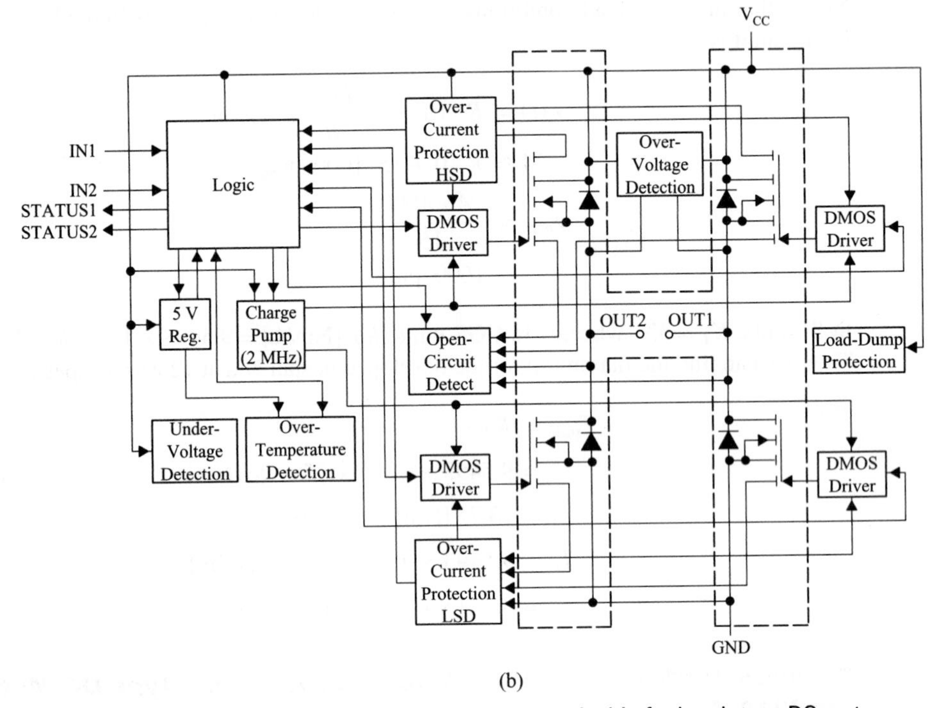


(b)

**FIGURE 8.35:** PWM controlled H-bridge integrated circuit chip for brush-type DC motors: TPIC01017B by Texas Instruments.

***Example: PM DC Motor*** Consider a permanent magnet DC motor. The armature resistance is measured to be $R_a = 0.5\ \Omega$. When $V_t = 120$ V is applied to the motor, it reaches 1200-rpm steady-state speed and draws 40-A current. Determine the back emf voltage, IR power losses, power delivered to the armature, and torque generated at that speed.

The basic relationship for the motor,

$$V_t(t) = L\frac{di(t)}{dt} + Ri(t) + V_{bemf} \tag{8.264}$$

$$120\ \text{V} = 0 + 0.5 \cdot 40\Omega\ A + V_{bemf} \tag{8.265}$$

$$V_{bemf} = 100\ \text{V} \tag{8.266}$$

where the transient effects of inductance is neglected in steady state. The IR power loss is

$$P_{IR} = R_a \cdot i^2 = 0.5 \cdot 40^2\ \Omega\ A^2 = 800\ \text{W} \tag{8.267}$$

and torque generated,

$$K_E = \frac{V_{bemf}}{\dot{\theta}} = \frac{100 \text{ V}}{1.2 \text{ krpm}} \tag{8.268}$$

$$K_T = 9.5493 \cdot 10^{-3} \cdot K_E[\text{Nm/A}] = 0.7958 \text{ [Nm/A]} \tag{8.269}$$

$$T = K_T \cdot i = 31.83 \text{ [Nm]} \tag{8.270}$$

and the electrical power converted to mechanical power,

$$P_m = T \cdot w \tag{8.271}$$

$$= 4000 \text{ W} \tag{8.272}$$

$$= P_e \tag{8.273}$$

$$P_e = V_{bemf} \cdot i_a \tag{8.274}$$

$$= 100 \text{ V} \cdot 40 \text{ A} = 4000 \text{ W} \tag{8.275}$$

$$\tag{8.276}$$

Because the motor speed is constant, the motor torque generated must be used by a total load toque of equal magnitude but in opposite direction.

## 8.4 AC INDUCTION MOTORS AND DRIVES

AC induction motors have been used in constant speed applications in huge quantities. They have been the work horse of the industrial world. In recent years, they are also used as closed-loop position servo motors with a sophisticated current commutation algorithm in the drive being the key to the increased capabilities. Three-phase AC motors are more common than single-phase AC motors in applications that require high efficiency and large power. The most common types of multiphase AC induction motors are as follows.

1. The squirrel-cage type AC induction motor is the most common type of AC motor. The stator has phase windings, and the rotor is a squirrel-cage type conductor (copper or aluminum conductor bars) placed in the rotor frame. The conductor bars are short-circuited with end rings. The rotor has no external electrical connection.
2. The wound-rotor AC induction motor differs from the squirrel-cage type in the construction of the rotor. The rotor has wound conductors. External electrical connections to the rotor winding are provided via slip rings.
3. Synchronous motors are used for constant speed applications. The motor design is such that the motor torque-speed curve in steady state provides a constant speed for a wide range of load torque. If the load torque exceeds a maximum value, motor speed decreases abruptly. Compared to the squirrel-cage motors, the synchronous motor speed shows much less variation as load torque varies.

Stator and rotor cores are made from thin steel disk laminations. The purpose of the laminated core is to reduce the Eddy current losses.

Conductors of the stator winding are covered with insulation material to protect them against high temperatures. Motor temperature raises primarily as a result of resistance losses, called IR or copper losses. There are four major insulation material classes: class A for temperatures up to 105°C, class B for temperatures up to 130°C, class F for temperatures up to 155°C, and class H for temperatures upto 180°C.

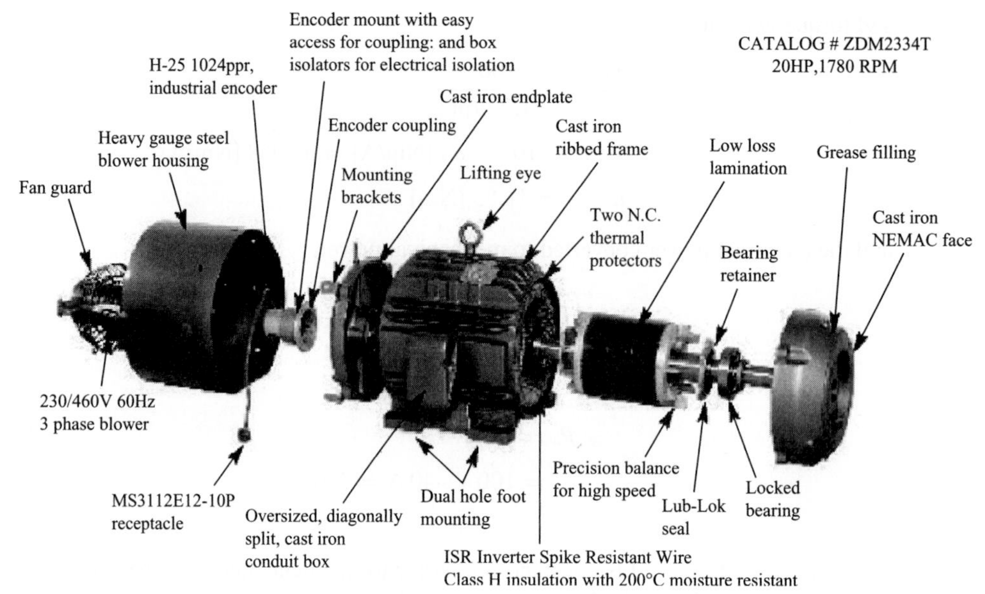


**FIGURE 8.36:** AC induction motor components and assembly view.

### 8.4.1 AC Induction Motor Operating Principles

The main components of an AC induction motor are the stator winding (single phase or three phase) and rotor with conductors (Fig. 8.36). The rotor of a squirrel-cage type AC induction motor is made of steel laminations with holes in its periphery. The holes are filled with conductors (copper or aluminum) and short-circuited to each other by conducting end rings (Fig. 8.37). The air-gap between the stator and rotor ranges from less than 1 mm (i.e., 0.25 mm) for motors up to 10 KW power to a few millimeters (i.e., 3 mm) for motors upto 100 KW (Fig. 8.36). The air-gap may be larger for motors which are designed for applications that have very large peak torque requirements.

The number of phases of the motor is determined by the number of independent windings connected to a separate AC line phase. Number of motor poles refers to the number of electromagnetic poles generated by the winding. Typical numbers of poles are $P = 2, 4$, or 8 (Fig. 8.38). The coil wire for each phase can be distributed over the periphery of the stator to shape the magnetic flux distribution.

The torque in any AC or DC electric motor is produced by the interaction of two magnetic fields, with one or both of these fields produced by electric currents. In an AC induction motor, the current in the stator generates a magnetic field which induces a current in the rotor conductors. This induction is a result of relative motion between stator magnetic

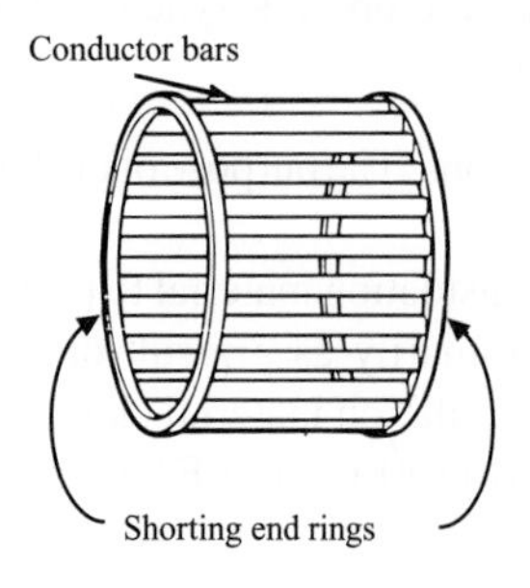


**FIGURE 8.37:** Rotor of a squirrel-cage type AC induction motor. The conductor bars are shorted together at both ends by two end rings.

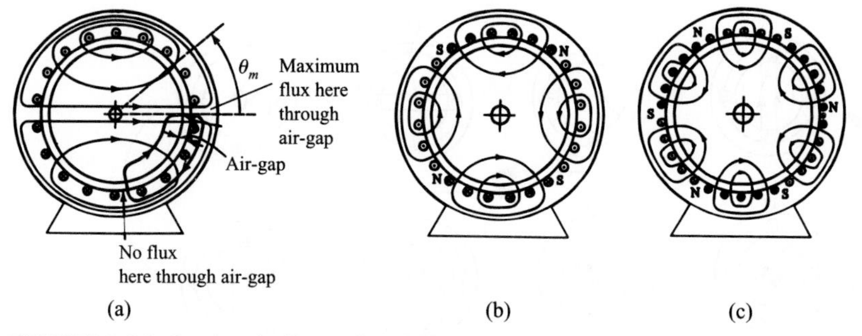


**FIGURE 8.38:** Stator windings of an AC induction motor: (a) two-pole ($P = 2$) configuration, (b) four-pole ($P = 4$) configuration, and (c) six-pole ($P = 6$) configuration.

field (rotating electrically due to AC current) and the rotor conductors (which is initially stationary). This is an example result of Faraday's induction law. Stator AC current sets up a rotating flux field. The changing magnetic field induces emf voltage, hence current, in the rotor conductors. The induced current in the rotor in turn generates its own magnetic field. The interaction of the two magnetic fields (the magnetic field of the rotor trying to keep up with the magnetic field of the stator) generates the torque on the rotor. When the rotor speed is identical to the electrical rotation speed of stator field, there is no induced voltage on the rotor, and hence the generated torque is zero. This is the main operating principle of an AC induction motor.

In order to draw a visual picture of how torque is generated, let us consider a two-phase AC induction motor (Fig. 8.39). The principle for other number of phases is similar. Let us consider the case where only phase 1 is energized and the current on the phase is a

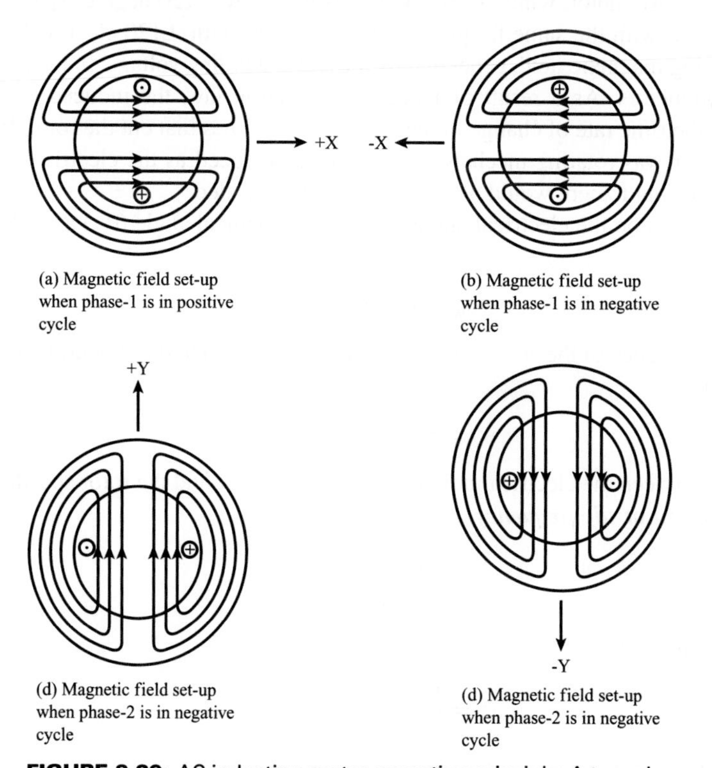

+X
-X
(a) Magnetic field set-up when phase-1 is in positive cycle
(b) Magnetic field set-up when phase-1 is in negative cycle
+Y
-Y
(d) Magnetic field set-up when phase-2 is in negative cycle
(d) Magnetic field set-up when phase-2 is in negative cycle


**FIGURE 8.39:** AC induction motor operating principle: A two-phase motor example.

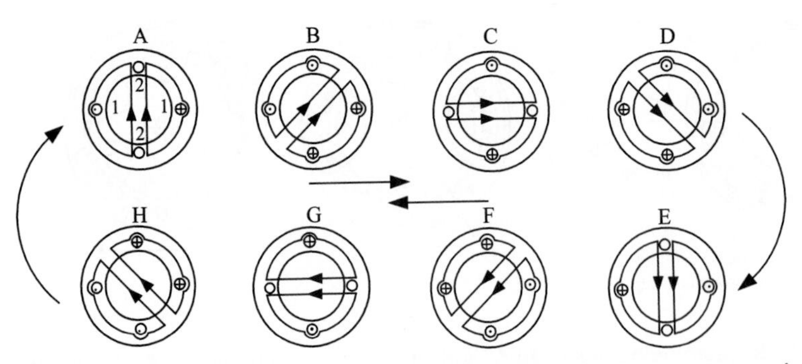


**FIGURE 8.40:** Progression of the magnetic field in a two-phase stator at eight different instants.

sinusoidal function of time. The induced magnetic field changes magnitude and direction as a function of current in the phase. It is basically a pulsating magnetic field in the X direction [Fig. 8.39(a-b)]. Next, let us consider the same for phase 2 which is spatially displaced by 90 degrees from the first phase. The same event occurs, except the direction of the magnetic field is in the Y direction [Fig. 8.39(c, d)]. Finally, if we consider the case where both phases are energized but by 90 electrical degrees apart and of the same frequency, the magnetic field would be the vector addition of two fields and rotate in space at the same frequency as the excitation frequency (Fig. 8.40). Therefore, we can think of the magnetic field as having a certain shape (that is distribution in space as a function of rotor angle) as a result of the winding distribution and current, and rotates in space as a result of the alternating current in time. In other words, the flux distribution is a rotating wave.

In a three-phase motor, windings would be displaced by $\pm 120$ degrees spatially and electrically excited with the same frequency source, except with $\pm 120$ electrical degrees apart. This rotating magnetic field, generated by the stator winding voltage, induces voltage in the rotor conductors. As a result of Faraday's induction law, the induced voltage is proportional to the time rate of change in the magnetic flux lines that cut the rotor. In other words, if the rotor was rotating mechanically at the same speed as the electrical rotating speed of the stator field, there would not be any torque generated.

In a P-pole motor, the electrical angle ($\theta_e$) and mechanical angle ($\theta_m$) are related by (Fig. 8.38):

$$\theta_m = \theta_e/(P/2) \tag{8.277}$$

Therefore, the frequency of the electrical excitation $w_e$, is related to the synchronous speed, $w_{syn}$ (Fig. 8.38):

$$w_{syn} = w_e/(P/2) \tag{8.278}$$

The difference between the electrical field rotation speed, $w_{syn}$, and the rotor speed, $w_{rm}$, is called the *slip speed* or *slip frequency*, $w_s$

$$w_s = w_{syn} - w_{rm} \tag{8.279}$$

$$= s \cdot w_{syn} \tag{8.280}$$

where $s$, slip, is defined as

$$s = \frac{w_{syn} - w_{rm}}{w_{syn}} \tag{8.281}$$

If we consider the case that the rotor is locked ($w_{rm} = 0$), then the slip is $s = 1$ and $w_s = w_{syn}$.

AC motor operating principle is similar to a transformer. Stator windings serve as the transformer primary winding. Stator structure serves as transformer "iron." Rotor serves as transformer secondary winding. The only difference is that the secondary winding is the rotor conductors, and it is mechanically rotating. The rotor sees an effective magnetic flux frequency of $w_s = w_{syn} - w_{rm}$, slip frequency, due to the relative motion between electrically rotating stator flux and mechanically rotating rotor. The induced voltage in rotor is analogous to induced voltage in the secondary winding of a transformer. The voltage in the primary winding generates a magnetic flux as follows. Let $P = 2$, hence $w_{syn} = w_e$. The stator AC voltage,

$$v_s(t) = V_s \, sin(w_e \, t) \tag{8.282}$$

The resulting flux is

$$\Phi = -\frac{V_s}{N_1 \cdot w_e} cos(w_e \, t) \tag{8.283}$$

where $N_1$ is the number of turns in the primary coil. The rotor sees a magnetic flux frequency of $w_s = w_{syn} - w_{rm}$. The induced output voltage as a result of the Faraday's law of induction is

$$v_r(t) = N_2 \frac{d\Phi}{dt} \tag{8.284}$$

$$= -\frac{N_2}{N_1} \frac{V_s}{w_e} \frac{d}{dt}[cos(w_e - w_{rm})t] \tag{8.285}$$

$$= \frac{N_2}{N_1} \frac{(w_e - w_{rm})}{w_e} V_s \, sin(w_s \, t) \tag{8.286}$$

The operating principle of an AC induction motor in terms of cause and effect relationship is illustrated below. The stator current is a result of the applied stator voltage ($v_s(t)$), and induced rotor current is a result of the induced rotor voltage ($v_r(t)$).

$$V_s(w_{syn}) \Longrightarrow i_s(w_{syn}) \Longrightarrow B_s(w_{syn}) \tag{8.287}$$

$$w_s \Longrightarrow V_r(t) \, induced \Longrightarrow i_r(t) \, induced \Longrightarrow B_r(w_{syn}) \tag{8.288}$$

$$B_s(w_{syn}) \,\&\, B_r(w_{syn}) \Longrightarrow T_m \, (torque) \tag{8.289}$$

Notice that the $B_s$ and $B_r$ rotate with the synchronous speed, $w_{syn}$. The rotor mechanically rotates close to the synchronous speed with the difference being the slip speed, $w_s = w_{syn} - w_{rm}$. When the mechanical speed of the rotor is smaller than the synchronous speed, torque is generated by the motor (motoring action). Whereas if the mechanical speed of rotor is larger than the synchronous speed, torque is consumed by the motor (it is in generating mode). In the vicinity of the rotor mechanical speed close to synchronous speed, the torque is proportional to the slip speed. When the slip is zero, generated torque is zero (Fig. 8.41). The steady-state torque-speed characteristics of an AC induction motor can be summarized as:

1. In the vicinity of small slip, the torque is proportional to the slip frequency for a given stator excitation. If the rotor speed is smaller than synchronous speed (slip is positive), then the torque is positive. The motor is in motoring mode. If the rotor speed is larger than the synchronous speed (slip is negative), then the torque is negative. The motor is in generator mode.
2. At a certain value of slip, torque reaches its maximum value. For slip frequencies larger than that, the inductance of the rotor becomes significant and current is limited at higher slip. As a result, torque drops after a certain magnitude of slip frequency.

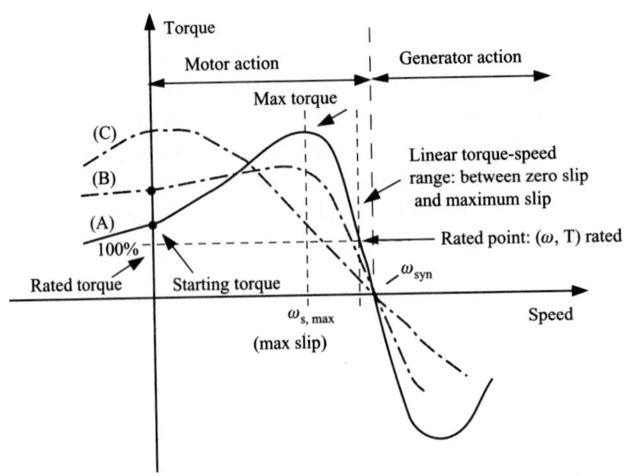


**FIGURE 8.41:** Open-loop torque-speed curve of an AC induction motor. Under a given constant AC voltage magnitude and frequency, the steady-state torque-speed relationship is nonlinear. This curve can be given different shapes by different electromagnetic designs. Curves (A), (B), and (C) show different AC motor torque-speed characteristics that can be achieved by varying designs.

3. The shape of the steady-state torque-speed curve can be modified for different applications by modifing the rotor conductor shape and stator winding distribution.

The torque generation in AC induction motor can be viewed as the result of the interaction between the magnetic flux distribution of the stator and the rotor. Let the stator flux density distribution be (assume that it is constant in time for simplicity)

$$B_s = B_{sm} \cdot cos(\theta_m) \tag{8.290}$$

and let the rotor flux density field be (as a result of the induction principle)

$$B_r = B_{rm} \cdot cos(\theta_m - \theta_{rs}) \tag{8.291}$$

where $B_{sm}$, $B_{rm}$ are the maximum value of the flux density distribution for stator and rotor, $\theta_m$ rotor mechanical angle, and $\theta_{rs}$ the angle between the $B_s$ and $B_r$, which is called the *power angle between stator and rotor*. From basic reasoning regarding the interaction of magnets and magnetic fields (Fig. 8.42), it can be concluded that

$$T_m = K_m \cdot \Phi_{sm} \cdot \Phi_{rm} \cdot sin(\theta_{rs}) \tag{8.292}$$

which states that the torque is proportional to the size and design parameters of the motor ($K_m$), the flux density in rotor and stator, and the angle between the two flux vectors. The *vector control* algorithm discussed later in this chapter attempts to maintain 90 degree angle between the stator and rotor flux density vectors in order to maximize torque production per unit current.

Open-loop, steady-state torque-speed characteristics of an AC induction motor is shown in Fig. 8.41 for the case of constant AC voltage magnitude and frequency supplied to the stator windings. Note that the shape of that curve can be changed by using different stator winding and slotting configurations. In fact, different motor designs are used to meet different steady-state torque-speed curve shapes needed by different applications. Most AC induction motors have maximum slip in the range of 5–20% of the synchronous speed. Maximum slip is defined as the slip speed at which maximum torque is reached. Clearly, the

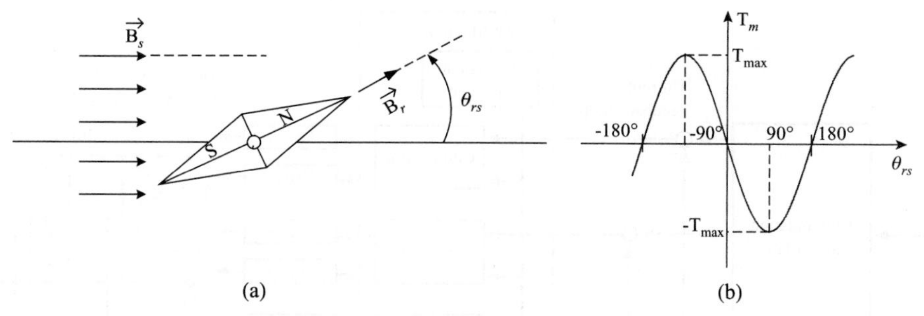


**FIGURE 8.42:** Torque production as a result of interaction between two magnetic fields. (a) A permanent magnet in an external magnetic field; (b) torque as function of the angle between the two flux vectors. In case of AC induction motors, the external field flux density ($B_s$) is set up by the stator current and the equivalent field of the magnet ($B_r$) is set up by the rotor conductors as a result of the induced voltage.

smaller the maximum slip, the smaller the speed variations of the motor under varying loads. Likewise, if large speed variation is desired as the load varies, a motor with large maximum slip should be used.

It should be quickly pointed out that these performance characteristics are only for the motor when controlled from a line voltage supply directly without any active commutation by a drive. The current-torque characteristics of an AC motor can be made to behave like a DC motor using the "field-oriented vector control" algorithm which is typically implemented in the drive. The performance of a motor should always be evaluated together with the drive that is used. Depending on the drive type used, the performance characteristics of the motor-drive combination can be quite different.

***Example*** Consider an AC induction motor driven by a line supply frequency of $w_e = 60$ Hz. Assume that it is a two-pole motor, $P = 2$. The motor design is such that at the maximum load, the slip is 20% of the synchronous speed, $s = 20\%$. Determine the speed of the motor for the following conditions: (a) no-load speed, (b) speed at maximum load, and (c) speed at a load that is 50% of the maximum load. Determine the speed variation (sensitivity) due to the variation of load as percentage of its maximum value. Let maximum load torque be 100 lb.in.

Referring to Fig. 8.41, let us assume that the curve connecting the no-load speed and maximum-load speed points is a linear line. In steady state, the actual motor speed is determined by the intersection of the torque-speed curve with the load torque. As long as the load torque is less than the maximum load torque, the motor speed will be somewhere between the no-load speed and the speed at maximum load. As the load varies up to the maximum load torque, the steady-state speed of the motor variation follows the linear torque-speed line.

The no-load speed of the motor is

$$w_{rm} = \frac{w_e}{P/2} = \frac{60\,\text{Hz}}{2/2} = 60\,\text{rev/sec} = 3600\,\text{rev/min} \tag{8.293}$$

At the maximum load, the motor specifications indicate that it has 20% slip,

$$s = \frac{w_{syn} - w_{rm}}{w_{syn}} = 0.2 \tag{8.294}$$

$$w_{rm} = w_{syn} - 0.2 \cdot w_{syn} = 0.8 \cdot w_{syn} = 2880\,\text{rev/min} \tag{8.295}$$

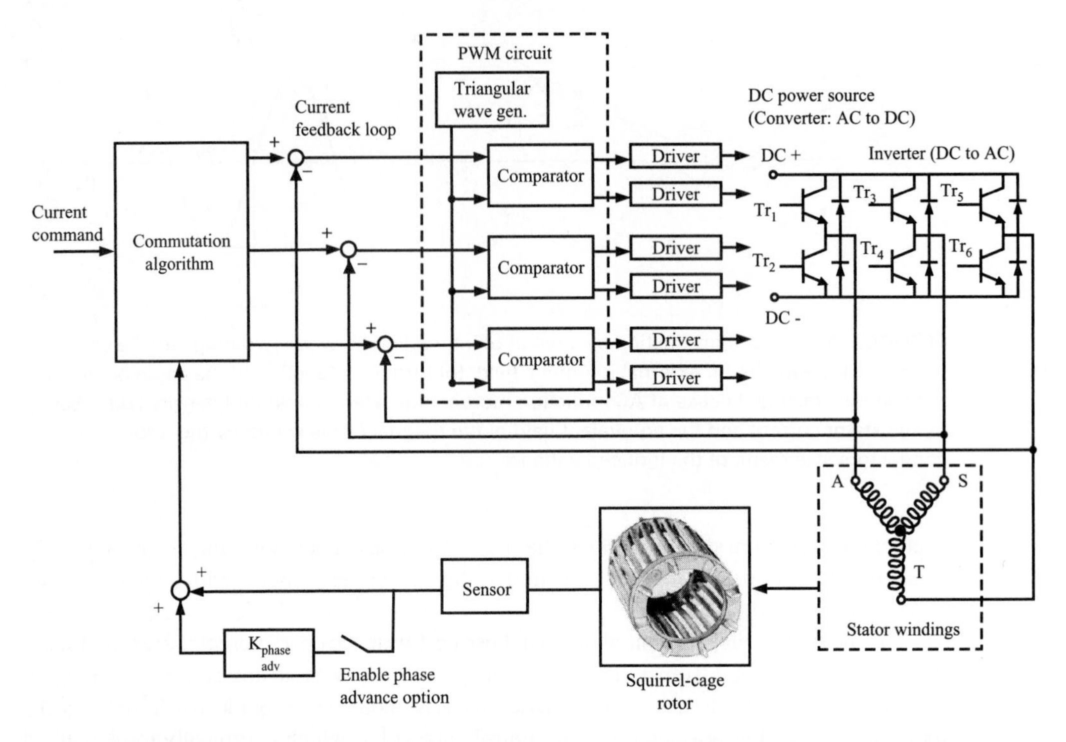


**FIGURE 8.43:** Current commutation and regulation in AC motor phases by a field-oriented vector control (FOVC) drive.

When the load is 50% of maximum load, the slip will be 50% of the maximum slip. Therefore, the steady-state rotor speed is

$$s = \frac{w_{syn} - w_{rm}}{w_{syn}} = 0.1 \tag{8.296}$$

$$w_{rm} = w_{syn} - 0.1 \cdot w_{syn} = 0.9 \cdot w_{syn} = 3240 \,\text{rev/min} \tag{8.297}$$

The speed varies from synchronous speed at no-load to 20% slip at maximum load, hence,

$$\frac{\Delta V}{\Delta T_l} = \frac{w_{syn} - ((1-s) \cdot w_{syn})}{100} \tag{8.298}$$

$$= \frac{s \cdot w_{syn}}{100} \tag{8.299}$$

$$= 7.2 \,[\text{rpm/lb–in.}] \tag{8.300}$$

## 8.4.2 Drives for AC Induction Motors

The drive controls the electrical variables, which are the voltage and current, in the stator windings of an AC induction motor in order to obtain the desired behavior in mechanical variables, which are torque, speed, and position. Frequency and magnitude of the voltage control are of particular interest. Four major drive types are discussed below (Fig. 8.44) where each drive type operates based on varying one or more of the electrical variables, that is voltage, frequency or current.

The steady-state torque-speed curve of an AC induction motor whose stator winding phases are fed directly from an AC line is shown in Fig. 8.41. The motor synchronous speed

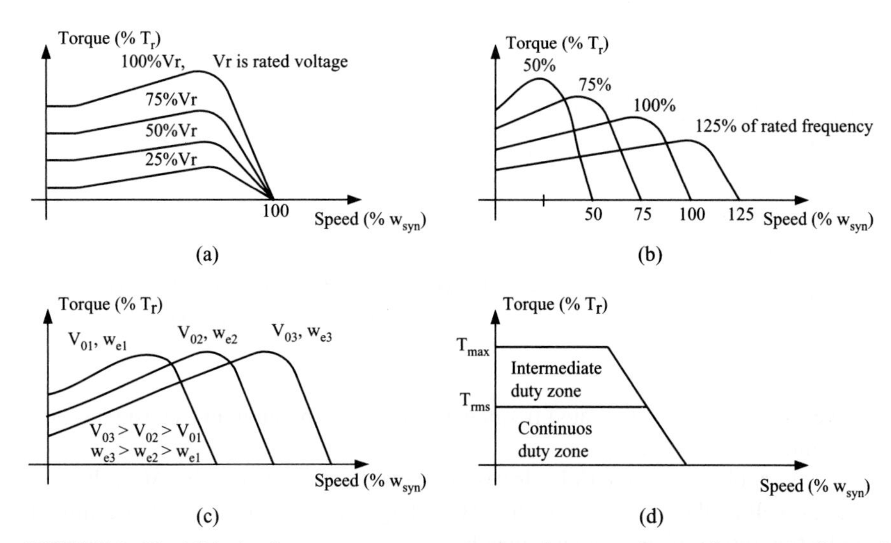


**FIGURE 8.44:** AC induction motor torque-speed performance in steady state under various control methods for varying voltage, frequency, and current in the motor stator phase windings: (a) variable voltage amplitude, fixed frequency; (b) variable frequency, fixed voltage; (c) variable voltage and variable freqency, while keeping voltage to freqency ratio constant for different frequency ranges (Volt/Hertz method), and (d) field-oriented vector control.

($w_{syn}$) is determined by the line voltage frequency ($w_e$). The actual speed of the rotor ($w_{rm}$) would be a little below that since there is a slip ($w_s$) between the synchronous speed and actual rotor speed in steady state. The slip speed depends on the load torque.

$$w_{syn} = \frac{w_e}{(P/2)} \tag{8.301}$$

$$w_{rm} = w_{syn} - w_s \tag{8.302}$$

The maximum torque characteristics are a function of the motor design and the line voltage magnitude. This basic relationship indicates that an AC motor driven directly from supply line has a speed that is largely determined by the frequency of the supply voltage. The slip frequency is a function of the load and the type of motor design (Fig. 8.42). The exact mechanical speed of the rotor is determined by the load around the synchronous speed.

***Scalar Control Drives*** If the drive varies the magnitude of voltage applied to the motor, while keeping the frequency constant, the torque-speed charateristics of the motor are as shown in Fig. 8.44(a). Notice that as the voltage magnitude decreases relative to the rated voltage ($V_r$), the torque gets smaller. It can be shown that the maximum torque is proportional to the square of the applied voltage magnitude. If the load is a constant torque load, by varying the amplitude of the voltage (variable voltage method), we can obtain some degree of variable speed control in the vicinity of syncronous speed of the motor.

The next method of control is to vary the frequency of the applied voltage while keeping the magnitude of the voltage constant. The steady-state torque-speed performance of an AC motor with such a drive is shown in Fig. 8.44(b). Notice that the synchronous speed of the motor is proportional to the applied frequency of the voltage; i.e., if the applied frequency is 50% of the base frequency, then the synchronous speed is also 50% of the original synchronous speed. However, the effective impedance of the motor is smaller at lower frequencies. This leads to large currents and results in magnetic saturation in the motor. Therefore, in order to improve the efficiency of the motor, it is determined that at

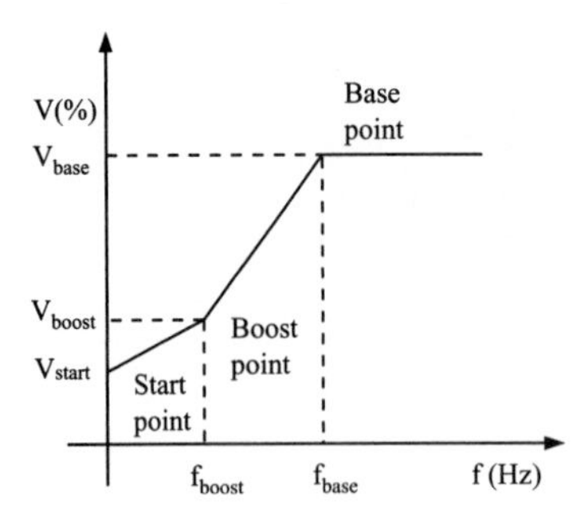


**FIGURE 8.45:** Look-up table relationship for the V/Hz algorithm used in adjustable speed drives.

lower freqencies than the rated frequency, it is better to maintain a constant ratio of voltage magnitude and frequency.

Variable frequency (VF) drives are capable of adjusting the AC voltage frequency, $w_e$, as well as the magnitude of the AC voltage of each phase, $V_0$. For a three-phase AC induction motor, phase voltages may be

$$V_a = V_0 \, sin(w_e t) \tag{8.303}$$

$$V_b = V_0 \, sin(w_e t + 2\pi/3) \tag{8.304}$$

$$V_a = V_0 \, sin(w_e t + 4\pi/3) \tag{8.305}$$

where the VF drive can control both $w_e$ and $V_0$ from zero to a maximum value. Hence, the steady-state torque-speed curve of the motor can be made as shown in Fig. 8.44(c). The power electronics of the VF drive are identical to that of a brushless motor drive, that is, a three-phase inverter (Fig. 8.43). The only essential difference is in the real-time control algorithm that operates the PWM circuit. The PWM circuit is controlled in such a way that the frequency and the magnitude of each phase voltage are changed depending on where on the torque-speed curve we want the motor to operate. Since both frequency and magnitude of voltage are controlled, such drives are also called variable frequency and variable voltage (VFVV) drives. Such a drive control method is also referred to as the *Volts/Hertz (V/Hz) method*. It is classified as one of the *scalar control* methods as opposed to *vector control* methods discussed next. The main objective of the V/Hz method is to maintain a constant air-gap flux. In steady state, the air-gap flux is proportional to the ratio of $V_0/w_e$. As the desired frequency increases, the voltage command is increased to maintain the contant ratio until the base frequency of the motor. After this point, the voltage command is saturated. The motor drive is operated in constant torque region until the base frequency, and in constant power region after the base frequency point. A typical open-loop look-up table for the V/Hz algorithm is shown in Fig. 8.45. The power stage of V/Hz drive for a three-phase AC induction motor is a three-phase inverter. Most common power switching devices are power MOSFETs and IGBTs. Power MOSFETs are voltage controlled transistors with low power loss, but they are sensitive to temperature. IGBTs (insulated gate bipolar transistors) are essentially bipolar transistors where the base is controlled by MOSFETs. IGBTs have higher switching frequency, but are less efficient than MOSFETs.

***Vector Control Drive: Field-Oriented Vector Control Algorithm***[1] The roles of the drive are the commutation and the amplification of current in the stator windings (Fig. 8.43). The only difference is that a DC brushless motor has permanent magnet rotor,

[1] This section can be skipped without loss of continuity.

whereas an AC induction motor has a squirrel-cage rotor with no magnets. Therefore the magnetic field of the rotor of AC motor is not physically locked to the rotor, unlike the DC brushless motor. In order to commutate the current in windings so that two magnetic fields are perpendicular for maximum torque generation per current unit, measuring rotor angle is not sufficient to know the relative angle between the magnetic field of the stator and the magnetic field of the rotor. The "field-oriented vector control algorithm" is the name used for AC motor current commutation where the angle between the magnetic fields (the magnetic field of the stator and the induced magnetic field of the rotor) is estimated based on the dynamic model of the motor.

Assuming that this angle between the two magnetic fields is known, an AC motor can be commutated to provide essentially the same torque-speed characteristics of a DC brushless motor [Fig. 8.44(d)]. The only difference may be in the transient response. The field-oriented vector control algorithm is a current commutation algorithm for AC induction motors. This current commutation algorithm attempts to make an AC induction motor behave like a DC motor, that is, to have a linear relationship between the torque and commutated current. The hardware components of a drive for AC motors which can implement the vector control commutation algorithm is identical to that of a drive for DC brushless motors. Both drives attempt the same thing: maintain a perpendicular relationship between the field magnetic flux vector and the controlled current vector.

The AC induction motor differs from DC brushless motors in two ways. First, the controlled current is on the stator which induces current in the conductors of the rotor. That induced current generates its own magnetic field in the rotor. The induced magnetic field is not locked to the rotor. There is a slip between the rotor and the induced field. Second, there are two components of the controlled current that are of interest: (1) The component that is parallel to the rotor field and (2) the component that is perpendicular to it. It can be shown mathematically [44, 49] that the parallel component (magnetization current) determines the torque gain of the motor, whereas the perpendicular component determines the current multiplier for torque generation. Let us express the torque–current relationship for a DC motor and the same desired torque–current relationship for a vector-controlled AC motor,

$$T_m^{DC} = K_T \cdot i \tag{8.306}$$

$$T_m^{AC} = K_T(i_{ds}) \cdot i_{qs} \tag{8.307}$$

Notice that the same linear relationship of DC motor between torque and current can be obtained. Let the 'dq'-coordinate frame be fixed to the field vector of the rotor. Hence, it rotates relative to the rotor with slip frequency (Fig. 8.46). The torque gain $K_T$ is a function

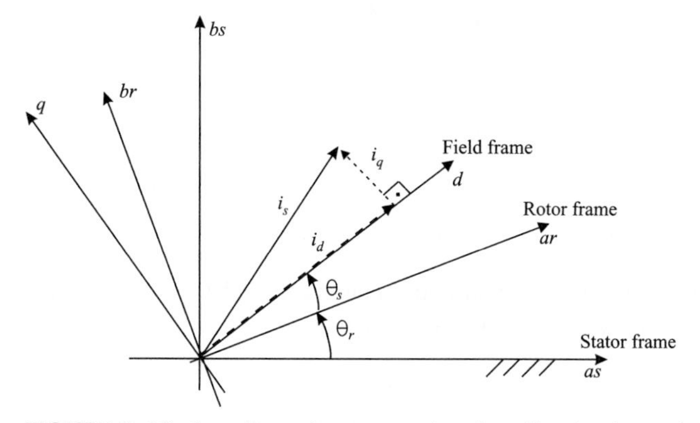


**FIGURE 8.46:** Coordinate frames used to describe the dynamics of an AC induction motor.

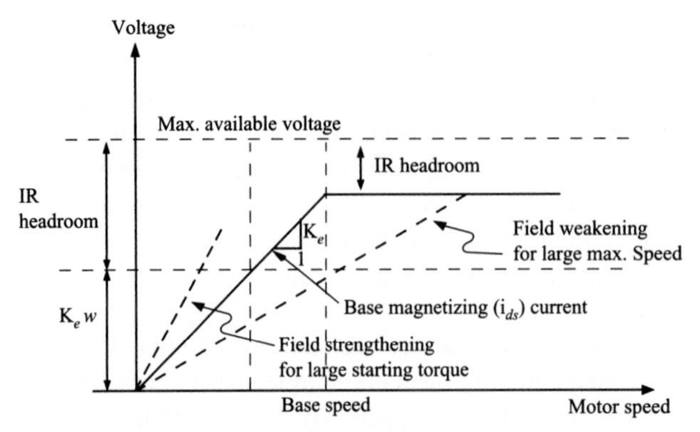


**FIGURE 8.47:** Field weakening and strengthening to change the torque-speed gain of a motor under a maximum available terminal voltage condition.

of the direct (parallel) component of the current in 'dq' coordinate frame, and the torque producing current is the 'q' (perpendicualar) component of the current [44].

From this viewpoint, a DC brushless motor can be treated as an AC induction motor with zero magnetization current and zero slip. In general, it is desirable to have a constant torque gain, hence the magnetization current (d-component of current, which is called parallel or direct component) is set to a constant value,

$$i_{ds} = i_{dso} \tag{8.308}$$

In special cases, the magnetization current can be modified to give the motor more specific performance. For instance, d-component of current can be increased at low speeds to give large torque gain at low speeds. It can also be reduced at high speeds so that the motor can run at higher speeds (at the cost of lower torque) without saturating the terminal voltage (Fig. 8.47) due to back emf voltage. This is called the field strengthening and field weakening, respectively.

Let us briefly show how we can achieve the linear current-torque relationship (equation 8.307). Consider the equations which describe the dynamics of an AC induction motor in the dq-coordinate frame. Let us consider the induced rotor voltages and the torque produced. Derivation of these equations is rather long and can be found in various references [44]. In dq-coordinates, the variable inductances of the AC motor can be expressed as constants which significantly simplify the dynamic equations [49].

$$V_{qr} = 0 = R_r\, i_{qr} + \frac{d\lambda_{qr}}{dt} + \omega_s \lambda_{dr} \tag{8.309}$$

$$V_{dr} = 0 = R_r\, i_{dr} + \frac{d\lambda_{dr}}{dt} - \omega_s \lambda_{qr} \tag{8.310}$$

$$T_m = K \cdot (\lambda_{dr}\, i_{qs} - \lambda_{qr}\, i_{ds}) \tag{8.311}$$

where $V_{qr}$, $V_{dr}$ are the rotor voltages in dq-coordinate frame and $T_m$ is the generated torque. The flux linkages in dq-coordinate frame are defined as

$$\lambda_{qr} = L_m\, i_{qs} + L_r\, i_{qr} \tag{8.312}$$

$$\lambda_{dr} = L_m\, i_{ds} + L_r\, i_{dr} \tag{8.313}$$

and the slip frequency is

$$w_s = w_{syn} - w_{rm} \tag{8.314}$$

Let us assume that the dq-frame is fixed to the magnetic field of the rotor. In other words, the dq-frame rotates with the magnetic field of the rotor. Then $\lambda_{qr} = 0$, and this simplifies equations as follows. The torque equation simplifies to

$$T_m = K \cdot \lambda_{dr} i_{qs} \tag{8.315}$$

Notice that if we can set the $K \cdot \lambda_{dr}$ to be a constant, it would act as the torque constant $K_T$ similar to the case in a DC motor. Hence, we would have a linear relationship between the torque generated and the perpendicular component of current in dq-coordinate frame, $i_{qs}$.

If $\lambda_{qr} = 0$, it can be shown that (equation 8.312)

$$i_{qr} = -\frac{L_m}{L_r} \cdot i_{qs} \tag{8.316}$$

Substitute this into equation 8.309, and solve for $w_s$, slip frequency, noting that $\lambda_{qr} = 0$ and $d\lambda_{dr}/dt = 0$,

$$w_s = -\frac{1}{\lambda_{dr}} \cdot R_r \cdot \left(-\frac{L_m}{L_r}\right) \cdot i_{qs} \tag{8.317}$$

From equation 8.310, we find

$$i_{dr} = -\frac{1}{R_r}\frac{d}{dt}(\lambda_{dr}) \tag{8.318}$$

Substitute $i_{dr}$ into equation 8.313,

$$\lambda_{dr} = L_m \cdot i_{ds} - \frac{L_r}{R_r} \cdot \frac{d}{dt}(\lambda_{dr}) \tag{8.319}$$

$$\lambda_{dr} = \frac{L_m}{\tau_r \frac{d}{dt}(\cdot) + 1} \cdot i_{ds} \tag{8.320}$$

where $\tau_r = L_r/R_r$, which is the electrical time constant of the rotor. Since resistance significantly varies with temperature (i.e., by a factor of two over typical operating temperatures), so does the electrical time constant of the rotor. If we keep the $i_{ds}$ constant, then the following relationship holds approximately,

$$\lambda_{dr} \approx L_m \cdot i_{ds} \tag{8.321}$$

If we substitute this relationship into the slip frequency relationship (equation 8.317), then

$$w_s = \frac{R_r}{L_r \cdot i_{ds}} \cdot i_{qs} \tag{8.322}$$

$$= K_s \cdot i_{qs} \tag{8.323}$$

The above relationship states that the slip frequency is proportional to the quadrature component of the current. The slip angle, $\theta_s$, is obtained by integrating the estimated slip frequency. This is the angle between the rotor's electrical field vector and the rotor itself. If we measure the position of the rotor relative to the stator, then we can add the two angles to find the angle of the rotor's field vector:

$$\theta_{ds} = \theta_r + \theta_s \tag{8.324}$$

where $\theta_r$ is the measured angular position of rotor relative to stator, $\theta_s$ is the estimated slip angle, and $\theta_{ds}$ is the total angle needed for current communtation (the angle between the stator frame and the $d$-axis of the dq-coordinate frame). Hence, the current commutation algorithm can attempt to maintain the desired vector relationship between the stator field and rotor field. Vector control algrorithm basically controls:

1. The direct (parallel) component of the current ($i_{ds}$) to control the magnetization, hence the torque gain
2. The perpendicular component of the current ($i_{qs}$), which is the torque-producing component in the current–torque relationship

The $K_s$ is the slip gain and is function of the direct (parallel) component of the current. Because $\lambda_{dr}$ is maintained constant through the direct component of the current, it determines the torque gain in the above equation:

$$K_T = K \cdot \lambda_{dr} = K \cdot L_m \cdot i_{ds} \tag{8.325}$$

and the desired linear relationship between torque and current is

$$T_m = K_T \cdot i_{qs} \tag{8.326}$$

In general, the direct component of the current is set to a constant value (unless field strengtening or weakening is done) and the quarature component is determined by the servo loop as being proportional to the torque command. The two components are combined in vector form to obtain the $i_s$ from which individual stator phase currents are calculated. The calculation requires the rotor angle relative to stator plus the slip angle of the field relative to the rotor. The algorithm described is called the *indirect field-oriented control (IFOC)* algorithm for the commutation of AC induction motors, which is well established among many other FOC algorithms [96].

Another practical problem to consider is the fact that the slip gain is a function of the rotor resistance and inductance. As the operating temperature changes, the resistance may change as much as by a factor of two. Therefore, accuracy estimation of slip angle can be improved by the temperature sensor and rotor resistance model as function of temperature.

It is interesting to notice that AC induction motor torque gain, hence the back EMF gain, can be adjusted by the direct component of the current. In a permanent magnet motor, these gains are constant and cannot be adjusted in real time. For instance, if large torque is needed at start up conditions under large load, the magnetizing component (direct component) of the current is increased to effectively increase the $K_T$ and $K_E$ of the motor. This is called the *field-strengthening* control (Fig. 8.47). If the maximum speed capacity of the motor is to be increased at the expense of reduced torque gain, the back EMF gain needs to be reduced. This is accomplished by reducing the magnetizing current component at higher speeds. This called the *field-weakening* control (Fig. 8.47).

A basic field-oriented vector control algorithm may take the following form, which must be repeated every servo sampling period:

1. Measure rotor angle ($\theta_r$), and estimate rotor speed ($\dot{\theta}_r(t)$).
2. Decide on the magnetizing current component, $i_{ds}$. Implement field-strengthening or field-weakening algorithm if desired depending on the speed range.
3. Estimate the slip frequency, $w_s$.
4. Integrate the slip frequency, and estimate the slip angle, $\theta_s$.

5. From desired torque (or current) command (which may be generated by position and velocity servo control loop), calculate the perpendicular component of current, $T_m = K_T(i_{ds}) \cdot i_{qs}$.
6. Vector sum $i_{ds}$ and $i_{qs}$ and determine the $\vec{i}_s$. In other words, determine the magnitude of $\vec{i}_s$ and its angle with the $d$-coordinate. $i_s = \sqrt{i_{ds}^2 + i_{qs}^2}$, $\phi = tan^{-1}(i_{qs}/i_{ds})$.
7. Determine the phase components of $\vec{i}_s$ in terms of $i_{as}$, $i_{bs}$, $i_{cs}$.

$$i_{as} = i_s \cdot cos(\theta_{cmd}) \tag{8.327}$$

$$i_{bs} = i_s \cdot cos(\theta_{cmd} + 120°) \tag{8.328}$$

$$i_{cs} = i_s \cdot cos(\theta_{cmd} - 120°) \tag{8.329}$$

where $\theta_{cmd} = \theta_r + \theta_s + \phi$.

If the dynamic lag, $\phi_{lag}$, between the actual current and commanded current is significant, i.e., at high-speed applications, the commanded current phase can be added with an estimated phase lead so that the actual current vector is at the desired 90° phase with the field vector,

$$\phi_{lag} = tan^{-1}(\omega \cdot \tau) \tag{8.330}$$

$$\theta_{cmd} = \theta_r + \theta_s + \phi + \phi_{lag} \tag{8.331}$$

where $\tau$ is the electrical time constant of the current regulation loop and $w$ is the speed of the rotor.

## 8.5 STEP MOTORS

Step motor, also called stepper motor, electromechanical construction is such that it moves in discrete mechanical steps. A change in phase current from one state to another creates a single step change in the rotor position. If the phase current state is not changed, the rotor position stays in that stable position. In contrast, a brush-type DC motor keeps accelerating for a fixed supply voltage condition until the back EMF voltage due a top speed balances the supply voltage.

Basic position control of a stepper motor does not require a position sensor. It can be position controlled in open loop (Fig. 8.48), whereas a DC motor must use a position sensor in order to be controlled in position mode.

The most significant advantages of step motors are the low cost, simplicity of the design, and ruggedness. The disadvantages are that step motors have mechanical resonance and step loss problems, although most of these drawbacks have been largely eliminated by the "microstepping drive" technology. Figure 8.49 shows a picture of a most common type of stepper motor (hybrid-stepper motor). Figure 8.50 shows the construction of the rotor, where laminations with teeth are assembled over an axially magnetized permanent magnet

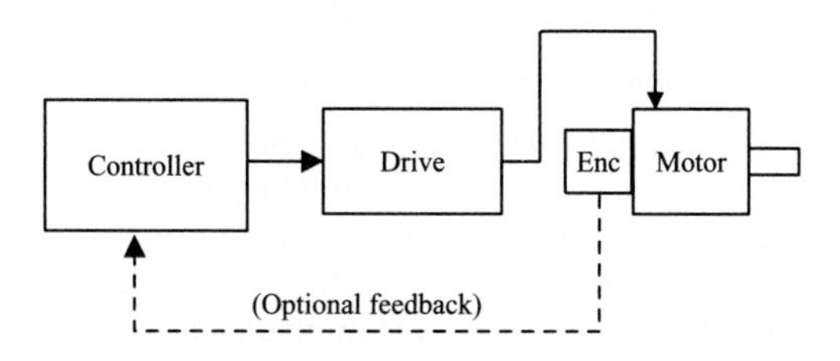


**FIGURE 8.48:** Stepper motor control system components. Position sensor feedback is optional.

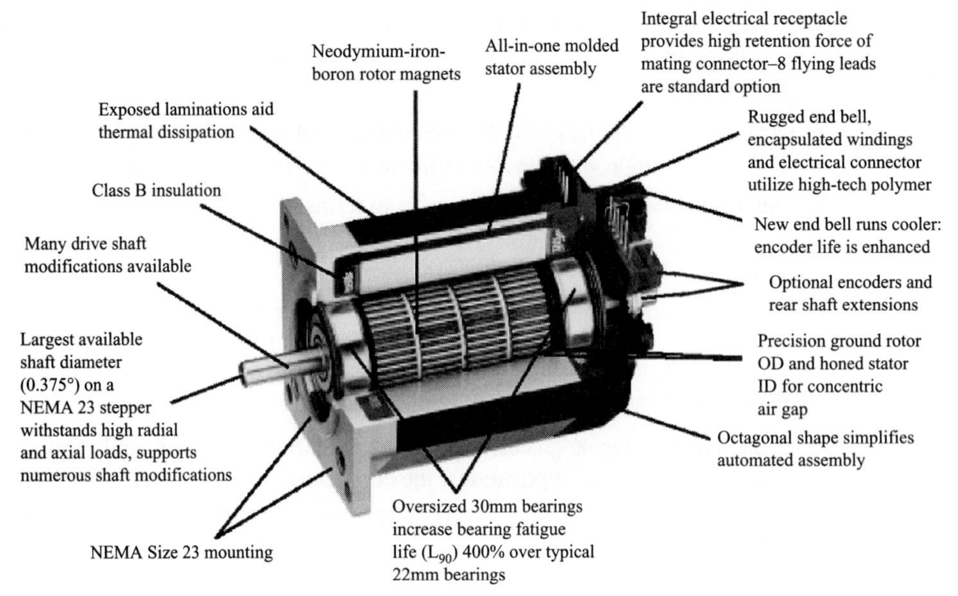
Integral electrical receptacle provides high retention force of mating connector–8 flying leads are standard option
All-in-one molded stator assembly
Neodymium-iron-boron rotor magnets
Exposed laminations aid thermal dissipation
Rugged end bell, encapsulated windings and electrical connector utilize high-tech polymer
Class B insulation
New end bell runs cooler: encoder life is enhanced
Many drive shaft modifications available
Optional encoders and rear shaft extensions
Largest available shaft diameter (0.375°) on a NEMA 23 stepper withstands high radial and axial loads, supports numerous shaft modifications
Precision ground rotor OD and honed stator ID for concentric air gap
Octagonal shape simplifies automated assembly
Oversized 30mm bearings increase bearing fatigue life ($L_{90}$) 400% over typical 22mm bearings
NEMA Size 23 mounting

**FIGURE 8.49:** Cross-sectional view of a four-stack hybrid type stepper motor.

pole pair. Furthermore, the north pole group of laminations is mounted on the rotor with one half-pitch of a tooth angular phase from the south pole group of the laminations. The permanent magnets are magnetized in place after the assembly. Therefore, disassembling a permanent magnet stepper motor by pulling the rotor out will result in loss of its magnetic strength by about 2/3.

A stepper motor has a rotor (permanent magnet in case of hybrid-stepper motors or soft iron in case of switched reluctance motors, Fig. 8.50) and stator winding (Fig. 8.49). The rotor and stator are made of laminations of soft iron material. Each stator pole has a concentrated coil. The stator winding and rotor have teeth. For a given state of stator

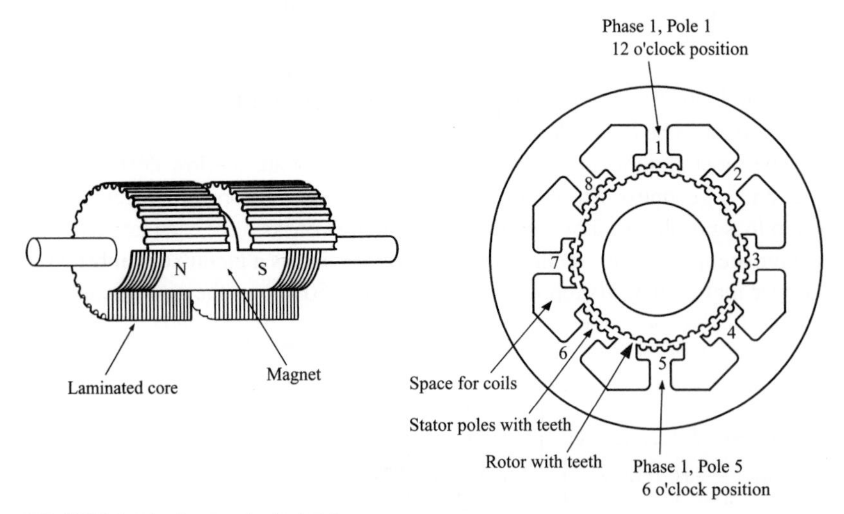
Phase 1, Pole 1
12 o'clock position
1
2
3
4
5
6
7
8
N
S
Laminated core
Magnet
Space for coils
Stator poles with teeth
Rotor with teeth
Phase 1, Pole 5
6 o'clock position

**FIGURE 8.50:** Rotor of a hybrid permanent magnet stepper motor: permanent magnet is polarized axially (North pole on one side, and South pole on the other side along the shaft) and laminated iron core with teeth. The laminated teeth group of the north pole is mounted with 1/2 of tooth pitch offset from the laminated teeth group over the South pole.

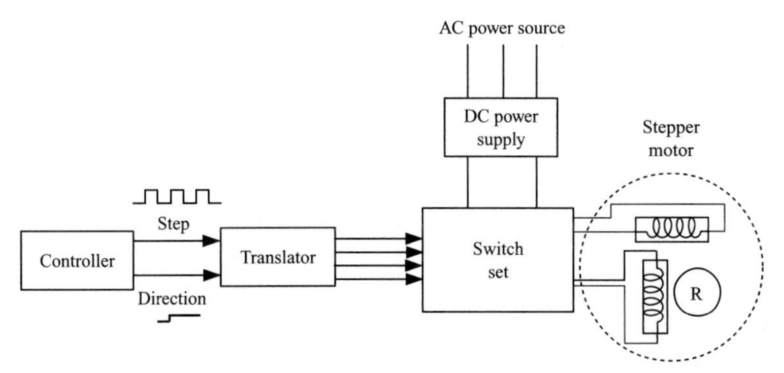


**FIGURE 8.51:** Stepper motor control system components: step motor, DC power supply, power switch set, translator, controller.

current, the rotor moves to align its teeth with those of the stator by the natural tendency to minimize reluctance to magnetic flux. The air-gap between the stator and rotor represents the resistance to the magnetic flux. The smaller the air-gap, the less the resistance to magentic flux, hence the higher the torque capacity of the motor. Typical air-gap in stepper motors is in the range of $30\,\mu$m to $125\,\mu$m. The components of a stepper motor control system is shown in Fig. 8.51. A given switch state of stator represents a field flux. There is a corresponding stable rotor position for each stator phase current condition. The "switch set" block represents the power transistors. The "translator" block represents the logic block which determines the order and time the power transistors should be switched based on a planned motion which is generated by the controller. It is the responsibility of the translator block to limit the maximum switching frequency so that the rotor is not left behind to the point of missing a step. Similarly, the translator block should minimize operating at switching frequency range, which might excite the natural resonance frequency of the stepper motor. Today, stepper motor sizes are standardized; the most common ones are NEMA 17, NEMA 23, NEMA 34, and NEMA 42.

A given phase current condition generates a certain stable rotor position, not continuous torque. Normally the drive or the controller does not need position feedback. But position feedback is commonly used as a way to detect possible step loss and compensate for it when necessary.

The translator, switch set, and DC power supply blocks are collectively called the "drive." Drive controls the current. Velocity or position feedback is used to control speed or position of the motor.

## 8.5.1 Basic Stepper Motor Operating Principles

Let us consider the operating principles of a basic stepper motor. In our basic stepper motor, the rotor has one north and one south pole permanent magnet. The stator has four-pole, two-phase winding with four switches (Fig. 8.52). At any given time either switch 1 or 2, and 3 or 4 can be ON to affect the polarity of electromagnets. For each switch state, there is a corresponding stable rotor position. In this concept figure, a stepper motor with unipolar winding is shown where each coil is center tapped.

Let us consider the switching sequence shown on the left four illustrations at the bottom of Fig. 8.52. In this case, at any given time all of the stator phases are energized. At all times, each rotor pole is attracted by two winding poles. Following the four switching sequence, the rotor would take the shown stable positions. The current pattern for these four discrete switch states are shown in Fig. 8.53(a). This type of phase current switching, where

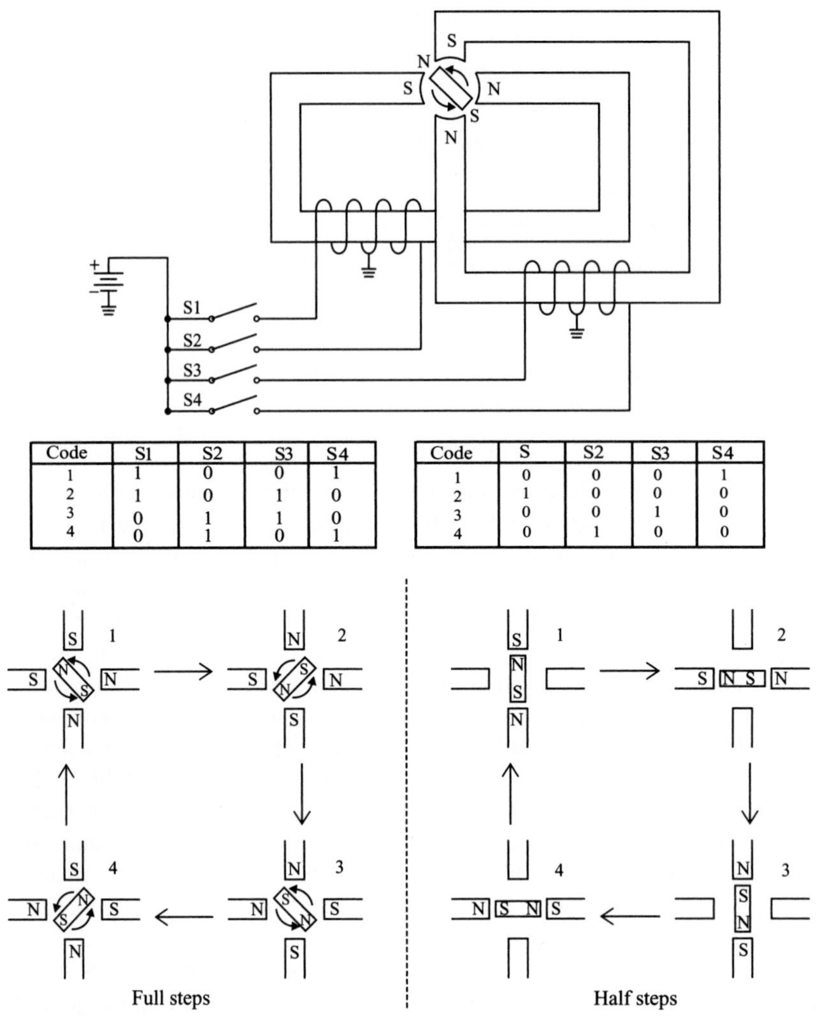


| Code | S1 | S2 | S3 | S4 |
|---|---|---|---|---|
| 1 | 1 | 0 | 0 | 1 |
| 2 | 1 | 0 | 1 | 0 |
| 3 | 0 | 1 | 1 | 0 |
| 4 | 0 | 1 | 0 | 1 |

| Code | S | S2 | S3 | S4 |
|---|---|---|---|---|
| 1 | 0 | 0 | 0 | 1 |
| 2 | 1 | 0 | 0 | 0 |
| 3 | 0 | 0 | 1 | 0 |
| 4 | 0 | 1 | 0 | 0 |

**FIGURE 8.52:** Operating principles of a stepper motor: A unipolar step motor winding with a unipolar drive model is shown.

both phases are energized is referred to as the "full-step" mode of operation. As ON/OFF states of switches are changed in the order shown in the figure, the magnetic field generated by the stator rotates in space. Hence, a permanent magnet (PM) rotor follows it.

Now, let us consider the four sequences of switch states shown on the right-hand side of Fig. 8.52. In this case, only one of the stator phases is energized while the other phase is OFF (i.e., both S1 and S2 OFF, or both S3 and S4 OFF). The corresponding stable rotor positions are shown in the figure. However, notice that since the magnetic force pulling the rotor is provided by only one phase, the holding torque of the motor at these switch states is less than (approximately 1/2) that of the holding torque at the full-step mode. This mode of switching power transistors at the drive is referred to as the "half-step" mode. We can, therefore, energize the motor in alternate modes (one full-step and one half-step) in every other step in order to increase its positioning resolution (from four stable full-step positions to eight half-step positions). The current pattern in the phase windings in full- and half-step modes are shown in Fig. 8.53. In order to make sure the torque capacity is similar at all steps, the current is increased during the half-step mode compared to the current during full-step mode [Fig. 8.53(c)].

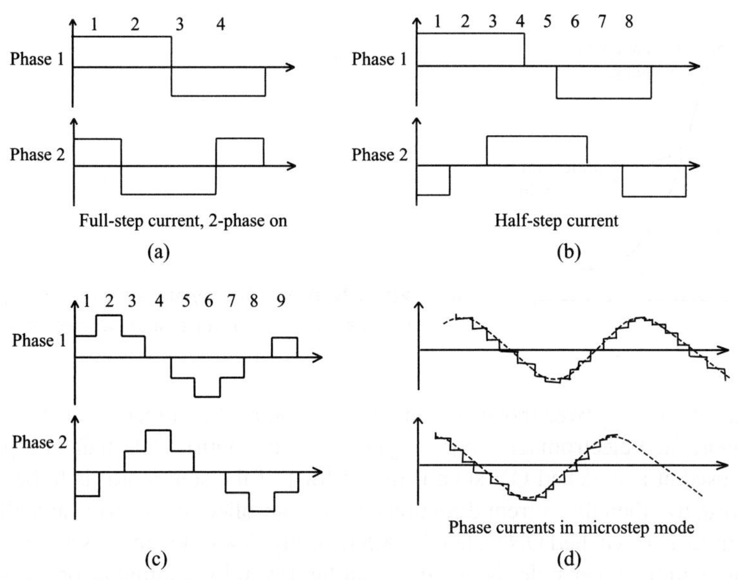


**FIGURE 8.53:** Phase currents in a stepper motor in different operating modes: (a) full-step mode, (b) half-step mode, (c) modified (increased current during alternate half-steps) half-step mode, and (d) micro-stepping mode.

The switching of power transistors from one state to another from ON-to-OFF and OFF-to-ON state instantaneously results in instantaneous change in the magnetic field. The motor behaves like a mass-spring system. We can take the concept of half-step mode further to ratio (smoothly change) the current in phases instead of making transitions from full ON to full OFF states. This will result in smoother motion and electronically controlled finer step sizes. This is the main operating principle of the "micro-stepping drives" (Fig. 8.53(d)). As a result of the smoother current switching between phases, the torque acting on the rotor shaft between steps is smoother, and the step motion of the rotor is less oscillatory. In addition, micro-stepping reduces the resonance and step-loss problems associated with step motors which are operated full-step and half-step current control drives.

The stator windings typically form two phases. If the step motor is to be operated by a *unipolar drive*, each winding must be center-tapped to the ground and positive voltage is connected to both ends [Fig. 8.54(a)]. Only one of the connections at a time per winding

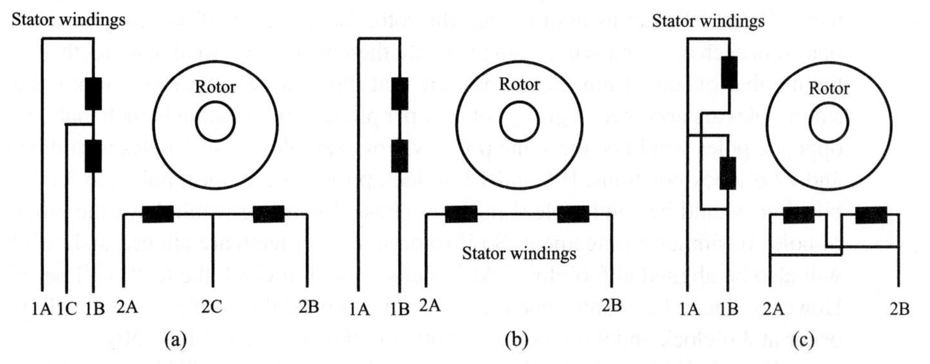


**FIGURE 8.54:** Stator winding connections of a two-phase, four-winding step motor: (a) unipolar drive configuration with center-tapped connection, (b) bipolar drive configuration with series connection, and (c) bipolar drive configuration with parallel connection.

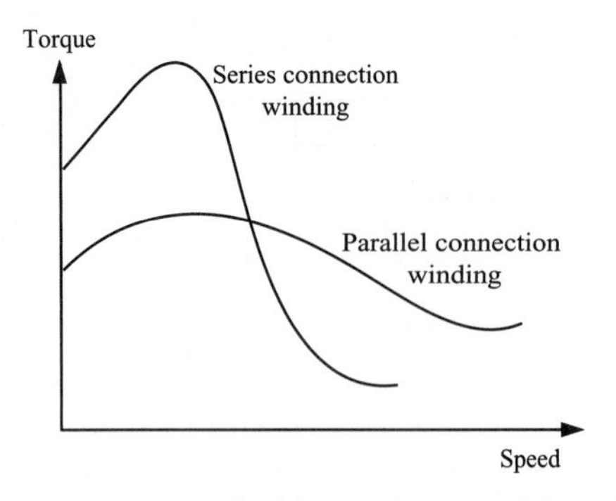


**FIGURE 8.55:** Steady-state torque and speed curves for series and parallel connected windings.

is switched ON (via a power transistor) in order to control the direction of the current—hence the generated electromagnetic pole type (north or south). Only half of a particular winding is used at a switched ON state [Fig. 8.54(a)]. If the step motor is to be operated by a *bipolar drive*, then the current direction can be controlled by the drive and all of each winding is used at a switched ON state [Fig. 8.54(b),(c)]. Some step motors are wound with two separate windings per pole; hence they can be driven by a unipolar or bipolar drive by appropriately terminating the winding ends. For unipolar operation, two windings are connected in series and center-tapped to ground. For bipolar operation, two windings can be either parallel or serial connected and current direction is controlled by the drive [Fig. 8.54(b, c)]. The implications of connecting the windings in series or in parallel in terms of the steady-state torque-speed characteristics of the motor are shown in Fig. 8.55.

Here, we consider two main step motor types:

1. Permanent magnet hybrid stepper motors
2. Switched (variable) reluctance (S.R. or V.R.) motors

The standard hybrid permanent magnet (PM) step motor operates in the same way as our simple model, but has a greater number of teeth on the rotor and stator, giving a smaller basic step size. The rotor is in two sections axially. The two sections of the rotor, both sections with teeth, are separated by a permanent magnet (Fig. 8.50). North and south poles are magnetized axially, so the N-pole is on one side and the S-pole is on the other side along the shaft. Hence, one side is magnetized as north pole, the other side is magnetized as south pole. Furthermore, there is a one-half tooth pitch angular displacement between the two sections (north and south sections) on the rotor. Let us consider an example hybrid step motor. The stator has eight poles each with five teeth on each, making a total of 40 teeth (Fig. 8.50). Let us assume that the rotor has 50 teeth. If we imagine that a tooth is placed in each of the gaps between the teeth, there would be a total of 48 teeth, two less than the number of rotor teeth. Let us assume that this is a two-phase step motor, hence eight stator poles are arranged in groups of four per phase. Coil winding is such that diametrically opposite poles would be the same polarity. For example, phase 1 poles would be at 3, 6, 9, and 12 o'clock positions. If 6 and 12 o'clock positions are north pole, the 3 and 9 o'clock positions would be south pole. Likewise, phase 2 winding would have the other four pair of poles in similar arrangement. So if rotor and stator teeth are aligned at 12 o'clock, they will also be aligned at 6 o'clock. At 3 o'clock and 9 o'clock the teeth will be misaligned. However, due to the displacement between the two axial sets of rotor teeth, alignment will occur at 3 o'clock and 9 o'clock at the other end of the rotor (Fig. 8.50).

By switching current to the second set of coils, the stator field pattern rotates through 45 degrees. However, to align with this new field, the rotor only has to turn through 1.8

degrees. This is equivalent to one quarter of a tooth pitch on the rotor, giving 200 full steps per revolution. The step angle of a hybrid PM stepper motor is determined by the number of electrical phases ($N_{ph}$) and the number of rotor teeth ($N_r$),

$$\theta_{step} = \frac{360^\circ}{2 \cdot N_r \cdot N_{ph}} \tag{8.332}$$

The number of steps of a PM stepper motor per revolution is the number of electrical phases times the number of rotor teeth times two,

$$N_{step} = 2 \cdot N_r \cdot N_{ph} \tag{8.333}$$

Let us look at a stepper motor condition where a set of phase switches is ON. This corresponds to a stable rotor position. As a result of the tendency to minimize the resistance to flux (minimize reluctance), the rotor-stator magnetic field interaction will try to maintain that stable position. If an external load torque is applied to the rotor and no change is made in the phase currents, torque is generated by the motor as a function of rotor displacement from neutral stable position. This is called the holding torque curve for a stepper motor. The steady-state holding torque is a function of the current and load angle of the rotor,

$$T = K \cdot i \cdot sin(\theta_r) \tag{8.334}$$

where $i$ is current and $K$ is the proportionality constant that is a function of the motor design (physical size and magnet strength). The angle, $\theta_r$, is called the *load angle* and is related to the rotor position angle $\theta$ as follows:

$$\theta_r = \frac{2\pi}{4\theta_{step}} \cdot \theta \tag{8.335}$$

where $\theta_{step}$ is one step angle of the motor. As the rotor position makes four-step angle displacement, the torque goes through one cycle of oscillation

If the load torque exceeds the maximum holding torque value, the rotor will move to a new position. The next stable position for a given switch state is two full steps apart. This static torque characteristic of stepper motor also helps explain why a stepper motor behaves like a mass-spring system in full stepping mode, and how micro-stepping effectively adds damping to the behavior.

Step loss and resonance are two fundamental problems inherent in stepper motors. If the phase currents are switched so fast that the rotor cannot keep up with it, then step loss occurs. The only way to correct the problem is to use a position sensor to detect the step loss and command additional steps to compensate for it. Therefore, the maximum value of switching frequency is limited. Higher switching frequencies can be used only if the rotor is accelerated to high rotational speeds under controlled acceleration profiles (Fig. 8.56). As the switching frequency increases, there is less time for the phase current to develop due to the electrical time constant of phase winding ($\tau = L/R$). Therefore, at higher switching frequency the torque capacity of the motor drops compared to the torque capacity at lower speeds.

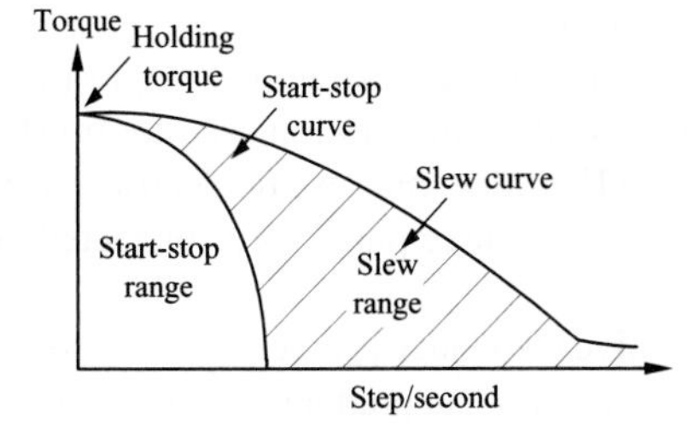


**FIGURE 8.56:** Torque capacity and step rate of stepper motors.

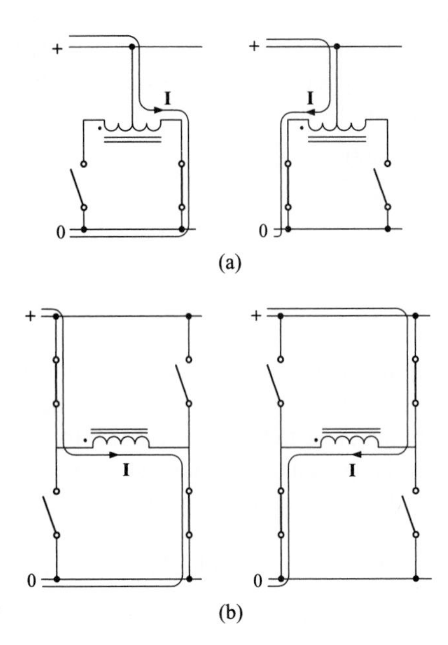


**FIGURE 8.57:** Drive types for step motor: (a) unipolar drive and (b) bipolar drive.

### 8.5.2 Step Motor Drives

Drive controls the direction and magnitude of current in each phase, hence it controls the direction and magnitude of torque. There are two types of drives which must be matched to the winding type of a step motor (Fig. 8.57),

1. Unipolar drive
2. Bipolar drive

Unipolar drive requires the motor windings to be center-tapped [Figs. 8.57(a) and 8.54(a)]. By turning ON the current at one end of the winding versus the other end, the direction of current in the winding is changed, hence the flux direction. When a phase winding is energized, only 50% of it is used by a unipolar drive. Unipolar drive requires two power transistor switches per phase. In a unipolar center-tapped motor, each phase has three leads: two sides and one center tap lead [Fig. 8.54(a)].

Bipolar drive uses an H-bridge (four power transistor switches) per phase. A bipolar drive for a two-phase step motor has two H-bridges ($2 \times 4 = 8$ transistors). The direction of the current is changed by controlling which two pairs of the H-bridge switches are turned ON. Although it requires twice as many power switches, it makes 100% use of the conductors of the energized coil. A stepper motor with bipolar winding has two leads per phase [Fig. 8.54(b) and (c)]. Some motors are wound such that they can be configured to be driven by either unipolar or bipolar drive. In this case, each winding has two identical windings with four leads. Depending on how the leads are connected, the motor can be made to operate with a unipolar- or a bipolar-type drive.

Three phase stepping motor may be driven by three separate H-bridges ($4 \times 3 = 12$ transistors) or by a six transistor inverter similar to the configuration used for brushless DC motors. If each phase is separately driven by an H-bridge, phases are electrically independent. If they are driven by a six-transistor inverter, the three phases are connected in star or delta configuration.

Micro-stepping drive was developed to solve resonance problems. In micro-stepping, current is not simply switched ON/OFF between phases, but gradually changed. The current vector does not change in sudden jumps, but instead it is smoothly changed. Resonance is greatly reduced, and resolution is sharply increased, because there are many more equilibrium points as current is ratioed between multiple phases. Micro-stepping is usually performed with two bidirectional phases. These two phases have similar torque equations as a brushless DC motor. Let us assume that the phase current torque gain is a sinusoidal function of the rotor position. The two phases are 90 degrees apart. The phase current and torque relationships are

$$T_a = K \cdot i_a \cdot sin(\theta_r) = K\, i\, cos(\theta_c) \cdot sin(\theta_r) \tag{8.336}$$

$$T_b = K \cdot i_b \cdot cos(\theta_r) = K\, i\, (-sin(\theta_c)) \cdot cos(\theta_r) \tag{8.337}$$

$$T_{total} = T_a + T_b = K\, i\, sin(\theta_c - \theta_r) \tag{8.338}$$

where $\theta_r$ is the real position and $\theta_c$ is the commanded position, $i_a$, $i_b$ are currents in phases a and b, $K$ is a proportionality constant. In equilibrium, these two positions are identical, so the total torque equation becomes

$$T_{total} = K\, i\, [sin(\theta)cos(\theta) - sin(\theta)cos(\theta)] = 0 \tag{8.339}$$

The step motor torque is developed as the rotor is forced to move away from the equilibrium position for a given state of winding current. The equilibrium position is the one that minimizes the magnetic reluctance (maximizes the inductance). When the rotor is at the exact position that minimizes the reluctance, the torque is zero. The holding torque is developed as the rotor position deviates from the ideal position due to load or current commutation in the windings.

***Example: Unipolar IC Drive for Step Motors*** An integrated circuit (IC) drive which can be used to drive a unipolar stepper motor is shown in Fig. 8.58. The SLA7051M (by Philips Semiconductors) integrates two blocks of Fig. 8.51, the translator and the power switch set. The translator section is made of low-power CMOS logic circuit and handles the logic for sequencing, direction, full- and half-step operation. Translator decides on the firing sequence of windings for full step (AB, BC, CD, DA in forward or reverse direction) or for half step (A, AB, B, BC, C, CD, D, DA in forward or reverse direction) as a function of the input signals at the terminals STEP (also named Clock), FULL/HALF, CW/CCW. At each low-to-high transition of the STEP input signal, the translator checks the state of the FULL/HALF pin to determine if *full-step mode* or *half-step mode* is commanded (high for full-step, low for half-step mode), and the CW/CCW input to determine the *direction* command. Then decide which one or two of the windings to be fired according to the sequence defined above (AB, BC, CD, DA or A, AB, B, BC, C, CD, D, DA).

The PWM current-controlled power stage uses FET output and can handle up to 2 A and 46 V per phase. The maximum current is controlled by the reference voltage and current sense resistors for each phase (pins REF, SENSA, SENSB). The output stages control the phase currents via terminals OUTA, OUTA\, OUTB, OUTB\ of the IC chip. Notice that this IC drive is capable of controlling only FULL- and HALF-step modes and does not support micro-stepping mode. For driving step motors which will require power dissipation more than 20 W, external heat sink must be added to the FET power transistor stage.

***Example: Bipolar IC Drive for Step Motors*** The LM18293 (by National Semiconductor) integrates two H-bridges (four-channel, push-pull driver) on the chip with 1 A continuous (2-A peak) current capacity per channel. Therefore, it can be used to drive a two-phase bipolar step motor (Fig. 8.59). It can also be used to drive two separate DC brush motors bidirectionally. The maximum DC supply voltage is 36 V. A separate enable pin ($V_{E1}$ and $V_{E2}$) controls one of the two H-bridges. When the ENABLE pin ($V_{E1}$ or $V_{E2}$) is low, the corresponding outputs are driven to tri-state condition. If the ENABLE pins

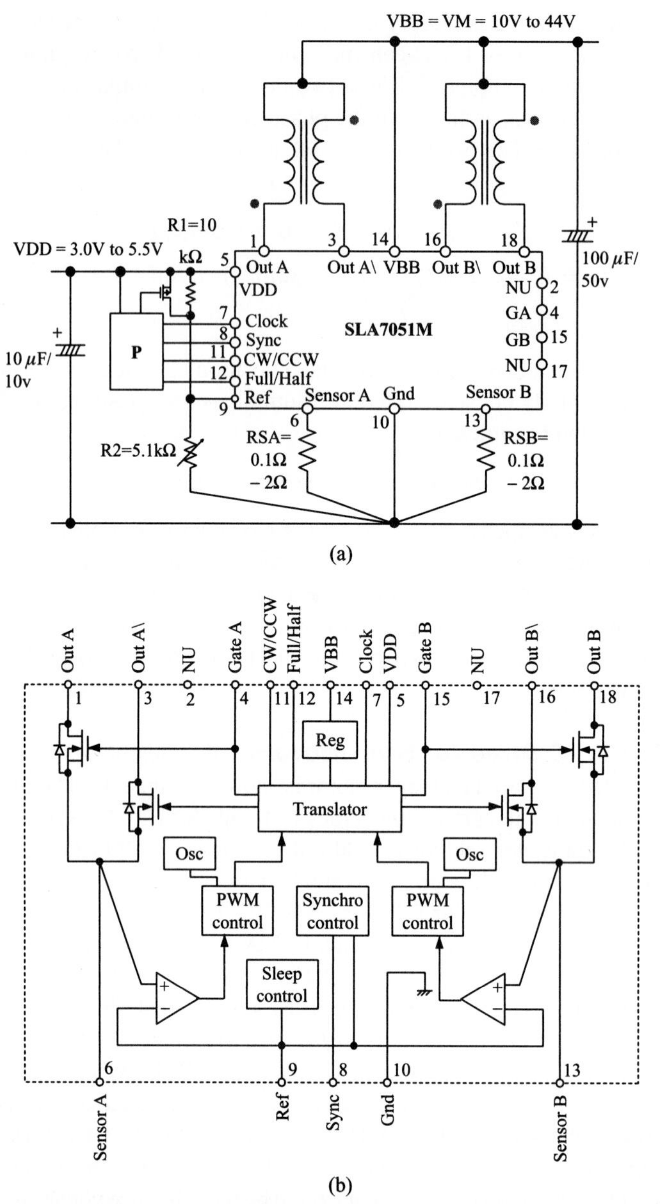


**FIGURE 8.58:** An example IC driver for unipolar step motors: SLA7051M by Philips Semiconductors. The unipolar step motor driver IC has the translator and output power stage integrated into one small IC package.

are floating (not connected), it is assumed that the H-bridges are enabled. Input channels (TTL level signals) INPUT1 (pin 2) and INPUT2 (pin 7) control the direction of current in one of the H-bridge. Similarly, INPUT3 and 4 (pin 9 and pin 15) control the current direction in the other H-bridge. The pairs INPUT1 and 2, and the INPUT3 and 4 are connected to be in opposite states via an inverter and driven by the same TTL level signal line for the step motor or DC motor control applications. Once an H-brige is enabled, the current direction is controlled by the pair of INPUT (1/2 or 3/4) pins for that H-bridge. By modulating the pulse width of the INPUT lines, the average current can be controlled by an external microcontroller.

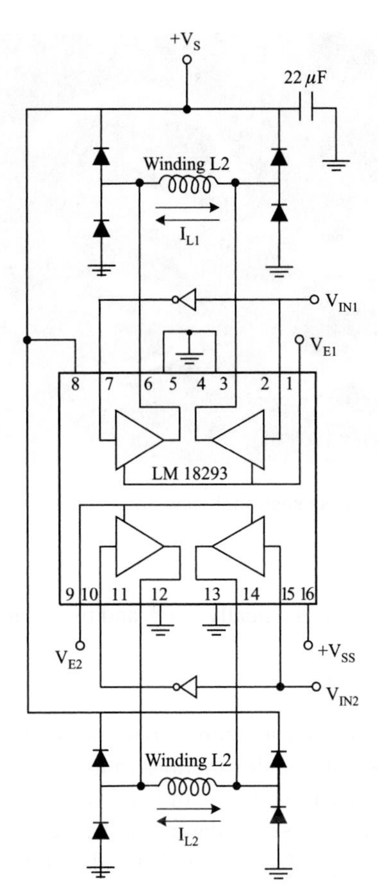


**FIGURE 8.59:** Integrated circuit drive chip (LM18293 by National Semiconductor) for bipolar step motors or brush-type DC motors.

# 8.6 SWITCHED RELUCTANCE MOTORS AND DRIVES

## 8.6.1 Switched Reluctance Motors

Switched reluctance (SR) motors, also known as variable reluctance motors or switched reluctance step motors, have been drawing increasing interest as servo motors in recent years (Fig. 8.60). The SR motors are potentially the cheapest and most rugged motors to manufacture because there are no permanent magnets or brushes in the motor. Currently, there are SR motor applications up to about the 500-HP range. The switched reluctance motor is essentially a stepper motor, except that the rotor is not a permanent magnet, but matter is made of soft ferromagnetic iron (also called soft iron). The magnetic field attraction works out in such a way to minimize the reluctance [47].

The SR motor has the following components:

1. A rotor that is simply a laminated stack of soft iron material with a number of teeth (salient poles).
2. A stator frame which is also a laminated stack of soft iron and concentrated stator windings, one for each stator pole (usually 6, 8, 12, or more poles). The stator windings are usually wired as three or four electrically independent phases.

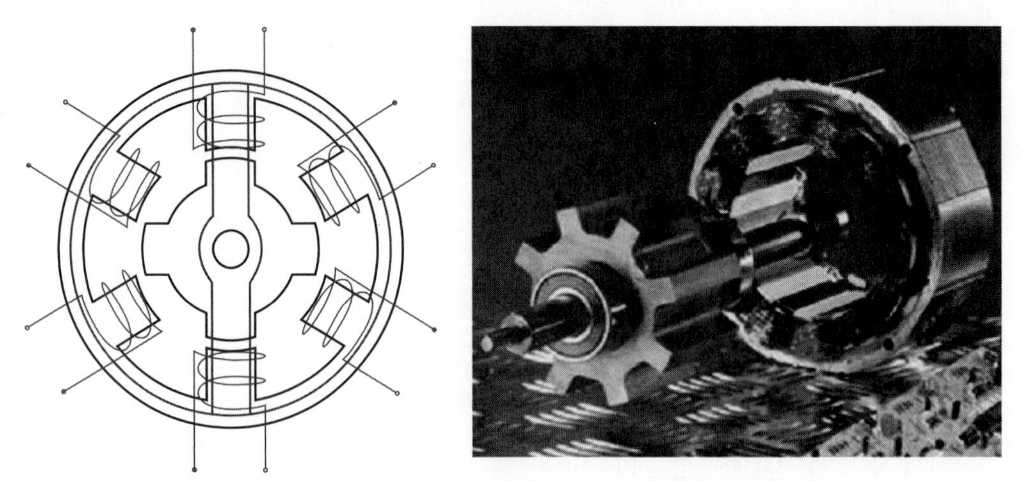

**FIGURE 8.60:** Switched reluctance (SR) motor (a) cross-sectional view and (b) picture of motor stator and rotor assembly.

3. Rotor shaft on which the rotor lamination is mechanically fixed, and two bearings on each end.
4. End plates and housing for the motor assembly.

The laminations used for the rotor and stator are die punched from a continuous roll of sheet metal. The sheet metal material is soft iron type with small variations in its material composition. Each lamination is coated with insulation. The purpose of lamination as opposed to solid metal design is to reduce the eddy current losses in the motor, hence reduce the generated heat. Stator windings are concentrated type, that is, a single winding is placed on one stator tooth instead of being spatially distributed over multiple teeth. Each concenrated winding is wound from a supply of insulated copper conductor using a winding machine. Then a winding is placed on each stator winding. Electrical connections between windings (two pairs, four pairs, etc.) are mechanically made by connecting the appropriate terminals. Then the spaces between the windings are filled with epoxy in vacuum chamber under heat. The epoxy filling closes the air-gaps between copper conductors in the winding, hence increasing heat conduction and helping insulate wires.

There are two basic principles which are at work for force generation in electric motors:

1. Force generated by the interaction between a magnetic field and a current-carrying conductor (i.e., the principle at work in DC, AC, and PM stepper motors).
2. Force generated by the attraction of ferrous metals in a magnetic field through temporary magnetization of the soft ferrromagnetic material (i.e., SR motor case). Attraction force is proportional to the variation in reluctance.

The basic principle of operation of an SR motor is based on the second principle. When an iron material is placed near a magnet, magnetic poles are induced on the surface of the iron. The surface of the iron close to the north pole develops a south pole, and the surface close to the south pole develops a north pole. When the magnet is moved away from the iron material, the induced magnetization on the iron material quickly disappears. When the stator windings have a certain current pattern, it generates a corresponding electromagnetic poles in the stator teeth. The teeth structure of the rotor provides a stable magnetic orientation for rotor under each condition of winding current in the stator. The rotor moves to minimize the magnetic reluctance between its teeth and the stator teeth. The torque is generated by the

**TABLE 8.3: Switched Reluctance Motor Design Parameters: Number of Phases, Number of Stator Teeth, and Number of Rotor Teeth**

| Number of phases and $N_s/N_r$ combinations | | | |
|---|---|---|---|
| 1-phase | 2-phase | 3-phase | 4-phase |
| 2/2 | 4/2 | 6/4 | 8/6 |
| 4/4 | 8/4 | 12/8 | 16/12 |
| 6/6 | 12/6 | 18/12 | 24/18 |
| 8/8 | 16/8 | 24/16 | 32/24 |

natural tendency of rotor position relative to stator to minimize the reluctance. As the rotor moves to decrease reluctance, it generates torque and is in "motoring" mode. As the rotor moves to increase reluctance, it requires torque from outside and is in "generating" mode.

The step angle of an SR motor is determined by the number of electrical phases ($N_{ph}$) and the number of rotor teeth ($N_r$),

$$\theta_{step} = \frac{360^\circ}{N_r \cdot N_{ph}} \tag{8.340}$$

Notice that the SR motor has half the resolution of a comparable PM hybrid stepper motor. The number of steps of an SR motor per revolution is the number of electrical phases times the number of rotor teeth,

$$N_{step} = N_r \cdot N_{ph} \tag{8.341}$$

Design parameters of an SR motor include the number of electrical phases, number of stator poles ($N_s$), and the number of rotor poles ($N_r$). Table 8.3 shows some of the possible phase, stator, and rotor pole number combinations. In Fig. 8.60, the number of stator and rotor teeth are not the same. As a result, the spatial angular spacing of stator pole and rotor teeth do not match up. When a particular stator phase current is turned ON, the closest rotor tooth moves to line up with the stator poles of that phase, minimizing the magnetic reluctance. Because the rotor has no permanent magnets, a field can be impressed on it in either direction (by directing stator current in either direction) and it will still turn in the same direction to line up with the stator field. This means that the SR motor torque is independent of the current direction in the windings. Therefore, there is no need to drive the phase currents bidirectionally.

As a result for a given state of the stator winding current (i.e., phase-1 is ON, phase-2 and phase-3 are OFF, Fig. 8.61), there is a stable position for the rotor. Depending on the angular position and the direction of speed of the rotor, a given stator current may make the motor act as a motor [Fig. 8.61(a) and (d)] or generator [Fig. 8.61(c) and (f)] or zero torque source [Fig. 8.61(b) and (e)]. Continuous torque in a particular direction (clockwise or counterclockwise) is obtained by turning ON/OFF the phase windings as a function of rotor position (Fig. 8.62). Each stator winding has a unidirectional current with varying magnitude. Current direction is not changed because it has no effect on the torque direction. It is the magnitude of the current relative to the rotor angular position that matters. The current in a phase should be OFF during the rotor position range where the inductance is constant, since the torque produced is zero due to constant inductance ($dL/dt = 0$) regardless of the value of the current. If positive torque is desired, then the current should be ON only during the rotor angle range where the inductance is increasing as the rotor moves, and it should be OFF in the rest of the cycle. If negative torque is desired, then the current should be ON during the rotor angle range where the inductance is decreasing as the rotor

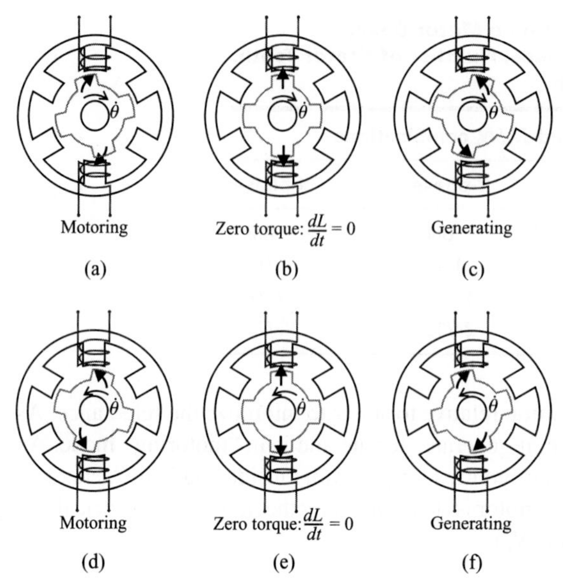


**FIGURE 8.61:** SR motor winding current, rotor position, direction of speed and torque relationship. One phase current is energized and constant. As rotor position changes, generated torque direction changes from positive (motoring) to zero and to negative (generating). If the motor motion is in the direction of reducing the reluctance (in this case same as better alignment of stator and rotor teeth), the motion generates torque, which means it is in "motoring" mode [(a) and (d) in the figure]. If the motion is the direction of increasing the reluctance (in this case, the alignment between stator and rotor teeth decreases), the motion requires torque, which means it is in "generating mode" [(c) and (f) in the figure]. If reluctance does not vary as a result of motion, no torque is generated or consumed [(b) and (e) in the figure].

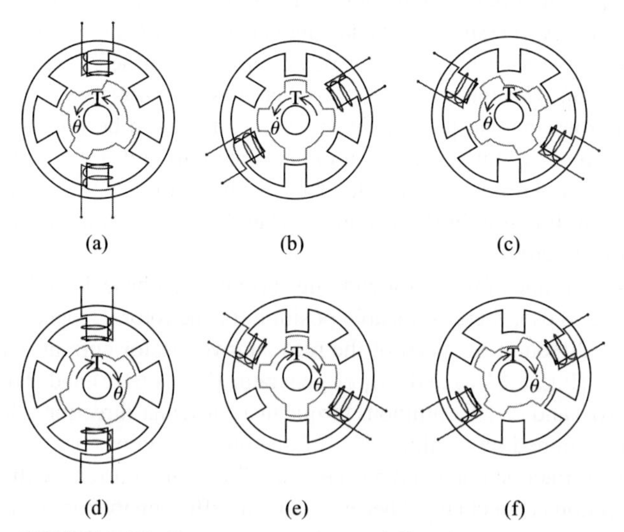


**FIGURE 8.62:** Torque generation and direction control by phase winding current. Torque direction is controlled by the phase current switching relative to rotor position. In (a) through (c), phase 1 through phase 3 winding current switching sequence generates torque in the counterclockwise direction. Notice that the current switching between phases must be done at the appropriate rotor position in order to generate the continuous counterclockwise torque. In (d) through (f), the sequence of stator winding phase current control is shown to generate continuous torque in a clockwise direction.

moves, hence negative torque is produced. Since at any given direction of speed (positive or negative), an SR motor can generate positive or negative torque by controlling the current as a function of rotor angular position, an SR motor can operate in all of the four quadrants of torque-speed plane. Therefore, a SR motor can operate both as motor and generator. As shown in Fig. 8.61(a) and (d), if the current in a phase is turned ON during a rotor angle range where the motion of rotor is in a direction that increases the inductance, it produces "motoring" torque or is said to be in "motoring" mode. Likewise, if the phase current is turned ON, during a rotor angle range where the motion of the rotor is in a direction that decreases the inductance, it consumes torque or is said to be in "generator" mode [Fig. 8.61(c) and (f)]. The direction of the torque is controlled by controlling the phase current relative to the rotor position and direction of speed. Magnitude of the current affects the magnitude of the torque generated, but the direction of the current has no affect on the torque. SR motors can be used with unipolar drives because the stator current does not change direction.

## 8.6.2 SR Motor Control System Components: Drive

The components of an SR motor control system are shown in Fig. 8.63. The power supply, power electronics, and switching control of the power transistors are similar to other stepper motor drives. The look-up table algorithm inverts the nonlinear torque-current relationship (equation 8.345) in order to obtain the desired torque from the motor using current regulation in stator windings as a function of rotor angle. The look-up table algorithm is used to determine current command levels and switching angle adjustments. The rotor speed is used to adjust the switching angles to compensate for the electrical time constant of windings. Switching controller operates the power transistors (Fig. 8.64) in order to regulate current in each phase as a function of rotor angle. The current is, therefore, controlled as a function of rotor angle and rotor speed.

Notice that the torque generated by one of the stator windings of an SR motor under constant current changes direction as a function of its rotor position. While the current is constant, the torque is positive for increasing inductance, zero for constant inductance, and negative for decreasing inductance as a function of rotor position [Fig. 8.61(a)–(c), (d)–(f) and Fig. 8.65]. In order to have a unidirectional torque contribution from a particular stator winding, current in that winding has to be turned OFF for about half of the cycle. Otherwise,

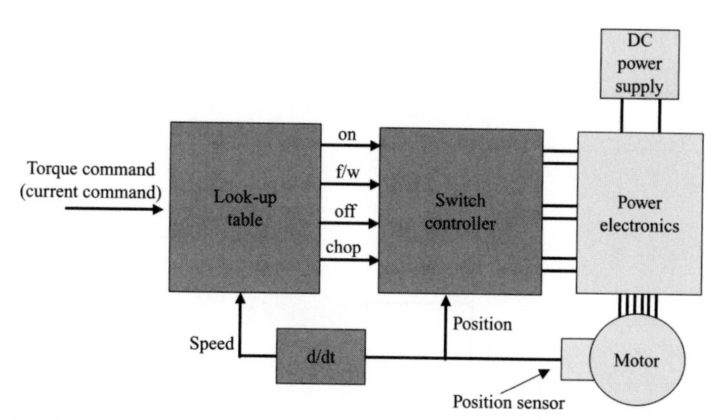


**FIGURE 8.63:** Components of a switched reluctance (SR) stepper motor control system. Two transistors and two diodes are used per phase current control in the power electronics block. Switch controller commutates the current in each phase as a function of rotor angular position and desired torque. The rotor speed information is used in high speed to compensate for transient delays and to modify current commutation algorithm to optimize performance.

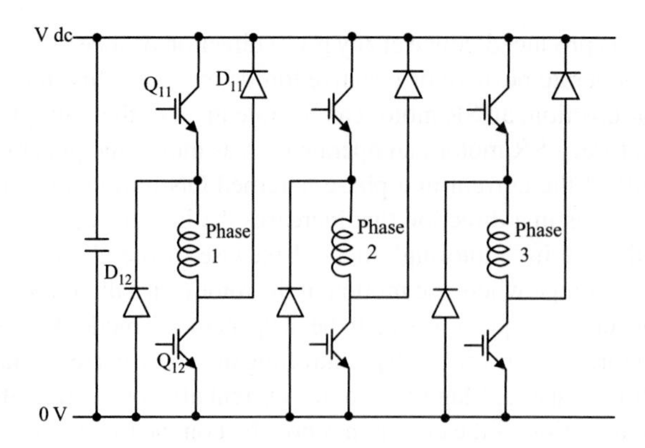


**FIGURE 8.64:** Inverter for switched reluctance motors.

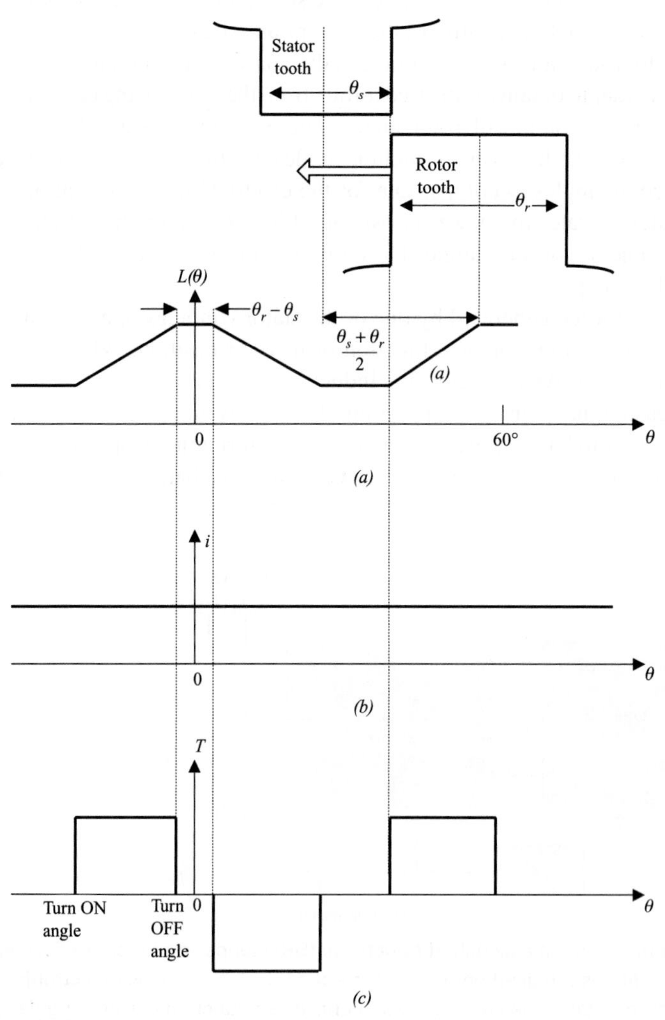


**FIGURE 8.65:** SR motor inductance, torque, and current relationship as a function of rotor angular position.

current contribution of the winding changes sign as a function of rotor position. A given phase can do nothing to contribute to torque in the direction desired for half of its cycle, so it should be turned OFF for this period.

By design, the inductance in a SR motor is not constant, but varies as a function of the rotor angle (Fig. 8.65). Inductance is a function of rotor position and current.

$$L = L(\theta, i) \tag{8.342}$$

The exact shape of the inductance as a function of the rotor position can be customized by specific rotor-stator teeth geometry,

$$L = L(\theta); \; for \; i \; = \; constant \tag{8.343}$$

Notice that $L(\theta)$ is a result of the motor design. The drive control can regulate the voltage or the current. In most applications, the drive is used in current regulation mode, and terminal voltages of the phases are PWM controlled to regulate the commanded current. In very high speed applications, the terminal voltage is switched between two extereme values only once. This is called the single-pulse mode current control. The switch ON and switch OFF time of the voltage relative to the rotor angle are adjusted as function of rotor speed in order to account for electrical time constant of each phase. In low-speed application, a typical current regulating circuit based on a PWM amplifier is sufficient for current loop control.

Let us consider one winding of an SR motor. Assume that the mutual flux linkage between phase windings is zero and that the $B$ versus $H$ characteristics of the ferromagnetic medium is linear. Furthermore, we assume that the inductance is a function of only rotor angular position, not the current. Torque produced by the excitation of a particular winding, $k$, is

$$T_k(t) = \left. \frac{\partial W_c}{\partial \theta} \right|_{i = constant} \tag{8.344}$$

$$= \frac{\partial}{\partial \theta} \left( \frac{1}{2} L_k(\theta) \cdot i_k^2 \right) \tag{8.345}$$

where $W_c$ is the generalized concept of *co-energy*,

$$W_c = \int_0^{i_k} \lambda \cdot di \tag{8.346}$$

$$= \int_0^{i_k} L_k(\theta) \cdot i_k \cdot di_k \tag{8.347}$$

$$= \frac{1}{2} L_k(\theta) \cdot i_k^2 \tag{8.348}$$

Let us assume that current is constant, and determine the variation of co-energy as a function of rotor position, which gives us the torque

$$T_k(t) = \frac{1}{2} \cdot \frac{\partial L_k(\theta)}{\partial \theta} \cdot i_k^2 \tag{8.349}$$

The voltage–current relationship of the phase winding is

$$V_k(t) = R \cdot i_k(t) + \frac{d}{dt} \lambda_k(t) \tag{8.350}$$

$$= R \cdot i_k(t) + \frac{d}{dt} (L_k(\theta) \cdot i_k(t)) \tag{8.351}$$

$$= R \cdot i_k(t) + L_k(\theta) \frac{di_k(t)}{dt} + \frac{dL_k(\theta)}{d\theta} \cdot i_k(t) \cdot \frac{d\theta}{dt} \tag{8.352}$$

Notice that the electrical model of a switched reluctance motor is similar to a permanent magnet DC motor in that the terminal voltage, $V_k(t)$, is balanced or consumed by three components:

1. Resistance times the current, $R \cdot i_k(t)$
2. Inductance times the time rate of change of current, $L_k(\theta)\frac{di_k(t)}{dt}$
3. And back EMF voltage, $\frac{dL_k(\theta)}{d\theta} \cdot i_k(t) \cdot \frac{d\theta}{dt}$

The back EMF voltage of an SR motor differs from the back EMF of a permanent magnet (PM) DC motor. In the case of PM DC motor, the back EMF voltage is proportional to the speed of the rotor and the proportionality constant is determined by the permanent magnets and the motor design,

$$V_{bemf} = K_E \cdot \frac{d\theta}{dt} \tag{8.353}$$

The back EMF voltage term in an SR motor is a function of current and change of inductance as function of rotor angle in addition to rotor speed. The back EMF voltage of an SR motor is

$$V_{bemf} = \frac{dL_k(\theta)}{d\theta} \cdot i_k(t) \cdot \frac{d\theta}{dt} \tag{8.354}$$

The back EMF voltage is zero if any one of the following is zero: current, change of inductance as a function of rotor position, rotor speed. We can think of back EMF of SR motors as having a variable gain $K_E(i, dL/d\theta)$ that is a function of current and inductance variation.

## 8.7 LINEAR MOTORS

In principle, all of the rotary electric motors can be designed and manufactured as linear electric motors by unrolling the cylindrical shape of stator and rotor into linear form. Such linear motors have rectangular cross sections and are called *flat linear motors*. In addition, the unrolled linear motor can be rolled back around the axis of linear motion to obtain *tubular* linear motor construction. Such linear motors have circular cross sections (Fig. 8.66).

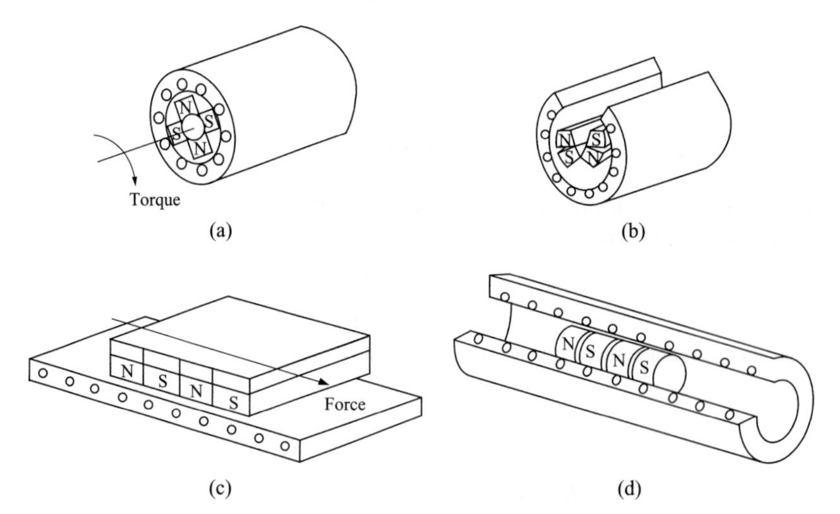


**FIGURE 8.66:** Principle of linear motor construction by unrolling a rotary motor: (a) a rotary permanent magnet type electric motor (i.e., brushless DC); (b) unrolling concept; (c) flat-type linear brushless DC motor; and (d) tubular-type linear motor by rolling again the flat-type linear motor concept around the axis of its force.

**TABLE 8.4: Comparison of Linear Brushless DC Motor Types in Terms of Cost, Cogging Force, Power Density, and Weight of the Moving Component**

| | Linear Brushless DC Motor Type | | |
|---|---|---|---|
| **Feature** | **Iron Core** | **Ironless (air) Corey** | **Slotless** |
| Attraction force | $F_{atr} \approx 10 \cdot F_{trust}$ | None | $F_{atr} \approx 5$ *to* $7 \cdot F_{trust}$ |
| Cost | Medium | High | Lowest |
| Force cogging | Highest | None | Medium |
| Power density | Highest | Medium | Medium |
| Forcer weight | Heaviest | Lighest | Moderate |

The electromagnetic force generation between the two components follows the same physical principles as in the case of their rotary counterparts. Figure 8.66 shows the basic principle of a linear motor by unrolling the cylindrical shape to flat shape. The number of actively controlled stator phases are the same. The commutation of the current in each phase is based on the cyclic linear distance (pitch) of the permanent magnet dimensions instead of the rotary angle. In a two-pole rotary brushless DC motor, the commutation cycle is 360 degrees, whereas the commutation cycle in a linear brushless motor is the distance between two consecutive pole pairs (i.e., distance between two north or two south pole magnets, total length of a north and a south pole magnet). In general, the same amplifiers used in the rotary version are used for the linear version of the motor. The feedback sensor (if used for commutation) is a linear displacement sensor instead of a rotary displacement sensor, i.e., hall effect sensors used to detect the relative position of stator with respect to rotor within a cyclic distance of pitch. Similarly, linear encoders may be used in place of rotary encoders for position sensing.

Linear brushless DC motors have three basic types: (1) iron core, (2) ironless (air core), and (3) slotless (Table 8.4). The basic design principles of each kind are shown in Fig. 8.67. In a brushless DC tubular linear motor, the permanent magnets are rolled into cylindrical shape around the axis perpendicular to the rotary motor axis. The stator has three phases and wound around the rotor. The controller and amplification stage of a linear motor are identical to those used in rotary version. The commutation in the amplifier is based on the cyclic pitch distance of the magnet pairs in the rotor, instead of the angular position in the case of rotary motors.

The dynamic model of a linear DC motor is identical to that of rotary DC motors. The electrical dynamics of the motor is

$$V_t(t) = R \cdot i(t) + L \cdot \frac{di(t)}{dt} + K_e \cdot \dot{x}(t) \tag{8.355}$$

where $V_t(t)$ is the terminal voltage, $R$ is the winding resistance, $L$ is the self-inductance, and $K_e$ is the back emf gain of the motor. The net electrical power converted to mechanical power is

$$\begin{aligned} P_e(t) &= V_{bemf}(t) \cdot i(t) && (8.356) \\ &= K_E \cdot \dot{x}(t) \cdot i(t) && (8.357) \\ &= K_T \cdot \dot{x}(t) \cdot i(t) && (8.358) \\ &= F(t) \cdot \dot{x}(t) && (8.359) \\ &= P_m(t) && (8.360) \end{aligned}$$

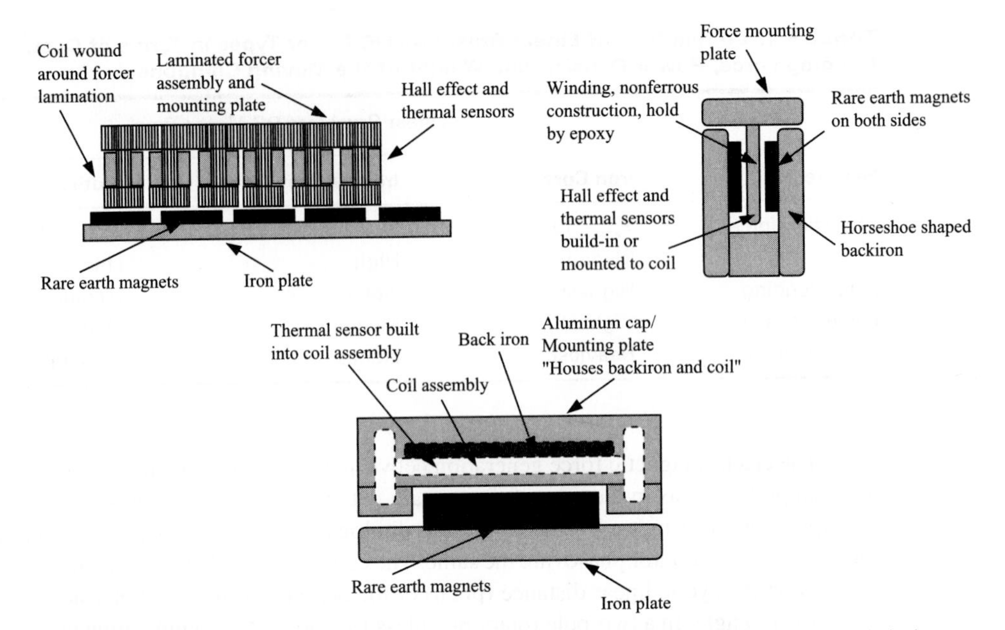


**FIGURE 8.67:** Linear brushless DC motor design types: iron core, ironless core, and slotless designs. Iron core has a large attraction force (up to 10 times the trust force) between rotor and stator which must be supported by the linear bearing. Ironless core does not have any attraction force. Slotless design has an attractive force somewhere in-between.

assuming 100% efficiency in converting the electrical power to mechanical power,

$$P_m(t) = F(t) \cdot \dot{x}(t) \tag{8.361}$$

$$= P_e(t) \tag{8.362}$$

$$= V_{bemf}(t) \cdot i(t) \tag{8.363}$$

where the force–current relationship is

$$F(t) = K_T \cdot i(t) \tag{8.364}$$

which also shows the force/torque gain is equal to the back emf gain. The gain $K_t$ is function of the magnetic flux (flux density times the cross-sectional area linking the magnetic field to the conductors) and the number of turns of the coil. In other words, it is a function of the magnetic flux at the operating point of the permanent magnet and its size, plus the number of coil turns that links the flux. For a practical motor, the solution of flux is best obtained by finite element based software tools. Therefore, the force gain is determined by the solution of FEA-based software instead of analytical solutions which are only possible for simple and idealized motor geometries. *Voice coil actuator* (also called moving coil actuator) is made of a tubular permanent magnet (PM) and a coil winding (Fig. 8.68). The interaction between the current-carrying coil assembly and PM generates the linear force. In principle, this is identical to the brush type DC motors. At any given instant in time, both motor action (force generation) and generator action (back EMF voltage) are in effect,

$$F = k \cdot l \cdot N \cdot B \cdot i = K_F \cdot i \tag{8.365}$$

$$V_{bemf} = k \cdot l \cdot N \cdot B \cdot \dot{x} = K_E \cdot \dot{x} \tag{8.366}$$

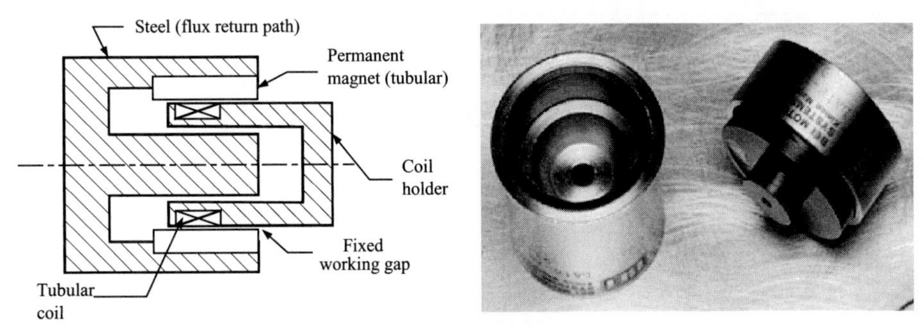


**FIGURE 8.68:** Voice coil actuator operating principle and components.

where $F$ is linear force, $V_{bemf}$ is the back emf voltage, $l$ is the length of the winding and $N$ is the number of turns of the coil, $B$ is the magnetic field strength across the air-gap between the rotor and stator, $i$ is the current in the coil, and $\dot{x}$ is the linear speed of the rotor. As long as the geometric overlap between the rotor and stator (coil and PM) is the same, the force is essentially independent of the displacement of the rotor and only function of the current. In order to provide such a force-current-displacement relationship, the axial length of the PM and the coil must be different (one longer than the other). The current in the coil winding is noncommutated. There are no commutator-brush components. Only the direction and magnitude of the current are controlled, i.e., using an H-bridge amplifier which is the same type used for a brush-type DC motor. Moving coil actuators are precision motion versions of the solenoid-plunger or audio-speaker designs. Voice coil actuators typically have small travel range (i.e., a micron to a few inches range) and very high bandwidth.

## 8.8 DC MOTOR: ELECTROMECHANICAL DYNAMIC MODEL

The most commonly used model for a direct current (d.c.) electric motor is shown in Fig. 8.69. This dynamic model includes electrical, electrical-to-mechanical power conversion, and mechanical dynamic relations.

The electrical relation between terminal voltage, current, and rotor speed is

$$V_t(t) = L_a \frac{di(t)}{dt} + R_a i(t) + K_e \cdot \dot{\theta}(t)$$

where $K_e\dot{\theta}(t)$ term is the voltage generated by the back electromotive force (back EMF) as a result of the generator action, and $L_a$, $R_a$ are inductance and resistance of the winding, respectively. The electrical-to-mechanical power conversion (current to torque) is given by

$$T_m(t) = K_T\, i(t)$$

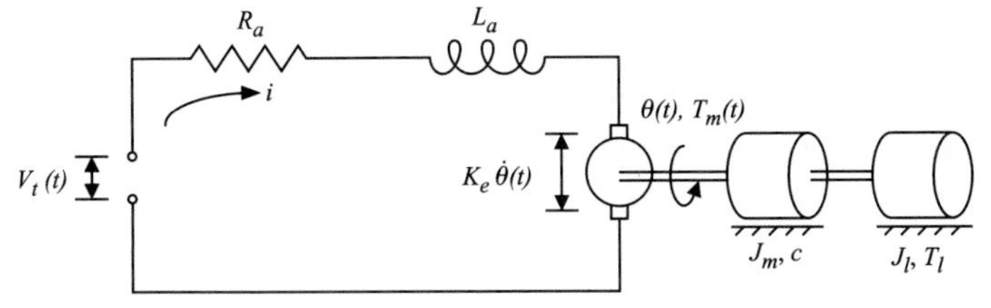


**FIGURE 8.69:** DC motor dynamic model.

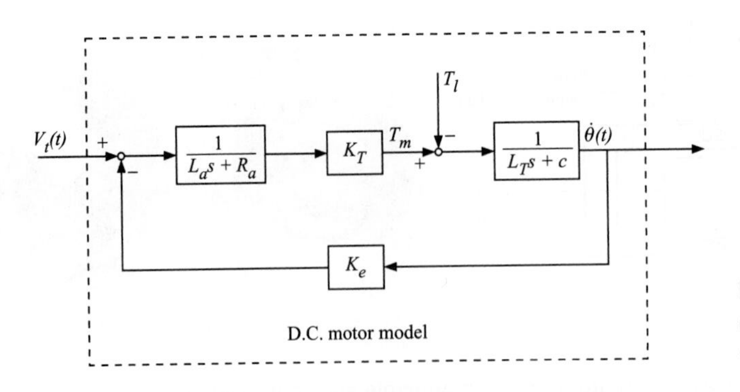


**FIGURE 8.70:** Block diagram of the DC motor dynamic model.

where $K_T$ is the torque gain and $T_m$ is the torque generated by the motor. Finally the mechanical relation between torque, inertia, and other load is given by

$$T_m(t) = (J_m + J_l)\,\ddot{\theta} + c\,\dot{\theta}(t) + T_l(t)$$

where $J_m$ is rotor inertia, $J_l$ is load inertia, $c$ is damping constant, and $T_l$ is load torque. From these three basic relationship, we can derive various transfer functions, i.e., the transfer function between terminal voltage to motor speed or motor position, or transfer function from armature current to motor speed. Physically, an amplifier manipulates the motor terminal voltage in order to control the motor motion. This voltage control may be based on current feedback, voltage feedback, or both.

Let us derive the transfer function from motor terminal voltage to motor speed. Taking the Laplace transform of the above equations for zero initial conditions (Fig. 8.70),

$$V_t(s) = (L_a s + R_a)i(s) + K_e\dot{\theta}(s) \tag{8.367}$$

$$\rightarrow i(s) = \frac{1}{L_a s + R_a}[V_t(s) - K_e\dot{\theta}(s)] \tag{8.368}$$

$$T_m(s) = K_T i(s)$$

$$(J_T s + c)\dot{\theta}(s) = T_m(s) - T_l(s)$$

where $J_T = J_m + J_l$.

The transfer function describing the effect of terminal voltage and load torque on the motor speed can be found as (Fig. 8.70)

$$\dot{\theta}(s) = \frac{K_T}{(J_t s + c)(L_a s + R_a) + K_T K_e}V_t(s) - \frac{(L_a s + R_a)}{(J_t s + c)(L_a s + R_a) + K_T K_e}T_l(s)$$

The transfer function from motor terminal voltage to motor speed is given by

$$\frac{\dot{\theta}(s)}{V_t(s)} = \frac{K_T}{(J_t s + c)(L_a s + R_a) + K_T K_e} \tag{8.369}$$

$$= \frac{K_T}{J_T L_a s^2 (L_a c + J_T R_a)s + (cR_a + K_T K_e)} \tag{8.370}$$

$$= \frac{K_T}{J_T L_a s^2 + \left(\frac{L_a c + J_T R_a}{J_T L_a}\right)s + \left(\frac{cR_a + K_T K_e}{J_T L_a}\right)} \tag{8.371}$$

The poles of the transfer function are given by

$$s^2 + \left(\frac{L_a c + J_T R_a}{J_T L_a}\right)s + \left(\frac{cR_a + K_T K_e}{J_T L_a}\right) = 0$$

Normally, this equation has two complex conjugate roots.

***Special Case: DC Servo Motors*** DC servo motors have very low inductance ($L$ small) and damping ($c$ small). Using these facts for the d.c. servo motors case, the transfer function can be approximated as

$$\frac{\dot{\theta}(s)}{V_t(s)} \simeq \frac{\frac{K_T}{J_T L_a}}{s^2 + \left(\frac{R_a}{L_a}\right)s + \left(\frac{K_T K_e}{J_T L_a}\right)}$$

where the poles are given by

$$p_{1,2} = -\frac{R_a}{2L_a} \pm \frac{\sqrt{\left(\frac{R_a}{L_a}\right)^2 - 4\left(\frac{K_T K_e}{J_T L_a}\right)}}{2} \tag{8.372}$$

$$= -\frac{R_a}{2L_a} \pm \frac{\sqrt{4K_T K_e J_T\left(\frac{R_a^2 J_T}{4K_T K_e} - L_a\right)}}{2L_a J_T} \tag{8.373}$$

$$\simeq -\frac{R_a J_T}{2L_a J_T} \pm \frac{R_a J_T\left(1 - \frac{2L_a K_T K_e}{R_a^2 J_T}\right)}{2L_a J_T} \tag{8.374}$$

where we used the approximation

$$\sqrt{1-x} \simeq 1 - \frac{x}{2}; \text{ for } x \ll 1$$

The poles are

$$p_1 = -\frac{K_T K_e}{J_T R_a} \tag{8.375}$$

$$p_2 = -\frac{2R_a J_T + \left(\frac{2L_a K_T K_e}{R_a}\right)}{2L_a J_T} \tag{8.376}$$

$$= -\frac{R_a}{L_a} + \frac{K_T K_e}{J_T R_a} \tag{8.377}$$

Because $\frac{R_a}{L_a} \gg \frac{K_T K_e}{J_T R_a}$, the second pole is approximately

$$p_2 \simeq -\frac{R_a}{L_a} \tag{8.378}$$

The motor terminal voltage to motor speed transfer function can be approximated as

$$\frac{\dot{\theta}(s)}{V_t(s)} = \frac{\left(\frac{K_T}{J_T L_a}\right)}{(s-p_1)(s-p_2)} \tag{8.379}$$

$$= \frac{\left(\frac{1}{K_e}\right)}{(\tau_m s + 1)(\tau_e s + 1)} \tag{8.380}$$

where

$$\tau_m = -\frac{1}{p_1} = \frac{J_T R_a}{K_T K_e} \quad \text{mechanical time constant} \tag{8.381}$$

$$\tau_e = -\frac{1}{p_2} = \frac{L_a}{R_a} \quad \text{electical time constant} \tag{8.382}$$

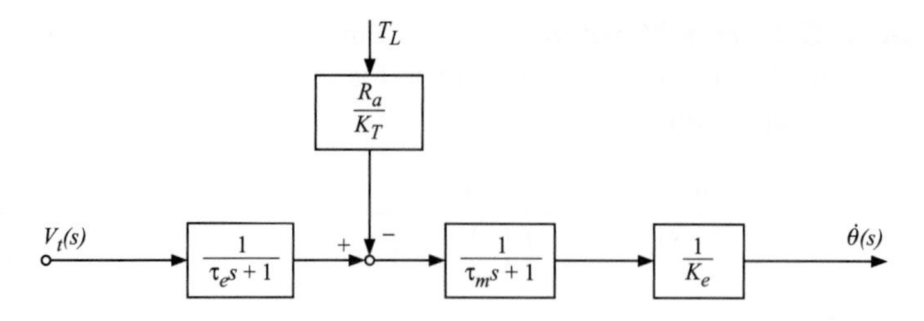


**FIGURE 8.71:** Block diagram of the relationship betweeen motor terminal voltage and load torque to motor speed for a typical DC servo motor.

and the mechanical time constant $\tau_m$ is much larger than the electrical time constant $\tau_e$ for most motors. The motor speed is also influenced by disturbance or load torque. The transfer function including the load torque can be derived as follows:

$$\dot{\theta}(s) = \frac{\frac{K_T}{J_T L_a}}{s^2 + \left(\frac{R_a}{L_a}\right)s + \left(\frac{K_T K_e}{J_T L_a}\right)} V_t(s) - \frac{\left(\frac{L_a s + R_a}{J_T L_a}\right)}{s^2 + \left(\frac{R_a}{L_a}\right)s + \left(\frac{K_T K_e}{J_T L_a}\right)} T_l(s) \tag{8.383}$$

$$= \frac{\frac{1}{K_e}}{(\tau_m s + 1)(\tau_e s + 1)} V_t(s) - \frac{(\tau_e s + 1) R_a \frac{1}{K_T} \frac{K_T}{J_T L_a}}{s^2 + \left(\frac{R_a}{L_a}\right)s + \left(\frac{K_T K_e}{J_T L_a}\right)} T_l(s) \tag{8.384}$$

$$= \frac{\frac{1}{K_e}}{(\tau_m s + 1)(\tau_e s + 1)} V_t(s) - \frac{\frac{1}{K_e}\frac{R_a}{K_T}(\tau_e s + 1)}{(\tau_m s + 1)(\tau_e s + 1)} T_l(s) \tag{8.385}$$

$$\dot{\theta}(s) = \frac{\frac{1}{K_e}}{(\tau_m s + 1)(\tau_e s + 1)} V_t(s) - \frac{\frac{1}{K_e}\frac{R_a}{K_T}}{(\tau_m s + 1)} T_l(s) \tag{8.386}$$

The transfer function from the terminal voltage and load torque to the motor speed in terms of time constants and dc gains is shown in block diagram form in Fig. 8.71.

## 8.8.1 Voltage Amplifier Driven DC Motor

If a voltage amplifier is used to drive the motor, the transfer function between voltage amplifier input and motor speed is given by (Fig. 8.72)

$$\frac{\dot{\theta}(s)}{V_{in}(s)} = \frac{\dot{\theta}(s)}{V_t(s)} \frac{V_t(s)}{V_{in}(s)} = G_{amp}(s) G_{motor}(s)$$

where the voltage amplifier is represented by a first-order filter model,

$$\frac{V_t(s)}{V_{in}(s)} = G_{amp}(s) = \frac{A_v}{(\tau_a s + 1)} \tag{8.387}$$

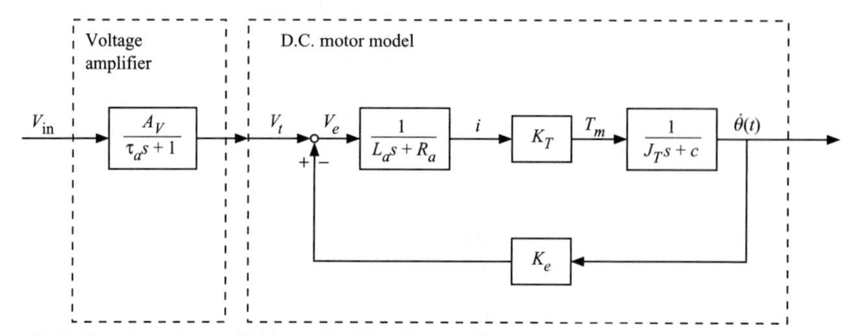


**FIGURE 8.72:** Block diagram of transfer functions for a DC servo motor driven by a voltage amplifier.

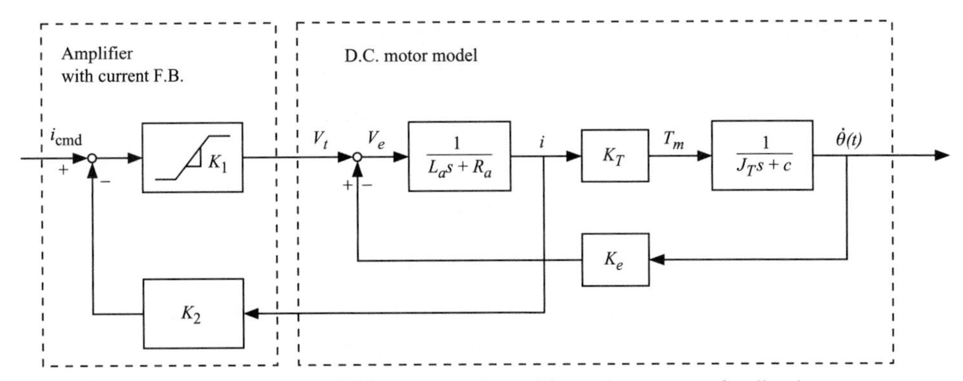


**FIGURE 8.73:** Block diagram of DC motor and amplifier using current feedback.

## 8.8.2 Current Amplifier Driven DC Motor

In the majority of cases, the d.c. motor terminal voltage is controlled by an amplifier which regulates armature current (Fig. 8.73). Directly regulating current is essentially directly regulating the torque generated by the motor because generated torque is proportional to the current. The transfer function of the current amplifier plus d.c. motor combination from amplifier input [commanded current, $i_{cmd}(t)$] to the motor speed can be derived as follows (neglecting the effect of amplifier terminal voltage saturation, Fig. 8.74):

$$\frac{\dot{\theta}(s)}{i_{cmd}(s)} = \frac{K_1 \frac{K_T}{(L_a s + R_a)(J_T s + c) + K_e K_T}}{1 + K_1 \frac{K_T}{(L_a s + R_a)(J_T s + c) + K_e K_T} K_2 \left(\frac{J_T s + c}{K_T}\right)} \tag{8.388}$$

$$= \frac{\frac{K_1 K_T}{L_a J_T}}{s^2 + \left(\frac{L_a c + R_a J_T s + K_1 K_2 J_T}{L_a J_T}\right) s + \left(\frac{K_e K_T + K_1 c + K_1 K_2 c}{L_a J_T}\right)} \tag{8.389}$$

$$= \frac{K}{s^2 + bs + c} \tag{8.390}$$

$$= \frac{K_a}{(\tau_a s + 1)} \frac{K_T}{(J_T s + c)} \tag{8.391}$$

If we consider the transfer function between armature current and motor speed,

$$T_m(t) = K_T i(t) \tag{8.392}$$

$$T_m(t) = J_T \ddot{\theta} + c\dot{\theta} - T_l(t) \tag{8.393}$$

$$\dot{\theta}(s) = \frac{K_T}{(J_T s + c)} i(s) - \frac{1}{(J_T s + c)} T_l(s) \tag{8.394}$$

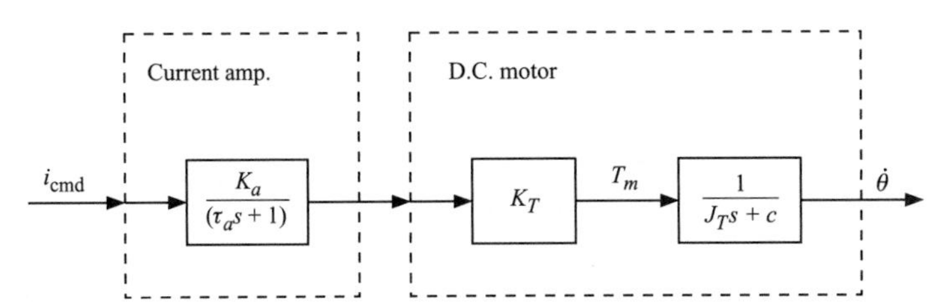


**FIGURE 8.74:** Current amplifier plus DC motor transfer function from commanded current to motor speed.

### 8.8.3 Steady-State Torque-Speed Characteristics of a DC Motor under Constant Terminal Voltage

Consider the electrical and electrical-to-mechanical power conversion relations for a d.c. motor:

$$V_t(t) = L_a \frac{di}{dt} + R_a i + K_e \dot{\theta}(t) \tag{8.395}$$

$$T_m(t) = K_T i(t) \tag{8.396}$$

In steady state the effect of $L_a$ will be zero. If we set $L_a = 0$ for steady-state analysis, the torque-speed terminal voltage relationship is given by

$$T_m(t) = \frac{K_T}{R_a} V_t(t) - \frac{K_T K_e}{R_a} \dot{\theta}(t) \tag{8.397}$$

This is a linear relationship of the type

$$y = -ax + b \tag{8.398}$$

Let us consider this equation for various constant terminal voltage values, $V_{ti}$ (Fig. 8.75). The steady-state torque-speed curve has a negative slope of $\frac{K_T K_e}{R_a}$ for a given constant terminal voltage. A given motor maximum torque at stall will saturate at the magnetic field saturation point, and the torque will not increase even if terminal voltage is increased beyond a certain voltage value. For a given constant terminal voltage, as the speed $(\Delta\dot{\theta})$ increases, the torque capacity of motor decreases by $\left(\frac{K_T K_e}{R_a}\right) \Delta\dot{\theta}$.

### 8.8.4 Steady-State Torque-Speed Characteristics of a DC Motor and Current Amplifier

When we consider a DC motor driven by a current amplifier, we need to consider the following additional relation,

$$V_t(t) = K_1(i_{cmd}(t) - K_2 i(t)); \quad V_t(t) \leq V_{tmax}$$

when the amplifier saturates, $V_t(t) > V_{tmax}$

$$V_t(t) = V_{tmax}$$

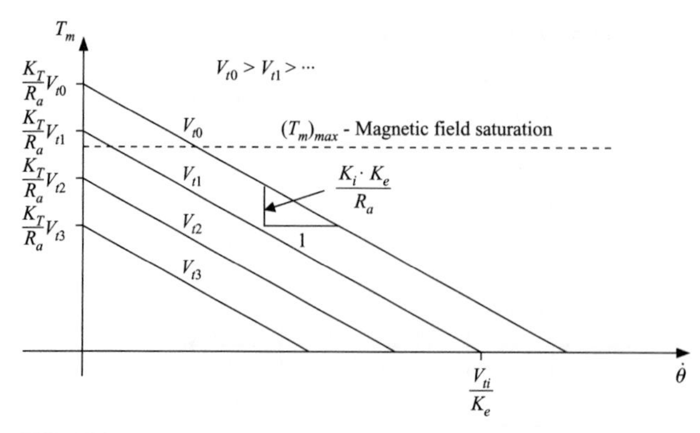


**FIGURE 8.75:** Steady-state torque-speed characteristics of a DC motor.

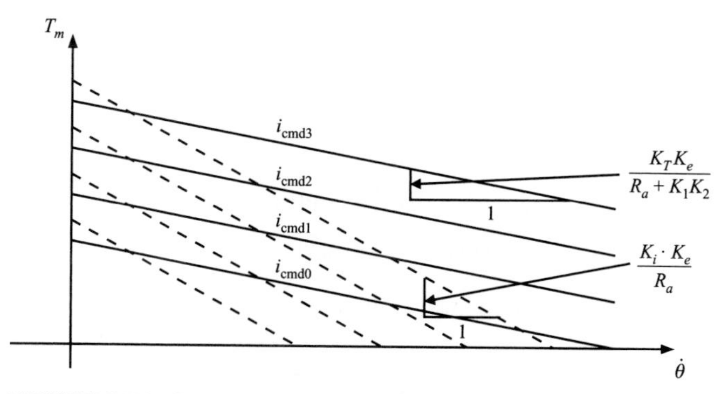


**FIGURE 8.76:** Speed-torque characteristics of a DC motor in steady state with current-controlled amplifier.

The steady-state torque-speed curve for the d.c. motor under current amplifier control,

$$T_m = \frac{K_T K_1}{R_a + K_1 K_2} i_{cmd} - \frac{K_T K_e}{R_a + K_1 K_2} \dot{\theta}$$

When the amplifier is commanded a constant current, ($i_{cmd}$), the speed torque characteristic is linear with negative slope. However, the slope is much flatter than the case of the d.c. motor with a constant terminal voltage (Fig. 8.76). In other words, the steady-state torque produced by a DC motor under constant current conditions (i.e., controlled by a current feedback amplifier) decreases with increasing speed at a much slower rate than the case when the motor is controlled by a voltage amplifier under constant terminal voltage condition.

***Example*** Consider a DC motor with the following parameters, $K_T = $ N m/A, $K_E = 6.7 \times 10^{-2}$ V/(rad/sec), $R_a = 0.5\ \Omega$, $L_a = 2$ mH, $J_m = 4.8 \times 10^{-5}$ kg m$^2$, $J_l = J_m$. Determine the mechanical and electrical time constants of the motor (consider the case in which the motor and the load inertia are connected). If the current loop control gains are $K_1 = 10.0$ and $K_2 = 1.0$, determine the speed-torque slopes under constant voltage and constant current control conditions.

The mechanical and electrical time constants of the motor with the load connected are

$$\tau_m = \frac{J_T R_a}{K_T K_E} = \frac{9.6 \times 10^{-5} \cdot 0.5}{6.7^2 \times 10^{-4}} \tag{8.399}$$

$$= 0.010 \, \text{sec} = 10 \, \text{msec} \tag{8.400}$$

$$\tau_e = \frac{L_a}{R_a} = \frac{2 \times 10^{-3} H}{0.5\ \Omega} = 4 \times 10^{-3} \, \text{sec} = 4 \, \text{msec} \tag{8.401}$$

The slope of the torque-speed curve in steady state under constant voltage and constant current conditions are

$$slope_v = \frac{K_T K_E}{R_a} \tag{8.402}$$

$$= \frac{6.7^2 \times 10^{-4}}{0.5} \tag{8.403}$$

$$= 89.7 \times 10^{-4} \, [\text{N m/(rad/sec)}] \tag{8.404}$$

$$= 0.0857 \, [\text{N m/rpm}] \tag{8.405}$$

$$slope_i = \frac{K_T K_E}{R_a + K_1 K_2} \tag{8.406}$$

$$= \frac{6.7^2 \times 10^{-4}}{0.5 + 10} \tag{8.407}$$

$$= 4.27 \times 10^{-4} \text{ [N m/(rad/sec)]} \tag{8.408}$$

$$= 0.00407 \text{ [N m/rpm]} \tag{8.409}$$

The ratio of the slope change in torque-speed curve is

$$\frac{slope_i}{slope_v} = \frac{4.27}{89.7} = \frac{1}{21} \tag{8.410}$$

For a given unit speed change, the torque variation under the current-controlled amplifier operation is 21 times less than that of under the voltage-controlled amplifier.

## 8.9 ENERGY LOSSES IN ELECTRIC MOTORS

An electric actuator is a device that converts energy from electrical to mechanical (Fig. 8.77). The conversion process is not 100% efficient; hence there are losses. These losses are in the form of heat. The energy losses can be categorized into three groups:

1. Resistance losses (also called copper losses): $R \cdot i^2$
2. Core losses (hysteresis and Eddy current losses)
3. Friction and windage losses

Resistance and core losses are electrical losses, whereas the friction and windage losses are mechanical losses.

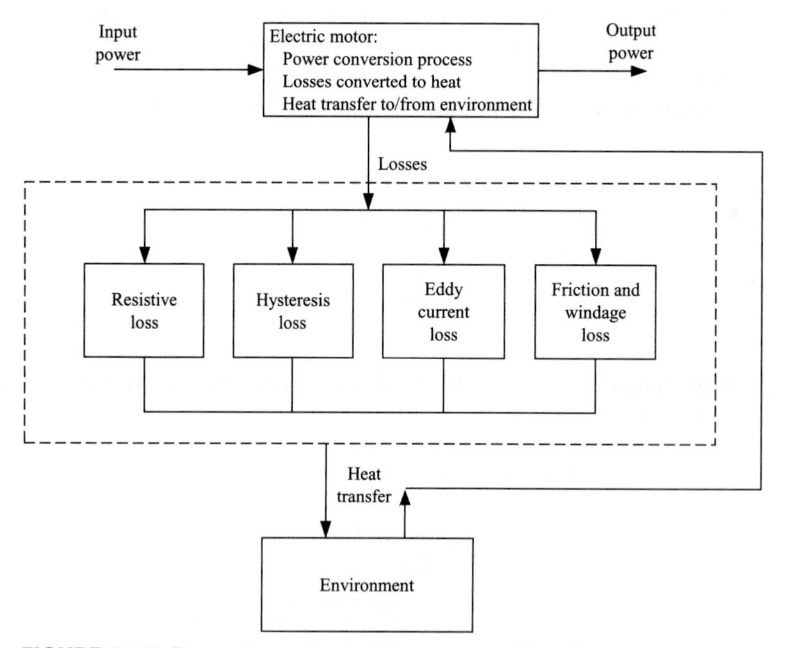


**FIGURE 8.77:** Energy losses in electric motors and its effect on the motor temperature.

The steady-state temperature of the actuator is determined by these losses, the heat transferred to the surrounding medium through conduction, convection, and radiation mechanisms. It is in general difficult to accurately predict the amount of heat transferred between the actuator and its surrounding medium. The energy balance ($Q_{net}$) is the difference between the generated heat due to the losses ($Q_{in}$) and the transferred heat from the actuator to its surroundings ($Q_{out}$). It is the energy that must be absorbed by the actuator in the form of temperature rise,

$$Q_{net} = Q_{in} - Q_{out} \tag{8.411}$$

$$= c_t \cdot m \cdot (T - T_0) \tag{8.412}$$

where $m$ is the mass of the actuator which absorbs heat, $c_t$ is the specific heat of the material of actuator in units of [Joule/(kg ·°C)], $T_0$ is the initial temperature, and $T$ is the steady-state temperature. The generated heat due to losses, $Q_{in}$,

$$Q_{in} = Q_R + Q_C + Q_F \tag{8.413}$$

where $Q_R$ is the resistive loss, $Q_C$ is the core loss, and $Q_F$ is the friction and windage losses.

The transferred heat is estimated with approximation as follows:

$$Q_{out} = c_{out} \cdot (T - T_{amb}) \cdot \Delta t \tag{8.414}$$

where $T_{amb}$ is the ambient temperature, and $c_{out}$ is the effective heat transfer coefficient between the actuator and its surroundings with units [(Joule/sec)/°C] = [Watts/°C], and $\Delta t$ is the time period.

The energy balance equation can also be expressed as power balance in differential equation form,

$$\frac{d}{dt}[Q_{net}] = \frac{d}{dt}[Q_{in} - Q_{out}] \tag{8.415}$$

$$P_{net} = P_{in} - P_{out} \tag{8.416}$$

$$c_t \cdot m \cdot \frac{dT}{dt} = P_{in} - c_{out}(T - T_{amb}) \tag{8.417}$$

$$c_t \cdot m \cdot \frac{dT}{dt} + c_{out}(T - T_{amb}) = P_{in} \tag{8.418}$$

This models the electric actuator thermal behavior like a first-order dynamic system. The difficulty in using such an analytical model is in the difficulty of accurately estimating $c_T$, $c_{out}$, and $P_{in}$. Note that

$$P_{in} = \frac{d}{dt}Q_{in} \tag{8.419}$$

$$= \frac{d}{dt}Q_R + \frac{d}{dt}Q_C + \frac{d}{dt}Q_F \tag{8.420}$$

$$= P_R + P_C + P_F \tag{8.421}$$

Next we discuss the nature of $P_R$, $P_C$, $P_F$ power losses.

### 8.9.1 Resistance Losses

Electric actuators have coils which are windings of current-carrying conductors. Coils act as current-controlled electromagnets. The conductor material, i.e., copper or aluminum, has finite electrical resistance. In order to generate an electromagnetic effect, we need to pass a certain amount of current. As the electrical potential pushes the current through the conductor, there is energy loss due to resistance. This is similar to the energy loss in pushing a certain fluid flow rate through a restriction. The power ($P_R$) lost as heat due to resistance ($R$) of a conductor when current $i$ is conducted,

$$P_R = R \cdot i^2 \tag{8.422}$$

Therefore, in order to minimize the heating of the motor, it is desirable to minimize the resistance and current. However, large current is desired to generate large force or torque. The energy ($Q_r$) that is converted to heat over a period of time $t_{cycle}$ is

$$Q_R = \int_o^{t_{cycle}} P_R \cdot dt \tag{8.423}$$

$$= \int_o^{t_{cycle}} R \cdot i(t)^2 \cdot dt \tag{8.424}$$

Resistive loss, $P_R = R \cdot i^2$, is fairly well estimated. The temperature dependence of resistance can also be taken into account for better accuracy of thermal prediction,

$$R(T) = R(T_0)[1 + \alpha_{cu}(T - T_0)] \tag{8.425}$$

where $R(T)$, $R(T_0)$ are resistances of coil at temperature $T$ and $T_0$, respectively, and $\alpha_{cu} = 0.00393$ for copper which is the property of the conductor material. Notice that for copper

$$R(125°\text{C}) \approx 1.4 \cdot R(25°\text{C}) \tag{8.426}$$

The resistance of the coil increases by 40% when its temperature increases by 100°C.

### 8.9.2 Core Losses

Core losses refer to the energy lost, in the form of heat, in electric actuators at their stator and rotor structure due to electromagnetic variations. There are two major physical phenomenon that contribute to core losses: *hysteresis losses* and *Eddy current losses.*

*Hysteresis loss* is due to the hysteresis loop in the B-H characteristics of the core (rotor and stator) material which is typically a soft ferromagnetic material. As the magnetic field changes, the hysteresis loop is traversed and energy is lost. The hysteresis loss is proportional to the magnitude of the magnetic field change and its frequency. This loss can be minimized by selecting core material that has small hysteresis loops, i.e., high silicon content in the steel used for the core material.

*Eddy currents and related losses* can be summarized as follows. If a bulk piece of metal (i.e., iron, copper, aluminum) moves through a magnetic field or a stationary in a changing magnetic field, a circulating current is induced on the metal. This current is called *Eddy current* and results in heat loss due to the resistance of the metal. The induced current is a result of Faraday's law of induction. That is, the changing magnetic field (whether due to change in $\vec{B}$ or motion of the metal conductor relative to $\vec{B}$ or both) induces a voltage on the conductor. This induced voltage generates the current. Power loss due to Eddy currents is proportional to the square of magnetic flux density and its frequency of variations. Therefore, it is expected that the hysteresis losses will be more dominant at low frequencies, whereas Eddy current losses will be more dominant at high frequencies.

Laminations of thin layers of metal which are insulated and stacked together, as opposed to bulk metal, are used to reduce the Eddy currents in motor applications. Hence, Eddy current-related heat losses are reduced, increasing the efficiency of the motor. The stator of DC and AC motors are built using laminated iron instead of bulk iron in order to reduce the Eddy current related loss of energy into heat. For different lamination materials, hysteresis and Eddy current losses are given based on measured data as a function of field intensity and frequency,

$$P_C^* = f(B_{max}, w) \tag{8.427}$$

where $B_{max}$ is the maximum flux density, and $w$ is the frequency of change in magnetic flux. The nonlinear funciton $f(B_{max}, w)$ is defined using numerical graphs for different lamination materials by manufacturers. Notice that the core loss data for different materials are given per unit mass of the material. Total core loss for a given mass of motor application,

$$P_C = m \cdot P_C{}^* \tag{8.428}$$

The designer can estimate the core losses for a given design using the manufacturer supplied data.

### 8.9.3 Friction and Windage Losses

These losses are significant at very high speeds due to the air resistance between the rotor and stator. The energy loss due to air resistance is called the windage loss. Energy loss due to bearing friction is called the friction loss. They are taken into account by increasing the resistive loss by a safety factor, because it is very difficult to accurately model friction and windage losses, i.e.,

$$P_F = 0.1 \cdot P_R \tag{8.429}$$

$$Q_F = 0.1 \cdot Q_R \tag{8.430}$$

## 8.10 PROBLEMS

**1.** Consider the solenoid shown in Fig. 8.18. Assume that the permeability of the iron core is much larger than the permeability of the air-gap ($x$). Hence, let us neglect the reluctance of the path in the iron core.

**(a)** Draw the equivalent electromagnetic circuit diagram.

**(b)** Derive the relationship for the inductance as function of the air-gap distance $x$, which is variable. Let air-gap cross-sectional area $A_g = 100$ cm$^2$, number of turns of the coil $N = 250$. Plot the inductance as a function of air-gap $x$ for 0.0 mm $\geq x \geq$ 10.0 mm.

**2.** Consider the linear motion mechanism shown in Fig. 3.3. Draw a block diagram of a closed-loop position control system components for this motion control system. Discuss the advantages and disadvantages of using (i) DC motor, (ii) stepper motor, (iii) switched reluctance motor, and (iv) vector-controlled AC motor as the actuator to provide the mechanical power to the ball-screw mechanism. Assume that the necessary power level is under 1.0 kW.

**3.** Consider a DC motor (or equivalent) motion control system. Assume that the DC bus voltage available is $V_s = 90$ VDC. The back EMF constant of the motor is $K_e = 20$ V/krpm. The nominal terminal resistance of the motor stator windings is $R = 10\ \Omega$.

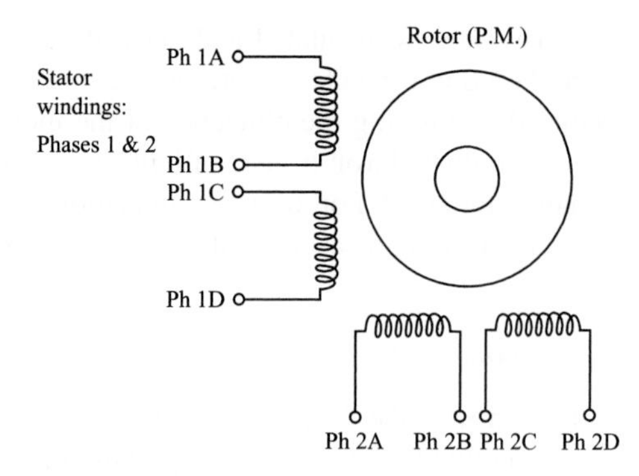


**FIGURE 8.78:** Two-phase stepper motor. Each phase is wound with two identical windings. Each phase has four terminal wires. The phase windings can be terminated to configure the motor for unipolar or bipolar operation.

**(a)** Determine the maximum no-load speed and the maximum torque capacity at stall (zero speed) of the motor in steady state. Discuss the factors that determine the peak torque and RMS torque capacity of the motor and predict peak and RMS torque capacity of the motor.

**(b)** Assume that the DC motor is controlled by a velocity mode amplifier and speed sensor feedback (i.e., tachometer). The amplifier sends a voltage command to the motor proportional to the speed error. Input speed is commanded with a 0 to 10 VDC voltage source. Determine the sensor gain and amplifier gain so that the closed-loop system has good closed-loop response and show that 0 to 10 VDC command signal results in proportional output speed in the speed range of the motor.

**4.** Consider a hybrid PM stepper motor.

**(a)** What is the difference between half-step and full-step mode. What is the main advantage and disadvantage of half stepping mode?

**(b)** Discuss how micro-stepping is accomplished and its advantages/disadvantages.

**(c)** Discuss the major performance differences between step motors and brushless DC motors.

**5.** Consider a stepper motor with two phases (Fig. 8.78). Phase 1 is wound with two identical coils. Phase 2 is also wound with two identical coils. Therefore, there are eight wires that come out from the stator winding, four for each phase.

**(a)** Connect the phase terminal wires in such a way that the motor can be used as a unipolar wound motor and draw a schematics for the amplifier circuit of the motor windings.

**(b)** Connect the phase terminal wires in such a way that the motor can be used as a bipolar wound motor with *series* connected windings (two identical windings for each phase are connected in series). Draw the amplifier circuit and its connections to motor windings.

**(c)** Connect the phase terminal wires in such a way that the motor can be used as a bipolar wound motor with *parallel* connected windings (two identical windings for each phase are connected in parallel). Draw the amplifier circuit and its connections to motor windings.

**(d)** What is the main performance difference between bipolar series and bipolar parallel configuration of the motor?

**6.** Consider the drive circuit for a three-phase brushless DC motor shown in Fig. 8.33. Assume that the rotor has one south and one north pole (like the one shown in Fig. 8.30(c)). Let us focus on the ON/OFF state of six power transistors ($Tr_1, \ldots, Tr_6$) in order to generate torque in forward and reverse directions. Determine the sequence of the ON/OFF conditions of the transistors for forward and reverse torque generations. Assume a nominal rotor position, hence the magnetic field of the rotor, as the beginning of the commutation cycle. *Hint*: Make a table with columns being the transistors 1 through 6 states, the position of the rotor at the beginning of each transistor state switching sequence, and the magnetic field vector generated by the stator at a particular switching pattern. The rows should be six different transistor states for forward torque generation in sequence and similar information for the reverse torque generation. In an actual motor, the power transistors would be controlled by a PWM signal to generate a sinusoidal or trapezoidal current as a function of rotor position as opposed to the

ON/OFF switching discussed in this problem. Nonetheless, the ON/OFF switching of transistors is still useful in understanding the current commutation in DC brushless motors.

**7.** Consider the torque versus rotor position under constant current conditions for a DC brush-type motor (Fig. 8.29). Assume that for each coil connected to the commutators, the torque is a sinusoidal function of the rotor position under constant current, as shown in the same figure. Plot the torque versus rotor position for one revolution, under constant current, for the following cases: (a) two coils only (and two commutators), (b) four coils (and four commutators), (c) eight coils, and (d) 16 coils. What is the benefit of having a large number of commutator segments? Assume that all coils are symmetrically distributed, i.e., in a four-coil case, the second coil is 180° electrically (1/4 revolution mechanical degrees) out of phase with the first coil; in the eight-coil case, each coil is 1/8 of a revolution out of phase from the previous one. Use of Matlab/Simulink for the plot and calculations is recommended.

**8.** Consider a DC motor, current amplifier, a closed-loop PD type controller, and a position feedback sensor (Figs. 8.69, 8.74, 2.35). Consider that the motor dynamics is described by its rotary intertia ($J_m$[kgm$^2$]) and current to torque gain ($K_t$[Nm/A]). Let the current amplifier be modeled by a static gain of voltage command to current gain of $K_a$[A/V]. The PD controller has gains $K_p$ and $K_d$ which defines the relationship between the position and velocity error and control signal (voltage command). Let the position feedback sensor also be represented by its gain and neglect any dynamic delays, $K_s$[counts/rad].

**(a)** Let $J_m = 10^{-4}$ kgm$^2$, $K_t = 0.10$ Nm/A, $K_a = 2.0$ A/V, $K_f = 2{,}000/(2\pi)$counts/rad, $K_p = 0.02$ V/counts, and $K_d = 10^{-4}$ V/(counts/sec). Find the loop transfer function (the transfer function from error signal to the output of the sensor (sensor signal)), then determine the frequency at which the magnitude of the loop transfer function is equal to unity. At that frequency, determine the phase angle of the loop transfer function.

**(b)** Similarly, determine the frequency at which the phase angle of the loop transfer function is equal to 180 degrees (if such a finite frequency exists) and find the magnitude of the loop transfer function at that frequency. In a real DC motor control system, such frequency does exist even if the above analysis may indicate that such frequency is not finite. Discuss why in real hardware such a frequency is finite. (Hint: Compare neglected filtering effects and time delays in the above analysis to real system. A pure time delay can be modelled in frequency domain as $e^{-jwt_d}$, where $t_d$ is the time delay. It adds phase lag to the loop transfer function.)

CHAPTER 9

# PROGRAMMABLE LOGIC CONTROLLERS

## 9.1 INTRODUCTION

Programmable logic controller (PLC) is the de-facto standard computer platform used in industrial control, factory automation, automated machine, and process control applications. PLCs were developed as a result of a need in automotive industry. In early 1960s, General Motors (GM) Corporation stated that the factory automation systems based on hardwired relay logic panels were not flexible enough for the changing needs of the industry. When a new car model required a different sequence of control logic, rewiring the control panels was taking too much time and the response time of the company was slow for new models. With hardwired relay logic panels, if there is a need to change the automation logic and functions of a line, the logic wiring between the input and output signals in the panel had to be physically changed. This is a time-consuming and costly process. It was requested in a design specification that a general-purpose wiring of all I/O devices be brought to a panel, but the logical relationship between the I/O be defined in software instead of being hardwired. In other words, it was desired that the logic be *softwired* instead of *hardwired*. That was the begining of the PLCs. PLCs play a very important role in the automated factories of the industrial world. The evolution of industrial automation can be viewed as shown in Fig. 9.1. The PLCs replaced the hardwired relay logic panels. The current trend is the large-scale networking between PLCs at factory and enterprise levels which may be physically distributed worldwide.

The physical shape of all PLCs made by many different companies all have the same form: it is rack mounted with standard-size slots to plug-in I/O interface units (Fig. 9.2). A typical PLC rack starts with a power supply and a CPU module plugged into a backplane of interface bus. A rack may have different number of slots, i.e., 4 slots, 7 slots, 10 slots, 15 slots. A PLC may support multiple I/O racks (main rack that has the CPU, and expansion I/O racks). Typical software development tools for a PLC include a notebook PC, a serial or ethernet communication interface and cable, and a software development environment for that particular PLC (Fig. 9.3). Depending on the I/O capabilities, CPU speed, and program functions, PLCs are categorized into four major sizes (Fig. 9.2). Any of the I/O interface modules can be plugged into the slots. All of the interface and unit power lines are provided by the snap-on connection to the rack. Typical I/O units supported by virtually every PLC platform include discrete input and output modules, analog input and output modules, high-speed counter and timer modules, serial communication modules, communication network interface modules (i.e., DeviceNet, CAN, ProfiBus), servo motor control modules and stepper motor control modules. For a given application, the necessary I/O units are selected and inserted into the slots. Furthermore, if the application needs change, additional I/O modules can be added by simply inserting them into the available slots on the main rack or the expansion racks. Each I/O module occupies a finite number of memory in the PLC's

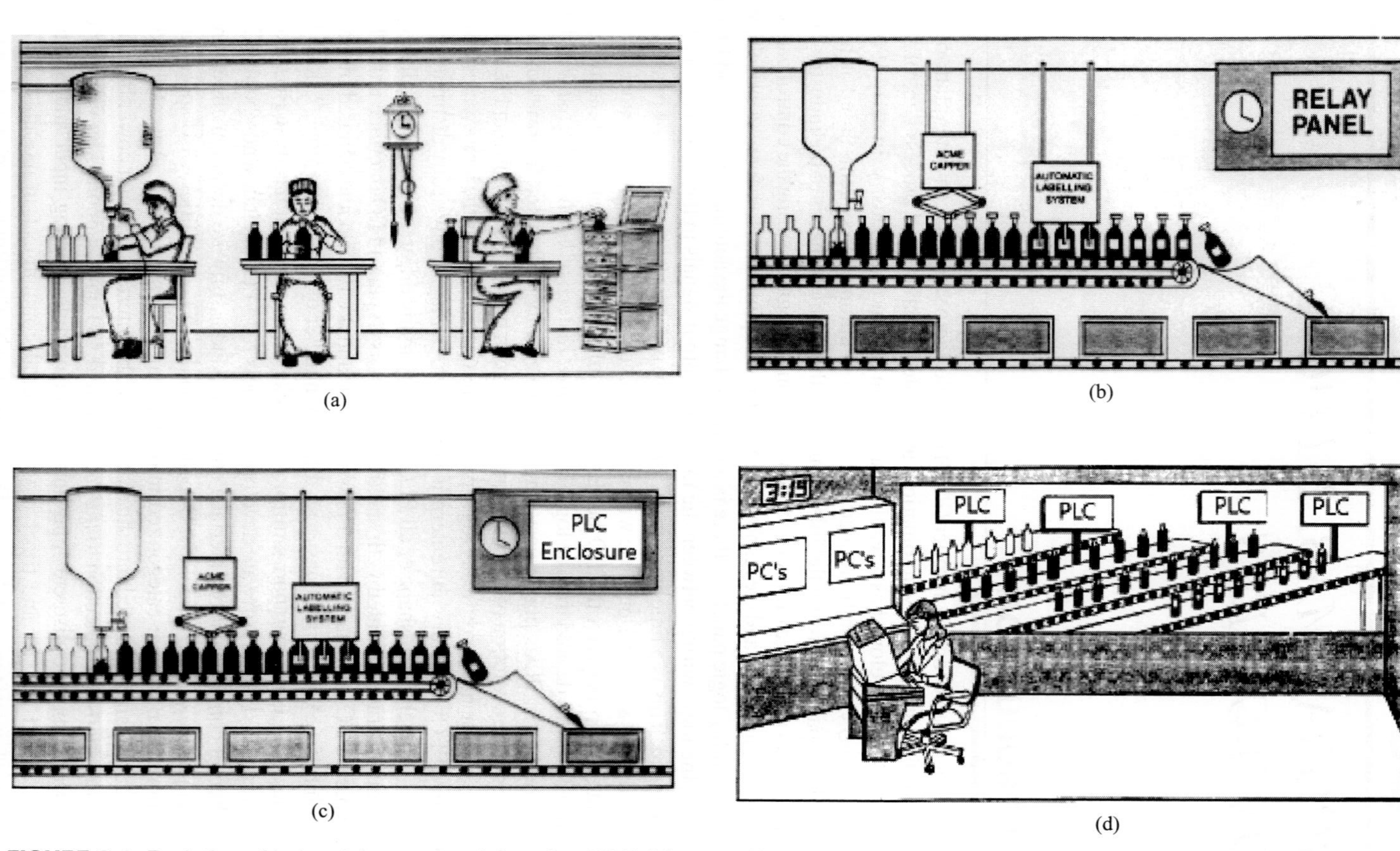


**FIGURE 9.1:** Evolution of industrial control and the role of PLC: (a) manual industrial production lines, (b) automated manufacturing using relay logic panels, (c) automated manufacturing using PLCs, and (d) networked automated manufacturing using PLCs and PCs.

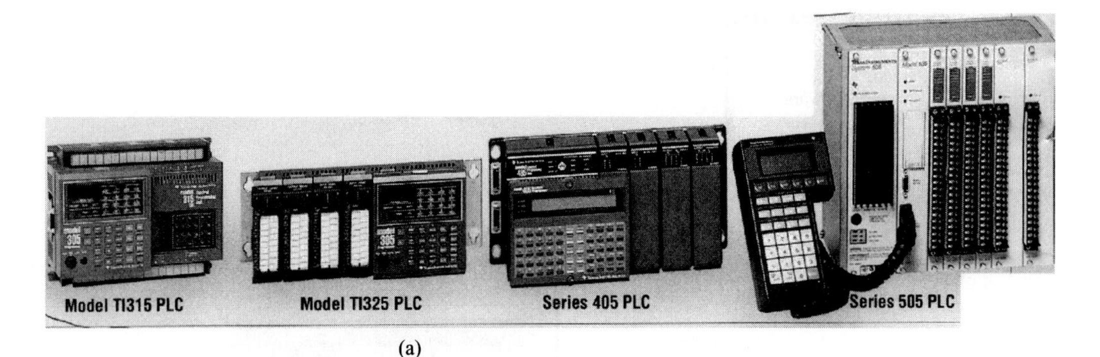


(a)

CPU Rack

CS1 Expansion I/O Rack

(b)

**FIGURE 9.2:** PLC hardware configuration and components: (a) different PLC categories in terms of their number of I/O capabilities and CPU speed, (b) a PLC CPU rack and three I/O expansion racks.

memory space. For instance, a 16-point discrete input module occupies a 2-byte space in the PLC's memory space. A four-channel 12-bit analog-to-digital (ADC) converter occupies $4 \times 12 = 48$ bit $= 6$ bytes of memory space. In other words, once the location of each module in the PLC rack is known, the I/O of all modules is memory mapped to the PLC's memory space. Then, in the PLC's logic, those memory locations are used.

Although there have been claims since the mid-1980s that PLCs would be a thing of the past and that personal computer (PC)-based control would take over the industrial control world, PLCs continue to be strong in the market. Let us compare the PLC-based control with the PC-based control.

1. PLCs have modular design. If a different type of I/O signal needs to be processed, all we need to do is add a different I/O interface module and modify software. Furthermore, the I/O modules include the terminals necessary for field wiring. In PC-based systems, for different I/Os, we need to insert a different PC card and provide a separate terminal block for wiring which connects to the PC card through a ribbon cable. This tends to be a messy and nonstandard wiring process.
2. PLCs have a rugged design suitable for harsh industrial environments against high temperature variations, dust, and vibrations.

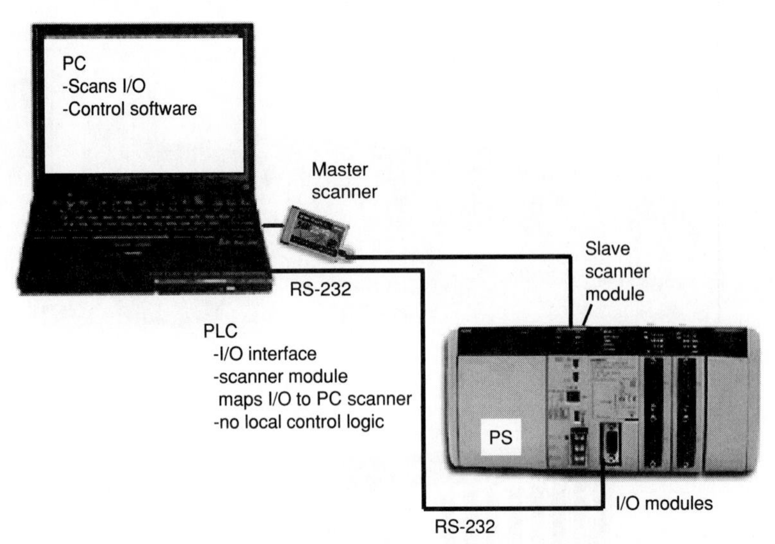


**FIGURE 9.3:** Typically, a PC is used as an off-line program development, as well as on-line debugging and monitoring tool (i.e., using the RS-232 serial interface). The PLC is used as the "brains" of the control system by implementing the control logic, and provides the means for I/O interface. In another mode (using master and slave scanner modules), the PC is involved in the real-time control logic implementation. The PLC merely acts as the I/O interface device and maps the I/O to the communication bus between the PLC and PC using a scanner (slave) card.

3. Programming of PLCs is mostly done using ladder logic diagrams which are understood by millions of technicians in the field. This proves to be one of the biggest advantages of PLCs. Even though ladder logic diagram (LLD) programming does not have the programming environment capablities provided for PCs, it is well established, proven to work well, and a large base of technical personnel can work with it. The reason it is called "ladder" logic diagram is because the logic program graphically looks like a ladder.

The real trend observed in industry is not a competition between PLCs and PCs in industrial control, but the complementary use of them. PCs are used in conjunction with PLCs at two different levels,

1. PCs are used as networking and user-interface devices [Fig. 9.1(d), Fig. 9.3].
2. PC implements the control logic, replacing the role of the CPU on the PLC while the PLC provides the I/O interface (Fig. 9.3). In this configuration, PLC has the I/O modules and a scanner module which updates the I/O between the PLC rack and the PC. The PC implements the control logic which may be developed using any of the programming tools under PC platform, i.e., using C, Basic, or PLC specific graphical program development tools. The key issue is to guarantee hard real-time performance in the PC using a real-time operating system. As real-time operating systems become more robust and low cost over time, this model of PC and PLC combination may be widely accepted in industry.

## 9.2 HARDWARE COMPONENTS OF PLCs

### 9.2.1 PLC: CPU and I/O Capabilities

Perhaps the biggest reason for the success of PLCs in an industrial control market is the fact that the hardware of almost all PLCs has a very similar design. The hardware design is based on a backplane which carries the power and communication bus. A snap-on input/output (I/O) module into any one of the slots makes the necessary electrical contacts for power as well as interface (Figs. 9.2, 9.3). Each PLC needs a power supply and a CPU (or a scanner card if PLC is used in conjunction with a master controller, i.e., PC). Then the I/O cards are inserted into the slots. The slots form the electrical interface between the I/O modules and the bus of the PLC in the backplane. The bus consists of four major groups of lines: power lines, address lines, data lines, and control lines. The end user does not need to be concerned with the details of the bus because the interface between the CPU and all of the I/O modules supported for a given PLC is already worked out and cannot be modified by the user. The real-time kernel on the PLC and the user program are stored in the memory. The memory can be ROM (read-only memory), EPROM (electrically programmable ROM), EEPROM (ereasable electrically programmable ROM), or battery backed RAM (random access memory) type.

Each I/O point on each unit must have a unique address on the PLC bus. The I/O address is generally determined based on the *rack number, slot number, and channel number.* Typically, there can be up to three to five racks supported by one PLC (Fig. 9.2). In each rack, there are 4 to 15 slots. In each slot, there can be a single I/O module. For instance address of a 16-point discrete input module on the main rack, slot number 3 would be determined by the address code Rack-Slot-IO Channel: 1 - 3 - $n$, where $n$ is 1 to 16 representing the 16 I/O channels on the module. Similarly, the I/O modules that have analog signal interface map their I/O values into the memory of the PLCs. Notice that a PLC with three racks, 12 slots in each rack, and support of 16 point discrete I/O module in each slot, can support a total of $3 \times 12 \times 16 = 576$ discrete I/O. Similarly, the same PLC can also support the following combination of discete and analog I/O:

1. Eight channels of 16-bit A/D converter ($8 \times 16 = 128$ bits)
2. Eight channels of 16-bit D/A converter ($8 \times 16 = 128$ bits)
3. Three hundred twenty discrete I/O channels (320 bits)

It was noted above that a main advantage of PLCs is the fact that a great variety of I/O modules are available in standard form. Hardware interface of a simple discrete I/O module has the same difficulty as the interface of a special purpose I/O module. They all snap on to one of the slots on the PLC rack. This standard hardware interface proves to be a very important asset. The I/O data associated with the I/O modules are memory-mapped to the CPU's address space. Below is a list of I/O interface modules available for most PLCs:

1. Discrete input modules for DC and AC type signals.
2. Discrete output modules for DC and AC type signals. Each discrete I/O module typically contains 8, 16, or 32 I/O points, and is rated for different voltages, i.e., DC modules for 5 V, 12 V, 24 V, AC modules for 120 VAC, 240 VAC.
3. Analog input modules (ADC with various voltage or current range and resolution, i.e., 0–5 VDC, 0–10 VDC, $-10$ to $+10$ VDC, 4–20 mAmp, 0–10 mAmp ranges, 10-bit, 12-bit, and 16-bit resolutions).
4. Analog output modules (DAC with various voltage and current ranges and resolutions).
5. Timer and counter modules (hardware timers, pulse and event counters). A hardware timer module can be programmed to generate an input to PLC as well as to generate an output to a device when a time period has elapsed. The beginning (trigger) of the timer may be program controlled, free running, or externally triggered by an input. Counter modules are used to count pulses. For instance, ON/OFF state transitions from a proximity sensor which counts the number of teeth on a gear can be used

by the counter module to count the number of teeth on a gear for quality-control purposes. The pulses from an optical encoder can be input to the counter module to measure displacement.

6. High-speed counter modules are used for counting high-frequency pulses and detecting very short periods of trigger signals (i.e., a high-resolution encoder signal input). For instance, this module can be used to measure position using high-resolution encoders for high-speed applications.
7. A programmable cam switch module is used to emulate the function of a mechanical cam switch set. A mechanical cam switch set turns ON/OFF a number of outputs as function of a master cam shaft position. In the mechanical system, the state of outputs is determined by the shape of each cam switch shape that is machined into the cam. In a programmable cam switch module, these functions are programmable as a function of a position sensor signal.
8. Thermocouple sensor interface module (and many other special sensor interface modules).
9. PID controller module (i.e., closed-loop temperature controller, closed-loop pressure regulator, closed-loop liquid-level regulator).
10. Motion control modules (servo motor, stepper motor, electrohydraulic valve control modules) used for the closed servo motion control. The actuators may be electric motor and drive or hydraulic valve and amplifiers. The motion control module sends out either the desired number of position pulses as the position command to the drive or sends a voltage command proportional to the desired speed or torque depending on the mode of the amplifier. For closed-loop operation, they have position sensor interface.
11. Most PLCs support a standard module called ASCII/BASIC module. This module provides RS-232 serial interface as well as a separate processor that supports BASIC programming language. The BASIC program is stored in the module on a battery-backed RAM. A PLC with an ASCII/BASIC module basically is a dual processor controller. The ladder logic running in the main CPU and the BASIC program running in the ASCII/BASIC module communicate with each other over a predefined memory for data exchange. Complicated mathematical calculations that may be difficult to code in ladder logic can be implemented in the ASCII/BASIC module.
12. Master and slave scanner modules when PLC is used as an I/O interface station and control logic is implemented by a master controller (another PLC or PC).
13. Network communication module (DeviceNet, CAN, ProfiBus, Ethernet, RS-232-C, etc.). PLCs are increasingly part of a larger networked control system. Many so-called fieldbus communication protocols are available.
14. Other special function modules such as fuzzy logic module. New special function modules are being added to the PLCs on an on-going basis.

Network communication protocols suitable for real-time control systems, such as DeviceNet, have been changing the hardware configuration of a PLC-controlled system in recent years (Fig. 9.4). More and more I/O devices (individual sensors, motor starters, closed-loop controllers) are being made available with a network interface (i.e., even a single proximity sensor or a motor starter with DeviceNet interface). Therefore, the I/O devices do not need to be wired into the modules on the PLC's I/O rack. Instead, each I/O device connects to a common communication bus. This reduces the amount of wiring needed between the I/O device and the PLC. The application programming does not change either way, other than the fact that there is a real-time network communication driver running in the background which is transparent to the application program developer. Networked PLC

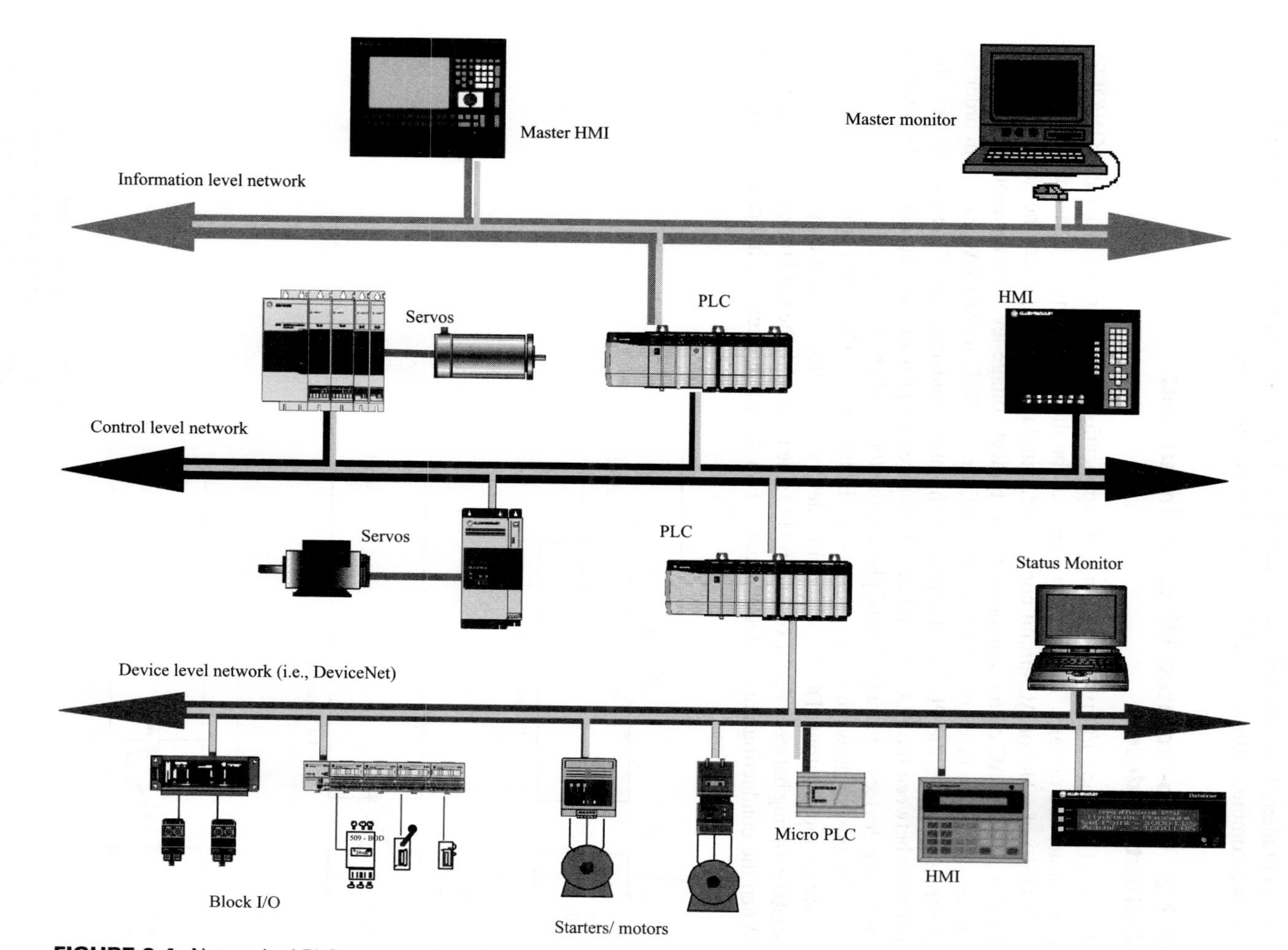


**FIGURE 9.4:** Networked PLC system: three layers of communication network. At the lowest level, simple I/O devices connected to a network (i.e., DeviceNet) using T-connection cables. Adding new I/O to the system simply requires the device and cables to connect to the common network cable. Second level is control network (i.e., Ethernet/IP). Finally the third level is enterprise-wide information network (i.e., Ethernet).

control reduces the wiring costs, distributes the intelligence to local devices, and makes the system I/O expansion easier (Fig. 9.4). When a new I/O device is added, the electical wires of the I/O device does not need to be run all the way to the physical location of the PLC rack, but rather, the communication wires of the I/O device simply need to be connected to the long communication bus cable using a T-type connector.

### 9.2.2 Opto-Isolated Discrete Input and Output Modules

The most common I/O types used in PLC applications are discrete (two state: ON/OFF) type inputs and outputs. The discrete input can be the conducting or nonconducting state (ON/OFF) of a DC or AC circuit component. Similarly, the output can be turning ON or OFF of a DC or AC circuit component. The typical voltage levels of the input and output circuits are in the order of 5 VDC to 48 VDC or 120 VAC to 240 VAC range. In order to electrically isolate the PLC hardware from the high voltage levels of the I/O devices, the interface between the PLC bus and the I/O devices are provided through optically coupled switching devices—namely, LEDs, phototransistors, and photo-triacs.

Figure 9.5 shows the four types of opto-isolated I/O modules that are used to interface two-state input/output (DC or AC circuit) devices to a PLC. Notice that in all cases, the signal coupling between the PLC side and I/O side is through the light (or optical coupling). When the conducting/nonconducting state of an AC input circuit is interfaced, a rectifier

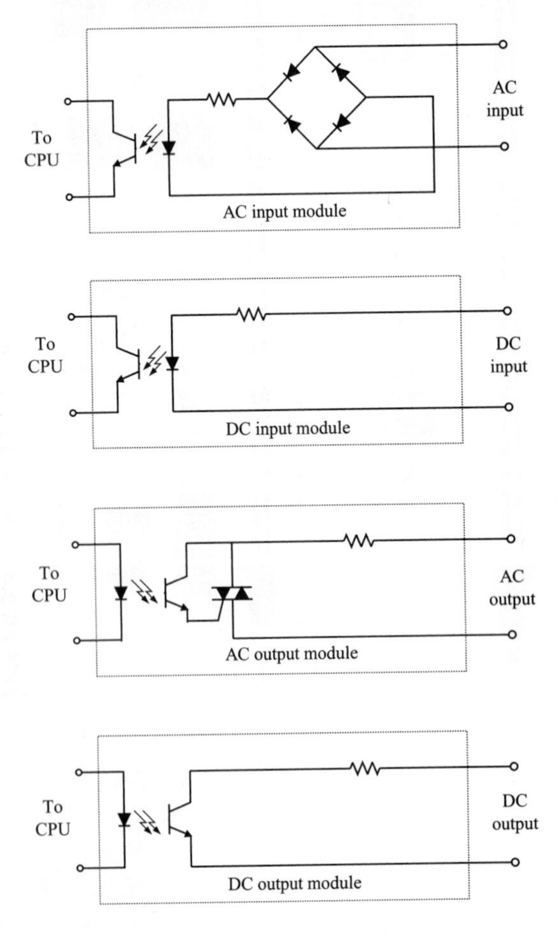


**FIGURE 9.5:** Opto-isolated input and output interface modules for two-state (ON/OFF, conducting/nonconducting) devices: DC input, AC input, DC output, and AC output types. Among DC input and output modules, those rated for 5 VDC are called TTL input and TTL output modules, respectively.

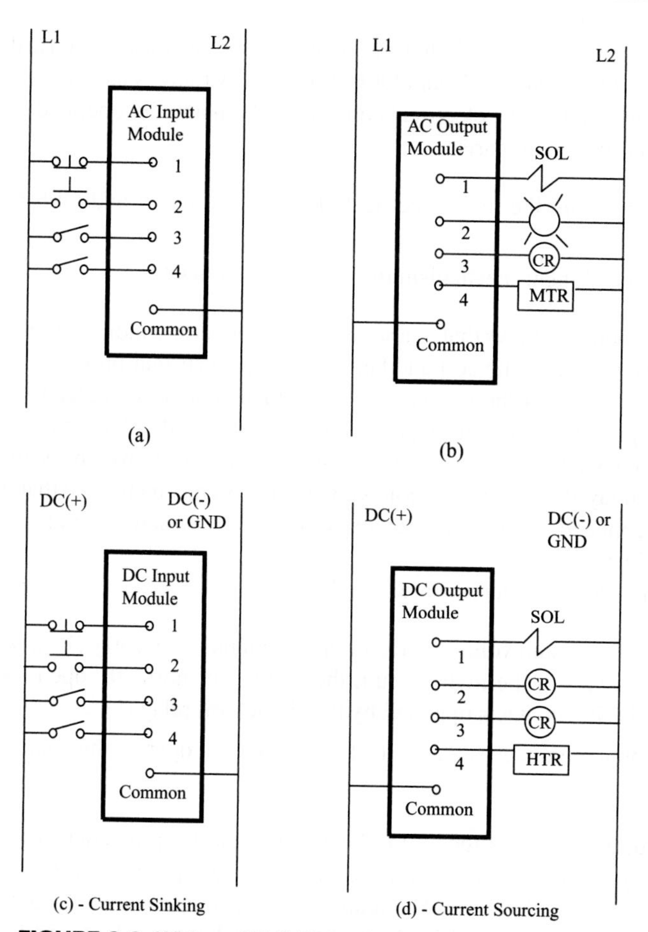


**FIGURE 9.6:** Wiring of PLC I/O modules to the field devices: (a) AC input module, (b) AC output module, (c) DC input module, (current sinking) (d) DC output module (current sourcing).

circuit is built into the opto I/O module to convert it to DC current and drive an LED. The resistor in series limits the amount of current flow through the LED. The light emitted by the LED turns ON a phototransistor on the PLC side. Similarly, the DC and AC outputs are turned ON/OFF by the PLC by driving an LED (phototransistor) on the PLC side. For DC outputs, the LED drives a phototransistor. For AC outputs, the LED drives a phototriac. Although not shown in the figures, it is important to point out that the solid-state opto-I/O interface modules also incorporate other functions useful in practical control systems such as switch debouncing circuits and inductive voltage surge protections.

The PLC I/O modules provide same type of I/O interface in groups of 8, 16, or 32 points, i.e., 8-point DC input module, 16-point AC output module. Fig. 9.6 shows the typical wiring of I/O modules. If a DC I/O module receives current from I/O device and connects it to common, it is called current sinking. If I/O module provides the current to the I/O device and the second terminal of the I/O device is connected to the common, it is called current sourcing.

### 9.2.3 Relays, Contactors, Starters

Relays, contactors, and starters are all electromechanical switches. Through an electromagnetic actuation principle, mechanical motion is obtained. The mechanical motion is used to open/close a switch which is used to control the continuity of an electric circuit.

*Relay* is an electromechanical switch. It is the most commonly used electrical switch in control circuits. It has two states, ON and OFF (Fig. 9.7). A relay is used to connect an electrical line with contacts. It is the electrical equivalent of a manually operated switch in electrical circuits. It has two main parts:

1. Contacts for connecting the main electrical circuit (to turn ON and OFF the main line)
2. Coil to operate the plunger mechanism that moves the contacts

The basic operating principle follows that of a solenoid. A conductor winding forms the coil. The control voltage is applied to the coil and hence develops a current proportional to the control voltage and resistance of the winding. The coil then generates an electromagnetic field. The plunger is pulled as a result of the electromagnetic field. When the coil is de-energized, the electromagnetic force goes to zero and the plunger moves back under the effect of a spring. In a relay, the plunger motion is connected to a mechanism so that it makes or breaks the contact connections. The plunger may be connected to activate multiple output contacts in the main line.

The performance ratings of a relay are as follows:

1. *Contact ratings*: maximum voltage and current the contacts can carry ($V_{max}$, $i_{max}$, i.e., 24 V to 600 V, 50 A) in the main circuit, the number of normally open (NO) and normally closed (NC) contacts operated by the single-coil relay (i.e., 6 NO, 6 NC).
2. *Coil ratings*: nominal voltage (i.e., 6 V to 120 V range) to operate the control circuit coil.

A basic relay design requires the coil to be energized to keep its contacts engaged. Variations in the designs include the *latching relay*. In this design, relay uses two coils: one for latching and one for unlatching. The relay is energized to engage the contacts and move the plunger. Then a mechanical latch mechanism locks it in place and keeps the contacts connected. Then, the coil does not need to be kept energized in order to keep the contacts connected. In order to disconnect the contacts and release the mechanical latching mechanism, the other coil is energized.

*Contactors* operate on the same principle and have a similar design as the relays. The main difference is in their mechanical components. The main line voltage and current capacity that are conducted by the contacts can be much larger than the contact ratings of a relay (Fig. 9.7).

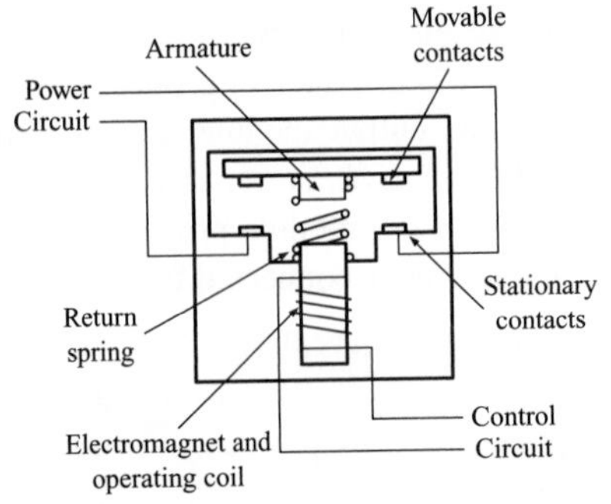

**FIGURE 9.7:** (a) Operating principles of relay, contactors, and starters, (b) Picture of a contactor.

*Starters* are similarly designed to relay principle except that they may have over-current protection as well as "smooth start" related current shaping and limiting mechanisms built in to the design. The overload protection is built-in based on the current–temperature relationship in contact material. When the current is too high for an extended period of time, a bimetallic material breaks the contact. Starters are used in motor control applications.

### 9.2.4 Counters and Timers

When a control logic requires that an action be taken after a certain delay, we use counters or timers. If the delay is based on counting something, a counter is used. If the delay is based on time, a timer is used. It should be emphasized that the counter and timer functions can be implemented in hardware as stand-alone modules or as PLC I/O modules (counter and timer modules) or in software by using general purpose I/O. Dedicated counter and timer hardware offers higher frequency counting and higher resolution timing functions than that can be accomplished in software implementation. Hardware counters and timers may also be used as part of a hardwired control circuit without any PLC software involvement. Stand-alone counter and timer hardware modules have three major circuits: (1) *power circuit*, (2) *control circuit,* and (3) *output circuit*. The power circuit is needed to power the counter and timer modules. The control circuit is the signal used to trigger the module. For a counter, it is the signal transition from one state to another state (OFF to ON or ON to OFF) that is to be counted. For a timer, it is the signal that triggers the start of the process of timing a period. Once it is triggered, the timer can run until the present time value, or the timing operation continues only while the control signal is ON and suspended when it is OFF. Both counter and timer have *preset values*. When the counter counts up to the preset value, the output circuit is turned ON. When the timer measures that the preset amount of time has passed since the control signal triggered it, the output circuit is turned ON. The output circuit of a timer and counter is similar to the output circuit of a relay. An electrical contact is made (ON) or broken (OFF) as output action.

## 9.3 PROGRAMMING OF PLCs

Every PLC includes a software development tool which allows the communication between a PC and the PLC, development of the PLC application software, debugs, downloads, and tests it (Fig. 9.3). A notebook PC is used only as development tool in this case, not as part of the real-time control. The development tools for different PLC manufacturers currently are not interchangable. The application development engineer must use the development tools supplied by the specific PLC vendor.

Assuming that we have the program development tools (i.e., a notebook PC and a PLC specific ladder logic program development environment), the software that runs on the PLC to control a specific industrial application is the issue discussed here.

There are three main types of programming languages available for PLCs:

1. ***Ladder logic diagrams (LLD)*** emulate the same structure of the hardwired relay logic diagrams. It has the widest use because most field technicans are familiar with hardware relay logic diagrams and understand LLD programs.
2. ***Boolean language*** is a statement list and similar to BASIC programming language.
3. ***Flowchart language*** uses graphical blocks. It is more intuitive than other two languages. Although the use of flowchart languages started to increase in recent years, LLD is still the dominant language.

The flowchart-type languages may eventually gain more widespread acceptance. Today every PLC has its own LLD language, and they are not compatible across different PLCs. The LLD programming environment for all PLCs looks very similar to the application program development point of view. In the rest of this chapter, only the LLD programs will be discussed, not the specific development environment for a particular PLC.

There is a fundamental difference between the way a PLC and a PC execute their programs. The program flow in a PC is controlled by the *flow control statements* such as *do-while, for* loops, *if-else if-else* blocks, function calls, jump or go-to statements. In high-level programming language, it is possible that the program can be limited to a local loop, whereas a PLC program runs in the *scan mode* (Fig. 9.8). The whole program logic, from the beginning to the end, is scanned every scan period of the PLC. Depending on the computational power of a PLC, typical scan times are on the order of a few millisecond or less per thousand lines of ladder logic code. The benchmark speed of scan time for per-thousand ladder logic code is generally limited to basic logic function such as AND, OR, NOT, and flip-flop. Special functions, trigonometric functions, and PID control algorithms implemented in the ladder logic take longer computation time. As a result, the scan time will be slower for ladder logic programs that include many special functions. Therefore, the scan time estimates given for a PLC by its manufacturer should be interpreted with this in mind. The actual scan time of a ladder program for an application can be exactly measured by the software development tools of the PLC. The software development tools installed and run on a PC for off-line program development and debugging purposes also include utilities that measure the scan time while the program is executing on the PLC. In the ladder logic programming, sections of the program execution can be conditional like the *if-else* blocks, and subroutines may be called or skipped based on the coded conditions. All of the PLC ladder logic program is scanned for execution every scan time. It is useful to imagine the PLC ladder logic as a circular instruction sequence where the CPU goes through the circle once every scan time period (Fig. 9.8).

From a programming point of view, once the racks and slots of a PLC are populated with I/O interface modules, all of the I/O is memory mapped. There is a one-to-one memory map between the I/O points on the PLC hardware and the CPU memory addresses (Fig. 9.9). The logic is implemented between the I/O (memory locations) using the logic functions provided by the ladder logic program. The ladder logic programming focuses on

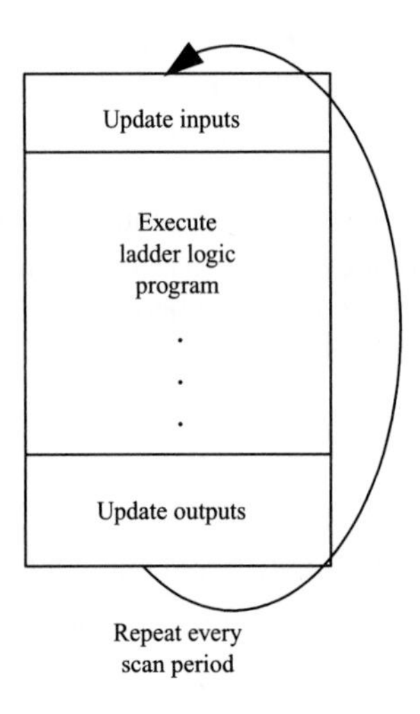


**FIGURE 9.8:** Ladder logic program execution model in a PLC. PLC executes its program in scan mode. All of the code is checked for execution every scan period.

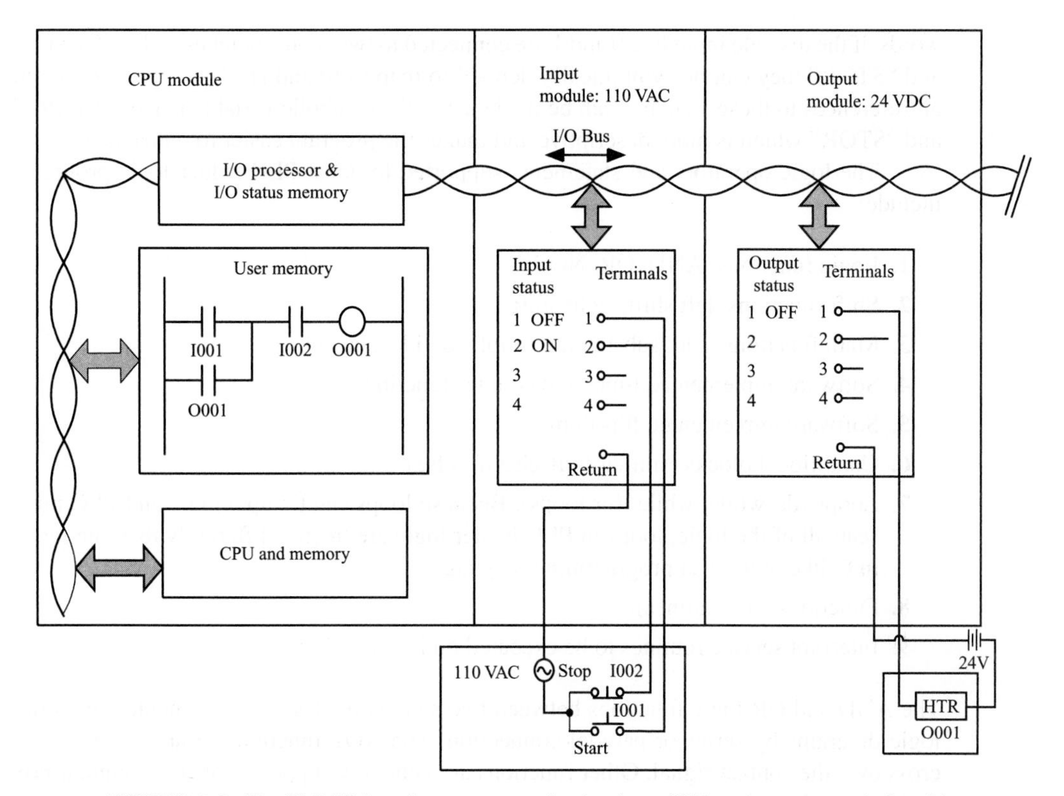


**FIGURE 9.9:** The PLC I/O interface, communication bus, CPU and memory relationship.

the industrial control and logic, not high-level, object-oriented data structures. Therefore, typical data structures supported include:

1. *Bits* for discrete I/O, and
2. *Bytes* and words for analog I/O and character data.

There are standard key words reserved for discrete input, output, timer, and counter funtions in the PLC. For instance, DIN 19239 standard specifies the following key words for PLC ladder logic programming concerning I/O and memory addresses:

1. I: for discrete input lines, for instance I0 through I1023
2. O: for discrete output lines, for instance O0 through O1023
3. T: for timer functions, for instance T0 through T15
4. C: for counter functions, for instance C0 through 31
5. F: for "Flag" to store and recall a bit of data, i.e., flip-flop operations

Control coils of relays, contactors, starters always appear on the output side of ladder logic diagram. Timers and counters also appear on the output side. The contacts that these devices operate appears as part of the logic on the input side of ladder logic. The contact name (letter and number designation) always matches the coil name designation of the output device.

Once the actual I/O terminal points on the I/O modules in the PLC slots are decided, then there is a memory map established between the I/O variables in memory and the actual I/O. Furthermore, most PLC program development tools support *symbolic names*. In other words, if the discrete input line 0 and 1 are connected to two input switches called "START" and "STOP," they can be symbolically defined to map to I0 and I1. Then, in the program all references to these switches can be made using the symbolic variable names "START" and "STOP," which is more descriptive and makes the program easier to understand.

The logic operators and statements supported by most PLC ladder logic programs include:

1. Logic functions: AND, OR, NOT.
2. Shift functions: left shift, right shift.
3. Math functions: add, substract, multiply, divide, sin, cos.
4. Software implemented timer and counter functions.
5. Software implemented flip-flops.
6. Conditional blocks (similar to if-else if - else).
7. Loops (do-while, while, for loops). Because loops can tie up a logic and PLC must scan all of the logic, loops in PLC ladder logic are treated differently than the loops in C-like high-level programming languages.
8. Functions (subroutines).
9. Interrupt service routines to be executed on interrupt inputs.

The AND and OR logic functions between two state variables are implemented in ladder logic diagrams by series or parallel connection. The NOT function is implemented by a cross over the contact signal. Other functions are generally supported by a rectangular box with appropriate input and output lines.

Some of the syntax rules common to all ladder language programs are as follows,

1. A ladder logic program is a sequence of rungs. Each rung must begin with an input or local data.
2. Output channels can appear at the end of a rung only once.
3. Discrete input channel state is represented by two vertical lines like a contact.
4. Discrete output channel state is represented as a circular symbol.
5. Timers, counter, and other function blocks are represented by a rectangular shape with appropriate input source and output lines in the rung.
6. Logic AND is implemented by putting two contacts in series, and logic OR is implemented by putting them in parallel (Fig. 9.9).

One of the standard logics in a ladder logic diagram (LLD) is a seal-in circuit. The idea is to turn ON an output channel (relay output) and keep it ON based on a momentary input (START switch). Then keep it ON until another momentary input (STOP Switch) is pressed. This is shown in Fig. 9.9.

The following software concepts are commonly used in PLC programs: *shift register, sequencer, and drums*. Shift register and sequencer are both registers. The individual bit locations in the register are used to indicate the location of a part in a multistation assembly line or the number of the operation currently performed from a sequence. A bit is associated with the current product or operation. As the bit is shifted in the register, its location on the assembly sequence is advanced. Similarly, the drum is the software analog of a mechanical cam. There is master shaft and multiple cams on it. Each cam actuates (turns ON and OFF) an output line as a function of the input cam shaft position. This functionality is duplicated

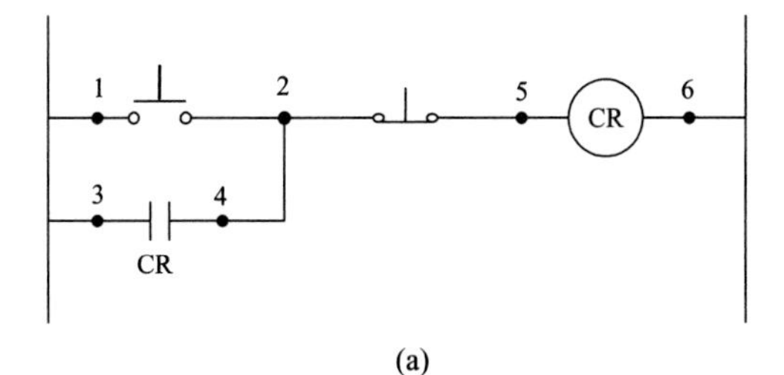


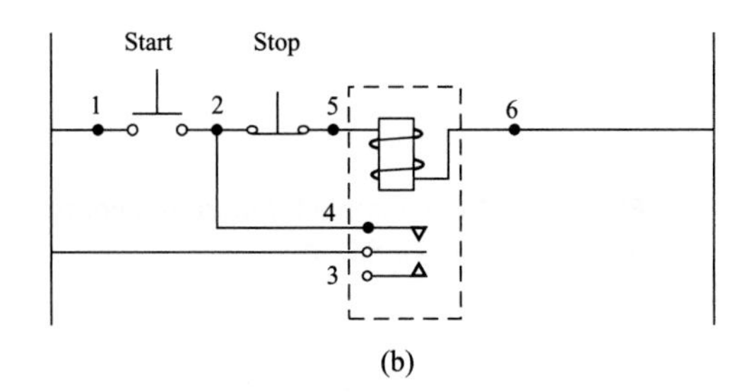


**FIGURE 9.10:** Hardwired seal-in circuit in a ladder logic diagram: (a) ladder logic diagram of connections, and (b) component connections between START button, STOP button and the RELAY.

in software with the software drum concept. The benefit of the software drum compared to the mechanical drum is that it is programmable.

## 9.3.1 Hardwired Seal-In Circuit

In original hardwired relay circuits, all of the control logic are physically wired into the electrical connections of input, output, and intermediate control devices. When programmable logic controllers replaced the hardwired relay logic, the logic between the inputs/outputs is software coded (programmed) in the PLC. Very little of the control logic is hardwired. Still, some of the safety-related functions are hardwired in case the software fails. Most emergency shutdown and cycle start functions are hardwired. One of the most often hardwired logic is the *seal-in circuit* used between START and STOP push buttons. The same diagram is used as the one shown in Fig. 9.10. The only difference is that the logic is hardwired instead of programmed in the PLC. A hardwired seal-in circuit includes the following components:

1. A START button, which is a momentary contact switch
2. A STOP button
3. A relay or starter coil for the output device (i.e., coil of a starter used to start a motor)

When the START button makes a momentary contact, it energizes the coil of the relay (or starter). Then one of the contacts from that coil maintain the flow of current, because it is wired in parallel with the START button, even when the START button no longer makes contact since it is a momentary button. Recall that a relay may operate multiple contact. The relay presumably have another contact that turns ON or OFF a separate power line to the motor. The power flow is cut off (stopped) any time the STOP button is pressed. This is called the *three-wire* control. If for some reason the power is lost to the circuit, the START button must be momentarily pressed in order to start the motor again. This is good for safety, but requires a human intervention if the power flow in the circuit is interrupted. In other words, the restart of the cycle is not automatic.

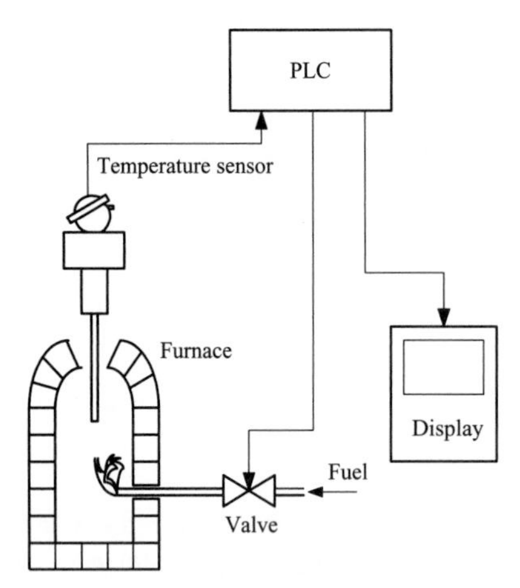


**FIGURE 9.11:** Example temperature control system with a PLC using a PID module.

## 9.4 PLC CONTROL SYSTEM APPLICATIONS

Figure 9.11 shows a closed-loop temperature control system in an oven using a PLC. The PLC-based control system uses a temperature sensor (i.e., a thermocouple interfaced to the PLC using a thermocouple sensor interface module), a proportional valve which regulates the fuel into the furnace proportional to the displacement of the valve spool, and a display to inform the operator of the process variables in real time (i.e., the desired temperature setting, actual temperature, fuel rate). The PLC controls the valve with an analog output module. If the current capacity of the analog drive is not large enough to drive the valve, then there would be a current amplifier between the analog output module and the valve. The closed-loop control algorithm is implemented in the ladder logic diagram using a PID algorithm function block where the parameters of the algorithm is to be tuned by the application engineer.

Conveyor speed control is a very common factory automation problem. Fig. 9.12 shows a motion control application using incremental encoder input and pulse output modules in the PLC rack. The PLC controls the speed (and position if desired) of two conveyors. The PLC program measures the speed of two conveyors using incremental encoders and commands the desired number of incremental position change pulses to the drive which controls the power to the motor. Typical applications include:

1. Independent speed control of two conveyors: The desired speed of each conveyor can be set by operator through a user-interface device, and PLC controls both conveyor speeds independent of each other.
2. Master–slave speed control: Speed of one of the conveyors may be set as the master speed and the other conveyor commanded speed is determined in proportion to the first one. The second conveyor is controlled in slave mode. The speed of the first conveyor may be programmed or determined based on a sensor (i.e., speed of another station in the line).
3. Master–slave speed control with position phase adjustment: In addition to making one conveyor follow the speed of the other conveyor (master–slave relationship), the position of one slave conveyor relative to the master conveyor can be adjusted based on position sensor information. This is very common in packaging applications where material is transferred from one conveyor to another. In steady state, two conveyors run in a certain gear ratio (i.e., 1:1 master–slave relationship). Due to the variation

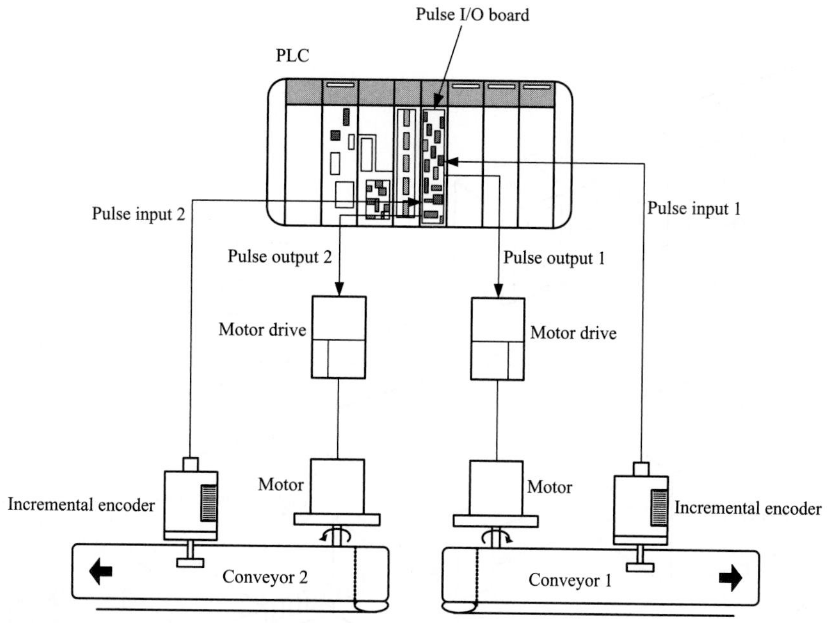


**FIGURE 9.12:** Example of two-conveyor speed control using PLC and pulse input and pulse output modules.

in material location on each conveyor (product on one and container on the other), in order to place the product on the container properly, the position phasing must be adjusted to correct for the placement variations.

Servo positioning using a lead screw (or ball screw) drive is a very common high-precision motion control application (Fig. 9.13). The lead-screw positioning system is used in all machine tools, gantry-type robots, printed circuit board (PCB) assembly machines, and others. This example shows the components of a PLC-based system to control the position of a linear stage using an electric motor (i.e., stepper motor, DC brushless motor, DC brush motor) and a rotary incremental encoder. Other real-world components are also shown such as the limit sensors to indicate the mechanical limits of motion, a proximity sensor to establish a reference position (home position) after power-up since an incremental encoder cannot measure absolute positions, and operator command buttons (START and STOP).

## 9.5 PLC APPLICATION EXAMPLE: CONVEYOR AND FURNACE CONTROL

Consider the conveyor and heat furnace shown in Fig. 9.14. The conveyor moves parts into a heating chamber. The part is kept inside the temperature-controlled furnace for a certain amount of time (programmable) and at a certain temperature range (also programmable).Then the conveyor moves to bring in the next part. In the process, the PLC controls the temperature of the furnace and speed of the conveyor. In addition, it controls the opening and closing of the furnace doors as well as the position of the part inside the furnace using a presence switch. The system inputs are:

1. START switch (ON/OFF)
2. STOP switch (ON/OFF)

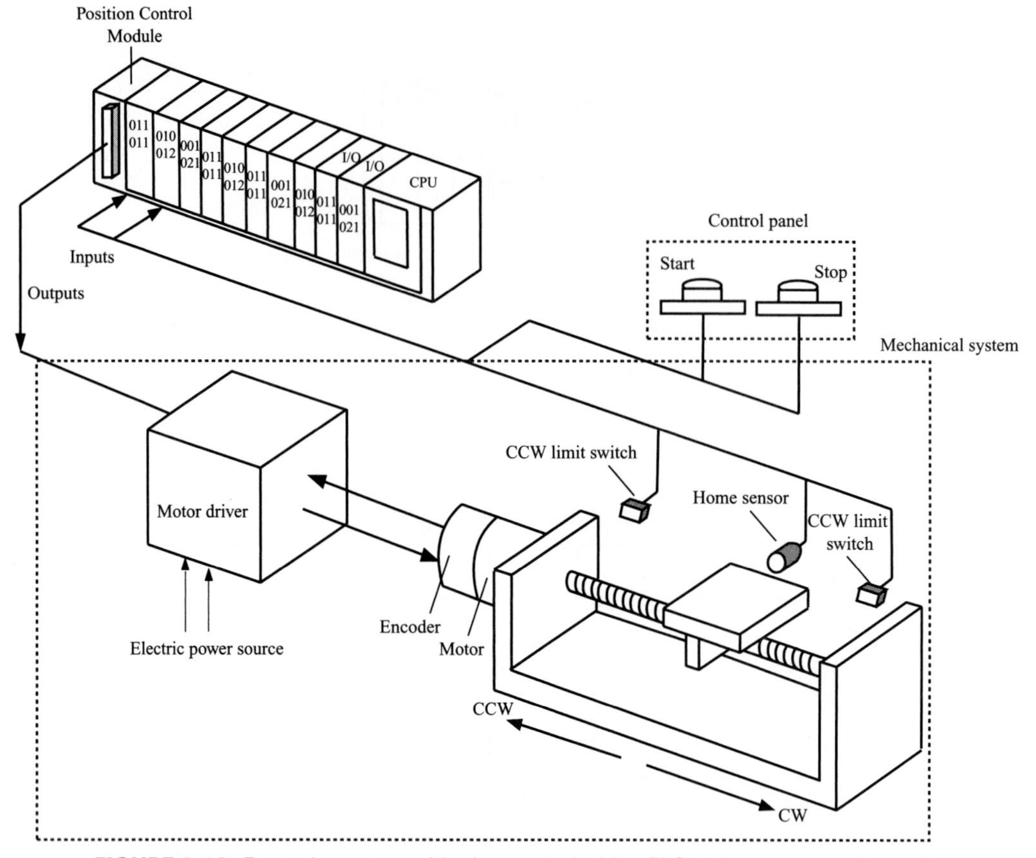


**FIGURE 9.13:** Example servo positioning control with a PLC and servo control module.

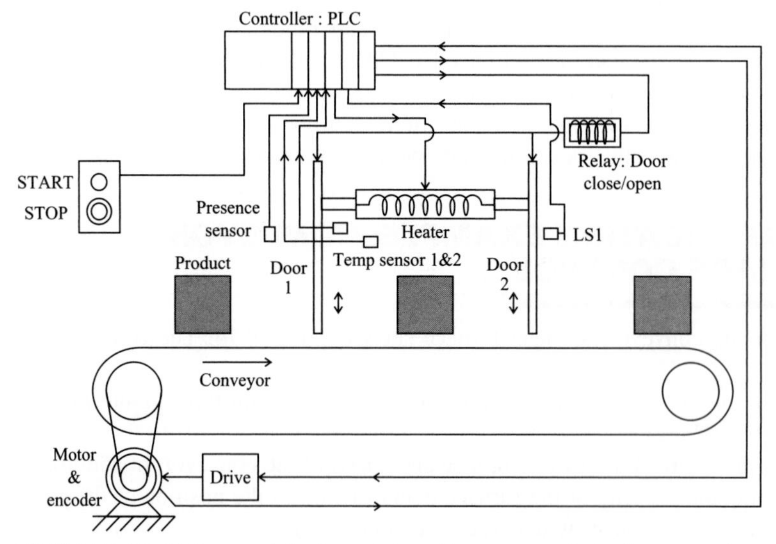


**FIGURE 9.14:** PLC control (logic and closed-loop control) of a heater and a conveyor motor.

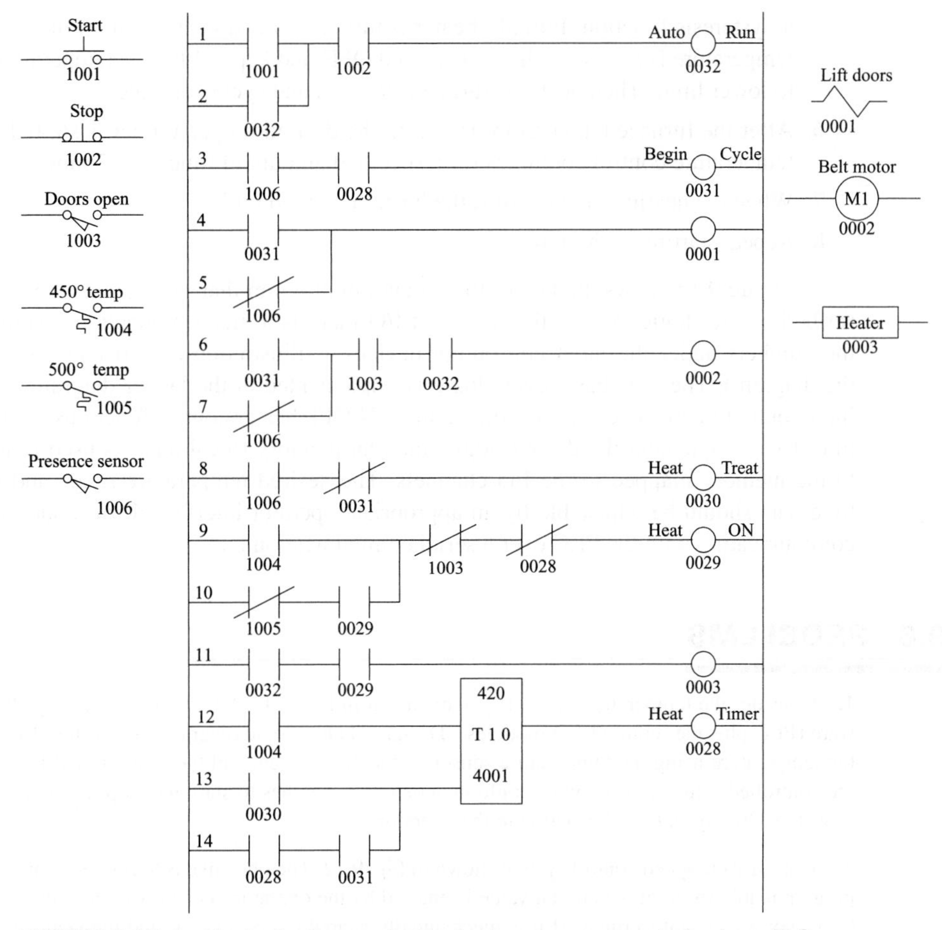


**FIGURE 9.15:** Ladder logic diagram for the heater-conveyor control using a PLC.

3. Part presence sensor (ON/OFF)
4. Door open switch LS1 (ON/OFF)
5. Temperature range sensor (analog)

The outputs are:

1. ON/OFF relay control of heater element
2. ON/OFF output to motor starter
3. ON/OFF output to door actuator relay control

The desired control logic is expressed in pseudo-code as follows:

1. The operation is started when the START button is pressed. When the STOP button is pressed, the cycle is stopped.
2. PLC opens the doors of the furnace and waits until the doors open switches are ON.
3. Then the motor is started. The motor is stopped when the part presence sensor is ON.
4. The furnace doors are closed.
5. The PLC controls the heater output based on the preprogrammed desired temperature range and temperature sensor feedback. The ON/OFF control of heater implements

a hysteresis function. Initially heater is turned ON and kept ON until upper limit of temperature is reached. Then it is turned OFF, and kept OFF until temperature drops to lower limit. Then heater is turned ON again and cycle continues.

6. After the furnace temperature is within the desired range, a timer is started and the temperature control continues for a specified amount of time (i.e., 2 min).
7. When the heating time expires, the heater is turned OFF.
8. Repeat starting with Step 2.

Figure 9.15 shows the ladder logic diagram program that implements the above described control logic. Notice that for each I/O there is a memory location assigned (see the numbers under the input and output devices as illustration on left and right side of the diagram). Then the basic cycle logic is implemented in the ladder diagram using the input–output memory references, timer, and ON/OFF heater control functions. Notice that in order to implement the desired logic, internal memory (local data) is used in addition to the memory mapped to the I/O channels. The desired temperature range and heating time data should be adjustable by an appropriate operator interface device, such as a PC communicating with the PLC over a serial communication bus.

## 9.6 PROBLEMS

**1.** Consider the temperature control system shown in Fig. 9.11. Specify the necessary PLC hardware (PLC plus the required I/O modules). Design a ladder logic diagram so that the PLC controls the temperature using the temperature sensor as feedback signal and the proportional fuel valve as the controlled actuator. The logic should include input switches to start and stop the control system as well as display status information to the operator.

**2.** Consider the speed control system shown in Fig. 9.12. The objective is to run the conveyor 1 at a programmable speed at which that value is entered by the operator via an operator interface device. Conveyor 2 is meant to run as if it is mechanically geared to conveyor 1, and the gear ratio is also programmable by the operator. Specify the necessary PLC hardware and design the LLD program.

**3.** Consider the position control system shown in Fig. 9.13. Our goal is that when the operator pushes the "Start" button, the automatic cycle of the positioning table starts. When the "Stop" button is pushed, the table is supposed to stop to a predefined home position. When the start button is pushed, the motor is to move in a direction to seek the "home" sensor. If it reaches one of the limit switches before reaching the home sensor, then it must reverse direction, go past the home sensor, stop, and come back to turn ON the home sensor from a predefined direction, then stop and establish a "home" reference. After that, the motor is supposed to make incremental forward and reverse motions, i.e., five in a row, 1.0-in. displacement motions separated by 100-msec delay between the completion and start of each motion. Specify the necessary PLC hardware, and design the program logic.

**4.** Consider the circuit shown in Fig. 9.10. What happens if the seal-in circuit at points 3 and 4 is not connected? If the seal-in circuit is not connected, and we change the "Start" switch to a latching switch (not momentary), how is the operation of the circuit different than the seal-in circuit operation? (Consider that we started the circuit, and stopped it with the "Stop" switch. How would you restart the system?)

CHAPTER 10

# PROGRAMMABLE MOTION CONTROL SYSTEMS

## 10.1 INTRODUCTION

Programmable motion control systems (PMCS) are used in all mechanical systems that involve computer-controlled motion. A robotic manipulator, an assembly machine, a CNC machine, XYZ table, and construction equipment control systems are all examples of application of PMCS. As the name indicates, PMCS are motion mechanisms where the motion is controlled by a digital computer, and hence programmable. PMCS are good examples of mechatronic systems in that they involve a mechanical motion system, various actuators and sensors, and computer control. Figure 10.1 shows the typical components of an electric motor-based programmable motion system, that is, a motor, amplifier and power supply (drive), and controller. The figure shows different types of rotary motors and drives (brushless and brush-type DC, AC induction) as well as linear motors.

In the past, coordinated motion control in automated machines was achieved by mechanically connecting various machine components with linkages, line shafts, and gears. Once the master line shaft is driven by a constant speed motor or engine, the rest of the motion axes derive their motion from it based on the mechanical linkage relation. This was what is called the "hard automation." The availability of low-cost microprocessors and digital signal processors (DSP) as well as their high reliability made it possible to control motion and coordinate various axes under computer control. The coordination between axes is not fixed by mechanical linkages, but coded in software. Hence, the coordination logic can be changed on the fly in software. The same machine can be used to perform with different coordination relations to make different products by simply changing software. In mechanically coordinated machines this may require a change of gears, linkages which may require very long set-up times. Some of the complicated coordination functions may not even be feasible to achieve by mechanical coordination while it can be easily coded in software. The "programmable" aspect of motion control comes from the fact that the control logic is programmed. Therefore, it is a "soft automation" or "flexible automation." The single most significant advantage of "soft automation" over "hard automation" is the significantly reduced set-up times for product change-overs. Figure 10.2 shows an example of a printing machine with both old mechanical automation and programmable automation versions. In the mechanical version, each station is coordinated with respect to the master shaft through mechanical gears. When a different product is required, the gear reducer ratios have to be mechanically changed. Therefore, for different products, different gear reducers have to be kept in the inventory. In addition, physically changing the gear reducers is labor intensive and time consuming. This results in long change-over and setup times. In the electronically coordinated version of the machine, the gear ratios between stations are defined in the application sofware. There is no physical gear between axes. The same

**FIGURE 10.1:** Components of an electric power-based programmable motion control system. The figure shows various electric motor types (brushless- and brush-type DC, AC induction, linear motors), drives (amplifier and power supply), and controllers which in some cases may be integrated on the drive.

functionality is achieved by software control of the motion of each axis. So, when a different product requires different gear ratios between axes, the only change needed is to select a different set of parameters in software from a database. The changeover time for different products is almost insignificant.

In a factory automation application, multiple PLCs and programmable motion controllers may be used to control the whole process. PLCs handle the general-purpose I/O and act as the higher level supervisory controller relative to the lower-level local controllers (Fig. 9.5). Programmable motion controllers handle the high-performance, coordinated motion control aspects of individual stations and communicate to the PLCs. In small automation applications, a standalone programmable motion controller may have enough I/O capability so that the machine control can be handled by it without the use of a PLC (Fig. 10.3).

Every programmable motion control system has the following components (Fig. 10.4):

1. Controller
2. Actuators (motor, amplifier, power supply)
3. Sensors (encoders, tachometers, tension sensors, etc.)
4. Motion transmission mechanisms (gears, lead screws)
5. Operator interface devices (also called human-machine interface (HMI))

Most of the technologies for the components of a PMCS is already covered in this text in various chapters. Here we will focus on the control software features specific to automation applications of PMCS.

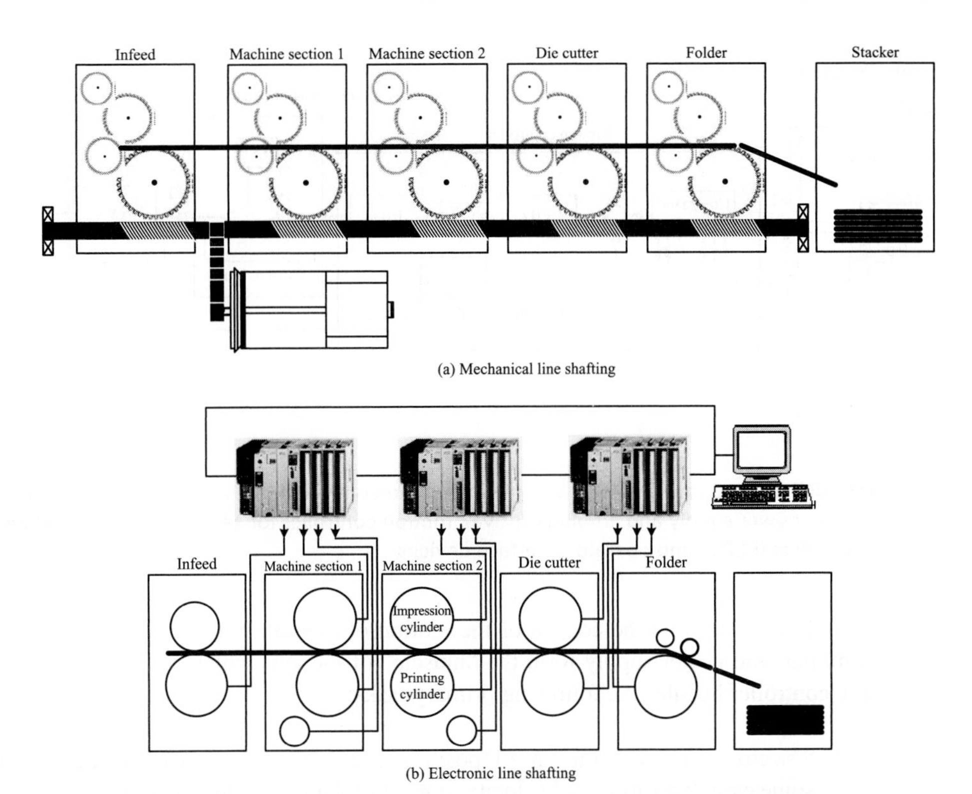


**FIGURE 10.2:** (a) An automated printing machine with mechanical coordination. A long master shaft is driven by a large electric motor. All other stations are geared to the master shaft mechanically. (b) An automated printing machine with electronic motion coordination. Each station is independently driven by a small actuator (i.e., electric servo motor with position sensors). The motion of each station is coordinated with a master station using the computer controller.

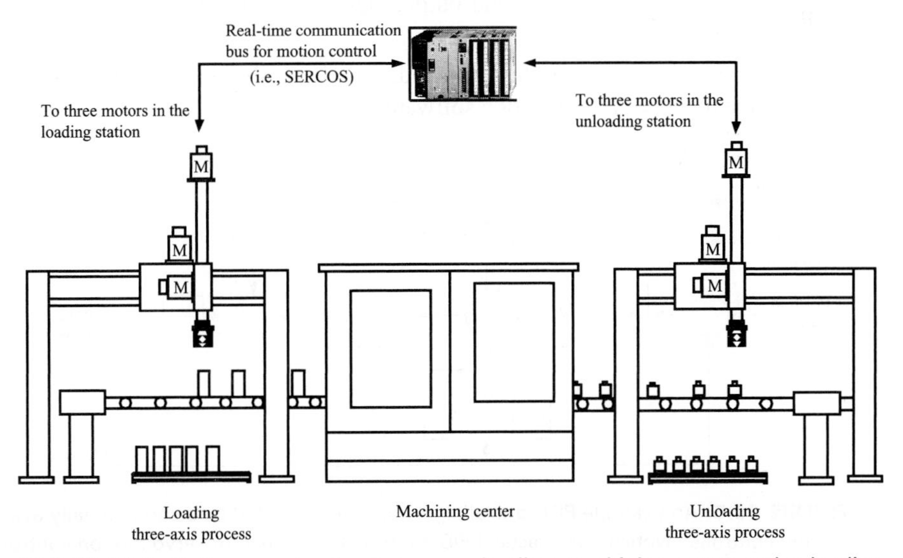


**FIGURE 10.3:** Motion coordination between a loading, machining center, and unloading station. Machining center typically has its own CNC controller. Custom loading and unloading stations need to be designed and controlled in coordination with the machine tool operation.

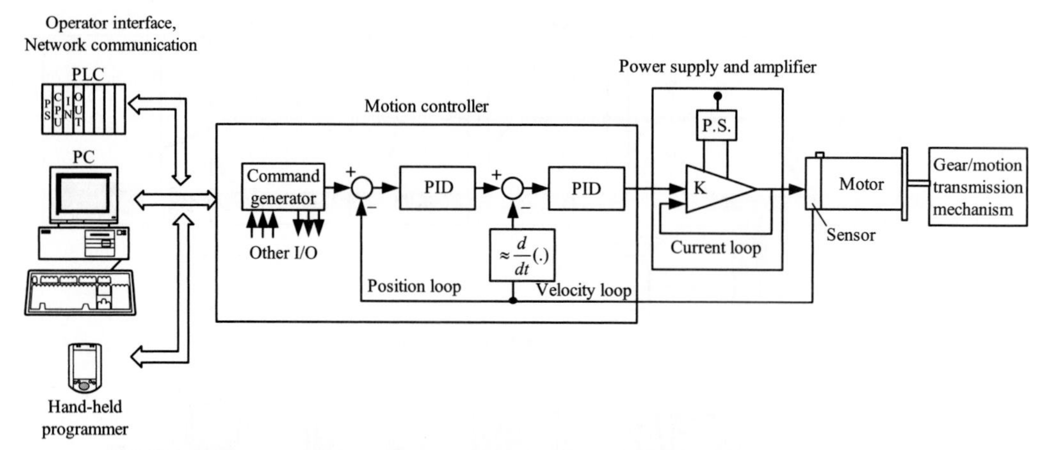


**FIGURE 10.4:** Components of a single-axis servo-controlled motion axis: motor, position sensor, power supply and amplifier (drive), motion controller for servo loop, machine level controllers (PLCs), and operator interface devices.

Let us focus on the control of the motion of one axis only (Fig. 10.4). The motor, amplifier, and power supply form the "muscle" of the axis which we call the "actuator." The controller handles the following primary tasks:

1. The closed-loop motion control (position loop, velocity loop, and even current loop in some cases): position and velocity loop control are typically implemented as a single closed-loop control algorithm. The most dominant servo position and velocity control algorithm is a form of PID control algorithm with various feedforward compensation terms, and limits on the contribution of each term. Figure 10.5 shows a servo position and velocity control algorithm used in industry, which is clearly a variation of the standard PID control algorithm.
2. It handles the other input and output signals associated with the machine control functions (Fig. 10.4).
3. The controller generates the commanded motion to the closed loop (servo loop) based on the logic of the application software (Fig. 10.4).

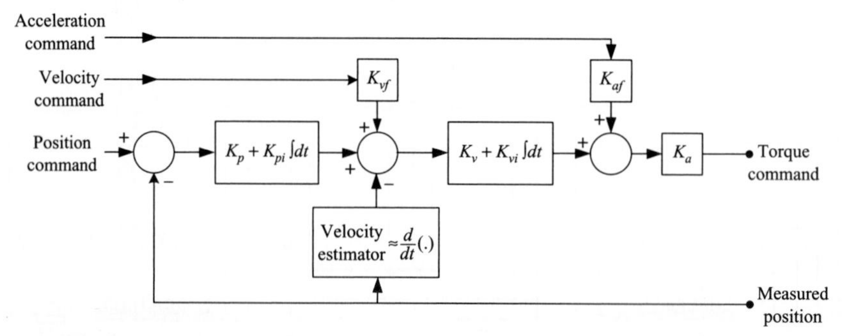


**FIGURE 10.5:** An example PID control algorithm implemented by a commercially available motion controller. Notice that practical PID control algroithms for servo motion control includes feedforward terms in addition to the textbook standard PID gains. Other modifications may include the addition of deadband, limits on integral control, activation and deactivation logic for integral action, integrator anti-windup, friction, and backlash compensation logic.

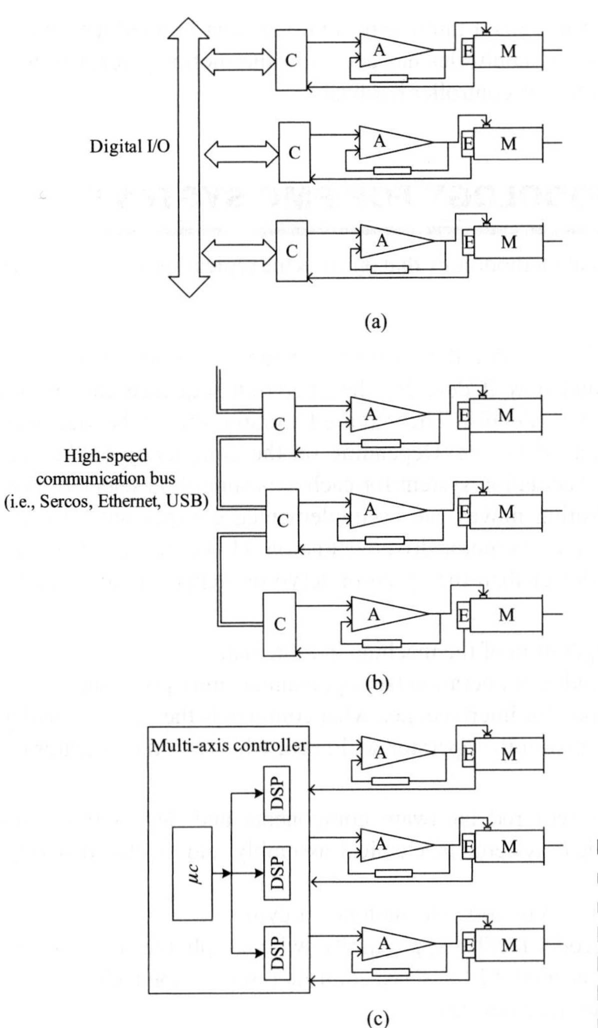


**FIGURE 10.6:** Coordination methods between multi-axis motion axes: (a) using digital I/O lines as dedicated triggers, (b) high-speed data-communication bus, and (c) central multi-axis motion controller.

4. The axis controller also communicates with the operator-interface devices and communication bus to coordinate its operations with the rest of the control system (Fig. 10.4).

In general, motion of one axis needs to be coordinated to other external machine events or to the motion of other axes. In order to coordinate motion between multiple axes, the controllers need to communicate with each other. Figure 10.6 shows different hardware options to establish the communication between multi-axis controllers. In the first case, there is a standalone controller for each axis, and the coordination data needed between axes are simple. In this case, the coordination signals between axes can be handled by digital I/O lines where one axis tells the other one that it is time to move [i.e., "Go" signal, Fig. 10.6(a)]. If more detailed data are needed for the motion coordination, i.e., more than just the "Go" signal such as the actual speed and position of another axis), each standalone axis controller must be on a common communication bus to exchange the necessary information in real time. Sercos is one of the serial communication standards adapted specifically for high-performance motion coordination applications

[Fig. 10.6(b)]. Finally, if there is a multi-axis controller that is interfaced with all of the tightly coordinated axes, the controller has access to the motion information of all axes [Fig. 10.6(c)] in one multi-axis controller hardware.

## 10.2 DESIGN METHODOLOGY FOR PMC SYSTEMS

The following is a practical methodology that outlines the typical steps involved in designing a PMC system.

*Step 1.* The first step is to define the purpose and functionalities of the machine, i.e., what it does and how it does it. The operation sequence can be roughly defined without fine details. The required motion should be decomposed into various coordinated moves. Depending on the complexity of the each motion axis, the proper actuation system for each axis should be selected. For instance, simple two-position moves can be implemented cheaper with two-position air cylinders than a servo motor driven actuation. However, the speed of air cylinders is much slower than the speed of servo or stepper motor actuated mechanisms.

*Step 2.* Describe the operation of the machine in more detail.

1. Required modes of operation (set-up, manual, auto, power-up sequence).
2. Required operator interface, i.e., what commands the operator will give, what status information the operator will need, what data the operator can change and monitor.

*Step 3.* Decide on the required hardware components and design integration of the electromechanical system (mechanical assembly and electrical wiring of components).

1. Actuators: servo motors, DC motors, air cylinders
2. Sensors: encoder feedback, proximity switches, photoelectric switches
3. Controllers needed: PLCs, servo controller, sensor controllers
4. Operator interface devices
5. Motion transmission mechanisms: gears, lead screws, timing belts

*Step 4.* Develop application software.

1. Top-down tree structure design of software
2. Pseudo-code
3. Code in specific programming language

*Step 5.* Set up each programmable motion axis.

1. Basic hardware check
2. Power-up test
3. Establish serial communication
4. Servo tuning ($K_p$, $K_v$, $K_i$, $K_{vf}$, $K_{af}$, ... values)
5. Default parameter set-ups (default acceleration, deceleration, jog speed, maximum motion planning parameters)
6. Simple moves: jog, home, single-index motion parameters

*Step 6.* Debug, test, verify performance.

*Step 7.* Documentation: document clearly so that later someone else can debug and/or modify the machine.

## 10.3 MOTION CONTROLLER HARDWARE AND SOFTWARE

The brain of a PMC system is the controller. The same controller can be used with different actuators such as electric servo, hydraulic servo, or pneumatic servo power. The degree of intelligence and sophistication that can be designed into the system is largely dependent on the capabilities of the controller. The custom application software in the controller uses the I/O hardware to define the logic between them and control the operation of the machine. The I/O hardware can be grouped into the following categories (Fig. 10.4):

1. Axis Level I/O:
   - (a) Servo control I/O:
     - Command signal to amplifier in the form of $\pm 10$ VDC range analog voltage or PWM signal
     - Feedback signal from the position sensor (encoder, resolver)
   - (b) Axis I/O:
     - Travel limits: positive and negative direction limits
     - Home sensor input
     - Amplifier fault input
     - Amplifier enable output
     - High-speed position capture input
     - High-speed position trigger output (cam function outputs)
2. Machine Level I/O:
   - (a) Discrete inputs (ON/OFF sensors: proximity, photoelectric, limit switches, operator buttons)
   - (b) Discrete outputs (ON/OFF output: relays, solenoids)
   - (c) Analog inputs ($\pm 10$ VDC, tension, temperature, force, position etc.)
   - (d) Analog outputs ($\pm 10$ VDC to other device amplifiers)
3. Communication:
   - (a) Serial RS 232/422 communication with HMI
   - (b) Network communication between higher-level intelligent devices

Let us review the software tasks of the motion controller in more detail. The real-time machine control software running in a motion controller generally performs many tasks. These tasks are run in some form of multitasking and priority order. Software task modules are:

1. Servo loop control task: $U_{DAC} = PID$ (desired and actual motion)
2. Trajectory generation task (desired motion profile): independent moves, coordinated moves (gearing, contouring)
3. Application program logic interpretation and execution
4. Communication with the host/higher-level controller and HMI
5. Check limits task: travel limits in software, torque limits, following error safety interlock checks, etc.
6. Other background tasks: PLC-type programs
7. Error handling routine called by other tasks with various flags
8. Interrupt handling: other application-specific interrupt-driven tasks

The tasks may be executed in multiple processors, i.e., servo loop may be closed by a DSP, and communication and planning tasks may be executed by another higher-level microprocessor or DSP which provides the command signal to the servo loop.

## 10.4 BASIC SINGLE-AXIS MOTIONS

The following motion types are so often encountered in automated machines that they are given specific industry standard names and are considered basic motion types: *home, jog, stop, index* motions. A typical motion is either a triangular or trapezoidal velocity profile as a function of time. Furthermore, it is also common to modifiy the trapezoidal velocity profiles with smoothed sine functions in order to reduce the jerk effect of the commanded motion. It is customary to define the motion profile in terms of velocity as a function of time.

**HOME:** Standard name used in the programmable motion control industry to describe the motion sequence used to establish a reference position for the motion axis. The objective of the homing motion is to bring the axis to a known reference position, called "home" position. All other positions are referenced to that position for the axis. For instance, the machine tool axes must be at a predefined position before the beginning of a cycle. Figure 10.7 shows examples of homing motion. If there is absolute position sensor for the motion axis, the homing can be performed from the current known position to the desired position in one move where the distance is known. Quite often, a motion axis uses an incremental position sensor. The position of the axis is not known at power-up. The axis must search for a reference position using a combination of sensors that indicates a known position. Typically an encoder index channel and/or an external ON/OFF sensor is used to establish the reference position. Depending on the accuracy required in starting at the known reference position, different homing motion sequences can be performed. Figure 10.7 shows three different types of common homing motion sequences. Others can be designed to meet the

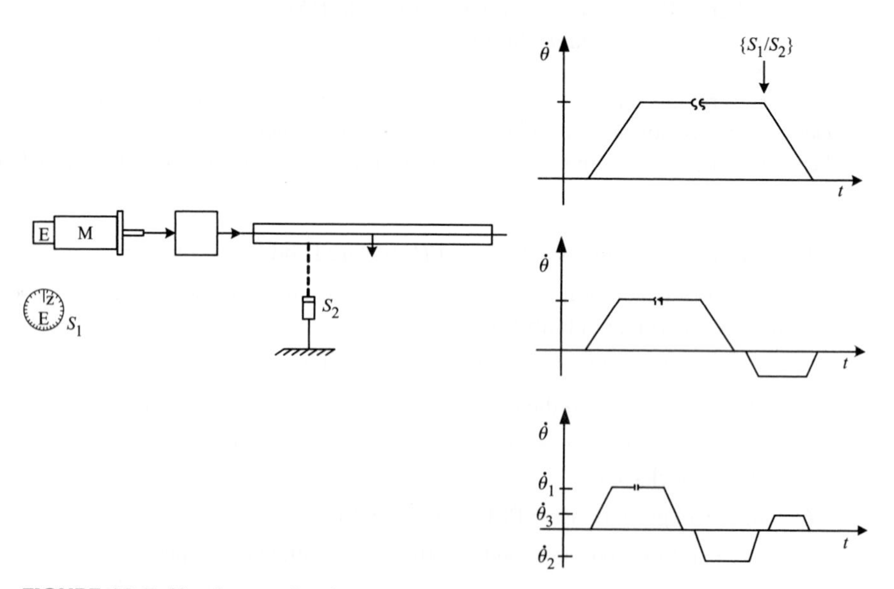


**FIGURE 10.7:** Homing routine (establishing a position reference) using motor and tool mounted sensors.

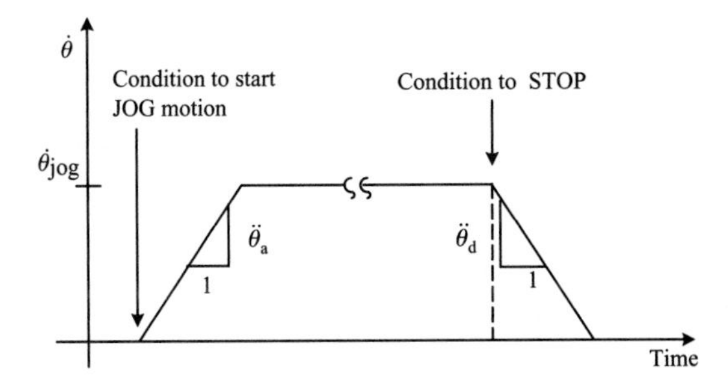


**FIGURE 10.8:** Motion executed on the JOG and STOP command.

needs of a particular application. The first home sequence simply moves at a constant speed until a sensor signal changes state (i.e., turns ON) and then stops the motion at a defined deceleration rate. The second home sequence does the same thing as the first home sequence, plus it moves backward by a predefined distance after stopping. The third home sequence adds another predefined motion distance in the forward direction. The small pre-defined movements at the end of the initial motion can also be defined based on external sensor input instead of a pre-defined distance. Notice the small wait periods between the stop and incremental moves. The purpose there is to allow the axis motion to settle to a complete stop. Parameters used to define a HOME motion are *home search speed, acceleration and decceleration rates, incremental move distances,* and *time periods to complete them.*

**JOG:** Standard name used to move at a certain speed while the operator holds an input switch on. Using the JOG function, the machine axes can be moved around to any position under the operator control. This function is needed in machine set-up and maintenance. JOG motion refers to a constant speed motion of an axis until a condition is met to stop it. The parameters used to define the jog motion are *jog speed, acceleration,* and *deceleration rates* (Fig. 10.8).

**STOP:** The rate at which the motion must be decelerated to a stop (zero speed) is defined by some PMC configuration parameters. When the STOP command (or equivalent) is executed, the motion is decelerated to zero speed from its current speed. The STOP motion may be initiated by an operator command or a sensor trigger or under program control. The only parameter needed to define the STOP motion is the *rate of deceleration* (Fig. 10.8).

**INDEX:** A common name used to describe a move which causes a predefined positional change in the axis. The described move position can be either in absolute or incremental units. Depending on the automation requirement, a typical automated cycle mode consists of various forms of INDEXes synchronized to other axis motion and machine I/O status (Fig. 10.9). Parameters used to describe the INDEX motion are *distance, acceleration and decceleration rates, top speed,* or *total index time period.* In some motion control applications an index motion may have multiple speeds. Such moves are called *compound indexes.* Compound indexes can be considered as a combination of various trapezoidal and triangular motion profiles. It is also very common to define an index motion with only distance and total time period for the motion. In this case, the time period is used in three equal portions: one third for the acceleration period, one third for the constant speed run period, and one third for the deceleration period.

***Example*** Consider a DC motor connected to a ball screw (Figs. 3.3, 9.13). Let the ball screw lead be 0.2 in/rev (or equivalently it has pitch of 5 rev/in). Assume that there are

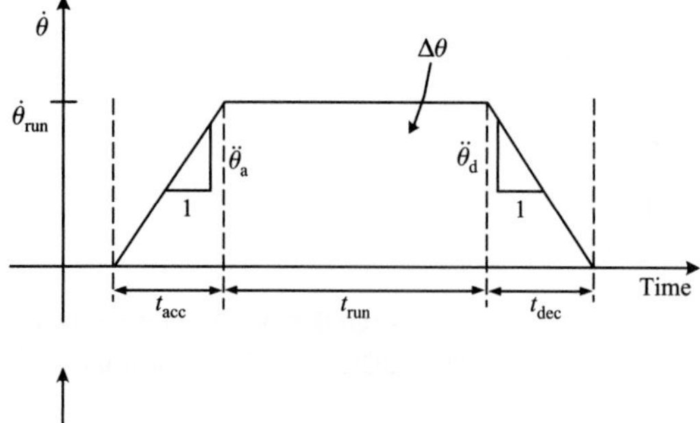


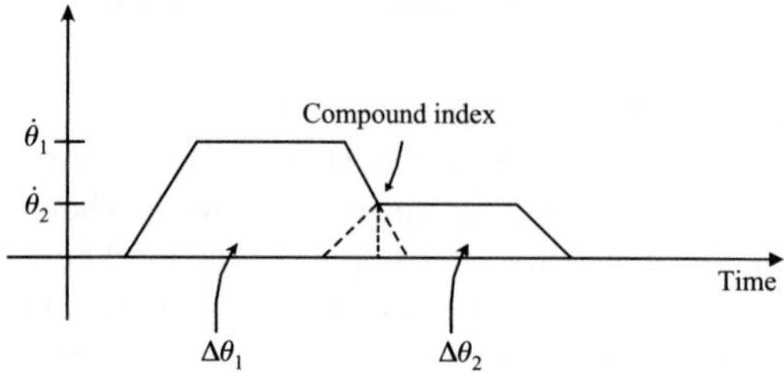


**FIGURE 10.9:** INDEX is a name given to a motion profile for a given distance (moving to an absolute position or moving for an incremental distance).

two limit switches, one at each end of the ball screw, to indicate the mechanical limits of travel. In addition, there is a presence sensor at the midpoint of ball screw's travel range to indicate a home position. Define a home motion sequence, a jog motion, a stop motion, and an incremental motion of 1.0 in distance movements.

A possible home motion may be defined as follows: The speed at which home sensor trigger will be sought is 0.5 in./sec. The acceleration rate is defined by the time allowed for the motor to reach the home speed, which is $t_{acc} = 0.5$ sec $= 500$ msec. Then when the home sensor tigger turns ON, the motion will stop at the deceleration rate defined by the stop motion. Then the axis will wait for 1 sec, move back 0.2 in. in 1.0 sec, wait for 1.0 sec, and move with a much smaller speed of 0.05 in./sec until the home sensor turns ON again. Then, it stops.

The jog motion can also be defined using two motion parameters: the acceleration rate (or the time period $t_{acc}$ to reach jog speed) and the jog speed value.

The stop motion definition requires only the deceleration rate and the condition on which the stop motion is initiated.

Finally, for index motion of 1.0 in. increments in either direction, we can define the acceleration, run, and deceleration times. Notice that

$$v = a \cdot t_{acc} \tag{10.1}$$

$$\Delta x_{acc} = \frac{1}{2} a \cdot t_{acc}^2; \text{ during constant acceleration motion phase} \tag{10.2}$$

$$\Delta x_{run} = v \cdot t_{run}; \text{ during constant speed motion phase} \tag{10.3}$$

$$\Delta x_{dec} = \frac{1}{2} d \cdot t_{dec}^2; \text{ during constant deceleration motion phase} \tag{10.4}$$

$$x = \Delta x_{acc} + \Delta x_{run} + \Delta x_{dec} \tag{10.5}$$

(10.6)

where $a$, $d$ are acceleration and deceleration values, $v$ is the top speed, and $t_{acc}$, $t_{dec}$, $t_{run}$ are the time periods for acceleration, deceleration, and constant speed run phases of the index motion (Fig. 10.9).

Let us assume that the index time period is equally divided between acceleration, constant speed run, and deceleration time periods,

$$t_{acc} = t_{run} = t_{dec} = \frac{1}{3} \cdot t_{index} \tag{10.7}$$

Let $t_{index} = 3.0$ sec. The motor rotation to make the 1.0 in linear displacement is 1.0 in · 5 rev/in = 5 rev. It can be shown that the top travel speed is

$$x = \frac{1}{2} \cdot v \cdot t_{acc} + v \cdot t_{run} + \frac{1}{2} \cdot v \cdot t_{dec} \tag{10.8}$$

$$= \left(\frac{1}{2} \cdot \frac{1}{3} + \frac{1}{3} + \frac{1}{2} \cdot \frac{1}{3}\right) \cdot v \cdot t_{index} \tag{10.9}$$

$$= \frac{2}{3} \cdot v \cdot t_{index} \tag{10.10}$$

$$v = \frac{3}{2} \frac{x}{t_{index}} \tag{10.11}$$

$$= \frac{3}{2} \frac{1.0 \text{ in}}{3.0 \text{ sec}} \tag{10.12}$$

$$= 0.5 \text{ in/sec} \tag{10.13}$$

The pseudo-code to define these motion sequences may look as follows. Exact syntax of the motion control software would depend on the particular motion controller.

```
% Assume we have two functions "MoveAt(...)" and "MoveFor
  (....) " to generate the desired motion.
% If these motion command (trajectory) generator functions
  are defined in C/C++, they can be
% overloaded to accept different argument list.
%
%
% Calculate or set parameters

   Home_Speed_1 = 2.5  % [rev/sec]
   Home_Speed_2 = 0.25 % [rev/sec]
   Home_Sensor = 1     % I/O channel number for the home
                         sensor
   Home_Index = 0.2
   Jog_Speed = 1.0     % [rev/sec]
   Jog_Stop = 2        % I/O channel to indicate to stop
                         motion.
   Stop_Rate = 10.0    % [rev]/[sec^2]
   Index_Value = 5.0   % [rev]
   t_acc = 1.0         % [sec]
   t_run = 1.0         % [sec]
   t_dec = 1.0         % [sec]

% Home motion

   MoveAt (Home_Speed_1)
   Wait until Home_Sensor = ON
```

```
        MoveAt (Zero_Speed, Stop_Rate)
        Wait for 1.0 sec
        MoveFor(Home_Index)
        Wait for 1.0 sec
        MoveAt (Home_Speed_2)
        Wait until Home_Sensor = ON
        MoveAt (Zero_Speed, Stop_Rate)

    % Jog Motion

        MoveAt (Jog_Speed, t_acc )
        Wait until Jog_Stop=ON
        MoveAt (0, Stop_Rate)

    % Stop motion

        MoveAt (0, Stop_Rate)

    % Index motion

        MoveFor (Index_Value, t_acc, t_run, t_dec )
    %
```

## 10.5 COORDINATED MOTION CONTROL METHODS

In PMCS that involve multiple actuators (also called *multi degrees of freedom* or *multi axis motion control system*), the motion coordination (also called motion synchronization) between different axes can be categorized as follows (Fig. 10.10),

1. Point-to-point control applications (insertion machines, assembly machines, pick-and-place machines)
2. Speed ratio (electronic gearing) applications (coil winding, packaging, printing, paper cutting, web handling machines)
3. Contouring applications (CNC machine tools, robots, laser cutting machines, knitting machines)
4. Sensor-based motion planning and autonomous motion control.

There are two variations of speed ratio-based motion coordination that are very common in web handling industry. They are:

1. Motion coordination using a "registration" signal from an external sensor (also called *registration application*)
2. Tension control applications (paper, plastic, wire handling machines)

Both of these applications use electronic gearing as the basis of motion coordination and add a motion modification to it based on the external sensor (registration sensor or tension sensor).

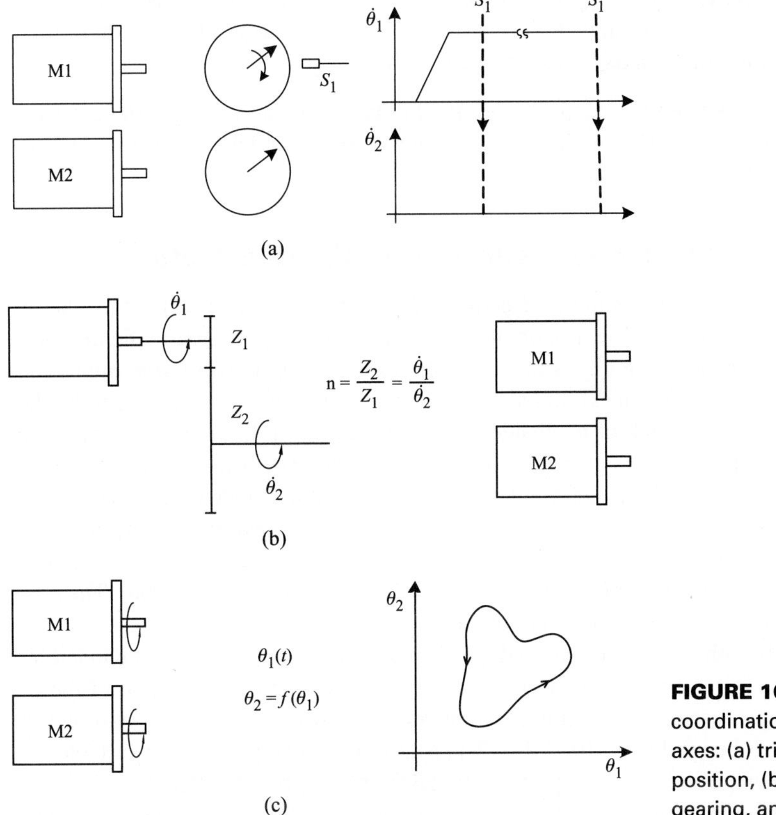


**FIGURE 10.10:** Motion coordination between two axes: (a) trigger at a certain position, (b) electronic gearing, and (c) camming.

## 10.5.1 Point-to-Point Synchronized Motion

Point-to-point position synchronization refers to the positioning of one axis or more with respect to another axis at a selected number of points during a cycle. In this type of application, there are a finite number of important points during each cycle where the relative positions are critical; i.e., in insertion applications, the pin and the housing must be properly positioned right before the insertion takes place. When an axis (master) is in a certain position, the other axis (slave) must be in a certain position relative to the master axis. The relative position of the slave with respect to the master is important at the synchronization reference points. During the motion between these points, the relative positions between axes are not critical. For instance, in pin-insertion applications, as long as the pin is in the correct position relative to the housing before the insertion starts, the motion profile of how each axis reaches to that point is not important [Fig. 10.10(a)].

Consider two axes motion control systems. Assume that axis 1 is the master and runs at a constant speed. During every cycle of axis 1 motion, axis 2 is supposed to make predefined motion when axis 1 is at a specific position in a given cycle relative to the beginning of the cycle. If the move of axis 2 is defined as a function of time, then the only information needed to coordinate the motion are:

- ***Axis 1:*** one cycle distance, zero reference at each cycle, current position within the current cycle in order to generate the trigger signal for axis 2.

- ***Axis 2:*** when triggered by axis 1 position, make a predefined move. In turn, at the completion of axis 2 motion (or at a certain position of axis 2 during its motion), it may trigger axis 1 to start a new cycle.

If the desired motion of axis 2 needs to be done within in a certain distance of axis 1 motion, the motion of axis 2 is defined as a function of axis 1, and that requires "electronic gearing" or "electronic contouring."

### 10.5.2 Electronic Gearing Coordinated Motion

In mechanically coordinated designs, one axis motion can be tied to the other axis through a mechanical gear. The motion of axis 1 is directly proportional to the motion of the other axis in position, velocity, and acceleration (neglecting transmission imperfections due to backlash and tolerances). In mechanical designs, the relationship is fixed by the gear ratio.

The drawback of mechanical gearing is that if a different gear ratio is needed for a different product, the gear ratio between the shafts needs to be mechanically changed by installing different gears for every different gear ratio. This will increase the set-up time. It may also be economically not feasible for applications involving many different gear ratios to keep many different gears in stock. The functionality needed here is to provide a motion to the second shaft which is proportional to the motion of the first shaft. The proportionality constant may vary for different products.

If both shafts had their independent actuation source, the commanded motion to the slave axis can be derived from the commanded or the actual motion of the master shaft. In the process of calculating the commanded motion for the slave shaft, any desired gear ratio can be used [Fig. 10.10(b)]. The gear ratio can be constant or variable. Furthermore, in certain positions in the cycle additional forward or backward moves can be implemented to change the phasing of slave shaft with respect to the master shaft. This type of motion profile generation is called the *electronic gearing* or *software gearing* because it is done electronically by software implementation as opposed to mechanical gears. The position tracking accuracy should be at least as good as the accuracy which would be obtained if the shafts were connected to each other through mechanical gears.

The gear ratio is set in software and easily changeable without any time-consuming mechanical disassembly and re-assembly. Electronic gearing is perhaps the most common type of programmable motion control application encountered in industry today. Let us consider two axes—each is driven by a separate motor. Axis 1 is the master. The axis 2 desired motion can be generated from the motion of axis 1 using a fixed (or variable) ratio. As long as servo loop delivers a good position tracking accuracy, the net result is a two-axis motion system making geared motion. The commanded motion to axis 2 can be generated from either the actual or the commanded motion of axis 1. Acceleration and deceleration rates of the slave axis can be defined as a function of the master axis position change instead of being defined as function of time. In all electronic gearing applications, the slave axis motion profile can be defined as a function of one of two independent variables:

1. Time
2. Master axis position

The accuracy of the electronic gear tracking is as good as the tracking accuracy of each slave axis, whatever method of control is used. Therefore, for the electronic gearing to be successful, the servo loops should be properly tuned. Otherwise, the tracking errors may deteriorate the performance to the point that it does not resemble the performance that would be achieved had the shafts been connected to each other by mechanical gears.

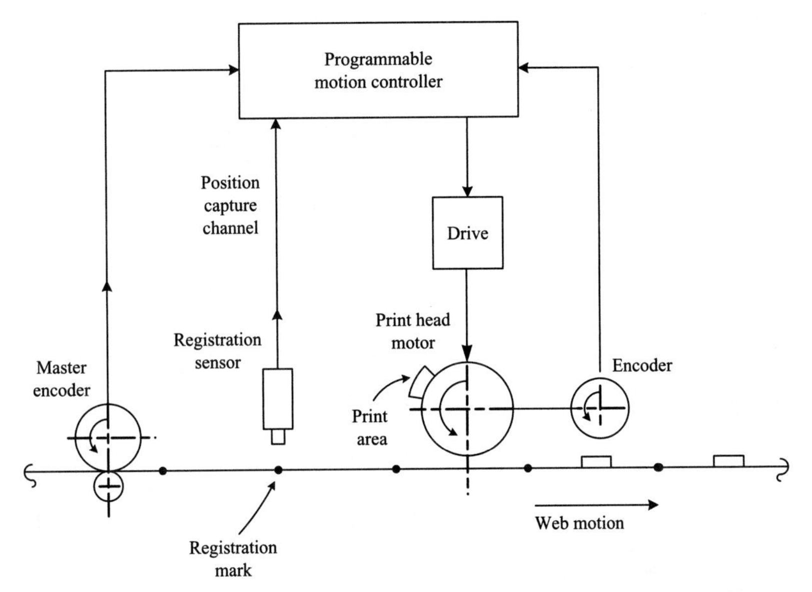


**FIGURE 10.11:** Web handling with registration application. A print head (or a cut head) follows the master encoder in order to match the speed of the web. It makes phase adjustments to its position based on a registration sensor. The position of the print head must be captured very accurately when the registration sensor triggers ON. The registration sensor must have good repeatability and fast response in order to have a good positioning accuracy for the print head.

***Fixed Ratio Gearing*** This is the simplest form of electronic gearing. It emulates exactly the relationship between two shafts connected to each other via a fixed gear ratio. During the cycle, the slave axis (or axes) are commanded to move at a certain gear ratio relative to the master axis motion. It is also very straightforward to use two or more fixed gear ratios between two axes where the gear ratio is changed from one fixed value to another at certain positions during a cycle—for instance, rotating knife applications where the web needs to be cut at a certain length (cut length is programmable). During the position range when the knife is in contact with the web, the linear motion speed of the knife must match the linear motion of the web (gear ratio $z_1$), and during the position range when knife is not in contact with the paper, the knife must speed up or slow down to a different gear ratio ($z_2$) to let the desired amount of web length pass-by [Figs. 10.11, 10.12(b)].

***Gearing with Registration*** Printing, cutting, and sealing applications in web handling processes require motion coordination which involves gearing plus registration. This very common form of motion coordination involves the following (Figs. 10.11, 10.12):

1. Electronic gearing of a slave axis to a master axis with a certain gear ratio.
2. For each cycle, after a sensor input condition (registration mark or registration signal) is true, the slave axis must make an incremental move during a certain distance of the master axis. This is called the phase adjustment based on registration mark sensing. A variation of this approach is to calculate a new gear ratio and to change the gear ratio in order to get the same amount of additional move.

There are many variations of the registration mark applications. The incremental motion distance may be either a predefined length or calculated based on the difference between actual position captured when the registration sensor triggers and a desired position.

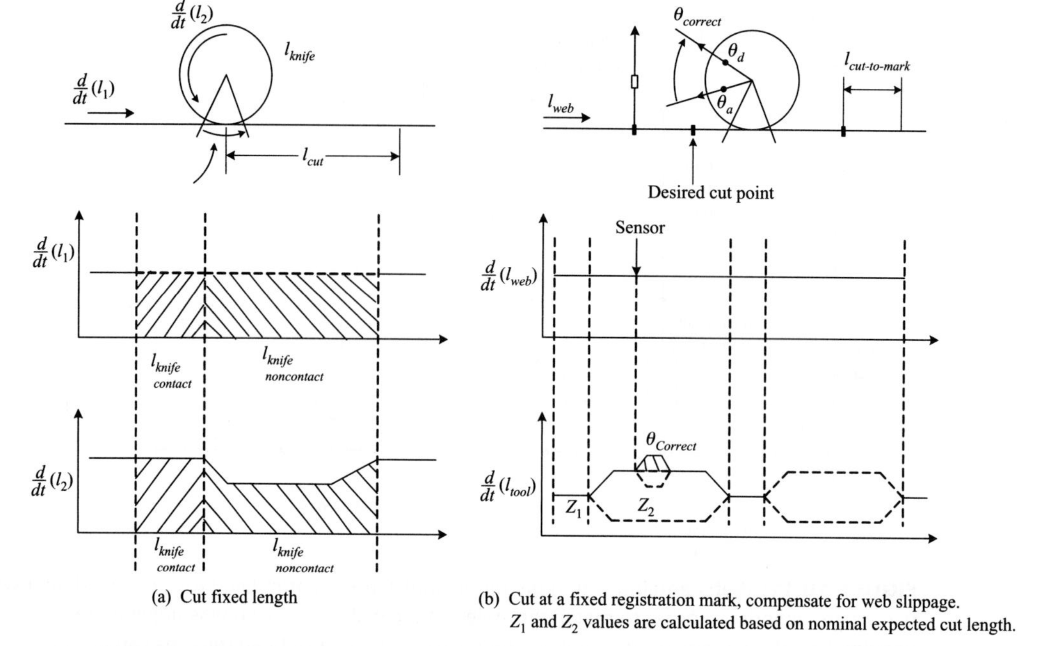


(a) Cut fixed length

(b) Cut at a fixed registration mark, compensate for web slippage. $Z_1$ and $Z_2$ values are calculated based on nominal expected cut length.

**FIGURE 10.12:** Web handling with registration mark application: (a) Cut a web every given length ($l_{cut}$) and there is no registration mark. (b) Cut web always at a certain distance from a registration mark.

The keys to achieving high accuracy in registration applications are to capture certain motion-related conditions very quickly (i.e., less than 25 $\mu$sec) and respond very quickly (i.e., a few milliseconds).

For instance, in paper-cutting machines, the flying knife can be geared to the master axis moving the paper (Figs. 10.11, 10.12). The paper must be cut at a certain offset distance from a registration mark. However, the registration mark may not be perfectly printed on the machine in exact distances from each other. The paper may slip in the master axis drive, hence the master encoder reading of paper movement cannot always be perfect. To overcome such unavoidable problems, a registration mark is always sensed on the paper at a fixed distance from the cutting point. Regardless of the printing or position accuracy loss due to slippage, there has to be a fixed phase relationship between the cutting knife and the location of registration mark. Therefore, the solution is:

1. Run the knife at nominal gear ratio relative to the master axis (i.e., using the master axis encoder position as the master position) feeding the paper.
2. At each registration mark sensor trigger, capture knife axis position, compare the actual position with the desired position, and correct the phase error, if any, with superimposed incremental moves on top of the current gearing motion.

The position capture on registration sensor input must be done very fast. This generally requires the use of axis level high-speed I/O lines instead of the general-purpose machine I/O. Let us consider a paper-cutting machine example. The paper is cut at a certain distance from a registration mark. Based on the nominal expected value of the cut length, a nominal gear ratio is set between the paper line speed (master axis) and the rotation knife axis (slave axis). When the controller receives the registration mark signal, it captures the knife position. Then it compares that captured position with the desired position that the knife should be at in order to cut the paper at the right location. Then the knife is commanded to make an incremental move to make the phase correction before the cutting action begins. Let us consider that the paper-cutting machine runs at 1200-ft/min line speed. When the registration mark is sensed, the controller will have to capture the knife axis position within a certain finite time. At 1200-ft/min rate, paper moves 20 ft/sec or 240 in/sec or 240/1000 in/msec. If the paper is to be cut with ±5/1000-in accuracy, the position must be accurately captured before that amount of paper passes. The 5/1000-in paper movement takes about 0.021 msec. Therefore, the position capture on registration mark must be done within 21 $\mu$sec. This type of response requires dedicated axis level high-speed I/O. Notice that the correction of position error does not (and cannot) need to be made within 21 $\mu$sec, rather within the period remaining in the cycle which is typically in the order of tens of milliseconds.

### 10.5.3 CAM Profile and Contouring Coordinated Motion

Consider a shaft driven by an independent actuation source (i.e., a motor-drive) and a mechanical cam on the shaft. The cam can be on the face of the shaft or around the shaft. A tool is connected to the cam profile through a cam follower. As the shaft rotates, the tool moves up and down as a function of the cam profile machined on the cam grove. For every point on one revolution of the shaft, there is a corresponding point on the cam; hence there is a corresponding position of the tool (Fig. 3.9). This relationship is mechanically determined by the cam design. For every revolution of the shaft, the cam relationship between the shaft and the tool is repeated. More than one cam can be connected to a shaft to drive more tools as slaved to the shaft motion. The slave motion synchronization for each tool is determined by the cam profile machined into each cam. This type of synchronization is more complicated than a constant gear-ratio relationship.

The functionality in cam synchronization is that the slave axis position is directly a function of the master axis position. The relationship repeats every revolution of the master axis output shaft.

The same functionality can be more flexibly achieved by *electronic cam* or *software cam* motion coordination. Both master and slave must have their independent actuation source (motor-amplifier). As the master axis moves, the desired (commanded) motion for the slave axis can be derived based on the position of the master axis. The desired relationship in this case happens to be a cam profile. The cam profile is defined only in software as a mathematical equation or a look-up table. As the master axis moves, the position of the master axis is periodically sampled. Usually, the sampling rate is the same as the servo loop update rate. Then the corresponding desired position of the slave axis is derived from the mathematical cam profile equation or from the loop-up table. Furthermore, in software implementation, the slave axis does not have to be driven by the cam relationship throughout the whole revolution or cycle. The cam following mode can be entered and exited at various points during a cycle based on some application dependent condition.

Contouring coordination is mostly used in machine tools and plasma cutting type of machines. The contour coordination is a more complicated form of cam coordination. The basic idea is that two or more axes must move in space in order to make a tool trace a certain path. In two-dimensional x-y motion, this corresponds to drawing an arbitrary curve in plane; in three-dimensional x-y-z motion, it corresponds to drawing an arbitrary curve in 3-d space. At any given time, all of the axes have a position that they must be at in order to trace the desired path. The way the commanded motion is generated for each axis can vary. One alternative is that the feed rate along the path (tangential linear speed) can be the indepedent variable. The feed rate can be planned based on application-specific requirements. Then as the tool travels along the path, the desired motion of each axis is calculated and commanded. In this case, the feed rate acts as the master (although it is not an axis), and all of the motion axes are slaved to it. Another alternative is to select one of the axes as master and define its motion, then derive the desired motion of other axes so that the tool traces the desired path. CNC machine tools are an example of the coordinated motion control application where contour coordination is most commonly implemented.

Consider a two-axis motion control application where the two axes is required to trace a contour, i.e., an X-Y stage holding a tool, and the tools are required to trace a path in the X-Y plane. Gearing coordination would generate a path that is a straight line. Contouring is a more general form of coordination and requires the ability to trace any path in the motion space.

The path can be defined as function of:

1. Time—each axis motion is separately defined as a function of time to generate the path.
2. One axis is set as master, and its motion for the path is defined; the other axis motion is defined as a function of the master axis position.
3. Path length or speed parameter—each axis motion is defined in terms of a path parameter.

***CNC Programming*** Computer numeric control (CNC) programming is a programming language used to define the desired motion of machine tool axes. The most common CNC language in use today is the G-code. The G-code standard is defined by Electronics Industry Association (EIA) standard 274-D. Although this standard exists, there are minor variations in the implementation of G-code from one manufacturer to another. The meaning of some codes also varies from one machine tool type to another. For example, G70 code means programming in units of inches in machining centers, whereas the same G70 code means edge finding in electric discharge machines (EDM).

### 10.5.4 Sensor-Based Real-Time Coordinated Motion

As the sophistication of computer-controlled machines increases, the type of motion demands placed on them gets more and more complicated. Particularly in robotic manipulator applications, the desired motion may not be known in advance. The machine is required to sense its environment (i.e., using vision systems) and decide on a motion strategy and generate the motion command profiles for each individual axis. The motion synchronization used may differ for different phases of its motion. Furthermore, this decision as to what strategy to use is determined on-line by the control software. Interpretation of sensory data and generating intelligent motion planning strategies in real-time is one of the current challenges of programmable motion control in robotics devices.

## 10.6 COORDINATED MOTION APPLICATIONS

### 10.6.1 Web Handling with Registration Mark

The following applications have very similar motion coordination requirements (Figs. 10.11, 10.12):

1. Rotating cutting knife to cut a web (i.e., paper, plastic) for fixed length and with reference to a registration mark
2. Rotating printing head to print over a web
3. Rotating sealing head to seal a web or bag

There are three key issues in general web handling applications with a registration mark:

1. Match linear speeds between the web and tool during contact phase
2. Adjust tool speed to meet the cycle web length requirements during noncontact phase
3. Adjust tool position (phase) with respect to the registration mark

The common key motion coordination requirement in these applications is that during a certain length of each cycle of web movement, the tool (cutting knife, print head, sealing head, etc.) must be moving at the same linear speed as the web. Usually, when the tool is in contact with the web (i.e., the knife is in the process of cutting the paper, the print head is in contact with the web), the linear speed of the tool and the web must be the same. Some printing heads print through the whole circumference of the print cylinder. In that case, the linear speeds must always be matched. The whole cycle is a contact phase. In such applications, only position phase corrections can be made in order to print at the correct position relative to the registration mark.

When the tool is not in contact with the web, the tool speed can be adjusted in order to let the proper amount of web distance to pass. If the design was not a programmable motion control system and it was mechanically geared, a rotating knife machine can only cut one paper length only which is the length of the knife circumference. The ability to program the motion of the knife so that while it is in contact with the web it matches the linear speed for a correct straigth cut, and speed up or slow down while it is not in contact with the web, provides us the capability to cut many different lengths of web with one rotary knife. The same discussion applies to print heads and sealing head applications where the cycle length for which a new print or seal to be made can be programmed.

If the web does not have a registration mark, the motion coordination problem is easier. There are only two phases to the coordination:

1. ***Contact phase***: Tool and web are in contact.
2. ***Noncontact phase***: Tool and web are not in contact.

The web axis motion is always the master because it is difficult to make the web do whatever we want. Rather, we need to follow and track the web with the tool axis. In contact phase, the tool axis gear ratio needs to be adjusted so that the linear speed between the web and tool is equal. If the gear ratio is programmed in linear motion units, it will be 1 to 1. During the noncontact phase, the gear ratio will be different than 1 to 1 linear ratio. The gear ratio will be such that the remaining length of web will pass during the time the tool moves over the noncontact phase. If the web cycle length happens to be equal to the circumference of the rotary tool, then the gear ratio will be 1:1 during the noncontact phase in linear distance units as well [Fig. 10.12(a)].

More often, a cut must be made relative to a registration mark [Figs. 10.11 and 10.12(b)]. A print must be made repeatedly but at a certain offset distance from a registration mark every cycle. Similar needs exist for sealing type applications. In other words, the tool will cut/print/seal the web every cycle where the web length per cycle is programmable, but it must be done relative to a registration mark. If the registration mark and length requirements are in conflict, the registration mark requirement will supercede. In other words, let us assume that we would like to cut a paper from a registration mark at 1.00 distance and 12.00 total length. When we receive the registration mark, we record that only 11.5 in. of paper has passed the knife. The conflict is this: If we cut the paper at 1.0-in. distance, the paper length will be 12.5 in. If we cut it at 12.0-in. length (0.5 in. after the registration mark), the offset distance from the registration mark will be 0.5 in. instead of 1.0 in. There is no solution to this problem because with one measure, we cannot satisfy two different requirements. Quite often, the offset distance from the registration mark takes priority over the cycle nominal length because the web can slip and the master axis reading may not be accurate. The registration mark provides a positive sensing mechanism for the actual web length passed.

From the electronic gearing motion coordination point of view, the registration mark requirement adds one more motion coordination problem. That is, the tool must be in correct positional phase relative to the mark in order to cut it in the right offset location. Assuming that there is no slip or variation in registration mark location, once the tool and paper are in phase, they will stay in phase for a fixed cycle length. However, to ensure accuracy in every cycle against web slippage or variations in web registration mark location, the tool-web position phase is checked and corrected if necessary.

Motion Coordination Algorithm: Rotating Tool (Cutting Knife, Print Head, Seal Head) with Registration Mark Application.

```
Main Program
{
    Initialize the rotating knife with registration mark
    application algorithm
    Verify HOME motion sequence is done
    Check Task 1
    Check Task 2
.....
   while (TRUE)
   {
     Check Print Phase and Registration Mark
     {
       When Entered Contact Phase, execute Contact_Phase
       Function once
       When Entered Noncontact Phase, execute
       Non_Contact_Phase Function once
       When Registration mark is received, execute
       Registration_Mark_Phase_Adjustment
     }
   }
}

Algorithm Initialize:
   web_cycle_length = ... % web length per cycle
   tool_circumference = .... % pi * diameter
   contact_percentage = ..... % 0 to 100
   gear_ratio_contact_num = 1
   gear_ratio_contact_den = 1
   gear_ratio_noncontact_num = web_cycle_length * (100-
   contact_percent)/100
   gear_ratio_noncontact_den = tool_circumference * (100-
   contact_percent)/100

   define desired position phase for tool on registration
   mark
   setup fast position capture (actual or commanded) of
   tool axis on registration mark

Return
```

```
Contact_Phase:
   change gear ratio of tool (slave) to web (master) to
   ratios defined by
   z = gear_ratio_contact_num / gear_ratio_contact_den
Return

Non_Contact_Phase:
   change gear ratio of tool (slave) to web (master) to
   ratios defined by
   z = gear_ratio_noncontact_num / gear_ratio_noncontact_den
Return

Registration_Mark_Phase_Adjustment:
   Get the position of the tool captured on registration
   mark with high speed capture
   Compare the actual position to the position desired for
   proper phase
   Command additional move (Position desired - Position
   captured) to complete fast
Return
```

### 10.6.2 Web Tension Control Using Electronic Gearing

Web handling is a generic name used to describe manufacturing processes where a continuous web must be moved and processed. The web material is generally one of the following (Figs. 10.13, 10.14):

1. Paper in printing machines, in paper-cutting machines
2. Plastic in packaging and in labeling machines

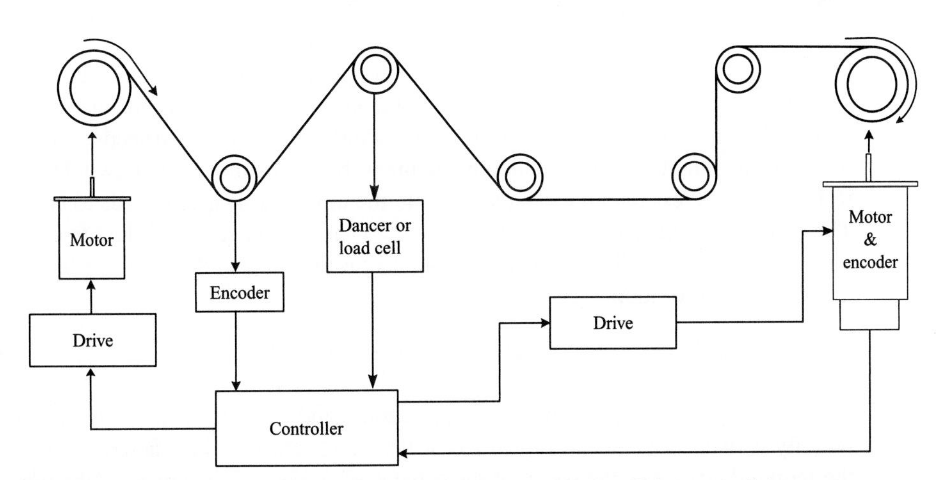


**FIGURE 10.13:** Web motion control with tension regulation. Either one of the two motors can be used to regulate tension, while the other motor is used to set the line speed. Encoder is used to determine the speed of the web. Dancer/load sensor is used to measure the web tension. The goal is to move the web at a desired speed while maintaining a desired tension during the motion.

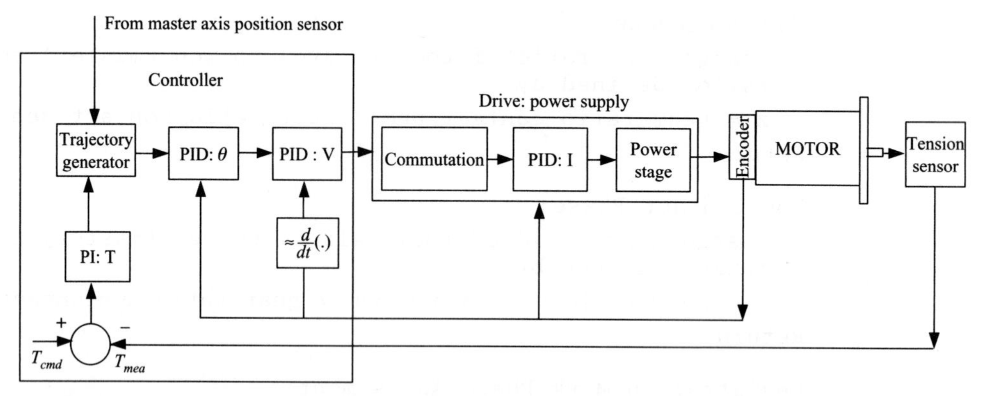


**FIGURE 10.14:** A servo control loop to control tension in the web. There are two different common implementations: (1) No master axis speed is provided. The commanded speed of this axis is modified (increased or decreased) based on the tension control loop: i.e., $\dot{\theta}_{cmd} = \dot{\theta}_{cmd0} + \Delta\dot{\theta}_{cmd0}(T_d - T_a)$. (2) a master axis speed is provided and the gear ratio between the master axis and this (slave) axis is modified based on the tension control loop: i.e., $\dot{\theta}_{cmd} = z \cdot \dot{\theta}_{master}$, $z = z_0 + \Delta z(T_d - T_a)$.

3. Wire in winding and wire drawing processes
4. Sheets of steel in steel mills

The web handling motion control problem has two components:

1. The web must be moved at a certain process speed, which is usually referred to as "line speed."
2. While being moved, web tension must be maintained at a desired level.

In general, the production rate dictates the line speed at which the web must be moving nominally. While the web is moving, the tension of the web must also be measured and corrections must be made to the feed in order to maintain the desired tension. In order to control both the speed and tension on a web, there needs to be two actuation sources: one to set the nominal line speed, the other to track the first nominally but to maintain the desired tension.

Consider that the web speed is set by the nip roller (or similar mechanism) which is specified by the required process speed. The unwind/rewind roll speed needs to be adjusted in order to maintain the desired tension while the web is being moved. In unwind applications, if the tension is less than the desired tension, the unwind roll speed needs to be decreased. The opposite is true in the rewind applications. That is, if the tension is less than the desired tension, the rewind roll speed needs to be increased. This is referred to as the polarity of the tension control loop

$$\Delta V = (sign) \cdot k \cdot (T_d - T_a) \tag{10.14}$$

where the sign is (+1) for unwind applications, and (−1) for rewind applications. The polarity is defined as the sign (positive or negative) ratios of the change in the tension to the required change in the speed of the drive used to control the tension. The polarities of unwind and rewind tension control are opposite (equation 10.14).

There are two main tension sensing methods: strain-gauge-based and dancer-arm-based tension sensors. The differences between using a strain-gauge-based tension sensor versus a dancer-arm-based tension sensor are as follows.

1. Strain-gauge-based tension sensor (also called load cell) results in larger loop gain because small changes in web in-feed and out-feed results in large tension changes compared to the dancer-arm-based tension sensor. On one hand, if the web is too stiff, the loop gain as a result of small differences in in-feed and out-feed web may be so large that large oscillations and closed-loop instability may occur. With sufficiently flexible web, on the other hand, this method would provide the fastest closed-loop response. The web stiffness and tension sensor sensitivity combination must be carefully judged to determine whether they will result in good, tight, closed-loop performance with large loop gain or they will result in too large loop gain that will create instability problems.
2. Dancer-arm-based tension sensor results in lower loop gain because the same change in web in-feed and out-feed will result in smaller tension variations. This effectively reduces the loop gain of the closed-loop system and instability problems are less likely to occur. The potential problem with this method is the inertia and spring effect of this type of tension sensor. The dancer mechanism has some inertia. It also usually has a preloaded spring. As the dancer moves under the web tension, it may tend to oscillate. The oscillations are more significant during high acceleration and deceleration. Therefore, the oscillations of the dancer arm itself causes fluctuations in actual tension. It may also excite natural resonance of the web and result in large oscillations. The only solution would be to move the web in such a way that the dancer arm is not displaced too fast, which means the variation in web in-feed and out-feed rates must be very slowly varying.

We will consider two different control ideas which differ only in their control algorithms. The only difference is in the way we control the unwind/rewind roll in order to regulate the tension. All of the following ideas are identically applicable to both unwind and rewind applications with the only exception of polarity difference. A tension control algorithm for an unwind tension control problem can be applied to a rewind tension control problem by changing the sign of the output of the tension control loop output.

- ***Approach 1***: Tension Control with Adjusted Velocity Command—web tension control by controlling the roll drive with inner velocity servo loop and outer PI tension servo loop.
- ***Approach 2***: Tension Control with Electronic Gearing—web tension control by controlling the unwind/rewind roll drive with inner position loop commanded by electronic gearing, and outer tension loop which modifies the electronic gear ratio.

***Approach 1: Tension Control with Adjusted Velocity Command*** The traditional approach is to control the rewind/unwind roll motor using two closed loops: 1. inner loop is a velocity PI type-loop (compare desired velocity and measured velocity and command the amplifier based on velocity error) which presumably makes the motor track the commanded velocity within its bandwidth capability; 2. Outer tension PI-loop which generates the commanded velocity based on the tension loop error. This approach has two limitations:

1. Closed-loop stability problem due to large loop gain
2. Slow acceleration/deceleration rate limitation

Let us explain these limitations by focusing on the tension control loop control system. Figure 10.14 shows the closed loop of the tension control from a different perspective. The key observation to be made in this block diagram is that the loop gain from web tension error to roll speed change is large. In other words, small changes in in-feed and out-feed rates will result in large tension errors, and that in turn will result in large commands and possibly transient oscillation on the roll speed. Both of the above-mentioned limitations are the result of large loop gain. Large loop gain not only causes stability problems, but also the acceleration/deceleration rates must be slow enough to avoid large oscillations. Large acceleration/deceleration rates will require large changes in commanded speed, which means large change in tension. This means that the system will not be able to accurately regulate tension (keep tension error small) under large acceleration/deceleration cases of motion. The result is that such an approach works only if the web is slowly accelerated and decelerated.

***Approach 2: Tension Control with Electronic Gearing*** Electronic gearing refers to a position servo controlled motion where the desired motion command is generated based on a gear ratio multiplied by a master motion source. In this case, the master is the encoder directly driven by the web line speed. In software, we monitor the motion of the web at the servo loop update rate, and command desired motion to the unwind roll proportional to the master motion speed. The proportional value is the gear ratio defined in the software.

The key difference in this approach is that the variation in tension is not allowed to get too large as seen by the feedback loop. The gear ratio in software is updated based on the tension error. The tension error is passed through a PI control algorithm, and the result is the updated gear ratio. The feedback loop sees much smaller tension error levels. Hence, it can have larger gain and larger acceleration/deceleration rates without causing closed-loop stability problems. Notice that the gear ratio initial starting value must be accurate for this approach to be more accurate than the first approach. This can be obtained by either directly sensing the roll diameter at the beginning of a cycle or always starting with a diameter known in advance or entered by the operator as set-up information.

```
Control Algorithm for Tension Control Using Electronic
Gearing

Initialize once:
  {Define/Input/Read} tension loop polarity (unwind/rewind)
                      sign = +1 or -1
  {Define/Input/Read} desired tension:     Td = ....
  {Define/Input/Read} master encoder roll diameter
                      d1 = ....
  {Define/Input/Read} current unwind/rewind roll diameter
                      d2 = .....
  Calculate initial gear ratio:    z =(d_1 / d_2)
  Initialize the parameters of the tension control
  algorithm (PI in this example):
    z_I = 0.0
    Kp  = 0.01
    Ki  = 0.001
End_Initialize

Update Periodically: (i.e. every 10 msec)

  Read tension sensor:             Ta
  Tension error:                   e_T = Td - Ta
  Integral portion of control      dz_i = dz_i + sign * Ki
                                   * e_T
```

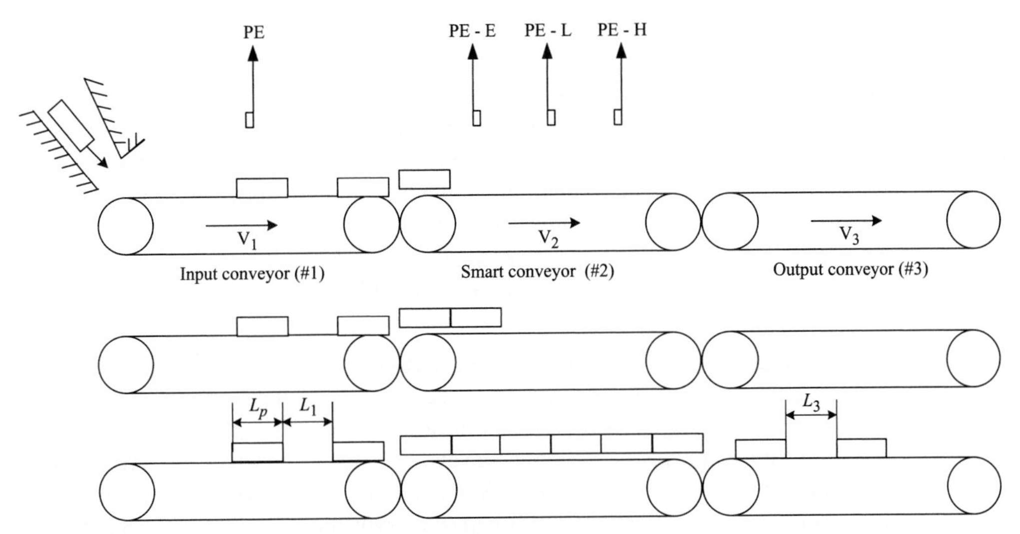


**FIGURE 10.15:** Smart conveyor operation.

```
        Proportional portion of control dz = sign * Kp * e_T
        P + I portion    dz = dz + dz_i
        Calculate the new gear ratio    z = z + dz
        Update the gear ratio 'z' in the electronically geared
        motion between master and slave axes.

    EndUpdate
```

### 10.6.3 Smart Conveyors

Conveyors are one of the most common mechanical systems used in industrial automation. They can be considered the "work-horse" of mass production. A typical conveyor runs at a constant speed and continuously moves material from one end to the other. The "smart" conveyor differs from the typical conveyors in that in addition to running at a certain speed, it makes position adjustments in order to space parts uniformly even though the in-feed rate into the conveyor may not be uniform.

A smart conveyor system typically involves three conveyors (Fig. 10.15): (1) input conveyor, (2) smart conveyor, and (3) output conveyor. In some cases, the smart conveyor and the output conveyor are the same conveyor. The generic automation problem is as follows. The parts come onto the input conveyor at random intervals. There is a fixed average feed rate, i.e., 300 parts per minute, but the spacing between them is not uniform. The objective is to adjust the spacing between the parts to a desired distance before they reach the output conveyor.

There are three basic questions about the application which affects the smart conveyor design:

1. Can the parts touch each other?
2. Does the speed of a conveyor need to match the other adjacent conveyor during the part transfer between them?
3. Is there a limit on the acceleration and deceleration rates due to part considerations?

Depending on how the first two questions are answered, the design can lead to one of two different approaches: (1) Constant gear ratio spacing conveyors, and (2) Position phase adjusting conveyors. The third question affects the number of smart conveyors to be used. If there is acceleration/deceleration limit, multiple smart conveyors would be used in order to give more time for the acceleration/deceleration moves. Soft or liquid products have acceleration/deceleration limits so that their shapes are not changed or spilled due to excessive inertial forces. If the parts can touch each other (i.e., bread doughs cannot touch each other), and the speeds of two adjacent conveyors do not have to match during the part transfer, constant speed spacing conveyors will be sufficient to handle the automation task.

***Constant Gear Ratio Spacing Conveyors with a Queue (Smart) Conveyor*** This design is appropriate when the parts can touch each other and when the speeds of two adjacent conveyors do not have to match during the transfer. In this case, there will be three conveyors: (1) input conveyor, (2) queue conveyor (the smart conveyor), and (3) output conveyor. Parts come to the input conveyor at a fixed average rate ($N_1$), but with nonuniform spacing. The queue (smart) conveyor runs slower than the input conveyor to provide a constant supply of continuous parts (Fig. 10.15). The output conveyor runs faster to adjust the proper spacing. If the queue conveyor is always full and parts are touching, the spacing will be proportional to the speed ratio between conveyor #2 and #3. Let us assume the part transfer rate is $N_1$. In continuous operation, the part rate in the conveyors must be the same. The length of each part is $L_p$, and the average spacing is $L_1$. The speed of the queue conveyor will be determined by the requirement that the part rate be the same in both conveyors:

$$N_1 = N_2 \tag{10.15}$$

$$\frac{V_1}{(L_p + L_1)} = \frac{V_2}{L_p} \tag{10.16}$$

$$V_2 = L_p * N_1 = (L_p/(L_p + L_1)) * V_1 \tag{10.17}$$

If the desired spacing between the parts is $L_3$, then the spacing conveyor speed needs to satisfy the same part rate requirement,

$$V_2/L_p = V_3/(L_p + L_3) \tag{10.18}$$

$$V_3 = ((L_p + L_3)/L_p) * V_2 \tag{10.19}$$

If the difference between $V_2$ and $V_3$ is so large that the acceleration rate is too high for the part integrity (i.e., soft parts or fluid content), then more than one spacing output conveyor would be used in order to reduce the maximum acceleration levels experienced by the part. From the queue conveyor point of view, the cycle time per part is

$$t_{cycle} = L_p/V_2 = \frac{1}{N_2} = \frac{1}{N_1} \tag{10.20}$$

Let's assume that $L_{pt}$ is the portion of the part length during which the part is in contact with both conveyors and changes speed from $V_2$ to $V_3$. The acceleration rate is then

$$t_{transfer} = \frac{L_{pt}}{L_p} \cdot t_{cycle} \tag{10.21}$$

$$A = (V_3 - V_2)/((L_{pt}/L_p) * t_{cycle}) \tag{10.22}$$

The number of spacing output conveyors needed is determined by rounding the following result to the next highest integer value

$$N_{spacing} = Int(A/A_{max}) \quad (10.23)$$

If the maximum allowed acceleration or deceleration is a limiting factor, the conveyor design parameters can be calculated to satisfy this constraint. Given $A_{max}$, calculate the number of spacing conveyors necessary following the queue conveyor. It can be shown that final part spacing $L_3$ is equal to (using eqn. 10.15–10.22)

$$L_3 = A \cdot \frac{L_{pt}}{L_p} \cdot \frac{1}{N_1^2} \quad (10.24)$$

In summary, the queue conveyor design is affected by the following parameters: $\{A_{max}, L_3, t_{transfer}, N_{spacing}\}$. If three of these parameters are specified, the fourth one can be calculated. The maximum allowed acceleration and deceleration rate and the desired spacing are given. Then, if we choose to use the $N_{spacing}$ conveyor (i.e., $N_{spacing} = 1$), that determines the minimum $t_{transfer}$, hence $t_{cycle}$, time in order not to exceed the acceleration limit. That sets the maximum throughput rate for a conveyor with $\{A_{max}, L_3, N_{spacing}\}$ specifications.

Let us consider the case that the input conveyor part feed rate may fluctuate. If it is desired that the conveyor system should be able to adjust to that condition, we need to add sensors. Assume we know $L_p$, $L_3$. First we need to measure the moving average part input rate, $N_1$. From that we can calculate the nominal speed of the queue conveyor and the output conveyor. In addition, we can add three ON/OFF sensors on the queue conveyor to indicate the following conditions: queue high, queue low, queue empty. If the queue is low, the conveyor speed $V_2$ can be reduced by a percentage; if the queue is high, it can be increased by a percentage. If the queue is empty, it should stop and go into a homing routine to fill the queue before continuing the normal cycle. As we modify the motion of queue conveyor, the output conveyor will track it with appropriate ratio because the output conveyor speed is gear ratioed to the queue conveyor as its master. A possible homing sequence before the automatic cycle begins is shown in Fig. 10.15.

```
Algorithm: Constant Gear Ratio Spacing Conveyors with a
Queue Conveyor

 I/O Required:

  Sensors (Inputs):
   One presence sensor (Photo electric eye or proximity
   sensor) and control logic to count the part rate input
   Three presence sensors to detect queue conveyor part
   level: high, low, empty

  Actuators (Outputs):
   Queue conveyor speed control,
   Output conveyor speed control by electronically
   gearing (slaved) to the speed of the queue conveyor

Control Algorithm Logic:

Initialize:

  Given the process parameters: L_p - part length, L_3-
                                desired part spacing,
```

```
   Given algorithmic parameter: V_percent (i.e. 0.9 = 90 %
                                speed reduction if queue is
                                full)

   Output conveyor speed is electronically geared to the
   queue conveyor speed by:    V_3 = ((L_p+L_3)/L_p ) * V_2
   Where the gear ratio between queue conveyor (master) and
   the output conveyor (slave) is   (L_p+L_3)/L_p.

Repeat Every Cycle:

   Measure, N_1, the part rate,
   Calculate queue conveyor speed: V_20 = L_p * N_1
   Check queue level and modify V_2 accordingly,
   Update speed command to the queue conveyor (V_2)
   If V_2 = 0, call homing routine to fill-up the queue
   conveyor.
Repeat Loop End

Return

Check queue level and modify queue speed algorithm:
    If (Queue is low (PRX-1 and PRX-2 are OFF) and PRX-3
    is ON)
       V_2 = V_20 / V_precent
    Else if (Queue is High (PRX-1 and PRX-2 are ON ) and
    PRX-3 is ON )
       V_2 = V_20 * V_precent
    Else if (Queue is Empty: PRX-3 is OFF)
       V_2 = 0.0
    Else
       V_2 = V_20
    End if
Return

Homing Routine:
' Queue conveyor is empty.
  Start input conveyor if it is stopped
  Disengage electronic gearing of output conveyor
  While (PRX-H is OFF)
    Wait until PRX-L triggers, then move for L_p
    distance.
  Repeat
  Engage gearing of output conveyor
' On return, queue conveyor is full and all parts are
  touching each other.
Return
```

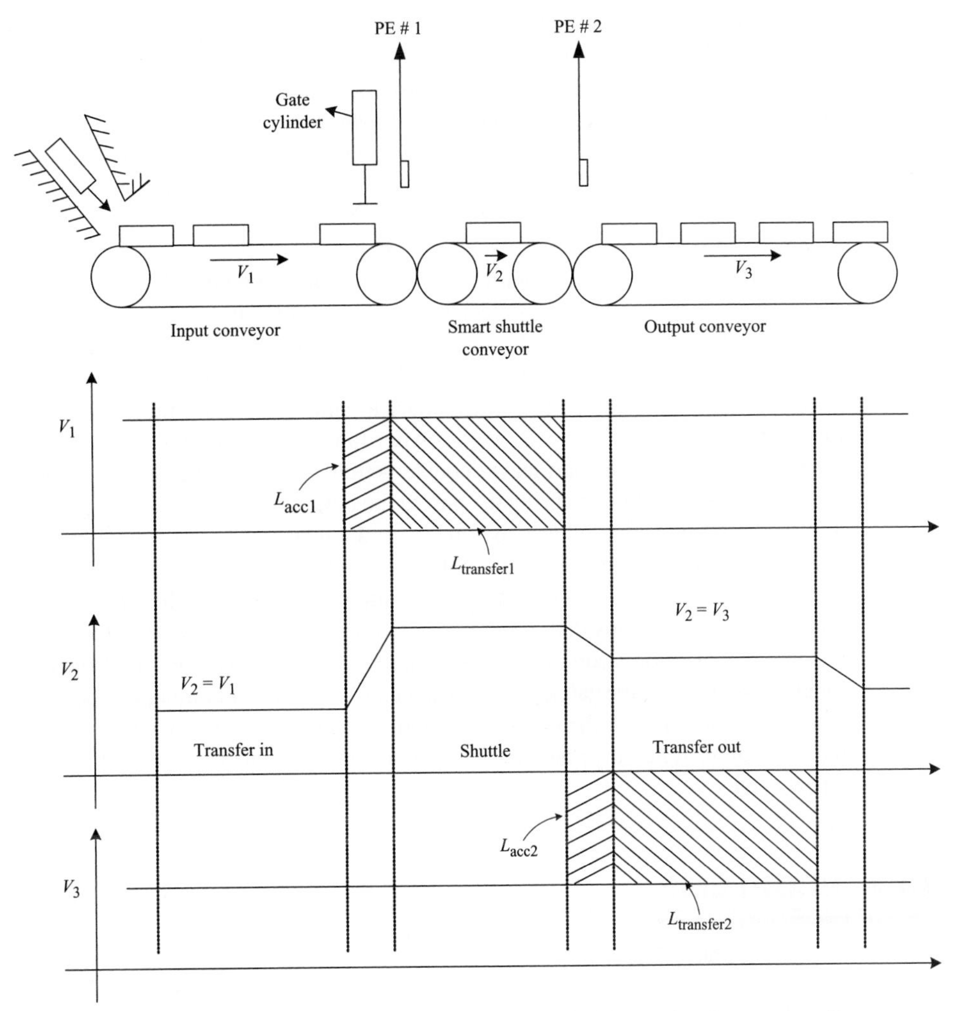


**FIGURE 10.16:** Another smart conveyor concept using a "smart shuttle conveyor."

***Position Phase Adjusting Conveyor: Smart Conveyor Slaved to Both the Input and the Output Conveyors*** In most general form, the smart (also called shuttle) conveyor must coordinate its motion both with input and output conveyors at different phases of the motion. In some applications, coordination only to input or output conveyor may be sufficient. The smart conveyor acts as a very fast slave between two masters (Fig. 10.16).

The cycle is divided into three phases: (1) Transfer phase from input to spacing conveyor (spacing conveyor matches linear speed to input conveyor by considering the input conveyor as the master); (2) nontransfer phase of spacing conveyor with no part contact to adjacent conveyors (spacing conveyor does not have to match linear speeds to any conveyor); and (3) transfer phase to output conveyor (spacing conveyor matches linear speed to output conveyor by considering the output conveyor as the master).

During each transfer (between input and smart conveyor, and smart conveyor and output conveyor), there are two phases of motion: (1) contact phase and (2) noncontact phase. During contact phase, a part is on both conveyors and the speed of both conveyors must be matched during that phase. During noncontact phase, part is on only the smart conveyor and it can move independent of the other conveyors. It is during this period that the smart conveyor makes the position corrections. In order for this approach to work, the time period during which input and smart converyors are coordinated should not have an overlap with the time period that the smart and output conveyors are coordinated. If the smart conveyor transfers one part at a time, this condition would be automatically met. The smart conveyor is the slave. There are two masters, input and output conveyors, for the spacing conveyor to match linear speeds with during two phases out of three of each cycle.

A possible design of a smart shuttle conveyor is shown in Fig. 10.16. Photoelectric sensor #1 (PE #1) is used to detect the beginning of the contact transfer phase of the part from the input conveyor to smart conveyor. This sensor must detect the incoming part before the physical contact between the part and smart conveyor. During a distance from the detection point to the part contact to smart conveyor, the smart conveyor must match the speed of the input conveyor. After that, the smart conveyor must maintain the speed matching until the part leaves contact with the input conveyor. Then smart conveyor speeds up or slows down to bring the part to the output conveyor. Another proximity sensor detects the trigger point when it is time to feed a new part to the output conveyor. This is done by another photoelectric sensor (PE #2). Again, within a defined distance, the smart conveyor brings the part in contact with the output conveyor, matches speed with the output conveyor, and maintains that speed until the transfer is complete. If before the transfer to the output converyor is completed another part is about to enter the smart conveyor from the input conveyor, then the gate cylinder is actuated to stop the incoming part because the smart conveyor cannot match the conflicting motion requirements of two different conveyors at the same time. When the new transfer cycle begins, the gate cylinder is deactivated to start a new cycle.

## 10.7 PROBLEMS

**1.** Consider an XY table where the X axis is stacked on top of the Y axis. Each axis is driven by a brushless DC motor. Each ball-screw has 2.0 rev/in. pitch. The motors have incremental encoders with 1024 lines/rev, and the controller has x4 encoder/decoder circuit.

**(a)** Define the external sensors necessary so that each axis can be stopped before hitting the travel limits as well as a physical home reference position can be established after power-up.

**(b)** Assume that the homing search speed is 1.0 in./sec speed. Write the pseudo-code of a homing motion.

**(c)** Calculate the parameters of an incremental move of 0.1 in. increments made in 300 msec. Command this motion in forward direction five times and in reverse direction five times and allow 200-msec wait period between each incremental motion.

**2.** Consider two independently driven rotary motion axes: axis 1 and axis 2. Our objective is to make axis 2 follow axis 1 as if they are connected with a mechanical gear ratio. Let the gear ratio be $z$.

**(a)** Draw the command motion generation block diagram for axis 2 for the case where the master motion it follows is the commanded motion of axis 1.

**(b)** Do the same; however, this time the master motion it follows is the actual position of the axis 1.

**(c)** Further consider the case that axis 1 is the drive for an unwind roll and axis 2 is the drive for a rewind roll for a web application. Axis 2 needs to modify the gear ratio as a function of the sensed tension. Draw the block diagram and write the pseudo-code to accomplish this.

**3.** Consider the registration application shown in Fig. 10.11. The rotating axis moves at a constant gear ratio relative to the speed of the web which is measured by an encoder. The objective is to make sure the rotating tool axis (print head or knife head) has to be at a certain position within one revolution when the registration mark passes under the registration sensor. When the registration mark sensor turns ON, we capture the position of the rotary axis and compare it with the desired position. Then the rotating axis makes a corrective index motion on top of the gear ratio motion. Assume that the web is moving at 2000 ft/min, and a cycle is repeated every 1.0 ft. It is desired that the accuracy of the positioning and operation done by the rotating tool axis relative to the web must be within $\pm 1/1000$ in.

**(a)** Determine the minimum psoition sensor resolution on the rotating axis.
**(b)** What happens to the accuracy if the registration mark sensor has 1.0-msec variation in its accuracy of responding to the mark?
**(c)** What is the maximum allowed variation in the registration mark sensors' response time?
**(d)** If the position capture is done under software polling and the application software may have 5-msec variation on when it may sample the position of the rotating axis relative to the instant the registration mark sensor came, what would be the positioning error?
**(e)** What is the viable position capturing method in such applications?

**4.** Consider an XY table. Our objective is to design a programmable control system such that the XY table makes various contour moves. Assume that the table X and Y axes have already established a home position.

**(a)** Design a pseudo-code so that the XY table draws a line with a given slope and length.
**(b)** Design a pseudo-code so the XY table draws a circle with given center coordinates and diameter.

**5.** Consider the tension control system shown in Figs. 1.6, B.30, 10.13, and 10.14. The second example on page 597 shows a web control method that changes the speed command to one of the motorized rolls in proportion to the tension error. Assume that Fig. 10.13 is a more complete and updated version of Fig. 1.6 where the analog op-amp controller circuit is replaced with a digital controller. In addition, the wind-off roll is a motorized drive, and there is a line position sensor (encoder). Update the control system using electronic gearing method to control the wind-up roll. Assume the wind-off roll is running at a constant speed. The control algorithm should use the line speed encoder in Fig. 10.13 as the master position to follow with a gear ratio, where the gear ratio is modified based on tension error through a PI type controller (Fig. 10.14). Show the benefits of electronic gearing based control method over the traditional direct speed control based on tension sensor method.

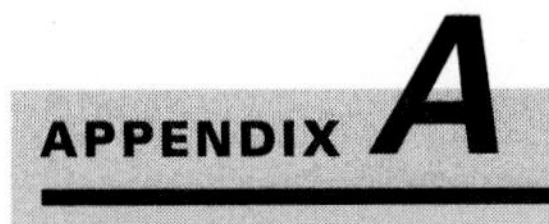

# APPENDIX A

# *TABLES*

**Unit Conversion Table**

| **Basic Units** | **SI Units** | **US Units** | **Conversions** |
|---|---|---|---|
| Time | second (s) | second (s) | 60 s = 1 min |
| | minute (min) | minutes (min) | 1 h = 60 min = 3600 s |
| | hour (h) | hour (h) | |
| Length | meter (m) | inch (in) | 1 m = 39.37 in |
| | | foot (ft) | 1 in = 0.0254 m |
| | | mile | 1 ft = 0.3048 m |
| | | | 1 mile = 1,609.344 m |
| Mass | kilogram (kg) | pound mass (lbm) | 1 slug = 14.59390 kg |
| | | slug | 1 lbm = 1 lbf/386 in/$s^2$ |
| Gravity | m/$s^2$ | in/$s^2$ | 9.80665 m/$s^2$ = 386.087 in/$s^2$ |
| | | ft/$s^2$ | 9.80665 m/$s^2$ = 32.174 ft/$s^2$ |
| **Derived Units** | **SI Units** | **US Units** | **Conversions** |
| Area | $m^2$ | $in^2$ | 1 $in^2$ = 6.4516 × $10^{-4}$ $m^2$ |
| | | $ft^2$ | 1 $ft^2$ = 0.09203 $m^2$ |
| Volume | $m^3$ | $in^3$ | 1 $in^3$ = 1.6387 × $10^{-5}$ $m^3$ |
| | | $ft^3$ | 1 $ft^3$ = 0.028317 $m^3$ |
| | liter (lt) = $10^{-3}$ $m^3$ | gallon (gal) | 1 gal = 3.785412 lt |
| Force | Newton (N) | pound (lb or lbf) | 1 N = 1 kg · 1 m/$s^2$ |
| | | | 1 lb = 4.448222 N |
| | | | 1 N = 0.2248 lb |
| | | | 1 lbf = 1 lbm · 32.176 ft/$s^2$ |
| | | | = 1 lbm · 386.087 in/$s^2$. |
| Torque | Nm | lb in | 1 lb ft = 1.355818 Nm |
| | | lb ft | 1 lb in = 0.112985 Nm |
| | | | 1 Nm = 8.850745 lb in |
| Pressure | N/$m^2$ = Pa | pound/$in^2$ | 1 psi = 6.894757 kPa |
| | | | 1 bar = $10^5$ Pa = 100 kPa |
| | | | 1 Pa = 0.145 × $10^{-3}$ psi |
| | | | 1 lbf/$ft^2$ = 47.88026 Pa |
| | | | 1 Pa = 0.020885434 lbf/$ft^2$ |
| | | | 1 atm = 1.01325 bar = 14.696 psi |
| | | | 1 bar = 14.504 psi |
| | | | 1 psi = 0.06894 bar |

(Continues)

**Unit Conversion Table (Cont.)**

| Derived Units | SI Units | US Units | Conversions |
|---|---|---|---|
| Flow rate | lt/s | gallon/s, | 1 gallon/s = 3.785412 lt/s |
| | lt/min | gallon/min | 1 gallon/min = 3.785412 lt/min |
| Power | Watt (W) | btu/hr | 1 W = 1 Nm/s |
| | | | 1 btu/hr = 0.293 W |
| | | | 1 HP = 550 lbft/s = 6600 lb in/s |
| | | | = 745.6999 W |
| | | | 1 lb ft/s = 1.3558 W |
| Energy | Joule (J) | British Thermal Unit (btu) | 1 J = 1 Nm = 1 Ws |
| | Calorie (cal) | Therm | 1 btu = 778.169 lbf ft |
| | | | = 9338.028 lbf in |
| | | | 1 btu = $1.055056 \times 10^3$ J |
| | | | 1 lbft = 1.3558 J |
| | | | 1 cal = 4.1868 J |
| | | | 1 Therm = $10^5$ btu |
| | | | 1 kW h = 3412.14 btu |
| One electron charge | Coulomb (C) | | $-1.60219 \times 10^{-19}$ [C] |
| One proton charge | Coulomb (C) | | $1.60219 \times 10^{-19}$ [C] |
| Coulomb constant | $k_e [\text{N} \cdot \text{m}^2/\text{C}^2]$ | | $8.9875 \times 10^9$ $[\text{N} \cdot \text{m}^2/\text{C}^2]$ |
| Permittivity of free space | $\epsilon_0 = \frac{1}{4\pi k_e}$ | | $8.8542 \times 10^{-12}$ $[\text{C}^2/(\text{N} \cdot \text{m}^2)]$ |
| Permeability of free space | $\mu_0$ | | $4\,\pi \cdot 10^{-7}$ [Tesla · m/A] |
| Flux density | Tesla (T) | Gauss (G) | $1[\text{T}] = 1\frac{\text{N}}{\text{Cm/s}} = 1\frac{\text{N}}{A \cdot m} = 10^4[\text{G}]$ |
| Flux | Weber | Maxwell | 1 Weber = 1 Tesla · $\text{m}^2$ |
| | | | = $10^8$ [Gauss · $\text{cm}^2$] |
| | | | = $10^8$ [Maxwell] |
| Other units | | | 1 [Gauss Oerstead] = 1 [GOe] |
| | | | $= \frac{1}{4\pi} 10^3 [\text{J}/\text{m}^3]$ |
| | | | 1 [Tesla][A/m] = 1 [Joule/$\text{m}^3$] |
| | | | 1 [A turn/m] = $4\pi \times 10^{-3}$ [Oe] |

APPENDIX **B**

# MODELING AND SIMULATION OF DYNAMIC SYSTEMS

## B.1 MODELING OF DYNAMIC SYSTEMS

A model of a system can be physical or mathematical. The model accuracy needed (closeness to the actual system) depends on the purpose. Generally, a simplified model is needed to study the main characteristics of the system. A detailed model is needed for accurate simulation and prediction studies. In this book, modeling refers to the mathematical model of a system. The mathematical model of a dynamic system is generally in the form of differential equations. Therefore, for our purposes, modeling means obtaining the differential equations of a dynamic system. First we will study the physical laws and the use of these laws in modeling various dynamic systems. Once we have a model of a system, we are interested in studying the behavior of it. The behavior of a dynamic system in time is described by the solution of its differential equations.

There are two different purposes for modeling a physical system (Fig. B.1).

- Develop a mathematical model in order to predict the dynamic behavior of the system as accurately as possible, using numerical solution methods. Such a model serves as a tool for extensive evaluation of system behavior without actually using or building the actual system.
- Develop models to gain insight into the behavior of dynamic system qualitatively instead of exact response prediction, i.e., knowledge of stability margins, controllability and observability of states, and sensitivity of response to parameter changes. Such models need not contain all the detail of an actual system, but only the most essential features so as to provide good insight from an engineering standpoint.

Therefore, we may develop simplified linear models for controller design and analysis purposes, and use more detailed, possibly nonlinear, models in testing and predicting the dynamic system response as accurately as possible. For instance, consider the robotic manipulator model shown in Fig. B.1. The dynamic model is a set of diffential equations which describes the relationship between the applied torques at the joints and motion of the joint angles in time. The set of nonlinear diffential equations can be used to predict the behavior of the robotic manipulator under various initial conditions and joint torque inputs.

Quite often, the dynamic models of physical systems are nonlinear. Most control system design methods and analytical methods are applicable only to linear systems. Therefore, for the sake of being able to analyze various controller alternatives, we need to obtain approximate linearized models from the nonlinear models.

In the following sections, we will discuss the basic complex variables and Laplace transforms which are the fundamental mathematical methods used in modeling and analysis

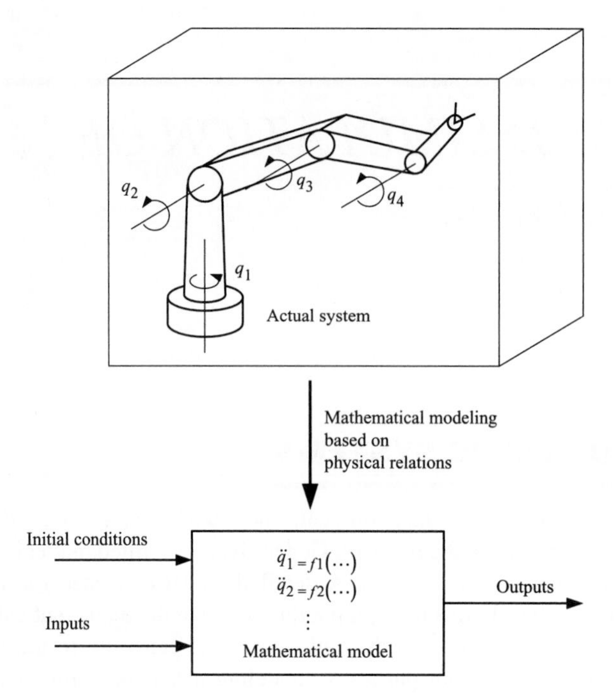


**FIGURE B.1:** Mathematical model and dynamic response prediction of a physical system.

of linear dynamic systems. Then we discuss ordinary differential equations which are used to represent the behavior of most dynamic systems we consider. Finally, we discuss analytical and well as numerical solutions of differential equations. Using Laplace transforms, we also establish the concept of transfer functions. Readers who are not familiar with *Matlab and Simulink* software should read the short review in the appendix.

## B.2 COMPLEX VARIABLES

The analysis and design of computer-controlled electromechanical systems rely to a great extent on the application of the theory of complex variables. A complex variable can be thought of as a point in two-dimensional space connected to the origin by a vector. The x-coordinate of the vector is called the *real* part, and the y-coordinate is called the *imaginary* part (Fig. B.2). Let $s$ be a complex variable defined as

$$s = x + jy \tag{B.1}$$

where $j$ is the imaginary number $\sqrt{-1}$. The complex variable $s$ can also be defined in terms

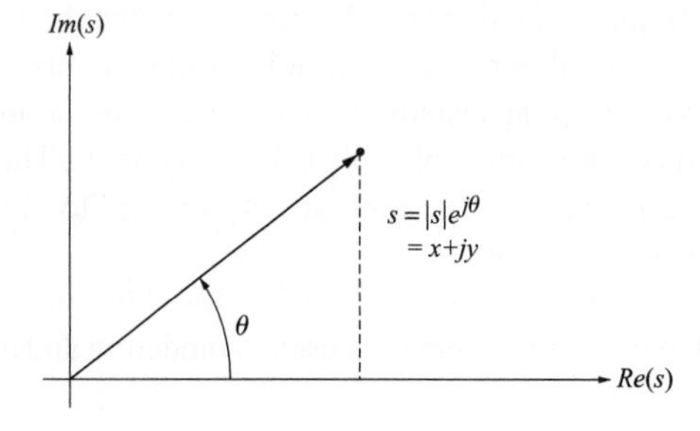


**FIGURE B.2:** Complex $s$-plane and a complex variable on the complex plane.

of its magnitude and direction, such that

$$s = |s|e^{j\theta} \tag{B.2}$$

$$= (x^2 + y^2)^{\frac{1}{2}} e^{jtan^{-1}(\frac{y}{x})} \tag{B.3}$$

with its complex conjugate defined as

$$\bar{s} = x - jy \tag{B.4}$$

The following exponential relations and series expansions are useful in the study of control theory:

$$sin\,\theta = \frac{e^{j\theta} - e^{-j\theta}}{2j} \tag{B.5}$$

$$cos\,\theta = \frac{e^{j\theta} + e^{-j\theta}}{2} \tag{B.6}$$

$$cos\,\theta = 1 - \frac{\theta^2}{2!} + \frac{\theta^4}{4!} - \frac{\theta^6}{6!} + \cdots \tag{B.7}$$

$$sin\,\theta = \theta - \frac{\theta^3}{3!} + \frac{\theta^5}{5!} - \frac{\theta^7}{7!} + \cdots \tag{B.8}$$

$$e^{\theta} = 1 + \theta + \frac{\theta^2}{2!} + \frac{\theta^3}{2!} + \frac{\theta^4}{4!} + \cdots \tag{B.9}$$

$$cos\,\theta + jsin\,\theta = 1 + (j\theta) - \frac{\theta^2}{2!} + (-j)\frac{\theta^3}{3!} + \cdots \tag{B.10}$$

$$= 1 + j\theta + \frac{(j\theta)^2}{2!} + \frac{(j\theta)^3}{3!} + \cdots \tag{B.11}$$

$$= e^{j\theta} \tag{B.12}$$

where

$$e^{j\theta} = cos\,\theta + jsin\,\theta \tag{B.13}$$

is known as *Euler's theorem.*

In the remainder of this section, we will discuss the basic *algebraic operations* on complex variables (addition, subtraction, multiplication, division). Let $s_1$ and $s_2$ be two complex variables defined as

$$s_1 = x_1 + jy_1 \tag{B.14}$$

$$s_2 = x_2 + jy_2 \tag{B.15}$$

The algebraic operations ( $+, -, *, /$ ) on complex variables can be defined as follows: The addition and subtraction of two complex variables are performed by adding or subtracting the real parts and imaginary parts of each variable. In multiplication and division operations, we need to keep in mind that $j = \sqrt{-1}$, $j^2 = -1$, $j^3 = -j$, $j^4 = 1$, $j^5 = j$, and so on.

$$\pm \qquad s_1 \mp s_2 = (x_1 \mp x_2) + j(y_1 \mp y_2) \tag{B.16}$$

$$*/ \qquad s_1 \cdot s_2 = (x_1x_2 - y_1y_2) + j(x_1y_2 + y_1x_2) \tag{B.17}$$

The multiplication of two complex variables may also be accomplished using the magnitude and phase representation,

$$s_1 \cdot s_2 = |s_1|e^{j\theta_1} \cdot |s_2|e^{j\theta_2} \tag{B.18}$$

$$= |s_1|\,|s_2|\,e^{j(\theta_1+\theta_2)} \tag{B.19}$$

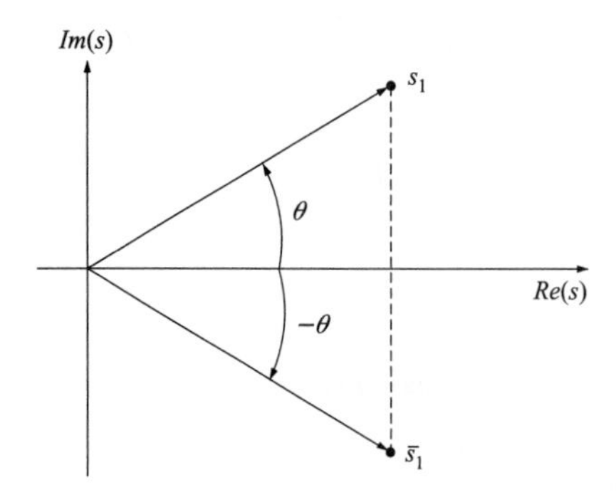


**FIGURE B.3:** Graphical description of complex conjugate, $\bar{s}_1$, of a complex number $s_1$.

Similarly, division of two complex variables can be achieved using the real and imaginary part representation or magnitude and phase representation. Note that multiplying a complex number by its complex conjugate always results in a real number.

$$\frac{s_1}{s_2} = \frac{x_1 + jy_1}{x_2 + jy_2} \tag{B.20}$$

$$= \frac{(x_1 + jy_1)(x_2 - jy_2)}{(x_2 + jy_2)(x_2 - jy_2)} \tag{B.21}$$

$$= \frac{(x_1 + jy_1)(x_2 - jy_2)}{x_2^2 + y_2^2} = \frac{s_1 \cdot \bar{s}_2}{|s_2|^2} \tag{B.22}$$

If the two complex numbers are expressed in magnitude and phase form, the division is accomplished as follows:

$$\frac{s_1}{s_2} = \frac{|s_1|e^{j\theta_1}}{|s_2|e^{j\theta_2}} \tag{B.23}$$

$$= \frac{|s_1|}{|s_2|}e^{j(\theta_1 - \theta_2)} \tag{B.24}$$

Note that the complex conjugate of $s_1$ is $\overline{s_1}$ and $s_2$ is $\overline{s_2}$ and are defined as (Fig. B.3)

$$s_1 = |s_1|e^{j\theta_1} \; \overline{s_1} = |s_1|e^{-j\theta_1} \tag{B.25}$$

$$s_2 = |s_2|e^{j\theta_2} \; \overline{s_2} = |s_2|e^{-j\theta_2} \tag{B.26}$$

Finally, the $n$th power of a complex variable can be obtained as

$$s_1{}^n = |s_1|^n e^{jn\theta} \tag{B.27}$$

# B.3 LAPLACE TRANSFORMS

## B.3.1 Definition of Laplace Transform

The Laplace transform is a mathematical tool which allows us to represent a function $f(t)$ as a continuous sum of generalized exponential functions with complex frequencies. Let the Laplace transform of a function $f(t)$ be $F(s)$, and the inverse Laplace transform of $F(s)$ be

$f(t)$. The definitions of the Laplace transform and the inverse Laplace transform are given as follows:

$$L\{f(t)\} = F(s) = \int_0^{\infty} f(t)e^{-st}dt \tag{B.28}$$

$$L^{-1}\{F(s)\} = f(t) = \frac{1}{2\pi j}\int_{\sigma-j\infty}^{\sigma+j\infty} F(s)e^{st}ds \tag{B.29}$$

where $s$ is a complex variable, defined as $s = \sigma + j\omega$ and $\sigma$ is a real constant greater than the real part of any singularity of $F(s)$. A singularity of a function is a point in the $s$-plane at which the function value goes to infinity. The condition for the Laplace transform to exist is given by

$$\int_{-\infty}^{\infty} |f(t)|e^{-\alpha t}dt < \infty \tag{B.30}$$

where $\alpha$ is a constant, real number with the property that $|f(t)|e^{-\alpha t}$ remains bounded as $t \to \infty$. Another way to state this is that there are constants $\alpha$, $M$, and $T$ such that $|f(t)|e^{-\alpha T} < M$ for all $t > T$.

Consider the function $e^{-at}u(t)$, where $u(t)$ is the unit step function, such that $u(t) = 0$, for $t < 0$, and 1 for $t > 0$

$$F(s) = \int_0^{\infty} e^{-at}u(t)e^{-st}dt \tag{B.31}$$

$$= \frac{-1}{(s+a)}e^{-(s+a)t}|_0^{\infty} \tag{B.32}$$

$$= \frac{1}{s+a} \tag{B.33}$$

for all $$(\sigma + a) > 0 \tag{B.34}$$

Note that if $(\sigma + a) < 0$, the value of the above integral goes to infinity and the transform does not exist. Hence, the region of convergence is given by $Re(s) > -a$.

Although the Laplace transform definition is used often in taking the Laplace transform of a function, the inverse Laplace transform equation is usually a difficult mathematical operation and is not normally used in engineering applications. An easier way to obtain the inverse Laplace transform is to use partial fraction expansions, which will be discussed later in this chapter.

***Existence Condition*** A function $f(t)$ must be defined for all $t > 0$, except at a finite set of points noted as discontinuities, in order for its Laplace transform to exist. Every such $f(t)$, which is piecewise continuous and of exponential order, has a unique Laplace transform.

In summary, for a function $f(t)$ to have Laplace transform, it is sufficient (but not necessary) for the function to be:

- Piecewise continuous.
- Exponential order, that is $\int_0^{\infty} e^{-\alpha t}|f(t)|dt < \infty$; there exists a finite $\alpha$ or equivalently $|f(t)| < Me^{\alpha t} \quad \forall t > T$.

For instance, $e^{2t}$ is an exponential order function, but $e^{(\alpha t)^2}$ is not an exponential order function.

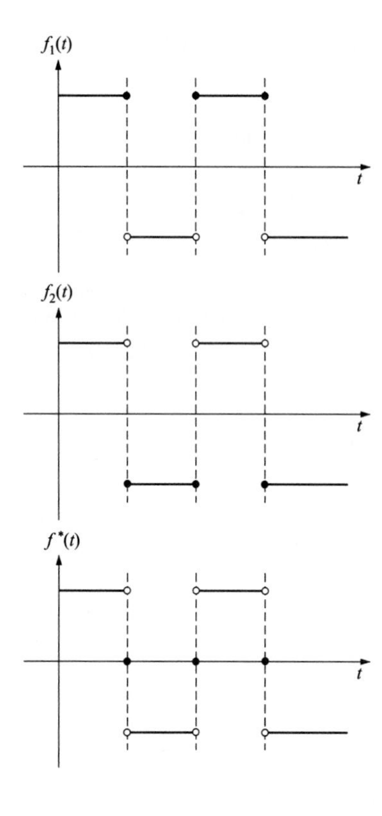


**FIGURE B.4:** Laplace transforms of functions that differ only at point discontinuities are the same. Inverse Laplace transformation gives a function which takes on the arithmetic average value at discontinuities.

If a function is exponential order, the derivative of it is not necessarily exponential order. Consider the following example:

$$f(t) = sin\,(e^{t^2}) \tag{B.35}$$

$$\frac{d}{dt}f(t) = (2te^{t^2})cos\,(e^{t^2}) \tag{B.36}$$

where $f(t)$ is exponential order, but not the first derivative of it.

If $f_1(t)$, $f_2(t)$ differ only at point discontinuities, never over a finite region, then they have the same Laplace transform. Laplace transform, $F(s)$, of a function, $f(t)$, is such that $F(s) \longrightarrow 0, sF(s) \longrightarrow Finite$, as $s \longrightarrow \infty$. Inverse Laplace transform of $F(s)$ gives a function $f^*(t)$ which takes on the average value at discontinuities (Fig. B.4).

## B.3.2 Properties of the Laplace Transform

We have defined the Laplace transforms and the sufficient conditions for their existence. Now we will discuss some of the properties of the Laplace transform. Keeping the properties of the Laplace transform in mind is very helpful in the analysis of control systems.

**Property 1** Linearity (also called the Superposition Property): Let the functions $f_1(t)$ and $f_2(t)$ be piecewise continuous and exponential order, and let $c_1$ and $c_2$ be constant scalars. The Laplace transform of the linear combination of two functions is equal to the linear combination of the Laplace transforms of each of the individual functions. This property is shown as follows:

$$L[c_1 f_1(t) + c_2 f_2(t)] = c_1 L[f_1(t)] + c_2 L[f_2(t)]$$

It is rather straightforward to show that this property is true by direct application of the Laplace transform definition,

$$\int_0^\infty [c_1 f_1(t) + c_2 f_2(t)]e^{-st}dt = c_1 \int_0^\infty f_1(t)e^{-st}dt + c_2 \int_0^\infty f_2(t)e^{-st}dt \quad (B.37)$$

$$= c_1 F_1(s) + c_2 F_2(s) \quad (B.38)$$

There is an analogy between differentiation and integration in the time domain, and multiplication and division by $s$ in the $s$-domain. Taking the derivative of a function in time domain is equivalent to multiplying its Laplace transform by $s$ in the $s$-domain. Similarly, taking the integral of a function in time domain is equivalent to dividing the Laplace transform of the function by $s$ in the $s$-domain. There are also initial conditions given as $f(0)$ involved in the relationships, which are given below as properties 2 and 3.

**Property 2** Let us assume that $f(t)$ is continuous and exponential order, and its first derivative $f'(t)$ is also piecewise continuous and exponential order. Then

$$L[f'(t)] = sL[f(t)] - f(0); \quad (B.39)$$

$$L[f^{(n)}(t)] = s^n L[f(t)] - s^{n-1} f(0) - s^{n-2} f'(0) \ldots - f^{(n-1)}(0) \quad (B.40)$$

where $f, f', \ldots, f^{n-1}$ are higher-order derivatives and are assumed continuous and exponential order. Thus $f^{(n)}$ is piecewise continuous and of exponential order.

Let us prove the above relations for the first derivative case.

$$L[f'(t)] = \int_o^\infty \frac{d}{dt}(f(t))e^{-st}dt \quad (B.41)$$

Notice that

$$\frac{d}{dt}(f(t)e^{-st}) = \frac{d}{dt}(f(t))e^{-st} - f(t)se^{-st} \quad (B.42)$$

Then, rearranging

$$\frac{d}{dt}(f(t))e^{-st} = \frac{d}{dt}(f(t)e^{-st}) + sf(t)e^{-st} \quad (B.43)$$

If we substitute this relationship in the above definition of the Laplace transform for $f'(t)$, and take the integration term by term, we will find that

$$L[f'(t)] = sL[f(t)] - f(0) \quad (B.44)$$

**Property 3** Let both $f(t)$ and $\int_0^t f(t)dt$ be piecewise continuous and exponential order; then it can be shown that

$$L\left[\int_0^t f(t)dt\right] = \frac{1}{s} L[f(t)] \quad (B.45)$$

**Property 4** Initial Value theorem: It can be shown that the initial value of a signal as $t \longrightarrow 0$ and its behavior as $s \longrightarrow \infty$ are related to each other as follows (Fig. B.5):

$$\lim_{t \to 0^+} f(t) = f(0^+) = \lim_{s \to \infty} sF(s) \quad (B.46)$$

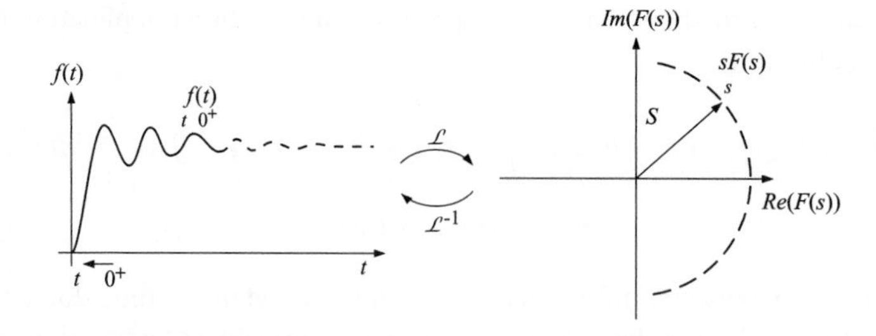


**FIGURE B.5:** Initial Value theorem describing the relation between time domain function and its Laplace transformation as time goes to zero.

**Property 5** Final Value theorem: The final value of a signal as $t \longrightarrow \infty$ in the time domain (in steady state) is related to the value of the Laplace transform of the function as $s \longrightarrow 0$ (Fig. B.6).

$$\lim_{t\to\infty} f(t) = \lim_{s\to 0} sF(s) \tag{B.47}$$

For the Final Value theorem to be applicable and meaningful, the function must have a steady-state value as time goes to infinity, such that

$$|f'(t)| < Me^{\alpha_0 t}; \quad \alpha_0 < 0 \tag{B.48}$$

where M is a positive finite value. This means that the function $f(t)$ is stable and converges to a steady-state constant value. The equivalent condition in the $s$-domain requires that the Laplace transform $F(s)$ at most can have one pole at $s = 0$; everything else must be on the left half of the $s$-plane. A *pole* is the value of $s$ which makes the denominator of $F(s)$ zero (a singularity). The Final Value theorem is a very useful term in control theory, as it is used to determine the final value of a time function by examining the behavior of its Laplace transform as $s$ tends to zero. However, the Final Value theorem is not valid if the denominator of $sF(s)$ contains any *pole* whose real part is zero or positive.

In other words, for $F(s)$ to be the Laplace transform of a signal which is stable and converges to a constant value as time goes to infinity it must have poles with negative real parts and at most can have one pole at $s = 0$.

The basic analogies between the time domain and $s$-domain operations can be stated as

$$s \longleftrightarrow \frac{d}{dt} \tag{B.49}$$

$$\frac{1}{s} \longleftrightarrow \int_0^t \ldots dt \tag{B.50}$$

**Property 6** Multiplying a time domain function by an exponential function is equivalent to the shifting of the Laplace transform of the function in $s$-domain by the exponential power. An analogous relationship exists between shifting a function in time domain and its

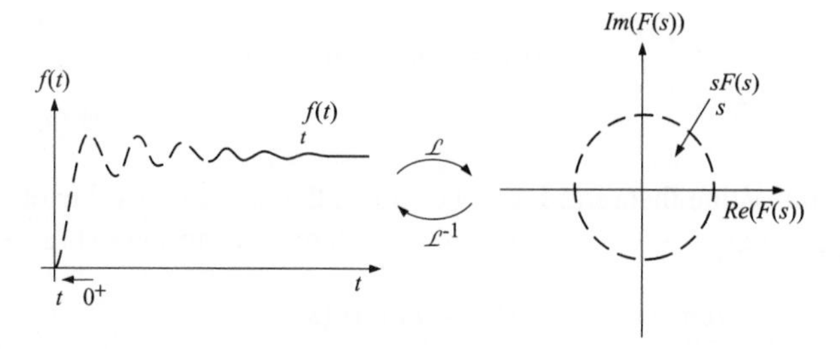


**FIGURE B.6:** Final Value theorem describing the relation between time domain function and its Laplace transformation as time goes to infinity.

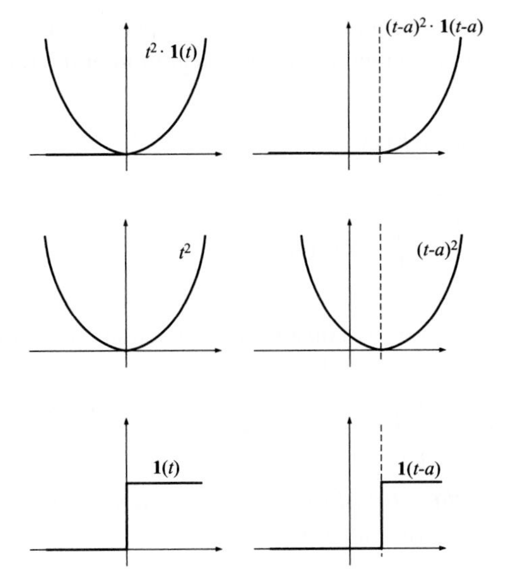


**FIGURE B.7:** Shifted functions in time domain.

corresponding effect in $s$-domain.

$$L[e^{-at} f(t)] = L[f(t)]_{s \to s+a} = F(s+a) \tag{B.51}$$

The following examples illustrate this property:

$$L[e^{-at} cos(bt)] = \frac{(s+a)}{(s+a)^2 + b^2} \tag{B.52}$$

$$L[e^{-at} sin(bt)] = \frac{b}{(s+a)^2 + b^2} \tag{B.53}$$

$$L[e^{-at} t^n] = \frac{n!}{(s+a)^{n+1}}; \quad n \text{ is positive integer} \tag{B.54}$$

**Property 7** Multiplying a function in $s$-domain by an exponential function, i.e, $e^{-as}$, has the effect of shifting the original function in time domain by $a$ units (Fig. B.7).

$$L[f(t-a)u(t-a)] = e^{-as} L[f(t)] \tag{B.55}$$

$$L^{-1}[e^{-as} F(s)] = f(t-a)u(t-a) \tag{B.56}$$

Note that in some cases we may need to take the Laplace transform of a function in the following form:

$$L[f(t)u(t-a)]$$

where $u(t-a)$ is the unit step function shifted in time from $t = 0$ to $t = a$. In order to take the Laplace transform using the above relations, we must make sure the arguments of the function and the shifted unit step function are the same.

$$f[(t+a) - a]u(t-a) = g(t-a)u(t-a) \tag{B.57}$$

$$L[g(t-a)u(t-a)] = e^{-as} L[g(t)] \tag{B.58}$$

$$= e^{-as} L[f(t+a)] \tag{B.59}$$

Therefore,

$$L[f(t)u(t-a)] = e^{-as} L[f(t+a)], \; a \geq 0 \tag{B.60}$$

**Property 8** The analog of properties 2 and 3—that is, multiplying and dividing a function in time domain by $t$, is related to differentiation, and integration of the function is the $s$-domain.

$$L[f(t)] = F(s) \tag{B.61}$$

$$L[tf(t)] = -F'(s) \tag{B.62}$$

$$L\left[\frac{f(t)}{t}\right] = \int_s^\infty F(s)d(s) \tag{B.63}$$

**Property 9** Convolution theorem: Let us consider two functions $f(t)$ and $g(t)$, which are both piecewise continuous and exponential order, then

$$L[f(t)]\, L[g(t)] = L\left[\int_0^t f(t-\tau)g(\tau)d(\tau)\right]$$

which states that the *product of the Laplace transforms of two functions is equal to the Laplace transform of the convolution of the two functions.*

Proof:

$$L\left[\int_0^t f(t-\tau)g(\tau)d\tau\right] = \int_0^\infty \left[\int_0^t f(t-\tau)g(\tau)d\tau\right] e^{-st}dt \tag{B.64}$$

$$= \int_0^\infty \int_0^\infty f(t-\tau)g(\tau)u(t-\tau)e^{-st}d\tau dt \tag{B.65}$$

$$= \int_0^\infty g(\tau)\left[\int_0^\infty f(t-\tau)u(t-\tau)e^{-st}dt\right] d\tau \tag{B.66}$$

$$= \int_0^\infty g(\tau)\left[\int_0^\infty f(\lambda)e^{-s\lambda}d\lambda\right] e^{-s\tau}d\tau \tag{B.67}$$

$$= \int_0^\infty g(\tau)e^{-s\tau}d\tau \int_0^\infty f(\lambda)e^{-s\lambda}d\lambda \tag{B.68}$$

$$= G(s)F(s) \tag{B.69}$$

Note that the following changes of variables and relations are used in the above derivation,

$$u(t-\tau) = \begin{cases} 1 & \tau < t \\ 0 & \tau > t \end{cases}$$

$$t - \tau = \lambda \tag{B.70}$$

$$dt = d\lambda \tag{B.71}$$

and integral limits on eqn. B.64 from $[0, t]$ is extended in eqn. B.65 to $[0, \infty]$ since for $\tau > t$, the $[t, \infty]$ range would be zero due to the definition of $u(t-\tau)$.

### B.3.3 Laplace Transforms of Some Common Functions

The Laplace transforms of a number of functions often encountered in control systems are studied below. The transforms are obtained by direct application of the definition and properties defined in this chapter. More complicated functions can be expressed as a linear summation of the elementary functions. Then the Laplace transform of the linear summation can be taken by application of the superposition (linearity) property (Figs. B.8 and B.9).

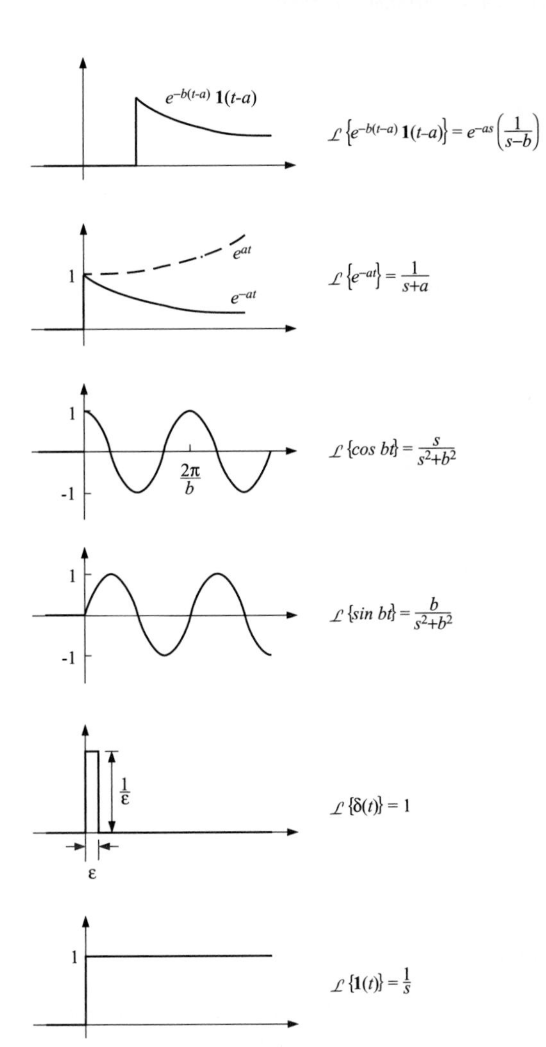


**FIGURE B.8:** Laplace transform of some common signals.

1. ***Unit Pulse:*** Consider the following unit pulse such that

$$u_1(t) = \begin{cases} lim_{\varepsilon\to 0} f(t); & t = 0 \\ 0; & t \neq 0 \end{cases} \tag{B.72}$$

   The Laplace transform of the unit pulse is

$$L[u_1(t)] = \int_0^{\infty} u_1(t)e^{-st}dt = \int_{0^-}^{0^+} e^{-st}dt = 1 \tag{B.73}$$

2. ***Unit Step:*** Consider the following unit step function such that

$$u_2(t) = \begin{cases} 1; & t \geq 0 \\ 0; & t \leq 0 \end{cases} \tag{B.74}$$

   where the Laplace transform is given by

$$F(s) = L[u_2(t)] = \int_0^{\infty} u_2(t)e^{-st}dt = -\frac{1}{s}[e^{-st}]_0^{\infty} = \frac{1}{s} \tag{B.75}$$

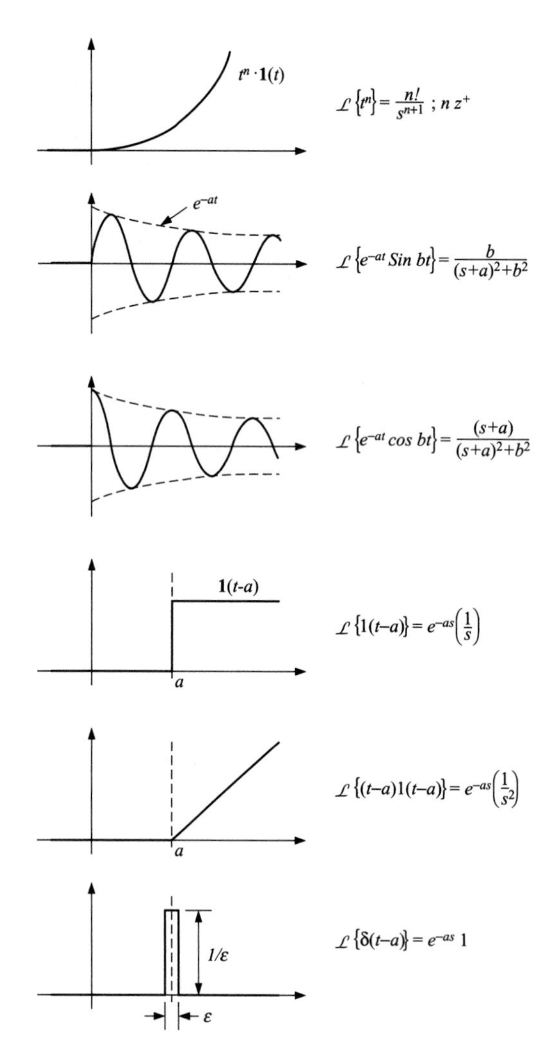


**FIGURE B.9:** Laplace transform of some common signals.

3. ***Exponential function:*** Consider the following function:

$$u_3(t) = e^{-at} \tag{B.76}$$

The Laplace transform follows as

$$F(s) = L\left[u_3(t)\right] = \int_0^\infty e^{-at}e^{-st}dt \tag{B.77}$$

$$= \left[\frac{e^{(s+a)t}}{s+a}\right]|_0^\infty \tag{B.78}$$

$$= \frac{1}{s+a} \tag{B.79}$$

for $Re(s) > -a$. Therefore, the Laplace transform of $f(t) = e^{-at}$ exists in the region to the right of the line $Re(s) = -a$ in the $s$-plane.

4. ***Sinusoid:*** Consider the function

$$u_4(t) = cos(\alpha t) \tag{B.80}$$

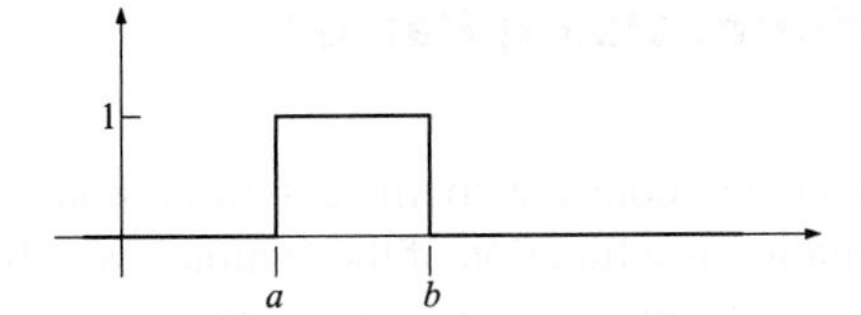

**FIGURE B.10:** Example signal.

With Laplace transform,

$$F(s) = L[cos(\alpha t)] = \int_0^{\infty} cos(\alpha t)e^{-st}dt \tag{B.81}$$

$$= \frac{1}{2}\int_0^{\infty} (e^{j\alpha t} + e^{-j\alpha t})e^{-st}dt \tag{B.82}$$

$$0 = \frac{1}{2}L[e^{j\alpha t}] + \frac{1}{2}L[e^{-j\alpha t}] \tag{B.83}$$

$$= \frac{1}{2}\left(\frac{1}{s - j\alpha} + \frac{1}{s + j\alpha}\right) \tag{B.84}$$

$$L[cos\,\alpha t] = \frac{s}{s^2 + \alpha^2} \tag{B.85}$$

It is not necessary to derive the Laplace transform of $f(t)$ each time. Laplace transform tables can conveniently be used to find the transform of a given function. Figures B.8 and B.9 show Laplace transforms of time functions that frequently appear in linear system analysis.

***Example*** Find the Laplace transform of the following function (Fig. B.10):

$$f(t) = u(t - a) - u(t - b)$$

This signal is comprised of a unit step function shifted in time by $b$ subtracted from a unit step function shifted in time by $a$ (Fig. B.11).

$$L[f(t)] = L[u(t - a)] - L[u(t - b)] \tag{B.86}$$

$$= e^{-as}\frac{1}{s} - e^{-bs}\frac{1}{s} \tag{B.87}$$

$$= (e^{-as} - e^{-bs})\frac{1}{s} \tag{B.88}$$

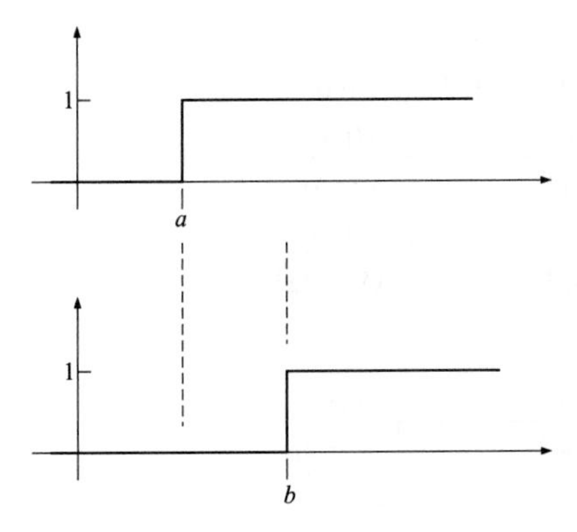

**FIGURE B.11:** Decomposition of the signal.

### B.3.4 Inverse Laplace Transform: Using Partial Fraction Expansions

We can use Laplace transforms to convert a linear constant coefficient differential equation to an algebraic equation. The algebraic equation is a function of the complex variable $s$ and can be expressed as the ratio of the Laplace transform of the output to the input. This function is called the transfer function. The Laplace transform of the response of a dynamic system is equal to the algebraic multiplication of the transfer function and the Laplace transform of the input function. In order to find the response in the time domain, we need to take the inverse Laplace transform of the response to return to the time domain. In practice, the difficulty of applying the inverse Laplace transform definition makes it impractical in most engineering applications. Instead, the method of partial fraction expansion (PFE) is commonly used in obtaining the inverse Laplace transform. The idea is to expand the rational function $F(s)$ into a summation of simpler components for which the inverse Laplace transforms are known.

Let us restate the definition of the Laplace and the inverse Laplace transforms

$$Lf(t) = F(s) = \int_0^{\infty} f(t)e^{-st}dt \tag{B.89}$$

$$L^{-1}[F(t)] = f(t) = \frac{1}{2\pi j}\int_{\sigma-j\omega}^{\sigma+j\omega} F(s)e^{st}dt \tag{B.90}$$

Consider a general rational function $F(s)$ of complex variable $s$, such that

$$F(s) = \frac{n(s)}{d(s)} \tag{B.91}$$

$$= F_1(s) + F_2(s) + \cdots + F_n(s) \tag{B.92}$$

where we assume that the function $F(s)$ can be expressed as a sum of simpler functions $F_i(s)$'s, and their corresponding inverse Laplace transforms $f_i(t)$ are known. Then the inverse Laplace transform can be obtained by the superposition (linearity) property,

$$f(t) = f_1(t) + f_2(t) + \cdots + f_n(t) \tag{B.93}$$

The mathematical derivation of the PFE is based on the Taylor and Laurent series expansions [56]. Let $F(s)$ be a rational polynomial of a complex variable $s$, and it has poles $s_1, s_2, \ldots, s_m$ with multiplicity $n_1, n_2, \ldots, n_m$. Also note that for most engineering systems the degree of the numerator is less than the degree of the denominator, $deg(n(s)) \leq deg(d(s))$. Then, $F(s)$ can be expanded into its partial fractions expansion as follows:

$$F(s) = \frac{B(s)}{A(s)} \tag{B.94}$$

$$= F_1(s) + \cdots + F_m(s) + a_0 \tag{B.95}$$

$$= \frac{a_{1,1}}{(s-s_1)^{n_1}} + \cdots + \frac{a_{1,n_1}}{s-s_1} \tag{B.96}$$

$$+ \frac{a_{2,1}}{(s-s_2)^{n_2}} + \cdots + \frac{a_{2,n_2}}{(s-s_2)} \tag{B.97}$$

$$\vdots$$

$$+ \frac{a_{m,1}}{(s-s_m)^{n_m}} + \cdots + \frac{a_{m,n_m}}{s-s_m} \tag{B.98}$$

$$+ a_0 \tag{B.99}$$

Given a rational function $F(s)$, we need to find the roots of its denominator and the multiplicity of each root. Then we can write out the PFE in the form of equation B.99 where only the constants of the numerators are unknown. Examining both sides of the expansion, it is easy to see that $a_{1,1}$ can be calculated as follows:

$$a_{1,1} = \lim_{s \to s_1} \left[ (s - s_1)_1^n F(s) \right] \tag{B.100}$$

Noting that roots of the expanded PFE in terms of the coefficients $a_{1,n_i}$ are repeated, $a_{1,2}$ can be found by taking the derivative of the expansion as calculated for $a_{1,1}$, such that

$$a_{1,2} = \lim_{s \to s_1} \frac{d}{ds} \left[ (s - s_1)_1^n F(s) \right] \tag{B.101}$$

This relationship can be generalized such that any of the coefficients of the expansion can be determined as

$$a_{i,j} = \lim_{s \to s_i} \frac{1}{(j-1)!} \frac{d^{j-1}}{ds^{j-1}} \left[ (s - s_i)_i^n F(s) \right] \tag{B.102}$$

Let us consider examples which illustrate applications of the PFE in obtaining inverse Laplace transforms. The constant term $a_0$ is obtained by direct division of the numerator by the denominator if $deg(n(s)) = deg(d(s))$. If $deg(n(s)) < deg(d(s))$, then $a_0$ is zero.

***Example*** Consider the following rational function of a complex variable, and let us take the inverse Laplace transform of it using PFE.

$$G(s) = \frac{s+2}{(s-1)^2(s+1)} \tag{B.103}$$

$$= \frac{a_{1,1}}{(s-1)^2} + \frac{a_{1,2}}{(s-1)} + \frac{a_{2,1}}{(s+1)} \tag{B.104}$$

The coefficients can be obtained using the formula given in equation B.102.

$$a_{1,1} = \lim_{s \to 1} [(s-1)^2 G(s)] = \frac{3}{2} \tag{B.105}$$

$$a_{1,2} = \lim_{s \to 1} \frac{d}{ds} [(s-1)^2 G(s)] = -\frac{1}{4} \tag{B.106}$$

$$a_{2,1} = \lim_{s \to -1} [(s+1) G(s)] = \frac{1}{4} \tag{B.107}$$

Now we can take the inverse Laplace transform of each component.

$$G(s) = \frac{\frac{3}{2}}{(s-1)^2} + \frac{-\frac{1}{4}}{s-1} + \frac{\frac{1}{4}}{s+1} \tag{B.108}$$

$$g(t) = \frac{3}{2} t e^t - \frac{1}{4} e^t + \frac{1}{4} e^{-t} \tag{B.109}$$

$$g(t) = \left( \frac{3}{2} t - \frac{1}{4} \right) e^t + \frac{1}{4} e^{-t} \tag{B.110}$$

***Example*** Consider the following complex function and its inverse Laplace transform using the PFE method:

$$Y(s) = \frac{1}{(s+1)(s+2)(s+3)} \tag{B.111}$$

$$= \frac{a_{1,1}}{s+1} + \frac{a_{2,1}}{s+2} + \frac{a_{2,1}}{s+3} \tag{B.112}$$

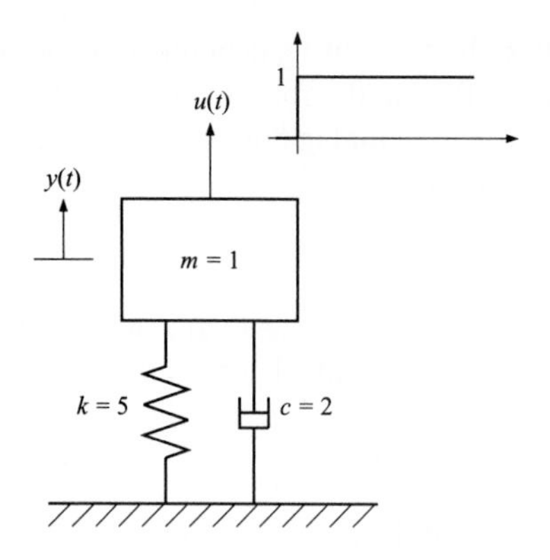


**FIGURE B.12:** A second-order linear system model: Mass, damper, and spring with external force.

The coefficients of PFE are found using equation B.102,

$$a_{1,1} = \lim_{s \to -1} [(s+1)Y(s)] = \frac{1}{2} \tag{B.113}$$

$$a_{2,1} = \lim_{s \to -2} [(s+2)Y(s)] = -1 \tag{B.114}$$

$$a_{3,1} = \lim_{s \to -3} [(s+3)Y(s)] = \frac{1}{2} \tag{B.115}$$

Hence, the inverse Laplace transform is easily found as follows:

$$Y(s) = \frac{\frac{1}{2}}{s+1} + \frac{-1}{s+2} + \frac{\frac{1}{2}}{s+3} \tag{B.116}$$

$$y(t) = \frac{1}{2}e^{-t} - e^{-2t} + \frac{1}{2}e^{-3t} \tag{B.117}$$

***Example*** Consider a linear dynamic system with transfer function, $G(s)$, input $u(t)$, and output $y(t)$. Let the Laplace transform of $u(t)$ be $U(s)$, and that of $y(t)$ be $Y(s)$. By application of the Convolution theorem, valid for linear time invariant systems, the Laplace transform of the response (Fig. B.12), $Y(s)$, is equal to the transfer function multiplied by the Laplace transform of the input, $U(s)$,

$$Y(s) = G(s) \cdot U(s) \tag{B.118}$$

By application of Newton's second law, the force–acceleration relationship of the mass at a displacement $y(t)$ and speed $\dot{y}(t)$ is

$$m \cdot \ddot{y}(t) = u_{net}(t) \tag{B.119}$$

$$= u(t) - c \cdot \dot{y}(t) - k \cdot y(t) \tag{B.120}$$

$$m \cdot \ddot{y}(t) + c \cdot \dot{y}(t) + k \cdot y(t) = u(t) \tag{B.121}$$

$$1.0 \cdot \ddot{y}(t) + 2.0 \cdot \dot{y}(t) + 5.0 \cdot y(t) = u(t) \tag{B.122}$$

If we take the Laplace transform of the differential equation, and assume zero initial conditions, that is $y(t_o) = \dot{y}(t_0) = 0$, then

$$1.0 \cdot s^2 Y(s) + 2.0 \cdot sY(s) + 5.0 \cdot Y(s) = U(s) \tag{B.123}$$

$$\frac{Y(s)}{U(s)} = G(s) \tag{B.124}$$

$$G(s) = \frac{1}{s^2 + 2s + 5} \tag{B.125}$$

In general, the transfer function between force (input) and displacement (output) of such a system has the following form:

$$\frac{Y(s)}{U(s)} = G(s) = \frac{1}{ms^2 + cs + k} \tag{B.126}$$

and when there is no damper or spring ($c = k = 0$), the transfer function reduces to

$$\frac{Y(s)}{U(s)} = G(s) = \frac{1}{ms^2} \tag{B.127}$$

The same transfer function relationship describes the dynamics of a rotary inertia-torque system. The analogy between the variables are as follows: displacement and angular displacement, force and torque, translational spring/damper and rotational spring/damper.

In particular, the input force function $u(t)$ is a step function with unit magnitude. Then we can detemine the response of the system to this particular input using the transfer function and inverse Laplace transform method. The Laplace transform of the step function is

$$U(s) = \frac{1}{s} \tag{B.128}$$

The response in the $s$-domain and its partial fraction expansion (PFE) can be expressed as

$$Y(s) = \frac{1}{s^2 + 2s + 5} \cdot \frac{1}{s} \tag{B.129}$$

$$= \frac{1}{s(s + 1 - 2j)(s + 1 + 2j)} \tag{B.130}$$

$$= \frac{a_{1,1}}{s} + \frac{a_{2,1}}{(s + 1 - 2j)} + \frac{a_{3,1}}{(s + 1 + 2j)} \tag{B.131}$$

$$a_{1,1} = \lim_{s \to 0} [sY(s)] = 1/5 \tag{B.132}$$

$$a_{2,1} = \lim_{s \to -1+2j} [(s + 1 - 2j)Y(s)] = -0.1 + 0.04j \tag{B.133}$$

$$a_{3,1} = \cdots = -0.1 - 0.04j \tag{B.134}$$

Note that $a_{3,1} = a^*_{2,1}$ because $s_2 = s^*_2$. This is always true. The residues of complex conjugate poles are complex conjugate of each other. Here we carried out the calculations in detail to illustrate the properties by example.

The time domain response can be obtained by taking the inverse Laplace transform of the PFE form of $Y(s)$

$$y(t) = L^{-1}\left\{\frac{\frac{1}{5}}{s} + \frac{-0.1 + 0.04j}{s + 1 - 2j} + \frac{-0.1 - 0.04j}{s + 1 + 2j}\right\}$$

$$= (1/5)u(t) + (-0.1 + 0.04j)e^{-t}e^{2jt} - (0.1 + 0.04j)e^{-t}e^{-2jt}$$

$$= (1/5)u(t) + e^{-t}[-0.1(e^{2jt} + e^{-2jt}) + 0.04j(e^{2jt} - e^{-2jt})]$$

$$= (1/5)u(t) - e^{-t}(0.2cos(2t) + 0.08sin(2t))u(t)$$

$$= (1/5)u(t) - (0.2cos(2t) + 0.08sin(2t))e^{-t}u(t)$$

$$= (1/5)u(t) - D\ sin(2t + \phi)e^{-t}u(t) \tag{B.135}$$

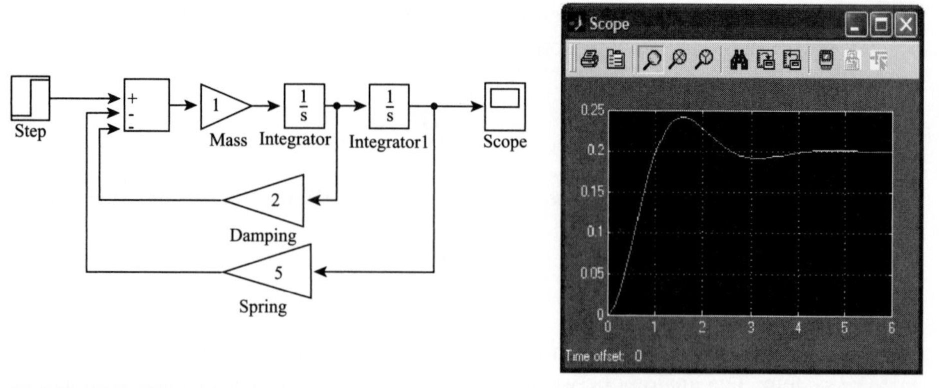


**FIGURE B.13:** Model and simulation of a mass, damper, and spring with an external force system using Simulink.

where

$$D = \sqrt{0.2^2 + 0.08^2} \tag{B.136}$$

$$\phi = +tan^{-1}\left(\frac{0.2}{0.08}\right) \tag{B.137}$$

Note that the following relations were used in the above example:

$$A\ cos\omega t + B\ sin\omega t = C\ sin(\omega t + \phi) \tag{B.138}$$

$$A = C\ sin\phi \tag{B.139}$$

$$B = C\ cos\phi \tag{B.140}$$

$$C = (A^2 + B^2)^{\frac{1}{2}} \tag{B.141}$$

$$\phi = tan^{-1}\left(\frac{A}{B}\right) \tag{B.142}$$

Notice that the response of the system can also be determined numerically by using a Matlab or Simulink environment. Figure B.13 shows the Simulink model and simulation result.

## B.4 FOURIER SERIES, FOURIER TRANSFORMS, AND FREQUENCY RESPONSE

Every signal can be viewed in terms of its frequency content. Fouries series is a mathematical expression of the fact that *any periodic function can be expressed as a series sum of sine and cosine functions which have frequencies that are integer multiples of the frequency of the periodic function*. Then, a periodic function can be viewed as having frequency content that includes its main frequency and the integer multiples of that frequency. Fourier transforms can be defined as the limiting case of Fourier series for nonperiodic functions where a nonperiodic function is viewed as a periodic function with infinite period. When the period is infinite, the fundamental frequency of the function is infinitesimally small. As a result, the *series summation* of Fourier series becomes an *integral* operation in Fourier transforms.

Let us consider a periodic function, $f(t)$, and its Fourier series expression and Fourier transform,

$$f(t) = f(t + T) \tag{B.143}$$

where $T > 0$ is the period of the function. If $T$ is finite, we can express the function as a Fourier series. Fourier series can be viewed as a frequency domain representation of the periodic function where only integer multiples of the fundamental frequency are involved in the function. We call fundamental frequency $w_1 = 1/T$ [Hz] or $w_1 = 2\pi/T$ [rad]. If $T$ is infinite (the function is not periodic), we can express the function in frequency domain as continuous function of frequency, which is referred to as its Fourier transform.

The following conditions are sufficient but not necessary for the existence of the Fourier series and transforms (first two conditions are for Fourier series, and all three conditions are for Fourier transforms):

1. Function must be piecewise continuous.
2. Function can have finite discontinuities, but the maximum and minimum values of the function are within a finite range.
3. $\int_{-\infty}^{+\infty} |f(t)| \cdot dt < M; \quad M -$ finite.

Any periodic function which satisfies the above two requirements can be experessed as an infinite series sum as follows:

$$f(t) = \frac{1}{2} \cdot a_o + \sum_{n=1}^{\infty} a_n \cdot cos\left(\frac{2\pi n}{T}t\right) + b_n \cdot sin\left(\frac{2\pi n}{T}t\right) \quad \text{(B.144)}$$

$$= \sum_{n=-\infty}^{\infty} c_n \cdot e^{j \cdot \frac{2\pi n}{T} t} \quad \text{(B.145)}$$

The second series expression is called the complex form, and it can be shown that it is equivalent to the first form. It can be shown that the coefficients in the above series are as follows (derivation is skipped here and can be found in any advanced mathematics book):

$$a_o = \frac{1}{T/2} \int_{t_0}^{t_o+T} f(t) \cdot dt \quad \text{(B.146)}$$

$$a_n = \frac{1}{T/2} \int_{t_0}^{t_o+T} f(t) \cdot cos\left(\frac{2\pi n}{T}t\right) \cdot dt \quad \text{(B.147)}$$

$$b_n = \frac{1}{T/2} \int_{t_0}^{t_o+T} f(t) \cdot sin\left(\frac{2\pi n}{T}t\right) \cdot dt \quad \text{(B.148)}$$

$$c_n = \frac{1}{T} \int_{t_0}^{t_o+T} f(t) \cdot e^{-j(\frac{2\pi n}{T}t)} \cdot dt \quad \text{(B.149)}$$

In the above integrations for coefficients, the $t_0$ is arbitrary. The important thing is that the integration should be taken over a complete period, $T$. Therefore, $t_0 = 0$ is often a convenient choice. It can be shown that $a_n$, $b_n$, and $c_n$ coefficients are related. In addition, $c_n$ coefficients appear in complex conjugate pairs. As a result, the complex form of the series expression still leads to a real function.

If we have a function that is nonperiodic, then we can view it as a limiting case of a periodic function where $T$ goes to infinity. In this case, the fundamental frequency, $w_1 = 2\pi/T$, is infinitesimally small, and integer multiples of it, $w_n = (2\pi/T)n$, are almost a continuous frequency spectrum. Figure B.14 shows a periodic function, its Fourier series frequency content, as the period of the function is extended to infinity, and its frequency content.

Fourier transform is a limiting case of Fourier series where the period of the original function is infinite, and the fundamental frequency of the function is very small. Hence, the integer multiples of the fundamental frequency actually form a continuous frequency

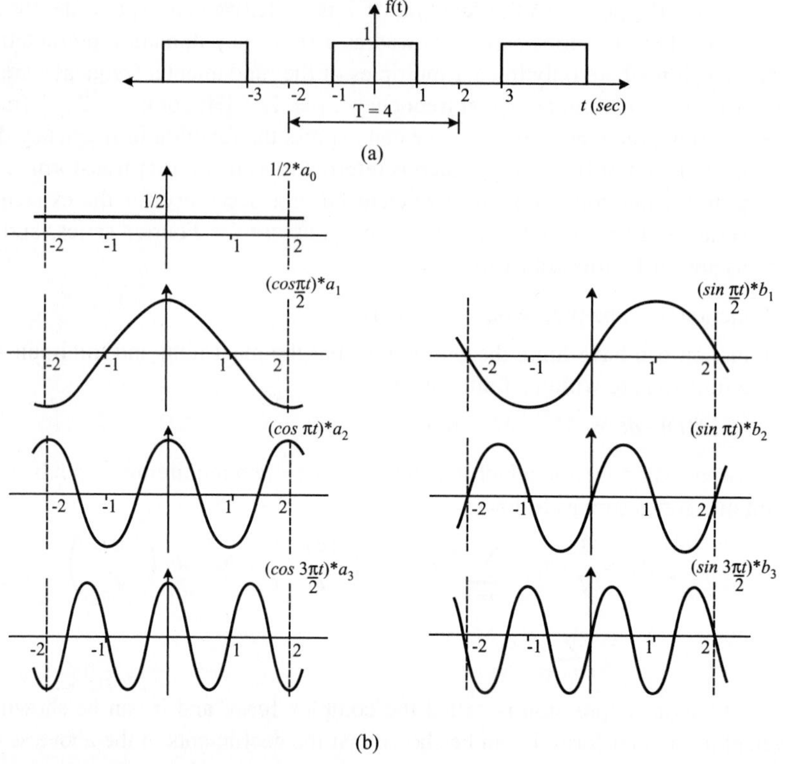


**FIGURE B.14:** Fourier series of a periodic function. (a) a periodic signal, and (b) its Fourier series components ($cos(\frac{2\pi n}{T}t)$, $sin(\frac{2\pi n}{T}t)$, $n = 1, 2, \ldots$) and magnitude of each components contribution ($a_i$, $b_i$) at integer multiples of the fundamental frequency.
$f(t) = f(t+T) = \frac{1}{2} \cdot a_o + \sum_{n=1}^{\infty}(a_n \cdot cos(\frac{2\pi n}{T}t) + b_n \cdot sin(\frac{2\pi n}{T}t))$.

spectrum. The series summation of discrete frequency contents becomes integral of continuous frequency content. Let $f(t)$ be the periodic function with period $T$ and its Fourier series experession be

$$f(t) = \sum_{n=-\infty}^{\infty} c_n \cdot e^{j \cdot \frac{2\pi n}{T} t} \tag{B.150}$$

where

$$c_n = \frac{1}{T} \int_0^T f(t) \cdot e^{-j(\frac{2\pi n}{T}t)} \cdot dt \tag{B.151}$$

Substitute the $c_n$ definition into the above series, and let $w_n = \frac{2\pi}{T} \cdot n = \Delta w \cdot n$,

$$f(t) = \sum_{n=-\infty}^{\infty} \frac{1}{2\pi} \cdot e^{jw_n t} \left( \int_0^T f(\tau) e^{-jw_n \tau} d\tau \right) \Delta w \tag{B.152}$$

$$= \sum_{n=-\infty}^{\infty} e^{jw_n t} \cdot F(jw_n) \cdot \Delta w \tag{B.153}$$

If we let $T \to \infty$, then $\Delta w \to dw$, $w_n \to w$, and the summation can be replaced by integration,

$$f(t) = \int_{-\infty}^{\infty} \frac{1}{2\pi} e^{jwt} \left( \int_{-\infty}^{\infty} f(\tau) e^{-jw\tau} d\tau \right) dw \tag{B.154}$$

where

$$F(jw) = \int_{-\infty}^{\infty} f(t) \cdot e^{-jwt} \cdot dt \quad \text{Fourier transform} \tag{B.155}$$

$$f(t) = \frac{1}{2\pi} \int_{-\infty}^{\infty} F(jw) \cdot e^{jwt} \cdot dt \quad \text{Inverse Fourier transform} \tag{B.156}$$

It should be noted that the constant $(1/2\pi)$ in the Fourier transform and inverse Fourier transform expression can be divided between the two expressions in any way desired depending on the convention used, as long as the multiplication of them equals $1/2\pi$.

Fourier transform is a linear operator and has the same properties as Laplace transforms, such as linearity, shift, convolution. Notice that Fourier transform is applied to functions that run from $-\infty$ to $\infty$ in the independent variable which is generally time in our case. Laplace transform is applied to functions that run from a finite time (i.e., zero time) to infinite. If we have the Laplace transform of a function, its Fourier transform can be obtained by

$$F(jw) = F(s)|_{s=jw} \tag{B.157}$$

***Example*** Consider the square function shown in Fig. B.14 with unit magnitude and period $T$[sec]. We will consider two cases of this function as follows:

1. Let $T = 4.0$ sec, and we are to determine the Fourier series description of it (Fig. B.14).
2. Let $T = \infty$ sec, and we are to determine the Fourier transform of it (Fig. B.15).

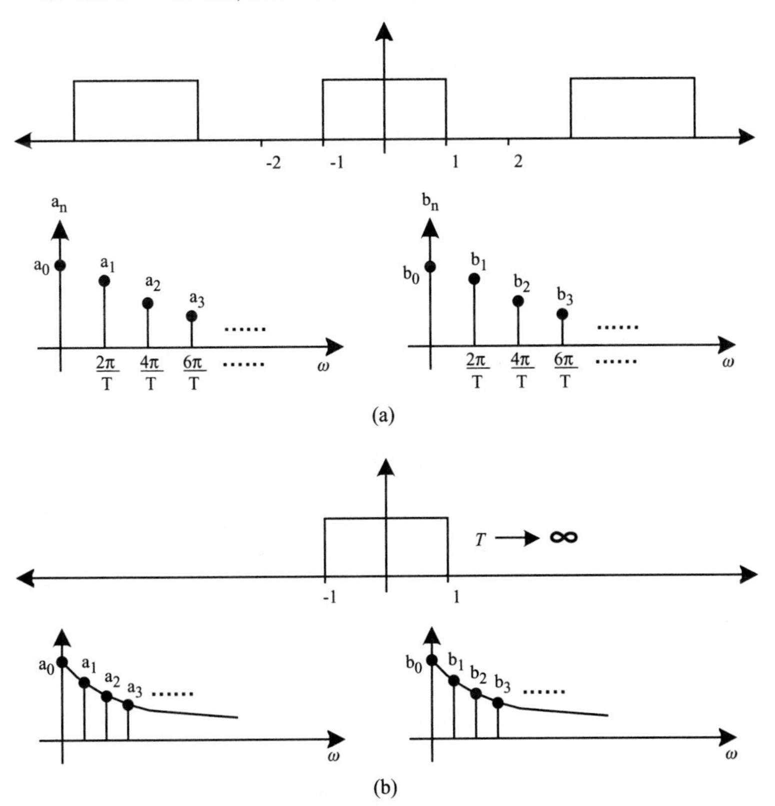


**FIGURE B.15:** Fourier transform of a nonperiodic function: (a) a nonperiodic signal as a limiting case of a periodic signal where the period $T \rightarrow \infty$, and (b) its Fourier transform function which can be interpreted as a limiting case of Fourier series. Fourier series represents the frequency content of a periodic signal with a function of discrete frequencies. Fourier transform represents the frequency content of a nonperiodic signal with a function of continuous frequency.

The fundamental frequency of the periodic function is

$$w_1 = \frac{2\pi}{T} = \frac{2\pi}{4} = \frac{\pi}{2} [\text{rad/sec}] \tag{B.158}$$

$$= \frac{1}{T} [\text{Hz}] = 0.25[\text{Hz}] \tag{B.159}$$

The frequency components of cosine and sine functions $[cos(w_n t), sin(w_n t)]$ contributing to the Fourier series description of the function are simply integer multiples of that fundamental frequency $w_1$,

$$w_n = n \cdot w_1 = n\frac{\pi}{2}; \quad n = 1, 2, 3, \ldots \tag{B.160}$$

$$= \frac{\pi}{2}, \pi, \frac{3\pi}{2}, \frac{4\pi}{2}, \frac{5\pi}{2}, \ldots. \tag{B.161}$$

The contribution of each sinusoidal function is determined by the coefficients $a_o, a_n, b_n$, $n = 1, 2, \ldots$ which can be determined by

$$a_o = \frac{1}{2}\int_{-2}^{2} f(t)\, dt = \frac{1}{2}\int_{-1}^{1} 1 \cdot dt = \frac{1}{2}(t)|_{-1}^{1} = 1 \tag{B.162}$$

$$a_n = \frac{1}{2}\int_{-2}^{2} f(t) \cdot cos\left(\frac{\pi}{2}nt\right) \cdot dt\, ; \quad n = 1, 2, 3, \ldots \tag{B.163}$$

$$= \frac{1}{2}\int_{-1}^{1} 1.0 \cdot cos\left(\frac{\pi}{2}nt\right) \cdot dt \tag{B.164}$$

$$= \frac{1}{2} \cdot \frac{2}{\pi n} \cdot sin\left(\frac{\pi}{2}nt\right)\Big|_{-1}^{1} \tag{B.165}$$

$$= \frac{1}{\pi n} \cdot (1 + 1) \tag{B.166}$$

$$= \frac{2}{\pi n} \tag{B.167}$$

$$b_n = \frac{1}{2}\int_{-2}^{2} f(t) \cdot sin\left(\frac{\pi}{2}nt\right) \cdot dt\, ; \quad n = 1, 2, 3, \ldots \tag{B.168}$$

$$= \frac{1}{2}\int_{-1}^{1} 1.0 \cdot sin\left(\frac{\pi}{2}nt\right) \cdot dt \tag{B.169}$$

$$= -\frac{1}{2} \cdot \frac{2}{\pi n} cos\left(\frac{\pi}{2}nt\right)\Big|_{-1}^{1} \tag{B.170}$$

$$= -\frac{1}{2} \cdot \frac{2}{\pi n}(0 - 0) \tag{B.171}$$

$$= 0 \tag{B.172}$$

If we think of the Fourier series as a summation of sinusioidal functions, it is clear that the periodic functions have discrete frequency content.

$$f(t) = \frac{1}{2} + \frac{1}{\pi}\sum_{n=1}^{\infty} \frac{1}{n} cos\left(\frac{\pi}{2}nt\right); \quad n = 1, 2, 3, \ldots \tag{B.173}$$

Notice that since this function is an even function, $b_n$ coefficients are all zero. For even functions, this is always the case. If the function were an odd periodic function instead, then $a_0, a_1, a_2, \ldots.$ would be zero. This is a property of Fourier series.

Let us now consider the nonperiodic function, which is only a pulse. The Fourier transform of the function can be obtained by evaluating the integral,

$$F(jw) = \int_{-\infty}^{\infty} f(t) \cdot e^{-jwt} \cdot dt \tag{B.174}$$

$$= \int_{-1}^{1} 1.0 \cdot e^{-jwt} \cdot dt \tag{B.175}$$

$$= -\frac{1}{jw} e^{-jwt}|_{-1}{}^{1} \tag{B.176}$$

$$= -\frac{1}{jw}\left(e^{-jw} - e^{jw}\right) \tag{B.177}$$

$$= \frac{1}{jw}\left(-e^{jw} + e^{jw}\right) \tag{B.178}$$

$$= \frac{2}{w} \cdot sin(w) \tag{B.179}$$

which indicates that the Fourier transform of a single-pulse function, which is a nonperiodic function, is a continuous function of frequency.

### B.4.1 Basics of Frequency Response: Meaning of Frequency Response

Consider a linear time invariant (LTI) stable dynamic system shown in Fig. B.16. If such a system is excited by a sinusoidal input signal, the response of the system in steady state is also sinusoidal with the same frequency. The only difference would be in the magnitude and phase of the steady-state response in relation to the input signal. Furthermore, the output-to-input magnitude ratio and the phase shift are functions of excitation frequency. This is a property of the LTI stable dynamic systems. For the input signal

$$u(t) = A\, sin\,(wt) \tag{B.180}$$

the steady-state response of the LTI system is

$$y(t) = B\, sin\,(wt + \psi) \tag{B.181}$$

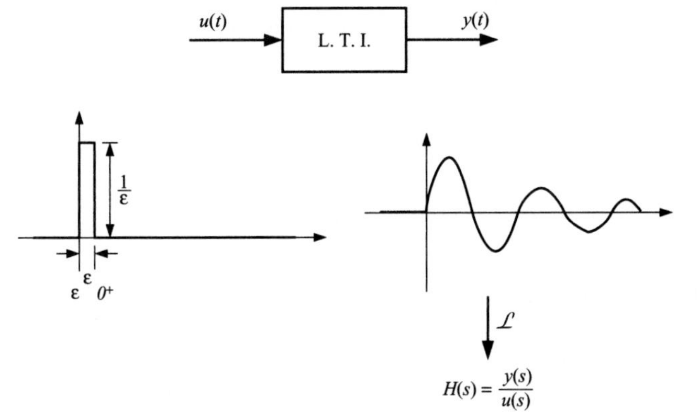


**FIGURE B.16:** For a linear time invariant dynamic system, Laplace transform of the impulse response is equal to the transfer function. Similarly, Fourier transform of the impulse response is equal to the frequency response relationship between input and output.

For nonlinear or time-varying systems, this relationship does not hold. For LTI systems, output-to-input magnitude ratio and the phase as functions of frequency completely characterize a specific LTI system steady-state response,

$$B/A = B/A(w) \tag{B.182}$$

$$\psi = \psi(w) \tag{B.183}$$

### B.4.2 Relationship Between the Frequency Response and Transfer Function

The magnitude ratio and phase difference in the steady-state response of an LTI system to a sinusoidal input, $\{B/A(w), \psi(w)\}$, are related to the transfer function of the LTI system. Let us derive this relationship. Let the transfer function of the LTI stable system be

$$G(s) = n(s)/d(s) \tag{B.184}$$

and note that because the dynamic system is assumed to be stable, all the roots of $d(s)$ have negative real parts. Hence, consider a transfer function

$$G(s) = n(s)/((s + p_1)(s + p_2)\ldots(s + p_n)) \tag{B.185}$$

where $Re(p_i) > 0$ for $i = 1, \ldots, n$. Let us calculate the response of the system to a sinusoidal input $Asin(wt)$ using partial fraction expansions and inverse Laplace transform,

$$y(t) = L^{-1}\{G(s)u(s)\} \tag{B.186}$$

$$= L^{-1}\{n(s)/((s + p_1)(s + p_2)\ldots(s + p_n)).Aw/(s^2 + w^2)\} \tag{B.187}$$

If we take the partial fraction expansion of the terms inside the bracket,

$$y(t) = L^{-1}\left\{\frac{k_1}{s + jw} + \frac{\bar{k}_1}{s - jw} + \frac{c_1}{s + p_1} + \frac{c_2}{s + p_2} + \cdots + \frac{c_n}{s + p_n}\right\} \tag{B.188}$$

$$= k_1 e^{-jwt} + \bar{k}_1 e^{jwt} + \sum_{i=1}^{p} c_i e^{-p_i t} \tag{B.189}$$

$$\lim_{t \longrightarrow \infty} y(t) = k_1 e^{-jwt} + \bar{k}_1 e^{jwt} \tag{B.190}$$

The summation terms associated with transient response go to zero because the LTI is stable and, therefore, has poles with negative real parts. It can be shown that the residues $k_1$ and the complex conjugate of it, $\bar{k}_1$, are

$$k_1 = \lim_{s \longrightarrow -jw} \{(s + jw)G(s)u(s)\} = -\frac{A}{2j}G(-jw) \tag{B.191}$$

$$\bar{k}_1 = \lim_{s \longrightarrow jw} \{(s - jw)G(s)u(s)\} = \frac{A}{2j}G(jw) \tag{B.192}$$

Let us note the following relations:

$$G(jw) = |G(jw)|e^{j\psi} \tag{B.193}$$

$$G(-jw) = |G(-jw)|e^{-j\psi} = |G(jw)|e^{-j\psi} \tag{B.194}$$

where $\psi = tan^{-}1\left(\frac{Im(G(jw))}{Re(G(jw))}\right)$. The steady-state response is

$$y_{ss}(t) = -\frac{A}{2j}G(-jw)e^{-jwt} + \frac{A}{2j}G(jw)e^{jwt} \tag{B.195}$$

$$= A|G(jw)|\left(-\frac{1}{2j}e^{-jwt}e^{-j\psi} + \frac{1}{2j}e^{jwt}e^{j\psi}\right) \tag{B.196}$$

$$= A|G(jw)|sin\,(wt + \psi) \tag{B.197}$$

$$= Bsin\,(wt + \psi) \tag{B.198}$$

From the above equation, the relationship between transfer function magnitude and its phase evaluated along $s = jw$ axis and the steady-state response to sinusoidal input which is characterized by the output-to-input magnitude ratio and phase shift between output and input,

$$B/A(w) = |G(jw)| \tag{B.199}$$

$$\psi(w) = tan^{-1}\left(\frac{Im(G(jw))}{Re(G(jw))}\right) \tag{B.200}$$

Therefore, the steady-state response of an LTI stable dynamic system to a sinusiodal input at various frequencies conveys the same information as the transfer function of the system evaluated along the imaginary axis. The transfer function evaluated along the imaginary axis conveys all the information conveyed by the transfer function over the s-plane, and it is called the *frequency response* of the LTI dynamic system.

### B.4.3 s-Domain Interpretation of Frequency Response

We can think of the frequency response as either the magnitude ratio and phase shift of a linear system in steady-state response to a sinusoidal stimulus or as the transfer function evaluated along the $s = jw$ axis on the complex s-plane. If the $G(s)$ is a stable transfer function, the $G(jw)$ completely describes the same information described by $G(s)$.

$$G(s)_{|s=jw} = G(jw) = |B/A(w)|e^{j\psi(w)} \tag{B.201}$$

Magnitude ratio at a frequency is equal to gain multiplied by the magnitude of product of the phasors from zeros divided by the the magnitude of product of the phasors from poles.

$$G(jw) = K_1 \left.\frac{\Pi(s + z_i)}{\Pi(s + p_i)}\right|_{s=jw} \tag{B.202}$$

$$= K_1 \left|\frac{\Pi|jw + z_i|}{\Pi|jw + p_i|}\right| e^{j(\sum\psi_{z_i} - \sum\psi_{p_i})} \tag{B.203}$$

$$= |G(jw)|e^{j\psi(w)} \tag{B.204}$$

Therefore, the following relations are obvious from the above:

$$|G(jw)| = K_1 \frac{\Pi|jw + z_i|}{\Pi|jw + p_i|} \tag{B.205}$$

$$\psi(w) = \left(\sum\psi_{z_i} - \sum\psi_{p_i}\right) \tag{B.206}$$

### B.4.4 Experimental Determination of Frequency Response

Consider the dynamic system shown in Fig. B.16. It is excited by an input signal in the range such that the dynamic system behaves as a LTI system. Let us assume that we can set the magnitude and phase of the input signal, and that we can measure the response magnitude and phase.

The experimantal procedure to determine the frequency response is:

1. Select $A$, and $w = w_0$ (i.e., $w_0 = 0.001$).
2. Apply input signal: $u(t) = Asin\,(wt)$.
3. Wait long enough so that the output reaches the steady-state response and so that transients die out.
4. Measure $B$ and $\psi$ of the response in $y(t) = B \cdot sin\,(wt + \psi)$.
5. Record $w$, $B/A$, $\psi$.
6. Repeat until $w = w_{max}$ where $w_{max}$ is the maximum frequency of interest, by incrementing $w = w + \Delta w$. $\Delta w$ is the increment of frequency as the experiment sweeps the freqeuncy range from $w_0$ to $w_{max}$.
7. Plot $B/A$, $\psi$ versus $w$.
8. Curve fit to $B/A$ and $\psi$ as function of $w$ and obtain a mathematical expression for the frequency response as a rational function.

### B.4.5 Graphical Representation of Frequency Response

The frequency response of a dynamic system is conveniently represented by a complex function of frequency. The complex function can be graphically represented in many different ways. In control system studies, three most commonly known representations are:

1. ***Bode plots***: plot $20log_{10}|G(jw)|$ v.s. $log_{10}(w)$ and $Phase(G(jw))$ v.s. $log_{10}(w)$.
2. ***Nyquist plots*** (polar plots): plot $Re(G(jw))$ v.s. $Im(G(jw))$ on the complex $G(jw)$ plane where $w$-frequency is parameterized along the curve.
3. ***Log magnitude versus phase plot***: plot the $20log_{10}(G(jw))$ ($y$-axis) vs. $phase(G(jw))$ ($x$-axis) and $w$-frequency is parameterized along the curve.

One can choose to graphically plot the complex frequency response function in many other ways. The above three representations are the most common ones. With the aid of CAD-tools, it is very simple to plot a given frequency response in any one of the above forms. However, the ability to plot basic building blocks of transfer functions by hand sketches still remains a powerful tool in design.

## B.5 TRANSFER FUNCTION AND IMPULSE RESPONSE RELATION

The equation that relates the Laplace transforms of input and output is called the transfer function, or input–output description of a linear time invariant system. The transfer function does not take the initial condition effects into account. One can define the relationship between the initial conditions and output as another transfer function which describes the way the initial conditions influence the output. Generally, the term transfer function alone refers to the transfer function from input to output.

We can take the Laplace transform of any linear constant coefficient ordinary differential equation and find out the transfer function between input–output as well as between initial conditions and output. Consider the following LTI system model:

$$\ddot{y} + 3\dot{y} + 2y = 2\dot{u} + u \tag{B.207}$$

and assume that initial conditions, $y_0 = y_0,\ \dot{y}(0) = \dot{y}_0,\ u(0) = u_0$, are given. The response of the system, $y(t)$, to any input, $u(t)$, plus the given initial conditions can be studied using Laplace transforms (Fig. B.23).

Note that

$$L\{y(t)\} = y(s) \tag{B.208}$$

$$L\{u(t)\} = u(s) \tag{B.209}$$

$$L\{\dot{y}(t)\} = sy(s) - y(0) \tag{B.210}$$

$$L\{\dot{u}(t)\} = su(s) - u(0) \tag{B.211}$$

$$L\{\ddot{y}(t)\} = s^2 y(s) - \dot{y}(0) - sy(0) \tag{B.212}$$

Using the above properties of Laplace transforms, we can take the Laplace transform of the O.D.E.,

$$(s^2 y(s) - \dot{y}(0) - sy(0)) + 3(sy(s) - y(0)) + 2y(s) = 2(su(s) - u(0)) + u(s) \tag{B.213}$$

$$(s^2 + 3s + 2)y(s) = (2s + 1)u(s) - 2u(0) + \dot{y}(0) + (s + 3)y(0) \tag{B.214}$$

and the Laplace transform of the response as a result of the input and the initial conditions can be found as

$$y(s) = \frac{2s + 1}{s^2 + 3s + 2}u(s) - \frac{2}{s^2 + 3s + 2}u_0 + \frac{1}{s^2 + 3s + 2}\dot{y}_0 + \frac{s + 3}{s^2 + 3s + 2}y_0 \tag{B.215}$$

The $y(s)$ can be obtained for any input and initial conditions, then by taking inverse Laplace transform (i.e., using partial fraction expansion), $y(t)$ can be obtained. Note that if the initial conditions are zero, the response will be only due to the input

$$y(s) = \frac{2s + 1}{s^2 + 3s + 2}u(s) \tag{B.216}$$

and in general, the relationship between $u(s)$ and $y(s)$ is described by a rational polynomial of $s$ called the transfer function,

$$G(s) = \frac{y(s)}{u(s)} \tag{B.217}$$

$$y(s) = G(s)u(s) \tag{B.218}$$

which describes the output due to input.

***Special Case*** If $u(t) = \delta(t)$ unit impulse function, the Laplace transform of it is $u(s) = 1.0$, then $y(s) = G(s)$, which means that the transfer function of an LTI system is the Laplace transform of its unit impulse response.

The transfer function and unit impulse response are related to each other as follows:

$$G(s) = L\{h(t)\} \tag{B.219}$$

$$h(t) = L^{-1}\{H(s)\} \tag{B.220}$$

Similarly, the Fourier transform of the impulse response is equal to the frequency response of the system,

$$G(jw) = F\{h(t)\} \tag{B.221}$$

$$= G(s)|_{s=jw} \tag{B.222}$$

$$h(t) = F^{-1}\{G(jw)\} \tag{B.223}$$

The transfer function of a LTI dynamic system consists of poles, zeros, and gain (constant),

$$\{A\ \ transfer\ function\ \} = \{\ \{\ poles\ \},\ \{\ zeros\ \},\ \{\ constant\ \ (gain)\ \}\ \}$$

$$G(s) = K\frac{\prod(s+z_i)}{\prod(s+p_i)} \tag{B.224}$$

$$= K_{dc}\frac{\prod(s/z_i+1)}{\prod(s/p_i+1)} \tag{B.225}$$

where it is clear that the DC gain $K_{dc}$ of the transfer function and the gain $K$ are related as

$$K_{dc} = G(0) = K\frac{\prod z_i}{\prod p_i} \tag{B.226}$$

For instance, consider the following transfer function:

$$G(s) = \frac{b(s)}{a(s)} = \frac{2s+1}{s^2+3s+2} \tag{B.227}$$

The *zeros* are these $s$ values for which $G(s)$ is zero, which means $b(s)$ is zero,

$$\{s \mid G(s) = 0\ :\rightarrow b(s) = 0\} \rightarrow 2s+1 = 0 \rightarrow s = -\frac{1}{2}\quad (zeros) \tag{B.228}$$

The *poles* are these $s$ values for which $G(s)$ goes to $\infty$, which means $a(s)$ is zero,

$$\{s \mid G(s) \rightarrow \infty\ :\rightarrow a(s) = 0\} \rightarrow s^2+3s+2 = 0 \rightarrow s = -1, -2\quad (poles) \tag{B.229}$$

Another convenient way of expressing $G(s)$ is poles, zeros, and DC gain. The DC gain is defined as the value of $G(s)$ at $s = 0$. Consider the same transfer function,

$$G(s) = \frac{2s+1}{s^2+3s+2} \tag{B.230}$$

$$= \frac{2\left(s+\frac{1}{2}\right)}{(s+1)(s+2)} \tag{B.231}$$

$$= \frac{\left(\frac{s}{\frac{1}{2}}+1\right)}{2\left(\frac{s}{1}+1\right)\left(\frac{s}{2}+1\right)} \tag{B.232}$$

$$= \frac{1}{2}\left[\frac{\left(\frac{s}{\frac{1}{2}}+1\right)}{\left(\frac{s}{1}+1\right)\left(\frac{s}{2}+1\right)}\right] \tag{B.233}$$

In general, a transfer function can be expressed as

$$G(s) = G(0)\frac{\prod_{i=1}^{m}\left(\frac{s}{z_i}+1\right)}{\prod_{i=1}^{n}\left(\frac{s}{p_i}+1\right)} \tag{B.234}$$

The unit magnitude impulse response of the system described by the above transfer function can be found:

$$h(t) = L^{-1}\left\{\frac{2s+1}{s^2+3s+2}\right\} = \frac{2(s+1/2)}{(s+2)(s+1)} \tag{B.235}$$

$$= L^{-1}\left\{\frac{-1}{s+1}+\frac{3}{s+2}\right\} \tag{B.236}$$

$$= L^{-1}\left\{\frac{2\cdot(s+1/2)}{(s+1)(s+2)}\right\} \tag{B.237}$$

$$h(t) = -e^{-t} + 3e^{-2t} \tag{B.238}$$

Notice that the contribution of each pole to the impulse response is determined by the residue associated with that pole when we expand the transfer function to its partial fraction expansion (P.F.E.). The graphical illustration of of the residue associated with each pole is shown in Fig. B.17. The residue associated with a particular pole is the ratio of vectors from zeros and poles to that pole location *multiplied by the gain* (in this example the gain is 2, not the d.c. gain which is 1/2) of the transfer function. Let us show the effect of poles and zeros on the response of an LTI system with an example.

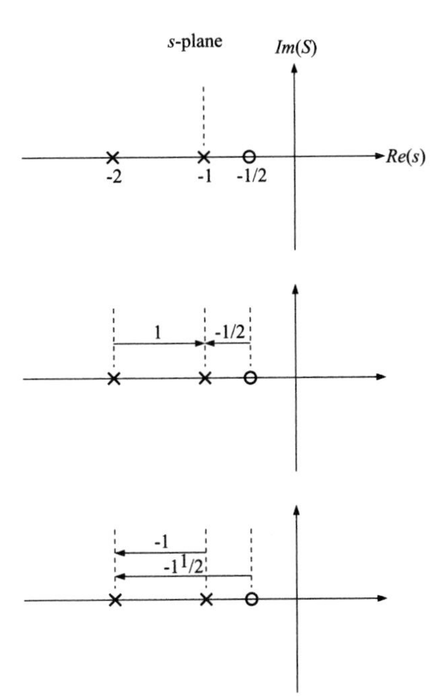


**FIGURE B.17:** Illustration of residue in the inverse Laplace transform and the effect of zeros on residue values.

***Example*** Consider the response of an LTI dynamic system due to a nonzero initial condition,

$$y(s) = \frac{s+3}{s^2+3s+2} y(0^+) \tag{B.239}$$

$$y(0^+) = 1 \tag{B.240}$$

We can obtain the time domain response by taking the inverse Laplace transform using the PFE method. The roots of the denominator are $-1$ and $-2$; hence, the PFE of $y(s)$ is

$$y(s) = \frac{k_1}{(s+1)} + \frac{k_2}{(s+2)} \tag{B.241}$$

where

$$k_1 = \lim_{s \to -1} (s+1) y(s) = \left. \frac{s+3}{s+2} \right|_{s=-1} = 2 \tag{B.242}$$

$$k_2 = \lim_{s \to -2} (s+2) y(s) = \left. \frac{s+3}{s+1} \right|_{s=-2} = -1 \tag{B.243}$$

Substituting the $k_1, k_2$ coefficients in the PFE,

$$y(s) = \frac{2}{s+1} - \frac{1}{s+2} \tag{B.244}$$

Taking the inverse Laplace transform,

$$y(t) = 2e^{-t} - e^{-2t} \tag{B.245}$$

Note that if a zero is too close to a pole, the residue ($k_i$) associated with that pole is very small; hence, the contribution of that pole to the response is small (Fig. B.18).

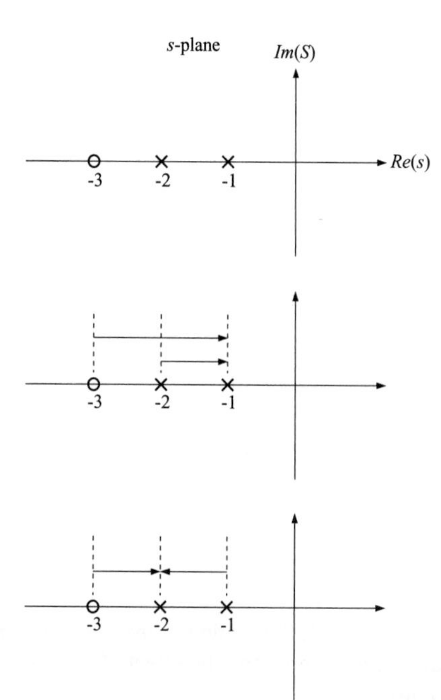


**FIGURE B.18:** Illustration of residue in the inverse Laplace transform.

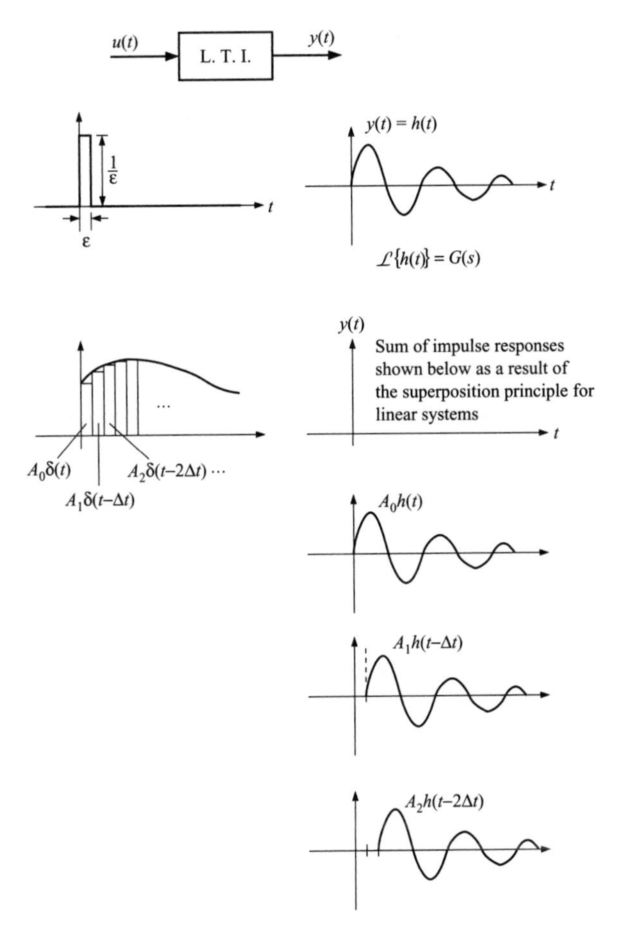


**FIGURE B.19:** Illustration of the convolution theorem for linear time invariant systems.

## B.6 CONVOLUTION

A linear system has superposition property. If the response of a linear system to an input $u_1(t)$ is $y_1(t)$, and to an input $u_2(t)$ is $y_2(t)$, then the response of the system to an input $c_1u_1(t) + c_2u_2(t)$ is $c_1y_1(t) + c_2y_2(t)$. The response of an LTI system to a unit impulse is called its impulse response and generally denoted as $h(t)$. The Laplace transform of impulse response is the transfer function of the system.

Any general input function (piecewise continuous, and exponential order) can be expressed as a sum of discrete impulses. Therefore, the response of the LTI system can be obtained as the sum of the responses to impulses (Fig. B.19). Let us describe a general input $u(t)$ as a sum of a series of impulses

$$u(t) = \sum_{i=0}^{\infty} A_i \delta(t - i.\Delta t) \tag{B.246}$$

The response is calculated by using the above superposition argument,

$$y(t) = \sum_{i=0}^{\infty} A_i h(t - i\,\Delta t) \tag{B.247}$$

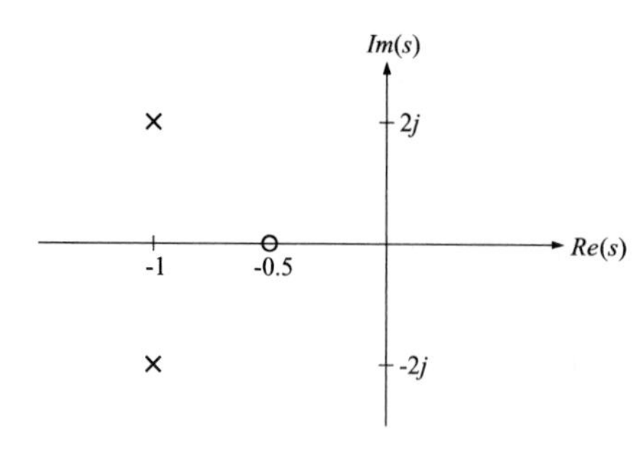


**FIGURE B.20:** Poles and zeros of the transfer function $H(s) = \frac{2s+1}{s^2+2s+5}$.

In limit, as $\Delta t$ converges to zero, we can replace the summation with integration as follows:

$$y(t) = \int_0^t h(t-\tau)u(\tau)d\tau \tag{B.248}$$

Notice that this is the same convolution property discussed in the study of Laplace transforms. If we take the Laplace transform of the above equation in both sides, we obtain (following the property number 9, Convolution theorem, of Laplace transforms)

$$y(s) = G(s)u(s) \tag{B.249}$$

***Example*** Consider the following transfer function (Fig. B.20):

$$G(s) = \frac{2s+1}{s^2+2s+5} \tag{B.250}$$

The zeros of the transfer function are $-\frac{1}{2}$. The poles are $-1 \pm 2j$. Therefore, the partial fraction expansion form of $G(s)$ can be expressed as

$$G(s) = \frac{k_1}{s+1-2j} + \frac{k_1^*}{s+1+2j} \tag{B.251}$$

where $k_1 = lim_{s \to -1+2j}[s-(-1+2j)]$

$$k_1 = \frac{-1+4j}{4j} = \sqrt{1+0.25^2}e^{jtan^{-1}(0.25)} \tag{B.252}$$

Note:

$$z = x + yj = \sqrt{x^2+y^2}e^{jtan^{-1}(\frac{y}{x})} \tag{B.253}$$

$$k_1^* = \sqrt{1+0.25^2}e^{-jtan^{-1}(0.25)} \tag{B.254}$$

The unit impulse response of the linear time invariant (LTI) system is given by

$$h(t) = L^{-1}\{H(s)\} \tag{B.255}$$

$$= \sqrt{1+0.25^2}(e^{jtan^{-1}(0.25)}e^{-t}e^{2tj} + e^{-jtan^{-1}(0.25)}e^{-t}e^{-2tj}) \tag{B.256}$$

$$= \sqrt{1+0.25^2}e^{-t}cos(2t+\theta) \tag{B.257}$$

$$= \sqrt{1+0.25^2}e^{-t}cos(2t+14.03^o); \quad \theta = tan^{-1}(0.25) = 14.03^o \tag{B.258}$$

## B.7 REVIEW OF DIFFERENTIAL EQUATIONS

### B.7.1 Definitions

Continuous time dynamic systems are described by differential equations. A differential equation is an equation involving derivatives of a dependent variable with respect to an independent variable, such as

$$\frac{dy(t)}{dt} + ay(t) = 0 \tag{B.259}$$

where $y$ is a dependent and $t$ is an independent variable. If there is only one independent variable, then the differential equation is called an *ordinary differential equation* (O.D.E.). If there are more than one independent variables, then it is called a *partial differential equation* (P.D.E.). If the dependent variables and their derivatives appear in nonlinear functions in the equations, then the differential equation (D.E.) is *nonlinear*; otherwise it is *linear*. Consider the following two O.D.E.s:

$$\frac{d^2y(t)}{dt^2} + \left(\frac{dy(t)}{dt}\right)^2 + ay(t) = 0 \tag{B.260}$$

This differential equation is nonlinear due to $\left(\frac{dy(t)}{dt}\right)^2$ term.

$$\frac{d^2y(t)}{dt^2} + 2\left(\frac{dy(t)}{dt}\right) + ay(t) = 0 \tag{B.261}$$

This is a linear differential equation because $y$, $dy/dy$, $d^2y/dt^2$ appear linearly in the equation.

The highest derivative in the equation is the order of the d.e. Solution of an $n$th-order d.e. contains $n$-arbitrary constants. These constants would be determined by $n$-conditions on dependent variables (i.e., the initial conditions).

### B.7.2 System of First-Order O.D.E.s

Any $n$th-order O.D.E. can be expressed as $n$ set of first-order O.D.E. There are infinitely many different equivalent ways of doing that. Especially in numerical and state space analysis, expressing differential equations as a set of first-order O.D.E. is very useful.

Consider an $n$th-order nonlinear O.D.E.,

$$\frac{d^nx(t)}{dt^n} = g(t, x(t), x(t)^{(1)}, x(t)^{(2)}, \ldots, x(t)^{(n-1)}) \tag{B.262}$$

Let's define $n$ new variables $x_1, x_2, \ldots, x_n$ as follows:

$$x_1(t) = x(t) \tag{B.263}$$

$$x_2(t) = x(t)^{(1)} = \frac{dx(t)}{dt} \tag{B.264}$$

$$x_3(t) = x(t)^{(2)} = \frac{d^2x(t)}{dt^2} \tag{B.265}$$

$$\vdots$$

$$x_n(t) = x(t)^{(n-1)} = \frac{dx(t)^{n-1}}{dt^{n-1}} \tag{B.266}$$

Then

$$\dot{x}_1 = x_2 \tag{B.267}$$

$$\dot{x}_2 = x_3 \tag{B.268}$$

$$\vdots$$

$$\dot{x}_{n-1} = x_n \tag{B.269}$$

$$\dot{x}_n = g(t, x, x^{(1)}, x^{(2)}, \ldots., x^{(n-1)}) \tag{B.270}$$

Hence the $n$th-order O.D.E. can be expressed as an $n$ set of first-order O.D.E.s in vector form as

$$\underline{\dot{x}} = \underline{f}(t, \underline{x})$$

where

$$\underline{x} = [x_1, x_2, \ldots., x_{n-1}, x_n]^T \tag{B.271}$$

$$\underline{f} = [x_2, x_3, \ldots., x_n, g(t, \underline{x})]^T \tag{B.272}$$

where superscript $T$ represents transpose of the vector.

Clearly, infinitely many other possible ways exist to obtain an equivalent $n$ set of first-order O.D.E.s, because any $T$ transformation ($\underline{x}^* = T\underline{x}$, $\underline{x} = T^{-1}\underline{x}^*$) describes a new equivalent first-order O.D.E. set.

Consider the following special case of an $n$-order O.D.E which is linear,

$$a_0(t)\frac{d^n x}{dt^n} + a_1(t)\frac{d^{n-1}x}{dt^{n-1}} + \cdots + a_{n-1}(t)\frac{dx}{dt} + a_n(t)x(t) = r(t)$$

Using the same approach, this O.D.E. can be expressed as an $n$-set of first-order O.D.E.s and it would have the following form,

$$\underline{\dot{x}}(t) = [A(t)]\underline{x}(t) + \underline{B}(t)\underline{r}(t)$$

where

$$A(t) = \begin{bmatrix} 0 & 1 & 0 \;\ldots.0 \\ 0 & 0 & 1 \;\;..0 \\ \ldots.. & & \\ -a_n(t) & -a_{n-1}(t) \;\ldots & -a_1(t) \end{bmatrix} \tag{B.273}$$

$$B(t) = \begin{bmatrix} 0 \\ 0 \\ .. \\ .. \\ 1/a_0(t) \end{bmatrix} \tag{B.274}$$

### B.7.3 Existence and Uniqueness of the Solution of O.D.E.s

***Nonlinear O.D.E.s*** Consider a general nonlinear O.D.E,

$$\frac{dx}{dt} = f(x, t); \quad x(t_0) = x_0 \quad \text{given} \tag{B.275}$$

If $f(x, t)$ continuous and has continuous partial derivative with respect to $x$ at each point of region $|t - t_0| < \delta_1$, $|x - x_0| < \delta_2$, there exists one unique solution in the region which passes through $(t_0, x_0)$. This theorem can be generalized to the $n$th order O.D.E.s, since an $n$th order O.D.E. can be expressed as $n$ set of first-order O.D.E. Therefore, the above statements are also valid for vector $\underline{f}$ function and vector $\underline{x}$ states.

***Linear O.D.E.s*** Consider an $n$th-order, linear, variable coefficient O.D.E.

$$a_0(t)\frac{d^n x}{dt^n} + a_1(t)\frac{d^{n-1}x}{dt^{n-1}} + \cdots + a_{n-1}(t)\frac{dx}{dt} + a_n(t)x = r(t) \quad \text{(B.276)}$$

$a_i(t)$, $r(t)$ continuous in region $|t - t_0| < \delta$ and $a_0(t) \neq 0$. Then there exists a unique solution which satisfies $n$ initial conditions $x(t_0), x'(t_0), \ldots, x^{(n-1)}(t_0)$.

## B.8 LINEARIZATION

We study the linearizarion in increasing order of complexity. First we study the linearization of nonlinear functions, then that of first-order nonlinear differential equations, and finally that of a set of first-order nonlinear differential equations.

### B.8.1 Linearization of Nonlinear Functions

Consider the function $y = ax$ or $y = ax + b$ which are both linear in $x$ (Fig. B.21). If the linear function does not pass through the origin, it may be convenient to define a new variable so that the input–output relationship of the static function passes through the origin,

$$y^* = y - b \quad \text{(B.277)}$$

$$= ax \quad \text{(B.278)}$$

If the function between independent variable and dependent variable (or input and output) is nonlinear,

$$y = y(x) \quad \text{(B.279)}$$

it can be approximated by a linear function about a nominal point, $(x_o, y_o)$. This approximation is accomplished by expanding the function to its Taylor series about the nominal point, and neglecting the second-order and higher-order terms. Let us define the total values of

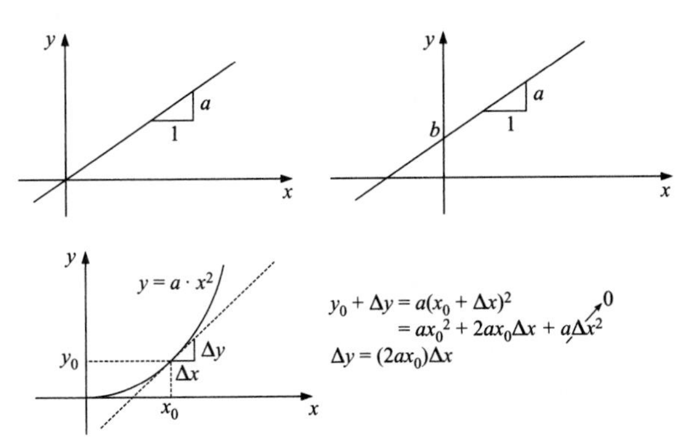


**FIGURE B.21:** Linearization of functions.

the independent and dependent variables as nominal values plus the small variations about that point,

$$x = x_o + \Delta x \tag{B.280}$$

$$y(x) = y(x_o) + \Delta y(x_o + \Delta x) \tag{B.281}$$

The Taylor series expansion of $y(x)$ about the nominal point $(x_o, y_o)$ is

$$y(x) = y(x_o) + \left.\frac{dy(x)}{dx}\right|_{x_o} (x - x_o) + \left.\frac{d^2y(x)}{dx^2}\right|_{x_o} \frac{(x - x_o)^2}{2!} + \cdots \tag{B.282}$$

If we neglect the second- and higher-order terms, assuming that the approximation will be used in the close vicinity of the nominal point $(x_o, y_o)$,

$$y(x) = y(x_o) + y'(x_o)(x - x_o) \tag{B.283}$$

The small variations around the $(x_o, y_o)$ nominal point have the following relation:

$$y(x) - y(x_o) = y'(x_o)(x - x_o) \tag{B.284}$$

$$\Delta y(x) = m \cdot \Delta x \tag{B.285}$$

where $m = y'(x_o)$.

Clearly, the closer the actual evaluation point to the nominal point, the more accurate the approximation. As the evaluation point gets farther away from the nominal point, the approximation gets poorer (Fig. B.21).

We can apply the same idea to the linearization of nonlinear differential equations. Every nonlinear function in the o.d.e. can be approximated by its first-order Taylor series expansion about the nominal point.

Notice that linearizing a nonlinear algebraic function about a nominal operating condition involves taking the first derivative of the nonlinear function and evaluating it at the nominal values of the independent variable. For multivariable, nonlinear functions, the same operation would be performed using the first-order partial derivatives of the nonlinear function with respect to each individual independent variable. Let $y = f(x)$ be a nonlinear albegraic function of single variable $x$, and $z = h(x_1, x_2, x_3)$ be a nonlinear algebraic multivariable function of variables $x_1, x_2, x_3$. Let us consider the linearized approximation of these functions about the nominal conditions of $x_0$, $(x_{10}, x_{20}, x_{30})$,

$$\Delta y = \left.\frac{df(x)}{dx}\right|_{x_0} \cdot \Delta x \tag{B.286}$$

which is a linearized approximation to $y = f(x)$ about nominal point $(x_o, y_o)$.

$$\Delta z = \left.\frac{\partial h(x_1, x_2, x_3)}{\partial x_1}\right|_{(x_{10}, x_{20}, x_{30})} \cdot \Delta x_1 + \left.\frac{\partial h(x_1, x_2, x_3)}{\partial x_2}\right|_{(x_{10}, x_{20}, x_{30})} \cdot \Delta x_2 + \left.\frac{\partial h(x_1, x_2, x_3)}{\partial x_3}\right|_{(x_{10}, x_{20}, x_{30})} \cdot \Delta x_3 \tag{B.287}$$

which is a linearized approximation to $z = h(x_1, x_2, x_3)$ about the nominal point $z_o = h(x_{10}, x_{20}, x_{30})$.

***Example*** Let us study the linearization of a set of nonlinear differential equations by an example. The key idea is that the system is operating about a nominal condition (a state or time varying trajectory), and the variations about that nominal condition are small such that the second-order and higher terms are negligible when all the nonlinearities are approximated by a Taylor series expansion about the nominal condition.

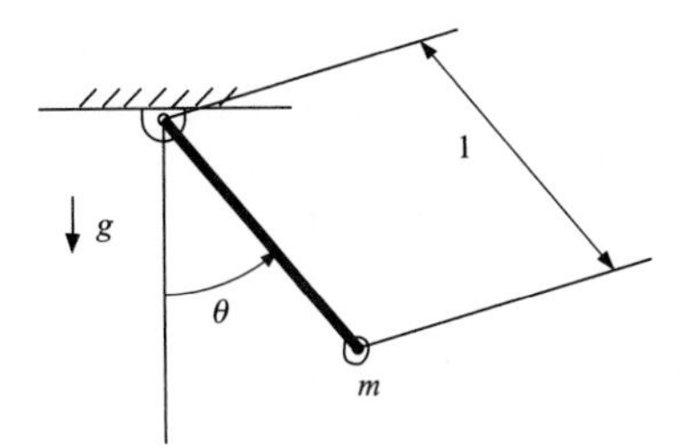


**FIGURE B.22:** Pendulum model.

Consider the dynamic model of a pendulum (Fig. B.22),

$$ml^2\ddot{\theta} + mgl\,sin\,\theta = 0 \tag{B.288}$$

$$\ddot{\theta} + \left(\frac{g}{l}\right) sin\,\theta = 0 \tag{B.289}$$

The differential equation is nonlinear due to the presence of the nonlinear function $sin(\theta)$ of the dependent variable $\theta$.

Let us consider the linearization of the nonlinear dynamic model of the pendulum about a nominal angle at rest. It is assumed that (1) The angular motion of the pendulum is small, and it stays in the neighborhood of the nominal angular position; and (2) position, speed, and acceleration changes are so small that the second-order terms are negligible.

$$\theta = \theta_0 + \Delta\theta \tag{B.290}$$

$$\dot{\theta} = \dot{\theta}_0 + \Delta\dot{\theta} \tag{B.291}$$

$$\ddot{\theta} = \ddot{\theta}_0 + \Delta\ddot{\theta} \tag{B.292}$$

$$sin\,\theta = sin\,(\theta_0 + \Delta\theta)|_{\theta_0} \tag{B.293}$$

$$= sin\,\theta_0 + (cos\,\theta_0)\Delta\theta \tag{B.294}$$

Let us consider the nominal angular position $\theta_0 = 0$ and, $\dot{\theta}_0 = 0$ , $\ddot{\theta}_0 = 0$:

$$sin\,\theta|_{\theta_0=0} = 0, \quad cos\,\theta|_{\theta_0=0} = 1 \tag{B.295}$$

Finally, the linearized dynamic model is

$$\Delta\ddot{\theta} + \left(\frac{g}{l}\right)\Delta\theta = 0 \tag{B.296}$$

Note that the linearization of $cos\,\theta$ and $sin\,\theta$ about the nominal angle $\theta_0$ for small variations of angle around that nominal value,

$$cos\,\theta = cos\,\theta_0 + (-sin\,\theta_0)\Delta\theta + \cdots \tag{B.297}$$

$$cos\,\theta = 1; \quad when \quad \theta_0 = 0 \tag{B.298}$$

$$sin\,\theta = sin(\theta_0 + \Delta\theta)|_{\theta_0} \tag{B.299}$$

$$= sin\,\theta_0 + (cos\,\theta_0)\Delta\theta \tag{B.300}$$

$$= \Delta\theta; \quad when \quad \theta_0 = 0 \tag{B.301}$$

### B.8.2 Linearization of Nonlinear First-Order Differential Equations

Consider the following generic form of a nonlinear first-order o.d.e:

$$\dot{x} = f(x, u) \tag{B.302}$$

We want to linearize about a nominal condition of state and inputs. If the state and input nominal conditions define a constant state, we call this linearization about a nominal state. If the nominal state and input is function of time, we call this linearization about a nominal trajectory. Clearly, linerization about a nominal state is a special case of linearization about a nominal trajectory.

1. If the nominal state and its derivative are specified $(x_0(t), \dot{x}_0(t))$, we can determine the necessary nominal input, $u_0(t)$,

$$x_0(t), \quad \dot{x}_0(t) \text{ specified; } \rightarrow \dot{x}_0(t) = f(x_0(t), u_0(t)), \qquad \text{solve for} \quad u_0(t).$$

2. If the nominal state and nominal input are specified $x_0(t), u_0(t)$, then we can calculate the corresponding nominal first derivative of the state,

$$\dot{x}_0(t) = f(x_0(t), u_0(t)), \quad \text{solve for} \quad \dot{x}_0(t)$$

Either way, the nominal state and nominal input condition $(x_0(t), u_0(t), \dot{x}_0(t))$ satisfy the nonlinear dynamic model. Otherwise, it could not be called a nominal state at which this particular dynamic system can be. Let us define the total state and input values as nominal values plus the small variations about the nominal state,

$$x = x_0 + \Delta x \quad \rightarrow \quad \dot{x} = \dot{x}_0 + \Delta \dot{x} \tag{B.303}$$

$$u = u_0 + \Delta u \tag{B.304}$$

The O.D.E. can be linearized about a nominal operating point or trajectory by substitution of the above relations and expanding the nonlinear function to its Taylor series up to the first-order terms,

$$\dot{x}(t) = f(x, u) \tag{B.305}$$

$$\dot{x}_o(t) + \Delta \dot{x}(t) = f(x_o(t), u_o(t)) + \left.\frac{\partial f}{\partial x}\right|_{[x_o(t), u_o(t)]} \Delta x + \left.\frac{\partial f}{\partial u}\right|_{[x_o(t), u_o(t)]} \Delta u \tag{B.306}$$

Because the nominal values would cancel out each other,

$$\Delta \dot{x}(t) = \left.\frac{\partial f}{\partial x}\right|_{[x_o(t), u_o(t)]} \Delta x + \left.\frac{\partial f}{\partial u}\right|_{[x_o(t), u_o(t)]} \Delta u \tag{B.307}$$

$$= a(t)\Delta x(t) + b(t)\Delta u(t) \tag{B.308}$$

It must be emphasized that this linearized equation is an approximation to the original nonlinear equation. It is accurate only within the close vicinity of the nominal conditions. As the operating conditions get farther away from the nominal conditions, the approximation gets poorer.

The same idea can be directly extended to the vector case which represents multi-dimensional dynamic systems.

### B.8.3 Linearization of Multidimensional Nonlinear Differential Equations

Consider the following $n$-set of first-order nonlinear differential equations (note that any $n$th-order differential equation can be re-expressed as an $n$ set of first-order differential equations),

$$\underline{\dot{x}} = \underline{f}(\underline{x}, \underline{u}) \tag{B.309}$$

$$\underline{x}^T = [x_1, x_2, \ldots, x_n] \tag{B.310}$$

$$\underline{u}^T = [u_1, u_2, \ldots, u_m] \tag{B.311}$$

$$\underline{f}^T = [f_1, \ldots, f_n] \tag{B.312}$$

$$\dot{x} + \Delta\dot{x} = f(x_0, u_0) + \left.\frac{\partial f}{\partial x}\right|_{x_0,u_0} \Delta x + \left.\frac{\partial f}{\partial x}\right|_{x_0,u_0} \Delta u + \cdots$$

$$\Delta\dot{x} = \left.\frac{\partial f}{\partial x}\right|_{x_0,u_0} \Delta x + \left.\frac{\partial f}{\partial u}\right|_{x_0,u_0} \Delta u \tag{B.313}$$

$$\Delta\dot{x} = [A]\Delta x + [B]\Delta u \tag{B.314}$$

where the elements of matrices are given by

$$\left[A_{ij}\right] = \left.\frac{\partial f_i}{\partial x_j}\right|_{x_0,u_0} \tag{B.315}$$

$$\left[B_{ij}\right] = \left.\frac{\partial f_i}{\partial u_j}\right|_{x_0,u_0} \tag{B.316}$$

Linearization is valid for operating conditions for which neglecting second- and higher-order terms are valid and accurate enough. Notice that if the nominal condition is an equilibrium point, which means $(x_o, u_o)$ are constant, the $A$ and $B$ matrices are constant. Dynamic systems represented by linear matrix differential equations where matrices $A$ and $B$ are constant are called *linear time invariant* (LTI) linear systems. If the nominal conditions define a nominal trajectory in time, not an equilibrium condition, then $(x_o(t), u_o(t))$ are functions of time; hence, the $A$ and $B$ matrices will be function of time, $A(t)$ and $B(t)$. Such systems are called the linear time varying (LTV) type.

```
% linear0.m
%
% Numerical linearizarion example using Matlab.
%

x0 = [ pi/2 0.0 ]' ;
     u0 = [ 0.0 ]  ;

     [A,B] = linear1('pendulum', x0, u0) ;


%
%

function [A,B] = linear1(Fname, x0, u0)
%
% Given: nonlinear function f(x,u)
%        nominal condition x0, u0
%
% Calculate linearized equation matrices A, B
%

n = size(x0) ;
     m = size(u0) ;

     delta = 0.0001 ;
     x = x0 ;
```

```
    u = u0 ;

    for j=1:n
      x(j) = x(j) + delta;
      A(:,j) = (feval(Fname,x,u)-feval(Fname,x0,u0))/delta;
      x(j) = x(j) - delta ;
    end

    for j=1:m
      u(j) = u(j) + delta;
      B(:,j) = (feval(Fname,x,u)-feval(Fname,x0,u0))/delta;
      u(j) = u(j) - delta ;
    end

%
%

function xdot=pendulum(x,u)
%
% Pendulum dynamic model: nonlinear model.
%

    g = 9.81 ;
    l = 1.0 ;
    xdot = [ x(2)
              - (g/l)*sin(x(1)) + u] ;

}
```

## B.9 NUMERICAL SOLUTION OF O.D.E.s AND SIMULATION OF DYNAMIC SYSTEMS

Analytical solution methods are available for only linear, constant coefficient differential equations (i.e., variation of parameters, Laplace transforms). Only a few special cases of first- or second-order nonlinear differential equations can be solved by analytical methods. For all practical purposes, the only viable engineering solution method for nonlinear differential equations is the numerical methods.

The behavior of a dynamic system in time can be predicted by the solution of its mathematical model which typically is a set of ordinary differential equations. Analytical solution of O.D.E.s is available for only linear O.D.E.s and very simple nonlinear O.D.E.s. Therefore, time domain response of any dynamic system model with reasonable complexity must be solved using numerical methods. The primary tool is the numerical integration of O.D.E.s in time. Numerical integration is performed by discretizing O.D.E.s using various approximations to differentiation (i.e., Euler's approximation, trapezoidal approximation, Runga–Kutta approximation). First we will study various numerical integration methods for a general nonlinear O.D.E. set of the form,

$$\dot{\underline{x}} = \underline{f}(\underline{x}, \underline{u}, t) \tag{B.317}$$

$$\underline{x}(t_0) = \underline{x}_0, \quad \underline{u}(t) \ \text{ given} \tag{B.318}$$

Then we will study the time domain simulation of a dynamic system that involves a digital controller in the determination of $u(t)$.

## B.9.1 Numerical Methods for Solving O.D.E.s

*The problem* is that given a nonlinear set of first-order O.D.E.s, with a specified initial condition and input, find the solution of the O.D.E. system.

$$\dot{\underline{x}} = \underline{f}(\underline{x}, \underline{u}; t) \tag{B.319}$$

$$\underline{x}(t_0) = \underline{x}_0, \qquad \text{initial condition} \tag{B.320}$$

$$\underline{u}(t) \quad \text{is given} \tag{B.321}$$

Numerical integration provides an approximate solution method for the differential equations. Although the solution is approximate, the error in the approximation can be controlled to the point of being negligible. The essential approximation is that the derivatives of the dependent variables are approximated by finite differences. Here we will consider various finite difference approximations.

## B.9.2 Numerical Solution of O.D.E.s

***Euler's Method*** The Euler's approximation is the simplest among others. In the following discussion, we drop the underline symbol from the $x$, $f$, and $u$ variables for simplicity in notation. The variables $x$, $f$, $u$ are still considered to be vector quantities. The derivative of the dependent variable at a given time is approximated by the difference between the values of the dependent variable at two consecutive samples divided by the sampling interval (Fig. B.23).

$$\dot{x} = \frac{dx}{dt} \simeq \lim_{\Delta t \to 0^+} \frac{x(t + \Delta t) - x(t)}{\Delta t} \tag{B.322}$$

$$x(t + \Delta t) = x(t) + \dot{x}\Delta t \tag{B.323}$$

$$x(t + \Delta t) = x(t) + f(x, u; t)\Delta t \tag{B.324}$$

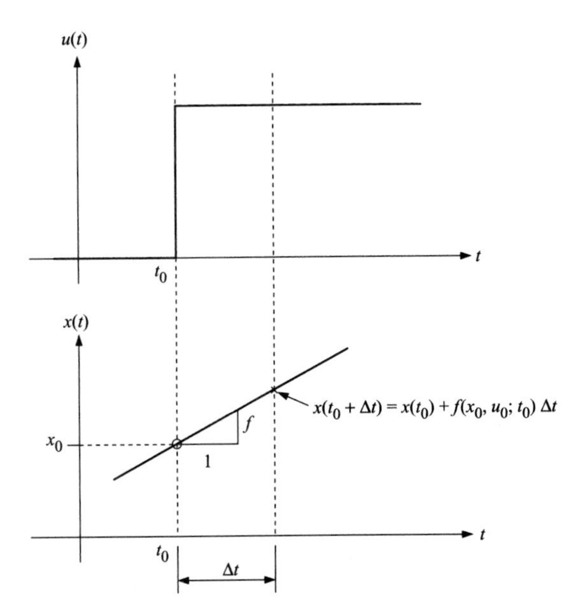


**FIGURE B.23:** Euler's method for solution of O.D.E.s.

where $\Delta t$ is the sampling interval. We can use this approach to solve any nonlinear O.D.E. (although it may not be the most efficient or accurate method) as follows: Starting with the initial condition, we know $x(t_0)$, $u(t_0)$, then we can evaluate $f(x_0, u_0; t_0)$, and calculate $x(t_0 + \Delta t)$ as

$$x(t_0 + \Delta t) = x(t_0) + f(x_0, u_0, t_0)\Delta t \tag{B.325}$$

Similarly, using the newfound values of $x(t_0 + \Delta t)$, we can find the $x(t_0 + 2\Delta t)$ as

$$x(t_0 + 2\Delta t) = x(t_0 + \Delta t) + f(x, u, t)\Delta t \tag{B.326}$$

where $f(x, u, t)$ is evaluated for those values at $(t_0 + \Delta t)$. This iteration can be continued until the desired solution time period is covered.

***Runga–Kutta Method: 4th Order*** Runga–Kutta's fourth-order numerical integration method may be the best compromise between the complexity and accuracy. The difference compared to Euler's method is that it uses a more accurate approximation for the differentiation. If the sampling period is taken very small, the difference between the two methods would be insignificant in terms of accuracy. Given the same o.d.e solution problem, Runga–Kutta's fourth-order approximation to the solution is given by (Fig. B.24).

$$x(t_i + \Delta t) = x(t_i) + \frac{1}{6}(k_1 + 2k_2 + 2k_3 + k_4) \tag{B.327}$$

where

$$k_1 = \Delta t \cdot f(t_i; x(t_i), u(t_i)) \tag{B.328}$$

$$k_2 = \Delta t \cdot f\left(t_i + \frac{\Delta t}{2}; x(t_i) + \frac{1}{2}k_1, u\left(t_i + \frac{\Delta t}{2}\right)\right) \tag{B.329}$$

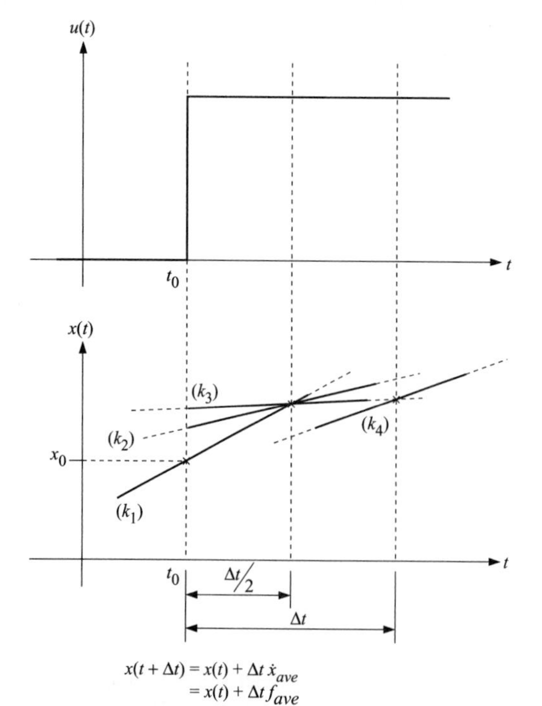


**FIGURE B.24:** Fourth-order Runga–Kutta finite difference approximation.

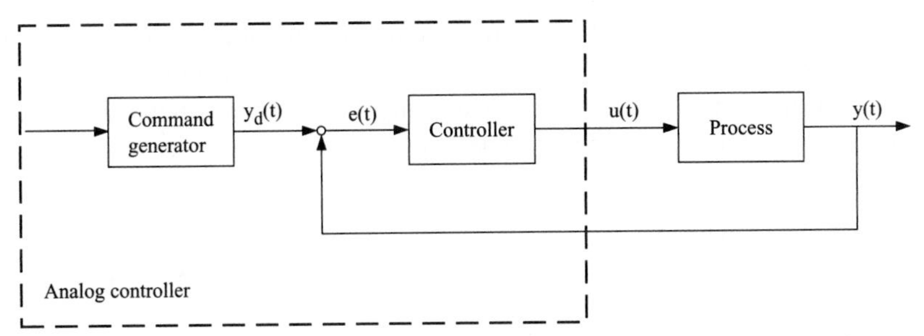


**FIGURE B.25:** A continuous time feedback control system—controller and process are both analog.

$$k_3 = \Delta t \cdot f\left(t_i + \frac{\Delta t}{2}; x(t_i) + \frac{1}{2}k_2, u\left(t_i + \frac{\Delta t}{2}\right)\right) \tag{B.330}$$

$$k_4 = \Delta t \cdot f(t_i + \Delta t; x(t_i) + k_3, u\,(t_i + \Delta t)) \tag{B.331}$$

## B.9.3 Time Domain Simulation of Dynamic Systems

We will consider the program structure for digital computer simulation of dynamic systems which has (a) analog controller (Fig. B.25), and (b) digital controller with a given control sampling period (Fig. B.26).

**(a)** ***Dynamic system with analog controller (Fig. B.25):*** The signals in such a system are all continuous in time, and no sampling is involved in reality. The numerical simulation will work on the samples of the signals in the system. For accuracy both the controller and process signals should be sampled at the same rate. Some numerical o.d.e. solution algorithms have automatic step size control in order to enssure local error in each integration step is smaller than a given value. In such a case, in order to enssure that the controller signals are sampled at the same rate as the process, thus simulate an analog control, the controller function should be called from the process dynamics. Hence, every time integration routine calls the process dynamic model, the controller routine will also be called.

**(b)** ***Dynamic system with digital controller (Fig. B.26):*** In this case there are two sampling periods—one is decided on by the numerical integration accuracy requirements for the purpose of accurate simulation of an actually continuous time process ($T_{int}$). The other is the sampling period of the digital controller which physically exists in real-time implementation, $T_{control}$. Almost all digital controllers are interfaced to a zero-order-hold (Z.O.H.) type D/A converter. Hence, between each sampling period, control action stays constant. In order to simulate the real-world behavior accurately, the controller function should be called to get an updated control signal only every, $T_{control}$, time period. The control signal value should be kept constant between the

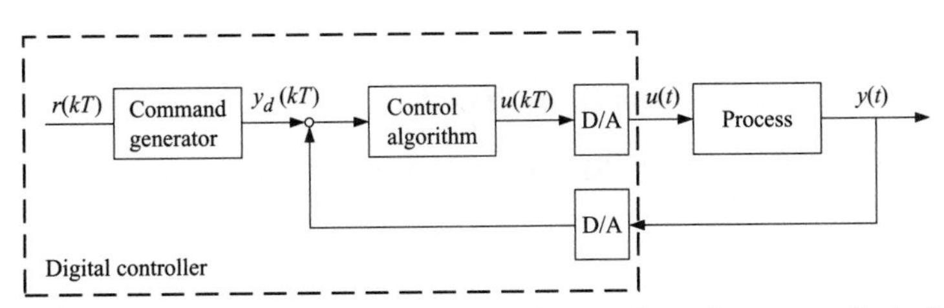


**FIGURE B.26:** A digital control system: Process is analog, controller is digital.

sampling instants to simulate a zero-order-hold D/A converter behavior. During that time period, the process dynamics may be called one or more time depending on the numerical accuracy requirements. Generally the integration step is integer multiples smaller than the control sampling period $\{T_{control} = n\ T_{int}\ ; \quad n = 1, 2, \ldots\}$.

- *Simulation program*
  - Initialize modules
  - If simulating a digital controller: run control sample loop every $T_{control}$ period call controller and get $u$
    - Simulation loop: $T_{int}$
      call o.d.e. solver
    - next
  - next (if simulating digital controller)
- *Controller*
  - desired response (command signal)
  - sensor dynamics: $y = g(x)$
  - calculate $u = \ldots..$
- *O.D.E. solver*
  - in : $t, x, u, \Delta t$, process dynamics
  - If simulating analog controller, call analog controller function here.
  - call process dynamics, get $\dot{x}$
  - out : $x(t + h)$
- *Process dynamics*
  - $\dot{x} = f(x, u; t)$

The structure of the simulation program is given below in more details.

- Initialize system parameters
  - $t_0, t_f$ - initial & final simulation time
  - $x_0$ - initial states
  - $t_{sample}, t_{int}$ - control loop update time (sampling time period), integration step size
- Initialize controller
  - controller parameters - i.e., gains, nonlinear compensation functions
  - controller initial condition - i.e., observer initial states.
- Loop for each control sampling period (assuming digital controller implementation)
  - begin simulation loop: $t_0, t_0 + t_{sampling}, \ldots, t_f$
  - control calculations: $u$
  - begin loop to simulate dynamic system during the control sampling period:
  - o.d.e. solver
  - loop end
  - loop end
- loop end
- output results

### *Simulation of a Dynamic System Using Matlab*

```
/* Implementation of dynamic system simulation program
   using MATLAB */


% mass_s.m
```

```
%
% simulates a continuous time dynamic system using
% 4 th order Runga-Kutta integration algorithm.
%
%     dynamic system: mass-force system
%     controller : PD control algorithm
%
% Initialize simulation...
%.....Dynamic system .....

   t0 = 0.0 ;
   tf = 4.0 ;
   t_sample = 0.01 ;
   t_int = 0.005 ;
   x = [ 0.0 0.0 ]' ;
   x_out = x' ;
   u_out = 0 ;

%.....initialize controller parameters....

     k_p = 16.0 ;
     k_v = 4.0 ;

% Start the simulation loop...

   for (t = t0 : t_sample : tf )
     mass_ct1 ;
     for (t1 = t : t_int : t+t_sample )
       x = rk4('mass_dyn',t1,t1+t_int, x, u) ;
     end
     x_out=[x_out ; x' ] ;
     u_out=[u_out ; u ] ;
    end

% ..Plot results....

   t_out=t0:t_sample:tf ;
   t_out = [t_out' ; tf+t_sample ] ;

   clg ;
   subplot(221)
     plot(t_out,x_out(:,1)) ;
     title('position vs time') ;
   subplot(222)
     plot(t_out,x_out(:,2)) ;
     title('velocity vs time ');
   subplot(223)
     plot(t_out,u_out) ;
     title('control vs time ');
```

```
    pause

% ... end......

% mass_ctl.m
%
% Implements a PD controller for a second order system...

%     get sensor measurements ....
%        in simulation this is readily available,
%        in hardware implementation that will require a
         call to sensor
%        device drivers i.e. A/D converters....
%         x(1) - measured position,
%         x(2) - measured velocity.

%     desired motion ...

       xd = [ 1.0
              0.0 ] ;

% PD control algorithm............

      u = k_p * (xd(1)-x(1)) + k_v * (xd(2) - x(2)) ;

%    Output to D/A converter in hardware
     implementation....
%    In simulation, the u is returned to the calling
     function...
% ..end....

function xdot=mass_dyn(t,x,u)
%
% describes the dynamic model: o.d.e's
% returns xdot vector.
%     xdot = [ x(2)
              u(1) ];
function x1=rk4(FuncName,t0,tf,x,u)
%
% implements Runga-Kutta 4th order ingetration algorithm
  on

% o.d.e's.
%
     h = tf - t0 ;
     h2= h/2 ;
     h6= h/6 ;
     th=t0+h2 ;
     xdot = feval(FuncName,t0,x,u) ;
     xt = x + h2 * xdot ;
```

```
        dxt = feval(FuncName,th,xt,u) ;
        xt = x + h2 * dxt ;

        dxm = feval(FuncName,th,xt,u) ;
        xt = x + h * dxm ;
        dxm = dxm + dxt ;

         dxt = feval(FuncName,tf,xt,u) ;

        x1 = x + h6 * (xdot + dxt + 2.0*dxm) ;

    % .... End.....
```

***Modeling and Simulation of a Dynamic System Using Simulink***
Simulink is a numerical modeling and time domain simulation software with a graphical user interface. It allows us to mix linear, nonlinear, analog, and digital systems in the model using interconnected blocks. In Simulink, the dynamic model and control algorithm are modeled using interconnected blocks. In order to simulate the system response for a given input conditions, inputs are connected to various source blocks (i.e., function generator block, step input block). The response is recorded by connecting various output signals to sink blocks (i.e., scope block). The simulation start time, stop time, sampling time, and integration method are selected in the Simulink set-up window. The same example (mass-force system and a PD controller) given earlier is simulated in Simulink as shown below. The Simulink model simulates an *analog PD controller* for the mass-force system. The controller is sampled at the same rate as the integration algorithm uses for the solution of the mass-force model. In Simulink models, the input–output relationships are described by interconnected blocks. The individual blocks may represent linear or nonlinear functions between its input and output, transfer functions in Laplace transform form for analog systems, and z-transforms form for digital systems. Input functions for the simulated case are represented in time domain form. The response of the system are also time domain functions.

***Example*** Consider the liquid level in a tank and its control system shown in (Fig. 1.4). Let us further consider a computer-controlled version of the system: The mechanical levers are replaced by a level sensor, a digital controller, and a valve that is actuated by a solenoid. In-flow rate to the tank is controlled by the valve. The valve is controlled by a solenoid. The input to the solenoid is the current signal from the controller, and output of the solenoid is proportional force. The force generated by the solenoid is balanced by a centering spring. Hence, the valve position or opening of the orifice is proportional to the current signal. The flow rate through the valve is proportional to the valve opening. Assuming simplified linear relationships, the input–output relationship for the valve can be expressed as

$$F_{valve}(t) = K_1 \cdot i(t) \tag{B.332}$$

$$= K_{spring} \cdot x_{valve}(t) \tag{B.333}$$

$$Q_{in}(t) = K_{flow} \cdot x_{valve}(t) \tag{B.334}$$

$$= K_{flow} \cdot \frac{1}{K_{spring}} \cdot K_1 \cdot i(t) \tag{B.335}$$

$$= K_{valve} \cdot i(t) \tag{B.336}$$

where $K_{valve} = K_{flow} \cdot K_1/K_{spring}$, which is the valve gain between current input and flow rate through the valve.

The liquid level in the tank is a function of the rate of in-flow, rate of out-flow, and the cross-sectional area of the tank. The time rate of change in the volume of the liquid in the tank is equal to the difference between in-flow rate and out-flow rate,

$$\frac{d(\textit{volume in tank})}{dt} = (\textit{in-flow rate}) - (\textit{out-flow rate}) \tag{B.337}$$

$$\frac{d(A \cdot y(t))}{dt} = Q_{in}(t) - Q_{out}(t) \tag{B.338}$$

$$A\frac{dy(t)}{dt} = Q_{in}(t) - Q_{out}(t) \tag{B.339}$$

The $Q_{in}$ is controlled by the valve to be between zero and maximum flow that can go through the valve, $[0, Q_{max}]$. The $Q_{out}$ is function of the liquid level and the orifice geometry at the outlet. Let us approximate the relationship as a linear one; that is, the higher the liquid height, the larger the out-flow rate,

$$Q_{out}(t) = \frac{1}{R} \cdot y(t) \tag{B.340}$$

where $R$ represents the orifice restriction as the resistance to flow. Then the tank dynamic model can be expressed as

$$A\frac{dy(t)}{dt} + \frac{1}{R} \cdot y(t) = Q_{in}(t) \tag{B.341}$$

Let us consider a practical ON/OFF type controller with hysteresis. The controller either fully turns ON or turns OFF the valve depending on the error between the actual and measured liquid level. In order to make sure the controller does not switch the valve ON/OFF at high frequency due to small changes in the liquid level, a small amount of hysteresis is added in the control function. This type of controller is called relay with hysteresis and is commonly used in many automatic control systems such as home temperature control and liquid level control. In Simulink, the controller function is implemented with a hysteresis block. In mathematical terms, the controller function is

$$e(t) = y_d(t) - y(t) \tag{B.342}$$

$$i(t) = Relay_{hysteresis}(e) \tag{B.343}$$

The relay control function with hysteresis where the hysteresis band is $[-e_{max}, e_{max}]$ range. Flow rate can vary linearly between zero and maximum flow rate as function of current signal. Because current signal is either zero or maximum value, flow rate will be either zero or maximum flow.

$$Q_{in}(t) = K_{valve} \cdot i(t) \tag{B.344}$$

$$= Q_{max}; \quad \text{when} \quad i(t) = i_{max} \tag{B.345}$$

$$= 0; \quad \text{when} \quad i(t) = 0 \tag{B.346}$$

Let us simulate the liquid level control system for the following conditions. The system parameters are $e_{max} = 0.05$, $i_{max} = 1.0\,A$, $Q_{max} = 1200$ liter/min $= 20$ liter/sec $=$ $0.02$ m$^3$/sec, $A = 0.01$ m$^2$, and $R = 500$ [m]/[m$^3$/sec]. Consider the case that the desired liquid height is $y_d(t) = 1.0$ m which is commanded as step function, and the initial height of the liquid is zero (empty tank). Figure B.27 shows the simulink model and simulation results.

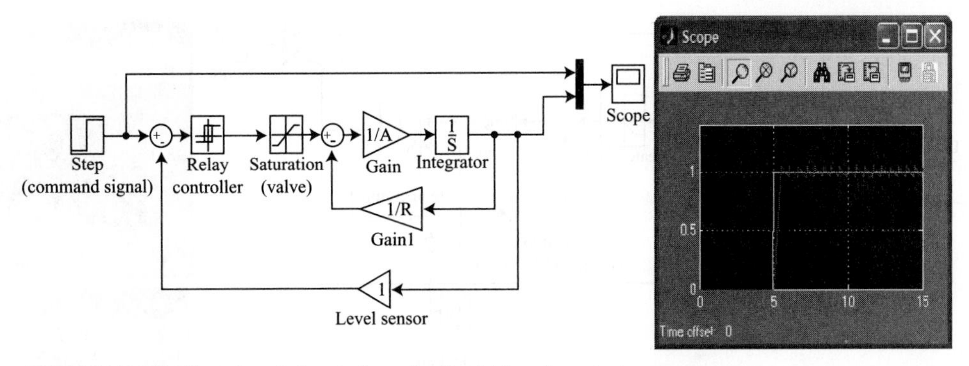


**FIGURE B.27:** Model and simulation of a liquid level control system.

***Example*** Consider the room or furnace temperature control system (Fig. 1.7). We need to consider room temperature, outside temperature (cold), and a heater. The heater is controlled by a relay type controller with hysteresis. The room temperature is initially at the same temperature as the outside temperature. The controller is set to increase the room temperature to a higher level. The heater is controlled to regulate the room temperature. As the room temperature increases and becomes larger than the outside temperature, there is a heat loss from room to outside. The net added heat rate to the room is the difference between the heat generated by the heater and the heat loss to the outside because outside temperature is colder. The temperature raise in the room is a function of this difference and the size of the room. The heat loss is linear function of the inside and outside room temperatures.

$$(\textit{net heat rate added to room}) = (\textit{heat-in rate}) - (\textit{heat-out rate}) \tag{B.347}$$

$$Q_{net} = Q_{in} - Q_{out} \tag{B.348}$$

$$mc\frac{dT}{dt} = Q_{in} - \frac{1}{R}(T - T_o) \tag{B.349}$$

where $mc$ is the heat capacitance of the room which is function of the room size and $R$ is the resistance of the heat transfer from walls due to the temperature difference. The effective resistance ($R$) to heat transfer is a function of the type of dominant mode of heat transfer (conduction, convection, radiation) as well as the size and insulation type of walls. $T$ and $T_o$ are inside and outside temperatures, respectively.

Let us simulate the room temperature control system for the following conditions: $T_d = 72°\text{F}$, $T_o = 42°\text{F}$, $e_{max} = 0.5°\text{F}$, $Q_{max} = 100$ , $R = 100$, $mc = 1.0$. Initially the room is assumed to be at the same temperature as the outside temperature. After entering the room, 1 sec later, the temperature is commanded to be $T_d = 72°\text{F}$. The relay controller is active if the temperature difference is beyond 2% of the commanded temperature. Figure B.28 shows the Simulink implementation of the model and simulation results.

***Example*** Consider the web tension control system shown in Fig. 1.6. The wind-off roll is driven by another part of the machinery where the speed $v_1(t)$ is dictated by other considerations. The wind-up roll is driven by an electric motor. This motor is required to run in such a way that the tension in the web ($F$) is maintained constant and equal to a desired value ($F_d$). So, if the wind-off roll speeds up, the wind-up roll is supposed to speed up. Similarly, if the wind-off roll slows down, the wind-up roll is suppose to slow down quickly. The wind-off roll speed is given as an external input and not under our control. The wind-up speed is our controlled variable. Our objective is to minimize the tension error, $e(t) = F_d(t) - F(t)$.

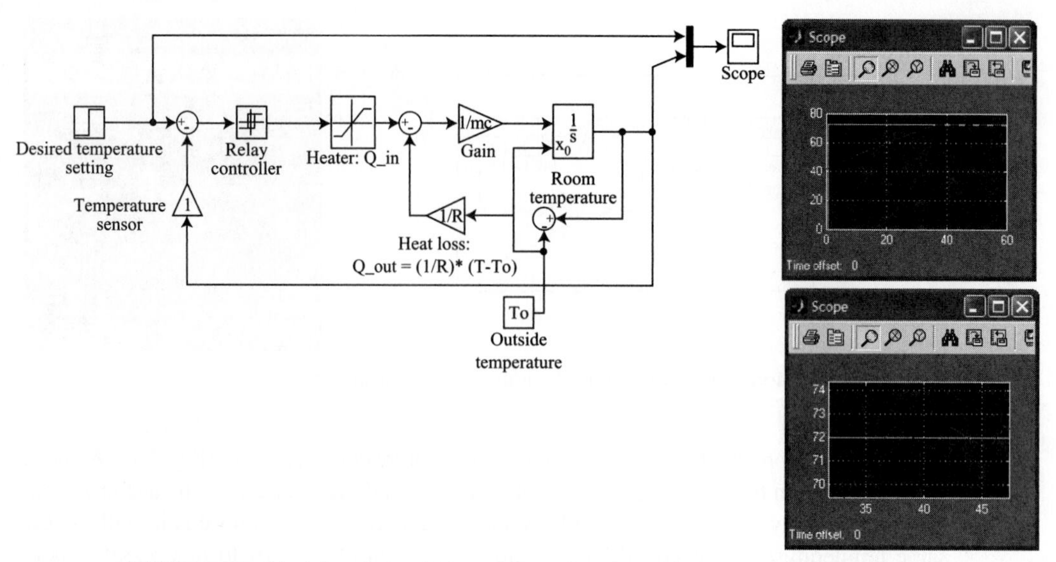


**FIGURE B.28:** Model and simulation of furnace or room temperature control system.

The tension in the web will be determined by the difference between the integral of $v_1(t)$ and $v_2(t)$,

$$y(t) = y(t_0) + \int_{t_o}^{t} (v_2(t) - v_1(t))\, dt \tag{B.350}$$

$$F(t) = F_o + k \cdot y(t) \tag{B.351}$$

If initially the web tension is adjusted so that when $y = y_0$, the tension $F = F_0 = 0$ by proper calibration, then we can express the tension as a function of change in $y(t)$,

$$\Delta y(t) = y(t) - y(t_0) \tag{B.352}$$

$$= \int_{t_o}^{t} (v_2(t) - v_1(t))\, dt \tag{B.353}$$

$$\Delta Y(s) = \frac{1}{s} \cdot (V_2(s) - V_1(s)) \tag{B.354}$$

$$F(t) = k \cdot \Delta y(t) \tag{B.355}$$

$$F(s) = \frac{k}{s} \cdot (V_2(s) - V_1(s)) \tag{B.356}$$

The control system that controls the $v_2(t)$ is a closed-loop control system and implemented using an analog controller [op-amp in the Fig. 5.31 (a)]. Let us consider the dynamics of the amplifier and motor as a first-order filter, that is, the transfer function between the commanded speed $w_{2,cmd}$ to actual speed $w_2$,

$$\frac{w_2(s)}{w_{2,cmd}(s)} = \frac{1}{\tau_m s + 1} \tag{B.357}$$

where $\tau_m$ is the first-order filter time constant for the amplifier and motor. The corresponding linear speeds are

$$v_{2,cmd}(t) = r_2 \cdot w_{2,cmd}(t) \tag{B.358}$$

$$v_2(t) = r_2 \cdot w_2(t) \tag{B.359}$$

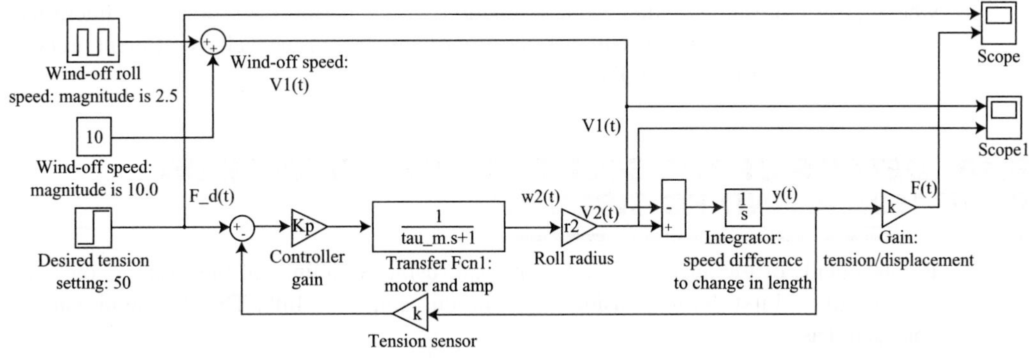


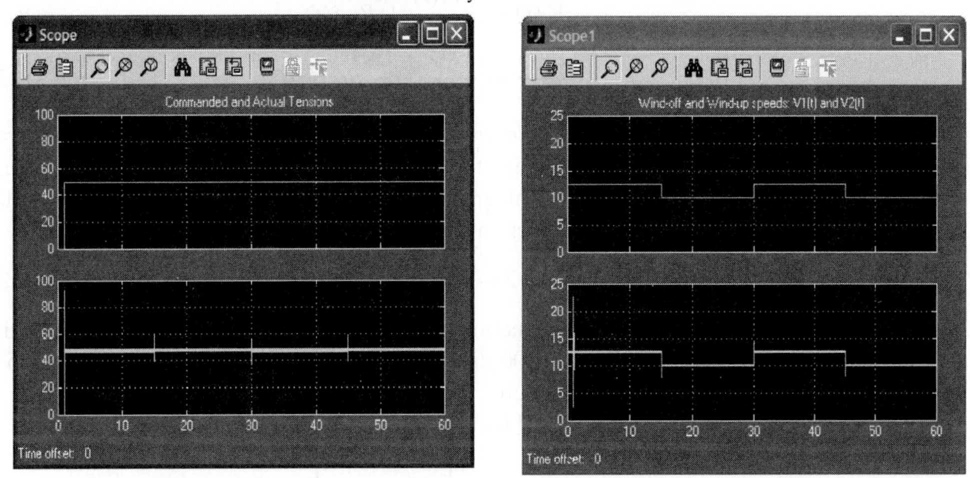


**FIGURE B.29:** Model and simulation of web tension control system. Top figure is the Simulink model of the tension control system. The left plot shows the commanded tension on the top and actual tension at the bottom. Right plot shows the wind-off and wind-up speeds.

Let us consider a proportional controller,

$$w_{2,cmd}(t) = K_p \cdot (F_d(t) - F(t)) \tag{B.360}$$

Figure B.29 shows the model and simulation conditions in Simulink. The parameters of the system used in the simulation are

$$k = 10000\,[\text{N/m}] \tag{B.361}$$

$$K_p = 10\,[\text{m/s/m}] \tag{B.362}$$

$$\tau_m = 0.01\,[\text{sec}] \tag{B.363}$$

$$r_2 = 0.5\,[\text{m}] \tag{B.364}$$

We will simulate a condition where $v_1(t)$ has a step change from its nominal value for a period of time,

$$v_1(t) = 10.0 + 2.5\,f_1(t) \tag{B.365}$$

$$F_d(t) = 50 \cdot step(t - 1.0); \quad \text{step function starts at 1.0 sec} \tag{B.366}$$

where $f_1(t)$ represents a square pulse function with a period of $T = 30$ sec. Interested reader can easily experiment with different control algorithms as well as different process parameters, i.e., different roll diameter values $r_2$.

## B.10 DETAILS OF THE SOLUTION FOR EXAMPLE ON PAGE 162: RL AND RC CIRCUITS

Below, details of the analytical and numerical (using Simulink) of the example on page 162 are given. First, let us consider the solution using Simulink. The RL circuit can be considered as

$$V_s(t) = L\frac{di(t)}{dt} + R \cdot i(t) \tag{B.367}$$

$$\frac{di(t)}{dt} = \frac{1}{L}(V_s(t) - R \cdot i(t)) \tag{B.368}$$

where we isolate the derivative of the current which will be presented as input to an integrator, whose output is the current. Then using the algebraic summation on the right-hand side, we define the quantity as input to the integrator. The whole case considered can be simulated by defining the $V_s(t)$ as a pulse function

$$V_s(t) = 24 \cdot (1(t - t_1) - 1(t - t_2)) \tag{B.369}$$

where $1(t)$ represents the unit step function, $1(t - t_1)$ represents the unit step function shifted in time by $t_1$. In this example, $t_1 = 100\ \mu\text{sec} = 0.0001$ sec and $t_2 = 500\ \mu\text{sec} = 0.0005$ sec. The initial current in the circuit is zero.

Similarly, the Simulink model for the RC circuit can be considered as

$$V_s(t) = R \cdot i(t) + \frac{1}{C}(Q_c(t_0) + \int_{t_0}^{t} i(\tau)d\tau) \tag{B.370}$$

$$i(t) = \frac{1}{R}(V_s(t) - \frac{1}{C}(Q_c(t_0) + \int_{t_0}^{t} i(\tau)d\tau)) \tag{B.371}$$

In this example, initial charge in the capacitor is zero, hence $Q_c(t_0) = 0$. The input to the integrator is the current $(i(t))$ and the output of the integrator is the charge (integral of the current). The Simulink models and the simulation results are shown in Figs. B.30 and B.31.

The analytical solution can be obtained by either treating each phase of switch (position A and position B) as (i) separate differential equations where the final state of one differential equation solution becomes the initial condition for the next differential equation solution, (ii) or use a single differential equation for each circuit and represent the switch state change by an equivalent change in the input voltage function. We will follow the first approach. The second approach is easier to solve using Laplace transforms and is left as an exercise to the reader.

The RL circuit for time period $t = 0$ to $t = t_1 = 100\ \mu$sec, the current and voltages are all zero since initial conditions and input are zero. For time period $t = t_1 = 100\ \mu$sec to $t = t_2 = 500\ \mu$sec, the voltage–current relationship is

$$V_s(t) = L\frac{di(t)}{dt} + R \cdot i(t) \tag{B.372}$$

Let us solve the differential equation for time period 0 to $t$ then, shift the solution by $t_1$ (replace $t$ by $(t - t_1)$ in the solution). Taking the Laplace transform, and noting that the

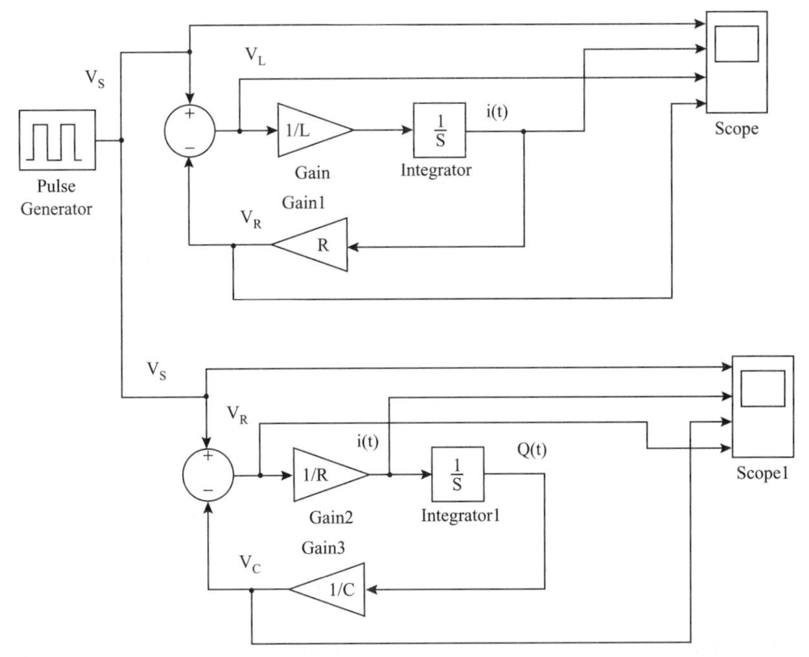


**FIGURE B.30:** Simulink model of an RL and RC circuit shown in Fig. 5.5.

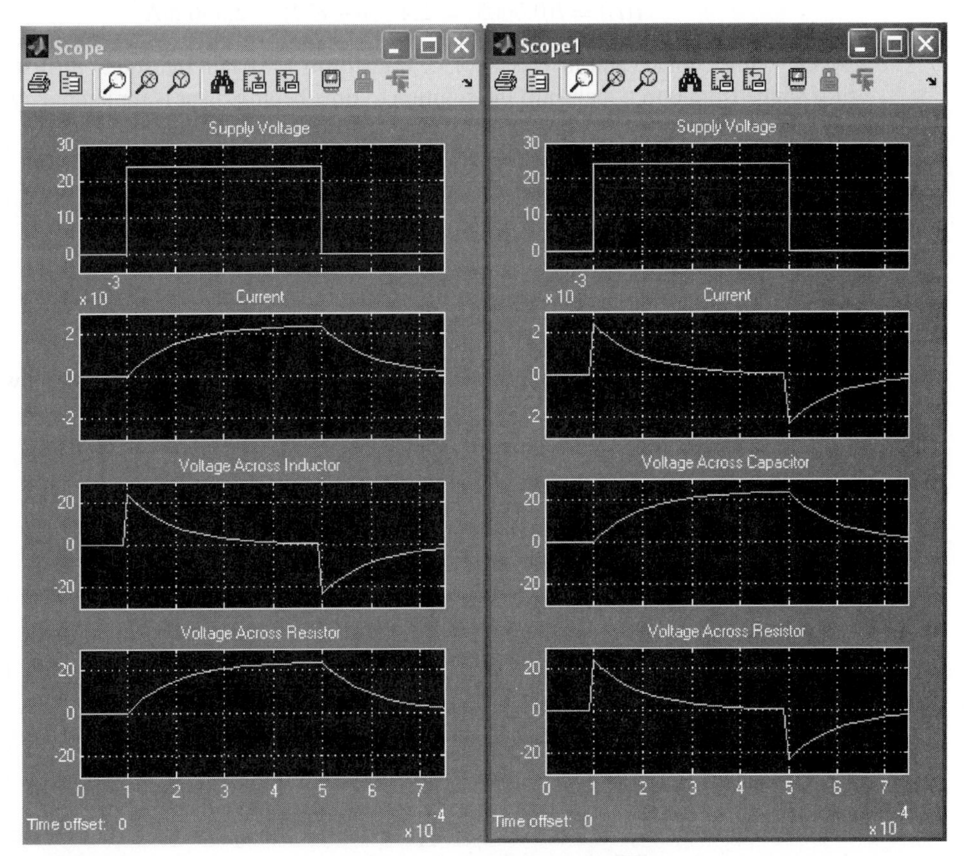


**FIGURE B.31:** Simulation results of the model shown in Fig. B.30.

input voltage is a step change,

$$(Ls + R)i(s) = V_s(s) \quad (B.373)$$

$$= \frac{V_0}{s} \quad (B.374)$$

$$i(s) = \frac{1/R}{(L/R \cdot s + 1)} \frac{V_0}{s} \quad (B.375)$$

By taking inverse Laplace transform to obtain and shift the solution by replacing $t$ with $t - t_1$, as we planned,

$$\bar{i}(t) = \frac{V_0}{R}(1 - e^{-t/(L/R)}) \cdot 1(t) \quad (B.376)$$

$$i(t) = \frac{V_0}{R}(1 - e^{-(t-t_1)/(L/R)}) \cdot 1(t - t_1); \quad \text{for } t_1 \leq t \leq t_2 \quad (B.377)$$

$$= 2.4 \cdot (1 - e^{-(t-0.0001)/0.0001}) \text{ mA}; \quad \text{for } t_1 \leq t \leq t_2 \quad (B.378)$$

For time period $t_2$ to $t_f$, the voltage–current relationship is governed by

$$0 = L\frac{di(t)}{dt} + Ri(t) \quad (B.379)$$

where the initial condition on current at $t = t_2$ is obtained from the solution of the previous period,

$$i(t_2) = i(0.0005) = 2.4 \cdot (1 - e^{-4}) = 2.356 \text{ mA} \quad (B.380)$$

Again, using Laplace and inverse Laplace transforms to obtain the solution (first assume time frame from 0 to $t$, then shift the solution to time beginning at $t_2$ for simplicity),

$$0 = L(si(s) - i(o)) + Ri(s) \quad (B.381)$$

$$i(s) = \frac{L}{Ls + R} i(0) \quad (B.382)$$

$$\bar{i}(t) = i(0) \cdot e^{-t/(L/R)} \quad (B.383)$$

$$i(t) = i(t_2) \cdot e^{-(t-t_2)/(L/R)} \cdot 1(t - t_2) \quad (B.384)$$

$$= 2.356 \cdot e^{-(t-0.0005)/(0.0001)} \cdot 1(t - t_2) \text{ mA}; \quad \text{for } t_2 \leq t \leq t_f \quad (B.385)$$

The voltage across resistor and inductor can be found in each time period as follows

$$V_R(t) = R \cdot i(t) \quad (B.386)$$

$$V_L(t) = L \cdot \frac{di(t)}{dt} \quad (B.387)$$

It is easy to show that

$$V_R(t) = 0.0; \quad \text{for } 0 \leq t \leq 0.0001 \text{ sec} \quad (B.388)$$

$$V_R(t) = 24 \cdot (1 - e^{-(t-0.0001)/0.0001}) \; V; \quad \text{for } 0.0001 \leq t \leq 0.0005 \text{ sec} \quad (B.389)$$

$$V_R(t) = 23.56 \cdot e^{-(t-0.0005)/(0.0001)} \; V; \quad \text{for } 0.0005 \leq t \leq 0.001 \text{ sec} \quad (B.390)$$

$$V_L(t) = 0.0; \quad \text{for } 0 \leq t \leq 0.0001 \text{ sec} \quad (B.391)$$

$$V_L(t) = 24.0 \cdot e^{-(t-0.0001)/0.0001} \; V; \quad \text{for } 0.0001 \leq t \leq 0.0005 \text{ sec} \quad (B.392)$$

$$V_L(t) = -23.56 \cdot e^{-(t-0.0005)/(0.0001)} \; V; \quad \text{for } 0.0005 \leq t \leq 0.001 \text{ sec} \quad (B.393)$$

For the RC circuit, we follow the same approach. The time period before switch connects the power supply to RC circuit, all voltages and current are zero since the initial condition (the current and initial charge in the capacitor) and input are zero. For time period from

$t = t_1 = 0.0001$ sec $= 100$ $\mu$sec to $t = t_2 = 0.0005$ sec $= 500$ $\mu$sec the voltage–current relationship is

$$V_s(t) = Ri(t) + V_c(t) \quad \text{(B.394)}$$

$$= Ri(t) + \frac{1}{C}(Q(t_1) + \int_{t_1}^{t} i(\tau)d\tau \quad \text{(B.395)}$$

Noting that $Q(t_1)$, the initial charge at the capacitor is zero, and taking Laplace transforms,

$$V_s(t) = Ri(t) + \frac{1}{C}\left(\int_{t_1}^{t} i(\tau)d\tau\right) \quad \text{(B.396)}$$

$$V_s(s) = Ri(s) + \frac{1}{Cs}i(s) \quad \text{(B.397)}$$

$$i(s) = \frac{Cs}{RCs+1} \cdot V_s(s) \quad \text{(B.398)}$$

$$= \frac{Cs}{RCs+1} \cdot \frac{V_0}{s} \quad \text{(B.399)}$$

$$\bar{i}(t) = \frac{V_0}{R} \cdot e^{-t/(RC)} \quad \text{(B.400)}$$

$$i(t) = \frac{V_0}{R} \cdot e^{-(t-t_1)/(RC)} \cdot 1(t - t_1) \quad \text{(B.401)}$$

$$= 2.4 \cdot e^{-(t-0.0001)/(0.0001)} \text{ mA}; \quad \text{for } 0.0001 \text{ sec} \le t \le 0.0005 \text{ sec} \quad \text{(B.402)}$$

The charge across the capacitor can be determined from

$$V_C(t) = V_s(t) + R \cdot i(t) \quad \text{(B.403)}$$

$$V_C(t_2) = V_s(t_2)r \cdot i(t_2) \quad \text{(B.404)}$$

$$= 23.56 \text{ V} \quad \text{(B.405)}$$

The $V_C(t_2)$ is the voltage across the capacitor due to accumulated charge at the end of this period. This will serve as an initial condition for the solution of the differential equations for the following phase, $t_2$ to $t_f$. The voltage–current relationship for this period is

$$0 = Ri(t) + \frac{1}{C}(Q_c(t_2)) + \int_{t_2}^{t} i(\tau)\, d\tau \quad \text{(B.406)}$$

$$= Ri(t) + \frac{1}{C}(Q_c(t_2)) + \frac{1}{C}\int_{t_2}^{t} i(\tau)\, d\tau \quad \text{(B.407)}$$

$$= Ri(t) + V_c(t_2) + \frac{1}{C}\int_{t_2}^{t} i(\tau)\, d\tau \quad \text{(B.408)}$$

Taking Laplace and inverse Laplace transforms to obtain the solution for current, it can be shown that

$$0 = R \cdot i(s) + \frac{V_c(t_2)}{s} + \frac{1}{Cs}i(s) \quad \text{(B.409)}$$

$$i(s) = -\frac{Cs}{RCs+1}\frac{V_c(t_2)}{s} \quad \text{(B.410)}$$

$$= -\frac{V_c(t_2)}{R}\frac{RC}{RCs+1} \quad \text{(B.411)}$$

$$\bar{i}(t) = -\frac{V_c(t_2)}{R}e^{-t/(RC)} \quad \text{(B.412)}$$

$$i(t) = -\frac{V_c(t_2)}{R}e^{-(t-t_2)/(RC)} \text{ A}; \quad \text{for } t_2 \le t \le t_f \quad \text{(B.413)}$$

$$= -2.356 \cdot e^{-(t-0.0005)/0.0001} \text{ mA} \quad \text{(B.414)}$$

The voltage across the resistor and capacitor can be found similarly,

$$V_R(t) = R \cdot i(t) \tag{B.415}$$

$$V_C(t) = V_s(t) - R \cdot i(t); \quad \text{for } t_1 \le t \le t_2 \tag{B.416}$$

$$= 24 \cdot (1 - e^{-(t-0.0001)/0.0001})\ V \tag{B.417}$$

$$V_C(t) = V_c(t_2) - R \cdot i(t); \quad \text{for } t_2 \le t \le t_f \tag{B.418}$$

$$= 23.24 - 23.24 \cdot (1 - e^{-(t-0.0001)/0.0001}) \tag{B.419}$$

$$= 23.24 \cdot e^{-(t-0.0001)/0.0001}\ V \tag{B.420}$$

If we plot the results of the current and voltages across each component that we obtained from the analytical solution, they should agree with the results obtained using numerical method (i.e., using Simulink).

## B.11 PROBLEMS

**1.** Consider the following complex number. Determine its magnitude and phase angle in complex plane and express it in magnitude times the exponential phase angle:

$$s = 2 + j\,4 \tag{B.421}$$

**2.** Consider the function shown in Fig. B.10. Let $a = 2.0$ sec, $b = 12.0$ sec.

**(a)** Determine the Laplace transform of the function.
**(b)** Determine the Fourier transform of the signal.

**3.** Consider that a first-order linear dynamic system is excited with the signal given in problem 2. Calculate the time domain response of the system using Laplace and inverse Laplace transform methods. The transfer function of the first-order system is

$$G(s) = \frac{10}{0.5\,s + 1} \tag{B.422}$$

**4.** Consider a nonlinear dynamic system. Write a Matlab program that linearizes the dynamic model about any given nominal operating condition ($\dot{y}(t_0)$, $y(t_0)$, $u(t_0)$) and obtains the result in first-order set of linear differential equations. Do the same analytically. The nonlinear dynamic system model is

$$\ddot{y}(t) = (\dot{y}(t))^3 + sin(y(t)) + u(t)^2 \tag{B.423}$$

Compare your Matlab results with your analytically obtained linearization results.

**5.** Consider the dynamics of a pendulum. Assume that the torque input is decided by an analog controller using the following relationship:

$$u(t) = 10.0 \cdot (\theta_d(t) - \theta(t)) \tag{B.424}$$

and the dynamics of the pendulum is given by

$$m\,l^2\,\ddot{\theta}(t) + m\,g\,l \cdot sin(\theta(t)) = u(t) \tag{B.425}$$

where $m = 1.0$ kg, $l = 1.0$ m, $g = 10$ m/s$^2$. Let the $\theta_d(t) = \pi/6$ rad. Simulate the response of the pendulum position under such a controller using Matlab or Simulink. Notice that the controller is an analog controller.

**6.** Repeat the problem 5 assignment for a digital controller. The same control algorithm is applied.

$$u(kT) = 10.0 \cdot (\theta_d(kT) - \theta(kT)) \tag{B.426}$$

where $T$ is the sampling period of the controller and $k$ is the sampling period number ($k = 0, 1, 2, \ldots$). Simulate the digital controller for two different sampling rates: (1) $T = 0.001$ sec (1 msec), and (2) $T = 1.0$ sec.

**7.** Consider the liquid level control system example given earlier in the chapter. Simulate the same system for two different digital controllers, each with 1.0-msec sampling period: (1) $u(kT) = 1.0 \cdot e(kT)$, and (2) $u(kT) = 1.0 \cdot e(kT) + 0.5 \cdot \dot{e}(kT)$.

**8.** Consider the temperature control system example given earlier in the chapter. Experiment with the effect of changing the relay hysteresis band on the control system. What are the effects of making the hysteresis band larger and smaller?

**9.** Consider the web tension control system shown in the example earlier in the chapter. Simulate the same conditions under two different radius values of the rewind roll: (1) $r_2 = 0.5$ m and (2) $r_2 = 5.0$ m. In particular, focus on the transient response around the sudden change in the speed of the unwind roll and magnify the response plot in that region in order to see the effect of the large changes in roll radius. What can be done in order to maintain the same dynamic response under the changing roll diameter conditions? Consider the dynamic bandwidth of the tension sensor. Assume that the sensor can be represented by a first-order filter. What is the effect of sensor dynamics on the overall system response if the time constant is $\tau_s = 0.001$ sec and $\tau_s = 0.1$ sec?

# BIBLIOGRAPHY


[1] Mori, T., "Mechatronics," Yasakawa Internal Trademark Application Memo, 21.121.01, July 12, 1969.

[2] Harashima, F., Tomizuka, M., and Fukuda, T., "Mechatronics—What Is It, Why and How?" *IEEE/ASME Trans. on Mechatronics*, Vol. 1, No. 1, 1996, pp. 1–4.

[2a] Grimheden, M. and Hansen, M., "Mechatronics—the evolution of an academic discipline in engineering education," *Int. Journal of Mechatronics*, Vol. 15 (2005), pp. 179–192.

[3] Brady, R. N., *Modern Diesel Technology,* Prentice Hall, Inc., Englewood Cliffs, NJ, 1996.

[4] Heywood, J. B., *Internal Combustion Engine Fundamentals*, McGraw Hill, New York, 1988.

[5] Heisler, H., *Advanced Engine Technology*, Society of Automotive Engineers, 1995, ISBN: 1-56091-734-2.

[6] Shigley, J. E., Mischke, C. R., and Budynas, R. G., *Mechanical Engineering Design*, 7th ed., McGraw-Hill, New York, 2004.

[7] Norton, R. L., *Design of Machinery*, 2nd ed., McGraw-Hill, New York, 1999, ISBN: 0-07-048395-7.

[8] Klafter, R. D., Chmielewski, T. A., and Negin, M., *Robotic Engineering: An Integrated Approach*, Prentice Hall, Englewood Cliffs, NJ, 1989, ISBN: 0-13-468752-3.

[9] Spong, M. W. and Vidyasagar, M., *Robot Dynamics and Control*, John Wiley & Sons, Hoboken, NJ, January 1989, ISBN:047161243X.

[10] Sciavicco, L. and Siciliano, B., *Modelling and Control of Robot Manipulators*, 2nd ed., Springer Verlag, New York, 2000, ISBN: 1852332212.

[11] Craig, J. J., *Introduction to Robotics: Mechanics and Control*, 3rd ed., Addison Wesley, 2004, ISBN:0-201-09528-9.

[12] Hartenberg, R. S. and Denavit, J., *Kinematic Synthesis of Linkages*, McGraw-Hill, New York, 1964.

[13] Paul, R., *Robot Manipulators*, The MIT Press, Cambridge, MA, 1981.

[14] Orin, D. E. and Schrader, W. W., *Efficient Computation of the Jacobian for Robot Manipulators*, International Journal of Robotics Research, Vol. 3, No. 4, 1984, pp. 66–75.

[15] Franklin, G. F., Powell, J. D., and Emami-Naeini, A., *Feedback Control of Dynamic Systems*, Addison Wesley, 2003.

[16] Franklin, G. F., Powell, J. D., and Workman, M., *Digital Control of Dynamic Systems*, Addison Wesley, 1998.

[17] Dorf, R. C. and Bishop, R. H., *Modern Control Systems*, Pearson-Prentice Hall, Englewood Cliffs, NJ, 2001.

[18] Ogata, K., *Modern Control Engineering*, Prentice Hall, Englewood Cliffs, NJ, 1990.

[19] Churchill, R. V., *Operational Mathematics*, 3rd ed., McGraw-Hill, New York, 1972.

[20] Figliola, R. S. and Beasley, D.E., *Theory and Design for Mechanical Measurements*, John Wiley & Sons, Hoboken, Ny, 1995.

[21] Cogdell, J. R., *Foundations of Electrical Engineering*, Prentice Hall, Englewood Cliffs, NJ, 1990.

[22] Fortney, L. R., *Principles of Electronics*, Harcourt, Brace, Jovanovich Publishers, 1987.

[23] Horowitz, P. and Hill, W., *The Art of Electronics*, 2nd ed., Cambridge Press, 1989.

[24] Jung, W. G., *IC Op-Amp Cookbook*, 3rd ed., Prentice Hall, Englewood Cliffs NJ, ISBN:0-13-889601-1.

[25] Bishop, R., *Basic Microprocessor and the 6800*, Hayden Book Co., 1979, ISBN:0-8104-0758-2.

[26] Valvano, J. W., *Embedded Microcomputer Systems:* Real Time Interfacing, Brooks/Cole, USA, 2000.

[27] Ford, W. and Topp, W., *MC 68000: Assembly Language and Systems Programming*, D. C. Heath and Company, Lexington, MA, 1987.

[28] Brey, B. B., *The Intel Microprocessors*, 6th ed., Prentice Hall, Englewood Cliffs, NJ, 2003.

[29] Iovine, J., *PIC Microcontroller Project Book*, McGraw-Hill, New York, 2000.

[30] Merritt, H. E., *Hydraulic Control Systems*, John Wiley & Sons, Hoboken, NJ, 1967.

[30a] Manring, N., *Hydraulic Control Systems*, John Wiley & Sons, 2005.

[31] *Graphical Symbols for Fluid Power Diagrams*, American National Standard, ANSI Standard y32. 10-1967, 1967.

[32] Pippenger, J. J. and Hicks, T. G., *Industrial Hydraulics*, McGraw-Hill, New York, 1970.

[33] *Design Engineers Handbook*, Vol. 1. Hydraulics, Motion Control Technology Series, Parker Hannifin Corp.

[34] Vickers Inc., Subsidiary of Eaton Corp., *Industrial Hydraulics Manual*, 1999. www.eatonhydraulics.com.

[35] *Making the Choice: Selecting and Applying Piston, Bladder and Diaphragm Accumulators*, Brochure 1660-USA, www.parker.com/accumulator.

[36] National Fluid Power Association, www.nfpa.com.

[37] International Standards Organization, www.iso.org.

[38] Deutsche Institute fur Normung, www2.din.org.

[39] Society of Automotive Engineers, www.sae.org.

[40] American Society of Mechanical Engineers, www.asme.org.

[41] Clark, D. C., "Selection and Performance Criteria for Electrohydraulic Servodrives," *Technical Bulletin 122*, Moog Inc., East Aurora, NY, www.moog.com.

[41a] Gamble, J. B. and Vaughan, N. D., "Comparison of Sliding Mode Control with State Feedback and PID Control Applied to a Proportional Solenoid Valve," *Trans. of the ASME, Journal of Dynamic Systems. Measurement and Control*, Vol. 118, September 1996, pp. 434–438.

[42] Neal, T. P., "Performance Estimation for Electrohydraulic Control Systems," *Technical Bulletin 126*, Moog Inc., East Aurora, NY, www.moog.com.

[42a] Geyer, L. H., "Controlled Damping Through Dynamic Pressure Feedback", Moog Technical Bulletin 101, June 1958, revised April 1972.

[42b] Clark, D. C., Selection and Performance Criteria for Electrohydraulic Servodrives, Technical Bulletin 122, Moog Inc.

[43] Vaughan, N. D. and Gamble, J. B., "*The Modelling and Simulation of a Proportional Solenoid Valve*," ASME Winter Annual Meeting, Nov. 25–30, 1990, 90-WA/FPST-11.

[43a] Lantto, B., Palmberg, J., Krus, P., Static and Dynamic Performance of Mobile Load-Sensing Systems with Two Different Types of Pressure-Compensated Valves, SAE paper #901552.

[44] Kenjo, T. and Nagamori, S., *Permanent-Magnet and Brushless DC Motors*, Oxford Science Publications, 1985.

[45] Kenjo, T., *Electric Motors and Their Controls*, Oxford Science Publications, 1991.

[45a] Haggag, S., Astrom, D., Egelja, A., Cetinkunt, S., "Modeling, Control and Validation of a Electro-hydraulic Steer-By-Wire System for Articulated Vehicle Applications", ASME/IEEE Trans. on Mechatronics, Vol. 10, No. 6, Dec. 2005, pp. 688–692.

[46] Kenjo, T. and Sugawara, A., *Stepping Motor and Their Microprocessor Controls*, Oxford Science Publications, 1994.

[47] Novotny, D. W., and Lipo, T. A., *Vector Control and Dynamics of AC Drives*, Clarendon Press, 2000.

[48] Hanselman, D., *Brushless Permanent Magnet Motor Design,* The Writers' Collective, 2003.

[49] DC Motors, Speed Controls, Servo Systems, Electro-Craft Corp, 1980.

[50] Miller, T. J. E., *Switched Reluctance Motors and Their Control*, Magna Physics Publishing and Clarendon Press, Oxford Science Publications, 1993.

[50a] Wylie, C. R. and Barrett, L. C., *Advanced Engineering Mathematics*, McGraw-Hill, New York, 1995.

[51] Ramshaw, R. and van Heeswijk, R. G., *Energy Conversion: Electric Motors and Generators*, Saunders College Publishing, 1990.

[52] Wilson, C. S., "Universal Commutation Algorithm Adapts Motion Controller for Multiple Motors," *Proceedings of PCIM 1989*, Intertec Communications, 1989, pp. 348–360.

[53] Uhlir, P. and Kubiczek, Z., "3-Phase AC Motor Control with V/Hz Speed Open Loop Using DSP56F80X," Motorola, Semiconductor Application Note, AN1911/D, 2001.

[53a] Cobo, M., Ingram, R., Reiners, E. A., Wiele, M. F. V., Positive Flow Control System, US Patent #5,873,244, Feb. 23, 1999.

[53b] Aardema, J., Koehler, D. W., System and Method for Controlling Independent Metering Valve, US Patent #5,947,140, Sep. 7, 1999.

[53c] Cetinkunt, S., Chen, C., Egelja, A., Muller, T., Ingram, R., Pinsopon, U., Method and System for Selecting Desired Response of an Electronic-Controlled Sub-System, US Patent #6,330,502, Dec. 11, 2001.

## Suppliers of Mechatronic Systems and Components

[54] Thompson Industries, Inc., 2 Channel Drive, Port Washington, NY 11050; Phone: 800-544-8466; www.thompsonindustries.com.
[55] Peerless-Winsmith Inc., 172 Eaton Street, PO Box 530, Springville, NY 14114; Phone: 716-592-9311; http://www.winsmith.com.
[56] Precision Industrial Components Corp., 86 Benson Road, PO Box 1004, Middlebury, CT 06762–1004; Phone: 800-243-6125; Fax: 203-758-8271; http://www.pic-design.com.
[57] MTS, Temposonic Sensor, 3001 Sheldon Dr., Cary, NC 27513; Phone: 919-677-0100; http://www.mtssensors.com.
[58] Omega, Sensors and Measurement Systems, http://www.omega.com.
[59] Encoders and various sensors, Dynapar Corp., http://www.dynapar-encoders.com.
[60] Piezoelectric accelerometers, PCB Piezotronics Inc., http://www.pcb.com.
[61] Sauer-Danfoss Co., 2800 E. 13th Street, Ames, IA 50010; Phone: 515-239-6000; Fax: 515-239-6618; www.sauer-danfoss.com.
[62] Bosch-Rexroth Corp., PO Box 25407, Lehigh Valley, PA 18002-5407; Phone: 610-694-8246; Fax: 610-694-8266; www.boschrexroth.com.
[63] Parker Hannfin Corp., Hydraulic Valve Division, 520 Ternes Ave., Elyria, OH 44035; Phone: 440-366-5200; Fax: 440-366-5253; www.parker.com/hydraulicvalve.
[64] Sun Hydraulics, 1500 West University Parkway, Sarasota, FL 34243; Phone: 941-362-1200; Fax: 941-355-4497; www.sunhydraulics.com.
[65] HYDAC Technology Corporation, HYDRAULIC Division, 445 Windy Point Drive, Glendale Heights, IL 60139; Phone: 630-545-0800; Fax: 630-545-0033; www.hydac.com.
[66] Moog Controls Inc., Industrial Division, East Aurora, NY 14052; Phone: 716-655-3000; Fax: 716-655-1803; http://www.moog.com.
[67] Hydraforce, Inc., 500 Barclay Blvd., Lincolnshire, IL 60069; Phone: 847-793-2300; Fax: 847-793-0086, http://www.hydraforce.com.
[68] The Oilgear Company, PO Box 343924, Milwaukee, WI 53234-3924; Phone: 414-327-1700; www.oilgear.com.
[69] Rockwell Automation Corporate Headquarters, Allen-Bradley US Bank Center, 777 East Wisconsin Avenue, Suite 1400, Milwaukee, WI 53202; Phone: 414-212-5200; http://www.rockwellautomation.com/, http://www.ab.com.
[70] Omron Electronics LLC, One East Commerce Drive, Schaumburg, IL 60173; Phone: 847-843-7900; Fax: 847-843-8081; http://oeiweb.omron.com/oei/.
[71] Mitsubishi Electric Automation, Inc., 500 Corporate Woods Parkway, Vernon Hills, IL 60061; Phone: 847-478-2100; Email: marcomm@meau.mea.com; http://www.mitsubishielectric.com.
[72] Siemens Energy and Automation, 1901 N. Roselle Rd., Suite 220, Schaumburg, IL 60195; Phone: 847-310-5900; Fax: 847-310-6570; http://www.sea.siemens.com.
[73] ABB Control Inc., 1206 Hatton Road, Wichita Falls, TX 76302; Phone: 940-397-7000; Fax: 940-397-7085; http://www.abb.com.
[74] Magnet Schultz Solenoids, Westmont, IL; www.magnetschultz.com.
[75] PC Based Data Acquisition Products, National Instruments, Austin, TX; www.ni.com.
[76] Analog Devices, Two Technology Way, PO Box 280, Norwood, MA 02062; www.analogdevices.com.
[77] Fairchild Semiconductor Corp., 313 Fairchild Drive, Mountain View, CA 94003; www.fairchild.com.
[78] Motorola Semiconductor Products, PO Box 20912, Phoenix, AZ 85036; www.motorola.com.
[79] National Semiconductor Products, Inc., 2900 Semiconductor Drive, PO Box 58090, Santa Clara CA 95052.
[80] Texas Instruments, PO Box 3640, Dallas, TX 75285; www.ti.com.

[81] Digi-Key, 701 Brooks Ave., PO Box 677, Tief River Falls, MN 56701-0677; Phone: 800-344-4539; Fax: 218-681-3380; http://www.digikey.com.

[82] Newark Electronics, 4801 N. Ravenswood, Chicago, IL 60640; Phone: 773-784-5100; Fax: 773-907-5339; http://www.newark.com.

[82a] Penny and Giles, Controls, 1 Airfield Rd., Christchurch, Dorset, RH23 3TH, UK, www.pennyandgiles.com.

[82b] ITT Industries, Engineered Valves, 1100 Bankhead Ave., Amory, MS 38821, USA, www.engvalves.com

## Suppliers of Industrial Robots

[83] Adept Technologies, Inc., 3011 Triad Drive, Livermore, CA 94551; Phone: 925-245-3400; Fax: 925-960-0452; www.adept.com.

[84] Fanuc Robotics (formerly GMF Robotics), 3900 W. Hamlin Road, Rochester Hills, MI 48309-3253; Phone: 800-47-ROBOT (or 1-800-477-6268); fanucrobotics.com.

[85] Motoman Inc., a Division of Yaskawa Electric Company, 805 Liberty Lane, West Carrollton, OH 45449; Phone: 937-847-6200; Fax: 937-847-6277; motoman.com.

[86] ABB, Inc. (Asea Robots), US Head Office, 501 Merritt 7, Norwalk, CT 06851; Phone: 203-750-2200; Fax: 203-750-2263; abb.com/robotics.

[87] EPSON Factory Automation/Robotics (formerly Seiko), 18300 Central Avenue, Carson, CA 90746; Phone: 562-290-5910; Fax: 562-290-5999; robots.epson.com.

[88] Mitsubishi Electric Automation Inc., a Mitsubishi Company, 500 Corporate Woods Parkway, Vernon Hills, IL 60061; Phone: 847-478-2100; Fax: 847-478-2396; www.meau.com.

[89] Kawasaki Robotics (USA), Inc., 28059 Center Oaks Court, Wixom, MI 48393; Phone: 248-305-7610; Fax: 248-305-7618; kawasakirobots.com.

[90] Volvo L120E wheel loader, Service Manual, 2004 Volvo Construction Equipment Co.

[91] Technical Manual 644H and 644HMH Loader, John Deere and Co., www.johndeere.com, 2004.

[92] Galeo WA 450-5L, WA 480-5L Wheel Loader Shop Manual, Komatsu America Int. Co., www.KomatsuAmerica.com, 2004.

[93] 950G Wheel Loader Technical Manual, 2004, Caterpillar Inc., www.cat.com.

# INDEX


**A**
A/D converter, 39, 499
Absolute encoders, 232
AC induction motor, 448
AC induction motors, 447
AC motors, 394
Acceleration Sensors, 248
Accumulator, 285
Accuracy, 217
Active Linear Region, 174
actuator, 393
Actuators, 3
ADC, 366
Address bus, 207
Admittance, 160
Aliasing, 36
Alnico, 416
Ampere's Law, 402
Amplifier, 163
Analog controller, 29
analog controller, 74
arithmetic logic unit, 130
assembler, 131

**B**
Back EMF, 407
Band-pass filter, 199
Band-reject filter, 199
Bandwidth, 358, 368
basic instruction set, 131
Beat Phenomenon, 38
Bernoulli's Equation, 294
BIBO Stable, 59, 60
bilinear transformation, 77
Biot-Savart law, 402
Bipolar drive, 468
BJT, 173
Borward difference approximation, 75
Bruhless DC motors, 408
Brush-type DC motor, 408
Brush-type DC motors, 430
Brushless, 394
Brushless DC motors, 430
Bulk Modulus, 297, 371
bus, 132
By-pass diode, 172

**C**
Cam synchronization, 531
CAN, 495
Capacitance, 155, 297
Capacitance Based Pressure Sensor, 262
Capacitive Gap Sensors, 238
capacitor, 154
carbon types, 154
Cartridge Valves, 350
Cavitation, 281, 310
Central processing unit, 30
Ceramic, 416
Charge Pump, 310
Check Valve, 298
clock, 130
Closed Circuit, 290
Closed Loop, 371, 377
Closed loop control, 29, 46, 56, 79
Closed Loop Control, 343
closed loop control system, 2
Closed-center, 320, 352, 381
Co-energy, 410
Comb function, 34
Commutation algorithm, 441
Compensator, 313
complementary metal oxide silicon, 201
Component Sizing, 360
Compressibility, 373
Computer numeric control, 532
conductance, 161
conductive part, 161
Contactor, 503
Contouring coordination, 531
Control bus, 207
Coriolis Flow Meters, 272
Coulomb constant, 400
Counterbalance Valve, 330
CPU, 130
Current, 153
Cut-off Region, 173
Cylinder, 300, 320

**D**
D/A converter, 39, 499
DAC, 366
Data bus, 207

DC motors, 394
DC servo motors, 430
Dead Head, 314
Decoders, 202
Deflection Method, 223
Derivative control, 64, 68
device, 1
DeviceNet, 495
Diamagnetic, 412
Dielectric constant, 155
Diesel Engine, 18
Diesel Engine Components, 14
Differential Pressure Flow Rate Sensors, 269
differential-ended signal, 185
Digital controller, 29, 30, 44, 76
digital controller, 74
Diode, 168
Direct Drive Valve, 351
Directional Flow Control, 335
Displacement Based Pressure Sensors, 260
Displacement Pistons, 313
Disturbances, 46
Doped silicon, 167
Drag Flow Meters, 268
Drive, 393
Drums, 508

E
EEPROM, 499
Efficiency, 299
EH, 381
Electric fields, 399
Electric flux, 400
Electric potential, 400
Electro-hydraulic, 294
Electromagnetic field, 401
electronic cam, 531
Electronic gearing, 528
Emission Issues, 26
Encoder, 434
Encoders, 232
Energy Efficient, 354
Engine Modelling, 22
EPROM, 499

F
feedback control, 29, 47, 48, 49, 64
Ferromagnetic, 412
Filter, 287
Filtering, 44
Filters, 44
filters, 52, 56, 76
Finite Difference Approximations, 76
Fixed Displacement, 289
Fixed Displacement Pump, 302
Fixed Ratio Gearing;, 529
Flapper, 351
Flip-flop circuits, 204
Float Position, 330
Flow Compensated, 315
Flow Control Valve, 324
flow gain, 353
Flow Rate Sensors, 267
Flowchart language, 505
Flux linkage, 406
Force and Torque Sensors, 256
Forward difference approximation, 75
Four-quadrant, 396
Free-wheeling diodes, 172
Fuzzy logic, 500

G
G-code, 532
gain margin, 50
Gearing with Registration, 529
Generator, 395
Generator action, 436
Germanium, 166

H
H-bridge amplifier, 438
Hall Effect Sensors, 237
Hall effect sensors, 434
Hard ferromagnetic, 413
Hard magnetic materials, 413
Hard-wired relay logic panel, 495
Hidden oscillations, 38
High pass filter, 198
Hole, 167
HOME, 522
Hot Wire Anemometer, 271
Humidity Sensors, 272
Hybrid-stepper motors, 462
Hydraulic Motor, 320
hydraulic power supply unit, 281

I
I/O devices, 130
I/O interface units, 495
IGBT, 177
Impedance, 160
IMV, 382
incremental encoder, 232

Independent Metered, 383
INDEX, 523
Indirect field oriented control, 460
Inductance, 298
inductive sensor, 244
Inductor, 156
Inertial Accelerometers, 249
Instrumentation op-amp, 199
Integral Control, 69
Internal combustion engine, 13
interrrupt service routine, 146
interrupt latency time, 146
Interrupt service routines, 508
Iron core armature, 431

**J**

JOG, 523

**K**

Kinectic Energy, 295

**L**

Ladder logic diagram, 508
Ladder logic diagrams, 498, 505
latching current, 170
Latching relay, 504
Leakage, 320
Linear electric motors, 478
Linear Models, 373
load cells, 256
Load Dynamics, 359
Load Sensing, 316
Loading errors, 165, 220
Low pass filter, 197
LVDT, 227, 348

**M**

machine instructions, 131
Magnetic fields, 399, 401
Magnetostriction Position Sensors, 239
Manifold, 351
Margin, 315
Mechanical Flow Rate Sensors, 267
Mechatronics, 1
Metering, 317, 350
Micro stepping drives, 465
MOSFET, 176
Motion Control, 359
Motion control modules, 500
Motion control system, 393
Motion coordination, 526
mounting plate, 352
Multiplexer, 202

**N**

N-type semiconductor, 168
Neodymium, 416
Norton's equivalent circuit, 157
Notch filter, 199
Nozzle, 351
Null Method, 223
Null Position, 377

**O**

Open Circuit, 290
Open Loop, 282, 371
Open loop control, 29, 49
Open-center, 289, 352, 381
Open-loop control, 46
Operational amplification, 183
Opto-couplers, 172
Opto-isolated I/O modules, 502
Orifice Equation, 333
Over-center, 307

**P**

P-type semiconductors, 167
Parallel, 318
Paramagnetic, 412
Pascal's Law, 294
passive components, 153
Permeance, 404
Permittivity, 155, 400
PFC, 318
Photoelectric Sensors, 241
PID, 367
PID controller module, 500
Piezoelectric Accelerometers, 252
Piezoelectric Based Pressure Sensor, 262
piezoelectric effect, 252
Pitot Tube, 269
PLC, 495
Point-to-point position synchronization, 527
Poppet Valve, 327
Position Sensors, 225
Positive Displacement, 301
Positive Displacement Flo meters, 267
Positive Flow Control, 318
Post-compensator, 333
Potential Energy, 295
Potentiometer, 225

Power transistors, 175
Pre-compensator, 333
Presence Sensors, 243
Pressure Compensated, 314
Pressure Control Valve, 324
Pressure Drop, 299
Pressure Feedback System, 358
pressure gain, 353
Pressure Intensifier, 321
Pressure Sensors, 259
Printed-disk armature, 431
Priority, 291
ProfiBus, 495
program counter register, 130
Programmable motion control systems, 515
Proportional control, 64
Proportional Control, 67
Proportional Valves, 324, 340, 344
Proportional-integral-derivative (PID), 29, 64, 73
  PD Control, 72
Proportional-integral-derivative (PID)
  PI Control, 70
pulse width modulated, 175
Pulse width modulation, 439
PWM, 366

## Q

Quality of response, 29, 49, 50

## R

RAM, 130
reduced instruction set computer, 131
Regenerative, 383
Registers, 130
Relay, 503
Relief Valve, 326
Relief Valves, 284, 381
relocatable code, 126
Reluctance, 404
Repeatability, 217
Residual magnetization, 413
resistance, 154
resistivity, 153
resistor, 153
Resolution, 217
Resolver, 330
Resolvers, 228
Restrictor, 333
Ride Control, 388
RISC, 131
Robotic manipulator, 8
Robustness, 29, 48, 49, 50
ROM, 130, 499
Root locus, 60, 61, 62, 63
Rotor, 393
RS-232, 500
RTD Temperature Sensors, 264

## S

Samarium cobalt, 416
sampling, 31
Sampling
  Sampling theorem, 36, 38, 44
  Shannon's sampling theorem, 34, 36
sandwitch style mounting plates, 352
Saturation Region, 174
Scan mode, 506
Schmitt trigger, 190
Seal-in circuit, 508
Semiconductor, 166
sensor, 217
Sensor calibration, 219
sensors, 2
Sensors, 29
Sequence Valve, 329
Sequencer, 508
Sercos, 519
Series, 318
Servo Valves, 340
Shell-armature DC motors, 431
Shift register, 508
Shuttle, 330
Shuttle Valves, 335
Silicon, 166
Silicon controller rectifier, 169
Simultaneous sample and hold circuit, 212
single-ended signal, 185
slip frequency, 450
snubber circuit, 170
Soft ferromagnetic, 413
Soft magnetic materials, 413
Soft-wired, 495
software cam, 531
Software gearing, 528
Solenoid, 281, 358, 423
Solid-state devices, 166
Sonic Distance Sensors, 240
Squirrel-cage, 447
Stability, 29
stability, 49, 50, 58
stage, 341
Starter, 503
Stationary charges, 401
Stator, 393
Steady state response, 50, 56

Step angle, 473
Step motor, 461
Stepper motors, 394
STOP, 523
Strain Gauge Based Pressure Sensor, 261
Strain Gauges, 254
sub-plate, 352
Superconductors, 154
susceptance, 161
susceptive part, 161
Swash Plate, 313
Switched reluctance, 471
Switching sequence, 463
Synchronous motors, 447
syncro, 231
system, 2

**T**
Tachometers, 245
Temperature Sensors, 263
Thermistors Temperature Sensors, 265
Thermocouple sensor, 500
Thermocouples, 265
Thevenin's equivalent circuit, 157
Time-delay, 42
Torque Motor, 281
Torque-speed performance, 455
Torque-speed plane, 396
transduction, 217
Transient response, 49, 74
Transistor, 172
Transistor-transistor logic, 201
Trapezoidal difference approximation, 75
Triac, 171
Turbine Flow Meters, 268
Tustin's method, 77

**U**
Unipolar drive, 468
Unloading Valve, 329

**V**
valve adaptor, 352
Valve Control, 358
Variable Displacement, 302
Velocity Sensors, 245
Viscosity, 288
Vision Systems, 273
Voltage surge protection, 172
Vortex flow meter, 269

**W**
Web handling, 535
Wheatstone Bridge Circuit, 222
wire wound, 154
Working Volume, 285
Wound-rotor, 447

**Z**
Zener diode, 169
Zero-Order-Hold, 39, 41